D0551714

Numerical and Statistical Methods for Bioengineering

This is the first MATLAB-based numerical methods textbook for bioengineers that uniquely integrates modeling concepts with statistical analysis, while maintaining a focus on enabling the user to report the error or uncertainty in their result. Between traditional numerical method topics of linear modeling concepts, non-linear root finding, and numerical integration, chapters on hypothesis testing, data regression, and probability are interweaved. A unique feature of the book is the inclusion of examples from clinical trials and bioinformatics, which are not found in other numerical methods textbooks for engineers. With a wealth of biomedical engineering examples, case studies on topical biomedical research, and the inclusion of end of chapter problems, this is a perfect core text for a one-semester undergraduate course.

Michael R. King is an Associate Professor of Biomedical Engineering at Cornell University. He is an expert on the receptor-mediated adhesion of circulating cells, and has developed new computational and in vitro models to study the function of leukocytes, platelets, stem, and cancer cells under flow. He has co-authored two books and received numerous awards, including the 2008 ICNMM Outstanding Researcher Award from the American Society of Mechanical Engineers, and received the 2009 Outstanding Contribution for a Publication in the International Journal *Clinical Chemistry*.

Nipa A. Mody is currently a postdoctoral research associate at Cornell University in the Department of Biomedical Engineering. She received her Ph.D. in Chemical Engineering from the University of Rochester in 2008 and has received a number of awards including a Ruth L. Kirschstein National Research Service Award (NRSA) from the NIH in 2005 and the Edward Peck Curtis Award for Excellence in Teaching from University of Rochester in 2004.

CAMBRIDGE TEXTS IN BIOMEDICAL ENGINEERING

Cambridge Texts in Biomedical Engineering provides a forum for high-quality accessible textbooks targeted at undergraduate and graduate courses in biomedical engineering. It covers a broad range of biomedical engineering topics from introductory texts to advanced topics including, but not limited to, biomechanics, physiology, biomedical instrumentation, imaging, signals and systems, cell engineering, and bioinformatics. The series blends theory and practice, aimed primarily at biomedical engineering students, it also suits broader courses in engineering, the life sciences and medicine.

Numerical and Statistical Methods for Bioengineering

Applications in MATLAB

Michael R. King and Nipa A. Mody
Cornell University

CAMBRIDGE
UNIVERSITY PRESS

University Printing House, Cambridge CB2 8BS, United Kingdom

One Liberty Plaza, 20th Floor, New York, NY 10006, USA

477 Williamstown Road, Port Melbourne, VIC 3207, Australia

4843/24, 2nd Floor, Ansari Road, Daryaganj, Delhi - 110002, India

79 Anson Road, #06-04/06, Singapore 079906

Cambridge University Press is part of the University of Cambridge.

It furthers the University's mission by disseminating knowledge in the pursuit of education, learning and research at the highest international levels of excellence.

www.cambridge.org
Information on this title: www.cambridge.org/9780521871587

First published 2011
Reprinted 2015

A catalogue record for this publication is available from the British Library

Library of Congress Cataloging in Publication data
King, Michael R.
Numerical and statistical methods for bioengineering : applications in MATLAB /
Michael R. King, Nipa A. Mody.
p. cm. – (Cambridge texts in biomedical engineering)
ISBN 978-0-521-87158-7 (hardback)
1. Biomedical engineering – Statistical methods. 2. Biomedical engineering – Mathematics.
3. MATLAB. I. Mody, Nipa A. II. Title. III. Series.
R857.M34K56 2010
610.28–dc22

2010025717

ISBN 978-0-521-87158-7 Hardback

Additional resources for this publication at www.cambridge.org/9780521871587

Contents

Preface *page* ix

1 Types and sources of numerical error 1
1.1 Introduction 1
1.2 Representation of floating-point numbers 4
1.2.1 How computers store numbers 7
1.2.2 Binary to decimal system 7
1.2.3 Decimal to binary system 9
1.2.4 Binary representation of floating-point numbers 10
1.3 Methods used to measure error 16
1.4 Significant digits 18
1.5 Round-off errors generated by floating-point operations 20
1.6 Taylor series and truncation error 26
1.6.1 Order of magnitude estimation of truncation error 28
1.6.2 Convergence of a series 32
1.6.3 Finite difference formulas for numerical differentiation 33
1.7 Criteria for convergence 39
1.8 End of Chapter 1: key points to consider 40
1.9 Problems 40
References 46

2 Systems of linear equations 47
2.1 Introduction 47
2.2 Fundamentals of linear algebra 53
2.2.1 Vectors and matrices 53
2.2.2 Matrix operations 56
2.2.3 Vector and matrix norms 64
2.2.4 Linear combinations of vectors 66
2.2.5 Vector spaces and basis vectors 69
2.2.6 Rank, determinant, and inverse of matrices 71
2.3 Matrix representation of a system of linear equations 75
2.4 Gaussian elimination with backward substitution 76
2.4.1 Gaussian elimination without pivoting 76
2.4.2 Gaussian elimination with pivoting 84
2.5 LU factorization 87
2.5.1 LU factorization without pivoting 88
2.5.2 LU factorization with pivoting 93
2.5.3 The MATLAB `lu` function 95
2.6 The MATLAB backslash (\) operator 96
2.7 Ill-conditioned problems and the condition number 97
2.8 Linear regression 101

2.9 Curve fitting using linear least-squares approximation 107
2.9.1 The normal equations 109
2.9.2 Coefficient of determination and quality of fit 115
2.10 Linear least-squares approximation of transformed equations 118
2.11 Multivariable linear least-squares regression 123
2.12 The MATLAB function `polyfit` 124
2.13 End of Chapter 2: key points to consider 125
2.14 Problems 127
References 139

3 Probability and statistics 141
3.1 Introduction 141
3.2 Characterizing a population: descriptive statistics 144
3.2.1 Measures of central tendency 145
3.2.2 Measures of dispersion 146
3.3 Concepts from probability 147
3.3.1 Random sampling and probability 149
3.3.2 Combinatorics: permutations and combinations 154
3.4 Discrete probability distributions 157
3.4.1 Binomial distribution 159
3.4.2 Poisson distribution 163
3.5 Normal distribution 166
3.5.1 Continuous probability distributions 167
3.5.2 Normal probability density 169
3.5.3 Expectations of sample-derived statistics 171
3.5.4 Standard normal distribution and the z statistic 175
3.5.5 Confidence intervals using the z statistic and the t statistic 177
3.5.6 Non-normal samples and the central-limit theorem 183
3.6 Propagation of error 186
3.6.1 Addition/subtraction of random variables 187
3.6.2 Multiplication/division of random variables 188
3.6.3 General functional relationship between two random variables 190
3.7 Linear regression error 191
3.7.1 Error in model parameters 193
3.7.2 Error in model predictions 196
3.8 End of Chapter 3: key points to consider 199
3.9 Problems 202
References 208

4 Hypothesis testing 209
4.1 Introduction 209
4.2 Formulating a hypothesis 210
4.2.1 Designing a scientific study 211
4.2.2 Null and alternate hypotheses 217
4.3 Testing a hypothesis 219
4.3.1 The p value and assessing statistical significance 220
4.3.2 Type I and type II errors 226
4.3.3 Types of variables 228
4.3.4 Choosing a hypothesis test 230

4.4 Parametric tests and assessing normality 231
4.5 The z test 235
4.5.1 One-sample z test 235
4.5.2 Two-sample z test 241
4.6 The t test 244
4.6.1 One-sample and paired sample t tests 244
4.6.2 Independent two-sample t test 249
4.7 Hypothesis testing for population proportions 251
4.7.1 Hypothesis testing for a single population proportion 256
4.7.2 Hypothesis testing for two population proportions 257
4.8 One-way ANOVA 260
4.9 Chi-square tests for nominal scale data 274
4.9.1 Goodness-of-fit test 276
4.9.2 Test of independence 281
4.9.3 Test of homogeneity 285
4.10 More on non-parametric (distribution-free) tests 288
4.10.1 Sign test 289
4.10.2 Wilcoxon signed-rank test 292
4.10.3 Wilcoxon rank-sum test 296
4.11 End of Chapter 4: key points to consider 299
4.12 Problems 299
References 308

5 Root-finding techniques for nonlinear equations 310
5.1 Introduction 310
5.2 Bisection method 312
5.3 Regula-falsi method 319
5.4 Fixed-point iteration 320
5.5 Newton's method 327
5.5.1 Convergence issues 329
5.6 Secant method 336
5.7 Solving systems of nonlinear equations 338
5.8 MATLAB function `fzero` 346
5.9 End of Chapter 5: key points to consider 348
5.10 Problems 349
References 353

6 Numerical quadrature 354
6.1 Introduction 354
6.2 Polynomial interpolation 361
6.3 Newton–Cotes formulas 371
6.3.1 Trapezoidal rule 372
6.3.2 Simpson's 1/3 rule 380
6.3.3 Simpson's 3/8 rule 384
6.4 Richardson's extrapolation and Romberg integration 387
6.5 Gaussian quadrature 391
6.6 End of Chapter 6: key points to consider 402
6.7 Problems 403
References 408

7 Numerical integration of ordinary differential equations 409
7.1 Introduction 409
7.2 Euler's methods 416
7.2.1 Euler's forward method 417
7.2.2 Euler's backward method 428
7.2.3 Modified Euler's method 431
7.3 Runge–Kutta (RK) methods 434
7.3.1 Second-order RK methods 434
7.3.2 Fourth-order RK methods 438
7.4 Adaptive step size methods 440
7.5 Multistep ODE solvers 451
7.5.1 Adams methods 452
7.5.2 Predictor–corrector methods 454
7.6 Stability and stiff equations 456
7.7 Shooting method for boundary-value problems 461
7.7.1 Linear ODEs 463
7.7.2 Nonlinear ODEs 464
7.8 End of Chapter 7: key points to consider 472
7.9 Problems 473
References 478

8 Nonlinear model regression and optimization 480
8.1 Introduction 480
8.2 Unconstrained single-variable optimization 487
8.2.1 Newton's method 488
8.2.2 Successive parabolic interpolation 492
8.2.3 Golden section search method 495
8.3 Unconstrained multivariable optimization 500
8.3.1 Steepest descent or gradient method 502
8.3.2 Multidimensional Newton's method 509
8.3.3 Simplex method 513
8.4 Constrained nonlinear optimization 523
8.5 Nonlinear error analysis 530
8.6 End of Chapter 8: key points to consider 533
8.7 Problems 534
References 538

9 Basic algorithms of bioinformatics 539
9.1 Introduction 539
9.2 Sequence alignment and database searches 540
9.3 Phylogenetic trees using distance-based methods 554
9.4 End of Chapter 9: key points to consider 557
9.5 Problems 558
References 558

Appendix A Introduction to MATLAB 560
Appendix B Location of nodes for Gauss–Legendre quadrature 576
Index for MATLAB commands 578
Index 579

Preface

Biomedical engineering programs have exploded in popularity and number over the past 20 years. In many programs, the fundamentals of engineering science are taught from textbooks borrowed from other, more traditional, engineering fields: statics, transport phenomena, circuits. Other courses in the biomedical engineering curriculum are so multidisciplinary (think of tissue engineering, Introduction to BME) that this approach does not apply; fortunately, excellent new textbooks have recently emerged on these topics. On the surface, numerical and statistical methods would seem to fall into this first category, and likely explains why biomedical engineers have not yet contributed textbooks on this subject. I mean ... math is math, right? Well, not exactly.

There exist some unique aspects of biomedical engineering relevant to numerical analysis. Graduate research in biomedical engineering is more often *hypothesis driven*, compared to research in other engineering disciplines. Similarly, biomedical engineers in industry design, test, and produce medical devices, instruments, and drugs, and thus must concern themselves with human clinical trials and gaining approval from regulatory agencies such as the US Food & Drug Administration. As a result, statistics and hypothesis testing play a bigger role in biomedical engineering and must be taught at the curricular level. This increased emphasis on statistical analysis is reflected in special "program criteria" established for biomedical engineering degree programs by the Accreditation Board for Engineering and Technology (ABET) in the USA.

There are many general textbooks on numerical methods available for undergraduate and graduate students in engineering; some of these use MATLAB as the teaching platform. A good undergraduate text along these lines is *Numerical Methods with Matlab* by G. Recktenwald, and a good graduate-level reference on numerical methods is the well-known *Numerical Recipes* by W. H. Press *et al.* These texts do a good job of covering topics such as programming basics, nonlinear root finding, systems of linear equations, least-squares curve fitting, and numerical integration, but tend not to devote much space to statistics and hypothesis testing. Certainly, topics such as genomic data and design of clinical trails are not covered. But beyond the basic numerical algorithms that may be common to engineering and the physical sciences, one thing an instructor learns is that *biomedical engineering students want to work on biomedical problems*! This requires a biomedical engineering instructor to supplement a general numerical methods textbook with a gathering of relevant lecture examples, homework, and exam problems, a labor-intensive task to be sure and one that may leave students confused and unsatisfied with their textbook investment. This book is designed to fill an unmet need, by providing a complete numerical and statistical methods textbook, tailored to the unique requirements of the modern BME curriculum and implemented in MATLAB, which is inundated with examples drawn from across the spectrum of biomedical science.

This book is designed to serve as the primary textbook for a one-semester course in numerical and statistical methods for biomedical engineering students. The level of the book is appropriate for sophomore year through first year of graduate studies, depending on the pace of the course and the number of advanced, optional topics that are covered. A course based on this book, together with later opportunities for implementation in a laboratory course or senior design project, is intended to fulfil the statistics and hypothesis testing requirements of the program criteria established by ABET, and served this purpose at the University of Rochester. The material within this book formed the basis for the required junior-level course "Biomedical computation," offered at the University of Rochester from 2002 to 2008. As of Fall 2009, an accelerated version of the "Biomedical computation" course is now offered at the masters level at Cornell University. It is recommended that students have previously taken calculus and an introductory programming course; a semester of linear algebra is helpful but not required. It is our hope that this book will also serve as a valuable reference for bioengineering practitioners and other researchers working in quantitative branches of the life sciences such as biophysics and physiology.

Format

As with most textbooks, the chapters have been organized so that concepts are progressively built upon as the reader advances from one chapter to the next. Chapters 1 and 2 develop basic concepts, such as types of errors, linear algebra concepts, linear problems, and linear regression, that are referred to in later chapters. Chapters 3 (Probability and statistics) and 5 (Nonlinear root-finding techniques) draw upon the material covered in Chapters 1 and 2. Chapter 4 (Hypothesis testing) exclusively draws upon the material covered in Chapter 3, and can be covered at any point after Chapter 3 (Sections 3.1 to 3.5) is completed. The material on linear regression error in Chapter 3 should precede the coverage of Chapter 8 (Nonlinear model regression and optimization). The following chapter order is strongly recommended to provide a seamless transition from one topic to the next:

Chapter 1 → Chapter 2 → Chapter 3 → Chapter 5 → Chapter 6 → Chapter 8.

Chapter 4 can be covered at any time once the first three chapters are completed, while Chapter 7 can be covered at any time after working through Chapters 1, 2, 3, and 5. Chapter 9 covers an elective topic that can be taken up at any time during a course of study.

The examples provided in each chapter are of two types: Examples and Boxes. The problems presented in the Examples are more straightforward and the equations simpler. Examples either illustrate concepts already introduced in earlier sections or are used to present new concepts. They are relatively quick to work through compared to the Boxes since little additional background information is needed to understand the example problems. The Boxes discuss biomedical research, clinical, or industrial problems and include an explanation of relevant biology or engineering concepts to present the nature of the problem. In a majority of the Boxes, the equations to be solved numerically are derived from first principles to provide a more complete understanding of the problem. The problems covered in Boxes can be more challenging and require more involvement by the reader. While the Examples are critical in mastering the text material, the choice of which boxed problems to focus on is left to the instructor or reader.

As a recurring theme of this book, we illustrate the implementation of numerical methods through programming with the technical software package MATLAB. Previous experience with MATLAB is not necessary to follow and understand this book, although some prior programming knowledge is recommended. The best way to learn how to program in a new language is to jump right into coding when a need presents itself. Sophistication of the code is increased gradually in successive chapters. New commands and programming concepts are introduced on a need-to-know basis. Readers who are unfamiliar with MATLAB should first study Appendix A, Introduction to MATLAB, to orient themselves with the MATLAB programming environment and to learn the basic commands and programming terminology. Examples and Boxed problems are accompanied by the MATLAB code containing the numerical algorithm to solve the numerical or statistical problem. The MATLAB programs presented throughout the book illustrate code writing practice. We show two ways to use MATLAB as a tool for solving numerical problems: (1) by developing a program (m-file) that contains the numerical algorithm, and (2) using built-in functions supplied by MATLAB to solve a problem numerically. While self-written numerical algorithms coded in MATLAB are instructive for teaching, MATLAB built-in functions that compute the numerical solution can be more efficient and robust for use in practice. The reader is taught to integrate MATLAB functions into their written code to solve a specific problem (e.g. the backslash operator).

The book has its own website hosted by Cambridge University Press at www.cambridge.org/kingmody. All of the m-files and numerical data sets within this book can be found at this website, along with additional MATLAB programs relevant to the topics and problems covered in the text.

Acknowledgements

First and foremost, M. R. K. owes a debt of gratitude to Professor David Leighton at the University of Notre Dame. The "Biomedical computation" course that led to this book was closely inspired by Leighton's "Computer methods for chemical engineers" course at Notre Dame. I (Michael King) had the good fortune of serving as a teaching assistant and later as a graduate instructor for that course, and it shaped the format and style of my own teaching on this subject. I would also like to thank former students, teaching assistants, and faculty colleagues at the University of Rochester and Cornell University; over the years, their valuable input has helped continually to improve the material that comprises this book. I thank my wife and colleague Cindy Reinhart-King for her constant support. Finally, I thank my co-author, friend, and former student Nipa Mody: without her tireless efforts this book would not exist.

N.A.M. would like to acknowledge the timely and helpful advice on creating bioengineering examples from the following business and medical professionals: Khyati Desai, Shimoni Shah, Shital Modi, and Pinky Shah. I (Nipa Mody) also thank Banu Sankaran and Ajay Sadrangani for their general feedback on parts of the book. I am indebted to the support provided by the following faculty and staff members of the Biomedical Engineering Department of the University of Rochester: Professor Richard E. Waugh, Donna Porcelli, Nancy Gronski, Mary

Gilmore, and Gayle Hurlbutt, in this endeavor. I very much appreciate the valiant support of my husband, Anand Mody, while I embarked on the formidable task of book writing.

We thank those people who have critically read versions of this manuscript, in particular, Ayotunde Oluwakorede Ositelu, Aram Chung, and Bryce Allio, and also to our reviewers for their excellent recommendations. We express many thanks to Michelle Carey, Sarah Matthews, Christopher Miller, and Irene Pizzie at Cambridge University Press for their help in the planning and execution of this book project and for their patience.

If readers wish to suggest additional topics or comments, please write to us. We welcome all comments and criticisms as this book (and the field of biomedical engineering) continue to evolve.

1 Types and sources of numerical error

1.1 Introduction

The job of a biomedical engineer often involves the task of formulating and solving mathematical equations that define, for example, the design criteria of biomedical equipment or a prosthetic organ or physiological/pathological processes occurring in the human body. Mathematics and engineering are inextricably linked. The types of equations that one may come across in various fields of engineering vary widely, but can be broadly categorized as: linear equations in one variable, linear equations with respect to multiple variables, nonlinear equations in one or more variables, linear and nonlinear ordinary differential equations, higher order differential equations of *n*th order, and integral equations. Not all mathematical equations are amenable to an analytical solution, i.e. a solution that gives an exact answer either as a number or as some function of the variables that define the problem. For example, the analytical solution for

(1) $x^2 + 2x + 1 = 0$ is $x = \pm 1$, and
(2) $dy/dx + 3x = 5$, with initial conditions $x = 0, y = 0$, is $y = 5x - 3x^2/2$.

Sometimes the analytical solution to a system of equations may be exceedingly difficult and time-consuming to obtain, or once obtained may be too complicated to provide insight.

The need to obtain a solution to these otherwise unsolvable problems in a reasonable amount of time and with the resources at hand has led to the development of **numerical methods**. Such methods are used to determine an approximation to the actual solution within some tolerable degree of error. A numerical method is an iterative mathematical procedure that can be applied to only certain types or forms of a mathematical equation, and under usual circumstances allows the solution to converge to a final value with a pre-determined level of accuracy or tolerance. Numerical methods can often provide exceedingly accurate solutions for the problem under consideration. However, keep in mind that the solutions are rarely ever exact. A closely related branch of mathematics is **numerical analysis**, which goes hand-in-hand with the development and application of numerical methods. This related field of study is concerned with analyzing the performance characteristics of established numerical methods, i.e. how quickly the numerical technique converges to the final solution and accuracy limitations. It is important to have, at the least, basic knowledge of numerical analysis so that you can make an informed decision when choosing a technique for solving a numerical problem. The accuracy and precision of the numerical solution obtained is dependent on a number of factors, which include the choice of the numerical technique and the implementation of the technique chosen.

Errors can creep into any mathematical solution or statistical analysis in several ways. Human mistakes include, for example, (1) entering incorrect data into a computer, (2) errors in the mathematical expressions that define the problem, or (3) bugs in the computer program written to solve the engineering or math problem, which can result from logical errors in the code. A source of error that we have less control over is the quality of the data. Most often, scientific or engineering data available for use are imperfect, i.e. the true values of the variables cannot be determined with complete certainty. Uncertainty in physical or experimental data is often the result of imperfections in the experimental measuring devices, inability to reproduce exactly the same experimental conditions and outcomes each time, the limited size of sample available for determining the average behavior of a population, presence of a bias in the sample chosen for predicting the properties of the population, and inherent variability in biological data. All these errors may to some extent be avoided, corrected, or estimated using statistical methods such as confidence intervals. Additional errors in the solution can also stem from inaccuracies in the mathematical model. The model equations themselves may be simplifications of actual phenomena or processes being mimicked, and the parameters used in the model may be approximate at best.

Even if all errors derived from the sources listed above are somehow eliminated, we will still find other errors in the solution, called **numerical errors**, that arise when using numerical methods and electronic computational devices to perform numerical computations. These are actually unavoidable! **Numerical errors**, which can be broadly classified into two categories – round-off errors and truncation errors – are an integral part of these methods of solution and preclude the attainment of an exact solution. The source of these errors lies in the fundamental approximations and/or simplifications that are made in the representation of numbers as well as in the mathematical expressions that formulate the numerical problem. Any computing device you use to perform calculations follows a specific method to store numbers in a memory in order to operate upon them. Real numbers, such as fractions, are stored in the computer memory in floating-point format using the binary number system, and cannot always be stored with exact precision. This limitation, coupled with the finite memory available, leads to what is known as round-off error. Even if the numerical method yields a highly accurate solution, the computer round-off error will pose a limit to the final accuracy that can be achieved.

You should familiarize yourself with the types of errors that limit the precision and accuracy of the final solution. By doing so, you will be well-equipped to (1) estimate the magnitude of the error inherent in your chosen numerical method, (2) choose the most appropriate method for solution, and (3) prudently implement the algorithm of the numerical technique.

The origin of round-off error is best illustrated by examining how numbers are stored by computers. In Section 1.2, we look closely at the floating-point representation method for storing numbers and the inherent limitations in numeric precision and accuracy as a result of using binary representation of decimal numbers and finite memory resources. Section 1.3 discusses methods to assess the accuracy of estimated or measured values. The accuracy of any measured value is conveyed by the number of significant digits it has. The method to calculate the number of significant digits is covered in Section 1.4. Arithmetic operations performed by computers also generate round-off errors. While many round-off errors are too small to be of significance, certain floating-point operations can produce large and unacceptable errors in the result and should be avoided when possible. In Section 1.5, strategies to prevent the inadvertent generation of large round-off errors are discussed. The origin of truncation error is examined in Section 1.6. In Section 1.7 we introduce useful termination

Box 1.1A Oxygen transport in skeletal muscle

Oxygen is required by all cells to perform respiration and thereby produce energy in the form of ATP to sustain cellular processes. Partial oxidation of glucose (energy source) occurs in the cytoplasm of the cell by an anaerobic process called glycolysis that produces pyruvate. Complete oxidation of pyruvate to CO_2 and H_2O occurs in the mitochondria, and is accompanied by the production of large amounts of ATP. If cells are temporarily deprived of an adequate oxygen supply, a condition called hypoxia develops. In this situation, pyruvate no longer undergoes conversion in the mitochondria and is instead converted to lactic acid within the cytoplasm itself. Prolonged oxygen starvation of cells leads to cell death, called necrosis.

The circulatory system and the specialized oxygen carrier, hemoglobin, cater to the metabolic needs of the vast number of cells in the body. Tissues in the body are extensively perfused with tiny blood vessels in order to enable efficient and timely oxygen transport. The oxygen released from the red blood cells flowing through the capillaries diffuses through the blood vessel membrane and enters into the tissue region. The driving force for oxygen diffusion is the oxygen concentration gradient at the vessel wall and within the tissues. The oxygen consumption by the cells in the tissues depletes the oxygen content in the tissue, and therefore the oxygen concentration is always lower in the tissues as compared to its concentration in arterial blood (except when O_2 partial pressure in the air is abnormally low, such as at high altitudes). During times of strenuous activity of the muscles, when oxygen demand is greatest, the O_2 concentrations in the muscle tissue are the lowest. At these times it is critical for oxygen transport to the skeletal tissue to be as efficient as possible.

The skeletal muscle tissue sandwiched between two capillaries can be modeled as a slab of length L (see Figure 1.1). Let $N(x, t)$ be the O_2 concentration in the tissue, where $0 \leq x \leq L$ is the distance along the muscle length and t is the time. Let D be the O_2 diffusivity in the tissue and let Γ be the volumetric rate of O_2 consumption within the tissue.

Performing an O_2 balance over the tissue produces the following partial differential equation:

$$\frac{\partial N}{\partial t} = D\frac{\partial^2 N}{\partial x^2} - \Gamma.$$

The boundary conditions for the problem are fixed at $N(0, t) = N(L, t) = N_o$, where N_o is the supply concentration of O_2 at the capillary wall. For this problem, it is assumed that the resistance to transport of O_2 posed by the vessel wall is small enough to be neglected. We also neglect the change in O_2 concentration along the length of the capillaries. The steady state or long-term O_2 distribution in the tissue is governed by the ordinary differential equation

$$D\frac{\partial^2 N}{\partial x^2} = \Gamma,$$

Figure 1.1

Schematic of O_2 transport in skeletal muscle tissue.

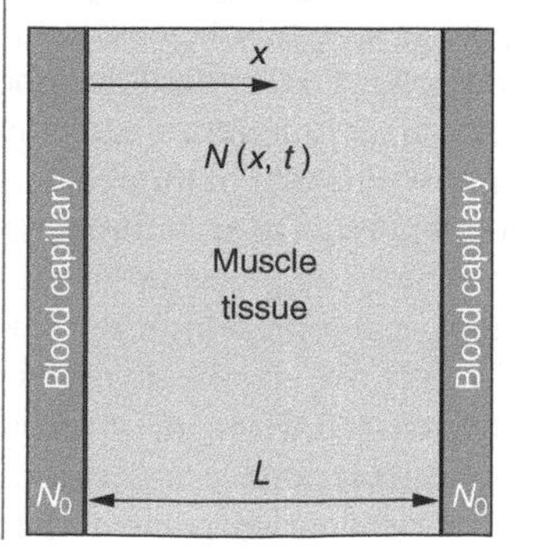

whose solution is given by

$$N_s = N_o - \frac{\Gamma x}{2D}(L - x).$$

Initially, the muscles are at rest, consuming only a small quantity of O_2, characterized by a volumetric O_2 consumption rate Γ_1. Accordingly, the initial O_2 distribution in the tissue is $N_o - \frac{\Gamma_1 x}{2D}(L - x)$.

Now the muscles enter into a state of heavy activity characterized by a volumetric O_2 consumption rate Γ_2. The time-dependent O_2 distribution is given by a Fourier series solution to the above partial differential equation:

$$N = N_o - \frac{\Gamma_2 x}{2D}(L - x) + \frac{4(\Gamma_2 - \Gamma_1)L^2}{D} \sum_{n=1,\ n \text{ is odd}}^{\infty} \left[\frac{1}{(n\pi)^3} e^{-(n\pi)^2 Dt/L^2} \sin\left(\frac{n\pi x}{L}\right) \right].$$

In Section 1.6 we investigate the truncation error involved when arriving at a solution to the O_2 distribution in muscle tissue.

criteria for numerical iterative procedures. The use of robust convergence criteria is essential to obtain reliable results.

1.2 Representation of floating-point numbers

The arithmetic calculations performed when solving problems in algebra and calculus produce exact results that are mathematically sound, such as:

$$\frac{2.5}{0.5} = 5, \quad \frac{\sqrt[3]{8}}{\sqrt{4}} = 1, \text{and} \quad \frac{d}{dx}\sqrt{x} = \frac{1}{2\sqrt{x}}.$$

Computations made by a computer or a calculator produce a true result for any integer manipulation, but have less than perfect precision when handling real numbers. It is important at this juncture to define the meaning of "precision." Numeric **precision** is defined as the exactness with which the value of a numerical estimate is known. For example, if the true value of $\sqrt{4}$ is 2 and the computed solution is 2.0001, then the computed value is precise to within the first four figures or four significant digits. We discuss the concept of significant digits and the method of calculating the number of significant digits in a number in Section 1.4.

Arithmetic calculations that are performed by retaining mathematical symbols, such as 1/3 or $\sqrt{7}$, produce exact solutions and are called symbolic computations. Numerical computations are not as precise as symbolic computations since numerical computations involve the conversion of fractional numbers and irrational numbers to their respective numerical or *digital* representations, such as $1/3 \sim 0.33333$ or $\sqrt{7} \sim 2.64575$. Numerical representation of numbers uses a finite number of digits to denote values that may possibly require an infinite number of digits and are, therefore, often inexact or approximate. The precision of the computed value is equal to the number of digits in the numerical representation that tally with the true digital value. Thus 0.33333 has a precision of five significant digits when compared to the true value of 1/3, which is also written as $0.\bar{3}$. Here, the overbar indicates infinite repetition of the underlying digit(s). Calculations using the digitized format to represent all real numbers are termed as **floating-point arithmetic**. The format

Box 1.2 Accuracy versus precision: blood pressure measurements

It is important that we contrast the two terms *accuracy* and *precision*, which are often confused. **Accuracy** measures how close the estimate of a measured variable or an observation is to its true value. **Precision** is the range or spread of the values obtained when repeated measurements are made of the same variable. A narrow spread of values indicates good precision.

For example, a digital sphygmomanometer consistently provides three readings of the systolic/diastolic blood pressure of a patient as 120/80 mm Hg. If the true blood pressure of the patient at the time the measurement was made is 110/70 mm Hg, then the instrument is said to be very precise, since it provided similar readings every time with few fluctuations. The three instrument readings are "on the same mark" every time. However, the readings are all inaccurate since the "correct target or mark" is not 120/80 mm Hg, but is 110/70 mm Hg.

An intuitive example commonly used to demonstrate the difference between accuracy and precision is the bulls-eye target (see Figure 1.2).

Figure 1.2

The bulls-eye target demonstrates the difference between accuracy and precision.

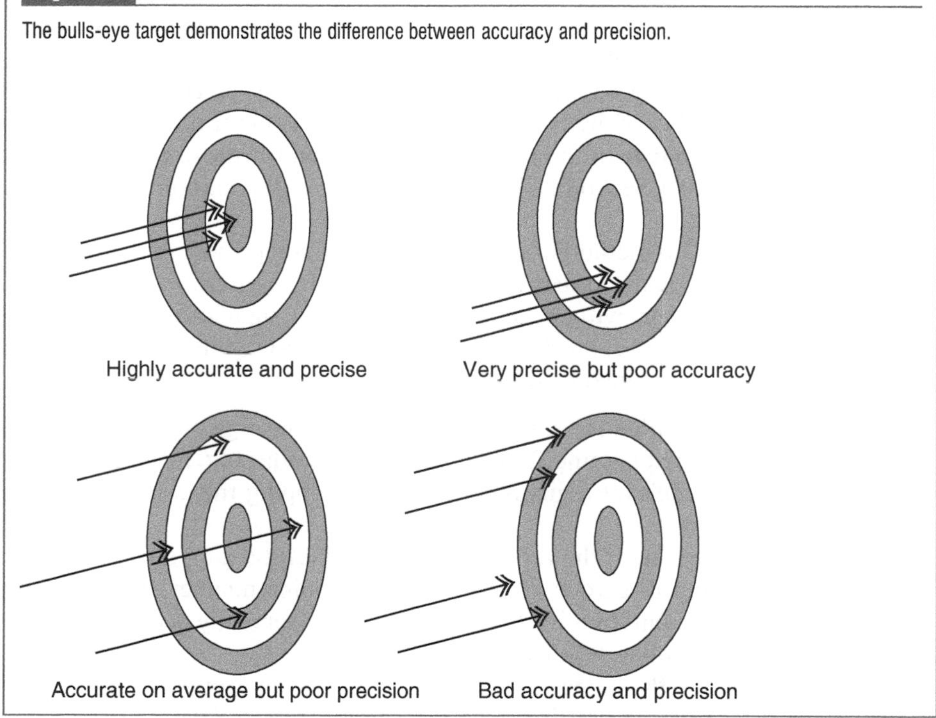

standards for **floating-point number representation** by computers are specified by the IEEE Standard 754. The binary or base-2 system is used by digital devices for storing numbers and performing arithmetic operations. The binary system cannot precisely represent all *rational numbers* in the decimal or base-10 system. The imprecision or error inherent in computing when using floating-point number representation is called **round-off error**. Round-off thus occurs when real numbers must be approximated using a limited number of significant digits. Once a round-off error is introduced in a floating-point calculation, it is carried over in all subsequent computations. However, round-off is of fundamental advantage to the efficiency of performing computations. Using a fixed and finite number of digits to represent each number ensures a reasonable speed in computation and economical use of the computer memory.

Round-off error results from a trade-off between the efficient use of computer memory and accuracy.

Decimal floating-point numbers are represented by a computer in standardized format as shown below:

$$\pm 0.f_1 f_2 f_3 f_4 f_5 f_6 \ldots f_{s-1} f_s \times 10^k,$$

```
|----------------------------|   |---|
              ↑                    ↑
         significand      10 raised to the power k
```

where f is a decimal digit from 0 to 9, s is the number of significant digits, i.e. the number of digits in the significand as dictated by the precision limit of the computer, and k is the exponent. The advantage of this numeric representation scheme is that the **range** of representable numbers (as determined by the largest and smallest values of k) can be separated from the degree of **precision** (which is determined by the number of digits in the significand). The power or exponent k indicates the **order of magnitude** of the number. The notation for the order of magnitude of a number is $O(10^k)$. Section 1.7 discusses the topic of "estimation of order of magnitude" in more detail.

This method of numeric representation provides the best approximation possible of a real number within the limit of s significant digits. The real number may, however, require an infinite number of significant digits to denote the true value with perfect precision. The value of the last significant digit of a *floating-point number* is determined by one of two methods.

(1) **Truncation** Here, the numeric value of the digit f_{s+1} is not considered. The value of the digit f_s is unaffected by the numeric value of f_{s+1}. The floating-point number with s significant digits is obtained by dropping the $(s + 1)$th digit and all digits to its right.

(2) **Rounding** If the value of the last significant digit f_s depends on the value of the digits being discarded from the floating-point number, then this method of numeric representation is called rounding. The generally accepted convention for rounding is as follows (Scarborough, 1966):
   (a) if the numeric value of the digits being dropped is greater than five units of the f_{s+1}th position, then f_s is changed to $f_s + 1$;
   (b) if the numeric value of the digits being dropped is less than five units of the f_{s+1}th position, then f_s remains unchanged;
   (c) if the numeric value of the digits being dropped equals five units of the f_{s+1}th position, then
      (i) if f_s is even, f_s remains unchanged,
      (ii) if f_s is odd, f_s is replaced by $f_s + 1$.

This last convention is important since, on average, one would expect the occurrence of the f_s digit as odd only half the time. Accordingly, by leaving f_s unchanged approximately half the time when the f_{s+1} digit is exactly equal to five units, it is intuitive that the errors caused by rounding will to a large extent cancel each other.

As per the rules stated above, the following six numbers are rounded to retain only five digits:

$$
\begin{aligned}
0.345903 &\rightarrow 0.34590,\\
13.85748 &\rightarrow 13.857,\\
7983.9394 &\rightarrow 7983.9,\\
5.20495478 &\rightarrow 5.2050,\\
8.94855 &\rightarrow 8.9486,\\
9.48465 &\rightarrow 9.4846.
\end{aligned}
$$

The rounding method is used by computers to store floating-point numbers. Every machine number is precise to within $5 \times 10^{-(s+1)}$ units of the original real number, where s is the number of significant figures.

Using MATLAB

A MATLAB function called `round` is used to round numbers to their nearest integer. The number produced by `round` does not have a fractional part. This function rounds down numbers with fractional parts less than 0.5 and rounds up for fractional parts equal to or greater than 0.5. Try using `round` on the numbers 1.1, 1.5, 2.50, 2.49, and 2.51. Note that the MATLAB function `round` does not round numbers to s significant digits as discussed in the definition above for rounding.

1.2.1 How computers store numbers

Computers store all data and instructions in binary coded format or the **base-2** number system. The machine language used to instruct a computer to execute various commands and manipulate operands is written in binary format – a number system that contains only two digits: 0 and 1. Every binary number is thus constructed from these two digits only, as opposed to the base-10 number system that uses ten digits to represent numbers. The differences between these two number systems can be further understood by studying Table 1.1.

Each digit in the binary system is called a bit (binary digit). Because a bit can take on two values, either 0 or 1, each value can represent one of two physical states – on or off, i.e. the presence or absence of an electrical pulse, or the ability of a transistor to switch between the on and off states. Binary code is thus found to be a convenient method of encoding instructions and data since it has obvious physical significance and can be easily understood by the operations performed by a computer.

The range of the magnitude of numbers, as well as numeric precision that a computer can work with, depends on the number of bits allotted for representing numbers. Programming languages, such as Fortran and C, and mathematical software packages such as MATLAB allow users to work in both single and double precision.

1.2.2 Binary to decimal system

It is important to be familiar with the methods for converting numbers from one base to another in order to understand the inherent limitations of computers in working with real numbers in our base-10 system. Once you are well-versed in the

Table 1.1. *Equivalence of numbers in the decimal (base-10) and binary (base-2) systems*

Decimal system (base 10)	Binary system (base 2)	Conversion of binary number to decimal number
0	0	$0 \times 2^0 = 0$
1	1	$1 \times 2^0 = 1$
2	1 0	$1 \times 2^1 + 0 \times 2^0 = 2$
3	1 1	$1 \times 2^1 + 1 \times 2^0 = 3$
4	1 0 0	$1 \times 2^2 + 0 \times 2^1 + 0 \times 2^0 = 4$
5	1 0 1	$1 \times 2^2 + 0 \times 2^1 + 1 \times 2^0 = 5$
6	1 1 0	$1 \times 2^2 + 1 \times 2^1 + 0 \times 2^0 = 6$
7	1 1 1	$1 \times 2^2 + 1 \times 2^1 + 1 \times 2^0 = 7$
8	1 0 0 0	$1 \times 2^3 + 0 \times 2^2 + 0 \times 2^1 + 0 \times 2^0 = 8$
9	1 0 0 1	$1 \times 2^3 + 0 \times 2^2 + 0 \times 2^1 + 1 \times 2^0 = 9$
10	1 0 1 0	$1 \times 2^3 + 0 \times 2^2 + 1 \times 2^1 + 0 \times 2^0 = 10$
	↑ ↑ ↑ ↑ $2^3\ 2^2\ 2^1 2^0$ binary position indicators	

ways in which round-off errors can arise in different situations, you will be able to devise suitable algorithms that are more likely to minimize round-off errors. First, let's consider the method of converting binary numbers to the decimal system. The *decimal number* 111 can be expanded to read as follows:

$$\begin{aligned} 111 &= 1 \times 10^2 + 1 \times 10^1 + 1 \times 10^0 \\ &= 100 + 10 + 1 \\ &= 111. \end{aligned}$$

Thus, the position of a digit in any number specifies the magnitude of that particular digit as indicated by the power of 10 that multiplies it in the expression above. The first digit of this base-10 integer from the right is a multiple of 1, the second digit is a multiple of 10, and so on. On the other hand, if 111 is a binary number, then the same number is now equal to

$$\begin{aligned} 111 &= 1 \times 2^2 + 1 \times 2^1 + 1 \times 2^0 \\ &= 4 + 2 + 1 \\ &= 7 \text{ in base 10}. \end{aligned}$$

The decimal equivalent of 111 is also provided in Table 1.1. In the binary system, the position of a binary digit in a binary number indicates to which power the multiplier, 2, is raised. Note that the largest decimal value of a binary number comprising n bits is equal to $2^n - 1$. For example, the binary number 11111 has 5 bits and is the binary equivalent of

$$\begin{aligned} 11111 &= 1 \times 2^4 + 1 \times 2^3 + 1 \times 2^2 + 1 \times 2^1 + 1 \times 2^0 \\ &= 16 + 8 + 4 + 2 + 1 \\ &= 31 \\ &= 2^5 - 1. \end{aligned}$$

What is the range of integers that a computer can represent? A certain fixed number of bits, such as 16, 32, or 64 bits, are allotted to represent every integer. This fixed maximum number of bits used for storing an integer value is determined by the computer hardware architecture. If 16 bits are used to store each integer value in binary form, then the maximum integer value that the computer can represent is $2^{16} - 1 = 65\,535$. To include representation of negative integer values, 32 768 is subtracted internally from the integer value represented by the 16-bit number to allow representation of integers in the range of [−32 768, 32 767].

What if we have a fractional binary number such as 1011.011 and wish to convert this binary value to the base-10 system? Just as a digit to the right of a radix point (decimal point) in the base-10 system represents a multiple of 1/10 raised to a power depending on the position or place value of the decimal digit with respect to the decimal point, similarly a binary digit placed to the right of a radix point (binary point) represents a multiple of 1/2 raised to some power that depends on the position of the binary digit with respect to the radix point. In other words, just as the fractional part of a decimal number can be expressed as a sum of the negative powers of 10 (or positive powers of 1/10), similarly a binary number fraction is actually the sum of the negative powers of 2 (or positive powers of 1/2). Thus, the decimal value of the binary number 1011.011 is calculated as follows:

$$
\begin{aligned}
&(1 \times 2^3) + (0 \times 2^2) + (1 \times 2^1) + (1 \times 2^0) + (0 \times (1/2)^1) \\
&\quad + (1 \times (1/2)^2) + (1 \times (1/2)^3) \\
&\quad = 8 + 2 + 1 + 0.25 + 0.125 \\
&\quad = 11.375.
\end{aligned}
$$

1.2.3 Decimal to binary system

Now let's tackle the method of converting decimal numbers to the base-2 system. Let's start with an easy example that involves the conversion of a decimal integer, say 123, to a binary number. This is simply done by resolving the integer 123 as a series of powers of 2, i.e. $123 = 2^6 + 2^5 + 2^4 + 2^3 + 2^1 + 2^0 = 64 + 32 + 16 + 8 + 2 + 1$. The powers to which 2 is raised in the expression indicate the positions for the binary digit 1. Thus, the binary number equivalent to the decimal value 123 is 1111011, which requires 7 bits. This expansion process of a decimal number into the sum of powers of 2 is tedious for large decimal numbers. A simplified and straightforward procedure to convert a decimal number into its binary equivalent is shown below (Mathews and Fink, 2004).

We can express a positive base-10 integer I as an expansion of powers of 2, i.e.

$$I = b_n \times 2^n + b_{n-1} \times 2^{n-1} + \cdots + b_2 \times 2^2 + b_1 \times 2^1 + b_0 \times 2^0,$$

where $b_0, b_1, \ldots, b_n$ are binary digits each of value 0 or 1. This expansion can be rewritten as follows:

$$I = 2(b_n \times 2^{n-1} + b_{n-1} \times 2^{n-2} + \cdots + b_2 \times 2^1 + b_1 \times 2^0) + b_0$$

or

$$I = 2 \times I_1 + b_0,$$

where

$$I_1 = b_n \times 2^{n-1} + b_{n-1} \times 2^{n-2} + \cdots + b_2 \times 2^1 + b_1 \times 2^0.$$

By writing I in this fashion, we obtain b_0.
Similarly,

$$I_1 = 2(b_n \times 2^{n-2} + b_{n-1} \times 2^{n-3} + \cdots + b_2 \times 2^0) + b_1,$$

i.e.

$$I_1 = 2 \times I_2 + b_1,$$

from which we obtain b_1 and

$$I_2 = b_n \times 2^{n-2} + b_{n-1} \times 2^{n-3} + \cdots + b_2 \times 2^0.$$

Proceeding in this way we can easily obtain all the digits in the binary representation for I.

Example 1.1 Convert the integer 5089 in base-10 into its binary equivalent. Based on the preceding discussion, 5089 can be written as

$$5089 = 2 \times 2544 + 1 \rightarrow b_0 = 1$$

$$2544 = 2 \times 1272 + 0 \rightarrow b_1 = 0$$

$$1272 = 2 \times 636 + 0 \rightarrow b_2 = 0$$

$$636 = 2 \times 318 + 0 \rightarrow b_3 = 0$$

$$318 = 2 \times 159 + 0 \rightarrow b_4 = 0$$

$$159 = 2 \times 79 + 1 \rightarrow b_5 = 1$$

$$79 = 2 \times 39 + 1 \rightarrow b_6 = 1$$

$$39 = 2 \times 19 + 1 \rightarrow b_7 = 1$$

$$19 = 2 \times 9 + 1 \rightarrow b_8 = 1$$

$$9 = 2 \times 4 + 1 \rightarrow b_9 = 1$$

$$4 = 2 \times 2 + 0 \rightarrow b_{10} = 0$$

$$2 = 2 \times 1 + 0 \rightarrow b_{11} = 0$$

$$1 = 2 \times 0 + 1 \rightarrow b_{12} = 1$$

Thus the binary equivalent of 5089 has 13 binary digits and is 1001111100001. This algorithm, used to convert a decimal number into its binary equivalent, can be easily incorporated into a MATLAB program.

1.2.4 Binary representation of floating-point numbers

Floating-point numbers are numeric quantities that have a significand indicating the value of the number, which is multiplied by a base raised to some power. You are familiar with the **scientific notation** used to represent real numbers, i.e. numbers with fractional parts. A number, say 1786.134, can be rewritten as 1.786134×10^3. Here, the **significand** is the number 1.786134 that is multiplied by the **base 10** raised to a power 3 that is called the **exponent** or **characteristic**. Scientific notation is one method of representing base-10 floating-point numbers. In this form of notation, only one digit to the left of the decimal point in the significand is retained, such as 5.64×10^{-3} or 9.8883×10^{67}, and the magnitude of the exponent is adjusted accordingly. The advantage of using the floating-point method as a convention for representing numbers is that it is concise, standardizable, and can be used to represent very large and very small numbers using a limited fixed number of bits.

Two commonly used standards for storing floating-point numbers are the 32-bit and 64-bit representations, and are known as the **single-precision** format and **double-precision** format, respectively. Since MATLAB stores all floating-point numbers by default using double precision, and since all major programming languages support the double-precision data type, we will concentrate our efforts on understanding how computers store numeric data as double-precision floating-point numbers.

64-bit digital representations of floating-point numbers use 52 bits to store the significand, 1 bit to store the sign of the number, and another 11 bits for the exponent. Note that computers store all floating-point numbers in base-2 format and therefore not only are the significand and exponent stored as binary numbers, but also the base to which the exponent is raised is 2. If x stands for 1 bit then a 64-bit floating-point number in the machine's memory looks like this:

x	$xxxxxxxxxxx$	$xxxxxxx\ldots xxxxxx$
↑	↑	↑
sign s	exponent k	significand d
1 bit	11 bits	52 bits

The single bit s that conveys the sign indicates a positive number when $s = 0$. The range of the exponent is calculated as $[0, 2^{11} - 1] = [0, 2047]$. In order to accommodate negative exponents and thereby extend the numeric range to very small numbers, 1023 is deducted from the binary exponent to give the range $[-1023, 1024]$ for the exponent. The exponent k that is stored in the computer is said to have a *bias* (which is 1023), since the stored value is the sum of the true exponent and the number 1023 $(2^{10} - 1)$. Therefore if 1040 is the value of the biased exponent that is stored, the true exponent is actually $1040 - 1023 = 17$.

Overflow and underflow errors

Using 64-bit floating-point number representation, we can determine approximately the largest and smallest exponent that can be represented in base 10:

- largest e(base 10) $= 308$ (since $2^{1024} \sim 1.8 \times 10^{308}$),
- smallest e(base 10) $= -308$ (since $2^{-1023} \sim 1.1 \times 10^{-308}$).

Thus, there is an upper limit on the largest magnitude that can be represented by a digital machine. The largest number that is recognized by MATLAB can be obtained by entering the following command in the MATLAB Command Window:

```
>> realmax
```

Box 1.3 Selecting subjects for a clinical trial

The fictitious biomedical company Biotektroniks is conducting a double-blind clinical trial to test a vaccine for sneazlepox, a recently discovered disease. The company has 100 healthy volunteers: 50 women and 50 men. The volunteers will be divided into two groups; one group will receive the normal vaccine, while the other group will be vaccinated with saline water, or placebos. Both groups will have 25 women and 25 men. In how many ways can one choose 25 women and 25 men from the group of 50 women and 50 men for the normal vaccine group?

The solution to this problem is simply $N = C_{25}^{50} \times C_{25}^{50}$, where $C_r^n = \frac{n!}{r!(n-r)!}$. (See Chapter 3 for a discussion on combinatorics.) Thus,

$$N = {}_{25}^{50}C \times {}_{25}^{50}C = \frac{50!50!}{(25!)^4}.$$

On evaluating the factorials we obtain

$$N = \frac{3.0414 \times 10^{64} \times 3.0414 \times 10^{64}}{(1.551 \times 10^{25})^4} = \frac{9.2501 \times 10^{128}}{5.784 \times 10^{100}}.$$

When using double precision, these extraordinarily large numbers will still be recognized. However, in single precision, the range of recognizable numbers extends from -3.403×10^{38} to 3.403×10^{38} (use the MATLAB function `realmax('single')` for obtaining these values). Thus, the factorial calculations would result in an overflow if one were working with single-precision arithmetic. Note that the final answer is 1.598×10^{28}, which is within the defined range for the single-precision data type.

How do we go about solving such problems without encountering overflow? If you know you will be working with large numbers, it is important to check your product to make sure it does not exceed a certain limit. In situations where the product exceeds the set bounds, divide the product periodically by a large number to keep it within range, while keeping track of the number of times this division step is performed. The final result can be recovered by a corresponding multiplication step at the end, if needed. There are other ways to implement algorithms for the calculation of large products without running into problems, and you will be introduced to them in this chapter.

MATLAB outputs the following result:

```
ans =
  1.7977e+308
```

Any number larger than this value is given a special value by MATLAB equal to infinity (`Inf`). Typically, however, when working with programming languages such as Fortran and C, numbers larger than this value are not recognized, and generation of such numbers within the machine results in an **overflow error**. Such errors generate a floating-point exception causing the program to terminate immediately unless error-handling measures have been employed within the program. If you type a number larger than `realmax`, such as

```
>> realmax + 1e+308
```

MATLAB recognizes this number as infinity and outputs

```
ans =
  Inf
```

This is MATLAB's built-in method to handle occurrences of overflow. Similarly, the smallest negative number supported by MATLAB is given by `-realmax` and numbers smaller than this number are assigned the value `-Inf`.

Similarly, you can find out the smallest positive number greater than zero that is recognizable by MATLAB by typing into the MATLAB Command Window

```
>> realmin
ans =
  2.2251e-308
```

Numbers produced by calculations that are smaller than the smallest number supported by the machine generate **underflow errors**. Most programming languages are equipped to handle underflow errors without resulting in a program crash and typically set the number to zero. Let's observe how MATLAB handles such a scenario. In the Command Window, if we type the number

```
>> 1.0e-309
```

MATLAB outputs

```
ans =
  1.0000e-309
```

MATLAB has special methods to handle numbers slightly smaller than `realmin` by taking a few bits from the significand and adding them to the exponent. This, of course, compromises the precision of the numeric value stored by reducing the number of significant digits.

The lack of continuity between the smallest number representable by a computer and 0 reveals an important source of error: floating-point numbers are not continuous due to the finite limits of range and precision. Thus, there are gaps between two floating-point numbers that are closest in value. The magnitude of this gap in the floating-point number line increases with the magnitude of the numbers. The size of the numeric gap between the number 1.0 and the next larger number distinct from 1.0 is called the **machine epsilon** and is calculated by the MATLAB function

```
>> eps
```

which MATLAB outputs as

```
ans =
  2.2204e-016
```

Note that `eps` is the minimum value that must be added to 1.0 to result in another number larger than 1.0 and is $\sim 2^{-52} = 2.22 \times 10^{-16}$, i.e. the incremental value of 1 bit in the significand's rightmost (52nd) position. The limit of precision as given by `eps` varies based on the magnitude of the number under consideration. For example,

```
>> eps(100.0)
ans =
  1.4211e-014
```

which is obviously larger than the precision limit for 1.0 by two orders of magnitude $O(10^2)$. If two numbers close in value have a difference that is smaller than their smallest significant digit, then the two are indistinguishable by the computer. Figure 1.3 pictorially describes the concepts of underflow, overflow, and discontinuity of double-precision floating-point numbers that are represented by a computer. As the order of magnitude of the numbers increases, the discontinuity between two floating-point numbers that are adjacent to each other on the floating-point number line also becomes larger.

Figure 1.3

Simple schematic of the floating-point number line.

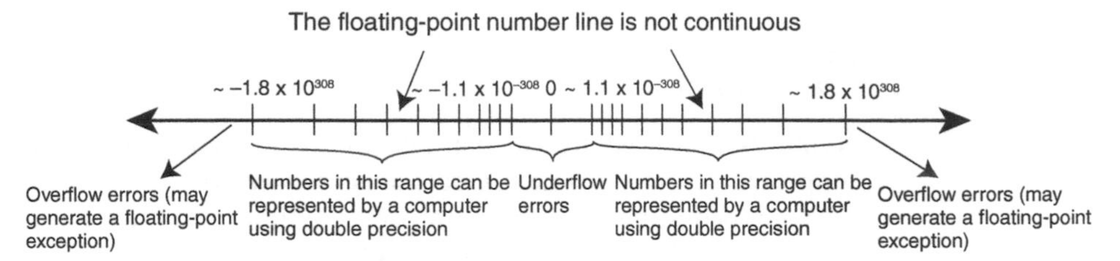

Binary significand – limits of precision

The number of bits allotted to the significand governs the limits of precision. A binary significand represents fractional numbers in the base-2 system. According to IEEE Standard 754 for floating-point representation, the computer maintains an implicitly assumed bit (termed a hidden bit) of value 1 that precedes the binary point and does not need to be stored (Tanenbaum, 1999). The 52 binary digits following the binary point are allowed any arbitrary values of 0 and 1. The binary number can thus be represented as

$$1.b_1b_2b_3b_4b_5b_6 \ldots b_{p-1}b_p \times 2^k, \tag{1.1}$$

where b stands for a binary digit, p is the maximum number of binary digits allowed in the significand based on the limits of precision, and k is the binary exponent. Therefore, the significand of every stored number has a value of $1.0 \leq \text{fraction} < 2.0$. The fraction has the exact value of 1.0 if all the 52 bits are 0, and has a value just slightly less than 2.0 if all the bits have a value of 1. How then is the value 0 represented? When all the 11 exponent bits and the 52 bits for the significand are of value 0, the implied or assumed bit value preceding the binary point is no longer considered as 1.0. A 52-bit binary number corresponds to *at least* 15 digits in the decimal system and *at least* 16 decimal digits when the binary value is a fractional number. Therefore, any double-precision floating-point number has a **maximum of 16 significant digits** and this defines the precision limit for this data type.

In Section 1.2.2 we looked at the method of converting binary numbers to decimal numbers and vice versa. These conversion techniques will come in handy here to help you understand the limits imposed by finite precision. In the next few examples, we disregard the implied or assumed bit of value 1 located to the left of the binary point.

Example 1.2 Convert the decimal numbers 0.6875 and 0.875 into its equivalent binary fraction or binary significand.

The method to convert fractional decimal numbers to their binary equivalent is similar to the method for converting integer decimal numbers to base-2 numbers. A fractional real number R can be expressed as the sum of powers of 1/2 as shown:

$$R = b_1 \times \left(\frac{1}{2}\right) + b_2 \times \left(\frac{1}{2}\right)^2 + b_3 \times \left(\frac{1}{2}\right)^3 + \cdots + b_n \times \left(\frac{1}{2}\right)^n + \cdots$$

such that

$$\underset{\substack{\uparrow \\ \textit{base-10 fraction}}}{R} \quad = \quad \underset{\substack{\uparrow \\ \text{binary fraction}}}{0.b_1b_2b_3 \ldots b_n \ldots}$$

Table 1.2. *Scheme for converting binary floating-point numbers to decimal numbers, where* $p = 4$ *and* $k = 0, -2$

For these conversions, the binary floating-point number follows the format $0.b_1\, b_2 \ldots b_{p-1}\, b_p \times 2^k$.

Binary significand	(p = 4)	Conversion calculations	Decimal number
0.1000	($k = 0$)	$(1\times(1/2)^1+0\times(1/2)^2+0\times(1/2)^3+0\times(1/2)^4)\times 2^0$	0.5
	($k = -2$)	$(1\times(1/2)^1+0\times(1/2)^2+0\times(1/2)^3+0\times(1/2)^4)\times 2^{-2}$	0.125
0.1010	($k = 0$)	$(1\times(1/2)^1+0\times(1/2)^2+1\times(1/2)^3+0\times(1/2)^4)\times 2^0$	0.625
	($k = -2$)	$(1\times(1/2)^1+0\times(1/2)^2+1\times(1/2)^3+0\times(1/2)^4)\times 2^{-2}$	0.156 25
0.1111	($k = 0$)	$(1\times(1/2)^1+1\times(1/2)^2+1\times(1/2)^3+1\times(1/2)^4)\times 2^0$	0.9375
	($k = -2$)	$(1\times(1/2)^1+1\times(1/2)^2+1\times(1/2)^3+1\times(1/2)^4)\times 2^{-2}$	0.234 375

where, $b_1, b_2, \ldots, b_n$ are binary digits each of value 0 or 1. One method to express a decimal fraction in terms of powers of 1/2 is to subtract successively increasing integral powers of 1/2 from the base-10 number until the remainder value becomes zero.

(1) $0.6875-(1/2)^1 = 0.1875$, which is the remainder. The first digit of the significand b_1 is 1. So

$0.1875 - (1/2)^2 < 0 \quad \rightarrow b_2 = 0;$

$0.1875 - (1/2)^3 = 0.0625 \quad \rightarrow b_3 = 1;$

$0.0625 - (1/2)^4 = 0.0 \quad \rightarrow b_4 = 1$ (this is the last digit of the significand that is not zero.)

Thus, the equivalent binary significand is 0.1011.

(2) $0.875 - (1/2)^1 = 0.375 \quad \rightarrow b_1 = 1;$

$0.375 - (1/2)^2 = 0.125 \quad \rightarrow b_2 = 1;$

$0.125 - (1/2)^3 = 0.0 \quad \rightarrow b_3 = 1$ (this is the last digit of the significand that is not zero.)

Thus, the equivalent binary significand is 0.111.

The binary equivalent of a number such as 0.7 is 0.1011 0011 0011 0011 ..., which terminates indefinitely. Show this yourself. A finite number of binary digits cannot represent the decimal number 0.7 exactly but can only approximate the true value of 0.7. Here is one instance of round-off error. As you can now see, round-off errors arise when a decimal number is substituted by an equivalent binary floating-point representation.

> Fractional numbers that may be exactly represented using a finite number of digits in the decimal system may not be exactly represented using a finite number of digits in the binary system.

Example 1.3 Convert the binary significand 0.10110 into its equivalent base-10 rational number.

The decimal fraction equivalent to the binary significand 0.10110 in base 2 is

$$1\times(1/2)^1+0\times(1/2)^2+1\times(1/2)^3+1\times(1/2)^4+0\times(1/2)^5 = 0.6875.$$

See Table 1.2 for more examples demonstrating the conversion scheme.

Now that we have discussed the methods for converting decimal integers into binary integers and decimal fractions into their binary significand, we are in a position to obtain the binary floating-point representation of a decimal number as described by Equation (1.1). The following sequential steps describe the method to obtain the binary floating-point representation for a decimal number a.

(1) Divide a by 2^n, where n is any integer such that 2^n is the largest power of 2 that is less than or equal to a, e.g. if $a = 40$, then $n = 5$ and $a/2^n = 40/32 = 1.25$. Therefore, $a = a/2^n \times 2^n = 1.25 \times 2^5$.

(2) Next, convert the decimal fraction (to the right of the decimal point) of the quotient into its binary significand, e.g. the binary significand of 0.25 is 0.01.

(3) Finally, convert the decimal exponent n into a binary integer, e.g. the binary equivalent of 5 is 101. Thus the binary floating-point representation of 40 is 1.01×2^{101}.

Using MATLAB

The numeric output generated by MATLAB depends on the display formation chosen, e.g. short or long. Hence the numeric display may include only a few digits to the right of the decimal point. You can change the format of the numeric display using the `format` command. Type `help format` for more information on the choices available. Regardless of the display format, the internal mathematical computations are always done in double precision, unless otherwise specified as, e.g., single precision.

1.3 Methods used to measure error

Before we can fully assess the significance of round-off errors produced by floating-point arithmetic, we need to familiarize ourselves with standard methods used to measure these errors. One method to measure the magnitude of error involves determining its absolute value. If m' is the approximation to m, the true quantity, then the **absolute error** is given by

$$\epsilon_{\mathrm{a}} = |m' - m|. \tag{1.2}$$

This error measurement uses the absolute difference between the two numbers to determine the precision of the approximation. However, it does not give us a feel for the **accuracy** of the approximation, since the absolute error does not compare the absolute difference with the magnitude of m. (For a discussion on accuracy vs. precision, see Box 1.2.) This is measured by the **relative error**,

$$\epsilon_{\mathrm{r}} = \frac{|m' - m|}{|m|}. \tag{1.3}$$

Example 1.4 Errors from repeated addition of a decimal fraction

The number 0.2 is equivalent to the binary significand $0.\overline{0011}$, where the overbar indicates the infinite repetition of the group of digits located underneath it. This fraction cannot be stored exactly by a computer when using floating-point representation of numbers. The relative error in the computer representation of 0.2 is practically insignificant compared to the true value. However, errors involved in binary floating-point approximations of decimal fractions are additive. If 0.2 is added to itself many times, the resulting error may become large enough to become significant. Here, we consider the **single** data type, which is a 32-bit representation format for floating-point numbers and has at most eight significant digits (as opposed to the 16 significant digits available for **double** data types (64-bit precision).

A MATLAB program is written to find the sum of 0.2 + 0.2 + 0.2 + ... 250 000 times and initially converts all variables to the single data type since MATLAB stores all numbers as **double** by default. The true final sum of the addition process is $0.2 \times 250\,000 = 50\,000$. We then check the computed sum with the exact solution. In the code below, the variables are converted to the single data type by using the `single` conversion function.

MATLAB program 1.1

```
% This program calculates the sum of a fraction added to itself n times

% Variable declaration and initialization
fraction = 0.2;      % fraction to be added
n = 250000;          % is the number of additions
summation = 0;       % sum of the input fraction added n times

% Converting double precision numbers to single precision
fraction = single (fraction);
summation = single(summation);

% Performing repeated additions
for l = 1:n  % l is the looping variable
    summation = summation + fraction;
end
summation
```

The result obtained after running the code is

```
sum =
     4.9879925e+004
```

The percentage relative error when using the single data type is 0.24%. The magnitude of this error is not insignificant considering that the digital floating-point representation of single numbers is accurate to the eighth decimal point. The percentage relative error when using double precision for representing floating-point numbers is 3.33×10^{-12} %.

> Double precision allows for much more accurate representation of floating-point numbers and should be preferred whenever possible.

In the real world, when performing experiments, procuring data, or solving numerical problems, you usually will not know the true or exact solution; otherwise you would not be performing these tasks in the first place! Hence, calculating the relative errors inherent in your numerical solution or laboratory results will not be as straightforward. For situations concerning numerical methods that are solved iteratively, the common method to assess the relative error in your solution is to compare the approximations obtained from two successive iterations. The relative error is then calculated as follows:

$$\epsilon_r = \frac{m_{i+1} - m_i}{m_{i+1}} \tag{1.4}$$

where i is the iteration count and m is the approximation of the solution that is being sought. With every successive iteration, the relative error is expected to decrease,

Box 1.4 Measurement of cells in a cell-counting device

Suppose that you are working in a clinical laboratory and you need to measure the number of stem cells present in two test tube solutions. From each test tube you make ten samples and run each of the ten samples obtained from the same test tube through a cell-counter device. The results are shown in Table 1.3.

Table 1.3.

	Test tube 1	Test tube 2
Mean count of ten samples	20 890	4 526 750
Maximum count	21 090	4 572 007
Minimum count	20 700	4 481 490
Difference (max. – min.)	390	90 517

At first glance, you might question the results obtained for test tube 2. You interpret the absolute difference in the minimum and maximum counts as the range of error possible, and that difference for test tube 2 is approximately 206 times the count difference calculated for test tube 1! You assume that the mean count is the true value of, or best approximation to, the actual stem cell count in the test tube solution, and you continue to do some number crunching.

You find that the maximum counts for both test tubes are not more than 1% larger than the values of the mean. For example, for the case of test tube 1: 1% of the mean count is $20\,890 \times 0.01 = 208.9$, and the maximum count (21 090) – mean count (20 890) = 200.

For test tube 2, you calculate

$$\frac{(\text{maximum count} - \text{mean count})}{\text{mean Count}} = \frac{4\,572\,007 - 4\,526\,750}{4\,526\,750} = 0.01.$$

The minimum counts for both test tubes are within 1% of their respective mean values. Thus, the accuracy of an individual cell count appears to be within ±1% of the true cell count. You also look up the equipment manual to find a mention of the same accuracy attainable with this device. Thus, despite the stark differences in the absolute deviations of cell count values from their mean, the accuracy to which the cell count is determined is the same for both cases.

thereby promising an incremental improvement in the accuracy of the calculated solution.

The importance of having different definitions for error is that you can choose which one to use for measuring and subsequently controlling or limiting the amount of error generated by your computer program or your calculations. As the magnitude of the measured value $|m|$ moves away from 1, the relative error will be increasingly different from the absolute error, and for this reason relative error measurements are often preferred for gauging the accuracy of the result.

1.4 Significant digits

The "significance" or "quality" of a numeric approximation is dependent on the number of digits that are significant. The **number of significant digits** provides a measure of how accurate the approximation is compared to its true value. For example, suppose that you use a cell counting device to calculate the number of bacterial cells in a suspended culture medium and arrive at the number 1 097 456.

However, if the actual cell count is 1 030 104, then the number of significant digits in the measured value is not seven. The measurement is not accurate due to various experimental and other errors. At least five of the seven digits in the number 1 097 456 are in error. Clearly, this measured value of bacterial count only gives you an idea of the order of magnitude of the quantity of cells in your sample. The number of significant figures in an approximation is not determined by the countable number of digits, but depends on the relative error in the approximation. The digit 0 is not counted as a significant figure in cases where it is used for the purpose of positioning the decimal point or when used for filling in one or more blank positions of unknown digits.

The method to determine the number of significant digits, i.e. the number of correct digits, is related to our earlier discussion on the method of "rounding" floating-point numbers. A number with s significant digits is said to be correct to s figures, and therefore, in *absolute terms*, has an error of within ± 0.5 units of the position of the sth figure in the number. In order to calculate the number of significant digits in a number m', first we determine the relative error between the approximation m' and its true value m. Then, we determine the value of s, where s is the largest non-negative integer that fulfils the following condition (proof given in Scarborough, 1966):

$$\frac{|m' - m|}{|m|} \leq 0.5 \times 10^{-s}, \tag{1.5}$$

where s is the number of significant digits, known with absolute certainty, with which m' approximates m.

Example 1.5 Number of significant digits in rounded numbers

You are given a number $m = 0.014\,682\,49$. Round this number to

(a) four decimal digit places after the decimal point,
(b) six decimal digits after the decimal point.

Determine the number of significant digits present in your rounded number for both cases.

The rounded number m' for (a) is 0.0147 and for (b) is 0.014 682. The number of significant digits in (a) is

$$\frac{|0.0147 - 0.01468249|}{|0.01468249|} = 0.0012 = 0.12 \times 10^{-2} < 0.5 \times 10^{-2}$$

The number of significant digits is therefore two. As you may notice, the preceding zero(s) to the left of the digits "147" are not considered significant since they do not establish any additional precision in the approximation. They only provide information on the order of magnitude of the number or the size of the exponent multiplier.

Suppose that we did not round the number in (a) and instead chopped off the remaining digits to arrive at the number 0.0146. The relative error is then $0.562 \times 10^{-2} < 0.5 \times 10^{-1}$, and the number of significant digits is one, one less than in the rounded case above. Thus, rounding improves the accuracy of the approximation.

For (b) we have

$$|0.014682 - 0.01468249|/|0.01468249| = 0.334 \times 10^{-4} < 0.5 \times 10^{-4}.$$

The number of significant digits in the second number is four.

When performing arithmetic operations on two or more quantities that have differing numbers of significant digits in their numeric representation, you will need to determine the acceptable number of significant digits that can be retained in your final result. Specific rules govern the number of significant digits that should be retained in the final computed result. The following discussion considers the cases of (i) addition and subtraction of several numbers and (ii) multiplication and division of two numbers.

Addition and subtraction

When adding or subtracting two or more numbers, all numbers being added or subtracted need to be adjusted so that they are multiplied by the same power of 10. Next, choose a number (operand) being added or subtracted whose last significant digit lies in the leftmost position as compared to the position of the last significant digit in all the other numbers (operands). The position of the last or rightmost significant digit of the final computed result corresponds to the position of the last or rightmost significant digit of the operand as chosen from the above step. For example:

(a) $10.343 + 4.56743 + 0.62 = 15.53043$. We can retain only two digits after the decimal point since the last number being added, 0.62 (shaded), has the last significant digit in the second position after the decimal point. Hence the final result is 15.53.
(b) $0.0000345 + 23.56 \times 10^{-4} + 8.12 \times 10^{-5} = 0.345 \times 10^{-4} + 23.56 \times 10^{-4} + 0.812 \times 10^{-4} = 24.717 \times 10^{-4}$. The shaded number dictates the number of digits that can be retained in the result. After rounding, the final result is 24.72×10^{-4}.
(c) $0.23 - 0.1235 = 0.1065$. The shaded number determines the number of significant digits in the final and correct computed value, which is 0.11, after rounding.

Multiplication and division

When multiplying or dividing two or more numbers, choose the operand that has the least number of significant digits. The number of significant digits in the computed result corresponds to the number of significant digits in the chosen operand from the above step. For example:

(a) $1.23 \times 0.3045 = 0.374\,535$. The shaded number has the smallest number of significant digits and therefore governs the number of significant digits present in the result. The final result based on the correct number of significant digits is 0.375, after rounding appropriately.
(b) $0.00\,234/1.2 = 0.001\,95$. The correct result is 0.0020. Note that if the denominator is written as 1.20, then this would imply that the denominator is precisely known to three significant figures.
(c) $301/0.045\,45 = 6622.662\,266$. The correct result is 6620.

1.5 Round-off errors generated by floating-point operations

So far we have learned that round-off errors are produced when real numbers are represented as binary floating-point numbers by computers, and that the finite available memory resources of digital machines limit the precision and range of the numeric representation. Round-off errors resulting from (1) lack of perfect precision in the floating-point representation of numbers and (2) arithmetic manipulation of floating-point numbers, are usually insignificant and fit well within the tolerance limits applicable to a majority of engineering calculations. However, due to precision limitations, certain arithmetic operations performed by a computer can produce large round-off errors that may either create problems during the evaluation of a mathematical expression, or generate erroneous results. Round-off errors are generated during a mathematical operation when there is a loss of significant digits, or the absolute error in the final result is large. For example, if two numbers of considerably disparate magnitudes are added, or two nearly equal numbers are subtracted from each other, there may be significant loss of information and loss of precision incurred in the process.

Example 1.6 Floating-point operations produce round-off errors

The following arithmetic calculations reveal the cause of spontaneous generation of round-off errors.
First we type the following command in the Command Window:

```
>> format long
```

This command instructs MATLAB to display all floating-point numbers output to the Command Window with 15 decimal digits after the decimal point. The command `format short` instructs MATLAB to display four decimal digits after the decimal point for every floating-point number and is the default output display in MATLAB.

(a) Inexact representation of a floating-point number

If the following is typed into the MATLAB Command Window

```
>> a = 32.8
```

MATLAB outputs the stored value of the variable `a` as

```
a =
  32.799999999999997
```

Note that the echoing of the stored value of a variable in the MATLAB Command Window can be suppressed by adding a semi-colon to the end of the statement.

(b) Inexact division of one floating-point number by another

Typing the following statements into the Command Window:

```
>> a = 33.0
>> b = 1.1
>> a/b
```

produces the following output:

```
a =
  33
b =
  1.100000000000000
ans =
  29.999999999999996
```

(c) Loss of information due to addition of two exceedingly disparate numbers

```
>> a = 1e10
>> b =1e-10
>> c = a+b
```

typed into the Command Window produces the following output:

```
a =
  1.000000000000000e+010
b =
  1.000000000000000e-010
c =
  1.000000000000000e+010
```

Evidently, there is no difference in the value of a and c despite the addition of b to a.

(d) Loss of a significant digit due to subtraction of one number from another

Typing the following series of statements

```
>> a = 10/3
>> b = a - 3.33
>> c = b*1000
```

produces this output:

```
a =
  3.333333333333334
b =
  0.003333333333333
c =
  3.333333333333410
```

If the value of c were exact, it would have the digit 3 in all decimal places. The appearance of a 0 in the last decimal position indicates the loss of a significant digit.

Example 1.7 Calculation of round-off errors in arithmetic operations

A particular computer has memory constraints that allow every floating-point number to be represented by a maximum of only four significant digits. In the following, the absolute and relative errors stemming from four-digit floating-point arithmetic are calculated.

(a) Arithmetic operations involving two numbers of disparate values

(i) $4000 + 3/2 = 4002$. Absolute error $= 0.5$; relative error $= 1.25 \times 10^{-4}$. The absolute error is $O(1)$, but the relative error is small.
(ii) $4000 - 3/2 = 3998$. Absolute error $= 0.5$; relative error $= 1.25 \times 10^{-4}$.

(b) Arithmetic operations involving two numbers close in value

(i) $5.0/7 + 4.999/7 = 1.428$. Absolute error $= 4.29 \times 10^{-4}$; relative error $= 3 \times 10^{-4}$.
(ii) $5.0/7 - 4.999/7 = 0.7143 - 0.7141 = 0.0002$. Absolute error $= 5.714 \times 10^{-5}$; relative error $= 0.4$ (or 40%)! There is a loss in the number of significant digits during the subtraction process and hence a considerable loss in the accuracy of the result. This round-off error manifests from **subtractive cancellation** of significant digits. This occurrence is termed as **loss of significance.**

Example 1.7 demonstrates that even a single arithmetic operation can result in either a large absolute error or a large relative error due to round-off. Addition of two numbers of largely disparate values or subtraction of two numbers close in value can cause arithmetic inaccuracy. If the result from subtraction of two nearly equal numbers is multiplied by a large number, great inaccuracies can result in subsequent computations. With knowledge of the types of floating-point operations prone to round-off errors, we can rewrite mathematical equations in a reliable and robust manner to prevent round-off errors from corrupting our computed results.

Example 1.8 Startling round-off errors found when solving a quadratic equation

Let's solve the quadratic equation $x^2 + 49.99x - 0.5 = 0$.

The analytical solution to the quadratic equation $ax^2 + bx + c = 0$ is given by

$$x_1 = \frac{-b + \sqrt{b^2 - 4ac}}{2a}; \tag{1.6}$$

$$x_2 = \frac{-b - \sqrt{b^2 - 4ac}}{2a}. \tag{1.7}$$

The exact answer is obtained by solving these equations (and also from algebraic manipulation of the quadratic equation), and is $x_1 = 0.01$; $x_2 = -50$.

Now we solve this equation using floating-point numbers containing no more than four significant digits. The frugal retainment of significant digits in these calculations is employed to emphasize the effects of round-off on the accuracy of the result.

Now, $\sqrt{b^2 - 4ac} = 50.01$ exactly. There is no approximation involved here. Thus,

$$x_1 = \frac{-49.99 + 50.01}{2} = 0.01;$$

$$x_2 = \frac{-49.99 - 50.01}{2} = -50.$$

We just obtained the exact solution.

Now let's see what happens if the equation is changed slightly. Let's solve $x^2 + 50x - 0.05 = 0$. We have $\sqrt{b^2 - 4ac} = 50.00$, using rounding of the last significant digit, yielding

$$x_1 = \frac{-50 + 50.00}{2} = 0.000;$$

$$x_2 = \frac{-50 - 50.00}{2} = -50.00.$$

The true solution of this quadratic equation to eight decimal digits is $x_1 = 0.000\,999\,98$ and $x_2 = -50.000\,999\,98$. The percentage relative error in determining x_1 is given by

$$\frac{0.000\,999\,98 - 0}{0.000\,999\,8} \times 100 = 100\%(!),$$

and the error in determining x_2 is

$$\frac{-(50.00099998 - 50.00)}{-50.00099998} \times 100 = 0.001999 \sim 0.002\%.$$

The relative error in the result for x_1 is obviously unacceptable, even though the absolute error is still small. Is there another way to solve quadratic equations such that we can confidently obtain a small relative error in the result? Look at the calculation for x_1 closely and you will notice that two numbers of exceedingly close values are being subtracted from each other. As discussed earlier in Section 1.5, this naturally has disastrous consequences on the result. If we prevent this subtraction step from occurring, we can obviate the problem of subtractive cancellation.

When b is a positive number, Equation (1.6) will always result in a subtraction of two numbers in the numerator. We can rewrite Equation (1.6) as

$$x_1 = \frac{-b+\sqrt{b^2-4ac}}{2a} \times \frac{b+\sqrt{b^2-4ac}}{b+\sqrt{b^2-4ac}}. \qquad (1.8)$$

Using the algebraic identity $(a+b)\cdot(a-b) = a^2 - b^2$ to simplify the numerator, we evaluate Equation (1.8) as

$$x_1 = \frac{-2c}{b+\sqrt{b^2-4ac}}. \qquad (1.9)$$

The denominator of Equation (1.9) involves the addition of two positive numbers. We have therefore successfully replaced the subtraction step in Equation (1.6) with an addition step. Using Equation (1.9) to obtain x_1, we obtain

$$x_1 = \frac{-2 \times -0.05}{50 + 50.00} = 0.001.$$

The percentage relative error for the value of x_1 is 0.002%.

Example 1.8 illustrates one method of algebraic manipulation of equations to circumvent round-off errors spawning from subtractive cancellation. Problem 1.5 at the end of this chapter uses another technique of algebraic manipulation for reducing round-off errors when working with polynomials.

Box 1.5 Problem encountered when calculating compound interest

You sold some stock of Biotektronics (see Box 1.3) and had a windfall. You purchased the stocks when they had been newly issued and you decided to part with them after the company enjoyed a huge success after promoting their artificial knee joint. Now you want to invest the proceeds of the sale in a safe place. There is a local bank that has an offer for ten-year certificates of deposit (CD). The scheme is as follows:

(a) Deposit x dollars greater than \$1000.00 for ten years and enjoy a return of 10% simple interest for the period of investment
OR
(b) Deposit x dollars greater than \$1000.00 for ten years and enjoy a return of 10% (over the period of investment of ten years) compounded annually, with a \$100.00 fee for compounding
OR
(c) Deposit x dollars greater than \$1000.00 for ten years and enjoy a return of 10% (over the period of investment of ten years) compounded semi-annually or a higher frequency with a \$500.00 fee for compounding.

You realize that to determine the best choice for your \$250 000 investment you will have to do some serious math. You will only do business with the bank once you have done your due diligence. The final value of the deposited amount at maturity is calculated as follows:

$$P\left(1+\frac{r}{100n}\right)^n,$$

where P is the initial principal or deposit made, r is the rate of interest per annum in percent, and n is based on the frequency of compounding over ten years.

Choice (c) looks interesting. Since you are familiar with the power of compounding, you know that a greater frequency in compounding leads to a higher yield. Now it is time to outsmart the bank. You decide that, since the frequency of compounding does not have a defined lower limit, you have the authority to make this choice. What options do you have? Daily? Every minute? Every second? Table 1.4 displays the compounding frequency for various compounding options.

A MATLAB program is written to calculate the amount that will accrue over the next ten years.

Table 1.4.

Compounding frequency	n
Semi-annually	20
Monthly	120
Daily	3650
Per hour	87 600
Per minute	5.256×10^6
Per second	3.1536×10^8
Per millisecond	3.1536×10^{11}
Per microsecond	3.1536×10^{14}

Table 1.5.

Compounding frequency	n	Maturity amount ($)
Semi-annually	20	276 223.89
Monthly	120	276 281.22
Daily	3650	276 292.35
Per hour	87 600	276 292.71
Per minute	5.256×10^6	276 292.73
Per second	3.1536×10^8	276 292.73
Per millisecond	3.1536×10^{11}	276 291.14
Per microsecond	3.1536×10^{14}	268 133.48

MATLAB program 1.2

```
% This program calculates the maturity value of principal P deposited in a
% CD for 10 years compounded n times

% Variables
p = 250000;     % Principal
r = 10;     % Interest rate
n = [20 120 3650 87600 5.256e6 3.1536e8 3.1536e11 3.1536e14 3.1536e17];
   % Frequency of compounding

% Calculation of final accrued amount at maturity
maturityamount = p*(1+r/100./n).^n;
fprintf('%7.2f\n',maturityamount)
```

In the MATLAB code above, the ./ and .^ operators (dot preceding the arithmetic operator) instruct MATLAB to perform element-wise operations on the vector n instead of matrix operations such as matrix division and matrix exponentiation.

The results of MATLAB program 1.2 are presented in Table 1.5. You go down the column to observe the effect of compounding frequency and notice that the maturity amount is the same for compounding every minute versus compounding every second.[1] Strangely, the maturity amount drops with even larger compounding frequencies. This cannot be correct.

This error occurs due to the large value of n and therefore the smallness of $r/100n$. For the last two scenarios, $r/100n = 3.171\times10^{-13}$ and 3.171×10^{-16}, respectively. Since floating-point numbers have at most 16 significant digits, adding two numbers that have a magnitude difference of 10^{16} results in the loss of significant digits or information as described in this section. To illustrate this point further, if the calculation were performed for a compounding frequency of every nanosecond, the maturity amount is calcuated by MATLAB to be $250 000 – the principal amount! This is the result of $1+(r/100n)=1$, as per floating-point addition. We have encountered a limit imposed by the finite precision in floating-point calculations.

How do round-off errors propagate in arithmetic computations? The error in the individual quantities is found to be additive for the arithmetic operations of addition, subtraction, and multiplication. For example, if a and b are two numbers to be added having errors of Δa and Δb, respectively, the sum will be $(a + b) + (\Delta a + \Delta b)$, with the error in the sum being $(\Delta a + \Delta b)$. The result of a calculation involving a number that has error ϵ, raised to the power of a, will have approximately an error of $a\epsilon$.

If a number that has a small error well within tolerance limits is fed into a series of arithmetic operations that produce small deviations in the final calculated value of the same order of magnitude as the initial error or less, the numerical algorithm or method is said to be **stable**. In such algorithms, the initial errors or variations in numerical values are kept in check during the calculations and the growth of error is either linear or diminished. However, if a slight change in the initial values produces a large change in the result, this indicates that the initial error is magnified through the course of successive computations. This can yield results that may not be of much use. Such algorithms or numerical systems are **unstable**. In Chapter 2, we discuss the source of ill-conditioning (instability) in systems of linear equations.

1.6 Taylor series and truncation error

You will be well aware that functions such as sin x, cos x, e^x, and log x can be represented as a series of infinite terms. For example,

$$e^x = 1 + \frac{x}{1!} + \frac{x^2}{2!} + \frac{x^3}{3!} + \cdots, \qquad -\infty < x < \infty. \tag{1.10}$$

The series shown above is known as the Taylor expansion of the function e^x. The **Taylor series** is an infinite series representation of a differentiable function $f(x)$. The series expansion is made about a point x_0 at which the value of f is known. The terms in the series are progressively higher-order derivatives of $f(x_0)$. The Taylor expansion in one variable is shown below:

$$f(x) = f(x_0) + f'(x_0)(x - x_0) + \frac{f''(x_0)(x - x_0)^2}{2!} + \cdots + \frac{f^n(x_0)(x - x_0)^n}{n!} + \cdots, \tag{1.11}$$

[1] The compound-interest problem spurred the discovery of the constant e. Jacob Bernoulli (1654–1705) noticed that compound interest approaches a limit as $n \to \infty$; $\lim_{n\to\infty}\left(1 + \frac{r}{n}\right)^n$, when expanded using the binomial theorem, produces the Taylor series expansion for e^r (see Section 1.6 for an explanation on the Taylor series). For starting principal P, continuous compounding (maximum frequency of compounding) at rate r per annum will yield $ Pe^r at the end of one year.

where x_0 and $x \in [a, b]$, $f'(x)$ is a first-order derivative of $f(x)$, $f''(x)$ is a second-order derivative of $f(x)$ and so on. The Taylor expansion for e^x can be derived by setting $f(x) = e^x$ and expanding this function about $x_0 = 0$ using the Taylor series. The Taylor series representation of functions is an extremely useful tool for approximating functions and thereby deriving numerical methods of solution. Because the series is infinite, only a finite number of initial terms can be retained for approximating the solution. The higher-order terms usually contribute negligibly to the final sum and can be justifiably discarded. Often, series approximations of functions require only the first few terms to generate the desired accuracy. In other words, the series is truncated, and the error in the approximation depends on the discarded terms and is called the **truncation error**. The Taylor series truncated to the nth order term can exactly represent an nth-order polynomial. However, an infinite Taylor series is required to converge exactly to a non-polynomial function.

Since the function value is known at x_0 and we are trying to obtain the function value at x, the difference, $x - x_0$, is called the **step size**, which we denote as the independent variable h. As you will see, the step size plays a key role in determining both the truncation error and the round-off error in the final solution. Let's rewrite the Taylor series expansion in terms of the powers of h:

$$f(x) = f(x_0) + f'(x_0)h + \frac{f''(x_0)h^2}{2!} + \frac{f'''(x_0)h^3}{3!} + \cdots + \frac{f^n(x_0)h^n}{n!} + \cdots. \quad (1.12)$$

As the step size h is gradually decreased when evaluating $f(x)$, the higher-order terms are observed to diminish much faster than the lower-order terms due to the dependency of the higher-order terms on larger powers of h. This is demonstrated in Examples 1.10 and 1.11. We can list two possible methods to reduce the truncation error:

(1) reduce the step size h, and/or
(2) retain as many terms as possible in the series approximation of the function.

While both these possibilities will reduce the truncation error, they can increase the round-off error. As h is decreased, the higher-order terms greatly diminish in value. This can lead to the addition operation of small numbers to large numbers or subtractive cancellation, especially in an alternating series in which some terms are added while others are subtracted such as in the following sine series:

$$\sin x = x - \frac{x^3}{3!} + \frac{x^5}{5!} - \frac{x^7}{7!} + \cdots. \quad (1.13)$$

One way to curtail loss in accuracy due to addition of large quantities to small quantities is to sum the small terms first, i.e. sum the terms in the series backwards.

Reducing the step size is usually synonymous with an increased number of computational steps: more steps must be "climbed" before the desired function value at x is obtained. Increased computations, either due to reduction in step size or increased number of terms in the series, will generally increase the round-off error.

> There is a trade-off between reducing the truncation error and limiting the round-off error, despite the fact that these two sources of error are independent from each other in origin.

We will explore the nature and implications of these two error types when evaluating functions numerically in the next few examples.

The Taylor series can be rewritten as a finite series with a remainder term R_n:

$$f(x_0 + h) = f(x_0) + f'(x_0)h + \frac{f''(x_0)h^2}{2!} + \frac{f'''(x_0)h^3}{3!} + \cdots + \frac{f^n(x_0)h^n}{n!} + R_n \quad (1.14)$$

and

$$R_n = \frac{f^{n+1}(\xi)h^{n+1}}{(n+1)!}, \qquad \xi\epsilon[x_0, x_0 + h]. \tag{1.15}$$

The remainder term R_n is generated when the $(n + 1)$th and higher-order terms are discarded from the series. The value of ξ is generally not known. Of course, if R_n could be determined exactly, it could be easily incorporated into the numerical algorithm and then there would unquestionably be no truncation error! When using truncated series representations of functions for solving numerical problems, it is beneficial to have, at the least, an order of magnitude estimate of the error involved, and the $(n + 1)$th term serves this purpose.

1.6.1 Order of magnitude estimation of truncation error

As mentioned earlier, the **order of magnitude** of a quantity provides a rough estimate of its size. A number p that is of order of magnitude 1, is written as $p \sim O(1)$. Knowledge of the order of magnitude of various quantities involved in a given problem highlights quantities that are important and those that are comparatively small enough to be neglected. The utility of an order of magnitude estimate of the truncation error inherent in a numerical algorithm is that it allows for comparison of the size of the error with respect to the solution.

If the function $|f^{n+1}(x)|$ in Equation (1.15) has a maximum value of K for $x_0 \leq x \leq x_0 + h$, then

$$R_n = \frac{|f^{n+1}(\xi)h^{n+1}|}{(n+1)!} \leq \frac{K|h|^{n+1}}{(n+1)!} \sim O\left(h^{n+1}\right). \tag{1.16}$$

The error term R_n is said to be *of order* h^{n+1} or $O\left(h^{n+1}\right)$. Although the multipliers K and $1/(n + 1)!$ influence the overall magnitude of the error, because they are constants they do not change as h is varied and hence are not as important as the "power of h" term when assessing the effects of step size or when comparing different algorithms. By expressing the error term as a power of h, the rate of decay of the error term with change in step size is conveyed. Truncation errors of higher orders of h decay much faster when the step size is halved compared with those of lower orders. For example, if the Taylor series approximation contains only the first two terms, then the error term, T_E, is $\sim O(h^2)$. If two different step sizes are used, h and $h/2$, then the magnitude of the truncation errors can be compared as follows:

$$\frac{T_{E,h/2}}{T_{E,h}} = \frac{(h/2)^2}{(h)^2} = \frac{1}{4}.$$

Note that when the step size h is halved the error is reduced to one-quarter of the original error.

Example 1.9 Estimation of round-off errors in the e^{-x} series

Using the Taylor series expansion for $f(x) = e^{-x}$ about $x_0 = 0$, we obtain

$$e^{-x} = 1 - x + \frac{x^2}{2!} - \frac{x^3}{3!} + \frac{x^4}{4!} - \frac{x^5}{5!} + \cdots. \tag{1.17}$$

For the purpose of a simplified estimation of the round-off error inherent in computing this series, assume that each term has associated with it an error ϵ of 10^{-8} times the magnitude of that term (single-precision arithmetic). Since round-off errors in addition and subtraction operations are additive, the total error due to summation of the series is given by

$$\text{error} = \epsilon \cdot 1 + \epsilon \cdot x + \epsilon \cdot \frac{x^2}{2!} + \epsilon \cdot \frac{x^3}{3!} + \epsilon \cdot \frac{x^4}{4!} + \epsilon \cdot \frac{x^5}{5!} + \cdots = \epsilon \cdot e^x.$$

The sum of the series inclusive of error is

$$e^{-x} = e^{-x} \pm \epsilon \cdot e^x = e^{-x}\left(1 \pm \epsilon \cdot e^{2x}\right).$$

If the magnitude of x is small, i.e. $0 < x < 1$, the error is minimal and therefore constrained. However, if $x \sim 9$, the error $\epsilon \cdot e^{2x}$ becomes $O(1)$! Note that, in this algorithm, no term has been truncated from the series, and, by retaining all terms, it is ensured that the series converges exactly to the desired limit for all x. However, the presence of round-off errors in the individual terms produces large errors in the final solution for all $x \geq 9$. How can we avoid large round-off errors when summing a large series? Let's rewrite Equation (1.17) as follows:

$$e^{-x} = \frac{1}{e^x} = \frac{1}{1 + \frac{x}{1!} + \frac{x^2}{2!} + \frac{x^3}{3!} + \cdots}.$$

The solution inclusive of round-off error is now

$$\frac{1}{e^x \pm \epsilon \cdot e^x} = \frac{e^{-x}}{1 \pm \epsilon}.$$

The error is now of $O(\varepsilon)$ for all $x > 0$.

Example 1.10 Exploring truncation errors using the Taylor series representation for e^{-x}

The Taylor series expansion for $f(x) = e^{-x}$ is given in Equation (1.17). A MATLAB program is written to evaluate the sum of this series for a user-specified value of x using n terms. The danger of repeated multiplications either in the numerator or denominator, as demonstrated in Box 1.3, can lead to fatal overflow errors. This problematic situation can be avoided if the previous term is used as a starting point to compute the next term. The MATLAB program codes for a function called `taylorenegativex` that requires input of two variables x and n. A MATLAB function that requires input variables must be run from the Command Window, by typing in the name of the function along with the values of the input parameters within parentheses after the function name. This MATLAB function is used to approximate the function e^{-x} for two values of x: 0.5 and 5. A section of the code in this program serves the purpose of detailing the appearance of the graphical output. Plotting functions are covered in Appendix A.

MATLAB program 1.3

```
function taylorenegativex(x, n)
% This function calculates the Taylor series approximation for e^-x for 2
% to n terms and determines the improvement in accuracy with the addition
% of each term.

% Input Variables
% x is the function variable
% n is the maximum number of terms in the Taylor series

% Only the positive value of x is desired
```

```
if (x < 0)
    x = abs(x);
end
% Additional Variables
term = 1; % first term in the series
summation(1) = term; % summation term
enegativex = exp(-x); % true value of e^-x
err(1) = abs(enegativex - summation(1));
fprintf(' n Series sum Absolute error Percent relative error\n')
for i = 2:n
     term = term*(-x)/(i-1);
     summation = summation + term;
     (% Absolute error after addition of each term)
     err(i) = abs(summation - enegativex);
     fprintf('%2d %14.10f %14.10f %18.10f\n',i, sum,err(i),err(i)/...
       enegativex*100)
end
plot([1:n], err, 'k-x','LineWidth',2)
xlabel('Number of terms in e^-^x series','FontSize',16)
ylabel('Absolute error','FontSize',16)
title(['Truncation error changes with number of terms, x =',num2str
   (x)],...'FontSize',14)
set(gca,'FontSize',16,'LineWidth',2)
```

We type the following into the Command Window:

```
>> taylorenegativex(0.5, 12)
```

MATLAB outputs:

n	Series sum	Absolute error	Percent relative error
2	2.0000000000	0.1065306597	17.5639364650
3	1.6000000000	0.0184693403	3.0450794188
4	1.6551724138	0.0023639930	0.3897565619
5	1.6480686695	0.0002401736	0.0395979357
6	1.6487762988	0.0000202430	0.0033375140
7	1.6487173065	0.0000014583	0.0002404401
8	1.6487215201	0.0000000918	0.0000151281
9	1.6487212568	0.0000000051	0.0000008450
10	1.6487212714	0.0000000003	0.0000000424
11	1.6487212707	0.0000000000	0.0000000019
12	1.6487212707	0.0000000000	0.0000000001

and draws Figure 1.4.

The results show that by keeping only three terms in the series, the error is reduced to ~3%. By including 12 terms in the series, the absolute error becomes less than 10^{-10} and the relative error is $O(10^{-12})$. Round-off error does not play a role until we begin to consider errors on the order of 10^{-15}, and this magnitude is of course too small to be of any importance when considering the absolute value of $e^{-0.5}$.

Repeating this exercise for $x = 5$, we obtain the following tabulated results and Figure 1.5.

Figure 1.4

Change in error with increase in number of terms in the negative exponential series for $x = 0.5$.

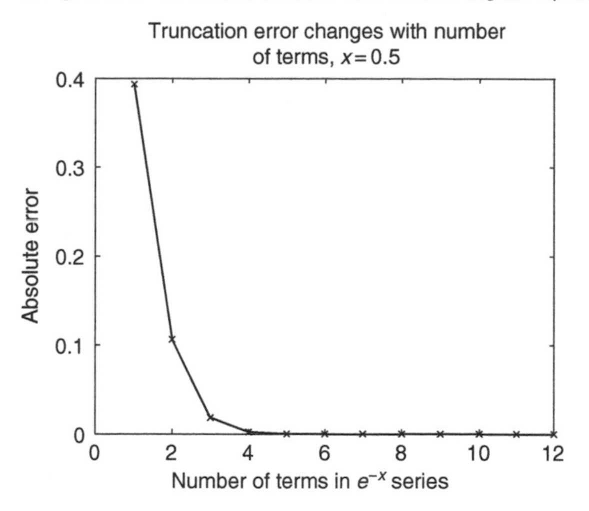

Figure 1.5

Change in error with increase in number of terms in the negative exponential series for $x = 5$.

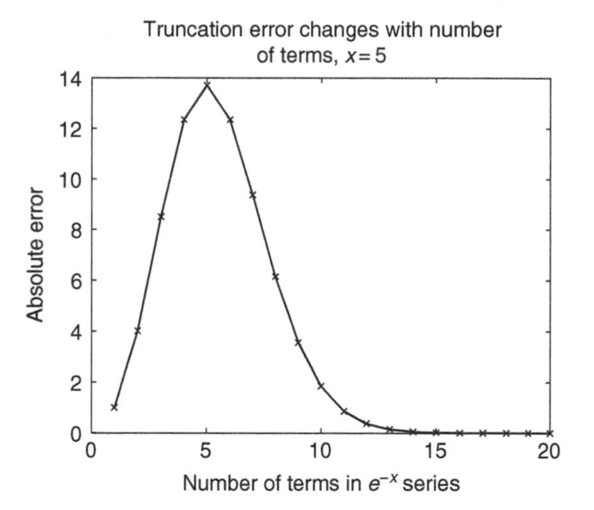


```
>> taylorenegativex(5,20)
 n    Series sum          Absolute error    Percent relative error
 2     -4.0000000000       4.0067379470      59465.2636410306
 3      8.5000000000       8.4932620530     126051.1852371901
 4    -12.3333333333      12.3400712803     183142.8962265111
 5     13.7083333333      13.7015953863     203349.7056031154
 6    -12.3333333333      12.3400712803     183142.8962265111
 7      9.3680555556       9.3613176086     138934.2719648443
 8     -6.1329365079       6.1396744549      91120.8481718382
 9      3.5551835317       3.5484455847      52663.6019135884
10     -1.8271053792       1.8338433262      27216.6481338708
11      0.8640390763       0.8573011293      12723.4768898588
12     -0.3592084035       0.3659463505       5431.1253936547
```

```
13     0.1504780464    0.1437400994    2133.2922244759
14    -0.0455552035    0.0522931505     776.0991671128
15     0.0244566714    0.0177187244     262.9691870261
16     0.0011193798    0.0056185672      83.3869310202
17     0.0084122834    0.0016743364      24.8493558692
18     0.0062673118    0.0004706352       6.9848461571
19     0.0068631372    0.0001251902       1.8579877391
20     0.0067063411    0.0000316059       0.4690738125
```

For $|x| > 1$, the Taylor series is less efficient in arriving at an approximate solution. In fact, the approximation is worse when including the second through the tenth term in the series than it is when only including the first term (zeroth-order approximation, i.e. the slope of the function is zero). The large errors that spawn from the inclusion of additional terms in the series sum are due to the dramatic increase in the numerator of these terms as compared to the growth of the denominator; this throws the series sum off track until there are a sufficient number of large terms in the sum to cancel each other. The relative error is $<5\%$ only after the inclusion of 19 terms! This example shows that a series such as this one which has alternating positive and negative terms takes longer to converge than a strictly positive term series. This brings us to the next topic: what is meant by the convergence of a series?

1.6.2 Convergence of a series

An **infinite series** S is defined as the sum of the terms in an infinite sequence $t_1, t_2, \ldots, t_n \ldots, t_\infty$. The sum of the first n terms of an infinite sequence is denoted by S_n. If S tends to a finite limit as n, the number of terms in the series, tends to ∞, then S is said to be a **convergent series**. A necessary (but not sufficient) condition for any convergent series is that $\lim_{n\to\infty} t_n = 0$, or that

$$\lim_{n\to\infty} S_n - S_{n-1} = 0.$$

There are a number of convergence tests that are formally used to determine whether a series is convergent or divergent. When the series limit is not known, a simple convergence condition used to determine if a convergent series has converged to within tolerable error is

$$|S_n - S_{n-1}| = |t_n| \le \delta,$$

where δ is a positive constant and specifies some tolerance limit.

Example 1.10 shows that the Taylor series expansion converges rapidly for $|x| < 1$ since the higher-order terms rapidly diminish. However, difficulties arise when we use the e^{-x} expansion for $|x| > 1$. Is there a way to rectify this problem?

Example 1.11 Alternate method to reduce truncation errors

As discussed earlier, it is best to minimize round-off errors by avoiding subtraction operations. This technique can also be applied to Example 1.10 to resolve the difficulty in the convergence rate of the series encountered for $|x| > 1$. Note that e^{-x} can be rewritten as

$$e^{-x} = \frac{1}{e^x} = \frac{1}{1 + \frac{x}{1!} + \frac{x^2}{2!} + \frac{x^3}{3!} + \cdots}.$$

After making the appropriate changes to MATLAB program 1.3, we rerun the function `taylorenegativex` for $x = 5$ in the Command Window, and obtain

```
>> taylorenegativex(5,20)
 n   Series sum     Absolute error   Percent relative error
 2   0.1666666667   0.1599287197     2373.5526517096
 3   0.0540540541   0.0473161071      702.2332924464
 4   0.0254237288   0.0186857818      277.3215909388
 5   0.0152963671   0.0085584201      127.0182166005
 6   0.0109389243   0.0042009773       62.3480318351
 7   0.0088403217   0.0021023747       31.2020069419
 8   0.0077748982   0.0010369512       15.3897201464
 9   0.0072302833   0.0004923363        7.3069180831
10   0.0069594529   0.0002215059        3.2874385120
11   0.0068315063   0.0000935593        1.3885433338
12   0.0067748911   0.0000369441        0.5482991168
13   0.0067515774   0.0000136304        0.2022935634
14   0.0067426533   0.0000047063        0.0698477509
15   0.0067394718   0.0000015248        0.0226304878
16   0.0067384120   0.0000004650        0.0069013004
17   0.0067380809   0.0000001339        0.0019869438
18   0.0067379835   0.0000000365        0.0005416368
19   0.0067379564   0.0000000094        0.0001401700
20   0.0067379493   0.0000000023        0.0000345214
```

Along with the table of results, we obtain Figure 1.6.

Observe the dramatic decrease in the absolute error as the number of terms is increased. The series converges much faster and accuracy improves with the addition of each term. Only ten terms are required in the series to reach an accuracy of $< 5\%$. For most applications, one may need to attain accuracies much less than 5%.

1.6.3 Finite difference formulas for numerical differentiation

In this section, the Taylor series is used to obtain the truncation errors associated with finite difference formulas used to approximate first-order

Figure 1.6

Rapidly decreasing error with increase in number of terms in the negative exponential series for $x = 5$ when using a more efficient algorithm.

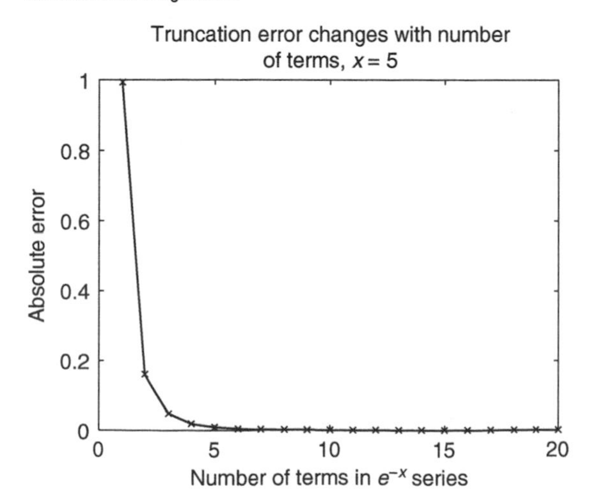

Box 1.1B Oxygen transport in skeletal muscle

Earlier we determined the concentration profile of O_2 in the muscle tissue as

$$N = N_0 - \frac{\Gamma_2 x}{2D}(L-x) + \frac{4(\Gamma_2 - \Gamma_1)L^2}{D} \sum_{n=1,\ n \text{ is odd}}^{n=\infty} \left[\frac{1}{(n\pi)^3} e^{-(n\pi)^2 Dt/L^2} \sin\left(\frac{n\pi x}{L}\right) \right]. \tag{1.18}$$

The third term on the right-hand side of Equation 1.18 contains an infinite series called the Fourier sine series. This particular sine series is a summation of only the terms with odd n. The terms corresponding to even numbers of n are all identically zero. Therefore the first term in the Fourier series expansion corresponds to $n = 1$, the second term corresponds to $n = 3$, and so on.

The diffusivity of O_2 in skeletal muscle is 1.5×10^{-5} cm^2/s (Fletcher, 1980). The partial pressure of O_2 in arterial blood is 95 mm Hg and in venous blood is 40 mm Hg. We assume an average O_2 partial pressure of 66 mm Hg in the capillaries surrounding the muscle tissue. The solubility of O_2 in skeletal muscle is 1.323 μM/mm Hg (Fletcher, 1980), from which we obtain the O_2 concentration N_0 at $x = 0, L$ as 87 μM. The resting volumetric O_2 consumption in tissue Γ_1 is 50 μM/s, while, after heavy exercise, volumetric O_2 consumption in tissue Γ_2 is 260 μM/s (Liguzinski and Korzeniewski, 2006). The distance between two capillaries is $L = 60$ μm.

The oxygen concentration will be least in the region of the tissue that is farthest from the O_2 supply zone, i.e. at $x = L/2$. We are interested in determining the change in O_2 concentration at $x = L/2$ with time. The time taken for the first exponential term in the series $e^{-(n\pi)^2 Dt/L^2}$ to decay to e^{-1} is called the **diffusion time**, i.e. $t_D = L^2/\pi^2 D$. It is the time within which appreciable diffusion has taken place. Note that when the first exponential term in the series has decayed to e^{-1}, the second exponential term has decayed to e^{-9}. After the diffusion time has passed, only the first term is important. In this example, the diffusion time $t_D = 0.243$ s.

Plotted in Figure 1.7 is the initial time-dependent O_2 concentration profile at $x = L/2$, after a sudden jump in O_2 demand in the muscle tissue from 50 μM/s to 260 μM/s. In Figure 1.7(a), N at $x = L/2$ remains greater than zero up to a period $t = 1$ s. At the new steady state corresponding to conditions of heavy exerise, $N_s = 9$ μM at $x = L/2$. We can say that the entire tissue remains well oxygenated even at times of strenuous exercise. Note that, for all t, the series converges quickly. Figure 1.7(b) is a close-up

Figure 1.7

Time-dependent oxygen concentration profile at the center of a slab of muscle tissue observed when the muscle tissue enters from rest into a state of heavy exercise for (a) initial time elapsed of 0 to 1 s and (b) initial time elapsed of 0 to 0.1 s.

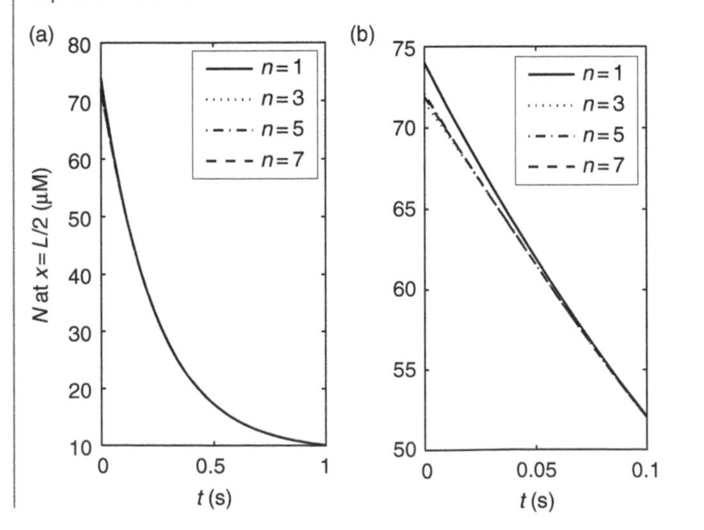

Figure 1.8

Estimated oxygen concentration N obtained by retaining n terms in the Fourier series. Estimates of N are plotted for odd n, since the terms corresponding to even n are equal to zero.

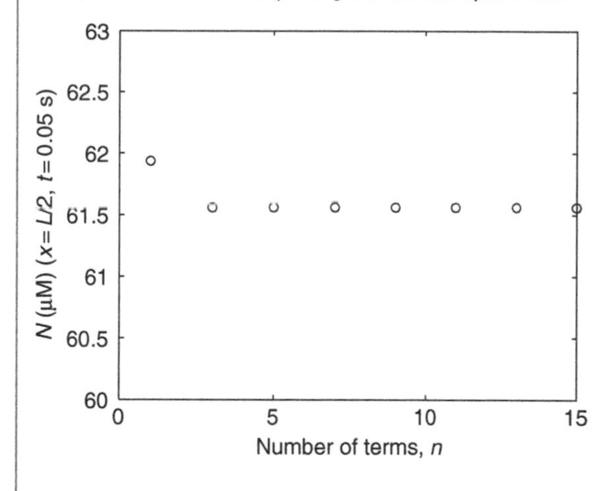

look at the O_2 concentration at $x = L/2$ for the first 0.1 s after the sudden increase in O_2 demand, and highlights the truncation errors involved as more terms are progressively retained in the Fourier series.

At $t = 0.01$ s, when only the first two terms are retained in the Fourier series, $N(n = [1, 3]) = 69.7363$. When the first three terms in the Fourier series are included then $N(n = [1, 3, 5]) = 69.9224$, which exhibits a relative difference of only 0.27%. Mathematically, we write

$$\frac{|N_{n=5} - N_{n=3}|}{|N_{n=3}|} \leq 0.0027.$$

For all practical purposes (see Figure 1.8), only the first two terms must be retained in the series for $t \geq 0.01$ s to obtain a reliable estimate. The efficient convergence is due to the $1/(n\pi)^3$ term in the series that forces the rapid decay of higher-order terms. When $t > t_D$, all terms in the Fourier series summation except for the first term can be neglected. The MATLAB code used to generate Figures 1.7 and 1.8 is available on the Cambridge University Press website.

derivatives. The first-order derivative of a function $f(x)$ that is differentiable at $x = x_0$ is defined as

$$f'(x) = \lim_{x \to x_0} \frac{f(x) - f(x_0)}{x - x_0}.$$

If x is very close to x_0, then the derivative of f with respect to x at x_0 can be approximated as

$$f'(x_0) \cong \frac{f(x_0 + h) - f(x_0)}{h}, \tag{1.19}$$

where $h = x - x_0$ is small. Equation (1.19) can also be derived from the Taylor series, with the advantage that we also get an estimate of the truncation error involved. Rearranging the Taylor series formula given in Equation (1.12) we get

$$f'(x_0) = \frac{f(x_0 + h) - f(x_0)}{h} - \frac{f''(x_0)h}{2!} - \frac{f'''(x_0)h^2}{3!} + \cdots$$

or

$$f'(x_0) = \frac{f(x_0 + h) - f(x_0)}{h} - \frac{f''(\xi)h}{2!}, \qquad \xi \in [x_0, x_0 + h]$$

or

$$f'(x_0) = \frac{f(x_0 + h) - f(x_0)}{h} + O(h), \tag{1.20}$$

with the truncation error of order h. The approximation given by Equation (1.20) for a first-order derivative can be rewritten as

$$f'(x_i) \cong \frac{f(x_{i+1}) - f(x_i)}{h}, \tag{1.21}$$

where $h = x_{i+1} - x_i$ is called the first **forward finite difference**, since each iteration proceeds forward by using the value of $f(x)$ from the current step i.

There are two other finite difference formulas that approximate the first-order derivative of a function: the backward finite difference and the central finite difference methods. The **backward finite difference** formula is derived by using the Taylor series to determine the function value at a previous step, as shown below:

$$f(x_0 - h) = f(x_0) - f'(x_0)h + \frac{f''(x_0)h^2}{2!} - \frac{f'''(x_0)h^3}{3!} + \cdots \tag{1.22}$$

or

$$f(x_{i-1}) = f(x_i) - f'(x_i)h + \frac{f''(x_i)h^2}{2!} + \cdots$$

or

$$f'(x_i) = \frac{f(x_i) - f(x_{i-1})}{h} + \frac{f''(x_i)h^2}{2!h} + \cdots,$$

from which we arrive at the formula for backward finite difference for a first-order derivative with an error of $O(h)$:

$$f'(x_i) \cong \frac{f(x_i) - f(x_{i-1})}{h}. \tag{1.23}$$

To obtain the **central finite difference** equation for a first-order derivative, Equation (1.22) is subtracted from Equation (1.12), and upon rearranging the terms it can be easily shown that

$$f'(x_0) = \frac{f(x_0 + h) - f(x_0 - h)}{2h} + O(h^2)$$

or

$$f'(x_i) \cong \frac{f(x_{i+1}) - f(x_{i-1})}{2h}. \tag{1.24}$$

Note that the central forward difference formula has a smaller truncation error than the forward and backward finite differences, and is therefore a better approximation method for a derivative. Box 1.6 addresses these three finite difference approximation methods and compares them.

Box 1.6 Monod growth kinetics: using finite difference approximations

Monod, in 1942, devised a mathematical relationship to relate the specific growth rate μ of a micro-organism to the concentration of a limiting nutrient in the growth medium. If μ_{max} is the maximum growth rate achievable when the substrate (nutrient) concentration s is large enough such that it is no longer limiting, and K_s is the substrate concentration when the specific growth rate is exactly one-half the maximum value, then the Monod equation[2] states that

$$\mu = \frac{\mu_{max} s}{K_s + s}.$$

We want to determine the rate of change in specific growth rate with substrate concentration s. Taking the first derivative of the Monod equation with respect to s, we obtain

$$\mu' = \frac{\mu_{max} K_s}{(K_s + s)^2}.$$

The unit of specific growth rate μ is per hour. The product of μ with the cell mass per unit volume (g/l) gives the rate of growth of cell mass (g/(l hr)). For $\mu_{max} = 1.0$ hr and $K_s = 0.5$ g/l for a fungal strain in a particular culture medium, determine μ' at $s = 1.0$ g/l.

The exact value of $\mu'(s = 1.0\ \text{g/l}) = 0.2222\ \text{l/(g hr)}$.

We use the finite difference approximations to determine the first-order derivative. Let the step size be 0.1 g/l. Then,

$$\mu\ (s = 1.0\ \text{g/l}) = 0.6666/\text{hr};$$

$$\mu\ (s = 1.1\ \text{g/l}) = 0.6875/\text{hr};$$

$$\mu\ (s = 0.9\ \text{g/l}) = 0.6428/\text{hr}.$$

Using a forward difference approximation to determine μ', we obtain

$$\mu' \cong \frac{0.6875 - 0.6666}{0.1} = 0.209\ \text{l/(g hr)};\ \text{relative error } \epsilon_r = 5.94\%.$$

Using a backward difference approximation to determine μ', we obtain

$$\mu' \cong \frac{0.6666 - 0.6428}{0.1} = 0.238\ \text{l/(g hr)};\ \text{relative error } \epsilon_r = 7.11\%.$$

Using a central difference approximation to determine μ', we obtain

$$\mu' \cong \frac{0.6875 - 0.6428}{2 \times 0.1} = 0.2235\ \text{l/(g hr)};\ \text{relative error } \epsilon_r = 0.585\%.$$

Note the smaller error in the approximation of μ' when using the central difference formula.
Observe what happens if we reduce the step size by half:

$$\mu\ (s = 1.05\ \text{g/l}) = 0.6774/\text{hr};$$

$$\mu\ (s = 0.95\ \text{g/l}) = 0.6552/\text{hr}.$$

Now recalculate the forward, backward, and central differences. Using a forward difference approximation we obtain

$$\mu' \cong \frac{0.6774 - 0.6666}{0.05} = 0.216\ \text{l/(g hr)};\ \text{relative error } \epsilon_r = 2.79\%.$$

[2] This model is mathematically identical to the well-known Michaelis–Menten kinetics for enzyme reactions.

Using a backward difference approximation we obtain

$$\mu' \cong \frac{0.6666 - 0.6552}{0.05} = 0.228 \ \mathrm{l/(g\ hr)}; \ \text{relative error } \epsilon_\mathrm{r} = 2.61\%.$$

Using a central difference approximation we obtain

$$\mu' \cong \frac{0.6774 - 0.6552}{2 \times 0.05} = 0.222 \ \mathrm{l/(g\ hr)}; \ \text{relative error } \epsilon_\mathrm{r} = 0.09\%.$$

By halving the step size, the truncation errors for the forward and backward difference approximations are reduced by approximately one-half. This is expected since the truncation error is of the same order as the step size. On the other hand, halving the step size resulted in an approximately four-fold decrease in the truncation error associated with the central difference approximation. The reason for the pronounced decrease in the truncation error should be clear to you.

Box 1.7 Monod growth kinetics: truncation error versus round-off error

We again use the forward finite difference approximation (Equation (1.21)) to estimate μ', the rate of change in specific growth rate at a substrate concentration $s = 1.0$ g/l, where

$$\mu' = \frac{\mu_{\mathrm{max}} K_\mathrm{s}}{(K_\mathrm{s} + s)^2}.$$

We progressively decrease the step size h by an order of magnitude and accordingly determine the relative error in the prediction of the first-order derivative. In other words, h assumes values $0.1, 0.01, 0.001, \ldots, 10^{-15}$.

Figure 1.9 shows that as h is decreased from 0.1 to 10^{-8}, the relative error is dominated by the truncation error and decreases with a slope of 1. (We know that the truncation error is dominant for this range of step-size values because the error reduces in a regular way with decrease in step size.) The relative error in the finite forward difference approximation is found to be linearly proportional to the step size. For $h < 10^{-8}$, the round-off error dominates the overall error in the estimated value of μ'. The large round-off error stems from the loss of significant digits (loss of numeric information) in the subtraction step. Note that the error at $h = 10^{-15}$ is even greater than the error at $h = 0.1$. Thus, there is an optimum step size at which we attain the least error in our estimation. On the right or left side of this optimum value, the truncation error and round-off error, respectively, constrain the attainable accuracy. The MATLAB code used to generate Figure 1.9 is available on the Cambridge University Press website.

Figure 1.9

An optimum step size h exists that yields minimum relative error in the estimation of a first-order derivative using a forward finite difference approximation.

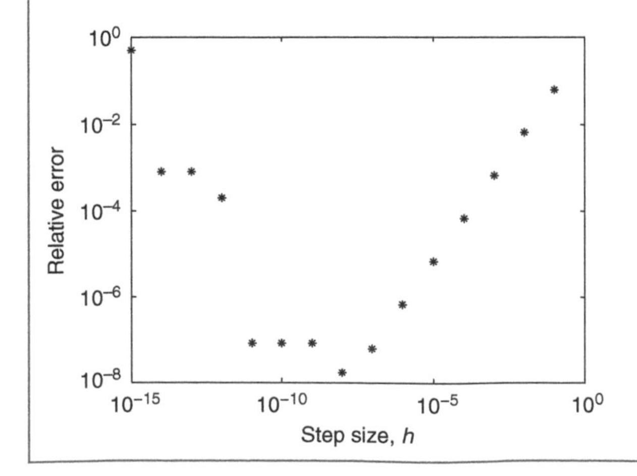

1.7 Criteria for convergence

In Sections 1.1–1.6 we discussed the types of errors generated as by-products of numerical computations and the different ways to assess the magnitude of the error. In this section, we introduce different criteria for determining whether the error in the solution has been reduced to a tolerable limit, or, in other words, whether a numerical algorithm has converged upon a solution. There are several convergence criteria that may be used, and you have the option of employing one or multiple criteria as you deem appropriate.

One condition that specifies whether convergence has been realized involves computing the change in the magnitude of the solution with every time step. The convergence condition may be based on the absolute difference of two successive iterative values of the numerical solution of x or their relative difference:

(a) $|x_{i+1} - x_i| \leq \epsilon$

or

(b) $\left|\dfrac{x_{i+1} - x_i}{x_{i+1}}\right| \leq \epsilon,$

where ϵ is the acceptable tolerance. In condition (a) ϵ is called **absolute tolerance**, while in condition (b) it is called **relative tolerance**. The stringency of condition (a) in comparison to condition (b) depends on the magnitude of x_{i+1}, if ϵ is the same in both cases. The first condition will allow termination of the iteration procedure only after $x_i - \epsilon \leq x_{i+1} \leq x_i + \epsilon$. In this case, the imposed tolerance governs the precision of the solution by specifying the rightmost digit of the number that must be correct or significant. Condition (b) will direct the algorithm to exit the iterative procedure when $x_i/(1+\epsilon) \leq x_{i+1} \leq x_i/(1-\epsilon)$. The imposed tolerance governs the total number of significant figures in the solution. If x_{i+1} is large, i.e. much greater than unity, then condition (a) is more stringent. If x_{i+1} is small, i.e. $\ll 1$, then condition (b) is a stricter convergence criterion than condition (a).

Another method to determine if convergence has been reached involves periodically monitoring the value of a function of the iteration variable(s). If a zero of the function is being sought, then the approach of the function towards zero can be the test for termination of the iterative method, such as

$$|f(x_i)| < \epsilon,$$

where ϵ is the tolerance limit close to zero.

Finally, other checks can be used to ensure that the algorithm is functioning properly and whether, in certain situations, convergence is not reached in an acceptable time limit, in which case the algorithm can be terminated. Iteration checks that are inserted into computational procedures to prevent an endless iterative loop include capping the total number of iterations allowed and placing bounds on the magnitude of the iteration variables or the function values being calculated. In the next few chapters, such convergence criteria will be suitably incorporated into the numerical algorithms introduced. It will become clear from the specific examples in subsequent chapters how the appropriate condition(s) may be chosen for terminating iterative procedures.

1.8 End of Chapter 1: key points to consider

(1) Numerical methods of solution usually possess two inherent sources of numerical error: **round-off errors** and **truncation errors**.

(2) Computers store numbers using binary floating-point representation and have finite memory constraints. As a result:
  (a) the **range** of numbers that a computer can represent is finite;
  (b) the **precision** by which real numbers can be represented by a computer is limited – many real numbers cannot be stored with exact precision;
  (c) the binary numbering system is incapable of exactly representing decimal fractions such as 0.1, 0.2, or 0.7 with a finite number of binary digits.

  This leads to discontinuity of the floating-point number line and the generation of round-off errors.

(3) The magnitude of round-off error can be calculated by determining the **absolute error** and **relative error** of the calculated result with respect to the true value, if known.

(4) In most cases, round-off error generated when using double-precision format is minimal and can be neglected. However, the magnitude of round-off error can become significant when performing arithmetic operations on floating-point numbers such as
  (a) addition of two numbers of disparate magnitudes;
  (b) subtraction of two numbers close in value that results in loss of significance;
  (c) multiplication by a large number (>1) and exponentiation, which can sometimes magnify a round-off error.

  Care should be taken to formulate numerical algorithms such that disruptive or fatal numerical errors, such as loss of significance due to subtractive cancellation, increase in round-off error from addition of disparate numbers, or generation of floating-point exceptions due to overflow error, are avoided.

(5) The number of **significant digits** present in a number determines the accuracy of the numeric quantity in question.

(6) Many numerical methods employ the Taylor series expansion for approximating a function $f(x)$. The inclusion of only a finite number of terms in the series leads to the origin of **truncation error**. The order of magnitude of the truncation error can be easily estimated from the Taylor series.

(7) While round-off error and truncation error arise independently of each other, there exists a trade-off between reducing truncation error and keeping the round-off error small. When using large step sizes or when retaining only a few terms of the infinite series to approximate the function of interest, the truncation error dominates. On the other hand, when the step size is very small or when many terms of the infinite series that represents the function are retained, round-off error can dominate.

1.9 Problems

1.1. Write a MATLAB program to convert a decimal fraction into its equivalent binary fraction. Run the program to determine the binary fraction representation for the decimal fractions 0.7 and 0.2. Verify the correctness of your algorithm from the answers in the chapter.

1.2. (i) Determine the 4-bit significand (binary fraction) for the following decimal fractions:
(a) 1/9,
(b) 2/3.
You can use the program you developed in Problem 1.1.
(ii) Add the 4-bit binary significands determined in (i) to find the binary equivalent for $(1/9) + (2/3)$. When performing binary addition, note these simple rules: $0 + 1 = 1$ and $1 + 1 = 10$. As an example, the addition of 1.01 and 0.11 would proceed as follows:

```
  1.01
+ 0.11
------
 10.00
```

(iii) Convert the 4-bit binary significand determined in (ii) to its decimal equivalent.
(iv) Determine the relative error in the binary addition of two decimal fractions, i.e. compare your answer in (iii) to the exact solution of the addition of the two decimal fractions, which is 7/9.

1.3. Write the binary floating-point representations ($1.b_1\, b_2\, b_3\, b_4\, b_5\, b_6 \ldots b_{p\text{-}1}\, b_p \times 2^e$) of the following numbers (you can use the program you developed in Problem 1.1 to extract the binary significand):
(a) 100,
(b) 80.042,
(c) 0.015 625. (Hint: Use a negative value of the power to which 2 is raised so that a "1" appears to the left of the decimal point.)

1.4. Evaluate the function $f(x) = \sqrt{x^2+1} - \sqrt{x^2}$ for $x = 300$ using only six-digit arithmetic. Calculate the relative error in your result. Find another way to evaluate $f(x)$ again using only six-digit arithmetic in order to obtain a better approximation of the value of the function at $x = 300$. (Hint: Algebraic rearrangement of $f(x)$.)

1.5. The equation $f(x) = x^3 + 5x^2 - 2x + 1 = 0$ can be rearranged to obtain an alternate nested form $g(x) = ((x+5)x - 2)x + 1 = 0$, both expressions being equivalent. Using four-digit rounding arithmetic, determine the value of $f(x)$ and $g(x)$ for $x = 1.17$ and check your results against the exact result of 7.106 113. Which expression produces a more accurate result, and can you explain why?

1.6. Using three-digit arithmetic, add the following numbers from left to right:

$$1.23 + 0.004 + 0.0036 + 5.6.$$

With every addition step, the intermediate result can only have three digits after rounding.

Check the precision of your final result by comparing with the exact answer. What is the absolute error?

Now rearrange the numbers so that the smallest numbers are summed first and this is followed by adding the larger numbers. Again, use three-digit arithmetic. Check the precision of your final result by comparing with the exact answer. What is the absolute error now?

Determine the number of significant digits in both calculated results. How has changing the order in which the numbers are added affected the precision of your result?

1.7. If b in the quadratic equation $ax^2 + bx + c = 0$ is negative, explain why the choice of formulas to obtain the solutions x_1 and x_2 should be chosen as

$$x_1 = \frac{-b + \sqrt{b^2 - 4ac}}{2a} \tag{1.6}$$

and

$$x_2 = \frac{-2c}{b - \sqrt{b^2 - 4ac}} \tag{P1.1}$$

in order to minimize round-off error. Show how the expression for x_2 is obtained.

Choose appropriately two out of the four Equations (1.6), (1.7), (1.9) and (P1.1) and solve the equations below. Justify your choice of equations each time.

(1) $x^2 - 75.01x + 1 = 0$,
(2) $x^2 + 100.01 + 0.1 = 0$.

1.8. Write an algorithm that accurately computes the roots of a quadratic equation even in situations when subtractive cancellation may severely corrupt the results, i.e. $|b| \approx \sqrt{b^2 - 4ac}$. Convert the algorithm into a MATLAB program and solve the quadratic equations given in Problem 1.7.

1.9. Use Equation (1.11) to derive the Taylor series expansion for $f(x) = \ln(1 + x)$. Use $x_0 = 0$. Show that

$$\ln(1 + x) = x - \frac{x^2}{2} + \frac{x^3}{3} - \frac{x^4}{4} + \cdots.$$

Calculate $\ln(1 + x)$ for $x = 0.5$ and 2. Estimate the relative error based on the exact solution if you include either one, two, three, or four terms. Construct and complete Table P1.1 to compare your answers. What conclusions can your draw from your solution?

Table P1.1. *Relative error in approximating ln(1 + x)*

Terms included	$x = 0.5$	$x = 2$
1		
2		
3		
4		

1.10. Write a MATLAB program to determine the value of sin x for values of x such that $0 < x < \pi$, using the sine series (Equation (1.13)). Find the answer to within a relative accuracy of 0.001 of the exact value of sin x, the criterion for convergence being

```
abs((sum - sin(x))/sin(x)) < 0.0001,
```

where `sum` is the summation of terms in the sin x series. Plot the number of terms that must be included in the sum to achieve this relative accuracy for each value of x.

Redetermine the value of sin x using the sine series for five values of x such that $0 < x < \pi$; however, this time use the following tolerance specification:

absolute value of (last term/summation of terms) < 0.0001

as the convergence criterion. Plot the number of terms that must be included in the sum to achieve this tolerance for each value of x.

How well does the tolerance specification for the second criterion match the first criterion of convergence?

1.11. Determine the first-order derivative of the function

$$f(x) = 4x^4 - 9x^3 + 1.5x^2 + 7x - 2$$

at $x = 2$, using forward difference, backward difference, and central difference (Equations (1.20), (1.22) and (1.23)). Use a step size of 0.1. Calculate the absolute and relative error for each case based on the exact derivative $f'(2)$.

1.12. **Novel hematopoietic stem cell separating device** Each year, more than 35 000 people in the United States are diagnosed with leukemia, lymphoma, or other blood, metabolic, or immune system disorders for which a bone marrow or adult blood stem cell transplant could be a cure. However, 70% of these patients do not receive a transplant due to both a lack of matched donors and the high, non-reimbursed costs incurred by treatment centers. For many patients, bone marrow or stem cell transplants provide a cure to an otherwise fatal or greatly debilitating condition. One of the most important steps in processing a bone marrow or peripheral blood sample for transplantation is separating out the hematopoietic stem cells. These cells are present in very low numbers and are responsible for restoring function to the patient's damaged bone marrow. Current separation technology is a labor-intensive and inefficient process. There is often a low yield of stem cells, and high expenses must be incurred by transplant centers.

An extracorporeal device to capture hematopoietic stem cells (HSCs) from blood quickly and efficiently is currently under design. This device is made of tubes that are coated internally with specific sticky molecules called *selectins* that preferentially bind HSCs (Wojciechowski *et al.*, 2008). Stem cells have a surface marker called CD34 that is absent on mature blood cells. Selectins can also bind white blood cells or leukocytes. If blood or bone marrow cells are perfused through such tubing, stem cells and some amount of white blood cells are selectively removed from the bulk of the fluid due to adhesion of these cells to the wall.

Rat experiments were performed to assess the feasibilty of purifying stem cells from blood by using this capture device. The inner surface area (SA) of a single tube is 35.34 mm^3. It was found that, on average, 1770 cells/mm^2 were found to be attached to the inner tubing surface following perfusion of blood for 60 min. The total number of stuck cells is calculated to be $1770 \times \text{SA} = 62\,550$ (rounded off to the nearest integer). However, the number of stuck cells per mm^2 was found to vary between 1605 and 1900 for different experiments (or 0.9068 to 1.073 times the average). Determine the range of the total number of stuck cells that are captured by a single tube in both absolute and relative terms. How many significant digits can you include in your answer?

1.13. In the experiment discusseed in Problem 1.12, the purity of the captured stem cells was found to be 18%; in other words, the fraction of captured cells that were CD34+ stem cells was 18%. Based on the range of the total number of stuck cells that are captured by a single tube, calculate the range of the number of captured stem cells by the coated tube. If the purity ranges from 12 to 25%, then determine the absolute lower and upper bounds of captured stem cells. (Take into account the variability in the total number of captured cells.)

1.14. **Separation of different populations of cancer cells** Some leukemic cancer cells exhibit stem cell-like properties and express the CD34 marker on their surface, a surface marker exhibited by immature stem cells. In this way, leukemic cancer cells can be loosely classified into two categories: CD34+ cancer cells and CD34– cancer cells. Current cancer therapies are not as effective in eradicating the cancer cells displaying CD34 on their surfaces. These CD34+ cancer cells are believed to be responsible for inducing relapse of the disease following treatment.

Table P1.2.

	CD34+ cancer cells		CD34– cancer cells	
Observation	Distance travelled (μm)	Time (s)	Distance travelled (μm)	Time (s)
1	147.27	61.40	124.77	58.47
2	129.93	46.12	162.74	60.90
3	42.29	51.87	173.99	39.67
4	142.54	61.40	163.20	60.90
5	59.70	61.40	167.78	60.90
6	163.94	61.40	127.36	60.90
7	83.99	61.40	112.08	43.67
8	121.9	61.40	159.32	60.90
9	89.85	27.43	123.76	60.90
10	83.35	61.40	176.41	60.90

A device for separating CD34+ cancer cells from CD34– cancer cells is currently being investigated. This device exploits the differences in the mechanical properties of cell rolling on a surface-bound glycoprotein called *selectin*. Cell rolling manifests from repeated bond formation and breakage between selectin molecules bound to a stationary wall and CD34 molecules bound to the cells' surface, when cells are flowing over a surface (see Figure P1.1).

A series of in vitro experiments were performed to measure the rolling velocities of primary (directly derived from patients) cancer cells (acute myeloid leukemia (AML)) to ascertain the differences in their rolling properties. Table P1.2 lists the observations for both CD34+ and CD34– cells.

Calculate the rolling velocities (in μm/s) using the appropriate number of significant figures. Determine the average rolling velocity and the minimum and maximum rolling velocities observed for CD34+ and CD34– cells. What are the maximum absolute and relative deviations from the average deviations in the measured rolling velocities for both CD34+ and CD34– cells?

Compare the average rolling velocites of CD34+ and CD34– cancer cells in absolute terms. How do you think they compare? Do you suggest the need for more data (observations)? The method used to determine whether the two rolling populations are significantly different using statistical tests is discussed in Chapter 4.

1.15. **Treatment of prostate cancer cells with vitamin D** A **tumor** (neoplasm) is a cluster of abnormally proliferating cells. As long as the tumor cells remain within the clustered mass and do not detach from it, the tumor is considered **benign**. A tumor is defined as **malignant** or a **cancer** if the abnormally growing and dividing (neoplastic) cells are capable of separating from the tumor of origin and invading other tissues. The process by which cancerous cells break away from the tumor growth and invade other surrounding normal tissues is called **metastasis**. During metastasis, these abnormally proliferating cells may even enter into the body's circulation (blood or lymphatic system), where they are transported to different regions of the body. It is important to note that a neoplastic cell cannot metastasize and colonize in

Table P1.3.

Observation	Vitamin D-treated cells		Untreated cells	
	Distance travelled (μm)	Time (s)	Distance travelled (μm)	Time (s)
1	114.865	11.50	87.838	19.70
2	121.622	18.18	165.541	44.53
3	114.865	46.73	351.416	57.93
4	108.108	25.17	442.568	41.27
5	334.528	57.93	362.259	57.90
6	233.108	27.60	375.380	57.90
7	212.865	57.93	432.485	57.90
8	209.568	57.93	523.649	56.73
9	331.512	57.93	496.725	57.90
10	479.777	57.93	317.639	57.90
11	446.266	57.93	375.243	57.90
12	321.017	57.93	351.368	57.90
13	127.297	57.93	425.796	57.90
14	320.964	57.93	270.355	57.90
15	321.106	57.93	459.472	57.90
16	517.289	57.93	341.216	57.90
17	391.906	53.47	365.256	57.90
18	328.138	57.93	405.631	57.90

This data set is available at the book's website www.cambridge.org/* as prostateCA.dat.
From K. P. Rana and M. R. King (unpublished data).

Figure P1.1

A CD34+ cancer cell rolling on selectin molecules bound to a wall while flowing through a flow chamber.

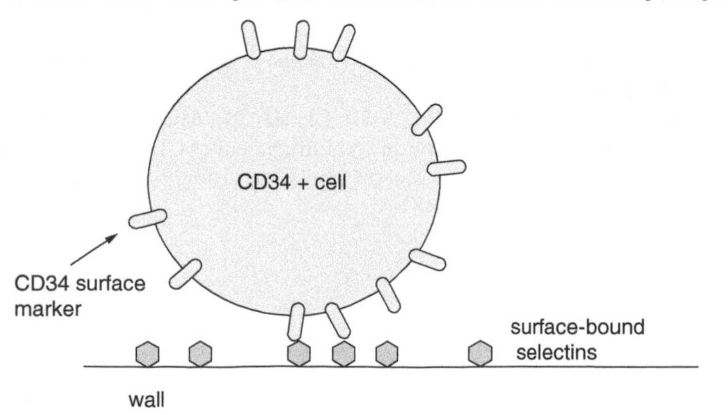

foreign tissues (tissues other than that of its origin) unless the cell has acquired certain properties that allow it to survive in an alien environment.

Prostate cancer cells are prone to metastasize to the bone and cause bone cancer. It has been shown that treatment of prostate cancer cells with vitamin D inhibits metastasis to the bone (Ali and Vaidya, 2007). In order for cancer cells to metastasize successfully, they must be able to contact and adhere to the vascular endothelium after they have entered the blood stream. Once adhered, the cancer cells must then find a way to squeeze through the vascular wall, escape from the blood stream, and enter new (foreign) tissues. In order for cells to adhere firmly to the endothelial lining of the vessel wall, they first contact the wall via weak non-covalent bonds that form and break under flow. The repeated formation of new cell–surface bonds and breakage translates macroscopically into cell rolling on the endothelial wall.

It is hypothesized that vitamin D may influence the rolling properties of cells on a selectin-coated surface. Selectins are glycoprotein molecules that are expressed by endothelial cells. Selectins reversibly bind the surface receptors of white blood cells and certain cancer cells and stem cells, the consequence of which is cell rolling on the surface. To test this hypothesis, the rolling velocities of vitamin D-treated prostate cancer cells are compared with the rolling velocities of non-treated prostate cancer cells (see Table P1.3).

Calculate the rolling velocities (in μm/s) using the appropriate number of significant figures. Determine the average rolling velocity and the minimum and maximum rolling velocities observed for vitamin D-treated and untreated cells. What are the maximum absolute and relative deviations from the average value in the measured rolling velocities for both cases? Compare the average rolling velocites of vitamin D-treated and untreated cancer cells in absolute terms. How do you think they compare?

References

Ali, M. M. and Vaidya, V. (2007) Vitamin D and Cancer. *J. Cancer. Res. Ther.*, **3**, 225–30.

Fletcher, J. E. (1980) On Facilitated Oxygen Diffusion in Muscle Tissues. *Biophys. J.*, **29**, 437–58.

Liguzinski, P. and Korzeniewski, B. (2006) Metabolic Control over the Oxygen Consumption Flux in Intact Skeletal Muscle: In Silico Studies. *Am. J. Physiol. Cell Physiol.*, **291**, C1213–24.

Mathews, J. H. and Fink, K. D. (2004) *Numerical Methods Using Matlab* (Upper Saddle River, NJ: Pearson Prentice Hall).

Scarborough, J. B. (1966) *Numerical Mathematical Analysis* (Baltimore, MD: The John Hopkins Press).

Tanenbaum, A. S. (1999) *Structured Computer Organization* (Upper Saddle River, NJ: Prentice Hall).

Wojciechowski, J. C., Narasipura, S. D., Charles, N., Mickelsen, D., Rana, K., Blair, M. L., and King, M. R. (2008) Capture and Enrichment of CD34-Positive Haematopoietic Stem and Progenitor Cells from Blood Circulation Using P-Selectin in an Implantable device. *Br. J. Haematol.*, **140**, 673–81.

2 Systems of linear equations

2.1 Introduction

A linear equation is one that has a simple linear dependency on all of the **variables** or **unknowns** in the equation. A generalized linear equation can be written as:

$$a_1x_1 + a_2x_2 + \cdots + a_nx_n = b.$$

The variables of the equation are $x_1, x_2, \ldots, x_n$. The coefficients of the variables $a_1, a_2, \ldots, a_n$, and the constant b on the right-hand side of the equation represents the known quantities of the equation. A **system of linear equations** consists of a set of two or more linear equations. For example, the system of equations

$$x_1 + x_2 + x_3 = 4,$$
$$2x_1 + 3x_2 - 4x_3 = 4$$

forms a set of two linear equations in three unknowns. A **solution** of a system of linear equations consists of the values of the unknowns that simultaneously satisfy each linear equation in the system of equations. As a *rule of thumb*, one requires n linear equations to solve a system of equations in n unknowns to obtain a unique solution. A problem is said to have a **unique solution** if *only one* solution exists that can exactly satisfy the governing equations. In order for a system of linear equations in three unknowns to yield a unique solution, we must have at least three linear equations. Therefore, the above set of two equations does not have a unique solution. If we complete the equation set by including one more equation, then we have three equations in three unknowns:

$$x_1 + x_2 + x_3 = 4,$$
$$2x_1 + 3x_2 - 4x_3 = 4,$$
$$6x_1 - 3x_2 = 0.$$

The solution to this system of equations is unique: $x_1 = 1, x_2 = 2, x_3 = 1$.

Sometimes one may come across a situation in which a system of n equations in n unknowns does not have a unique solution, i.e. more than one solution may exist that completely satisfies all equations in the set. In other cases, the linear problem can have no solution at all. Certain properties of the coefficients and constants of the equations govern the type of solution that is obtained.

Many standard engineering problems, such as force balances on a beam or truss, flow of a reagent or chemical through a series of stirred tanks, or flow of electricity through a circuit of wires, can be modeled using a system of linear equations. Solution methods for simultaneous linear equations (that contain two or more unknowns) are in most cases straightforward, intuitive, and usually easier to implement than those for nonlinear systems of equations. Linear equations are solved using mathematical techniques derived from the foundations of linear algebra theory. Mastery of key concepts of linear algebra is important to apply solution methods developed for systems of linear equations, and understand their limitations.

Box 2.1A Drug development and toxicity studies: animal-on-a-chip

Efficient and reliable drug screening methods are critical for successful drug development in a cost-effective manner. The FDA (Food and Drug Administration) mandates that the testing of all proposed drugs follow a set of protocols involving (expensive) *in vivo* animal studies and human clinical trials. If the trials are deemed successful, only then may the approved drug be sold to the public. It is most advantageous for pharmaceutical companies to adopt *in vitro* methods that pre-screen all drug candidates in order to select for only those that have the best chances of success in animal and human trials. Key information sought in these pre-clinical trial experiments include:

- absorption rate into blood and tissues, and extent of distribution of drug in the body,
- metabolism of the drug,
- toxicity effects of the drug and drug metabolites on the body tissues, and
- routes and rate of excretion of the drug.

The biochemical processing of drug by the body can be represented by **physiologically based** mathematical models that describe the body's response to the drug. These models are "physiological" because the parameters of the equations are obtained from *in vivo* animal experiments and therefore reliably predict the physiological response of the body when the drug is introduced. Mathematical models that predict the distribution (concentration of drug in various parts of the body), metabolism (breakdown of drug by enzymatic reactions), covalent binding of drug or drug metabolites with cellular components or extracellular proteins, and elimination of the drug from the body are called **pharmacokinetic models**. See Box 2.3 for a discussion on "pharmacokinetics" as a branch of pharmacology.

A **physiologically based pharmacokinetic model (PBPK)** using experimental data from *in vitro* cell-based assays was successfully developed by Quick and Shuler (1999) at Cornell University. This model could predict many of the drug (toxicant: naphthalene) responses observed *in vivo*, i.e. distribution, enzymatic consumption of drug, and non-enzymatic binding of drug metabolites to tissue proteins, in several different organs of the rat and mouse body (e.g. lung, liver, and fatty tissue). The merit in developing *in vitro*-based physiological models stems from the fact that data from *in vitro* studies on animal cells are relatively abundant; *in vivo* animal studies are considerably more expensive and are subject to ethical issues. Performing experimental studies of drug effects on humans is even more difficult. Instead, if an overall human body response to a drug can be reliably predicted by formulating a PBPK model based on samples of human cells, then this represents advancement by leaps and bounds in medical technology.

A novel microfabricated device called **microscale cell culture analog (μCCA)** or '**animal-on-a-chip**' was recently invented by Shuler and co-workers (Sin *et al*., 2004). The μCCA uses microchannels to transport media to different compartments that mimic body organs and contain representative animal cells (hence "animal-on-a-chip"). The flow channels and individual organ chambers are etched onto a silicon wafer using standard lithographic methods. The μCCA serves as a physical replica of the PBPK model and approximates the distribution, metabolism, toxic effects, and excretion of drug occurring within the body. The advantages of using a μCCA as opposed to a macro CCA apparatus (consisting of a set of flasks containing cultured mammalian cells, tubes, and a pump connected by tubing) are several-fold:

(1) physiological compartment residence times, physiological flowrates (fluid shear stresses), and liquid-to-cell ratios are more easily realized;

(2) rate processes occur at biological length scales (on the order of 10 μm), which is established only by the microscale device (Sin *et al.*, 2004);
(3) smaller volumes require fewer reagents, chemicals and cells (especially useful when materials are scarce); and
(4) the device is easily adaptable to large-scale automated drug screening applications.

Figure 2.1 is a simplified block diagram of the μCCA (Sin *et al.*, 2004) that has three compartments: lung, liver, and other tissues.

The effects of the environmental toxicant naphthalene ($C_{10}H_8$) – a volatile polycyclic aromatic hydrocarbon – on body tissues was studied by Shuler and co-workers using the μCCA apparatus (Viravaidya and Shuler, 2004; Viravaidya *et al.*, 2004). Naphthalene is the primary component of moth balls (a form of pesticide), giving it a distinctive odor. This hydrocarbon is widely used as an intermediate in the production of various chemicals such as synthetic dyes. Exposure to large amounts of naphthalene is toxic to humans, and can result in conditions such as hemolytic anemia (destruction of red blood cells), lung damage, and cataracts. Naphthalene is also believed to have carcinogenic effects on the body.

On exposure to naphthalene, cytochrome P450 monooxygenases in liver cells and Clara lung cells catalyze the conversion of **naphthalene** into **naphthalene epoxide** (naphthalene oxide). The circulating epoxides can bind to **glutathione (GSH)** in cells to form glutathione conjugates. GSH is an antioxidant and protects the cells from free radicals, oxygen ions, and peroxides, collectively known as **reactive oxygen species (ROS)**, at the time of cellular oxidative stress. Also, GSH binds toxic metabolic by-products, as part of the cellular detoxification process, and is then eliminated from the body in its bound state (e.g. as bile). Loss of GSH functionality can lead to cell death.

Naphthalene epoxide is consumed in several other pathways: binding to tissue proteins, spontaneous rearrangement to **naphthol**, and conversion to **naphthalene dihydrodiol** by the enzyme **epoxide hydrolase**. Both naphthol and dihydrodiol are ultimately converted to **naphthalenediol**, which then participates in a reversible redox reaction to produce **naphthoquinone**. ROS generation occurs in this redox pathway. The build up of ROS can result in oxidative stress of the cell and cellular damage. Also, naphthoquinone binds GSH, as well as other proteins and DNA. The loss and/or alteration

Figure 2.1

Block diagram of the μCCA (Sin *et al.*, 2004). The pump that recirculates the media is located external to the silicon chip. The chip houses the organ compartments and the microchannels that transport media containing drug metabolites to individual compartments.

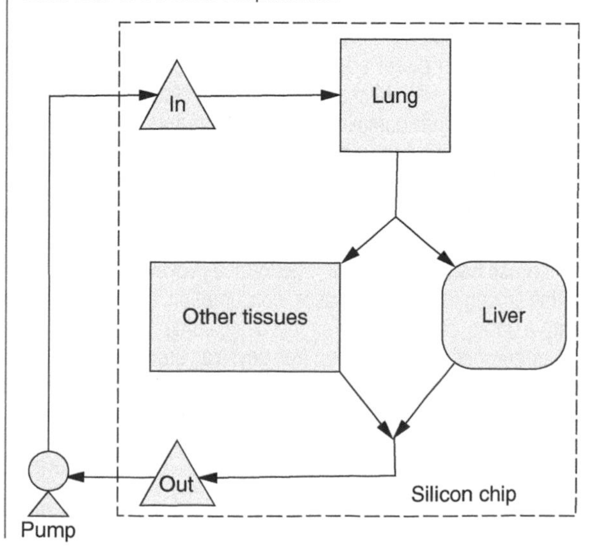

Figure 2.2

Naphthalene metabolic pathway in mice (Viravaidya *et al.*, 2004) .

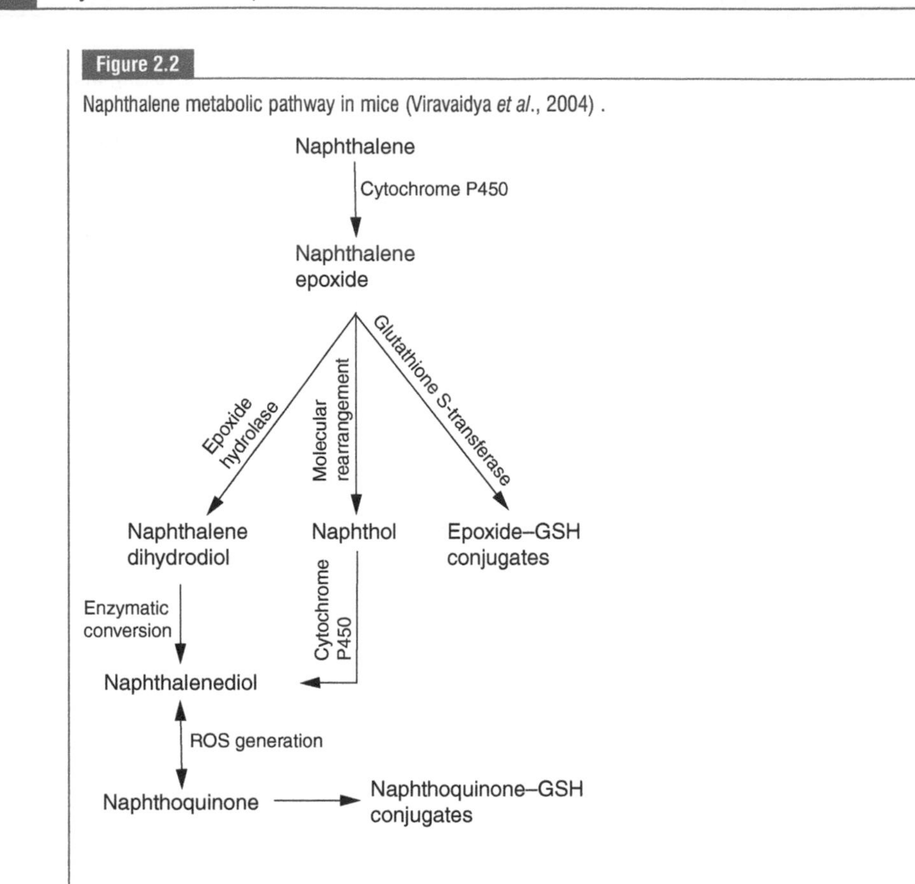

of protein and DNA function can result in widespread havoc in the cell. Figure 2.2 illustrates the series of chemical reactions that are a part of the metabolic processing of naphthalene into its metabolites.

Naphthalene is excreted from the body in the form of its metabolites: mainly as GSH conjugates, naphthalene dihydrodiol, and naphthol. Quick and Shuler (1999) developed a comprehensive PBPK model for exposure to naphthalene in mice. In mice, exposure to naphthalene results in cell death in the lung.

Here, we envision the steady state operation of Shuler's μCCA that might mimic continuous exposure of murine cells to naphthalene over a period of time. Figure 2.3 is a material balance diagram for naphthalene introduced into the μCCA depicted in Figure 2.1. The feed stream contains a constant concentration of naphthalene in dissolved form. Conversion of naphthalene to epoxide occurs in the lung and liver in accordance with Michaelis–Menten kinetics (see Box 4.3 for a discussion on Michaelis–Menten kinetics). Both compartments are treated as well-mixed reactors. Therefore, the prevailing concentration of naphthalene within these compartments is equal to the exit concentration. No conversion of naphthalene occurs in the other tissues (ot). Naphthalene is not absorbed anywhere in the three compartments. The purge stream mimics the loss of the drug/toxicant from the circuit via excretory channels or leakage into other body parts (see Figure 2.3).

In engineering, a **mass balance** is routinely performed over every compartment (unit) to account for transport, distribution among immiscible phases, consumption, and generation of a particular product, drug, or chemical. A mass balance is a simple accounting process of material quantities when phenomena such as convective transport and/or physical processes (diffusion, absorption) and/or chemical processes (reaction, binding) are involved. A generic mass or material balance of any component A over a unit, equipment, or compartment is represented as:

$$(\text{flow in of } A - \text{flow out of } A) + (\text{generation of } A - \text{consumption of } A) \\ = \text{accumulation of } A \text{ in unit.}$$

Figure 2.3

Material balance of naphthalene in the μCCA.

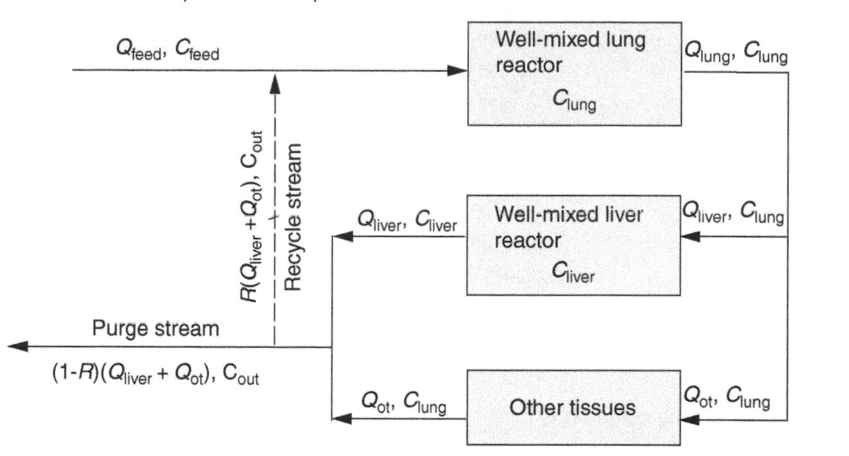

In **steady state processes**, the concentrations of components as well as the volumetric flowrates do not change with time, and thus there is no accumulation of material anywhere in the system. For a steady state process, the material balance is:

$$(\text{flow in of } A - \text{flow out of } A) + (\text{generation of } A - \text{consumption of } A) = 0.$$

Mass balances over individual compartments (units) as well as an overall mass balance over the entire system of units provide a set of equations that are solved to obtain the unknowns such as the outlet concentrations of the species of interest A from each unit.

On performing a mass balance of naphthalene over the two chambers – lung and liver – we obtain two equations for the unknowns of the problem, C_{lung} and C_{liver}.

Lung compartment

$$Q_{\text{feed}}C_{\text{feed}} + R\left(Q_{\text{liver}}C_{\text{liver}} + Q_{\text{ot}}C_{\text{lung}}\right) - \frac{v_{\text{max,P450-lung}}C_{\text{lung}}}{K_{\text{m,P450-lung}} + C_{\text{lung}}}V_{\text{lung}} - Q_{\text{lung}}C_{\text{lung}} = 0. \tag{2.1}$$

Liver compartment

$$Q_{\text{liver}}C_{\text{lung}} - \frac{v_{\text{max,P450-liver}}C_{\text{liver}}}{K_{\text{m,P450-liver}} + C_{\text{liver}}}V_{\text{liver}} - Q_{\text{lung}}C_{\text{lung}} = 0 \tag{2.2}$$

The mass balance of naphthalene over the other tissues compartment is trivial since no loss or conversion of mass occurs in that compartment.

The values for the parameters in Equations (2.1) and (2.2) are provided below.

Q_{lung}: flowrate through lung compartment = 2 μl/min;
Q_{liver}: flowrate through liver compartment = 0.5 μl/min;

Q_{ot}: flowrate through other tissues compartment = 1.5 μl/min;
C_{feed}: concentration of naphthalene in feed stream = 390 μM;
Q_{feed}: flowrate of feed stream = $(1 - R)Q_{lung}$;
$v_{max,P450-lung}$: maximum reaction velocity for conversion of naphthalene into napthalene epoxide by cytochrome P450 in lung cells = 8.75 μM/min;
$K_{m,P450-lung}$: Michaelis constant = 2.1 μM;
$v_{max,P450-liver}$: maximum reaction velocity for conversion of naphthalene into napthalene epoxide by cytochrome P450 in liver cells = 118 μM/min;
$K_{m,P450-liver}$: Michaelis constant = 7.0 μM;
V_{lung}: volume of lung compartment = 2 mm × 2 mm × 20 μm = 8 × 10^{-8} l = 0.08μ
V_{liver}: volume of liver compartment = 3.5 mm × 4.6 mm × 20 μm = 0.322 μl
R: specifies the fraction of the exiting stream that re-enters the microcircuit.

Substituting these values into Equations (2.1) and (2.2), we obtain the following.

Lung compartment

$$780(1 - R) + R(0.5C_{liver} + 1.5C_{lung}) - \frac{8.75C_{lung}}{2.1 + C_{lung}}0.08 - 2C_{lung} = 0.$$

Liver compartment

$$0.5C_{lung} - \frac{118C_{liver}}{7.0 + C_{liver}}0.322 - 0.5C_{liver} = 0.$$

Here, we make the assumption that $C_{lung} \gg K_{m,P450-lung}$ and $C_{liver} \gg K_{m,P450-liver}$. As a result, the Michaelis–Menten kinetics tends towards zeroth-order kinetics. We then have the following.

Lung compartment

$$780(1 - R) + R(0.5C_{liver} + 1.5C_{lung}) - 0.7 - 2C_{lung} = 0.$$

Liver compartment

$$0.5C_{lung} - 38 - 0.5C_{liver} = 0,$$

which can be simplified further as follows.

Lung compartment

$$(2 - 1.5R)C_{lung} - 0.5RC_{liver} = 779.3 - 780R. \tag{2.3}$$

Liver compartment

$$C_{lung} - C_{liver} = 76. \tag{2.4}$$

We have two simultaneous linear equations in two variables. Equations (2.3) and (2.4) will be solved using solution techniques discussed in this chapter.

Some *key assumptions* made in setting up these equations are:

(1) all reactions are irreversible;
(2) all compartments are well-mixed;
(3) no density changes occur in the system;
(4) The tissue : medium partition coefficient is 1, i.e. there is an equal distribution between tissue and medium.

2.2 Fundamentals of linear algebra

This section covers topics on linear algebra related to the solution methods for simultaneous linear equations. In addition to serving as a review of fundamental concepts of linear algebra, this section introduces many useful MATLAB operators and functions relevant to matrix construction and manipulation. Application of mathematical tools from linear algebra is common in many fields of engineering and science.

MATLAB or "MATrix LABoratory" is an ideal tool for studying and performing operations involving matrix algebra. The MATLAB computing environment stores each data set in a matrix or multidimensional array format. For example, a scalar value of `a = 1` is stored as a 1×1 matrix. A character set such as `'computer'` is treated as a 1×8 matrix of eight character elements. Dimensioning of arrays prior to use is not required by MATLAB. Any algebraic, relational, or logical operation performed on any data or variables is treated as a matrix or array operation. MATLAB contains many built-in tools to analyze and manipulate matrices.

2.2.1 Vectors and matrices

In linear algebra, a **vector** is written as a set of numbers arranged in sequential fashion either row-wise or column-wise and enclosed within brackets. The individual entries contained in a vector are the elements. For example, a vector $\mathbf{v}$ that contains three elements 1, 2 and 3 is written as:

$$\mathbf{v} = [1\,2\,3] \quad \text{or} \quad \mathbf{v} = \begin{bmatrix} 1 \\ 2 \\ 3 \end{bmatrix}.$$

1×3 matrix $\qquad$ 3×1 matrix

It is not necessary to include a comma to delineate each entry within the brackets. The vector $\mathbf{v}$ written row-wise is called a **row vector** and can be visualized as a matrix of size 1×3, where 1 is the number of rows and 3 is the number of columns. On the other hand, $\mathbf{v}$ is a **column vector** when its elements are arranged in a single column and can be regarded as a matrix of size 3×1, where 3 is the number of rows and 1 is the number of columns. An n-element column vector $\mathbf{v}$ is represented as follows:

$$\mathbf{v} = \begin{bmatrix} v_1 \\ v_2 \\ \cdot \\ \cdot \\ \cdot \\ v_n \end{bmatrix},$$

where v_i are the individual elements or components of $\mathbf{v}$ and $i = 1, 2, \ldots, n$. Each element has a single subscript that specifies its position in the sequence. Throughout this text, variables representing vectors are denoted by lower-case bold roman letters, and elements of a vector will be represented by subscripted lower-case roman letters. Henceforth, **every vector is assumed to be a column vector** (unless noted otherwise) as per common convention used by many standard linear algebra textbooks.

A **matrix** is defined as a two-dimensional rectangular array of numbers. Examples of matrices are shown below:

$$\mathbf{A} = \begin{bmatrix} 4 & 7 & 2 \\ 5 & 1 & 8 \end{bmatrix}, \quad \mathbf{B} = \begin{bmatrix} 7 & 7 \\ 1 & 2 \end{bmatrix}, \quad \mathbf{C} = \begin{bmatrix} \frac{3}{2} & \frac{5}{6} \\ \frac{10}{3} & \frac{6}{7} \\ 2 & 3 \end{bmatrix}.$$

Matrix variables are denoted by bold upper-case roman letters in order to provide a contrast with vectors or one-dimensional matrices. The **size** of a matrix is determined by the number of rows and columns that the matrix contains. In the examples above, $\mathbf{A}$ and $\mathbf{C}$ are rectangular matrices of sizes 2×3 and 3×2, respectively, and $\mathbf{B}$ is a square matrix of size 2×2, in which the number of rows equal the number of columns. An $m \times n$ matrix $\mathbf{A}$, where m is the number of rows and n is the number of columns, can be written in a generalized format as shown below, in which every element has two subscripts to specify its row and column position. The first subscript specifies the row location and the second subscript indicates the column position of an element:

$$\mathbf{A} = \begin{bmatrix} a_{11} & a_{12} & a_{13} & \ldots & a_{1n} \\ a_{21} & a_{22} & a_{23} & \ldots & a_{2n} \\ & & \cdot & & \\ & & \cdot & & \\ & & \cdot & & \\ a_{m1} & a_{m2} & a_{m3} & \ldots & a_{mn} \end{bmatrix}.$$

An element in the ith row and jth column of $\mathbf{A}$ is represented as a_{ij}.

Using MATLAB

A vector or matrix can be created by using the matrix constructor operator, [ ]. Elements of the matrix are entered within square brackets and individual rows are distinguished by using a semi-colon. Elements within a row can be separated using either a comma or space. Here, we construct a 3×3 matrix $\mathbf{A}$ in MATLAB:

```
>> A = [1 2 3; 4 5 6; 7 8 9]
```

MATLAB provides the following output (since we have not suppressed output generation by adding a semi-colon at the end of the statement):

```
A =
    1   2   3
    4   5   6
    7   8   9
```

There are several MATLAB functions available to create special matrices for general use. The `zeros` function produces a matrix consisting entirely of zeros of specified size. Similarly, the `ones` function produces a matrix consisting entirely of ones of specified size. For example,

```
>> Z = zeros(2,3)
Z =
     0     0     0
     0     0     0
>> M = ones(4,4)
M =
     1     1     1     1
     1     1     1     1
     1     1     1     1
     1     1     1     1
```

Two other important matrix generating functions are the `eye` and `diag` functions. The `eye` function produces the identity matrix I, in which the elements are ones along the diagonal and zeros elsewhere. The `diag` function creates a diagonal matrix from an input vector by arranging the vector elements along the diagonal. Elements along the diagonal of a matrix can also be extracted using the `diag` function. Examples of the usage of `eye` and `diag` are shown below.

```
>> I = eye(3)
I =
     1     0     0
     0     1     0
     0     0     1
>> D = diag([1 2 3])
D =
     1     0     0
     0     2     0
     0     0     3
>> b = diag(I)
b =
     1
     1
     1
```

To access a particular element of a matrix, the row and column indices need to be specified along with the matrix name as shown below. To access the element in the second row and second column of matrix

```
A =
     1     2     3
     4     5     6
     7     8     9
```

we type

```
>> A(2,2)
ans =
     5
```

If multiple elements need to be accessed simultaneously, the **colon operator** (:) is used. The colon operator represents a series of numbers; for example, $1{:}n$ represents a series of numbers from 1 to n in increments of 1:

```
>> A(1:3, 2)
ans =
      2
      5
      8
```

If access to an entire row or column is desired, the colon by itself can be used as a subscript. A stand-alone colon refers to all the elements along the specified column or row, such as that shown below for column 2 of matrix $\boldsymbol{A}$:

```
>> A(:,2)
ans =
      2
      5
      8
```

MATLAB provides the `size` function that returns the length of each dimension of a vector or matrix:

```
>> [rows, columns] = size(A)
rows =
      3
columns =
      3
```

2.2.2 Matrix operations

Matrices (and vectors) can undergo the usual arithmetic operations of addition and subtraction, and also scalar, vector, and matrix multiplication, subject to certain rules that govern matrix operations. Let $\boldsymbol{A}$ be an $m \times n$ matrix and let $\boldsymbol{u}$ and $\boldsymbol{v}$ be column vectors of length $n \times 1$.

Addition and subtraction of matrices

Addition or subtraction of two matrices is carried out in element-wise fashion and can only be performed on matrices of identical size. Only elements that have the same subscripts or position within their respective matrices are added (or subtracted). Thus,

$$\boldsymbol{A} + \boldsymbol{B} = \left[a_{ij} + b_{ij}\right]_{m \times n}, \qquad \text{where } 1 \le i \le m \text{ and } 1 \le j \le n.$$

Scalar multiplication

If a matrix $\boldsymbol{A}$ is multiplied by a scalar value or a single number p, all the elements of the matrix $\boldsymbol{A}$ are multiplied by p:

$$p\boldsymbol{A} = \left[pa_{ij}\right]_{m \times n}, \text{ where } 1 \le i \le m \text{ and } 1 \le j \le n.$$

Using MATLAB

If we define

```
>> A = 3*eye(3)
```

```
A =
     3   0   0
     0   3   0
     0   0   3
>> B = 2*ones(3,3) + diag([4 4 4])
B =
     6   2   2
     2   6   2
     2   2   6
```

then,

```
>> C = B - A
C =
     3   2   2
     2   3   2
     2   2   3
```

If we define

```
>> p = 0.5
>> D = p*B
```

we get the MATLAB output

```
p =
    0.5000
D =
     3   1   1
     1   3   1
     1   1   3
```

Transpose operations

The **transpose** of an $m \times n$ matrix $\boldsymbol{A}$ is an $n \times m$ matrix that contains the same elements as that of $\boldsymbol{A}$. However, the elements are "transposed" such that the rows of $\boldsymbol{A}$ are converted into the corresponding columns, and vice versa. In other words, the elements of $\boldsymbol{A}$ are "reflected" across the diagonal of the matrix. Symbolically the transpose of a matrix $\boldsymbol{A}$ is written as $\boldsymbol{A}^{\mathrm{T}}$. If $\boldsymbol{B} = \boldsymbol{A}^{\mathrm{T}}$, then $b_{ij} = a_{ji}$ for all $1 \le i \le n$ and $1 \le j \le m$.

For a 3×3 matrix $\boldsymbol{C} = \begin{bmatrix} c_{11} & c_{12} & c_{13} \\ c_{21} & c_{22} & c_{23} \\ c_{31} & c_{32} & c_{33} \end{bmatrix}$, its transpose is $\boldsymbol{C}^{\mathrm{T}} = \begin{bmatrix} c_{11} & c_{21} & c_{31} \\ c_{12} & c_{22} & c_{32} \\ c_{13} & c_{23} & c_{33} \end{bmatrix}$.

Since a vector can also be viewed as a matrix, each element of a vector can be written using double subscript notation, and thus the vector $\boldsymbol{v}$ can also be represented as

$$\boldsymbol{v} = \begin{bmatrix} v_{11} \\ v_{21} \\ \cdot \\ \cdot \\ \cdot \\ v_{n1} \end{bmatrix}.$$

By taking the transpose of $\boldsymbol{v}$ we obtain

$$\mathbf{v}^{\mathrm{T}} = [v_{11}\ v_{12}\ \ldots\ v_{1n}].$$

Thus, a column vector can be converted to a row vector by the transpose operation.

Using MATLAB

The **colon operator** can be used to represent a series of numbers. For example, if

```
>> a = [5:10]
```

then MATLAB creates a row vector $\boldsymbol{a}$ that has six elements as shown below:

```
a =
    5   6   7   8   9   10
```

Each element in $\boldsymbol{a}$ is greater than the last by an increment of one. The increment can be set to a different value by specifying a step value between the first and last values of the series, i.e.

```
vector = [start value : increment : end value] .
```

For example

```
>> b = [1:0.2:2]
b =
    1.0000   1.2000   1.4000   1.6000   1.8000   2.0000
```

The transpose operation is performed by typing an apostrophe after the vector or matrix variable. At the command prompt, type

```
>> v = a'
```

The MATLAB output is

```
v =
     5
     6
     7
     8
     9
    10
```

At the command prompt, type

```
>> v = b'
```

The MATLAB output is

```
v =
    1.0000
    1.2000
    1.4000
    1.6000
    1.8000
    2.0000
```

The transpose of the matrix $\boldsymbol{A}$ defined at the command prompt

```
>> A = [1 3 5; 2 4 6]
```

```
A =
     1   3   5
     2   4   6
```

is equated to B:

```
>> B = A'
B =
     1     2
     3     4
     5     6
```

Vector–vector multiplication: inner or dot product

The dot product is a scalar quantity obtained by a multiplication operation on two vectors. When a column vector $\boldsymbol{v}$ is pre-multiplied by a row vector $\boldsymbol{u}^{\mathrm{T}}$, and both vectors $\boldsymbol{u}$ and $\boldsymbol{v}$ have the same number of elements, then the dot product (where $\cdot$ is the dot product operator) is defined as follows:

$$\boldsymbol{u} \cdot \boldsymbol{v} = \boldsymbol{u}^{\mathrm{T}} \boldsymbol{v} = [u_1 \; u_2 \; \ldots \; u_n] \begin{bmatrix} v_1 \\ v_2 \\ \cdot \\ \cdot \\ \cdot \\ v_n \end{bmatrix} = \sum_{i=1}^{i=n} u_i v_i. \tag{2.5}$$

It is straightforward to show that $\boldsymbol{u}^{\mathrm{T}} \boldsymbol{v} = \boldsymbol{v}^{\mathrm{T}} \boldsymbol{u}$. Note that to obtain the dot product of two vectors, the vector that is being multiplied on the left must be a row vector and the vector being multiplied on the right must be a column vector. This is in accordance with matrix multiplication rules (see the subsection on "Matrix–matrix multiplication" below); i.e., if two matrices are multiplied, then their inner dimensions must agree. Here, a $1 \times n$ matrix is multiplied with an $n \times 1$ matrix. The inner dimension in this case is "n."

If a vector $\boldsymbol{u}$ is dotted with itself, we obtain

$$\boldsymbol{u} \cdot \boldsymbol{u} = \boldsymbol{u}^{\mathrm{T}} \boldsymbol{u} = [u_1 \; u_2 \; \ldots \; u_n] \begin{bmatrix} u_1 \\ u_2 \\ \cdot \\ \cdot \\ \cdot \\ u_n \end{bmatrix} = \sum_{i=1}^{i=n} u_i^2. \tag{2.6}$$

Using MATLAB

The scalar multiplication operator $*$ also functions as the dot product operator. We define

```
>> u = [0:0.5:2.5]'
u =
          0
     0.5000
     1.0000
     1.5000
     2.0000
     2.5000
>> v = [0:2:10]'
```

```
v =
     0
     2
     4
     6
     8
    10
```

Then, the following expressions are evaluated by MATLAB as

```
>> u'*v
ans =
    55
>> v'*u
ans =
    55
>> v'*v
ans =
    220
>> u'*u
ans =
    13.7500
```

Three-dimensional Euclidean space is a mathematical representation of our physical three-dimensional space. Vectors in this space are commonly expressed in terms of the mutually perpendicular unit (orthonormal) vectors $\boldsymbol{i}$, $\boldsymbol{j}$, and $\boldsymbol{k}$. For example, $\boldsymbol{u} = u_1\boldsymbol{i} + u_2\boldsymbol{j} + u_3\boldsymbol{k}$ and $\boldsymbol{v} = v_1\boldsymbol{i} + v_2\boldsymbol{j} + v_3\boldsymbol{k}$, where $\boldsymbol{u}$ and $\boldsymbol{v}$ are expressed as a linear combination of the orthonormal vectors $\boldsymbol{i}$, $\boldsymbol{j}$, $\boldsymbol{k}$ (see Section 2.2.4). The dot product of two three-dimensional vectors $\boldsymbol{u} \cdot \boldsymbol{v}$ is calculated as

$$\boldsymbol{u} \cdot \boldsymbol{v} = u_1 v_1 + u_2 v_2 + u_3 v_3 = \|u\| \|v\| \cos \theta,$$

where

$$\|\boldsymbol{u}\| = \sqrt{u_1^2 + u_2^2 + u_3^2} = \sqrt{\boldsymbol{u}^{\mathrm{T}}\boldsymbol{u}},$$

$$\|\boldsymbol{v}\| = \sqrt{v_1^2 + v_2^2 + v_3^2} = \sqrt{\boldsymbol{v}^{\mathrm{T}}\boldsymbol{v}},$$

and θ is the angle between the two vectors ($\|\boldsymbol{u}\|$ is the length of vector $\boldsymbol{u}$; see Section 2.2.3 on vector norms). If $\theta = 90°$, then $\cos\theta = 0$, and $\boldsymbol{u}$ and $\boldsymbol{v}$ are said to be **perpendicular** or **normal** to each other.

Using MATLAB

If

```
>> a = [1 0 1];
>> b = [0; 1; 0];
>> v = a*b
```

MATLAB evaluates the dot product as

```
v =
    0
```

Since the dot product of $\boldsymbol{a}$ and $\boldsymbol{b}$ is zero, vectors $\boldsymbol{a}$ and $\boldsymbol{b}$ are said to be **orthogonal** (or perpendicular in three-dimensional space) with respect to each other.

Matrix–vector multiplication

The outcome of a matrix–vector multiplication is another vector whose dimension and magnitude depend on whether the vector is multiplied on the right-hand side or the left-hand side of the matrix. The product of two matrices or a matrix and a vector is usually not commutative, i.e. $\boldsymbol{AB} \neq \boldsymbol{BA}$ or $\boldsymbol{Av} \neq \boldsymbol{vA}$. In this chapter, we will be dealing only with the multiplication of a vector on the right-hand side of a matrix. A matrix–vector multiplication, $\boldsymbol{Av}$, involves a series of dot products between the rows of the matrix $\boldsymbol{A}$ and the single column of vector $\boldsymbol{v}$. It is *absolutely essential* that the number of columns in $\boldsymbol{A}$ equal the number of rows in $\boldsymbol{v}$, or, in other words, that the number of elements in a row of $\boldsymbol{A}$ equal the number of elements contained in $\boldsymbol{v}$. If the elements of each row in $\boldsymbol{A}$ are viewed as components of a row vector, then $\boldsymbol{A}$ can be shown to consist of a single column of row vectors:

$$\boldsymbol{A} = \begin{bmatrix} a_{11} & a_{12} & a_{13} & \ldots & a_{1n} \\ a_{21} & a_{22} & a_{23} & \ldots & a_{2n} \\ & & \cdot & & \\ & & \cdot & & \\ & & \cdot & & \\ a_{m1} & a_{m2} & a_{m3} & \ldots & a_{mn} \end{bmatrix} = \begin{bmatrix} \boldsymbol{a}_1^{\mathrm{T}} \\ \boldsymbol{a}_2^{\mathrm{T}} \\ \cdot \\ \cdot \\ \cdot \\ \boldsymbol{a}_m^{\mathrm{T}} \end{bmatrix},$$

where $\boldsymbol{a}_i^{\mathrm{T}} = [a_{i1}\ a_{i2}\ a_{i3}\ \ldots\ a_{in}]$ and $1 \leq i \leq m$.

If $\boldsymbol{b}$ is the product of the matrix–vector multiplication of $\boldsymbol{A}$ and $\boldsymbol{v}$, then

$$\boldsymbol{b} = \boldsymbol{Av} = \begin{bmatrix} a_{11} & a_{12} & a_{13} & \ldots & a_{1n} \\ a_{21} & a_{22} & a_{23} & \ldots & a_{2n} \\ & & \cdot & & \\ & & \cdot & & \\ & & \cdot & & \\ a_{m1} & a_{m2} & a_{m3} & \ldots & a_{mn} \end{bmatrix} \begin{bmatrix} v_1 \\ v_2 \\ \cdot \\ \cdot \\ \cdot \\ v_n \end{bmatrix} = \begin{bmatrix} \boldsymbol{a}_1^{\mathrm{T}} \\ \boldsymbol{a}_2^{\mathrm{T}} \\ \cdot \\ \cdot \\ \cdot \\ \boldsymbol{a}_m^{\mathrm{T}} \end{bmatrix} \boldsymbol{v}$$

$$= \begin{bmatrix} \boldsymbol{a}_1^{\mathrm{T}}\boldsymbol{v} \\ \boldsymbol{a}_2^{\mathrm{T}}\boldsymbol{v} \\ \cdot \\ \cdot \\ \cdot \\ \boldsymbol{a}_m^{\mathrm{T}}\boldsymbol{v} \end{bmatrix} = \begin{bmatrix} \boldsymbol{a}_1 \cdot \boldsymbol{v} \\ \boldsymbol{a}_2 \cdot \boldsymbol{v} \\ \cdot \\ \cdot \\ \cdot \\ \boldsymbol{a}_m \cdot \boldsymbol{v} \end{bmatrix} = \begin{bmatrix} a_{11}v_1 & a_{12}v_2 & a_{13}v_3 & \ldots & a_{1n}v_n \\ a_{21}v_1 & a_{22}v_2 & a_{23}v_3 & \ldots & a_{2n}v_n \\ & & \cdot & & \\ & & \cdot & & \\ & & \cdot & & \\ a_{m1}v_1 & a_{m2}v_2 & a_{m3}v_3 & \ldots & a_{mn}v_n \end{bmatrix}. \tag{2.7}$$

Here, $\boldsymbol{a}_i^{\mathrm{T}}\boldsymbol{v} = \boldsymbol{a}_i \cdot \boldsymbol{v} = a_{i1}v_1 + a_{i2}v_2 + \cdots + a_{in}v_n$ for $1 \leq i \leq m$.

Matrix–matrix multiplication

The end result of the multiplication of two matrices is another matrix. Matrix multiplication of $\boldsymbol{A}$ with $\boldsymbol{B}$ is defined if $\boldsymbol{A}$ is an $m \times n$ matrix and $\boldsymbol{B}$ is an $n \times p$ matrix. The resulting matrix $\boldsymbol{C} = \boldsymbol{AB}$ has the dimension $m \times p$. When two matrices are multiplied, the row vectors of the left-hand side matrix undergo inner product or dot product vector multiplication[1] with the column vectors of the right-hand side matrix. An element c_{ij} in the resulting matrix $\boldsymbol{C}$ in row i and column j is the dot product of the elements in the ith row of $\boldsymbol{A}$ and the elements in the jth column of $\boldsymbol{B}$. Thus, the number of elements in the ith row of $\boldsymbol{A}$ *must equal* the number of elements in the jth column of $\boldsymbol{B}$ for the dot product operation to be valid. In other words, the number of columns in $\boldsymbol{A}$ must equal the number of rows in $\boldsymbol{B}$. It is *absolutely essential* that the inner dimensions of the two matrices being multiplied are the same. This is illustrated in the following:

$$\begin{array}{c} \boldsymbol{A} * \boldsymbol{B} \quad = \boldsymbol{C} \\ (m \times n) * (n \times p) = (m \times p) \\ \bigvee \\ \text{same inner dimension} \end{array}$$

The matrix multiplication of $\boldsymbol{A}$ and $\boldsymbol{B}$ is defined as

$$\boldsymbol{C} = \boldsymbol{AB} = \begin{bmatrix} a_{11} & a_{12} & a_{13} & \ldots & a_{1n} \\ a_{21} & a_{22} & a_{23} & \ldots & a_{2n} \\ & & \cdot & & \\ & & \cdot & & \\ & & \cdot & & \\ a_{m1} & a_{m2} & a_{m3} & \ldots & a_{mn} \end{bmatrix} \begin{bmatrix} b_{11} & b_{12} & b_{13} & \ldots & b_{1p} \\ b_{21} & b_{22} & b_{23} & \ldots & b_{2p} \\ & & \cdot & & \\ & & \cdot & & \\ & & \cdot & & \\ b_{n1} & b_{n2} & b_{n3} & \ldots & b_{np} \end{bmatrix}$$

$$= \begin{bmatrix} c_{11} & c_{12} & c_{13} & \ldots & c_{1p} \\ c_{21} & c_{22} & c_{23} & \ldots & c_{2p} \\ & & \cdot & & \\ & & \cdot & & \\ & & \cdot & & \\ c_{m1} & c_{m2} & c_{m3} & \ldots & c_{mp} \end{bmatrix}, \qquad (2.8)$$

where

$$c_{ij} = \boldsymbol{a}_i^{\mathrm{T}} \boldsymbol{b}_j \ (\text{or } \boldsymbol{a}_i \cdot \boldsymbol{b}_j) = a_{i1}b_{1j} + a_{i2}b_{2j} + \cdots + a_{in}b_{nj} \text{ for } 1 \leq i \leq m \text{ and } 1 \leq j \leq p.$$

Note that because matrix–matrix multiplication is not commutative, the order of the multiplication affects the result of the matrix multiplication operation.

A matrix can be interpreted as either a single column of row vectors (as discussed above) or as a single row of column vectors as shown in the following:

$$\boldsymbol{B} = \begin{bmatrix} b_{11} & b_{12} & b_{13} & \ldots & b_{1p} \\ b_{21} & b_{22} & b_{23} & \ldots & b_{2p} \\ & & \cdot & & \\ & & \cdot & & \\ & & \cdot & & \\ b_{n1} & b_{n2} & b_{n3} & \ldots & b_{np} \end{bmatrix} = [\boldsymbol{b}_1\ \boldsymbol{b}_2\ \ldots\ \boldsymbol{b}_p],$$

where

[1] There is another type of vector product called the "outer product" which results when a column vector on the left is multiplied with a row vector on the right.

$$\boldsymbol{b}_i = \begin{bmatrix} b_{1i} \\ b_{2i} \\ \cdot \\ \cdot \\ \cdot \\ b_{ni} \end{bmatrix} \quad \text{for } 1 \le i \le p.$$

If $\boldsymbol{A}$ is represented by a column series of row vectors and $\boldsymbol{B}$ is represented by a row series of column vectors, then the matrix multiplication of $\boldsymbol{A}$ and $\boldsymbol{B}$ can also be interpreted as

$$\boldsymbol{C} = \boldsymbol{AB} = \begin{bmatrix} a_{11} & a_{12} & a_{13} & \dots & a_{1n} \\ a_{21} & a_{22} & a_{23} & \dots & a_{2n} \\ & & \cdot & & \\ & & \cdot & & \\ & & \cdot & & \\ a_{m1} & a_{m2} & a_{m3} & \dots & a_{mn} \end{bmatrix} \begin{bmatrix} b_{11} & b_{12} & b_{13} & \dots & b_{1p} \\ b_{21} & b_{22} & b_{23} & \dots & b_{2p} \\ & & \cdot & & \\ & & \cdot & & \\ & & \cdot & & \\ b_{n1} & b_{n2} & b_{n3} & \dots & b_{np} \end{bmatrix}$$

$$= \begin{bmatrix} \boldsymbol{a}_1^{\mathrm{T}} \\ \boldsymbol{a}_2^{\mathrm{T}} \\ \cdot \\ \cdot \\ \cdot \\ \boldsymbol{a}_m^{\mathrm{T}} \end{bmatrix} \begin{bmatrix} \boldsymbol{b}_1 \; \boldsymbol{b}_2 \dots \boldsymbol{b}_p \end{bmatrix} = \begin{bmatrix} \boldsymbol{a}_1^{\mathrm{T}}\boldsymbol{b}_1 & \boldsymbol{a}_1^{\mathrm{T}}\boldsymbol{b}_2 & \dots & \boldsymbol{a}_1^{\mathrm{T}}\boldsymbol{b}_p \\ \boldsymbol{a}_2^{\mathrm{T}}\boldsymbol{b}_1 & \boldsymbol{a}_2^{\mathrm{T}}\boldsymbol{b}_2 & \dots & \boldsymbol{a}_2^{\mathrm{T}}\boldsymbol{b}_p \\ & \cdot & & \\ & \cdot & & \\ & \cdot & & \\ \boldsymbol{a}_m^{\mathrm{T}}\boldsymbol{b}_1 & \boldsymbol{a}_m^{\mathrm{T}}\boldsymbol{b}_2 & \dots & \boldsymbol{a}_m^{\mathrm{T}}\boldsymbol{b}_p \end{bmatrix}$$

(outer product)

Using MATLAB

The matrix multiplication operator is the same as the scalar multiplication operator, $*$. We define

```
>> A = [1 1; 2 2]; % 2 x 2 matrix
>> B = [1 2;1 2]; % 2 x 2 matrix
>> P = [3 2 3; 1 2 1]; % 2 x 3 matrix
```

Next we evaluate the matrix products:

```
>> C = A*B
C =
    2   4
    4   8
>> C = B*A
C =
    5   5
    5   5
>> C = A*P
C =
    4   4   4
    8   8   8
>> C = P*A
??? Error using ==> mtimes
Inner matrix dimensions must agree.
```

Can you determine why $\boldsymbol{P} * \boldsymbol{A}$ is undefined?

2.2.3 Vector and matrix norms

A **norm** of a vector or matrix is a real number that provides a measure of its length or magnitude. There are numerous instances in which we need to compare or determine the length of a vector or the magnitude (norm) of a matrix. In this chapter, you will be introduced to the concept of ill-conditioned problems in which the "condition number" measures the extent of difficulty in obtaining an accurate solution. The condition number depends on the norm of the associated matrix that defines the system coefficients.

The norm of a vector or matrix has several important properties as discussed in the following.

(1) The norm of a vector $\boldsymbol{v}$, denoted $\|\boldsymbol{v}\|$, and that of a matrix $\boldsymbol{A}$, denoted $\|\boldsymbol{A}\|$, is always positive and is equal to zero only when all the elements of the vector or matrix are zero. The double vertical lines on both sides of the vector that denote the scalar value of the vector length *should not be confused with* the single vertical lines used to denote the absolute value of a number. Only a zero vector or a zero matrix can have a norm equal to zero.

(2) The norm of the product of two vectors, a vector and a matrix, or two matrices is equal to or less than the product of the norms of the two quantities being multiplied; i.e., if $\boldsymbol{u}$ and $\boldsymbol{v}$ are vectors, and $\boldsymbol{A}$ and $\boldsymbol{B}$ are matrices, and the operations $\boldsymbol{uv}$, $\boldsymbol{Av}$, and $\boldsymbol{AB}$ are all defined, then
(a) $\|\boldsymbol{uv}\| \leq \|\boldsymbol{u}\|\|\boldsymbol{v}\|$,
(b) $\|\boldsymbol{Av}\| \leq \|\boldsymbol{A}\|\|\boldsymbol{v}\|$,
(c) $\|\boldsymbol{AB}\| \leq \|\boldsymbol{A}\|\|\boldsymbol{B}\|$.

(3) For any two vectors or matrices, $\|\boldsymbol{u}+\boldsymbol{v}\| \leq \|\boldsymbol{u}\| + \|\boldsymbol{v}\|$, which is also known as the **triangle inequality** or the **Cauchy–Schwarz inequality**.

Norm of a vector

In Euclidean vector space,[2] the norm of an $n \times 1$ vector $\boldsymbol{v}$ is defined as

$$\|\boldsymbol{v}\| = \sqrt{\boldsymbol{v}^{\mathrm{T}}\boldsymbol{v}} = \left(\sum_{i=1}^{i=n} v_i^2\right)^{1/2}. \tag{2.9}$$

The norm is thus the square root of the inner product of a vector with itself. This method of calculating the vector length may be familiar to you since this is the same method used to determine the length of vectors in our three-dimensional physical space.

A p-norm is a general method used to determine the vector length, and is defined as

$$\|\boldsymbol{v}\|_p = \left\{\sum_{i=1}^{n} |v_i|^p\right\}^{1/p}, \tag{2.10}$$

where p can take on any value. When $p = 2$, we obtain Equation (2.9), which calculates the Euclidean norm. Two other important norms commonly used are the $\boldsymbol{p} = \mathbf{1}$ **norm** and the $\boldsymbol{p} = \boldsymbol{\infty}$ **norm**:

$p = 1$ norm: $\|\boldsymbol{v}\|_1 = |v_1| + |v_2| + |v_3| + \cdots + |v_n|$;
$p = \infty$ norm: $\|\boldsymbol{v}\|_\infty = \max_{1 \leq i \leq n} |v_i|$.

[2] Euclidean vector space is a finite real vector space and is represented as R^n, i.e. each vector in this space is defined by a sequence of n real numbers. In Euclidean space, the inner product is defined by Equation (2.5). Other vector/function spaces may have different definitions for the vector norm and the inner product.

Norm of a matrix

A matrix is constructed to perform a **linear transformation** operation on a vector. The matrix operates or transforms the elements of the vector in a linear fashion to produce another vector (e.g. by rotating and/or stretching or shrinking the vector). If a vector of n elements is transformed into another vector of the same number of elements, i.e. the size remains unchanged, then the operating matrix is square and is called a **linear operator**. If $\boldsymbol{A}$ is the matrix that linearly transforms a vector $\boldsymbol{x}$ into a vector $\boldsymbol{b}$, this linear transformation operation is written as

$$\boldsymbol{A}\boldsymbol{x} = \boldsymbol{b}.$$

A **matrix norm** measures the maximum extent to which a matrix can lengthen a **unit vector** (a vector whose length or norm is one). If $\boldsymbol{x}$ is a unit vector the matrix norm is defined as

$$\|\boldsymbol{A}\| = \max_{\|\boldsymbol{x}\|=1} \|\boldsymbol{A}\boldsymbol{x}\|.$$

As we saw for vector norms, there are three commonly used matrix norms, the 1, 2, and ∞ norm. In this book we will concern ourselves with the 1 and ∞ norms, which are calculated, respectively, as

$$\|\boldsymbol{A}\|_1 = \max_{1 \leq j \leq n} \sum_{i=1}^{m} |a_{ij}| \tag{2.11}$$

or the maximum sum of column elements, and

$$\|\boldsymbol{A}\|_\infty = \max_{1 \leq i \leq n} \sum_{j=1}^{n} |a_{ij}| \tag{2.12}$$

or the maximum sum of row elements.

Using MATLAB

The `norm` function calculates the p-norm of a vector for any $p \geq 1$, and the norm of a matrix for $p = 1$, 2, and ∞. The syntax is

```
norm(x, p)
```

where $\boldsymbol{x}$ is a vector or matrix and p is the type of norm desired. If p is omitted, the default value of $p = 2$ is used. We have

```
>> u = [1 1 1];
>> norm(u)
ans =
    1.7321
>> norm (u,1)
ans =
     3
>> norm(u,inf)
ans =
     1
```

Note that infinity in MATLAB is represented by `inf`.

```
>> A = [3 2; 4 5];
>> norm(A,1)
```

```
ans =
      7
>> norm(A, inf)
ans =
      9
```

2.2.4 Linear combinations of vectors

A linear transformation operation is denoted by $\boldsymbol{Ax} = \boldsymbol{b}$; however, this equation is also a compact notation used to represent a system of simultaneous linear equations. How is this possible? If we assume that $\boldsymbol{A}$ is a 3×3 matrix and $\boldsymbol{x}$ and $\boldsymbol{b}$ are 3×1 vectors, then

$$\boldsymbol{Ax} = \begin{bmatrix} a_{11} & a_{12} & a_{13} \\ a_{21} & a_{22} & a_{23} \\ a_{31} & a_{32} & a_{33} \end{bmatrix} \begin{bmatrix} x_1 \\ x_2 \\ x_3 \end{bmatrix} = \begin{bmatrix} b_1 \\ b_2 \\ b_3 \end{bmatrix} = \boldsymbol{b}.$$

On performing the matrix–vector multiplication of $\boldsymbol{Ax}$, and equating the inner products between the three rows of $\boldsymbol{A}$ and the column vector $\boldsymbol{x}$ with the appropriate element of $\boldsymbol{b}$, we obtain

$$a_{11}x_1 + a_{12}x_2 + a_{13}x_3 = b_1,$$
$$a_{21}x_1 + a_{22}x_2 + a_{23}x_3 = b_2,$$
$$a_{31}x_1 + a_{32}x_2 + a_{33}x_3 = b_3.$$

If $\boldsymbol{x} = [x_1\ x_2\ x_3]$ is the only set of unknowns in this algebraic equation, then we have a set of three simultaneous linear equations in three variables x_1, x_2, and x_3. The properties of the matrix $\boldsymbol{A}$, which stores the constant coefficients of the equations in a systematic way, play a key role in determining the outcome and type of solution available for a system of linear equations. A system of linear equations can have *only one of three outcomes*:

(1) a unique solution;
(2) infinitely many solutions;
(3) no solution.

The outcome of the solution solely depends on the properties of $\boldsymbol{A}$ and $\boldsymbol{b}$. If a system of linear equations yields either a unique solution or infinitely many solutions, then the system is said to be **consistent**. On the other hand, if a system of linear equations has no solution, then the set of equations is said to be **inconsistent** (See Problem 2.1).

The topics "linear independence of vectors" and "vector spaces" are crucial to the understanding of the properties of $\boldsymbol{A}$, and are discussed next.

If $V = (\boldsymbol{v}_1, \boldsymbol{v}_2, \ldots, \boldsymbol{v}_n)$ is a set of vectors, and if another vector $\boldsymbol{u}$ can be expressed as

$$\boldsymbol{u} = k_1\boldsymbol{v}_1 + k_2\boldsymbol{v}_2 + \cdots + k_n\boldsymbol{v}_n,$$

where k_i for $1 \leq i \leq n$ are scalars, then $\boldsymbol{u}$ is said to be a **linear combination** of the given set of vectors V. For example, if

$$\boldsymbol{v}_1 = \begin{bmatrix} 2 \\ 3 \end{bmatrix}, \qquad \boldsymbol{v}_2 = \begin{bmatrix} 1 \\ 1 \end{bmatrix},$$

and

$$\boldsymbol{u} = \begin{bmatrix} 3 \\ 5 \end{bmatrix},$$

Figure 2.4

Three-dimensional rectilinear coordinate system and the unit vectors (length or norm = 1) ***i***, ***j***, and ***k*** along the *x*-, *y*-, and *z*-axes. This coordinate system defines our three-dimensional physical space.

Unit vectors along the *x*-, *y*-, and *z*-axes

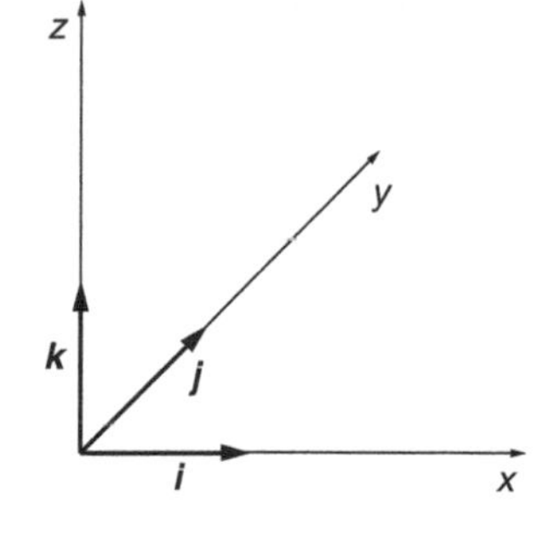

then, since we can write

$$\boldsymbol{u} = 2\boldsymbol{v_1} - \boldsymbol{v_2},$$

$\boldsymbol{u}$ can be expressed as a linear combination of vectors $\boldsymbol{v_1}$ and $\boldsymbol{v_2}$.

As another example, in three-dimensional space, every vector can be written as a linear combination of the mutually perpendicular orthonormal vectors (see Figure 2.4):

$$\boldsymbol{i} = \begin{bmatrix} 1 \\ 0 \\ 0 \end{bmatrix}, \boldsymbol{j} = \begin{bmatrix} 0 \\ 1 \\ 0 \end{bmatrix}, \boldsymbol{k} = \begin{bmatrix} 0 \\ 0 \\ 1 \end{bmatrix},$$

such as

$$\boldsymbol{a} = 4\boldsymbol{i} - 3\boldsymbol{j} + 7\boldsymbol{k}; \ \boldsymbol{b} = \boldsymbol{i} + \boldsymbol{k}; \ \boldsymbol{c} = 2\boldsymbol{i} + \boldsymbol{j}.$$

In fact, the matrix–vector multiplication operation $\boldsymbol{Ax}$ involves a linear combination of the matrix column vectors to produce another column vector $\boldsymbol{b}$. If a matrix $\boldsymbol{A}$ is written as a series of column vectors, then the following demonstrates how the transformation of the $(n \times 1)$ vector $\boldsymbol{x}$ by the $(m \times n)$ matrix $\boldsymbol{A}$ can be viewed as an operation involving a linear combination of the matrix columns:

$$\boldsymbol{b} = \boldsymbol{Ax} = \begin{bmatrix} a_{11} & a_{12} & a_{13} & \dots & a_{1n} \\ a_{21} & a_{22} & a_{23} & \dots & a_{2n} \\ & & \cdot & & \\ & & \cdot & & \\ & & \cdot & & \\ a_{m1} & a_{m2} & a_{m3} & \dots & a_{mn} \end{bmatrix} \begin{bmatrix} x_1 \\ x_2 \\ \cdot \\ \cdot \\ \cdot \\ x_n \end{bmatrix} = [\boldsymbol{a_1} \ \boldsymbol{a_2} \ \dots \ \boldsymbol{a_n}] \begin{bmatrix} x_1 \\ x_2 \\ \cdot \\ \cdot \\ \cdot \\ x_n \end{bmatrix},$$

or

$$\boldsymbol{b} = x_1\boldsymbol{a_1} + x_2\boldsymbol{a_2} + \cdots + x_n\boldsymbol{a_n},$$

or

$$\boldsymbol{b} = x_1 \begin{bmatrix} a_{11} \\ a_{21} \\ \cdot \\ \cdot \\ \cdot \\ a_{m1} \end{bmatrix} + x_2 \begin{bmatrix} a_{12} \\ a_{22} \\ \cdot \\ \cdot \\ \cdot \\ a_{m2} \end{bmatrix} + \cdots + x_n \begin{bmatrix} a_{1n} \\ a_{2n} \\ \cdot \\ \cdot \\ \cdot \\ a_{mn} \end{bmatrix},$$

where

$$\boldsymbol{a_i} = \begin{bmatrix} a_{1i} \\ a_{2i} \\ \cdot \\ \cdot \\ \cdot \\ a_{mi} \end{bmatrix} \text{ for } 1 \leq i \leq n,$$

and $\boldsymbol{b}$ is another column vector of size $m \times 1$.

In order for a vector $\boldsymbol{u}$ to be a linear combination of the vectors in set V, i.e.

$$\boldsymbol{u} = k_1\boldsymbol{v_1} + k_2\boldsymbol{v_2} + \cdots + k_n\boldsymbol{v_n},$$

it is necessary that at least one non-zero scalar k_i for $1 \leq i \leq n$ exists. If $\boldsymbol{u}$ cannot be constructed from the vectors contained in the vector set V, then no combination exists of one or more vectors in the set that can produce $\boldsymbol{u}$. In other words, there are no non-zero values for the scalar multipliers that permit $\boldsymbol{u}$ to be expressed as a linear combination of vectors $\boldsymbol{v_1}, \boldsymbol{v_2}, \ldots, \boldsymbol{v_n}$.

Now let's explore the properties of the vectors in the set V. Within a set of vectors $V = (\boldsymbol{v_1}, \boldsymbol{v_2}, \ldots, \boldsymbol{v_n})$, it may be possible that we can express one or more vectors as a linear combination of the other vectors in the set. We can write the vector equation

$$k_1\boldsymbol{v_1} + k_2\boldsymbol{v_2} + \cdots + k_n\boldsymbol{v_n} = 0,$$

where again k_i for $1 \leq i \leq n$ are scalars. The values of the scalar quantities for which this equation holds true determine whether a vector in the set can be expressed as a linear combination of one or more of the other vectors in the same set.

- If one or more scalars in this equation can take on non-zero values, then we say that the vectors in the set V are **linearly dependent**, i.e. a linear combination of one or more vectors in the set can produce another vector in the same set.
- If this equation holds true only when all $k_i = 0$ for $1 \leq i \leq n$, then the set of vectors is said to be **linearly independent**. In other words, none of the vectors in the set can be expressed as a linear combination of any of the other vectors in the same set.

For example, the mutually perpendicular orthonormal vectors:

$$\boldsymbol{i} = \begin{bmatrix} 1 \\ 0 \\ 0 \end{bmatrix}, \boldsymbol{j} = \begin{bmatrix} 0 \\ 1 \\ 0 \end{bmatrix}, \boldsymbol{k} = \begin{bmatrix} 0 \\ 0 \\ 1 \end{bmatrix}$$

are linearly independent. If we write the equation

$$k_1\boldsymbol{i} + k_2\boldsymbol{j} + k_3\boldsymbol{k} = 0,$$

we obtain

$$\begin{bmatrix} k_1 \\ k_2 \\ k_3 \end{bmatrix} = 0,$$

which is valid only for $k_1 = k_2 = k_3 = 0$.

2.2.5 Vector spaces and basis vectors

The space S that contains the set of vectors V and includes all vectors that result from vector addition and scalar multiplication operations on the set V, i.e. all possible linear combinations of the vectors in V is called a **vector space**. If $\boldsymbol{u}$ and $\boldsymbol{v}$ are two vectors, and if W is a real vector space containing $\boldsymbol{u}$ and $\boldsymbol{v}$, then W must also contain $p\boldsymbol{u} + q\boldsymbol{v}$, where p and q are any real scalar quantities. Commutative and associative properties of vector addition and subtraction as well as scalar multiplication must also hold in any vector space.

If the set of vectors $V = (\boldsymbol{v}_1, \boldsymbol{v}_2, \ldots, \boldsymbol{v}_n)$ and all linear combinations of the vectors contained in V completely define the size and extent of the vector space S, then S is said to be **spanned** by the set of vectors V. For example, the set of vectors

$$\boldsymbol{i} = \begin{bmatrix} 1 \\ 0 \\ 0 \end{bmatrix}, \boldsymbol{j} = \begin{bmatrix} 0 \\ 1 \\ 0 \end{bmatrix}, \boldsymbol{k} = \begin{bmatrix} 0 \\ 0 \\ 1 \end{bmatrix}$$

span the three-dimensional real space (R^3).

If the vectors contained in set V are *linearly independent*, and if V *spans* a vector space S, then the set of vectors V is called a **basis** for the vector space S, and the individual vectors in V are called **basis vectors** of S. There can be many basis sets for any vector space S; however, all basis sets for the vector space S contain the same number of linearly independent vectors. The number of vectors in a basis set determines the **dimension** of the vector space S. If there are n vectors in a basis set V for vector space S, then the dimension of the vector space is n. Any set of vectors that span S of dimension n and has more than n vectors *must be* a linearly dependent set of vectors. If R is the set of real numbers, then R^n is a real space of dimension n and has n vectors in any of its basis sets. For example, R^3 represents the three-dimensional physical space that we live in, and an example of a basis vector set for this vector space is $V = (\boldsymbol{i}, \boldsymbol{j}, \boldsymbol{k})$.

A matrix $\boldsymbol{A}$ of size $m \times n$ has a set of n column vectors:

$$\boldsymbol{A} = \begin{bmatrix} a_{11} & a_{12} & a_{13} & \ldots & a_{1n} \\ a_{21} & a_{22} & a_{23} & \ldots & a_{2n} \\ & & \cdot & & \\ & & \cdot & & \\ & & \cdot & & \\ a_{m1} & a_{m2} & a_{m3} & \ldots & a_{mn} \end{bmatrix} = [\boldsymbol{a}_1 \; \boldsymbol{a}_2 \; \ldots \; \boldsymbol{a}_n],$$

where

$$\boldsymbol{a}_i = \begin{bmatrix} a_{1i} \\ a_{2i} \\ \cdot \\ \cdot \\ \cdot \\ a_{mi} \end{bmatrix} \quad \text{for } 1 \leq i \leq n.$$

The vector space spanned by the n column vectors is called the **column space** of A. If the column vectors are linearly independent, then the dimension of the column space of A is n. If the column vectors are linearly dependent on one another, i.e. at least one column vector can be expressed in terms of a linear combination of one or more other column vectors, then the dimension of the column space is $<n$. (The vector space spanned by the m row vectors of the $m \times n$ matrix A is called the **row space**[3] of A.)

Example 2.1 Column space of a matrix

The 3×4 matrix

$$\mathbf{D} = \begin{bmatrix} 1 & 0 & 0 & 5 \\ 0 & 1 & 0 & 8 \\ 0 & 0 & 1 & 9 \end{bmatrix}$$

has four column vectors

$$\mathbf{d}_1 = \begin{bmatrix} 1 \\ 0 \\ 0 \end{bmatrix}, \mathbf{d}_2 = \begin{bmatrix} 0 \\ 1 \\ 0 \end{bmatrix}, \mathbf{d}_3 = \begin{bmatrix} 0 \\ 0 \\ 1 \end{bmatrix}, \text{ and } \mathbf{d}_4 = \begin{bmatrix} 5 \\ 8 \\ 9 \end{bmatrix}.$$

The column space for this matrix is that vector space that is spanned by the four column vectors. Each column vector can be expressed as a linear combination of the other three column vectors. For example,

$$\mathbf{d}_4 = 5\mathbf{d}_1 + 8\mathbf{d}_2 + 9\mathbf{d}_3$$

and

$$\mathbf{d}_1 = \frac{1}{5}(\mathbf{d}_4 - 8\mathbf{d}_2 - 9\mathbf{d}_3).$$

Thus, the column vectors are *linearly dependent*. However, any vector set obtained by choosing three out of the four column vectors can be shown to be *linearly independent*. The *dimension* of the column space of $\mathbf{D}$ is therefore 3. In other words, a set of any three linearly independent vectors (each of size 3×1) can completely define the column space of $\mathbf{D}$.

Now that you are familiar with the concepts of column space of a matrix and linear combination of vectors, we introduce the simple criterion for the existence of a solution to a given set of linear equations $Ax = b$, where A is called the coefficient matrix. The matrix–vector multiplication Ax, which forms the left-hand side of the linear equations, produces a column vector b (right-hand side of the equations) that is a linear combination of the column vectors of A.

A solution $\mathbf{x}$ to a set of linear equations $\mathbf{Ax} = \mathbf{b}$ can be found if and only if the column vector $\mathbf{b}$ lies in the column space of $\mathbf{A}$, in which case the system of linear equations is said to be **consistent**. If the column vector $\mathbf{b}$ does not lie in the column space of $\mathbf{A}$ (then $\mathbf{b}$ cannot be formed from a linear combination of the column vectors of $\mathbf{A}$), no solution to the system of linear equations exists, and the set of equations is said to be **inconsistent**.

[3] If the row vectors are linearly independent, then the dimension of the row space of A is m; otherwise, the dimension of the row space is $< m$. For any matrix A, the row space and the column space will always have the *same* dimension.

2.2.6 Rank, determinant, and inverse of matrices

Matrix rank

The dimension of the column space of $\boldsymbol{A}$ (or equivalently the row space of $\boldsymbol{A}$) is called the **rank** of the matrix $\boldsymbol{A}$. The rank of a matrix indicates the number of linearly independent row vectors or number of linearly independent column vectors contained in the matrix. It is always the case that the number of linearly independent column vectors equals the number of linearly independent row vectors in any matrix. Since $\boldsymbol{A}$ has a set of n column vectors and a set of m row vectors, if $n > m$ the column vectors must be linearly dependent. Similarly, if $m > n$ the row vectors of the matrix must be linearly dependent. The rank of an $m \times n$ matrix $\boldsymbol{A}$ is always less than or equal to the minimum of m and n. If rank($\boldsymbol{A}$) < minimum of (m, n), then the column vectors form a linearly dependent set and so do the row vectors. Such a matrix $\boldsymbol{A}$ is said to be **rank deficient**.

The rank of the coefficient matrix $\boldsymbol{A}$ is of primary importance in determining whether a unique solution exists, infinitely many solutions exist, or no solution exists for a system of linear equations $\boldsymbol{Ax} = \boldsymbol{b}$. The test for consistency of a system of linear equations involves the construction of an augmented matrix [$\boldsymbol{A}$ $\boldsymbol{b}$], in which the column vector $\boldsymbol{b}$ is added as an additional column to the right of the columns of the $m \times n$ coefficient matrix $\boldsymbol{A}$. The vertical dotted line in the matrix representation shown below separates the coefficients a_{ij} of the variables from the constants b_i of the equations:

$$\textbf{augmented matrix} \rightarrow \boldsymbol{aug\ A} = [\boldsymbol{A}:\boldsymbol{b}]$$

$$= \begin{bmatrix} a_{11} & a_{12} & a_{13} & \ldots & a_{1n} & : & b_1 \\ a_{21} & a_{22} & a_{23} & \ldots & a_{2n} & : & b_2 \\ & \cdot & & & & : & \\ & \cdot & & & & : & \\ & \cdot & & & & : & \\ a_{m1} & a_{m2} & a_{m3} & \ldots & a_{mn} & : & b_m \end{bmatrix}.$$

Consistency of the system of linear equations is guaranteed if rank($\boldsymbol{A}$) = rank(***aug*** $\boldsymbol{A}$). Why? If the rank remains unchanged even after augmenting $\boldsymbol{A}$ with the column vector $\boldsymbol{b}$, then this ensures that the columns of ***aug A*** are linearly dependent, and thus $\boldsymbol{b}$ lies within the column space of $\boldsymbol{A}$. In Example 2.1, if $\boldsymbol{D}$ is an augmented matrix, i.e.

$$\boldsymbol{A} = \begin{bmatrix} 1 & 0 & 0 \\ 0 & 1 & 0 \\ 0 & 0 & 1 \end{bmatrix}, \ \boldsymbol{b} = \begin{bmatrix} 5 \\ 8 \\ 9 \end{bmatrix}, \text{ and } \boldsymbol{D} = [\boldsymbol{A}\ \boldsymbol{b}],$$

then rank($\boldsymbol{A}$) = 3, rank($\boldsymbol{D}$) = 3, and the system of linear equations defined by $\boldsymbol{D}$ is consistent.

If rank(***aug*** $\boldsymbol{A}$) > rank($\boldsymbol{A}$), then, necessarily, the number of linearly independent columns in the augmented matrix is greater than that in $\boldsymbol{A}$, and thus $\boldsymbol{b}$ cannot be expressed as a linear combination of the column vectors of $\boldsymbol{A}$ and thus does not lie in the column space of $\boldsymbol{A}$.

For example, if

$$\boldsymbol{D} = [\boldsymbol{A}\ \boldsymbol{b}] = \begin{bmatrix} 1 & 0 & 0 & 5 \\ 0 & 1 & 0 & 8 \\ 0 & 0 & 0 & 9 \end{bmatrix}$$

where

$$A = \begin{bmatrix} 1 & 0 & 0 \\ 0 & 1 & 0 \\ 0 & 0 & 0 \end{bmatrix}, \quad \boldsymbol{b} = \begin{bmatrix} 5 \\ 8 \\ 9 \end{bmatrix},$$

you should show yourself that rank($\boldsymbol{A}$)$\neq$rank($\boldsymbol{D}$). The system of equations defined by $\boldsymbol{D}$ is inconsistent and has no solution.

The next question is: how does the rank of a matrix determine whether a unique solution exists or infinitely many solutions exist for a *consistent* set of linear equations?

- If rank($\boldsymbol{A}$) = the number of column vectors n of the coefficient matrix $\boldsymbol{A}$, then the coefficient matrix has n linearly independent row vectors and column vectors, which is also equal to the number of unknowns or the number of elements in $\boldsymbol{x}$. Thus, there exists a unique solution to the system of linear equations.
- If rank($\boldsymbol{A}$) < the number of column vectors n of the coefficient matrix $\boldsymbol{A}$, then the number of linearly independent row vectors (or column vectors) of $\boldsymbol{A}$ is less than the number of unknowns in the set of linear equations. The linear problem is underdefined, and there exist infinitely many solutions.

In summary, the test that determines the type of solution that a system of linear equations will yield is defined below.

Unique solution:

(1) rank($\boldsymbol{A}$) = rank([$\boldsymbol{A}$ $\boldsymbol{b}$]), and
(2) rank($\boldsymbol{A}$) = number of column vectors of $\boldsymbol{A}$.

Infinitely many solutions:

(1) rank($\boldsymbol{A}$) = rank([$\boldsymbol{A}$ $\boldsymbol{b}$]), and
(2) rank($\boldsymbol{A}$) < number of column vectors of $\boldsymbol{A}$.

No solution:

(1) rank($\boldsymbol{A}$) < rank([$\boldsymbol{A}$ $\boldsymbol{b}$]).

In this chapter we are concerned with finding the solution for systems of equations that are either

- consistent and have a unique solution, or
- inconsistent but overdetermined (or overspecified – a greater number of equations than unknowns), i.e. rank($\boldsymbol{A}$) = n and $m > n$.

Using MATLAB

The `rank` function is provided by MATLAB to determine the rank of any matrix. The use of `rank` is demonstrated below:

```
>> A = 2*ones(3,3) + diag([1 1], -1)
```

(Note that the second parameter in the `diag` function indicates the position of the diagonal of interest, i.e. the main diagonal or diagonals below or above the main diagonal.)

```
A =
       2   2   2
       3   2   2
       2   3   2
>> rank(A)
ans =
       3
>> B = [1 0 2; 2 6 4; 3 4 6]
B =
     1   0   2
     2   6   4
     3   4   6
>> rank(B)
ans =
       2
```

Can you see why the matrix $\boldsymbol{B}$ does not have three linearly independent rows/columns?

Determinant of a matrix

The **determinant** is a scalar quantity that serves as a useful metric, and is defined for square $n \times n$ matrices. The determinant of a matrix $\boldsymbol{A}$ is denoted by $\det(\boldsymbol{A})$ or $|\boldsymbol{A}|$. Note that $|\boldsymbol{A}|$ is non-zero for square matrices $\boldsymbol{A}$ of rank $= n$ and is zero for all square matrices of rank $< n$. The determinant has properties that are different from its corresponding matrix.

If $\boldsymbol{A}$ is an $n \times n$ matrix, then

$$\boldsymbol{A} = \begin{bmatrix} a_{11} & a_{12} & a_{13} & \ldots & a_{1n} \\ a_{21} & a_{22} & a_{23} & \ldots & a_{2n} \\ & & \cdot & & \\ & & \cdot & & \\ & & \cdot & & \\ a_{n1} & a_{n2} & a_{n3} & \ldots & a_{nn} \end{bmatrix},$$

and the determinant of $\boldsymbol{A}$ is written as

$$\det(\boldsymbol{A}) = \begin{vmatrix} a_{11} & a_{12} & a_{13} & \ldots & a_{1n} \\ a_{21} & a_{22} & a_{23} & \ldots & a_{2n} \\ & & \cdot & & \\ & & \cdot & & \\ & & \cdot & & \\ a_{n1} & a_{n2} & a_{n3} & \ldots & a_{nn} \end{vmatrix}.$$

Note the different styles of brackets used to distinguish a matrix from a determinant.

In order to calculate the determinant of a matrix, it is first necessary to know how to calculate the **cofactor** of an element in a matrix. For any element a_{ij} of matrix $\boldsymbol{A}$, the **minor** of a_{ij} is denoted as M_{ij} and is defined as the determinant of the **submatrix** that is obtained by deleting from $\boldsymbol{A}$ the ith row and the jth column that contain a_{ij}. For example, the minor of a_{11} is given by

$$\boldsymbol{M}_{11} = \begin{vmatrix} a_{22} & a_{23} & \ldots & a_{2n} \\ a_{32} & a_{33} & \ldots & a_{3n} \\ & & \cdot & \\ & & \cdot & \\ a_{n2} & a_{n3} & \ldots & a_{nn} \end{vmatrix}.$$

The **cofactor** of a_{ij} is defined as $A_{ij} = (-1)^{i+j} M_{ij}$. The determinant of a matrix $\boldsymbol{A}$ can be calculated using either of two formulas.

(1) Row-wise calculation of the determinant

$$|\boldsymbol{A}| = \sum_{j=1}^{n} a_{ij}A_{ij} \qquad \text{for any single choice of } i \in \{1, 2, \ldots, n\}. \tag{2.13}$$

(2) Column-wise calculation of the determinant

$$|\boldsymbol{A}| = \sum_{i=1}^{n} a_{ij}A_{ij} \qquad \text{for any single choice of } j \in \{1, 2, \ldots, n\}. \tag{2.14}$$

Note that the determinant of a single 1×1 matrix $|a_{11}| = a_{11}$.

A 2×2 determinant is calculated by expanding along the first row:

$$\begin{vmatrix} a_{11} & a_{12} \\ a_{21} & a_{22} \end{vmatrix} = a_{11}|a_{22}| - a_{12}|a_{21}| = a_{11}a_{22} - a_{12}a_{21}.$$

Similarly, a 3×3 determinant is calculated by expanding along the first row:

$$\begin{vmatrix} a_{11} & a_{12} & a_{13} \\ a_{21} & a_{22} & a_{23} \\ a_{31} & a_{32} & a_{33} \end{vmatrix} = a_{11}\begin{vmatrix} a_{22} & a_{23} \\ a_{32} & a_{33} \end{vmatrix} - a_{12}\begin{vmatrix} a_{21} & a_{23} \\ a_{31} & a_{33} \end{vmatrix} + a_{13}\begin{vmatrix} a_{21} & a_{22} \\ a_{31} & a_{32} \end{vmatrix}$$
$$= a_{11}(a_{22}a_{33} - a_{23}a_{32}) - a_{12}(a_{21}a_{33} - a_{23}a_{31}).$$
$$+ a_{13}(a_{21}a_{32} - a_{22}a_{31})$$

If the determinant of a matrix is non-zero, the matrix is said to be **non-singular**; if the determinant is zero, the matrix is said to be **singular**.

The calculation of determinants, especially for matrices of size $n \geq 4$, is likely to involve numerical errors involving scaling (overflow error) and round-off errors, and requires a formidable number of computational steps. Expressing a matrix as a product of diagonal or triangular matrices (LU factorization) enables faster and simpler evaluation of the determinant and requires minimal number of computations. MATLAB provides the `det` function to calculate the determinant of a matrix.

Inverse of a matrix

Unlike matrix multiplication, matrix division is not formally defined. However, inversion of a matrix is a well-defined procedure that is equivalent to matrix division. Only square matrices that are non-singular ($\det(\boldsymbol{A}) \neq 0$) can be inverted. Such matrices are said to be **invertible**, and the inverted matrix $\boldsymbol{A}^{-1}$ is called the **inverse** of $\boldsymbol{A}$. A number divided by itself is equal to one. Similarly, if a matrix is multiplied by its inverse, the identity matrix $\boldsymbol{I}$ is the result. Remember that the identity matrix is a diagonal matrix that has ones on the diagonal and zeros everywhere else. Thus,

$$\boldsymbol{A}\boldsymbol{A}^{-1} = \boldsymbol{A}^{-1}\boldsymbol{A} = \boldsymbol{I}. \tag{2.15}$$

It can be shown that only one inverse of an invertible matrix exists that satisfies Equation (2.15). The method used to calculate the inverse of a matrix is derived from first principles in many standard linear algebra textbooks (see, for instance, Davis and Thomson (2000)) and is not discussed here. A brief demonstration of the method used to calculate the inverse is shown in the following for a 3×3 square matrix $\boldsymbol{A}$. If

$$A = \begin{bmatrix} a_{11} & a_{12} & a_{13} \\ a_{21} & a_{22} & a_{23} \\ a_{31} & a_{32} & a_{33} \end{bmatrix}$$

then

$$A^{-1} = \frac{1}{|A|} \begin{bmatrix} A_{11} & A_{21} & A_{31} \\ A_{12} & A_{22} & A_{32} \\ A_{13} & A_{23} & A_{33} \end{bmatrix}, \tag{2.16}$$

where A_{ij} are cofactors of the elements of A. Note that the ijth element of the inverse A^{-1} is equal to the cofactor of the jith element of A divided by the determinant of A. MATLAB provides the `inv` function to calculate the inverse of a matrix.

For the matrix equation $Ax = b$, the solution x can be written as $x = A^{-1}b$, where A^{-1} is the inverse of A. However, computation of the solution by calculating the inverse is less efficient since it involves evaluating the determinant and requires many more computational steps than solution techniques that are specially designed for efficiently solving a system of linear equations. Some of these solution techniques will be discussed in subsequent sections in this chapter.

An excellent introductory text on linear algebra by Anton (2004) is recommended for further reading.

2.3 Matrix representation of a system of linear equations

A linear equation is defined as an equation that has a linear dependency on the variables or unknowns in the equation. When the problem at hand involves more than one variable, a system of coupled equations that describe the physical or theoretical problem is formulated and solved simultaneously to obtain the solution. If all the equations in the system are linear with respect to the $n > 1$ unknowns, then we have a system of linear equations that can be solved using linear algebra theory. If we have a system of n linear equations in n variables,

$$\begin{aligned} a_{11}x_1 + a_{12}x_2 + a_{13}x_3 + \cdots + a_{1n}x_n &= b_1, \\ a_{21}x_1 + a_{22}x_2 + a_{23}x_3 + \cdots + a_{2n}x_n &= b_2, \\ &\vdots \\ a_{n1}x_1 + a_{n2}x_2 + a_{n3}x_3 + \cdots + a_{nn}x_n &= b_n, \end{aligned}$$

then this system can be represented in the following compact form:

$$Ax = b,$$

where

$$A = \begin{bmatrix} a_{11} & a_{12} & a_{13} & \ldots & a_{1n} \\ a_{21} & a_{22} & a_{23} & \ldots & a_{2n} \\ & & \cdot & & \\ & & \cdot & & \\ & & \cdot & & \\ a_{n1} & a_{n2} & a_{n3} & \ldots & a_{nn} \end{bmatrix}, \quad x = \begin{bmatrix} x_1 \\ x_2 \\ \cdot \\ \cdot \\ \cdot \\ x_n \end{bmatrix}, \quad \text{and } b = \begin{bmatrix} b_1 \\ b_2 \\ \cdot \\ \cdot \\ \cdot \\ b_n \end{bmatrix}.$$

For this system of n equations to have a unique solution, it is necessary that the determinant of $\boldsymbol{A}$, $|\boldsymbol{A}| \neq 0$. If this is the case, the $n \times (n + 1)$ augmented matrix ***aug A*** = [***A b***] will have the same rank as that of $\boldsymbol{A}$. The solution of a consistent set of n linear equations in n unknowns is now reduced to a matrix problem that can be solved using standard matrix manipulation techniques prescribed by linear algebra. Sections 2.4–2.6 discuss popular algebraic techniques used to manipulate either $\boldsymbol{A}$ or ***aug A*** to arrive at the desired solution. The pitfalls of using these solution techniques in the event that $|\boldsymbol{A}|$ is close to zero are also explored in these sections.

2.4 Gaussian elimination with backward substitution

Gaussian elimination with backward substitution is a popular method used to solve a system of equations that is consistent and has a coefficient matrix whose determinant is non-zero. This method is a **direct** method. It involves a pre-defined number of steps that only depends on the number of unknowns or variables in the equations. Direct methods can be contrasted with iterative techniques, which require a set of guess values to start the solution process and may converge upon a solution after a number of iterations that is not known *a priori* and depends on the initial guess values supplied. As the name implies, this solution method involves a two-part procedure:

(1) **Gaussian elimination**, to reduce the set of linear equations to an upper triangular format, and
(2) **backward substitution**, to solve for the unknowns by proceeding step-wise from the last reduced equation to the first equation (i.e. "backwards") and is suitable for a system of equations in "upper" triangular form.

2.4.1 Gaussian elimination without pivoting

Example 2.2

The basic technique of Gaussian elimination with backward substitution can be illustrated by solving the following set of three simultaneous linear equations numbered (E1)–(E3):

$$2x_1 - 4x_2 + 3x_3 = 1, \tag{E1}$$

$$x_1 + 2x_2 - 3x_3 = 4, \tag{E2}$$

$$3x_1 + x_2 + x_3 = 12. \tag{E3}$$

Our first goal is to reduce the equations to a form that allows the formation of a quick and easy solution for the unknowns. Gaussian elimination is a process that uses linear combinations of equations to eliminate variables from equations. The theoretical basis behind this elimination method is that the solution (x_1, x_2, x_3) to the set of equations (E1), (E2), (E3) satisfies (E1), (E2) and (E3) individually *and* any linear combinations of these equations, such as (E2) – 0.5(E1) or (E3) + (E2).

Our first aim in this elimination process is to remove the term x_1 from (E2) and (E3) using (E1). If we multiply (E1) with the constant 0.5 and subtract it from (E2), we obtain a new equation for (E2):

$$4x_2 - 4.5x_3 = 3.5. \tag{E2$'$}$$

If we multiply (E1) with the constant 1.5 and subtract it from (E3), we obtain a new equation for (E3):

$$7x_2 - 3.5x_3 = 10.5 \tag{E3$'$}$$

or

$$x_2 - 0.5x_3 = 1.5 \quad \text{(E3')}$$

We now have an **equivalent** set of equations that have different coefficients and constants compared to the original set, but which have the same exact solution:

$$2x_1 - 4x_2 + 3x_3 = 1, \quad \text{(E1)}$$

$$4x_2 - 4.5x_3 = 3.5, \quad \text{(E2')}$$

$$x_2 - 0.5x_3 = 1.5. \quad \text{(E3')}$$

Our next aim is to remove the term x_2 from Equation (E3′) using (E2′). This is done by multiplying (E2′) by 0.25 and subtracting the resulting equation from (E3′). We obtain a new equation for (E3′):

$$0.625x_3 = 0.625. \quad (E3')$$

Our new set of equivalent linear equations is now in triangular form:

$$2x_1 - 4x_2 + 3x_3 = 1, \quad \text{(E1)}$$

$$4x_2 - 4.5x_3 = 3.5, \quad \text{(E2')}$$

$$0.625x_3 = 0.625. \quad \text{(E3')}$$

Note that the Gaussian elimination procedure involves modification of only the coefficients of the variables and the constants of the equation. If the system of linear equations is expressed as an augmented matrix, similar transformations can be carried out on the row elements of the 3×4 augmented matrix.

Back substitution is now employed to solve for the unknowns. Equation (E3″) is the easiest to solve since the solution is obtained by simply dividing the constant term by the coefficient of x_3. Once x_3 is known, it can be plugged into (E2′) to solve for x_2. Then x_1 can be obtained by substituting the values of x_2 and x_3 into (E1). Thus,

$$x_3 = \frac{0.625}{0.625} = 1,$$

$$x_2 = \frac{3.5 + 4.5 * 1}{4} = 2,$$

$$x_1 = \frac{1 + 4 * 2 - 3 * 1}{2} = 3.$$

Next, we generalize this solution method for any system of four linear equations as shown below:

$$\begin{aligned}
a_{11}x_1 + a_{12}x_2 + a_{13}x_3 + a_{14}x_4 &= b_1, \\
a_{21}x_1 + a_{22}x_2 + a_{23}x_3 + a_{24}x_4 &= b_2, \\
a_{31}x_1 + a_{32}x_2 + a_{33}x_3 + a_{34}x_4 &= b_3, \\
a_{41}x_1 + a_{42}x_2 + a_{43}x_3 + a_{44}x_4 &= b_4.
\end{aligned}$$

A system of four linear equations can be represented by a 4×5 augmented matrix that contains the coefficients a_{ij} of the variables and the constant terms b_i on the right-hand side of the equations:

$$\textbf{\textit{aug}}\ A = \begin{bmatrix} a_{11} & a_{12} & a_{13} & a_{14} & : & b_1 \\ a_{21} & a_{22} & a_{23} & a_{24} & : & b_2 \\ a_{31} & a_{32} & a_{33} & a_{34} & : & b_3 \\ a_{41} & a_{42} & a_{43} & a_{44} & : & b_4 \end{bmatrix}.$$

The elements of a matrix can be modified based on certain rules or allowable operations called **elementary row transformations**, which permit the following:

(1) interchanging of any two rows;
(2) multiplication of the elements of a row by a non-zero scalar quantity;
(3) addition of a non-zero scalar multiple of the elements of one row to corresponding elements of another row.

A matrix that is modified by one or more of these elementary transformations is called an **equivalent matrix** since the solution of the corresponding linear equations remains the same. The purpose of Gaussian elimination when applied to the augmented matrix is to reduce it to an **upper triangular matrix**. An upper triangular matrix has non-zero elements on the diagonal and above the diagonal and has only zeros located below the diagonal. It is the opposite for the case of a **lower triangular matrix**, which has non-zero elements only on and below the diagonal.

The elimination process can be broken down into three overall steps:

(1) use a_{11} to zero the elements in column 1 (x_1 coefficients) in rows 2, 3, and 4;
(2) use a_{22} to zero the elements below it in column 2 (x_2 coefficients) in rows 3 and 4;
(3) use a_{33} to zero the element below it in column 3 (x_3 coefficient) in row 4.

Step 1

In step 1, the first row is multiplied by an appropriate scalar quantity and then subtracted from the second row so that $a_{21} \to 0$. The first row is then multiplied by another pre-determined scalar and subtracted from the third row such that $a_{31} \to 0$. This procedure is again repeated for the fourth row so that $a_{41} \to 0$. The first row is called the **pivot row**. The pivot row itself remains unchanged, but is used to change the rows below it. Since a_{11} is the element that is used to zero the other elements beneath it, it is called the **pivot element**. It acts as the pivot (a stationary point), about which the other elements are transformed. The column in which the pivot element lies, i.e. the column in which the zeroing takes place, is called the **pivot column**.

To convert a_{21} to zero, the elements of the first row are divided by a_{11}, multiplied by a_{21}, and then subtracted from corresponding elements of the second row. This is mathematically expressed as

$$a'_{2j} = a_{2j} - \frac{a_{21}}{a_{11}} a_{1j} \qquad 1 \le j \le 5,$$

where the prime indicates the new or modified value of the element. Hence, when $j = 1$ we obtain

$$a'_{21} = a_{21} - \frac{a_{21}}{a_{11}} a_{11} = 0.$$

To create a zero at the position of a_{31}, the first row is multiplied by the quantity a_{31}/a_{11} and subtracted from the third row. This is mathematically expressed as

$$a'_{3j} = a_{3j} - \frac{a_{31}}{a_{11}} a_{1j} \qquad 1 \le j \le 5.$$

Similarly, for the fourth row, the following elementary row transformation is carried out:

$$a'_{4j} = a_{4j} - \frac{a_{41}}{a_{11}} a_{1j} \qquad 1 \leq j \leq 5.$$

The *algorithm* for step 1 is thus given by

for $i = 2$ to 4
  for $j = 1$ to 5
    $a'_{ij} = a_{ij} - \frac{a_{i1}}{a_{11}} a_{1j}$
  end j
end i

The resulting equivalent augmented matrix is given by

$$\textbf{\textit{aug A}} = \begin{bmatrix} a_{11} & a_{12} & a_{13} & a_{14} & : & b_1 \\ 0 & a'_{22} & a'_{23} & a'_{24} & : & b'_2 \\ 0 & a'_{32} & a'_{33} & a'_{34} & : & b'_3 \\ 0 & a'_{42} & a'_{43} & a'_{44} & : & b'_4 \end{bmatrix},$$

where the primes indicate that the elements have undergone a single modification.

Step 2

In step 2, the second row is now the pivot row and a'_{22} is the pivot element, which is used to zero the elements a'_{32} and a'_{42} below it. Again, no changes occur to the pivot row elements in this step. The elementary row transformation that results in $a'_{32} \rightarrow 0$, and the modification of the other elements in row 3, is mathematically expressed as

$$a''_{3j} = a'_{3j} - \frac{a'_{32}}{a'_{22}} a'_{2j} \quad 2 \leq j \leq 5,$$

where the double prime indicates the new or modified value of the element resulting from step 2 transformations. These row operations do not change the value of the zero element located in column 1 in row 3, and therefore performing this calculation for $j = 1$ is unnecessary.

To zero the element a'_{42}, the second row is multiplied by the quantity a'_{42}/a'_{22} and is subtracted from the fourth row. This is mathematically expressed as

$$a''_{4j} = a'_{4j} - \frac{a'_{42}}{a'_{22}} a'_{2j} \quad 2 \leq j \leq 5.$$

The algorithm for step 2 is thus given by

for $i = 3$ to 4
  for $j = 2$ to 5
    $a''_{ij} = a'_{ij} - \frac{a'_{i2}}{a'_{22}} a'_{2j}$
  end j
end i

The resulting equivalent augmented matrix is given by

$$aug\ A = \begin{bmatrix} a_{11} & a_{12} & a_{13} & a_{14} & : & b_1 \\ 0 & a'_{22} & a'_{23} & a'_{24} & : & b'_2 \\ 0 & 0 & a''_{33} & a''_{34} & : & b''_3 \\ 0 & 0 & a''_{43} & a''_{44} & : & b''_4 \end{bmatrix},$$

where a single prime indicates that the corresponding elements have had a single change in value (in step 1), while double primes indicate that the elements have been modified twice (in steps 1 and 2).

Step 3

The pivot row shifts from the second to the third row in this step. The new pivot element is a''_{33}. Only the last row is modified in this step in order to zero the a''_{43} element. The third row is multiplied by the quantity a''_{43}/a''_{33} and subtracted from the fourth row. This is mathematically expressed as

$$a'''_{4j} = a''_{4j} - \frac{a''_{43}}{a''_{33}} a''_{3j} \quad 3 \leq j \leq 5.$$

The algorithm for step 3 is thus

for $i = 4$ to 4
 for $j = 3$ to 5

$$a'''_{ij} = a''_{ij} - \frac{a''_{i3}}{a''_{33}} a''_{3j}$$

 end j
end i

The resulting equivalent augmented matrix is given by

$$aug\ A = \begin{bmatrix} a_{11} & a_{12} & a_{13} & a_{14} & : & b_1 \\ 0 & a'_{22} & a'_{23} & a'_{24} & : & b'_2 \\ 0 & 0 & a''_{33} & a''_{34} & : & b''_3 \\ 0 & 0 & 0 & a'''_{44} & : & b'''_4 \end{bmatrix}, \qquad (2.17)$$

where a triple prime indicates that the elements have been modified three times (in steps 1, 2, and 3).

The algorithms for the three individual steps discussed above can be consolidated to arrive at an overall algorithm for the Gaussian elimination method for a 4×5 augmented matrix, which can easily be extended to any $n \times (n + 1)$ augmented matrix. In the algorithm, the $a_{i,n+1}$ elements correspond to the b_i constants for $1 \leq i \leq n$.

Algorithm for Gaussian elimination without pivoting

(1) Loop through $k = 1$ to $n - 1$; k specifies the pivot row (and pivot column).
(2) Assign pivot $= a_{kk}$.
(3) Loop through $i = (k + 1)$ to n; i specifies the row being transformed.
(4) Calculate the multiplier $m = a_{ik}/\text{pivot}$.
(5) Loop through $j = k$ to $n + 1$; j specifies the column in which the individual element is modified.
(6) Calculate $a_{ij} = a_{ij} - m a_{kj}$.

(7) End loop j.
(8) End loop i.
(9) End loop k.

Written in MATLAB programming style, the algorithm appears as follows:

```
for k = 1:n-1
    pivot = a(k,k)
    for row = k+1:n
        m = a(row,k)/pivot;
        a(row,k+1:n+1) = a(row,k+1:n+1) - m*a(k,k+1:n+1);
    end
end
```

Note that the calculation of the zeroed element a(i,k) is excluded in the MATLAB code above since we already know that its value is zero and is inconsequential thereafter.

Now that the elimination step is complete, the back substitution step is employed to obtain the solution. The equivalent upper triangular 4×5 augmented matrix given by Equation (2.17) produces a set of four linear equations in triangular form.

The solution for x_4 is most obvious and is given by

$$x_4 = \frac{b_4'''}{a_{44}'''}.$$

With x_4 known, x_3 can be solved for using the equation specified by the third row of ***aug A*** in Equation (2.17):

$$x_3 = \frac{b_3'' - a_{34}''x_4}{a_{33}''}.$$

Proceeding with the next higher row, we can solve for x_2:

$$x_2 = \frac{b_2' - a_{24}'x_4 - a_{23}'x_3}{a_{22}'}.$$

Finally, x_1 is obtained from

$$x_1 = \frac{b_1 - a_{14}x_4 - a_{13}x_3 - a_{12}x_2}{a_{11}}.$$

The algorithm for back substitution can be generalized for any system of n linear equations once an upper triangular augmented matrix is available, and is detailed below.

Algorithm for back substitution

(1) Calculate $x_n = b_n/a_{nn}$.
(2) Loop backwards through $i = n - 1$ to 1.
(3) Calculate $x_i = \frac{1}{a_{ii}}\left(b_i - \sum_{j=i+1}^{j=n} a_{ij}x_j\right)$.
(4) End loop i.

Program 2.1 solves a system of n linear equations using the basic Gaussian elimination with back substitution.

MATLAB program 2.1

```
function x = Gaussianbasic(A, b)
% Uses basic Gaussian elimination with backward substitution to solve a set
% of n linear equations in n unknowns.

% Input Variables
% A is the coefficient matrix
% b is the constants vector

% Output Variables
% x is the solution

Aug = [A b];
[n, ncol] = size(Aug);
x = zeros(n,1); %initializing the column vector x

% Elimination steps
for k = 1:n-1
    pivot = Aug(k,k);
    for row = k + 1:n
            m = Aug(row,k)/pivot;
            Aug(row,k+1:ncol) = Aug(row,k+1:ncol) - m*Aug(k,k+1:   ncol);
      end
end

% Backward substitution steps
x(n)=Aug(n,ncol)/Aug(n,n);
for k = n - 1:-1:1
    x(k) = (Aug(k,ncol) - Aug(k,k+1:n)*x(k+1:n))/Aug(k,k);
      % Note the inner product
end
```

Let's use `Gaussianbasic` to solve the system of three equations (E1)–(E3) solved by hand earlier in this section. For this example

$$A = \begin{bmatrix} 2 & -4 & 3 \\ 1 & 2 & -3 \\ 3 & 1 & 1 \end{bmatrix}, \quad b = \begin{bmatrix} 1 \\ 4 \\ 12 \end{bmatrix}.$$

The following is typed at the MATLAB prompt:

```
>> A = [2 -4 3; 1 2 -3; 3 1 1];
>> b = [1; 4; 12]; % Column vector
>> x = Gaussianbasic(A, b)
x =
    3
    2
    1
```

During the elimination process, if any of the elements along the diagonal becomes zero, either the elimination method or the backward substitution procedure will fail due to fatal errors arising from "division by zero." In such cases, when a diagonal element becomes zero, the algorithm cannot proceed further. In fact, significant problems arise even if any of the diagonal or pivot elements are not exactly zero but

Box 2.1B Drug development and toxicity studies: animal-on-a-chip

Gaussian elimination is used to solve the system of equations specified by Equations (2.3) and (2.4), which are reproduced below:

$$(2 - 1.5R)C_{\text{lung}} - 0.5RC_{\text{liver}} = 779.3 - 780R; \quad (2.3)$$

$$C_{\text{lung}} - C_{\text{liver}} = 76. \quad (2.4)$$

Here, $R = 0.6$, i.e. 60% of the exit stream is recycled back into the chip.

Equation (2.3) becomes

$$1.1C_{\text{lung}} - 0.3C_{\text{liver}} = 311.3.$$

For this problem,

$$\boldsymbol{A} = \begin{bmatrix} 1.1 & -0.3 \\ 1 & -1 \end{bmatrix}; \quad \boldsymbol{b} = \begin{bmatrix} 311.3 \\ 76 \end{bmatrix}.$$

We use `Gaussianbasic` to solve the system of two equations:

```
>> A = [1.1 -0.3; 1 -1];
>> b = [311.3; 76];
>> x = Gaussianbasic(A, b)
x =
  360.6250
  284.6250
```

The steady state naphthalene concentration in the lung (360.625 μM) is 92.5% that of the feed concentration. In the liver, the steady state concentration is not as high (284.625 μM; approx. 73% that of the feed concentration) due to the faster conversion rate in the liver cells of naphthalene to epoxide. If R is increased, will the steady state concentrations in the liver and lung increase or decrease? Verify your answer by solving Equations (2.3) and (2.4) for $R = 0.8$.

close to zero. Division by a small number results in magnification of round-off errors which are carried through the arithmetic calculations and produce erroneous results.

Example 2.3

Let's solve the example problem

$$0.001x_1 + 6.4x_2 = 4.801,$$
$$22.3x_1 - 8x_2 = 16.3,$$

using the Gaussian elimination method. As in some of the examples in Chapter 1, we retain only four significant figures in each number in order to emphasize the limitations of finite-precision arithmetic. The exact solution to this set of two equations is $x_1 = 1$ and $x_2 = 0.75$. The augmented matrix is

$$\begin{bmatrix} 0.001 & 6.4 & : & 4.801 \\ 22.3 & -8 & : & 16.3 \end{bmatrix}.$$

In this example, the pivot element 0.001 is much smaller in magnitude than the other coefficients. A single elimination step is performed to place a zero in the first column of the second row. The corresponding matrix is

$$\begin{bmatrix} 0.001 & 6.4 & : & 4.801 \\ 0 & -142700 & : & -107000 \end{bmatrix}.$$

The matrices are not exactly equivalent because the four-digit arithmetic results in a slight loss of accuracy in the new values. Next, performing back substitution, we obtain

$$x_2 = \frac{-107000}{-142700} = 0.7498$$

and

$$x_1 = \frac{4.801 - 6.4 \times 0.7498}{0.001} = 2.28.$$

While the solution for x_2 may be a sufficiently good approximation, the result for x_1 is quite inaccurate. The round-off errors that arise during calculations due to the limits in precision produce a slight error in the value of x_2. This small error is greatly magnified through loss of significant digits during subtractive cancellation and division by a small number (the pivot element). Suitable modifications of the elimination technique to avoid such situations are discussed in Section 2.4.2.

2.4.2 Gaussian elimination with pivoting

The procedure of exchanging rows in the augmented matrix either to remove from the diagonal zeros or small numbers is called **partial pivoting** or simply **pivoting**. To achieve greater accuracy in the final result, it is highly recommended that pivoting is performed for every pivot row. The strategy involves placing the element with the largest magnitude in the pivot position, such that the pivot element is the largest of all elements in the pivot column, and using this number to zero the entries lying below it in the same column. This way the multiplier used to transform elements in the rows beneath the pivot row will always be less than or equal to one, which will keep the propagation of round-off errors in check.

Algorithm for partial pivoting

(1) The pivot element is compared with the other elements in the same column. In MATLAB, the `max` function is used to find the largest number in the column.
(2) The row in the augmented matrix that contains the largest element is swapped with the pivot row.

Using MATLAB

(1) The syntax for the `max` function is `[Y I] = max(X)`, where `X` is the vector within which the maximum element is searched, `Y` is the maximum element of `X`, and `I` is the index or location of the element of maximum value.
(2) **Array indexing** enables the use of an array as an index for another array. For example, if $\boldsymbol{a}$ is a row vector that contains even numbers in sequential order from 2 to 40, and $\boldsymbol{b}$ is a vector containing integers from 1 to 10, and if we use $\boldsymbol{b}$ as the index into $\boldsymbol{a}$, the numeric values of the individual elements of $\boldsymbol{b}$ specify the position of the desired elements of $\boldsymbol{a}$. In our example the index $\boldsymbol{b}$ chooses the first ten elements of $\boldsymbol{a}$. This is demonstrated below.

```
>> a = 2:2:30
a =
  2 4 6 8 10 12 14 16 18 20 22 24 26 28 30
>> b = 1:10
b =
  1 2 3 4 5 6 7 8 9 10
```

```
>> c = a(b)
c =
  2 4 6 8 10 12 14 16 18 20
```

Program 2.1 is rewritten to incorporate partial pivoting. The program uses the `max` function to determine the largest element in the pivot column, and also uses array indexing to swap rows of the matrix.

MATLAB program 2.2

```
function x = Gaussianpivoting(A, b)
% Uses Gaussian elimination with pivoting to solve a set
% of n linear equations in n unknowns.

% Input Variables
% A is the coefficient matrix
% b is the constants vector
% Output Variables
% x is the solution

Aug = [A b];
[n, ncol] = size(Aug);
x = zeros(n,1); % initializing the column vector x

% Elimination with Pivoting
for k = 1:n-1
   [maxele, l] = max(abs(Aug(k:n,k)));
     % l is the location of max value between k and n
   p = k + l - 1; % row to be interchanged with pivot row
   if (p ~= k) % if p is not equal to k
      Aug([k p],:) = Aug([p k],:); % array index used for efficient swap
   end
   pivot = Aug(k,k);
       for row = k + 1:n
              m = Aug(row,k)/pivot;
              Aug(row,k+1:ncol) = Aug(row,k+1:ncol) - m*Aug (k,k+1:ncol);
       end

end

% Backward substitution steps
x(n)=Aug(n,ncol)/Aug(n,n);
for k = n - 1:-1:1
    x(k) = (Aug(k,ncol) - Aug(k,k+1:n)*x(k+1:n))/ Aug(k,k);
      % Note the inner product
end
```

Example 2.3 (continued)

Let's now swap the first and second rows so that the pivot element is now larger than all the other elements in the pivot column. The new augmented matrix is

$$\begin{bmatrix} 22.3 & -8 & : & 16.3 \\ 0.001 & 6.4 & : & 4.801 \end{bmatrix}.$$

After using the pivot row to transform the second row, we obtain

$$\begin{bmatrix} 22.3 & -8 & : & 16.3 \\ 0 & 6.400 & : & 4.800 \end{bmatrix}.$$

Again, due to finite precision, the original augmented matrix and the transformed matrix are not exactly equivalent. Performing back substitution, we obtain

$$x_2 = \frac{4.800}{6.400} = 0.75$$

and

$$x_1 = \frac{16.3 + 8 \times 0.75}{22.3} = 1$$

We have arrived at the exact solution!

Although partial pivoting is useful in safeguarding against corruption of the results by round-off error is some cases, it is not a foolproof strategy. If elements in a coefficient matrix are of disparate magnitudes, then multiplication operations with the relatively larger numbers can also be problematic. The number of multiplication/division operations (floating-point operations) carried out during Gaussian elimination of an $n \times n$ matrix $\boldsymbol{A}$ is $O(n^3/3)$, and the number of addition/subtraction operations carried out is also $O(n^3/3)$. This can be compared to the number of floating-point operations that are required to carry out the backward substitution procedure, which is $O(n^2)$, or the number of comparisons that are needed to perform partial pivoting, which is also $O(n^2)$. Gaussian elimination entails many more arithmetic operations than that performed by either backward substitution or the row comparisons and swapping steps. Gaussian elimination is therefore the most computationally intensive step of the entire solution method, posing a bottleneck in terms of computational time and resources. For large matrices, the number of arithmetic operations that are carried out during the elimination process is considerable and can result in sizeable accumulation of round-off errors.

Several methods have been devised to resolve such situations. One strategy is **maximal** or **complete pivoting**, in which the maximum value of the pivot element is searched amongst all rows and columns (except rows above the pivot row and columns to the left of the pivot column). The row and column containing the largest element are then swapped with the pivot row and pivot column. This entails a more complicated procedure since column interchanges require that the order of the variables be interchanged too. **Permutation matrices**, which are identity matrices that contain the same row and column interchanges as the augmented matrix, must be created to keep track of the interchanges performed during the elimination operations.

Another method that is used to reduce the magnitude of the round-off errors is called **scaled partial pivoting**. In this method every element in the pivot column is first scaled by the largest element in its row. The element of largest absolute magnitude in the pivot row p and in each row below the pivot row serves as the scale factor for that row, and will be different for different rows. If $s_i = \max_{1 \leq j \leq n} |a_{ij}|$ for $p \leq i \leq n$ are the scale factors, then the the pivot element, and therefore the new pivot row, is chosen based on the following criteria:

$$\frac{|a_{kp}|}{s_k} = \max_{p \leq i \leq n} \frac{|a_{ip}|}{s_i}.$$

The kth row is then interchanged with the current pivot row.

For large matrices, the number of multiplication/division operations and addition/subtraction operations required to carry out the elimination routine is rather formidable. For large **sparse matrices**, i.e. matrices that consist of many zeros, traditional Gaussian elimination results in many wasteful floating-point operations since addition or subtraction of any number with zero or multiplication of two zeros are unnecessary steps. **Tridiagonal matrices** are sparse matrices that have non-zero elements only along the main diagonal, and one diagonal above and below the main, and are often encountered in engineering problems (see Problems 2.7–2.9). Such matrices that have non-zero elements only along the main diagonal and one or more other smaller diagonals to the left/right of the main diagonal are called **banded matrices**. Special elimination methods, such as the Thomas method, which are faster than $O(n^3)$ are employed to reduce tridiagonal (or banded) matrices to the appropriate triangular form for performing back substitution.

2.5 LU factorization

Analysis of the number of floating-point operations (**flops**) carried out by a numerical procedure allows one to compare the efficiency of different computational methods in terms of the computational time and effort expended to achieve the same goal. As mentioned earlier, Gaussian elimination requires $O(n^3/3)$ multiplication/division operations and $O(n^3/3)$ addition/subtraction operations to reduce an augmented matrix that represents n equations with n unknowns to a upper triangular matrix (see Section 1.6.1 for a discussion on "order of magnitude"). Back substitution is an $O(n^2)$ process and is thus achieved much more quickly than the elimination method for large matrices. If we wish to solve systems of equations whose coefficient matrix $\boldsymbol{A}$ remains the same but have different $\boldsymbol{b}$ vectors using the Gaussian elimination method, we would be forced to carry out the same operations to reduce the coefficients in $\boldsymbol{A}$ in the augmented matrix every time. This is a wasteful process considering that most of the computational work is performed during the elimination process.

This problem can be addressed by decoupling the elimination steps carried out on $\boldsymbol{A}$ from that on $\boldsymbol{b}$. $\boldsymbol{A}$ can be converted into a product of two matrices: a lower triangular matrix $\boldsymbol{L}$ and an upper triangular matrix $\boldsymbol{U}$, such that

$$\boldsymbol{A} = \boldsymbol{LU}.$$

The matrix equation $\boldsymbol{Ax} = \boldsymbol{b}$ now becomes

$$\boldsymbol{LUx} = \boldsymbol{b}. \tag{2.18}$$

Once $\boldsymbol{A}$ has been factored into a product of two triangular matrices, the solution is straightforward. First, we substitute

$$\boldsymbol{y} = \boldsymbol{Ux} \tag{2.19}$$

into Equation (2.18) to obtain

$$\boldsymbol{Ly} = \boldsymbol{b}. \tag{2.20}$$

In Equation (2.20), $\boldsymbol{L}$ is a lower triangular matrix, and $\boldsymbol{y}$ can be easily solved for by the **forward substitution method**, which is similar to the backward substitution process except that the sequence of equations used to obtain the values of the unknowns starts with the first equation and proceeds sequentially downwards to the next equation. Once $\boldsymbol{y}$ is known, Equation (2.19) can easily be solved for $\boldsymbol{x}$ using the backward substitution method. This procedure used to solve linear systems of equations is called the **LU factorization method** or **LU decomposition method**, and is well-suited to solving problems with similar coefficient matrices. There are three steps involved in the LU factorization method as compared to the two-step Gauss elimination method.

(1) Factorization of $\boldsymbol{A}$ into two triangular matrices $\boldsymbol{L}$ (lower triangular) and $\boldsymbol{U}$ (upper triangular).
(2) Forward substitution method to solve $\boldsymbol{Ly} = \boldsymbol{b}$.
(3) Backward substitution method to solve $\boldsymbol{Ux} = \boldsymbol{y}$.

There are two important questions that arise here: first, how can $\boldsymbol{A}$ be factored into $\boldsymbol{L}$ and $\boldsymbol{U}$, and second how computationally efficient is the factorization step? The answer to both questions lies with the Gaussian elimination method, for it is this very procedure that points us in the direction of how to obtain the matrices $\boldsymbol{L}$ and $\boldsymbol{U}$. As discussed in Section 2.4, Gaussian elimination converts $\boldsymbol{A}$ into an upper triangular matrix $\boldsymbol{U}$. The technique used to facilitate this conversion process, i.e. the use of multipliers by which the pivot row is added to another row, produces the $\boldsymbol{L}$ matrix. Thus, factoring $\boldsymbol{A}$ into $\boldsymbol{L}$ and $\boldsymbol{U}$ is as computationally efficient as performing the Gaussian elimination procedure on $\boldsymbol{A}$. The advantage of using LU factorization stems from the fact that once $\boldsymbol{A}$ is factorized, the solution of any system of equations for which $\boldsymbol{A}$ is a coefficient matrix becomes a two-step process of $O(n^2)$.

There are several different methods that have been developed to perform the LU factorization of $\boldsymbol{A}$. Here, we discuss the **Doolittle method**, a technique that is easily derived from the Gaussian elimination process. In this factorization method, $\boldsymbol{U}$ is similar to the upper triangular matrix that is obtained as the end result of Gaussian elimination, while $\boldsymbol{L}$ has ones along its diagonal, and the multipliers used to carry out the elimination are placed as entries below the diagonal. Before we learn how to implement the Doolittle method, let's familiarize ourselves with how this method is derived from the Gaussian elimination method. We will first consider the case in which pivoting of the rows is not performed.

2.5.1 LU factorization without pivoting

Recall that the Gaussian elimination method involves zeroing all the elements below the diagonal, starting from the first or leftmost to the last or rightmost column. When the first row is the pivot row, the elements in the first column below the topmost entry are zeroed, and the other elements in the second to nth rows are transformed using the following formula:

$$a'_{ij} = a_{ij} - \frac{a_{i1}}{a_{11}} a_{1j} \qquad 1 \leq j \leq n, \qquad 2 \leq i \leq n,$$

where the prime indicates the new or modified value of the element. In this formula

$$m_{i1} = \frac{a_{i1}}{a_{11}}$$

is called the **multiplier** since its products with the elements of the pivot row are correspondingly subtracted from the elements of the other rows located below. The multiplier has two subscripts that denote the row and the column location of the element that is converted to zero after applying the transformation.

The transformation operations on the elements of the first column of the coefficient matrix can be compactly and efficiently represented by a single matrix multiplication operation. Let $\boldsymbol{A}$ be a 4×4 non-singular matrix. We create a matrix $\boldsymbol{M}_1$ that contains ones along the diagonal (as in an identity matrix) and negatives of the multipliers situated in the appropriate places along the first column of $\boldsymbol{M}_1$. Observe that the product of $\boldsymbol{M}_1$ and $\boldsymbol{A}$ produces the desired transformation:

$$\boldsymbol{M}_1\boldsymbol{A} = \begin{bmatrix} 1 & 0 & 0 & 0 \\ -m_{21} & 1 & 0 & 0 \\ -m_{31} & 0 & 1 & 0 \\ -m_{41} & 0 & 0 & 1 \end{bmatrix} \begin{bmatrix} a_{11} & a_{12} & a_{13} & a_{14} \\ a_{21} & a_{22} & a_{23} & a_{24} \\ a_{31} & a_{32} & a_{33} & a_{34} \\ a_{41} & a_{42} & a_{43} & a_{44} \end{bmatrix} = \begin{bmatrix} a_{11} & a_{12} & a_{13} & a_{14} \\ 0 & a'_{22} & a'_{23} & a'_{24} \\ 0 & a'_{32} & a'_{33} & a'_{34} \\ 0 & a'_{42} & a'_{43} & a'_{44} \end{bmatrix}.$$

Similarly, matrices $\boldsymbol{M}_2$ and $\boldsymbol{M}_3$ can be created to zero the elements of the coefficient matrix in the second and third columns. The matrix multiplications are sequentially performed to arrive at our end product – an upper triangular matrix:

$$\boldsymbol{M}_2(\boldsymbol{M}_1\boldsymbol{A}) = \begin{bmatrix} 1 & 0 & 0 & 0 \\ 0 & 1 & 0 & 0 \\ 0 & -m_{32} & 1 & 0 \\ 0 & -m_{42} & 0 & 1 \end{bmatrix} \begin{bmatrix} a_{11} & a_{12} & a_{13} & a_{14} \\ 0 & a'_{22} & a'_{23} & a'_{24} \\ 0 & a'_{32} & a'_{33} & a'_{34} \\ 0 & a'_{42} & a'_{43} & a'_{44} \end{bmatrix} = \begin{bmatrix} a_{11} & a_{12} & a_{13} & a_{14} \\ 0 & a'_{22} & a'_{23} & a'_{24} \\ 0 & 0 & a''_{33} & a''_{34} \\ 0 & 0 & a''_{43} & a''_{44} \end{bmatrix},$$

$$\boldsymbol{M}_3(\boldsymbol{M}_2\boldsymbol{M}_1\boldsymbol{A}) = \begin{bmatrix} 1 & 0 & 0 & 0 \\ 0 & 1 & 0 & 0 \\ 0 & 0 & 1 & 0 \\ 0 & 0 & -m_{43} & 1 \end{bmatrix} \begin{bmatrix} a_{11} & a_{12} & a_{13} & a_{14} \\ 0 & a'_{22} & a'_{23} & a'_{24} \\ 0 & 0 & a''_{33} & a''_{34} \\ 0 & 0 & a''_{43} & a''_{44} \end{bmatrix}$$

$$= \begin{bmatrix} a_{11} & a_{12} & a_{13} & a_{14} \\ 0 & a'_{22} & a'_{23} & a'_{24} \\ 0 & 0 & a''_{33} & a''_{34} \\ 0 & 0 & 0 & a'''_{44} \end{bmatrix} = \boldsymbol{U}.$$

Thus, we have

$$\boldsymbol{U} = \boldsymbol{M}_3\boldsymbol{M}_2\boldsymbol{M}_1\boldsymbol{A}.$$

Left-multiplying with the inverse of $\boldsymbol{M}_3$ on both sides of the equation, we get

$$(\boldsymbol{M}_3^{-1}\boldsymbol{M}_3)\boldsymbol{M}_2\boldsymbol{M}_1\boldsymbol{A} = \boldsymbol{M}_3^{-1}\boldsymbol{U}.$$

Again, left-multiplying with the inverse of $\boldsymbol{M}_2$ on both sides of the equation ($\boldsymbol{M}_3^{-1}\boldsymbol{M}_3 = \boldsymbol{I}$) (see Equation (2.15)), we get

$$\boldsymbol{M}_2^{-1}\boldsymbol{I}\boldsymbol{M}_2\boldsymbol{M}_1\boldsymbol{A} = \boldsymbol{M}_2^{-1}\boldsymbol{M}_3^{-1}\boldsymbol{U}.$$

Since $\boldsymbol{I}\boldsymbol{M}_2 = \boldsymbol{M}_2$ (see Equation (2.15))

$$\boldsymbol{I}\boldsymbol{M}_1\boldsymbol{A} = \boldsymbol{M}_2^{-1}\boldsymbol{M}_3^{-1}\boldsymbol{U}.$$

Finally, left-multiplying with the inverse of $\boldsymbol{M}_1$ on both sides of the equation, we get

$$\boldsymbol{M}_1^{-1}\boldsymbol{M}_1\boldsymbol{A} = \boldsymbol{A} = \boldsymbol{M}_1^{-1}\boldsymbol{M}_2^{-1}\boldsymbol{M}_3^{-1}\boldsymbol{U} = \boldsymbol{L}\boldsymbol{U},$$

where

$$\boldsymbol{L} = \boldsymbol{M}_1^{-1}\boldsymbol{M}_2^{-1}\boldsymbol{M}_3^{-1}$$

Now let's evaluate the product of the inverses of M_i to verify if L is indeed a lower triangular matrix. Determining the product $M_1^{-1}M_2^{-1}M_3^{-1}$ will yield clues on how to construct L from the multipliers used to perform the elimination process.

The inverse of

$$M_1 = \begin{bmatrix} 1 & 0 & 0 & 0 \\ -m_{21} & 1 & 0 & 0 \\ -m_{31} & 0 & 1 & 0 \\ -m_{41} & 0 & 0 & 1 \end{bmatrix}$$

is simply (calculate this yourself using Equation (2.16))

$$M_1^{-1} = \begin{bmatrix} 1 & 0 & 0 & 0 \\ m_{21} & 1 & 0 & 0 \\ m_{31} & 0 & 1 & 0 \\ m_{41} & 0 & 0 & 1 \end{bmatrix}.$$

Similarly,

$$M_2^{-1} = \begin{bmatrix} 1 & 0 & 0 & 0 \\ 0 & 1 & 0 & 0 \\ 0 & m_{32} & 1 & 0 \\ 0 & m_{42} & 0 & 1 \end{bmatrix}, \quad M_1^{-1} = \begin{bmatrix} 1 & 0 & 0 & 0 \\ 0 & 1 & 0 & 0 \\ 0 & 0 & 1 & 0 \\ 0 & 0 & m_{43} & 1 \end{bmatrix}.$$

Now,

$$M_1^{-1}M_2^{-1} = \begin{bmatrix} 1 & 0 & 0 & 0 \\ m_{21} & 1 & 0 & 0 \\ m_{31} & 0 & 1 & 0 \\ m_{41} & 0 & 0 & 1 \end{bmatrix} \begin{bmatrix} 1 & 0 & 0 & 0 \\ 0 & 1 & 0 & 0 \\ 0 & m_{32} & 1 & 0 \\ 0 & m_{42} & 0 & 1 \end{bmatrix} = \begin{bmatrix} 1 & 0 & 0 & 0 \\ m_{21} & 1 & 0 & 0 \\ m_{31} & m_{32} & 1 & 0 \\ m_{41} & m_{42} & 0 & 1 \end{bmatrix}$$

and

$$(M_1^{-1}M_2^{-1})M_3^{-1} = \begin{bmatrix} 1 & 0 & 0 & 0 \\ m_{21} & 1 & 0 & 0 \\ m_{31} & m_{32} & 1 & 0 \\ m_{41} & m_{42} & 0 & 1 \end{bmatrix} \begin{bmatrix} 1 & 0 & 0 & 0 \\ 0 & 1 & 0 & 0 \\ 0 & 0 & 1 & 0 \\ 0 & 0 & m_{43} & 1 \end{bmatrix}$$

$$= \begin{bmatrix} 1 & 0 & 0 & 0 \\ m_{21} & 1 & 0 & 0 \\ m_{31} & m_{32} & 1 & 0 \\ m_{41} & m_{42} & m_{43} & 1 \end{bmatrix} = L.$$

Thus, L is simply a lower triangular matrix in which the elements below the diagonal are the multipliers whose positions coincide with their subscript notations. A multiplier used to zero the entry in the second row, first column is placed in the second row and first column in L. While carrying out the transformation $A \to U$, no extra effort is expended in producing L since the multipliers are calculated along the way.

As mentioned earlier, the determinant of any triangular matrix is easily obtained from the product of all the elements along the diagonal. Since L has only ones on its diagonal, its determinant, $|L| = 1$, and U will have a non-zero determinant only if all the elements along the diagonal are non-zero. If even one element on the diagonal is zero, then $|U| = 0$. Since $|A| = |L||U|$, the determinant of A can be obtained by simply multiplying the elements along the diagonal of U. If $|A| = 0$, a unique solution of the existing set of equations cannot be determined.

Example 2.4

We use the LU factorization method to solve the system of equations given by Equations (E1)–(E3). The coefficient matrix is

$$A = \begin{bmatrix} 2 & -4 & 3 \\ 1 & 2 & -3 \\ 3 & 1 & 1 \end{bmatrix}.$$

LU factorization/decomposition

We begin the procedure by initializing the L and U matrices:

$$L = \begin{bmatrix} 1 & 0 & 0 \\ 0 & 1 & 0 \\ 0 & 0 & 1 \end{bmatrix}, \quad U = \begin{bmatrix} 2 & -4 & 3 \\ 1 & 2 & -3 \\ 3 & 1 & 1 \end{bmatrix}.$$

We divide the factorization procedure into two steps. In the first step we zero the elements in the first column below the pivot row (first row), and in the second step we zero the element in the second column below the pivot row (second row).

Step 1: The multipliers m_{21} and m_{31} are calculated to be

$$m_{21} = \frac{1}{2}, \quad m_{31} = \frac{3}{2}.$$

The transformation

$$u'_{ij} = u_{ij} - m_{i1}u_{1j}, \qquad 1 \le j \le 3, \qquad 2 \le i \le 3,$$

produces the following revised matrices:

$$L = \begin{bmatrix} 1 & 0 & 0 \\ \frac{1}{2} & 1 & 0 \\ \frac{3}{2} & 0 & 1 \end{bmatrix} \quad \text{and} \quad U = \begin{bmatrix} 2 & -4 & 3 \\ 0 & 4 & -\frac{9}{2} \\ 0 & 7 & -\frac{7}{2} \end{bmatrix}.$$

Step 2: The multiplier m_{32} is calculated as

$$m_{32} = \frac{7}{4}.$$

The transformation

$$u'_{ij} = u_{ij} - m_{i2}u_{2j}, \qquad 2 \le j \le 3, \qquad i = 3,$$

produces the following revised matrices:

$$L = \begin{bmatrix} 1 & 0 & 0 \\ \frac{1}{2} & 1 & 0 \\ \frac{3}{2} & \frac{7}{4} & 1 \end{bmatrix} \quad \text{and} \quad U = \begin{bmatrix} 2 & -4 & 3 \\ 0 & 4 & -\frac{9}{2} \\ 0 & 0 & \frac{35}{8} \end{bmatrix}.$$

The LU factorization of A is complete. If the matrix product LU is computed, one obtains the original matrix A. Next, we use the forward and backward substitution methods to obtain the solution.

Forward substitution

First, we solve the equation $\mathbf{Ly} = \mathbf{b}$:

$$\begin{bmatrix} 1 & 0 & 0 \\ \frac{1}{2} & 1 & 0 \\ \frac{3}{2} & \frac{7}{4} & 1 \end{bmatrix} \begin{bmatrix} y_1 \\ y_2 \\ y_3 \end{bmatrix} = \begin{bmatrix} 1 \\ 4 \\ 12 \end{bmatrix}.$$

This matrix equation represents the following three equations:

$$y_1 = 1, \tag{E4}$$

$$\frac{1}{2}y_1 + y_2 = 4, \tag{E5}$$

$$\frac{3}{2}y_1 + \frac{7}{4}y_2 + y_3 = 12. \tag{E6}$$

Equation (E4) gives us y_1 directly. Substituting the value of y_1 into Equation (E5), we get $y_2 = 3.5$. Equation (E6) then gives us $y_3 = 35/8$.

Backward substitution

Next we solve the equation $\mathbf{Ux} = \mathbf{y}$:

$$\begin{bmatrix} 2 & -4 & 3 \\ 0 & 4 & -\frac{9}{2} \\ 0 & 0 & \frac{35}{8} \end{bmatrix} \begin{bmatrix} x_1 \\ x_2 \\ x_3 \end{bmatrix} = \begin{bmatrix} 1 \\ \frac{7}{2} \\ \frac{35}{8} \end{bmatrix}.$$

This matrix equation represents the following three equations:

$$2x_1 - 4x_2 + 3x_3 = 1, \tag{E7}$$

$$4x_2 - \frac{9}{2}x_3 = \frac{7}{2}, \tag{E8}$$

$$\frac{35}{8}x_3 = \frac{35}{8}. \tag{E9}$$

Equation (E9) directly yields $x_3 = 1$. Working in the upwards direction, we use Equation (E8) to solve for $x_2 = 2$, and then use Equation (E7) to obtain $x_1 = 3$.

In the above example, $\boldsymbol{L}$ and $\boldsymbol{U}$ were created as separate matrices. The storage and manipulation of the elements of $\boldsymbol{L}$ and $\boldsymbol{U}$ can be achieved most efficiently by housing the elements in a single matrix $\boldsymbol{A}$. Matrix elements of $\boldsymbol{A}$ can be transformed *in situ* as in Gaussian elimination to yield $\boldsymbol{U}$. The multipliers of $\boldsymbol{L}$ can be stored beneath the diagonal of $\boldsymbol{A}$, with the ones along the diagonal being implied. For a 4×4 non-singular matrix $\boldsymbol{A}$, the factorized matrices $\boldsymbol{L}$ and $\boldsymbol{U}$ can be stored in a single matrix given by

$$\begin{bmatrix} a_{11} & a_{12} & a_{13} & a_{14} \\ m_{21} & a'_{22} & a'_{23} & a'_{24} \\ m_{31} & m_{32} & a''_{33} & a''_{34} \\ m_{41} & m_{42} & m_{43} & a'''_{44} \end{bmatrix}.$$

2.5.2 LU factorization with pivoting

Situations can arise in which row interchanges are necessary either to limit the round-off error generated during the factorization or substitution process, or to remove a zero from the diagonal of $\mathbf{U}$. If pivoting is performed during the factorization process, information on the row swapping performed must be stored so that elements of the $\mathbf{b}$ vector can be permuted accordingly. **Permutation matrices**, which are identity matrices with permuted rows (or columns), are used for this purpose. A permutation matrix has only one non-zero entry equal to one in any column and in any row, and has a determinant equal to one. Pre-multiplication of $\mathbf{A}$ or $\mathbf{b}$ by an appropriate permutation matrix $\mathbf{P}$ effectively swaps rows in accordance with the structure of $\mathbf{P}$. (Post-multiplication of any matrix with a permutation matrix will result in swapping of columns.)

For example, if permutation matrix $\mathbf{P}$ is an identity matrix in which the first two rows have been interchanged, then

$$\mathbf{P} = \begin{bmatrix} 0 & 1 & 0 & 0 \\ 1 & 0 & 0 & 0 \\ 0 & 0 & 1 & 0 \\ 0 & 0 & 0 & 1 \end{bmatrix}$$

and

$$\mathbf{PA} = \begin{bmatrix} 0 & 1 & 0 & 0 \\ 1 & 0 & 0 & 0 \\ 0 & 0 & 1 & 0 \\ 0 & 0 & 0 & 1 \end{bmatrix} \begin{bmatrix} a_{11} & a_{12} & a_{13} & a_{14} \\ a_{21} & a_{22} & a_{23} & a_{24} \\ a_{31} & a_{32} & a_{33} & a_{34} \\ a_{41} & a_{42} & a_{43} & a_{44} \end{bmatrix} = \begin{bmatrix} a_{21} & a_{22} & a_{23} & a_{24} \\ a_{11} & a_{12} & a_{13} & a_{14} \\ a_{31} & a_{32} & a_{33} & a_{34} \\ a_{41} & a_{42} & a_{43} & a_{44} \end{bmatrix}.$$

Here, the pre-multiplication operation with $\mathbf{P}$ interchanged the first two rows of $\mathbf{A}$.

The permutation matrix $\mathbf{P}$ is generated during the factorization process. Prior to any row swapping operations, $\mathbf{P}$ is initialized as an identity matrix, and any row swaps executed while creating $\mathbf{L}$ and $\mathbf{U}$ are also performed on $\mathbf{P}$.

Example 2.5

Here Equations (E1)–(E3) are solved again using LU factorization with pivoting. Matrices $\mathbf{P}$ and $\mathbf{A}$ are initialized as follows:

$$\mathbf{P} = \begin{bmatrix} 1 & 0 & 0 \\ 0 & 1 & 0 \\ 0 & 0 & 1 \end{bmatrix}, \quad \mathbf{A} = \begin{bmatrix} 2 & -4 & 3 \\ 1 & 2 & -3 \\ 3 & 1 & 1 \end{bmatrix}.$$

The largest entry is searched for in the first column or pivot column of $\mathbf{A}$. Since $|a_{31}| > |a_{11}|$, rows 1 and 3 are swapped. We also interchange the first and third rows of $\mathbf{P}$ to obtain

$$\mathbf{P} = \begin{bmatrix} 0 & 0 & 1 \\ 0 & 1 & 0 \\ 1 & 0 & 0 \end{bmatrix}, \quad \mathbf{A} = \begin{bmatrix} 3 & 1 & 1 \\ 1 & 2 & -3 \\ 2 & -4 & 3 \end{bmatrix}.$$

Elimination is carried out using the first row, which is the pivot row. The multipliers are stored in the matrix $\mathbf{A}$ itself below the diagonal in place of the zeroed elements. The multipliers used to zero the two entries below the diagonal in the first column are $m_{21} = 1/3$ and $m_{31} = 2/3$.

We obtain

$$\mathbf{A} = \begin{bmatrix} 3 & 1 & 1 \\ \frac{1}{3} & \frac{5}{3} & -\frac{10}{3} \\ \frac{2}{3} & -\frac{14}{3} & \frac{7}{3} \end{bmatrix}.$$

Next, the pivot column shifts to the second column and the largest entry is searched for in this column. Since $|a_{32}| > |a_{22}|$, rows 2 and 3 are interchanged in both $\mathbf{A}$ and $\mathbf{P}$ to obtain

$$\mathbf{P} = \begin{bmatrix} 0 & 0 & 1 \\ 1 & 0 & 0 \\ 0 & 1 & 0 \end{bmatrix} \qquad \mathbf{A} = \begin{bmatrix} 3 & 1 & 1 \\ \frac{2}{3} & -\frac{14}{3} & \frac{7}{3} \\ \frac{1}{3} & \frac{5}{3} & -\frac{10}{3} \end{bmatrix}.$$

Note that m_{32} is calculated as $-5/14$. Transformation of the third row using the pivot (second) row yields

$$\mathbf{A} = \begin{bmatrix} 3 & 1 & 1 \\ \frac{2}{3} & -\frac{14}{3} & \frac{7}{3} \\ \frac{1}{3} & -\frac{5}{14} & -\frac{5}{2} \end{bmatrix}.$$

The LU factorization process is complete and the matrices $\mathbf{L}$ and $\mathbf{U}$ can be extracted from $\mathbf{A}$:

$$\mathbf{L} = \begin{bmatrix} 1 & 0 & 0 \\ \frac{2}{3} & 1 & 0 \\ \frac{1}{3} & -\frac{5}{14} & 1 \end{bmatrix} \quad \text{and} \quad \mathbf{U} = \begin{bmatrix} 3 & 1 & 1 \\ 0 & -\frac{14}{3} & \frac{7}{3} \\ 0 & 0 & -\frac{5}{2} \end{bmatrix}.$$

Since $\mathbf{A}$ is permuted during the factorization process, the LU factorization obtained does not correspond to $\mathbf{A}$ but to the product $\mathbf{PA}$. In other words,

$$\mathbf{LU} = \mathbf{PA}$$

and

$$\mathbf{LUx} = \mathbf{PAx} = \mathbf{Pb}$$

Here,

$$\mathbf{Pb} = \begin{bmatrix} 12 \\ 1 \\ 4 \end{bmatrix}.$$

We use the forward substitution process to solve $\mathbf{Ly} = \mathbf{Pb}$, which represents the following set of equations:

$$\begin{aligned} &y_1 = 12, \\ &\frac{2}{3}y_1 + y_2 = 1, \\ &\frac{1}{3}y_1 - \frac{5}{14}y_2 + y_3 = 4, \end{aligned}$$

which has the solution $y_1 = 12$, $y_2 = -7$, and $y_3 = -5/2$.

The equation set $\mathbf{Ux} = \mathbf{y}$ is solved via backward substitution:

$$\begin{aligned} &3x_1 + x_2 + x_3 = 12, \\ &-\frac{14}{3}x_2 + \frac{7}{3}x_3 = -7, \\ &-\frac{5}{2}x_3 = -\frac{5}{2}, \end{aligned}$$

which has the solution $x_1 = 3$, $x_2 = 2$, $x_3 = 1$.

2.5.3 The MATLAB `lu` function

MATLAB provides a built-in function to resolve a matrix A into two triangular matrices: a lower triangular matrix L and an upper triangular matrix U. The function `lu` is versatile; it can factorize both square and rectangular matrices, and can accept a variety of parameters. A commonly used syntax for the `lu` function is discussed here. For more information on the `lu` function, type `help lu` in the Command Window. The `lu` function uses partial pivoting and therefore outputs the permutation matrix P along with the factorized matrices L and U.

One syntax form is

```
[L,U,P] = lu(A)
```

where $LU = PA$. Let

```
>> A = [ 2 -4 3; 1 2 -3; 3 1 1]
A =
    2   -4    3
    1    2   -3
    3    1    1
```

Applying the `lu` function to factorize A,

```
>> [L, U, P] = lu(A)
```

gives the following result:

```
L =
    1.0000         0         0
    0.6667    1.0000         0
    0.3333   -0.3571    1.0000
U =
    3.0000    1.0000    1.0000
         0   -4.6667    2.3333
         0         0   -2.5000
P =
    0   0   1
    1   0   0
    0   1   0
```

Another syntax form for the `lu` function is

```
[L,U] = lu(A)
```

where $A = L_pU$, and $L_p = P^{-1}L$.

```
>> [L,U] = lu(A)
```

gives the following result:

```
L =
    0.6667    1.0000         0
    0.3333   -0.3571    1.0000
    1.0000         0         0
U =
    3.0000    1.0000    1.0000
         0   -4.6667    2.3333
         0         0   -2.5000
```

Box 2.1C Drug development and toxicity studies: animal-on-a-chip

LU decomposition is used to solve the system of equations specified by Equations (2.3) and (2.4), which are reproduced below:

$$(2 - 1.5R)C_{\text{lung}} - 0.5RC_{\text{liver}} = 779.3 - 780R; \tag{2.3}$$

$$C_{\text{lung}} - C_{\text{liver}} = 76. \tag{2.4}$$

We choose $R = 0.6$, i.e. 60% of the exit stream is recycled back into the chip.

Equation (2.3) becomes

$$1.1C_{\text{lung}} - 0.3C_{\text{liver}} = 311.3.$$

For this problem,

$$\boldsymbol{A} = \begin{bmatrix} 1.1 & -0.3 \\ 1 & -1 \end{bmatrix}; \; \boldsymbol{b} = \begin{bmatrix} 311.3 \\ 76 \end{bmatrix}.$$

```
>> A = [1.1 -0.3; 1 -1]; b= [311.3; 76];
>> [L, U, P] = lu(A)
L =
    1.0000         0
    0.9091    1.0000
U =
    1.1000   -0.3000
         0   -0.7273
P =
    1    0
    0    1
```

On performing forward ($\boldsymbol{Ly} = \boldsymbol{b}$) and backward ($\boldsymbol{Ux} = \boldsymbol{y}$) substitutions by hand (verify this yourself), the solution

$$C_{\text{lung}} = 360.6\,\text{uM},$$
$$C_{\text{liver}} = 284.6\,\text{uM}$$

is obtained.

2.6 The MATLAB backslash (\) operator

Solutions to systems of linear equations can be obtained in MATLAB with the use of the backslash operator. If we desire the solution of the matrix equation $\boldsymbol{Ax} = \boldsymbol{b}$, where $\boldsymbol{A}$ is an $n \times n$ matrix, we can instruct MATLAB to solve the equation by typing

```
x = A\b.
```

The backslash operator, also known as the "matrix left-division operator," can be interpreted in several ways, such as "$\boldsymbol{b}$ divided by $\boldsymbol{A}$" or "$\boldsymbol{b}$ multiplied on the left by the inverse of $\boldsymbol{A}$." However, MATLAB does not explicitly compute $\boldsymbol{A}^{-1}$ when solving a linear system of equations, but instead uses efficient algorithms such as Gaussian elimination for square $n \times n$ matrices. MATLAB assesses the structural properties of $\boldsymbol{A}$ before deciding which algorithm to use for solving $\boldsymbol{A} \backslash \boldsymbol{b}$. If $\boldsymbol{A}$ is a triangular matrix, then backward or forward substitution is applied directly to obtain $\boldsymbol{x}$. If $\boldsymbol{A}$ is singular or nearly singular, MATLAB issues a warning message so that the result produced may be interpreted by the user accordingly. The function `mldivide` executes the left-division operation that is represented by the \ operator. Type `help mldivide` for more information on the backslash operator.

We use the backslash operator to solve Equations (E1)–(E3):

```
>> A = [2 -4 3; 1 2 -3; 3 1 1];
>> b = [1; 4; 12];
>> x = A\b
```

The output produced is

```
x =
     3
     2
     1
```

The backslash operator is capable of handling overdetermined and underdetermined systems as well, e.g. systems described by a rectangular $m \times n$ matrix $\boldsymbol{A}$.

2.7 Ill-conditioned problems and the condition number

One may encounter a system of equations whose solution is considerably sensitive to slight variations in the values of any of the coefficients or constants. Such systems exhibit large fluctuations in the solutions despite comparatively small changes in the parametric values. Any solution of these equations is usually not reliable or useful since it is often far removed from the true solution. Such systems of equations are said to be **ill-conditioned**. Often coefficient values or constants of equations are derived from results of experiments or from estimations using empirical equations and are only approximations to the true values. Even if the problem at hand contains an approximate set of equations, we wish the solution to be as close as possible to the true solution corresponding to the set of equations having the correct coefficients and constants. Equations that show little variability in the solution despite modest changes in the parameters are said to be **well-conditioned**. Even in cases where the coefficients and constants of the equations are known exactly, when performing numerical computations finite precision will introduce round-off errors that will perturb the true value of the parameters. Such small but finite perturbations will produce little change in the solutions of well-conditioned problems, but can dramatically alter the result when solving ill-conditioned systems of equations. Therefore, when working with ill-conditioned problems, or with problems involving numbers of disparate magnitudes, it is always recommended to carry out numerical computations in double precision, since powerful computational resources – high speeds and large memory – are easily available.

Ill-conditioning is an inherent property of the system of equations and cannot be changed by using a different solution algorithm or by using superior solution techniques such as scaling and pivoting. The value of the determinant of the coefficient matrix indicates the **conditioning** of a system of linear equations. A set of equations with a coefficient matrix determinant close to zero will always be ill-conditioned. The effect of the conditioning of the coefficient matrix on the solution outcome is demonstrated below.

Example 2.6

Let's consider a 2×2 well-conditioned coefficient matrix and 2×1 constants vector:

```
>> A = [2 3; 4 1];
>> b = [5; 5];
>> x = A\b
x =
    1
    1
```

Now let's change the constants vector to

```
>> bnew = [5.001; 5];
>> xnew = A\bnew
xnew =
    0.9999
    1.0004
```

The relative change in the first element of $\boldsymbol{b} = 0.001/5 = 0.0002$. The relative change in the solutions is calculated as 0.0001 and 0.0004 for both unknowns, which is of the same order of magnitude as the slight perturbation made to the constants vector. This is representative of the behavior of well-conditioned coefficient matrices.

Now let's study the behavior of an ill-conditioned coefficient matrix:

```
>> A = [4 2; 2.001 1];
>> b = [6; 3.001];
>> x = A\b
x =
    1.0000
    1.0000
```

The answer is correct. Let's introduce a slight modification in the first element of $\boldsymbol{b}$:

```
>> bnew = [6.001; 3.001];
>> xnew = A\bnew
xnew =
    0.5000
    2.0005
```

A relative perturbation of $0.001/6 = 0.000167$ (0.0167%) in only one of the elements of $\boldsymbol{b}$ resulted in a relative change of 0.5 (50%) and 1.0005 (100%) in the magnitudes of the two variables. The changes observed in the resulting solution are approximately 3000–6000 times the original change made in the first element of $\boldsymbol{b}$!

Next, we calculate how well `xnew` satisfies the original system of equations ($\boldsymbol{b} = [6; 3.001]$). To determine this, we calculate the **residual** (see Section 2.9 for a discussion on "residuals"), which is defined as $\|\boldsymbol{r}\| = \|\boldsymbol{b} - \boldsymbol{Ax}\|$. The residual is an indicator of how well the left-hand side equates to the right-hand side of the equations, or, in other words, how well the solution $\boldsymbol{x}$ satisfies the set of equations:

```
>> r = b - A*xnew
r =
    -0.0010
          0
```

The solution to the perturbed set of equations satisfies the original system of equations well, though not exactly. Clearly, working with ill-conditioned equations is most troublesome since numerous solutions of a range of magnitudes may satisfy the equation within tolerable error, yet the true solution may remain elusive.

In order to understand the origin of ill-conditioning, we ask the following question: how is the magnitude of the slight perturbations in the constants in $\boldsymbol{b}$ or the coefficients in $\boldsymbol{A}$ related to the resulting change in the solution of the equations? Consider first small variations occurring in the individual elements of $\boldsymbol{b}$ only. Consider the original equation $\boldsymbol{Ax} = \boldsymbol{b}$. Let $\Delta\boldsymbol{b}$ be a small change made to $\boldsymbol{b}$, such that the solution changes by $\Delta\boldsymbol{x}$. The new system of equations is given by

$$\boldsymbol{A}(\boldsymbol{x} + \Delta\boldsymbol{x}) = \boldsymbol{b} + \Delta\boldsymbol{b}.$$

Subtracting off $\boldsymbol{Ax} = \boldsymbol{b}$, we obtain

$$\boldsymbol{A}\Delta\boldsymbol{x} = \Delta\boldsymbol{b}$$

or

$$\Delta\boldsymbol{x} = \boldsymbol{A}^{-1}\Delta\boldsymbol{b}.$$

Taking the norms of both sides yields

$$\|\Delta\boldsymbol{x}\| = \|\boldsymbol{A}^{-1}\Delta\boldsymbol{b}\| \leq \|\boldsymbol{A}^{-1}\|\|\Delta\boldsymbol{b}\|. \tag{2.21}$$

Now we take the norm of the original equation $\boldsymbol{Ax} = \boldsymbol{b}$ to obtain

$$\|\boldsymbol{b}\| = \|\boldsymbol{Ax}\| \leq \|\boldsymbol{A}\|\|\boldsymbol{x}\|$$

or

$$\frac{1}{\|\boldsymbol{A}\|\|\boldsymbol{x}\|} \leq \frac{1}{\|\boldsymbol{b}\|}. \tag{2.22}$$

Mutliplying Equation (2.21) with (2.22), we obtain

$$\frac{\|\Delta\boldsymbol{x}\|}{\|\boldsymbol{A}\|\|\boldsymbol{x}\|} \leq \frac{\|\boldsymbol{A}^{-1}\|\|\Delta\boldsymbol{b}\|}{\|\boldsymbol{b}\|}.$$

Rearranging,

$$\frac{\|\Delta\boldsymbol{x}\|}{\|\boldsymbol{x}\|} \leq \|\boldsymbol{A}\|\|\boldsymbol{A}^{-1}\|\frac{\|\Delta\boldsymbol{b}\|}{\|\boldsymbol{b}\|}. \tag{2.23}$$

Here, we have a relationship between the relative change in the solution and the relative change in $\boldsymbol{b}$. Note that $\|\Delta\boldsymbol{x}\|/\|\boldsymbol{x}\|$ is dependent on two factors:

(1) $\|\Delta\boldsymbol{b}\|/\|\boldsymbol{b}\|$, and
(2) $\|\boldsymbol{A}\|\|\boldsymbol{A}^{-1}\|$.

The latter term defines the **condition number**, and is denoted as cond($\boldsymbol{A}$). The condition number is a measure of the conditioning of a system of linear equations. The minimum value that a condition number can have is one. This can be proven as follows:

$$\|\boldsymbol{A}\cdot\boldsymbol{A}^{-1}\| = \|\boldsymbol{I}\| = 1.$$

Now,

$$\|\boldsymbol{A}\cdot\boldsymbol{A}^{-1}\| \leq \|\boldsymbol{A}\|\|\boldsymbol{A}^{-1}\| = \mathrm{cond}(\boldsymbol{A}).$$

Thus,

$$\mathrm{cond}(\boldsymbol{A}) \geq 1.$$

The condition number can take on values from one to infinity. A well-conditioned matrix will have low condition numbers close to one, while large condition numbers indicate an ill-conditioned problem. The error in the solution $\|\Delta x\|/\|x\|$ will be of the order of $\|\Delta b\|/\|b\|$, only if cond(A) is of $O(1)$. Knowledge of the condition number of a system enables one to interpret the accuracy of the solution to a perturbed system of equations. The exact value of the condition number depends on the type of norm used ($p = 1, 2$, or ∞), but ill-conditioning will be evident regardless of the type of norm. A fixed cut-off condition number that defines an ill-conditioned system does not exist partly because the ability to resolve an ill-conditioned problem is dependent on the level of precision used to perform the arithmetic operations. It is always recommended that you calculate the condition number when solving systems of equations.

The goal of the Gaussian elimination method of solution (and LU decomposition method) is to minimize the residual $\|\boldsymbol{r}\| = \|\boldsymbol{b} - \boldsymbol{Ax}\|$. When numerical procedures are carried out, round-off errors creep in and can corrupt the calculated values. Such errors are carried forward to the final solution. If $\boldsymbol{x} + \Delta\boldsymbol{x}$ is the numerically calculated solution to the system of equations $\boldsymbol{Ax} = \boldsymbol{b}$, and $\boldsymbol{x}$ is the exact solution to $\boldsymbol{Ax} = \boldsymbol{b}$, then the residual is given by

$$\boldsymbol{r} = \boldsymbol{b} - \boldsymbol{A}(\boldsymbol{x} + \Delta\boldsymbol{x}) = -\boldsymbol{A}\Delta\boldsymbol{x}$$

or

$$\|\Delta\boldsymbol{x}\| = \|\boldsymbol{A}^{-1}\boldsymbol{r}\| \leq \|\boldsymbol{A}^{-1}\|\|\boldsymbol{r}\|. \tag{2.24}$$

Multiplying Equation (2.22) by Equation (2.24) and rearranging, we arrive at

$$\frac{\|\Delta\boldsymbol{x}\|}{\|\boldsymbol{x}\|} \leq \|\boldsymbol{A}\|\|\boldsymbol{A}^{-1}\|\frac{\|\boldsymbol{r}\|}{\|\boldsymbol{b}\|}. \tag{2.25}$$

The magnitude of error in the solution obtained using the Gaussian elimination method also depends on the condition number of the coefficient matrix.

Let's look now at the effect of perturbing the coefficient matrix $\boldsymbol{A}$. If $\boldsymbol{x}$ is the exact solution to $\boldsymbol{Ax} = \boldsymbol{b}$, and perturbations of magnitude $\Delta\boldsymbol{A}$ change the solution $\boldsymbol{x}$ by an amount $\Delta\boldsymbol{x}$, then our corresponding matrix equation is given by

$$(\boldsymbol{A} + \Delta\boldsymbol{A})(\boldsymbol{x} + \Delta\boldsymbol{x}) = \boldsymbol{b}.$$

After some rearrangement,

$$\Delta\boldsymbol{x} = -\boldsymbol{A}^{-1}\Delta\boldsymbol{A}(\boldsymbol{x} + \Delta\boldsymbol{x}).$$

Taking norms,

$$\|\Delta\boldsymbol{x}\| \leq \|\boldsymbol{A}^{-1}\|\|\Delta\boldsymbol{A}\|\|(\boldsymbol{x} + \Delta\boldsymbol{x})\|.$$

Multiplying and dividing the right-hand side by $\|\boldsymbol{A}\|$, we obtain

$$\frac{\|\Delta\boldsymbol{x}\|}{\|(\boldsymbol{x} + \Delta\boldsymbol{x})\|} \leq \|\boldsymbol{A}^{-1}\|\|\boldsymbol{A}\|\frac{\|\Delta\boldsymbol{A}\|}{\|\boldsymbol{A}\|}. \tag{2.26}$$

Again we observe that the magnitude of error in the solution that results from modification to the elements of $\boldsymbol{A}$ is dependent on the condition number. Note that the left-hand side denominator is the solution to the perturbed set of equations.

Using MATLAB

MATLAB provides the cond function to calculate the condition number of a matrix using either the 1, 2, or ∞ norms (the default norm that MATLAB uses is $p = 2$).

Let's use this function to calculate the condition numbers of the coefficient matrices in Example 2.6:

```
>> A = [2 3; 4 1];
>> C1 = cond(A,1)
C1 =
     3
>> Cinf = cond(A,inf)
Cinf =
    3.0000
>> A = [4 2; 2.001 1];
>> C1 = cond(A,1)
C1 =
  1.8003e+004
>> Cinf = cond(A,inf)
Cinf =
  1.8003e+004
```

The calculation of the condition number involves determining the inverse of the matrix which requires $O(n^3)$ floating-point operations or flops. Calculating A^{-1} is an inefficient way to determine cond(A), and computers use specially devised $O(n^2)$ algorithms to estimate the condition number.

2.8 Linear regression

Experimental observations from laboratories, pilot-plant studies, mathematical modeling of natural phenomena, and outdoor field studies provide us with a host of biomedical, biological, and engineering data, which often specify a variable of interest that exhibits certain trends in behavior as a result of its dependency on another variable. Finding a "best-fit" function that reasonably fits the given data is valuable. In some cases, an established theory provides a functional relationship between the dependent and independent variables. In other cases, when a guiding theory is lacking, the data itself may suggest a functional form for the relationship. Sometimes, we may wish to use simple or uncomplicated functions, such as a linear or quadratic equation to represent the data, as advised by *Ockham's (Occam's) razor* – a heuristic principle posited by William Ockham, a fourteenth century logician. Occam's razor dictates the use of economy and simplicity when developing scientific theories and explanations and the elimination of superfluous or "non-key" assumptions that make little difference to the characteristics of the observed phenomena.

The technique used to derive an empirical relationship between two or more variables based on data obtained from experimental observations is called **regression**. The purpose of any **regression analysis** is to determine the parameters or coefficients of the function that best relate the dependent variable to one or more independent variables. A regression of the data to a "best-fit" curve produces a function that only approximates the given data to within a certain degree, but does not usually predict the exact values specified by the data. Experimentally obtained data contain not only inherent errors due to inaccuracies in measurements and underlying assumptions, but also inherent variability or randomness that is characteristic of all natural processes (see Section 3.1 for a discussion on "variability in nature and biology"). Hence, it is desirable to obtain an approximate function that agrees with the data in an overall fashion rather than forcing the function to pass through every experimentally derived data point since the exact magnitude of error in the data is usually unknown. The parameters of the fitted curve are adjusted according to an error minimization criterion to produce a best-fit.

Linear regression models involve a *strict linear dependency* of the functional form on the parameters to be determined. For example, if y depends on the value of x, the independent variable, then a linear regression problem is set up as

$$y = \beta_1 f_1(x) + \beta_2 f_2(x) + \cdots + \beta_n f_n(x),$$

where f can be any function of x, and $\beta_1, \beta_2, \ldots, \beta_n$ are constants of the equation that are to be determined using a suitable regression technique. Note that the individual functions $f_i(x)$ $(1 \leq i \leq n)$ may depend nonlinearly on x, but the unknown parameter β_i must be separable from the functional form specified by $f_i(x)$. The **linear least-squares method of regression** is a popular curve-fitting method used to approximate a line or a curve to the given data. The functional relationship of the dependent variable on *one or more* independent variables can be linear, exponential, quadratic, cubic, or any other nonlinear dependency.

In any linear regression problem, the number of linear equations m is greater than the number of coefficients n whose values we seek. As a result, the problem most often constitutes an **overdetermined system of linear equations**, i.e. a set of equations whose coefficient matrix has a rank equal to the number of columns n, but unequal to the rank of the augmented matrix, which is equal to $n + 1$, resulting in inconsistency of the equation set and therefore the lack of an exact solution. Given a functional form that relates the variables under consideration, the principle of the least-squares method is to minimize the error or "scatter" of the data points about the fitted curve. Before proceeding, let's consider a few examples of biomedical problems that can be solved using linear regression techniques.

Box 2.2A Using hemoglobin as a blood substitute: hemoglobin–oxygen binding

Hemoglobin (Hb) is a protein present in red blood cells that is responsible for the transport of oxygen (O_2) from lungs to individual tissues throughout the body and removal of carbon dioxide (CO_2) from the tissue spaces for transport to the lungs. The hemoglobin molecule is a tetramer and consists of four subunits, two α chains, and two β chains. Each α or β polypeptide chain contains a single iron atom-containing heme group that can bind to one O_2 molecule. Thus, a hemoglobin molecule can bind up to four O_2 molecules. The subunits work cooperatively with each other (allosteric binding), such that the binding of one O_2 molecule to one of the four subunits produces a conformational change within the protein that makes O_2 binding to the other subunits more favorable. The binding equation between hemoglobin and oxygen is as follows:

$$\mathrm{Hb(O_2)}_n \leftrightarrow \mathrm{Hb} + n\mathrm{O_2},$$

where $n = 1, \ldots, 4$.

The exchange of O_2 and CO_2 gases between the lungs and tissue spaces via the blood occurs due to prevailing differences in the partial pressures of pO_2 and pCO_2, respectively. The atmospheric air contains 21% O_2. The O_2 in inspired air exerts a partial pressure of 158 mm Hg (millimeters of mercury), which then reduces to 100 mm Hg when the inspired air mixes with the alveolar air, which is rich in CO_2. Venous blood that contacts the alveoli contains O_2 at a partial pressure of 40 mm Hg. The large difference in partial pressure drives diffusion of O_2 across the alveolar membranes into blood. Most of the oxygen in blood enters into the red blood cells and binds to hemoglobin molecules to form **oxyhemoglobin**. The oxygenated blood travels to various parts of the body and releases oxygen from oxyhemoglobin. The partial pressure of oxygen in the tissue spaces depends on the activity level of the tissues and is lower in more active tissues. The high levels of CO_2 in the surrounding tissue drive entry of CO_2 into blood and subsequent reactions with water to produce bicarbonates (HCO_3^-). Some of the bicarbonates enter the red blood cells and bind to hemoglobin to form carbaminohemoglobin. At the lungs, the oxygenation of blood is responsible for the transformation of carbaminohemoglobin to

Table 2.1. *Fractional saturation of hemoglobin as a function of partial pressure of oxygen in blood*

pO_2 (mm Hg)	10	20	30	40	50	60	70	80	90	100	110	120
S	0.18	0.40	0.65	0.80	0.87	0.92	0.94	0.95	0.95	0.96	0.96	0.97

Figure 2.5

A plot of the fractional saturation of hemoglobin S as a function of the partial pressure of oxygen in plasma.

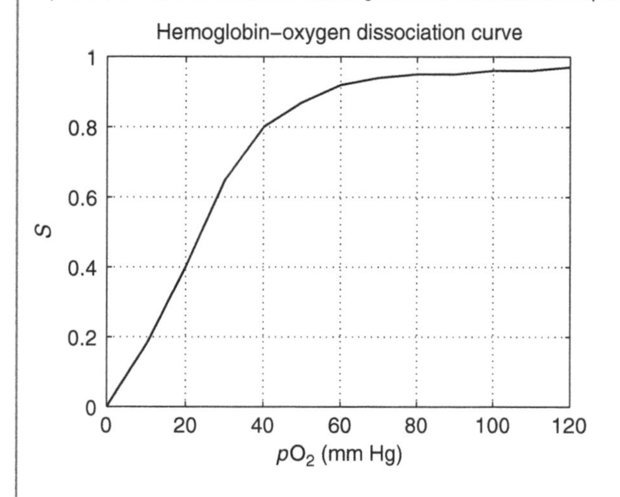

oxyhemoglobin or the dissociation of CO_2 from hemoglobin, and conversion of bicarbonates into CO_2 and H_2O, resulting in CO_2 leaving the blood and entering into the lungs.

The **Hill equation** describes a mathematical relationship between the extent of oxygen saturation of hemoglobin and the partial pressure of O_2 in blood. This equation is derived from the application of the law of mass action to the state of chemical equilibrium of the reaction

$$\mathrm{Hb(O_2)}_n \leftrightarrow \mathrm{Hb} + n\mathrm{O_2},$$

$$S = \frac{[\mathrm{Hb(O_2)}_n]}{[\mathrm{Hb(O_2)}_n] + [\mathrm{Hb}]} = \frac{(p\mathrm{O_2})^n}{P_{50}^n + (p\mathrm{O_2})^n}. \tag{2.27}$$

Equation (2.27) is the Hill equation, and accounts for the observed cooperative binding that produces the characteristic sigmoidal shape of the hemoglobin–oxygen dissociation curve.

The partial pressure of oxygen that results in occupancy of half of the oxygen binding sites of hemoglobin is denoted by P_{50}.

Biochemical engineers at several universities have explored the use of polymerized bovine hemoglobin as a possible blood substitute (Levy *et al.*, 2002). Monomeric hemoglobin is small enough to leak out of capillary pores; however, by chemically attaching several hemoglobin units together, the end product is large enough to be retained within the circulatory system. Data collected for tetrameric bovine hemoglobin binding to oxygen are given in Table 2.1.

We first examine the shape of the oxygen dissociation curve by plotting the data in Figure 2.5.

We wish to determine the best-fit values of P_{50} and n in the Hill equation for this data set and compare with values measured for human hemoglobin. Equation (2.27) is nonlinear but can be converted to linear form by first rearranging to

$$\frac{S}{1-S} = \frac{(pO_2)^n}{P_{50}^n}$$

and then taking the logarithm of both sides to obtain

$$\ln\frac{S}{1-S} = n\ln(pO_2) - n\ln P_{50}. \tag{2.28}$$

Plotting $\ln(S/1-S)$ as a function of $\ln(pO_2)$ produces a line with slope n and intercept $-n\ln P_{50}$.

Equation (2.28) is the functional form for the dependency of oxygen saturation of hemoglobin as a function of the oxygen partial pressure, and is linear in the regression parameters.

We will be in a position to evaluate the data once we learn how to fit the line described by Equation (2.28) to the data presented above.

Box 2.3A Time course of drug concentration in the body

Drugs can be delivered into the body by many different routes, a few examples of which are ingestion of drug and subsequent absorption into blood via the gastrointestinal tract, inhalation, topical application on the skin, subcutaneous penetration, and intravenous injection of drug directly into the circulatory system. The rate at which the drug enters into the vascular system, the extravascular space, and/or subsequently into cells, depends on a variety of factors, namely the mode of introduction of the drug into the body, resistance to mass transfer based on the chemical properties of the drug and inert material (such as capsules) associated with the drug, and the barriers offered by the body's interior. Once the drug enters into the blood, it is transported to all parts of the body and may enter into the tissue spaces surrounding the capillaries. The amount of drug that succeeds in entering into the extravascular space depends on the resistance faced by the drug while diffusing across the capillary membranes. While the active form of the drug distributes throughout the body, it is simultaneously removed from the body or chemically decomposed and rendered inactive either by liver metabolism, removal by kidney filtration, excretion of feces, exhalation of air, and/or perspiration. Thus, a number of processes govern the time-dependent concentrations of the drug in blood and various parts of the body. The study of the processes by which drugs are absorbed in, distributed in, metabolized by, and eliminated from the body is called **pharmacokinetics**, and is a branch of the field of pharmacology. In other words, pharmacokinetics is the study of *how the body affects the transport and efficacy of the drug*. A parallel field of study called **pharmacodynamics**, which is another branch of pharmacology, deals with the *time course of action and effect of drugs on the body.*

Pharmacokinetic studies of drug distribution and elimination aim at developing mathematical models that predict concentration profiles of a particular drug as a function of time. A simplified model of the time-dependent concentration of a drug that is injected all at once (bolus injection) and is cleared from the blood by one or more elimination processes is given by

$$C(t) = C_0 e^{-kt}, \tag{2.29}$$

where C_0 is the initial well-mixed blood plasma concentration of the drug immediately following intravenous injection and is calculated as $C_0 = D/V_{app}$, where D is the dosage and V_{app} is the apparent distribution volume. Note that k is the elimination rate constant with units of time^{-1} and is equal to the volumetric flowrate of plasma that is completely cleared of drug, per unit of the distribution volume of the drug in the body.

Voriconazole is a drug manufactured by Pfizer used to treat invasive fungal infections generally observed in patients with compromised immune systems. The use of this drug to treat fungal endophthalmitis is currently being explored. Shen *et al.* (2007) studied the clearance of this drug following injection into the vitreous humor (also known as "the vitreous") of the eyes of rabbits. The vitreous is a stagnant clear gel situated between the lens and the retina of the eyeball. The rate of

Table 2.2. *Time-dependent concentrations of a drug called Voriconazole in the vitreous and aqueous humor of rabbit eyes following intravitreal bolus injection of the drug*

Time, t (hr)	Vitreous concentration, $C_v(t)$ (μg/ml)	Aqueous concentration, $C_a(t)$ (μg/ml)
1	18.912 ± 2.058	0.240 ± 0.051
2	13.702 ± 1.519	0.187 ± 0.066
4	7.406 ± 1.783	0.127 ± 0.008
8	2.351 ± 0.680	0.000 ± 0.000
16	0.292 ± 0.090	0.000 ± 0.000
24	0.000 ± 0.000	0.000 ± 0.000
48	0.000 ± 0.000	0.000 ± 0.000

exchange of compounds between the vitreous and the systemic circulation is very slow. The aqueous humor is contained between the lens and the cornea and is continuously produced and drained from the eye at an equal rate.

The time-dependent concentration of Voriconazole in the vitreous and aqueous humor is given in Table 2.2 following intravitreal injection (Shen *et al.*, 2007).

We use Equation (2.29) to model the clearance of the drug from rabbit eyes following a bolus injection of the drug. If we take the logarithm of both sides of the equation, we get

$$\ln C(t) = \ln C_0 - kt. \tag{2.30}$$

Equation (2.30) is linear when $\ln C(t)$ is plotted against t, and has slope $-k$ and intercept equal to $\ln C_0$.

We want to determine the following:

(a) the initial concentration C_0,
(b) the elimination rate constant k,
(c) the half life of the drug, and
(d) the drug clearance rate from the vitreous of the eye relative to the clearance rate of drug from the aqueous fluid surrounding the eye.

Box 2.4A Platelet flow rheology

A computational three-dimensional numerical model called "multiparticle adhesive dynamics" (MAD) uses state-of-the-art fluid mechanical theory to predict the nature of flow of blood cells (e.g. white blood cells and platelets) in a blood vessel (King and Hammer, 2001). White blood cells play an indispensable role in the inflammatory response and the immune system, and one of their main roles is to search for and destroy harmful foreign molecules that have entered the body. Platelets patrol the blood vessel wall for lesions. They aggregate to form clots to curtail blood loss and promote healing at injured sites.

The flow behavior of a platelet-shaped cell (flattened ellipsoid) is dramatically different from that of a spherical-shaped cell (Mody and King, 2005). It was determined from MAD studies that platelets flowing in linear shear flow demonstrate three distinct regimes of flow near a plane wall. Platelets of dimension $2 \times 2 \times 0.5\ \mu m^3$ (see Figure 2.6) are observed to rotate periodically while moving in the direction of flow at flow distances greater than 0.75 μm from the wall. Platelets at these heights also contact the surface at regular intervals. When the platelet centroid height is less than 0.75 μm, the platelet cell simply glides or cruises over the surface in the direction of flow without rotating, and does not contact the wall.

Table 2.3.

Initial tilt α (degrees)	10.5	10.0	8.5	7	4.5
Height H (μm)	0.50	0.55	0.60	0.65	0.70

Figure 2.6

Three-dimensional view of a simulated platelet-shaped cell of dimension $2 \times 2 \times 0.5$ μm^3.

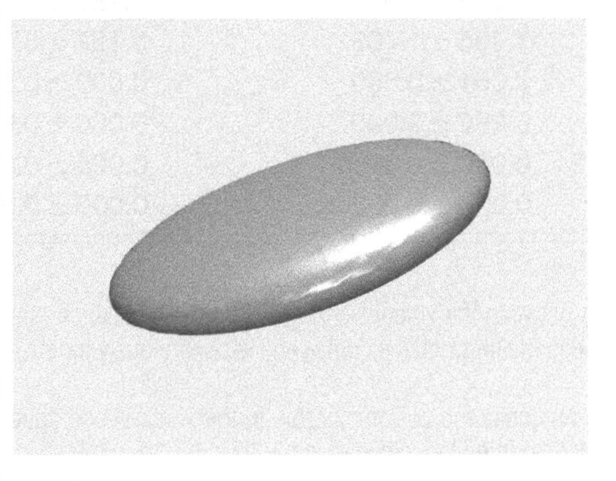

Figure 2.7

Platelet modeled as an oblate spheroid (flattened ellipsoid) that is flowing (translating and rotating) in linear shear flow over an infinite planar surface.

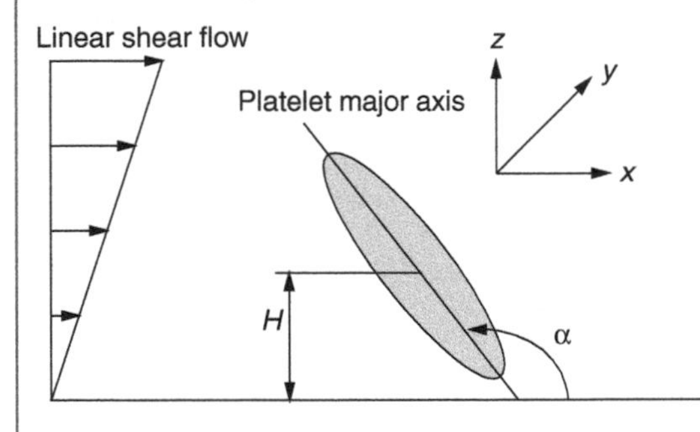

If a platelet flowing near a wall is subjected to linear shear flow, the relevant flow geometry is depicted in Figure 2.7.

A platelet-shaped cell dramatically changes its flow behavior from a non-tumbling, non-wall-contacting regime to a tumbling, wall-contacting flow regime under specific conditions of initial cell orientation and distance of cell from the vessel wall (Mody and King, 2005). The starting orientation of a platelet that forces a qualitative change in motion depends on the platelet proximity to the wall. The initial platelet orientations α that produce a regime shift are given in Table 2.3 as a function of the platelet height H from the surface and plotted in Figure 2.8.

If we wish to relate the initial orientation angle of the platelet α that produces a regime change with initial platelet height H using a quadratic function, we require only three equations to determine the three unknown parameters. The data in Table 2.3 contain five sets of (H, α) data points. Linear regression must be employed to obtain a quadratic function that relates the dependency of α on H as prescribed by these five data points.

Figure 2.8

Transition from rotating to gliding motion. Plot of the initial angles of tilt about the y-axis that result in a transition of platelet flow regime as a function of initial platelet centroid height from the surface.

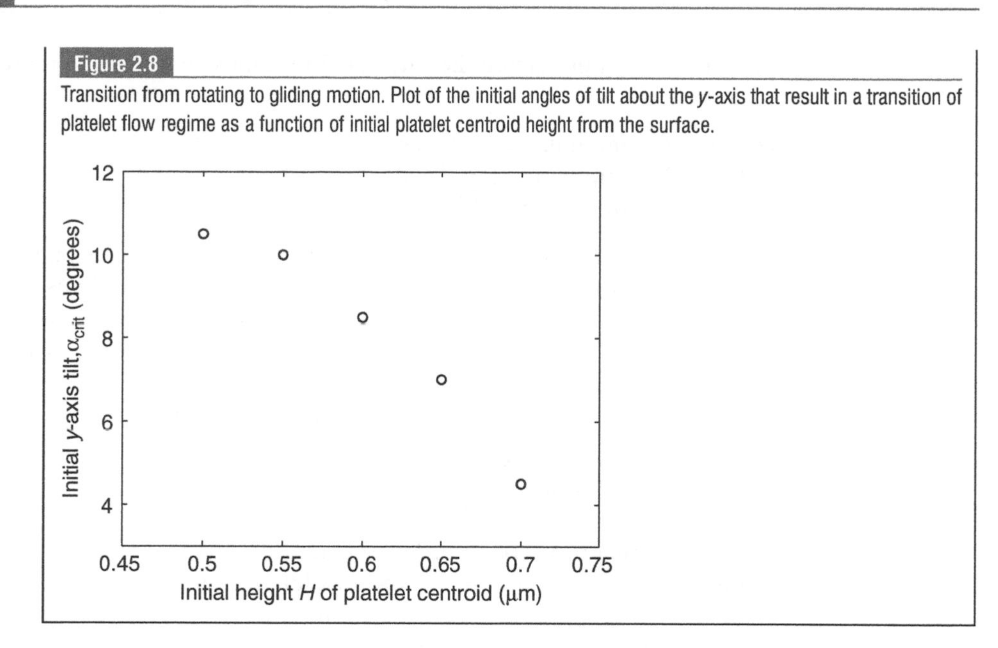

2.9 Curve fitting using linear least-squares approximation

For any set of data that can be represented by a straight line, the dependent variable y must vary with the n independent variables $x_1, x_2, \ldots, x_n$ in a linear fashion:

$$y = \beta_0 + \beta_1 x_1 + \beta_2 x_2 + \cdots + \beta_n x_n. \tag{2.31}$$

The dependency on the n variables is defined by the values of the $n + 1$ constant coefficients $\beta_0, \beta_1, \ldots, \beta_n$. Most often, situations arise in which the number of data points $(x_1, x_2, \ldots, x_n, y)$ m is greater than the number of undetermined coefficients $n + 1$. The system of linear equations consists of m equations in $n + 1$ variables (the unknown coefficients are the variables to be determined). Since $m > n + 1$, such a system is **overdetermined**. No exact solution can be obtained. Let us consider the simplest linear relationship $y = \beta_0 + \beta_1 x$, where the association of y is with only one independent variable x. The number of unknowns in this equation is two. If the m number of data points (x_i, y_i) $1 \leq i \leq m$ that describe the linear trend is greater than two, then one single line cannot pass through all these points unless all the data points actually lie on a single line. Our goal is to determine the values of β_0 and β_1 of a linear equation that produce the best approximation of the linear trend of the data (x_i, y_i) $1 \leq i \leq m$.

The quality of the fit of the straight-line approximation is measured by calculating the residuals. A **residual** is the difference between the actual data value y_i at x_i and the corresponding approximated value predicted by the fitted line $\hat{y}_i = \beta_0 + \beta_1 x_i$, where the hat symbol represents an approximation that is derived from regression. Thus, the residual at x_i is calculated as

$$r_i = y_i - (\beta_0 + \beta_1 x_i). \tag{2.32}$$

It is assumed in the following development of the regression equations that the x_i values (independent variable) are exactly known without error. The best-fit can be determined by minimizing the calculated residuals according to a pre-defined criterion such as

criterion 1: minimum of $\sum_{i=1}^{m} r_i$;
criterion 2: minimum of $\sum_{i=1}^{m} |r_i|$;
criterion 3: minimum of $\sum_{i=1}^{m} r_i^2$.

The flaw with criterion 1 is obvious. Residual values can be either positive or negative. Even for a badly fitted line, large positive residuals can cancel large negative residuals, producing a sum nearly equal to zero and thereby masking a bad approximation. Figure 2.9(a) shows a line that is clearly a bad fit, yet gives $\sum_{i=1}^{m} r_i = 0$, thus satisfying the criterion.

The problem with criterion 2 becomes obvious when we try to take the derivatives

$$\frac{d\sum_{i=1}^{m}|r_i|}{d\beta_0} \text{ and } \frac{d\sum_{i=1}^{m}|r_i|}{d\beta_1}$$

and equate them to zero in order to determine the value of the coefficients at the point of minimum of the sum of the absolute residuals. A unique solution does not exist, i.e. multiple lines can satisfy this criterion. This is clearly evident from Figure 2.9(b), which shows three lines, each equally satisfying criterion 2.

Figure 2.9

llustration of the difficulties faced in obtaining a best-fit for a line when using (a) minimization criterion 1 and (b) minimization criterion 2.

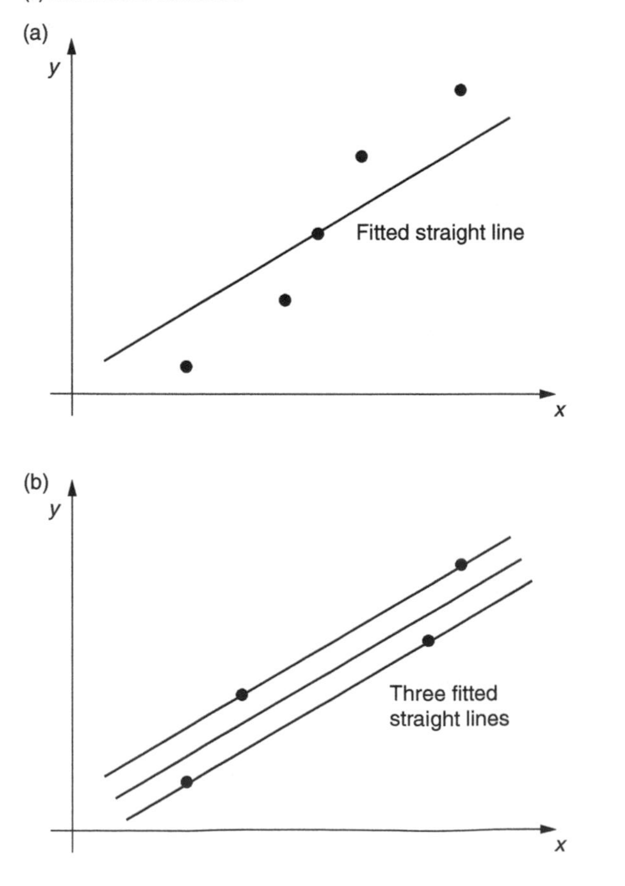

Criterion 3 is called the **least-squares fit**. The goal involves minimizing the sum of the squared residuals, or **SSR**. The least-squares criterion provides a *unique* best-fit solution. In fact, in Euclidean R^n space, where the norm of a vector is defined by Equation (2.9), determination of the least-squares of $\boldsymbol{r}$ is synonymous with the minimization of $\|\boldsymbol{r}\|$, and therefore specifies the best possible fit.[4]

2.9.1 The normal equations

Now we delve into the mathematical derivation of the linear regression equations based on the least-squares principle. The residual is defined by Equation (2.32).

Criterion 3 is rewritten as

$$SSR = \sum_{i=1}^{m} (y_i - (\beta_0 + \beta_1 x_i))^2. \tag{2.33}$$

Minimization of the SSR is achieved by differentiating Equation (2.33) with respect to each unknown coefficient. The solution of the following system of equations yields coefficient values that provide the unique least-squares fit to the data:

$$\frac{dSSR}{d\beta_0} = \sum_{i=1}^{m} -2(y_i - (\beta_0 + \beta_1 x_i)) = 0,$$

$$\frac{dSSR}{d\beta_1} = \sum_{i=1}^{m} -2x_i(y_i - (\beta_0 + \beta_1 x_i)) = 0.$$

After simplification of the above two equations, we get

$$m\beta_0 + \beta_1 \sum_{i=1}^{m} x_i = \sum_{i=1}^{m} y_i, \tag{2.34}$$

$$\beta_0 \sum_{i=1}^{m} x_i + \beta_1 \sum_{i=1}^{m} x_i^2 = \sum_{i=1}^{m} x_i y_i. \tag{2.35}$$

Equations (2.34) and (2.35) are referred to as the **normal equations**. This name derives from the geometric interpretation of these equations, which is discussed shortly.

This set of two equations can be expressed in the form of a single matrix equation:

$$\begin{bmatrix} m & \sum_{i=1}^{m} x_i \\ \sum_{i=1}^{m} x_i & \sum_{i=1}^{m} x_i^2 \end{bmatrix} \begin{bmatrix} \beta_0 \\ \beta_1 \end{bmatrix} = \begin{bmatrix} \sum_{i=1}^{m} y_i \\ \sum_{i=1}^{m} x_i y_i \end{bmatrix} \tag{2.36}$$

Solving Equations (2.34) and (2.35) simultaneously, we obtain

$$\beta_1 = \frac{m \sum_{i=1}^{m} x_i y_i - \sum_{i=1}^{m} x_i \sum_{i=1}^{m} y_i}{m \sum_{i=1}^{m} x_i^2 - \left(\sum_{i=1}^{m} x_i\right)^2} \tag{2.37}$$

and

$$\beta_0 = \bar{y} - \beta_1 \bar{x}, \tag{2.38}$$

[4] A linear least-squares fit is the best-fit only if the phenomenon which the data describes adheres to certain behaviors, such as independence of data values, normality of y distribution about its mean, and constant variance of y over the range of x studied. These assumptions are discussed in detail in Chapter 3.

where the means of x and y are defined as

$$\bar{x} = \frac{1}{m}\sum_{i=1}^{m} x_i \quad \text{and} \quad \bar{y} = \frac{1}{m}\sum_{i=1}^{m} y_i.$$

In fact, β_1 can be expressed in somewhat simpler form as follows:

$$\beta_1 = \frac{\sum_{i=1}^{m}(x_i - \bar{x})(y_i - \bar{y})}{\sum_{i=1}^{m}(x_i - \bar{x})^2}. \tag{2.39}$$

You are asked to verify the equivalence. (Hint: Expand Equation (2.39) and use the definition of the mean to simplify the terms.)

To illustrate the use of Equation (2.36) in determining the coefficients of a regression line, let's consider a simple example.

Example 2.7

Determine the linear trend line for the following data set:

$$\begin{aligned} \mathbf{x} &= 1.0,\ 2.0,\ 3.0,\ 4.0,\ 5.0,\ 6.0, \\ \mathbf{y} &= 1.6,\ 4.2,\ 6.6,\ 8.7,\ 11.5,\ 13.7. \end{aligned}$$

While a MATLAB program can be written to perform the necessary calculations, we will perform them by hand for demonstration purposes.

First, we calculate the summations of x_i, x_i^2, y_i and x_iy_i which can be conveniently represented in tabular form (see Table 2.4).

Using Equation (2.36) we obtain the matrix equation

$$\begin{bmatrix} 6 & 21.0 \\ 21.0 & 91.0 \end{bmatrix} \begin{bmatrix} \beta_0 \\ \beta_1 \end{bmatrix} = \begin{bmatrix} 46.3 \\ 204.3 \end{bmatrix}.$$

MATLAB can be used to solve the above linear problem:

```
>> A = [6 21.0; 21.0 91.0];
>> b = [46.3; 204.3];
>> c = A\b
c =
    -0.7333
     2.4143
```

Thus, $\beta_0 = -0.7333$ and $\beta_1 = 2.4143$.

Next we wish to assess the quality of the fit by calculating the residuals. The values $\hat{y}$ as predicted by the regression line at the x values specified in the data set are determined using the `polyval` function. The

Table 2.4.

x	**y**	$\mathbf{x}^2$	**xy**
1.0	1.6	1.0	1.6
2.0	4.2	4.0	8.4
3.0	6.6	9.0	19.8
4.0	8.7	16.0	34.8
5.0	11.5	25.0	57.5
6.0	13.7	36.0	82.2
$\sum\mathbf{x} = 21.0$	$\sum\mathbf{y} = 46.3$	$\sum\mathbf{x}^2 = 91.0$	$\sum\mathbf{xy} = 204.3$

polyval function accepts as its arguments the coefficients of a polynomial (specified in descending order of powers of x) and the x values at which the polynomial is to be determined:

```
>> x = [1.0 2.0 3.0 4.0 5.0 6.0];
>> y = [1.6 4.2 6.6 8.7 11.5 13.7];
>> r = y - polyval([2.4143 -0.7333], x)
r =
    -0.0810    0.1047    0.0904   -0.2239    0.1618   -0.0525
```

These residuals will be used to obtain a single number that will provide a measure of the quality of the fit.

Equation (2.36) is compact and easy to evaluate. However, a drawback in the method used for obtaining the individual equations (2.34) and (2.35) is that the process must be repeated for every different function that relates y to its independent variables. To emphasize this disadvantage, we now derive the normal equations for a polynomial fitted to data that relates x and y.

Often a polynomial function is used to describe the relationship between y and x. The nature of the fitted polynomial curve, i.e. the number of bends or kinks in the curve, is a function of the order of the polynomial. A data set of $n + 1$ points can be exactly represented by an nth-order polynomial (the nth-order polynomial will pass through all $n + 1$ data points). If many data points are known to exceedingly good accuracy, then it may be useful to use higher-order polynomials. However, in the majority of situations, the data set itself is only approximate, and forcing the fitted curve to pass through every point is not justified. As the order of the polynomial increases, the fitted curve may not necessarily connect the points with a smooth curve, but may display erratic oscillatory behavior between the data points. Unless the biological phenomenon or engineering application that the data describes suggests such complicated dependencies, it is recommended that one avoids using higher-order polynomials (fourth-order and higher) unless one is justified in doing so. With higher-order polynomials, the related normal equations are prone to ill-conditioning and round-off errors, which corrupt the accuracy of the solution.

We derive the normal equations for fitting a second-order polynomial to an (x, y) data set. The quadratic dependence of y on x is expressed as

$$y = \beta_0 + \beta_1 x + \beta_2 x^2.$$

The undetermined **model parameters** that we seek are β_0, β_1, and β_2. The **objective function** to be minimized is

$$SSR = \sum_{i=1}^{m} \left(y_i - (\beta_0 + \beta_1 x_i + \beta_2 x_i^2)\right)^2. \tag{2.40}$$

The partial derivatives of Equation (2.40) are taken with respect to each of the three unknown model parameters and are set equal to zero to determine the equations that simultaneously represent the minimum of the SSR:

$$\frac{dSSR}{d\beta_0} = \sum_{i=1}^{m} -2\left(y_i - (\beta_0 + \beta_1 x_i + \beta_2 x_i^2)\right) = 0,$$

$$\frac{dSSR}{d\beta_1} = \sum_{i=1}^{m} -2x_i\left(y_i - (\beta_0 + \beta_1 x_i + \beta_2 x_i^2)\right) = 0,$$

$$\frac{dSSR}{d\beta_2} = \sum_{i=1}^{m} -2x_i^2\left(y_i - (\beta_0 + \beta_1 x_i + \beta_2 x_i^2)\right) = 0.$$

The final form of the above three equations is as follows:

$$m\beta_0 + \beta_1 \sum_{i=1}^{m} x_i + \beta_2 \sum_{i=1}^{m} x_i^2 = \sum_{i=1}^{m} y_i, \tag{2.41}$$

$$\beta_0 \sum_{i=1}^{m} x_i + \beta_1 \sum_{i=1}^{m} x_i^2 + \beta_2 \sum_{i=1}^{m} x_i^3 = \sum_{i=1}^{m} x_i y_i, \tag{2.42}$$

$$\beta_0 \sum_{i=1}^{m} x_i^2 + \beta_1 \sum_{i=1}^{m} x_i^3 + \beta_2 \sum_{i=1}^{m} x_i^4 = \sum_{i=1}^{m} x_i^2 y_i, \tag{2.43}$$

which can be compactly represented as

$$\begin{bmatrix} m & \sum_{i=1}^{m} x_i & \sum_{i=1}^{m} x_i^2 \\ \sum_{i=1}^{m} x_i & \sum_{i=1}^{m} x_i^2 & \sum_{i=1}^{m} x_i^3 \\ \sum_{i=1}^{m} x_i^2 & \sum_{i=1}^{m} x_i^3 & \sum_{i=1}^{m} x_i^4 \end{bmatrix} \begin{bmatrix} \beta_0 \\ \beta_1 \\ \beta_2 \end{bmatrix} = \begin{bmatrix} \sum_{i=1}^{m} y_i \\ \sum_{i=1}^{m} x_i y_i \\ \sum_{i=1}^{m} x_i^2 y_i \end{bmatrix}. \tag{2.44}$$

Equation (2.44) constitutes the **normal equations for a quadratic function** in x, and has different terms compared to the normal equations (Equation (2.36)) for a linear function in x. In fact, the normal equations can be derived for any overdetermined linear system, as long as the form of dependence of y on x involves a *linear combination* of functions in x. In other words, any functional relationship of the form

$$y = \beta_1 f_1(x) + \beta_2 f_2(x) + \cdots + \beta_n f_n(x)$$

can be fitted to data using a linear least-squares approach. The individual functions $f_1, f_2, \ldots, f_n$ can involve polynomial functions, trignometric functions, logarithms, and so on. For example, using any of the following fitting functions, a given data set is amenable to linear least-squares regression:

$$y = \frac{\beta_1}{x} + \beta_2 \tan x,$$
$$y = \sqrt{x}(\beta_1 \sin x + \beta_2 \cos x),$$
$$y = \beta_1 \ln x + \beta_2 x^3 + \beta_3 x^{1/3}.$$

Rederiving the regression or normal equations for every new model is tedious and cumbersome. A generalized method allows us to perform least-squares linear regression much more efficiently. Assume that the functional dependence of y on x can be expressed as

$$y = \beta_0 + \beta_1 f_1(x) + \cdots + \beta_n f_n(x). \tag{2.45}$$

If we have a data set that consists of m (x, y) pairs, then using Equation (2.45) we can write m equations that are linear in $n + 1$ coefficients. We can write this set of m equations in compact matrix form. If we set

$$\mathbf{A} = \begin{bmatrix} 1 & f_1(x_1) & \ldots & f_n(x_1) \\ 1 & f_1(x_2) & \ldots & f_n(x_2) \\ & & \vdots & \\ 1 & f_1(x_m) & \ldots & f_n(x_m) \end{bmatrix}, \quad \mathbf{y} = \begin{bmatrix} y_1 \\ y_2 \\ \vdots \\ y_m \end{bmatrix},$$

and $\boldsymbol{c}$ as the vector of the coefficients $\beta_0, \beta_1, \ldots, \beta_n$ whose values we seek, then we obtain the equation $\boldsymbol{Ac} = \boldsymbol{y}$. Here the number of equations (one equation for each data point) is greater than the number of unknown coefficients, i.e. $m > n + 1$. The system of equations is overdetermined if the rank of $\boldsymbol{A}$ is less than the rank of the augmented matrix $[\boldsymbol{A}\ \boldsymbol{y}]$. In this case an exact solution does not exist since $\boldsymbol{y}$ is located outside the column space of $\boldsymbol{A}$. As a result, no vector of coefficients $\boldsymbol{c}$ exists that allows $\boldsymbol{Ac}$ to equal $\boldsymbol{y}$ exactly. Thus, $\boldsymbol{y} - \boldsymbol{Ac} \neq \boldsymbol{0}$, for any $\boldsymbol{c}$.

In order to determine a best-fit, we must minimize the norm of the residual $\|\boldsymbol{r}\|$, where $\boldsymbol{r} = \boldsymbol{y} - \hat{\boldsymbol{y}} = \boldsymbol{y} - \boldsymbol{Ac}$. From the definition of the norm (Euclidean norm) given by Equation (2.9), $\|\boldsymbol{r}\| = \sqrt{\boldsymbol{r} \cdot \boldsymbol{r}} = \sqrt{\boldsymbol{r}^{\mathrm{T}}\boldsymbol{r}}$. We wish to find the value of $\boldsymbol{c}$ that minimizes $\boldsymbol{r}^{\mathrm{T}}\boldsymbol{r} = \sum_{i=1}^{m} r_i^2$, or the smallest value of the sum of squares of the individual residuals corresponding to each data point. The linear least-squares regression equations (normal equations) can be derived in two different ways.

Geometric method

The product $\boldsymbol{Ac}$ produces a vector that is a linear combination of the column vectors of $\boldsymbol{A}$, and thus lies in the column space of $\boldsymbol{A}$. To minimize $\boldsymbol{r}$, we need to determine the value of $\boldsymbol{Ac}$ that is closest to $\boldsymbol{y}$. It can be shown (in Euclidean space) that the orthogonal projection of $\boldsymbol{y}$ in the column space of $\boldsymbol{A}$ lies closest in value to $\boldsymbol{y}$ than to any other vector in the column space of $\boldsymbol{A}$. If we denote the orthogonal projection of $\boldsymbol{y}$ in the column space of $\boldsymbol{A}$ as $\hat{\boldsymbol{y}}$, and if

$$\boldsymbol{y} = \hat{\boldsymbol{y}} + \boldsymbol{y}^{\perp} \quad \text{and} \quad \hat{\boldsymbol{y}} \perp \boldsymbol{y}^{\perp}$$

then the residual

$$\boldsymbol{r} = \boldsymbol{y}^{\perp} = \boldsymbol{y} - \hat{\boldsymbol{y}}.$$

Figure 2.10 depicts $\hat{\boldsymbol{y}}$ in the column space of $\boldsymbol{A}$. Any other vector equal to $\boldsymbol{y} - \boldsymbol{y}'$, where $\boldsymbol{y}'$ is a non-orthogonal projection of $\boldsymbol{y}$ in the column space of $\boldsymbol{A}$, as shown in Figure 2.10, will necessarily be greater than $\boldsymbol{y}^{\perp}$.

We seek to minimize

$$\boldsymbol{r} = \boldsymbol{y}^{\perp} = \boldsymbol{y} - \hat{\boldsymbol{y}} = \boldsymbol{y} - \boldsymbol{Ac}.$$

Figure 2.10

Geometrical interpretation of the least-squares solution, illustrating the orthogonal projection of $\boldsymbol{y}$ in the column space of $\boldsymbol{A}$.

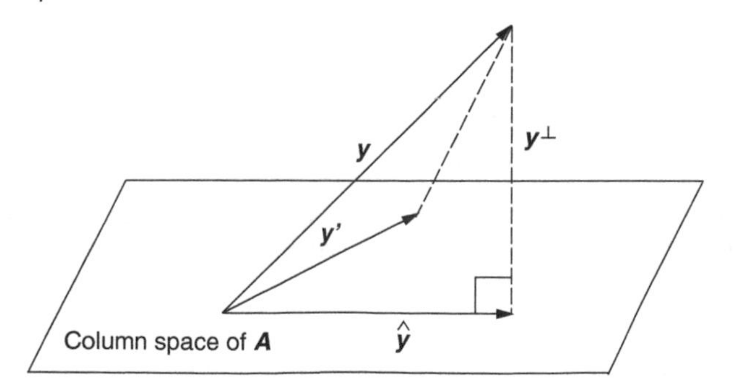

Now, $\boldsymbol{y}^{\perp}$ is normal or orthogonal to the column space of $\boldsymbol{A}$ (see Figure 2.10). The *column space* of $\boldsymbol{A}$ = *row space* of $\boldsymbol{A}^{\mathrm{T}}$. Therefore, $\boldsymbol{y}^{\perp}$ is orthogonal to the *row space* of $\boldsymbol{A}^{\mathrm{T}}$. Mathematically, we write

$$\boldsymbol{A}^{\mathrm{T}} \cdot \boldsymbol{y}^{\perp} = 0 \text{ or } \boldsymbol{A}^{\mathrm{T}}(\boldsymbol{y} - \boldsymbol{Ac}) = 0$$

or

$$\boldsymbol{A}^{\mathrm{T}}\boldsymbol{Ac} = \boldsymbol{A}^{\mathrm{T}}\boldsymbol{y}. \tag{2.46}$$

Equation (2.46) is called the **normal system**, and the resulting set of equations is called the **normal equations**. Why are these linear equations referred to as the "normal equations"? The matrix equation given by Equation (2.46) describes the condition that $\boldsymbol{y} - \boldsymbol{Ac}$ must be *normal* to the column space of $\boldsymbol{A}$ in order to obtain the set of values $\boldsymbol{c}$ that yields the best least-squares approximation of $\hat{\boldsymbol{y}}$ to $\boldsymbol{y}$. The normal equations have a unique solution if $|\boldsymbol{A}^{\mathrm{T}}\boldsymbol{A}| \neq 0$ and infinitely many solutions if $|\boldsymbol{A}^{\mathrm{T}}\boldsymbol{A}| = 0$.

Equation (2.46) is exactly the same as Equation (2.36) for the case of regression of a straight line to data. If $y = a_0 + a_1x$ and

$$\boldsymbol{A} = \begin{bmatrix} 1 & x_1 \\ 1 & x_2 \\ \cdot & \\ \cdot & \\ \cdot & \\ 1 & x_m \end{bmatrix},$$

show that $\boldsymbol{A}^{\mathrm{T}}\boldsymbol{A}$ and $\boldsymbol{A}^{\mathrm{T}}\boldsymbol{y}$ produce the corresponding matrices given in Equation (2.36).

Algebraic Method

We seek the minimum of

$$\|\boldsymbol{r}\|^2 = \boldsymbol{r}^{\mathrm{T}}\boldsymbol{r} = (\boldsymbol{y} - \hat{\boldsymbol{y}})^{\mathrm{T}}(\boldsymbol{y} - \hat{\boldsymbol{y}}) = (\boldsymbol{y} - \boldsymbol{Ac})^{\mathrm{T}}(\boldsymbol{y} - \boldsymbol{Ac}),$$
$$\|\boldsymbol{r}\|^2 = \boldsymbol{y}^{\mathrm{T}}\boldsymbol{y} - \boldsymbol{y}^{\mathrm{T}}\boldsymbol{Ac} - \boldsymbol{c}^{\mathrm{T}}\boldsymbol{A}^{\mathrm{T}}\boldsymbol{y} + \boldsymbol{c}^{\mathrm{T}}\boldsymbol{A}^{\mathrm{T}}\boldsymbol{Ac}.$$

Each term in the above equation is a scalar number. Note that the transpose operation follows the rule $(\boldsymbol{AB})^{\mathrm{T}} = \boldsymbol{B}^{\mathrm{T}}\boldsymbol{A}^{\mathrm{T}}$.

Determining the minimum is equivalent to finding the gradient of $\boldsymbol{r}^{\mathrm{T}}\boldsymbol{r}$ with respect to $\boldsymbol{c}$ and equating it to zero:

$$\nabla_c(\boldsymbol{r}^{\mathrm{T}}\boldsymbol{r}) = 0 - \boldsymbol{A}^{\mathrm{T}}\boldsymbol{y} - \boldsymbol{A}^{\mathrm{T}}\boldsymbol{y} + \boldsymbol{A}^{\mathrm{T}}\boldsymbol{Ac} + \boldsymbol{A}^{\mathrm{T}}\boldsymbol{Ac} = 0,$$

where

$$\nabla_c = \left(\frac{\partial}{\partial \beta_0}, \frac{\partial}{\partial \beta_1}, \ldots, \frac{\partial}{\partial \beta_n}\right)$$

and each term in the above equation is a column vector. The final form of the gradient of the squared residual norm is given by

$$\boldsymbol{A}^{\mathrm{T}}\boldsymbol{Ac} = \boldsymbol{A}^{\mathrm{T}}\boldsymbol{y}.$$

We obtain the same system of normal equations as before.

Two main steps are involved when applying the normal equations (Equation 2.46):

(1) construction of $\mathbf{A}$, and
(2) solution of $\mathbf{c} = (\mathbf{A}^T\mathbf{A})^{-1}\mathbf{A}^T\mathbf{y}$ in MATLAB as `c = (A'*A)\A'*y`.

2.9.2 Coefficient of determination and quality of fit

The **mean value** $\bar{y}$ can be used to represent a set of data points y_i, and is a useful statistic if the scatter can be explained by the stochastic nature of a process or errors in measurement. If the data show a trend with respect to a change in a property or condition (i.e. the independent variable(s)), then the mean value will not capture the nature of this trend. A model fitted to the data is expected to capture the observed trend either wholly or partially. The outcome of the fit will depend on many factors, such as the type of model chosen and the magnitude of error in the data. A "best-fit" model is usually a superior way to approximate data than simply stating the mean of all y_i points, since the model contains *at least* one adjustable coefficient *and* a term that represents the nature of dependency of y on the independent variable(s).

The difference $(y_i - \bar{y})$ (also called **deviation**) measures the extent of deviation of each data point from the **mean**. The sum of the squared difference $\sum(y_i - \bar{y})^2$ is a useful quantitative measure of the spread of the data about the mean. When a fitted function is used to approximate the data, the sum of the squared residuals is $\|\mathbf{r}\|^2 = \sum(y_i - \hat{y})^2$. The **coefficient of determination** R^2 is popularly used to determine the quality of the fit; R^2 conveys the improvement attained in using the model to describe the data, compared to using a horizontal line that passes through the mean. Calculation of R^2 involves comparing the deviations of the y data points from the model prediction, i.e. $\sum(y_i - \hat{y})^2$, with the deviations of the y data points from the mean, i.e. $\sum(y_i - \bar{y})^2$. Note that R^2 is a number between zero and one and is calculated as shown in Equation (2.47):

$$R^2 = 1 - \frac{\sum(y_i - \hat{y}_i)^2}{\sum(y_i - \bar{y})^2} = 1 - \frac{\|\mathbf{r}\|^2}{\sum(y_i - \bar{y})^2}. \tag{2.47}$$

The summation subscripts have been omitted to improve readability.

If $\sum(y_i - \hat{y})^2$ is much less than $\sum(y_i - \bar{y})^2$, then the model is a better approximation to the data compared to the mean of the data. In that case, R^2 will be close to unity. If the model poorly approximates the data, the deviations of the data points from the model will be comparable to the data variation about the mean and R^2 will be close to zero.

Example 2.7 (continued)

We readdress the problem of fitting the data set to a straight line, but this time Equation (2.46) is used instead of Equation (2.36). For this system of equations

$$\mathbf{A} = \begin{bmatrix} 1 & 1.0 \\ 1 & 2.0 \\ 1 & 3.0 \\ 1 & 4.0 \\ 1 & 5.0 \\ 1 & 6.0 \end{bmatrix} \text{ and } \mathbf{y} = \begin{bmatrix} 1.6 \\ 4.2 \\ 6.6 \\ 8.7 \\ 11.5 \\ 13.7 \end{bmatrix}.$$

Now, evaluating in MATLAB

```
>> A = [1 1; 1 2; 1 3; 1 4; 1 5; 1 6]; y = [1.6; 4.2; 6.6; 8.7; 11.5; 13.7];
>> c = (A'*A)\(A'*y)
c =
    -0.7333
     2.4143
```

The elements of $\boldsymbol{c}$ are exactly equal to the slope and intercept values determined in Example 2.7. Next, we calculate the coefficient of determination using MATLAB to determine how well the line fits the data:

```
>> ybar = mean(y)
ybar =
    7.7167
>> x = [1 2 3 4 5 6];
>> yhat = A*c
yhat' =
    1.6810  4.0952  6.5095  8.9238  11.3381  13.7524
>> SSR = sum((y - yhat(:)).^2) % sum of squared residuals
SSR =
    0.1048
>> R2 = 1 - SSR/sum((y - ybar).^2 % coefficient of determination
R2 =
    0.9990
```

The straight-line fit is indeed a good approximation.

Using MATLAB

(1) Common statistical functions such as `mean` and `std` (standard deviation) are included in all versions of MATLAB.

(2) The element-wise operator `.^` calculates the square of each term in the vector, i.e. the exponentiation operation is performed element-by-element.

(3) Calling a vector or matrix `y(:)` or `A(:)` retrieves all of the elements as a single column. For matrices, this usage stacks the columns vertically.

Box 2.4B Platelet flow rheology

We wish to fit the function

$$\alpha_{\text{crit}} = \beta_2 H^2 + \beta_1 H + \beta_0,$$

where α_{crit} is the critical angle of tilt of the platelet at which the transition in flow regime occurs. Using the data, we define the matrices

$$\boldsymbol{A} = \begin{bmatrix} H_1^2 & H_1 & 1 \\ H_2^2 & H_2 & 1 \\ H_3^2 & H_3 & 1 \\ H_4^2 & H_4 & 1 \\ H_5^2 & H_5 & 1 \end{bmatrix} \text{ and } \boldsymbol{y} = \begin{bmatrix} \alpha_{\text{crit},1} \\ \alpha_{\text{crit},2} \\ \alpha_{\text{crit},3} \\ \alpha_{\text{crit},4} \\ \alpha_{\text{crit},5} \end{bmatrix}.$$

We then apply the normal equations to solve for the coefficients of the quadratic function. In MATLAB, we type

```
>> H = [0.5; 0.55; 0.6; 0.65; 0.7];
>> A = [H.^2 H ones(size(H))];
>> alpha = [10.5; 10.0; 8.5; 7.0; 4.5];
>> c = (A'*A)\(A'*alpha)
```

MATLAB outputs

```
c =
    -114.2857
     107.1429
      14.4714
```

Now we determine the coefficient of determination:

```
>> alphabar = mean(alpha)
alphabar =
    8.1000
>> alphahat = polyval(c,H)
alphahat =
    10.5286
     9.8857
     8.6714
     6.8857
     4.5286
>> SSR = sum((alpha-alphahat).^2)
SSR =
    0.0571
>> R2 = 1 - SSR/sum((alpha-alphabar).^2)
R2 =
    0.9976
```

This value of R^2 tells us that the quadratic fit accounts for 99.76% of the variation in the data. The remaining 0.34% is assumed to be due to errors in obtaining the data, inherent variability in the data, or due to higher-order dependencies. The fitted quadratic curve is plotted in Figure 2.11 along with the data points to inspect the quality of the fit.

Figure 2.11

Transition from gliding to rotating motion. Quadratic curve fitted to data points (H, α_{crit}).

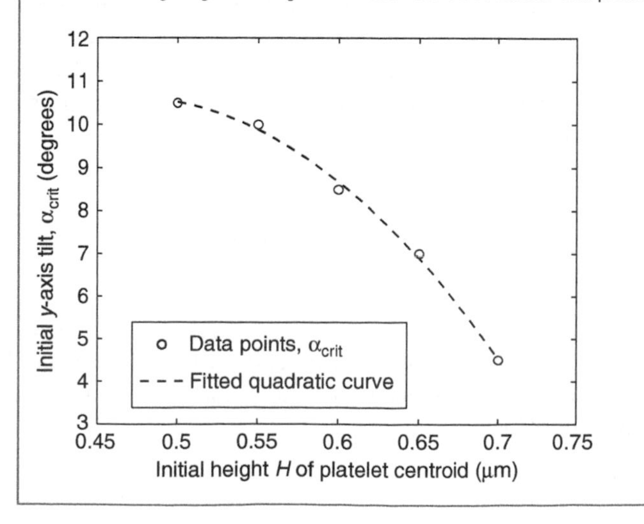

It is a good idea to plot the residuals with respect to the independent variable and visually inspect the residual plot to see if there is any observable trend in the unexplained error. If the distribution of the residuals about zero appears to be entirely random, then a trend in the residual error does not exist and the variations in the data are approximately random (errors or fluctuations are usually assumed to be normally distributed; see Chapter 3). If a distinct trend is noticeable in the residual values with respect to the independent variable, this suggests that the model is incomplete and does not fully account for the observed trend in the data.

Note on the MATLAB backslash (\) operator

The MATLAB backslash operator can solve overdetermined linear equations $\boldsymbol{Ac} = \boldsymbol{y}$ in which the number of equations m is greater than the number of variables n. One simply needs to type

```
c = A\y
```

to obtain the least-squares fit. MATLAB does not employ the normal equations discussed in this chapter, but uses another method called QR factorization with pivoting. QR factorization expresses the coefficient matrix $\boldsymbol{A}$ as a product of two matrices: an orthogonal matrix and an upper triangular matrix. Either method will, under usual circumstances, produce the same result. The methods differ in terms of (i) computational efficiency and (ii) stability in the face of minor fluctuations in data values (conditioning of the system). An analysis of the differences between various methods employed for linear least-squares regression can be found in Demmel (1997).

2.10 Linear least-squares approximation of transformed equations

In the problems described in Boxes 2.2 and 2.3, the dependent variable does not have a linear relationship with the undetermined coefficients. In both cases, the nonlinear equation is amenable to rearrangement such that the new form of the equation is a straight line. These nonlinear equations are said to be *transformed* to a linear equation. Such transformed equations have a linear relationship with respect to their undetermined coefficients, and can therefore be fitted to data sets using the linear least-squares method of regression. For example, many natural phenomena, such as radioactive decay, population growth, chemical reaction rate, and forward/backward chemical rate constants, exhibit an exponential dependence on some factor or environmental condition (temperature, concentrations of chemicals etc.). A nonlinear exponential dependency of m on t such as

$$m = m_0 e^{-kt}$$

is linearized by taking the natural logarithm of both sides:

$$\ln m = \ln m_0 - kt.$$

The dependency of $\ln m$ on t is linear, and a straight line can be used to approximate data that can be described by such a relationship. The linear least-squares line estimates the two constants, m_0 and k, that characterize the process. If we make the following substitutions:

$$y = \ln m, \qquad x = t, \qquad \beta_1 = -k, \qquad \beta_0 = \ln m_0,$$

then the model simplifies to $y = \beta_1 x + \beta_0$. The data points used to obtain a best-fit line that approximates the data are (x_i, y_i) or $(t, \ln m)$. The linear least-squares technique calculates values of β_0 and β_1 that represent the best line fit to data points $(\ln m, t)$ and not (m, t). Therefore, it is important to keep in mind that a best-fit solution for the transformed data may not be as good a fit for the original data. That is why it is important to extract the constants m_0 and k from β_0 and β_1 and determine the errors in the approximation with respect to the original data. Plotting the original data and the fitted curve to check the accuracy of the fit is recommended. Table 2.5 contains several additional examples of commonly encountered forms of nonlinear equations and strategies to linearize them.

Table 2.5. *Examples of nonlinear equations that can be linearized*

Nonlinear equation	Transformed equation of form $y = a_1x + a_0$	Transformed data points	Transformed coefficients
$m = m_0 t^k$ (power-law model)	$\ln m = k \ln t + \ln m_0$	$y = \ln m; x = \ln t$	$\beta_1 = k; \beta_0 = \ln m_0$
$m = m_0 10^{kt}$	$\log m = kt + \log m_0$	$y = \log m; x = t$	$\beta_1 = k; \beta_0 = \log m_0$
$m = m_0 \frac{1}{k+t}$	$\frac{1}{m} = \frac{t}{m_0} + \frac{k}{m_0}$	$y = \frac{1}{m}; x = t$	$\beta_1 = \frac{1}{m_0}; \beta_0 = \frac{k}{m_0}$
$m = m_0 \frac{t}{k+t}$	$\frac{1}{m} = \frac{k}{m_0 t} + \frac{1}{m_0}$	$y = \frac{1}{m}; x = \frac{1}{t}$	$\beta_1 = \frac{k}{m_0}; \beta_0 = \frac{1}{m_0}$

Box 2.2B Using hemoglobin as a blood substitute: hemoglobin–oxygen binding

We seek the best-fit values for the coefficients P_{50} and n using the transformed linear equation

$$\ln \frac{S}{1-S} = n \ln(pO_2) - n \ln P_{50}. \tag{2.28}$$

Let

$$y = \ln \frac{S}{1-S}, \quad x = \ln(pO_2), \quad \beta_1 = n, \text{ and } \beta_0 = -n \ln P_{50}.$$

Equation (2.28) is rewritten as

$$y = \beta_1 x + \beta_0.$$

Table 2.1 provides the data points (pO_2, S), where pO_2 is given in units of mm Hg. At the command line, we create the two variables x and y. Note that in MATLAB the `log` function calculates the natural logarithm of a scalar or the individual terms in a vector.

```
>> pO2 = [10:10:120]';
>> x = log(pO2);
>> S = [0.18; 0.40; 0.65; 0.80; 0.87; 0.92; 0.94; 0.95; 0.95;
        0.96; 0.96; 0.97];
>> y = log(S./(1-S));
```

The matrix $\boldsymbol{A}$ for this problem is

```
>> A = [x ones(size(x))];
```

The normal system is solved as shown below.

```
>> c = (A'*A)\A'*y
c =
    2.0906
    -6.3877
```

Matrix $\boldsymbol{c} = \begin{bmatrix} \beta_1 \\ \beta_0 \end{bmatrix}$ and the model coefficients are extracted from $\boldsymbol{c}$ as follows:

```
>> n = c(1)
n =
  2.0906
>> P50 = exp(-c(2)/n)
P50 =
    21.2297
```

The x, y data are plotted in Figures 2.12 and 2.13 to check the fit. The code is presented below:

```
yhat = A*c;
plot(x,y,'k*',x,yhat,'k-','LineWidth',2)
set(gca,'FontSize',16,'LineWidth',2)
xlabel('log({\itp}O_2)')
ylabel('log({\itS}/(1-{\itS})')
legend('transformed data','fitted line')
S_hat = (pO2.^n)./(P50^n+(pO2.^n));
plot(pO2, S, 'k*',pO2,S_hat,'k-','LineWidth',2)
set(gca,'FontSize',16,'LineWidth',2)
xlabel('{\itp}O_2 in mmHg')
ylabel('{\itS}')
legend('original data','fitted line')
```

Figure 2.12

Transformed oxygen dissociation curve.

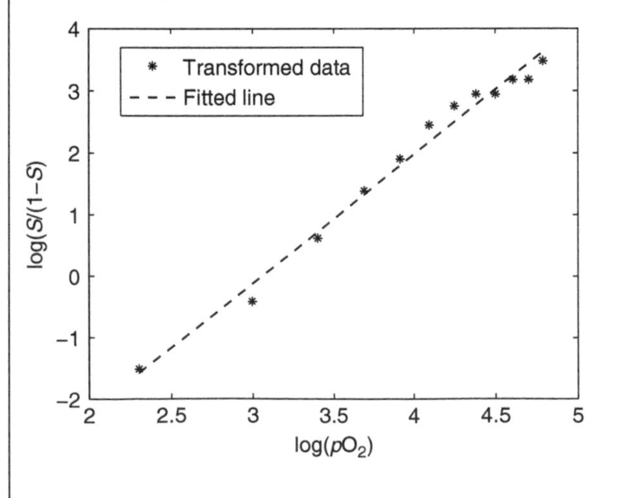

Figure 2.13

Original oxygen dissociation curve with model prediction.

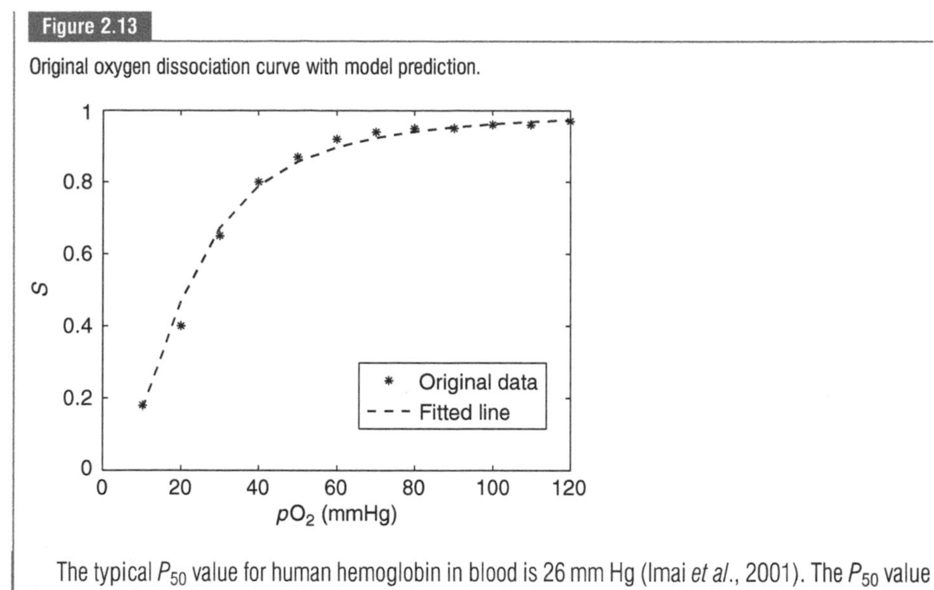

The typical P_{50} value for human hemoglobin in blood is 26 mm Hg (Imai *et al.*, 2001). The P_{50} value for tetrameric bovine hemoglobin is lower than that for the normal case, indicating higher affinity of bovine hemoglobin for oxygen. Such values are typically found in diseased (hemoglobinopathic) conditions and may not be suitable for *in vivo* use as a blood substitute (Imai *et al.*, 2001).

Box 2.3B Time course of drug concentration in the body

The transformed linear equation that we use to determine the "best-fit" values for the coefficients C_0 and k is as follows:

$$\ln C(t) = \ln C_0 - kt. \tag{2.30}$$

Let $y = \ln C$, $x = t$, $\beta_1 = -k$, and $\beta_0 = \ln C_0$. Equation (2.30) is rewritten as

$$y = \beta_1 x + \beta_0.$$

Table 2.2 lists the data points (t, $\ln C_v$) and (t, $\ln C_a$). In a script file, we create three variables x, y_v, and y_a as shown below. Equation (2.30) is not defined for C_v or $C_a = 0$ and therefore we do not include the zero concentration data points.

```
t = [1 2 3 8 16 24 48]';
Cv = [18.912 13.702 7.406 2.351 0.292]';
Ca = [0.240 0.187 0.127]';
x = t; yv = log(Cv); ya = log(Ca);
```

For the linear equation $y = \beta_1 x + \beta_0$, the matrix $\boldsymbol{A}$ is given by

```
A = [x ones(size(x))];
Av = A(1:length(Cv),:);
Aa = A(1:length(Ca),:);
```

The normal system is solved as follows:

```
cv = (Av'*Av)\(Av'*yv);
ca = (Aa'*Aa)\(Aa'*ya);
```

Since $c = \begin{bmatrix} \beta_1 \\ \beta_0 \end{bmatrix}$, the model coefficients are extracted as shown below:

```
kv = -cv(1)
C0v = exp(cv(2))
ka = -ca(1)
C0a=exp(ca(2))
```

The MATLAB output of the script is

```
kv =
    0.2713
C0v =
    21.4204
ka =
    0.3182
C0a =
    0.3376
```

To determine whether the drug clearance rates k are similar or different for the vitreous and aqueous humor, we would need to carry out a hypothesis test. Hypothesis tests are statistical tests that determine with some confidence level (probability) if the apparent differences between two means (or parameters) obtained from two distinct samples are due to chance or whether they actually signify a legitimate difference in the populations from which the samples have been derived (see Chapter 4). The transformed data points and the best-fit models are plotted in Figure 2.14 and the relevant code of the script is presented below.

```
% Plotting least-squares fit for drug elimination from the Vitreous
subplot(1,2,1)
plot(x(1:length(Cv)),yv,'k*',x(1:length(Cv)),Av*cv,'k-',...
  'LineWidth',2)
xlabel('{\it t} in hours')
ylabel('log({\itC_v})')
legend('transformed data','best-fit model')

% Plotting least-squares fit for drug elimination from the Aqueous
subplot(1,2,2)
```

Figure 2.14

Drug elimination profile using transformed coordinates.

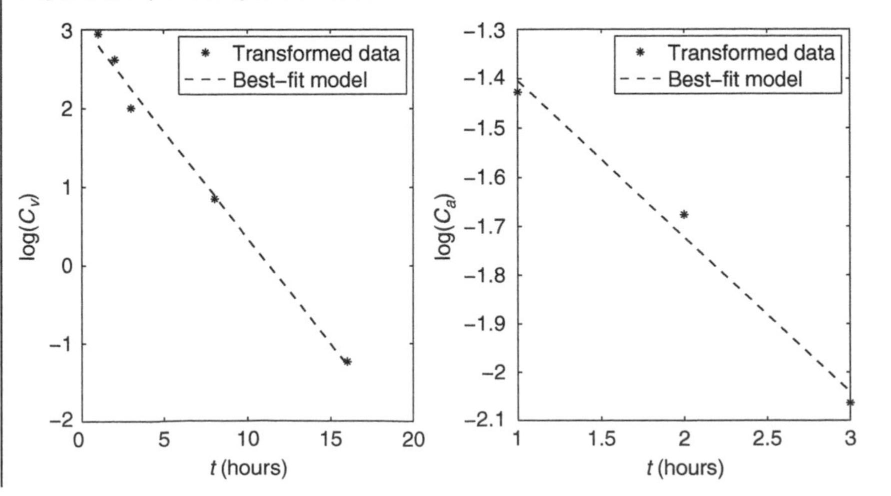

```
plot(x(1:length(Ca)),ya,'k*',x(1:length(Ca)),Aa*ca,'k-',...
  'LineWidth',2)
xlabel('{\it t} in hours')
ylabel('log({\itC_a})')
legend('transformed data', 'best-fit model')
```

Exponential, power-law, logarithmic, and rational functions are all subject to certain specific limitations in the data, such as:

- a reciprocal function in $1/t$ cannot include the point $t = 0$;
- logarithmic functions in x are valid strictly for $x > 0$;
- an exponential function in y does not allow a sign change in y.

Polynomial functions are, on the other hand, more flexible and do not have such limitations. These are important limitations to keep in mind when selecting a fitting function.

2.11 Multivariable linear least-squares regression

At times, we may model the simultaneous dependency of one variable on two or more independent variables. Equation (2.31) is an example of the linear dependency of y on more than one independent variable. Of course, the dependency need not be linear for any physical, chemical, or biological process, and linear least-squares regression can still be applicable in many cases. For example, the biochemical reaction below states that enzyme E must combine with substrates A and B in order to catalyze the irreversible reaction to form product P:

$$E + a\mathrm{A} + b\mathrm{B} \rightarrow \mathrm{E} + \mathrm{P}.$$

An example of an enzyme reaction that acts on two substrates is L-asparagine hydrolysis, which is catalyzed by the enzyme L-asparaginase via the deamination reaction

$$\text{L-asparagine} + \mathrm{H_2} \rightarrow \text{L-aspartate} + \mathrm{NH_3}.$$

Normal cells are capable of synthesizing asparagine, whereas certain leukemia cells are unable to do so, although they require L-asparagine for their growth and survival. L-asparaginase has been found to act as an anti-cancer agent and acts to remove all asparagine externally available to cells. As the internal supply of asparagine becomes depleted, the cancer cells die (Livingston and Krakoff, 1970).

The reaction rate $r = kC_\mathrm{E}{C_\mathrm{A}}^a{C_\mathrm{B}}^b$, where k is the reaction rate constant, C_E is the concentration of enzyme in solution, C_A is the concentration of substrate A in solution, and C_B is the concentration of substrate B in solution. Rate r is a function of three independent concentration variables, and k, a, and b are the coefficients that are calculated from experiments that measure a range of reaction rates as a function of concentration values of each reacting substance. How do we approach this regression problem?

An equation of the form $m = kt_1^p t_2^q t_3^r$ is called a **multivariable power-law equation**. This equation can be linearized by taking the logarithm of both sides to obtain

$$\ln m = \ln k + p \ln t_1 + q \ln t_2 + r \ln t_3. \tag{2.48}$$

If we substitute $y = m; x_1 = \ln t_1; x_2 = \ln t_2; x_3 = \ln t_3$, and $\beta_0 = \ln k; \beta_1 = p$; $\beta_2 = q; \beta_3 = r$, Equation (2.48) becomes

$$y = \beta_0 + \beta_1 x_1 + \beta_2 x_2 + \beta_3 x_3. \tag{2.49}$$

Equation (2.49) has four unknowns: $\beta_0, \beta_1, \beta_2, \beta_3$. The normal equations $\boldsymbol{A}^{\mathrm{T}}\boldsymbol{A}\boldsymbol{c} = \boldsymbol{A}^{\mathrm{T}}\boldsymbol{y}$ apply, where

$$\boldsymbol{A} = \begin{bmatrix} 1 & x_{1,1} & x_{2,1} & x_{3,1} \\ 1 & x_{1,2} & x_{2,2} & x_{3,2} \\ & & \cdot & \\ & & \cdot & \\ & & \cdot & \\ 1 & x_{1,m} & x_{2,m} & x_{3,m} \end{bmatrix}, \quad \boldsymbol{c} = \begin{bmatrix} \beta_0 \\ \beta_1 \\ \beta_2 \\ \beta_3 \end{bmatrix}, \text{ and } \boldsymbol{y} = \begin{bmatrix} y_1 \\ y_2 \\ \cdot \\ \cdot \\ \cdot \\ y_m \end{bmatrix}.$$

The second subscript in the individual terms of $\boldsymbol{A}$ indicates the $1 \leq i \leq m$ data points.

Equation (2.49) presents the simple dependence of y on the independent variables. Remember that, while the model must be linear in the model parameters $\boldsymbol{\beta}$, the relationships between the dependent and independent variables need not be linear. The normal system can be applied for linear regression if the model is, for example, $y = a_0 + a_1 \ln x_1 + a_2 x_1^2 x_2 + a_3 \sin x_3$.

2.12 The MATLAB function `polyfit`

MATLAB provides a built-in function called `polyfit` that uses the linear least-squares method to fit a polynomial in x of degree n to a set of data y. The syntax for this function is

```
p = polyfit(x, y, n)
```

where x is the independent variable vector, y is the dependent variable vector, n is the degree of the polynomial, and p is a vector containing the $n+1$ coefficients of the polynomial in x. The coefficients of the polynomial in p are arranged in descending order of the powers of x. The MATLAB `polyfit` function uses the QR factorization method to solve overdetermined systems using the least-squares method.

Box 2.4C Platelet Flow Rheology

We wish to fit the function $\alpha_{\mathrm{crit}} = \beta_2 H^2 + \beta_1 H + \beta_0$, where α_{crit} is the critical angle of tilt of the platelet at which the transition in flow behavior occurs.

The vectors $\boldsymbol{H}$ and $\boldsymbol{\alpha}$ contain the data to be fitted by a second-order polynomial:

```
>> H = [0.5 0.55 0.6 0.65 0.7];
>> alpha = [10.5 10 8.5 7.0 4.5];
>> c = polyfit(H, alpha, 2)
c =
-114.2857  107.1429  -14.4714
```

The least-squares fitted polynomial is

$$\alpha_{\mathrm{crit}} = -114.2857H^2 + 107.1429H - 14.4714.$$

2.13 End of Chapter 2: key points to consider

Concepts from linear algebra and MATLAB programming

(1) A one-dimensional array is called a **vector** and a two-dimensional array is called a **matrix**. A 1×1 array is a **scalar**.

(2) The **dot product** of two vectors is defined as $\boldsymbol{u} \cdot \boldsymbol{v} = \sum_{i=1}^{i=n} u_i v_i$, where both vectors have the same number of elements n. To obtain the dot product of two vectors, a column vector on the right must be pre-multiplied by a row vector on the left. The dot product is an example of matrix–matrix multiplication in which a $1 \times n$ matrix is multiplied with an $n \times 1$ matrix.

(3) **Matrix–matrix multiplication** of $\boldsymbol{A}$ with $\boldsymbol{B}$ is defined if $\boldsymbol{A}$ is an $m \times n$ matrix and $\boldsymbol{B}$ is an $n \times p$ matrix, i.e. their inner dimensions must agree. The resulting matrix $\boldsymbol{C} = \boldsymbol{AB}$ has the dimensions $m \times p$. Every element c_{ij} of $\boldsymbol{C}$ is equal to the dot product of the ith row of $\boldsymbol{A}$ and the jth column of $\boldsymbol{B}$.

(4) The **norm of a vector** is a real number that specifies the length of the vector. A p-norm is a general method used to calculate the vector length and is defined, for vector $\boldsymbol{v}$, as $\|\boldsymbol{v}\|_p = \{\sum_{i=1}^{n} |v_i|^p\}^{1/p}$. The $p = 2$ norm (Euclidean norm) is defined as $\|\boldsymbol{v}\| = \sqrt{\boldsymbol{v} \cdot \boldsymbol{v}} = \left(\sum_{i=1}^{i=n} v_i^2\right)^{1/2}$.

(5) The **norm of a matrix** $\boldsymbol{A}$ is a measure of the maximum degree to which $\boldsymbol{A}$ can stretch a unit vector $\boldsymbol{x}$. Mathematically, we write $\|\boldsymbol{A}\| = \max_{\|\boldsymbol{x}\|=1} \|\boldsymbol{Ax}\|$.

(6) **A linear combination of vectors** is the sum of one or more vectors of equal dimension, each multiplied by a scalar, and results in another vector of the same dimension. If $V = (\boldsymbol{v}_1, \boldsymbol{v}_2, \ldots, \boldsymbol{v}_n)$ is a set of vectors, and if $\boldsymbol{u} = k_1\boldsymbol{v}_1 + k_2\boldsymbol{v}_2 + \cdots + k_n\boldsymbol{v}_n$, where k_i for $1 \leq i \leq n$ are scalars, then $\boldsymbol{u}$ is said to be a linear combination of the given set of vectors V.

(7) If $V = (\boldsymbol{v}_1, \boldsymbol{v}_2, \ldots, \boldsymbol{v}_n)$ is a set of vectors, and the vector equation $k_1\boldsymbol{v}_1 + k_2\boldsymbol{v}_2 + \cdots + k_n\boldsymbol{v}_n = 0$ holds true only when all $k_i = 0$ for $1 \leq i \leq n$, then the vectors are said to be **linearly independent**. If it holds true for non-zero values of the scalars, i.e. there exists multiple solutions to the vector equation, then the vectors in set V are **linearly dependent**.

(8) A **vector space** S that contains a set of vectors V must also contain all vectors that result from vector addition and scalar multiplication operations on the set V, i.e. all possible linear combinations of the vectors in V. The physical three-dimensional space that we live in is a vector space. If the vectors in the set V are linearly independent, and all possible linear combinations of these vectors fully describe the vector space S, then these vectors are called **basis vectors** for the space S. The number of basis vectors that are required to describe the vector space S fully is called the **dimension** of the vector space.

(9) The vector space spanned by the columns of a matrix is called the **column space**.

(10) The **determinant** of a matrix is defined only for square matrices and is a scalar quantity. The determinant of an $n \times n$ matrix is the sum of all possible products of n elements of matrix $\boldsymbol{A}$, such that no two elements of the same row or column are multiplied together. A matrix that has a non-zero determinant is called **non-singular**, whereas a matrix whose determinant is equal to zero is **singular**.

(11) The **rank** of a matrix $\boldsymbol{A}$ is the dimension of the largest submatrix within $\boldsymbol{A}$ that has a non-zero determinant. Equivalently, the rank of a matrix is equal to the dimension of the column space of the matrix $\boldsymbol{A}$, i.e. the number of linearly independent columns in $\boldsymbol{A}$.

(12) The **inverse** of a matrix $\boldsymbol{A}$ is denoted as $\boldsymbol{A}^{-1}$ and is only defined for non-singular square matrices. When a matrix is multiplied by its inverse, the identity matrix $\boldsymbol{I}$ is obtained: $\boldsymbol{A}\boldsymbol{A}^{-1} = \boldsymbol{A}^{-1}\boldsymbol{A} = \boldsymbol{I}$.

Solution techniques for systems of linear equations

(1) A system of linear equations can be compactly represented by the matrix equation $\boldsymbol{A}\boldsymbol{x} = \boldsymbol{b}$, where $\boldsymbol{A}$ is the coefficient matrix, $\boldsymbol{b}$ is the vector of constants, and $\boldsymbol{x}$ is the vector containing the unknowns.

(2) The rank of the coefficient matrix $\boldsymbol{A}$ and the augmented matrix $[\boldsymbol{A}\ \boldsymbol{b}]$ tells us whether
(a) a unique solution exists,
(b) infinitely many solutions exist, or
(c) no solution exists
for a system of linear equations $\boldsymbol{A}\boldsymbol{x} = \boldsymbol{b}$.

(3) **Gaussian elimination** is a direct method that is used to solve consistent systems of linear equations $\boldsymbol{A}\boldsymbol{x} = \boldsymbol{b}$ for which the determinant of the coefficient matrix $|\boldsymbol{A}| \neq 0$.
(a) Gaussian elimination reduces the augmented matrix $[\boldsymbol{A}\ \boldsymbol{b}]$ to **upper triangular form**.
(b) The **pivot row** is used to zero the elements in rows below it. The column in which the elements are being zeroed is called the **pivot column**. The **pivot element** is located in the pivot row and pivot column and is used to zero the elements beneath it.

(4) **Backward substitution** works backwards from the last equation of the linear system, specified by the upper triangular form of the augmented matrix $[\boldsymbol{A}\ \boldsymbol{b}]$ that results from Gaussian elimination, to the first, to obtain the solution $\boldsymbol{x}$. For a system of n equations in n unknowns, x_n is obtained first using the last (bottom-most) equation specified by $[\boldsymbol{A}\ \boldsymbol{b}]$. The knowledge of x_n is used to obtain x_{n-1} using the second-last equation from $[\boldsymbol{A}\ \boldsymbol{b}]$. This procedure continues until x_1 is obtained.

(5) Gaussian elimination fails if any of the elements along the main diagonal of the augmented matrix is either exactly zero or close to zero. **Partial pivoting** is used to remove zeros or small numbers from the diagonal. Before any zeroing of elements is performed, the pivot element in the pivot row is compared with all other elements in the pivot column. Row swapping is carried out to place at the pivot row position, the row containing the element with the largest absolute magnitude.

(6) **LU factorization** is a direct method to solve systems of linear equations and is most useful when the same coefficient matrix $\boldsymbol{A}$ is shared by several systems of equations. In this method, $\boldsymbol{A}$ is factorized into two matrices: a lower triangular matrix $\boldsymbol{L}$ and an upper triangular matrix $\boldsymbol{U}$, i.e. $\boldsymbol{A} = \boldsymbol{L}\boldsymbol{U}$. Note that $\boldsymbol{U}$ is the same as the upper triangular matrix that results from Gaussian elimination, and $\boldsymbol{L}$ contains the multipliers that are used to zero the below-diagonal entries in the augmented matrix. In this process, we decouple the transformations carried out on $\boldsymbol{A}$ from that on $\boldsymbol{b}$. Two sequential steps are required to extract the solution:
(a) **forward substitution** to solve $\boldsymbol{L}\boldsymbol{y} = \boldsymbol{b}$;
(b) **backward substitution** to solve $\boldsymbol{U}\boldsymbol{x} = \boldsymbol{y}$.

(7) Pivoting or exchanging of rows is recommended when performing LU factorization in order to limit round-off errors generated during elimination steps. A **permutation matrix $\boldsymbol{P}$** must be used to track row swaps so that $\boldsymbol{b}$ can be transformed accordingly prior to forward substitution. Mathematically, $\boldsymbol{L}\boldsymbol{U} = \boldsymbol{P}\boldsymbol{A}$ and $\boldsymbol{L}\boldsymbol{U}\boldsymbol{x} = \boldsymbol{P}\boldsymbol{b}$.

(8) A problem is said to be **ill-conditioned** if the solution is hypersensitive to small changes in the coefficients and constants of the linear equations. On the other

hand, equations that exhibit stability (change in solution is on the same order of magnitude as fluctuations in the equation parameters) constitute a **well-conditioned** problem. The conditioning of a system of equations depends on the determinant of the coefficient matrix.

(9) The **condition number** is a measure of the conditioning of a system of linear equations and is defined as $\text{cond}(\boldsymbol{A}) = \|\boldsymbol{A}\|\|\boldsymbol{A}^{-1}\|$, where $\boldsymbol{A}$ is the coefficient matrix. Note that $\text{cond}(\boldsymbol{A})$ can take on values from one to infinity. Low condition numbers close to one indicate a well-conditioned coefficient matrix. An ill-conditioned problem has a large condition number.

Linear regression

(1) Experimental studies yield quantitative values that describe a relationship between a dependent variable y and one or more independent variables x. It is useful to obtain a mathematical model that relates the variables in the data set. The technique of determining the model parameters that best describe the relationship between the variables is called **regression.**

(2) The "best-fit" curve obtained on regressing data to the model approximates the data to within some degree but does not predict the original data points exactly. The **residual** is defined as the difference between the experimentally obtained value of y and that predicted by the model. The **residual** (error) minimization criterion used to obtain the "best-fit" curve is called the **objective function.**

(3) In **linear regression**, the functional dependence of the dependent variable y on the model parameters is strictly linear. Since the number of data points is greater than the number of unknowns, the system of linear equations is **overdetermined** and an exact solution does not exist.

(4) The **linear least-squares criterion** minimizes the sum of the squared residuals for an overdetermined system. This criterion produces the normal equations $\boldsymbol{A}^{\mathrm{T}}\boldsymbol{A}\boldsymbol{c} = \boldsymbol{A}^{\mathrm{T}}\boldsymbol{y}$, which on solving yields $\boldsymbol{c}$, the vector containing the model parameters. Note that $\boldsymbol{A}$ is the coefficient matrix, and its size and structure depends on the model and the number of data points in the data set.

(5) The **coefficient of determination,** R^2, measures the quality of the fit between the model and the data set; R^2 compares the sum of the squared residuals from the least-squares fit, $\sum(y - \hat{y})^2$, to the sum of the squared deviations, $\sum(y - \bar{y})^2$.

2.14 Problems

Solving systems of linear equations

2.1. Types of solutions for systems of linear equations.

(a) Each equation below represents a straight line. If two lines intersect at one point, then the system of two lines is said to possess a "unique solution." What type of solution do the following systems of equations (lines) have? Determine by solving the equations by hand. Also plot the equations (lines) using MATLAB (`plot` function) to understand the physical significance. In two-dimensional space, how are lines oriented with respect to each other when no solution exists, i.e. the system is inconsistent? What about when there exist infinitely many solutions?

(i) $2x + 3y = 3; x + 2y = 1;$
(ii) $x + 2y = 5; 3x + 6y = 15;$
(iii) $2x + 3y = 5; 4x + 6y = 11.$

(b) What value(s) should p take for the system of the following two linear equations to be consistent? (Hint: Determine the rank of the coefficient matrix and the augmented matrix. For what value of p is the rank of the augmented matrix equal to that of the coefficient matrix? You may find it useful to use elementary row transformations to simplify the augmented matrix.)

$$2x - 3y = 4,$$
$$4x - 6y = p.$$

What value(s) should p take for the system of two linear equations given below to be inconsistent (no solution)? Does a unique solution exist for any value of p? Why or why not? Explain the situation graphically by choosing three different numbers for p.

2.2. Linear independence Show that the vectors

$$\boldsymbol{v}_1 = \begin{bmatrix} 3 \\ 4 \\ 5 \end{bmatrix}, \quad \boldsymbol{v}_2 = \begin{bmatrix} 1 \\ 6 \\ 7 \end{bmatrix}, \quad \boldsymbol{v}_3 = \begin{bmatrix} -2 \\ 1 \\ 3 \end{bmatrix}$$

are linearly independent. To do this, rewrite the vector equation

$$k_1\boldsymbol{v}_1 + k_2\boldsymbol{v}_2 + \cdots + k_n\boldsymbol{v}_n = 0$$

as $\boldsymbol{Ak} = 0$, where $\boldsymbol{A}$ consists of $\boldsymbol{v}_1$, $\boldsymbol{v}_2$, and $\boldsymbol{v}_3$ as column vectors and k is the coefficients vector, and solve for k.

2.3. Use of finite-precision in solving systems of equations You are given the following system of equations:

$$10^{-5}x_1 + 10^{-5}x_2 + x_3 = 0.99998,$$
$$-10^{-5}x_1 + 10^{-5}x_2 - 4x_3 = 6 \times 10^{-5},$$
$$x_1 + x_2 + 3x_3 = 1.$$

(a) Does this set of equations have a solution? (Hint: Check to see if rank($\boldsymbol{A}$) = rank ($\boldsymbol{aug\ A}$).)
(b) Solve the above system of equations by hand using four-digit arithmetic and Gaussian elimination without pivoting. Substitute the solution back into the equation. Does the solution satisfy the equations? Re-solve by including partial pivoting. Are the solutions different?
(c) Repeat (b) but perform six-digit arithmetic. What is the solution now? Does it exactly satisfy the equations?
(d) What is the condition number of this system? Change the right-hand side constant of the first equation, 0.9998, to 1. Do you get an exact solution using only six-digit arithmetic? What is the relative change in solution compared to the relative change made in the right-hand-side constant? Note the importance of using greater precision in numerical computations. Often by using greater precision, problems having large condition numbers can remain tractable.

2.4. Pharmacokinetic modeling for "animal-on-a-chip" – 1 A material balance is performed for naphthalene epoxide (NO) generation, consumption, and transport in the μCCA device described in Figure 2.1; NO is an intermediate formed during the metabolism of naphthalene.

Routes of generation of naphthalene epoxide:

(1) conversion of naphthalene into its epoxide.

Figure P2.1

Material balance diagram for naphthalene epoxide.

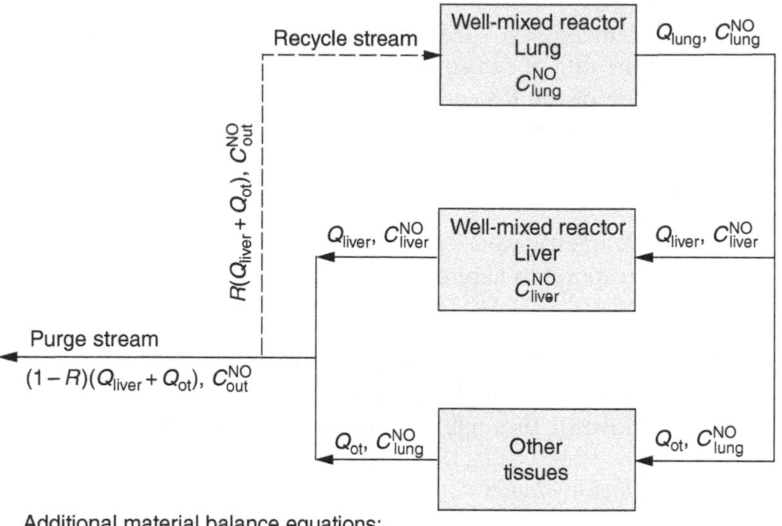

Routes of consumption of naphthalene epoxide:

(1) conversion of epoxide to naphthalene dihydrodiol;
(2) binding to GSH to form epoxide–GSH conjugates;
(3) rearrangement to naphthol.

The material balance diagram for naphthalene epoxide (NO) is shown in Figure P2.1.

Since we are dealing with a multicomponent system, we use superscripts N, NO, and NOH for naphthalene, naphthalene epoxide, and naphthol, respectively, to differentiate between the concentration terms in various compartments.

A mass balance of NO is performed over the two chambers – lung and liver. This yields two linear equations in the unknowns C_{lung}^{NO} and C_{liver}^{NO}. Note that simplifications have been made to the original equations (Quick and Shuler, 1999) by assuming that C_{lung}^{NO} and C_{liver}^{NO} are small in comparison to relevant constants present in the equations.

Lung compartment

$$R\left(Q_{liver}C_{liver}^{NO} + Q_{ot}C_{lung}^{NO}\right) + v_{max,P450\text{-}lung}V_{lung} - \frac{v_{max,EH\text{-}lung}C_{lung}^{NO}}{K_{m,EH\text{-}lung}}V_{lung}$$
$$- V_{lung}\frac{v_{max,GST}C_{lung}^{NO}C_{lung}^{GSH}}{K1_{lung} + K2_{lung}C_{lung}^{GSH}} - k_{NOH}\exp(l_{NOH}TP_{lung})C_{lung}^{NO}V_{lung} - Q_{lung}C_{lung}^{NO} = 0 \tag{P2.1}$$

Liver compartment

$$Q_{liver}C_{lung}^{NO} + v_{max,P450\text{-}liver}V_{liver} - \frac{v_{max,EH\text{-}liver}C_{liver}^{NO}}{K_{m,EH\text{-}liver}}V_{liver}$$
$$- V_{liver}\frac{v_{max,GST}C_{liver}^{NO}C_{liver}^{GSH}}{K1_{liver} + K2_{liver}C_{liver}^{GSH}} - k_{NOH}\exp(l_{NOH}TP_{liver})C_{liver}^{NO}V_{liver} - Q_{liver}C_{liver}^{NO} = 0. \tag{P2.2}$$

NO balance assumptions

(1) Binding of naphthalene epoxide to proteins is comparatively less important and can be neglected.
(2) The concentration of GSH in cells is constant. It is assumed that GSH is resynthesized at the rate of consumption.
(3) Production of the RS enantiomer of the epoxide (compared to SR oxide) is dominant, and hence reaction parameters pertaining to RS production only are used.
(4) The total protein content in the cells to which the metabolites bind remains fairly constant.

The parametric values and definitions are provided below. The modeling parameters correspond to naphthalene processing in mice.

Flowrates

Q_{lung}: flowrate through lung compartment = 2 μl/min;
Q_{liver}: flowrate through liver compartment = 0.5 μl/min;
Q_{ot}: flowrate through other tissues compartment = 1.5 μl/min.

Compartment volumes

V_{lung}: volume of lung compartment = $2\,\text{mm} \times 2\,\text{mm} \times 20\,\mu\text{m} = 8 \times 10^{-8}\,\text{l} = 0.08\,\mu\text{l}$;
V_{liver}: volume of liver compartment = $3.5\,\text{mm} \times 4.6\,\text{mm} \times 20\,\mu\text{m} = 3.22 \times 10^{-7}\,\text{l} = 0.322\,\mu\text{l}$.

Reaction constants

(1) **Naphthalene → naphthalene epoxide**

$v_{\text{max,P450-lung}}$: maximum reaction velocity for conversion of naphthalene into napthalene epoxide by cytochrome P450 monooxygenases in lung cells = 8.75 μM/min;
$v_{\text{max,P450-liver}}$: maximum reaction velocity for conversion of naphthalene into napthalene epoxide by cytochrome P450 monooxygenases in liver cells = 118 μM/min.

(2) **Naphthalene epoxide → naphthalene dihydrodiol**

$v_{\text{max,EH-lung}}$: maximum reaction velocity for conversion of naphthalene epoxide to dihydrodiol by epoxide hydrolase in the lung = 26.5 μM/min;
$K_{\text{m,EH-lung}}$: Michaelis constant = 4.0 μM;
$v_{\text{max,EH-liver}}$: maximum reaction velocity for conversion of naphthalene epoxide to dihydrodiol by epoxide hydrolase in the liver = 336 μM/min;
$K_{\text{m,EH-lung}}$: Michaelis constant = 21 μM

(3) **Naphthalene epoxide → naphthol**

k_{NOH}: rate constant for rearrangement of epoxide to naphthol = 0.173 μM/μM of NO/min;
l_{NOH}: constant that relates naphthol formation rate to total protein content = −20.2 ml/g protein

(4) **Naphthalene epoxide → epoxide–GSH conjugates**

$v_{\text{max,GST}}$: maximum reaction velocity for binding of naphthalene epoxide to GSH catalyzed by GST (glutathione S-transferase) = 2750 μM/min;
$K1_{\text{lung}}$: constant in epoxide–GSH binding rate = 310 000 μM^2;
$K2_{\text{lung}}$: constant in epoxide–GSH binding rate = 35 μM;
$K1_{\text{liver}}$: constant in epoxide–GSH binding rate = 150 000 μM^2;
$K2_{\text{liver}}$: constant in epoxide–GSH binding rate = 35 μM.

Protein concentrations

TP_{lung}: total protein content in lung compartment = 92 mg/ml;
TP_{liver}: total protein content in liver compartment = 192 mg/ml;
C_{lung}^{GSH}: GSH concentration in lung compartment = 1800 μM;
C_{liver}^{GSH}: GSH concentration in liver compartment = 7500 μM;
R: fraction of the exiting stream that reenters the microcircuit.

Your goal is to vary the recycle fraction from 0.6 to 0.95 in increasing increments of 0.05 in order to study the effect of reduced excretion of toxicant on circulating concentration values of naphthalene and its primary metabolite naphthalene epoxide. You will need to write a MATLAB program to perform the following.

(a) Use the Gaussian elimination method to determine the napthalene epoxide concentrations at the outlet of the lung and liver compartments of the animal-on-a-chip for the range of R specified.
(b) Plot the concentration values of epoxide in the liver and lung chambers as a function of R.
(c) Verify your preceding answers by repeating your calculations using the LU decomposition method. (You can use the MATLAB backslash operator to perform the forward and backward substitutions.)

2.5. Pharmacokinetic modeling for "animal-on-a-chip" – 2 Repeat problem 2.4 for naphthalene concentrations in lung and liver for recycle ratios 0.6 to 0.9. The constants and relevant equations are given in Box 2.1.What happens when we increase R to 0.95? Why do you think this occurs? (Hint: Look at the assumptions made in deriving the Michaelis–Menten equation.) Chapter 5 discusses solution techniques for nonlinear equation sets. See Problem 5.10.

2.6. Pharmacokinetic modeling for "animal-on-a-chip" – 3

(a) Are the sets of equations dealt with in Problems 2.4 and 2.5 always well-conditioned? What happens to the conditioning of the system of Equations (2.3) and (2.4) as $R \to 1$? Calculate the condition numbers for the range of R specified.
(b) What happens to the nature of the solution of the two simultaneous linear Equations (2.3) and (2.4) when R is exactly equal to one?

Figure P2.2

Banded matrix.

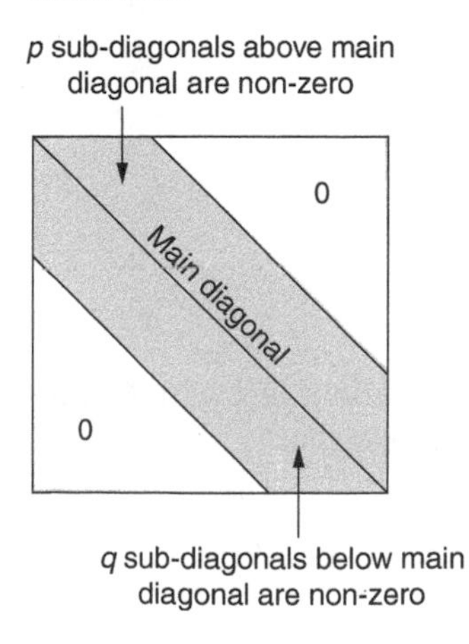

(c) Do you see a similar situation for Equations (P2.1) and (P2.2) as $R \rightarrow 1$? Why or why not? If Equations (P2.1) and (P2.2) produce a solution even when $R = 1$, explain why the steady state assumption is still not valid.

2.7. LU decomposition of tridiagonal matrices In banded matrices, the non-zero elements are located in a specific part of the matrix as shown in Figure P2.2. The non-zero part of the matrix consists of a set of diagonals and includes the main diagonal. Everywhere else, the matrix elements are all zero.

The **bandwidth** of a banded matrix is equal to $p + q + 1$, where

p is the number of non-zero sub-diagonals located above the main diagonal,
q is the number of non-zero sub-diagonals located below the main diagonal, and
1 is added to include the main diagonal.

Large banded matrices are **sparse matrices**. The minimum number of operations necessary to obtain a solution to a system of linear equations whose coefficient matrix is a large banded matrix is much less than the $O(n^3)$ number of operations required by Gaussian elimination or LU decomposition.

Consider the tridiagonal matrix $\boldsymbol{A}$:

$$\boldsymbol{A} = \begin{bmatrix} a_1 & c_1 & 0 & 0 & 0 & & \ldots & 0 \\ b_2 & a_2 & c_2 & 0 & 0 & & \ldots & 0 \\ 0 & b_3 & a_3 & c_3 & 0 & & \ldots & 0 \\ & & \cdot & \cdot & \cdot & & & \\ & & & \cdot & \cdot & \cdot & & \\ & & & & \cdot & \cdot & \cdot & \\ 0 & \ldots & & & 0 & b_{n-1} & a_{n-1} & c_{n-1} \\ 0 & \ldots & & & 0 & 0 & b_n & a_n \end{bmatrix}.$$

$\boldsymbol{A}$ is an example of a banded matrix. We seek an efficient elimination method to resolve a banded matrix to a triangular matrix.

A banded matrix has a convenient property. If LU decomposition is performed on a banded matrix via the Doolittle method without pivoting, the resulting $\boldsymbol{U}$ and $\boldsymbol{L}$ matrices have narrower bandwidths than the original matrix. $\boldsymbol{A}$ can be broken down into the product of two triangular banded matrices as shown below:

$\boldsymbol{A} = \boldsymbol{LU}$

$$= \begin{bmatrix} 1 & 0 & 0 & & 0 & \ldots & 0 \\ \beta_2 & 1 & 0 & & 0 & \ldots & 0 \\ 0 & \beta_3 & 1 & & 0 & \ldots & 0 \\ & & & \cdot\ \cdot & & & \\ & & & & \cdot\ \cdot & & \\ & & & & \cdot\ \cdot & & \\ \ldots & & 0 & & \beta_n & 1 & 0 \\ 0 & \ldots & & & 0 & \beta_n & 1 \end{bmatrix} \begin{bmatrix} \alpha_1 & \delta_1 & 0 & 0 & & \ldots & 0 \\ 0 & \alpha_2 & \delta_2 & 0 & & \ldots & 0 \\ 0 & 0 & \alpha_3 & \delta_3 & 0 & \ldots & 0 \\ & & & \cdot\ \cdot & & & \\ & & & \cdot\ \cdot & & & \\ & & & \cdot\ \cdot & & & \\ 0 & \ldots & & 0 & \alpha_{n-1} & & \delta_{n-1} \\ 0 & \ldots & & 0 & 0 & & \alpha_n \end{bmatrix}.$$

You are asked to perform matrix multiplications of $\boldsymbol{L}$ with $\boldsymbol{U}$, and compare the result with the corresponding elements of $\boldsymbol{A}$. Show that

$$\alpha_1 = a_1; \quad \alpha_k = a_k - \beta_k c_{k-1}; \qquad \beta_k = \frac{b_k}{\alpha_{k-1}}, \qquad \text{for } 2 \leq k \leq n$$

$$\delta_k = c_k, \qquad \text{for } 1 \leq k \leq n.$$

Note that the number of addition/subtraction, multiplication, and division operations involved in obtaining the elements of the $\boldsymbol{L}$ and $\boldsymbol{U}$ matrices is $3(n-1)$. This method is much more efficient than the traditional $O(n^3)$ operations required by standard Gaussian elimination for full matrices.

2.8. **Thomas method for tridiagonal matrices** Tridiagonal matrices are commonly encountered in many engineering and science applications. Large tridiagonal matrices have many zero elements. Standard methods of solution for linear systems are not well-suited for handling tridiagonal matrices since use of these methods entails many wasteful operations such as addition and multiplication with zeros. Special algorithms have been formulated to handle tridiagonal matrices efficiently. The Thomas method is a Gaussian elimination procedure tailored specifically for tridiagonal systems of equations as shown below:

$$\boldsymbol{Ax} = \begin{bmatrix} d_1 & u_1 & 0 & 0 & 0 \\ l_2 & d_2 & u_2 & 0 & 0 \\ 0 & l_3 & d_3 & u_3 & 0 \\ 0 & 0 & l_4 & d_4 & u_4 \\ & & \cdot & & \\ & & \cdot & & \\ 0 & 0 & 0 & l_n & d_n \end{bmatrix} \begin{bmatrix} x_1 \\ x_2 \\ x_3 \\ x_4 \\ \cdot \\ \cdot \\ x_n \end{bmatrix} = \begin{bmatrix} b_1 \\ b_2 \\ b_3 \\ b_4 \\ \cdot \\ \cdot \\ b_n \end{bmatrix} = \boldsymbol{b}$$

The Thomas algorithm converts the tridiagonal matrix into a upper triangular matrix with ones along the main diagonal. In this method, the three diagonals containing non-zero elements are stored as three vectors, $\boldsymbol{d}$, $\boldsymbol{l}$, and $\boldsymbol{u}$, where

$\boldsymbol{d}$ is the main diagonal;
$\boldsymbol{l}$ is the sub-diagonal immediately below the main diagonal; and
$\boldsymbol{u}$ is the sub-diagonal immediately above the main diagonal.

The Thomas method is illustrated below by performing matrix manipulations on a full matrix. Keep in mind that the operations are actually carried out on a set of vectors.

Step 1 The top row of the augmented matrix is divided by d_1 to convert the first element in this row to 1. Accordingly,

$$u'_1 = \frac{u_1}{d_1}, \quad b'_1 = \frac{b_1}{d_1}.$$

Step 2 The top row is multiplied by l_2 and subtracted from the second row to eliminate l_2. The other non-zero elements in the second row are modified accordingly as

$$d'_2 = d_2 - l_2 u'_1, \qquad b'_2 = b_2 - l_2 b'_1.$$

The augmented matrix becomes

$$\boldsymbol{augA} = \begin{bmatrix} 1 & u'_1 & 0 & 0 & 0 & : & b'_1 \\ 0 & d'_2 & u_2 & 0 & 0 & : & b'_2 \\ 0 & l_3 & d_3 & u_3 & 0 & : & b_3 \\ 0 & 0 & l_4 & d_4 & u_4 & : & b_4 \\ & & & \cdot & & & \\ & & & \cdot & & & \\ & & & \cdot & & & \\ 0 & 0 & 0 & l_n & d_n & : & b_n \end{bmatrix}.$$

Step 3 The second row is divided by d_2' to make the new pivot element 1. Thus, the non-zero elements in the second row are again modified as follows:

$$u_2' = \frac{u_2}{d_2'} = \frac{u_2}{d_2 - l_2 u_1'}, \qquad b_2'' = \frac{b_2'}{d_2'} = \frac{b_2 - l_2 b_1'}{d_2 - l_2 u_1'}.$$

The augmented matrix is now

$$\textbf{aug}\boldsymbol{A} = \begin{bmatrix} 1 & u_1' & 0 & 0 & 0 & : & b_1' \\ 0 & 1 & u_2' & 0 & 0 & : & b_2'' \\ 0 & l_3 & d_3 & u_3 & 0 & : & b_3 \\ 0 & 0 & l_4 & d_4 & u_4 & : & b_4 \\ & & & \cdot & & & \\ & & & \cdot & & & \\ & & & \cdot & & & \\ 0 & 0 & 0 & l_n & d_n & : & b_n \end{bmatrix}.$$

Steps 2 and 3 are repeated for all subsequent rows $i = 3, \ldots, n$. The only elements that need to be calculated at each step are u_i and b_i. Therefore, for all remaining rows,

$$u_i' = \frac{u_i}{d_i - l_i u_{i-1}'}, \quad b_i'' = \frac{b_i - l_i b_{i-1}''}{d_i - l_i u_{i-1}'}.$$

For the nth row, only b_n needs to be calculated, since u_n does not exist. The solution is recovered using back substitution.

Write a MATLAB code to implement the Thomas method of solution for tri-diagonal systems. Use the `diag` function to extract the matrix diagonals. Remember that the vector l must have length n in order to use the mathematical relationships presented here.

2.9. Countercurrent liquid extraction of an antibiotic Product recovery operations downstream of a fermentation or enzymatic process are necessary to concentrate and purify the product. The fermentation broth leaving the bioreactor is first filtered to remove suspended particles. The aqueous filtrate containing the product undergoes a number of separation operations such as solvent extraction, adsorption, chromatography, ultrafiltration, centrifugation, and crystallization (Bailey and Ollis, 1986). In the liquid extraction process, the feed stream containing the solute of interest is contacted with an immiscible solvent. The solvent has some affinity for the solute, and, during contacting between the two liquid streams, the solute distributes between the two phases. The exiting solvent stream enriched in solute is called the **extract**. The residual liquid from which the desired component has been extracted is called the **raffinate**. A single liquid–liquid contact operation is called a **stage**. Ideally, during a stagewise contact the distribution of the solute between the two liquids reaches equilibrium. A stage that achieves maximum separation as dictated by equilibrium conditions is called a **theoretical stage**. A scheme of repeated contacting of two immiscible liquid streams to improve separation is called a multistage process. An example of this is multistage countercurrent extraction. Figure P2.3 shows a schematic of a countercurrent extractor that contains n stages. The concentration of solute in the entering feed stream is denoted by x_F or x_{n+1}, and that in the entering solvent stream is denoted by y_S or y_0. We simplify the analysis by assuming that the mass flowrates of the liquid streams change very little at each stage. If the raffinate stream dissolves significant quantities of solvent, or vice versa, then this assumption is invalid. A mass balance performed on any stage $i = 2, \ldots, n-1$ under steady state operation yields

Figure P2.3

Countercurrent liquid extraction.

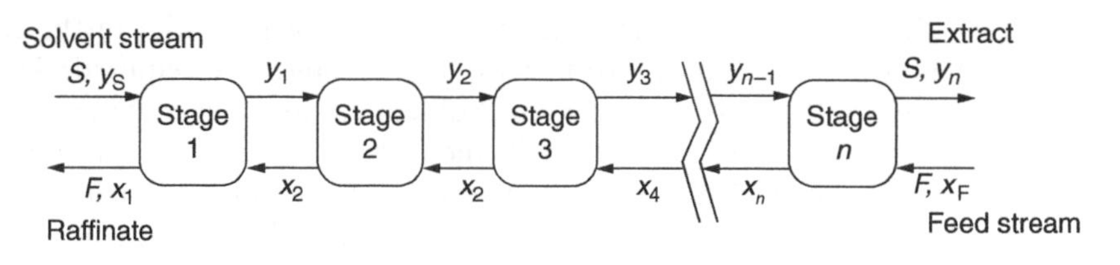

$$Sy_i + Fx_i = Sy_{i-1} + Fx_{i+1}. \tag{P2.3}$$

The equilibrium distribution between the two phases is specified by the distribution coefficient,

$$K = y/x. \tag{P2.4}$$

Combining Equations (P2.3) and (P2.4), we get

$$SKx_i + Fx_i = SKx_{i-1} + Fx_{i+1}.$$

Rearranging, the above expression becomes

$$SKx_{i-1} - (SK + F)x_i + Fx_{i+1} = 0.$$

We are interested in determining the concentration of the solute (product) in the final extract and in the exiting raffinate. For this purpose a mass balance is performed on each of the n stages. The set of simultaneous linear equations can be expressed in matrix form $\boldsymbol{Ax} = \boldsymbol{b}$, where

$$\boldsymbol{A} = \begin{bmatrix} -(SK+F) & F & 0 & 0 & \dots & 0 \\ SK & -(SK+F) & F & 0 & \dots & 0 \\ 0 & SK & -(SK+F) & F & \dots & 0 \\ & & \cdot & & & \\ & & \cdot & & & \\ & & \cdot & & & \\ 0 & \dots & 0 & 0 & SK & -(SK+F) \end{bmatrix},$$

$$\boldsymbol{x} = \begin{bmatrix} x_1 \\ x_2 \\ x_3 \\ \cdot \\ \cdot \\ \cdot \\ x_n \end{bmatrix}, \quad \text{and} \quad \boldsymbol{b} = \begin{bmatrix} -Sy_S \\ 0 \\ 0 \\ \cdot \\ \cdot \\ \cdot \\ -Fx_F \end{bmatrix}.$$

A countercurrent solvent extraction of penicillin from broth filtrate uses the organic solvent amyl acetate, and is defined by the following process parameters:

$F = 280$ kg/hr,
$S = 130$ kg/hr,
$x_F = 0.035$ kg penicillin/kg broth,

$y_S = 0$, and
$K = 2.0$.

It is desired that the final concentration of penicillin in the raffinate be $x_1 = 0.005$ kg penicillin/kg broth. How many stages are required? Use the MATLAB program that you developed in Problem 2.8 to solve tridiagonal matrix equations. Start with a guess value of $n = 3$, and increase n in steps of one until the desired value of x_1 is achieved. You will need to create new $\boldsymbol{A}$ and $\boldsymbol{b}$ for each n considered. You can use the MATLAB function `diag` to create three diagonal matrices from three vectors and sum them up to obtain $\boldsymbol{A}$. Draw a plot in MATLAB of the exiting pencillin concentration versus the number of theoretical stages required to achieve the corresponding separation.

Linear regression

2.10. If a straight line

$$\hat{y}_i = \beta_0 + \beta_1 x_i \tag{P2.5}$$

is fitted to a data set (x, y), then the intercept is given by Equation (2.38) (repeated here) as

$$\beta_0 = \bar{y} - \beta_1 \bar{x}. \tag{2.38}$$

If we substitute β_0 in Equation (P2.5) with the expression given by Equation (2.38), we obtain

$$\hat{y}_i - \bar{y} = \beta_1 (x_i - \bar{x}). \tag{P2.6}$$

Using Equation (P2.6), show that the point $(\bar{x}, \bar{y})$ lies on a straight line obtained by regression of y on x.

2.11. Use Equation (P2.6) to show that the sum of the residuals obtained from a least-squares regression of y on x is equal to zero, i.e.

$$\sum_{i=1}^{m} y_i - \hat{y}_i = 0. \tag{P2.7}$$

Use Equation (P2.7) to show that the average of $\hat{y}_i$ is equal to the mean of y_i.

2.12. **Power-law prediction of animal metabolism** Find a power-law relationship between an animal's mass and its metabolism. Table P2.1 gives animals' masses in kilograms and metabolism rates in kilocalories per day (Schmidt-Nielsen, 1972).

You will need to take the natural log of both the independent and dependent variables to determine the power-law exponent and coefficient (i.e. log (metabolic

Table P2.1.

Animal	Mass (kg)	Metabolism (kcal/day)
Horse	700	11 760
Human	70	1700
Sheep	55	1450
Cat	15	720
Rat	0.4	30

Table P2.2.

Time (s)	Tether length (μm)	Model 1 prediction (μm)	Model 2 prediction (μm)
0.65	1.0	2.0	1.0
1.30	3.0	4.0	3.0
1.95	4.0	5.0	6.0

rate) = $a \cdot \log$ (body weight) + k). Write a program in MATLAB and use the normal equations to perform your least-squares regression.

2.13. Data regression of two competing models Table P2.2 gives a measurement of microvillus tether extraction from a live B-cell as a function of time, along with the predictions of two different theoretical models. Evaluate which of the two models provides the best overall fit to the data.

2.14. Transformation of nonlinear equations For the second-order irreversible reaction

$$A + A \xrightarrow{k_2} A_2,$$

the decrease in reactant concentration C_A as a function of time t can be expressed as

$$\frac{C_A}{C_{A0}} = \frac{1}{1 + k_2 C_{A0} t},$$

where C_{A0} is the initial concentration of A. Rewrite the above equation so that it is linear in the two modeling parameters C_{A0} and k_2.

2.15. Will women's sprint time beat the men's in the Olympics someday? In a famous (and controversial) article in *Nature*, Tatem and coworkers performed linear regressions on the winning Olympic 100 meter sprint times for men and women (see Table P2.3) and used these results to predict that the winning women's sprint time would surpass the winning men's time in the summer Olympics of 2156 (Tatem *et al.*, 2004).[5] Your task is to recreate the analysis of Tatem *et al.* by performing linear regressions of the winning women's and men's 100 meter sprint times *using the normal equations*. Show graphically that the two best-fit lines intersect in the year 2156. Be careful not to include zeros (gaps in the data due to the World Wars and the absence of women competitors in Olympics prior to 1928) because these spurious points will bias your regression.

This data set is available at the book's website www.cambridge.org/* as sprint-times.dat.

2.16. Number of eyelash mites correlate with host's age Andrews (1976) hypothesized that the number of follicle mites living on human eyelashes would correlate with the age of the host organism, and collected the data given in Table P2.4 (McKillup, 2005).

[5] Although Tatem and coworkers carefully calculated the 95% confidence intervals in their model predictions (see Chapter 3 for an in-depth discussion on confidence intervals), many researchers have criticized their *extrapolation* of the model far beyond the range of available data. Also, various assumptions must be made, such as neglecting the possible influence of performance-enhancing drugs and the existence of a future plateau in human abilities.

Table P2.3. *Winning times of the men's and women's Olympic 100 m sprints from 1900*

Olympic year	Men's winning time (s)	Women's winning time (s)
1900	11.00	
1904	11.00	
1908	10.80	
1912	10.80	
1916		
1920	10.80	
1924	10.60	
1928	10.80	12.20
1932	10.30	11.90
1936	10.30	11.50
1940		
1944		
1948	10.30	11.90
1952	10.40	11.50
1956	10.50	11.50
1960	10.20	11.00
1964	10.00	11.40
1968	9.95	11.08
1972	10.14	11.07
1976	10.06	11.08
1980	10.25	11.06
1984	9.99	10.97
1988	9.92	10.54
1992	9.96	10.82
1996	9.84	10.94
2000	9.87	10.75
2004	9.85	10.93

Source: Rendell, M. (ed.) *The Olympics: Athens to Athens 1896–2004* (London: Weidenfeld and Nicolson, 2004), pp. 338–340.

Fit these data to a line using the MATLAB function `polyfit`, and then plot the data points (as symbols) along with the best-fit line (as a solid line). Also determine the quality of the fit.

2.17. Lung volume scales with body weight In mammals, the lung volume is found to scale with the body weight. The lung volume is measured by the **vital capacity**, which is the amount of air exhaled when one first breathes in with maximum effort and then breathes out with maximum effort. A higher vital capacity translates into a greater amount of respiratory gas exchange. People who engage in strenuous work or exercise tend to have a higher vital capacity. Given the data in Table P2.5 (Schmidt-Nielsen, 1972), show that a power law $V = aW^b$ fits these data reasonably well by performing a linear regression using the transformed equation (base 10): log

Table P2.4.

Age (years)	Number of mites
3	5
6	13
9	16
12	14
15	18
18	23
21	20
24	32
27	29
30	28

Table P2.5.

W (kg)	0.3 (rat)	3 (cat)	4 (rabbit)	15 (dog)	35 (pig)	80 (man)	600 (cow)	1100 (whale)
V (liter)	0.01	0.3	0.4	2	6	7	10	100

$V = b\log W + \log a$. Evaluate the constants a and b and show that $b \sim 1.0$. Perform the linear regression again, this time assuming a linear dependency of vital capacity V on body weight W. This time the slope you obtain for the straight line will equal the constant a from the power law.

The **tidal volume** (T) is the volume of air that a person breathes in and out effortlessly. It is approximately one-tenth of the vital capacity. With this information only, obtain the relationship between T and W.

2.18. For the problem stated in Box 2.4, plot the residuals of the α_{crit} values obtained from the quadratic fit as a function of height H of the platelet from the wall. Do you observe a trend in the residuals or simply random scatter?

2.19. Determine the coefficient of determination R for fitted curves obtained in Boxes 2.2 and 2.3. What can you say about the quality of the fit of the models?

References

Andrews, M. (1976) *The Life That Lives on Man* (London: Faber & Faber).

Anton, H. A. (2004) *Elementary Linear Algebra* (New York: Wiley, John & Sons, Inc.).

Bailey, J. E. and Ollis, D. F. (1986) *Biochemical Engineering Fundamentals* The Mc-Graw Hill Companies.

Davis, T. H. and Thomson, K. T. (2000) *Linear Algebra and Linear Operators in Engineering* (San Diego: Academic Press).

Demmel, J. W. (1997) *Applied Numerical Linear Algebra* (Philadelphia: Society for Industrial and Applied Mathematics (SIAM)).

Imai, K., Tientadakul, P., Opartkiattikul, N., Luenee, P., Winichagoon, P., Svasti, J., and Fucharoen, S. (2001) Detection of Haemoglobin Variants and Inference of Their

Functional Properties Using Complete Oxygen Dissociation Curve Measurements. *Br. J. Haematol.*, **112**, 483–7.

King, M. R. and Hammer, D. A. (2001) Multiparticle Adhesive Dynamics: Hydrodynamic Recruitment of Rolling Leukocytes. *Proc. Natl. Acad. Sci. USA*, **98**, 14919–24.

Levy, J. H., Goodnough, L. T., Greilich, P. E. *et al.* (2002) Polymerized Bovine Hemoglobin Solution as a Replacement for Allogeneic Red Blood Cell Transfusion after Cardiac Surgery: Results of a Randomized, Double-Blind Trial. *J. Thorac. Cardiovasc. Surg.*, **124**, 35–42.

Livingston, B. M. and Krakoff, I. H. (1970) L-Asparaginase; a New Type of Anticancer Drug. *Am. J. Nurs.*, **70**, 1910–15.

McKillup, S. (2005) *Statistics Explained: An Introductory Guide for Life Scientists* (Cambridge: Cambridge University Press).

Mody, N. A. and King, M. R. (2005) Three-Dimensional Simulations of a Platelet-Shaped Spheroid near a Wall in Shear Flow. *Phys. Fluids*, **17**, 1432–43.

Quick, D. J. and Shuler, M. L. (1999) Use of In Vitro Data for Construction of a Physiologically Based Pharmacokinetic Model for Naphthalene in Rats and Mice to Probe Species Differences. *Biotechnol. Prog.*, **15**, 540–55.

Schmidt-Nielsen, K. (1972) *How Animals Work* (Cambridge: Cambridge University Press).

Shen, Y.-C., Wang, M.-Y., Wang, C.-Y., Tsai, T.-C., Tsai, H.-Y., Lee, Y.-F., and Wei, L.-C. (2007) Clearance of Intravitreal Voriconazole. *Invest. Ophthalmol. Vis. Sci.*, **48**, 2238–41.

Sin, A., Chin, K. C., Jamil, M. F., Kostov, Y., Rao, G., and Shuler, M. L. (2004) The Design and Fabrication of Three-Chamber Microscale Cell Culture Analog Devices with Integrated Dissolved Oxygen Sensors. *Biotechnol. Prog.*, **20**, 338–45.

Tatem, A. J., Guerra, C. A., Atkinson, P. M., and Hay, S. I. (2004) Athletics: Momentous Sprint at the 2156 Olympics? *Nature*, **431**, 525.

Viravaidya, K. and Shuler, M. L. (2004) Incorporation of 3t3-L1 Cells to Mimic Bioaccumulation in a Microscale Cell Culture Analog Device for Toxicity Studies. *Biotechnol. Prog.*, **20**, 590–7.

Viravaidya, K., Sin, A., and Shuler, M. L. (2004) Development of a Microscale Cell Culture Analog to Probe Naphthalene Toxicity. *Biotechnol. Prog.*, **20**, 316–23.

3 Probability and statistics

3.1 Introduction

Variability abounds in our environment, such as in daily weather patterns, the number of accidents occurring on a certain highway per week, the number of patients in a hospital, the response of a group of patients to a particular type of drug, or the birth/death rate in any town in a given year. The "observed" occurrences in nature or in our man-made environment can be quantitatively described using "statistics" such as mean value, standard deviation, range, percentiles, and median. In the fields of biology and medicine, it is very difficult, and often impossible, to obtain data from every individual that possesses the characteristic of interest that we wish to observe. Say, for instance, we want to assess the success rate of a coronary drug-eluting stent. To do this, we must define our population of interest – patients who underwent coronary stent implantation. It is an almost impossible task to track down and obtain the medical records of every individual in the world who has been treated with a drug-eluting stent, not to mention the challenges faced in obtaining consent from patients for collection and use of their personal data. Therefore we narrow our search method and choose only a part of the population of interest (called a **sample**), such as all Medicare patients of age >65 that received a coronary stent from May 2008 to April 2009. The statistics obtained after studying the performance of a drug-eluting stent exactly describe the sample, but can only serve as estimates for the population, since they do not fully describe the varied experiences of the entire population. Therefore, it is most critical that a sample be chosen so that it exhibits the variability inherent in the population without systematic bias. Such a sample that possesses the population characteristics is called a **representative sample**. A sample can be a series of measurements, a set of experimental observations, or a number of trials.

A statistic is calculated by choosing a random sample from the entire population and subsequently quantifying data obtained from this sample. If the individuals that make up the sample exhibit sufficient variation, the sample-derived statistic will adequately mimic the behavior of the whole population. A common method to generate a sample is to conduct an extensive survey. In this case, the sample includes all the individuals who respond to the survey, or from whom data are collected. For example, a health insurance company may request that their customers provide feedback on the care received during a doctor's visit. Depending on the number of responses, and the characteristics of the people who responded, the sample may either (1) constitute a sizably heterogeneous group of individuals that can reliably represent the entire population from which the sample was drawn, or (2) contain too few responses or consist of a homogeneous set of individuals that does not mimic the variability within the population. The method of sampling is an important step in obtaining reliable statistical estimates and is discussed in Section 3.3. In the event that an inadequate sample is used to obtain information from which we wish to infer the behavior of the entire population, the derived estimates may be misleading and lie far from the actual values.

We are specifically interested in understanding and quantifying the variability observed in biological data. What do we mean by variability in biology? Due to varying environmental and genetic factors, different individuals respond in a dissimilar fashion to the same stimulus. Suppose 20 16-year-old boys are asked to run for 20 minutes at 6 miles/hour. At the end of this exercise, three variables – the measured heart rate, the blood pressure, and the breathing rate – will vary from one boy to another. If we were to graph the measured data for each of the three variables, we would expect to see a distribution of data such that most of the values lie close to the average value. A few values might be observed to lie further away from the average. It is important to realize that the differences in any of these measurements from the average value are not "errors." In fact, the differences, or "spread," of the data about the average value are very important pieces of information that characterize the differences inherent in the population. Even if the environmental conditions are kept constant, variable behavior cannot be avoided. For example, groups of cells of the same type cultured under identical conditions exhibit different growth curves. On the other hand, identical genetically engineered plants may bear different numbers of flowers or fruit, implying that environmental differences such as conditions of soil, light, or water intake are at work.

The methods of collection of numerical data, its analysis, and *scientifically sound* and *responsible* interpretation, for the purposes of generating descriptive numbers that characterize a population of interest, are collectively prescribed by the field of **statistics**, a branch of mathematics. The need for statistical evaluations is pervasive in all fields of biomedicine and bioengineering. As scientists and engineers we are required to calculate and report statistics frequently. The usefulness of published statistics depends on the methods used to choose a representative (unbiased) sample, obtain unbiased data from the sample, and analyze and interpret the data. Poorly made decisions in the data collection phase, or poor judgment used when interpreting a statistical value, can render its value useless. The statistician should be well aware of the ethical responsibility that comes with designing the experiment that yields the data and the correct interpretation of the data. As part of the scientific community that collects and utilizes statistical values on a regular basis, you are advised to regard all published statistical information with a critical and discerning eye.

Statistical information can be broadly classified into two categories: descriptive statistics and inferential statistics. **Descriptive statistics** are numerical quantities that condense the information available in a set of data or sample into a few numbers, and exactly specify the characteristics exhibited by the sample, such as average weight, size, mechanical strength, pressure, or concentration. **Inferential statistics** are numerical estimations of the behavior of a population of individuals or items based on the characteristics of a smaller pool of information, i.e. a representative sample or data set. Some very useful and commonly employed descriptive statistics are introduced in Section 3.2. The role of probability in the field of statistics and some important probability concepts are discussed in Section 3.3. In Section 3.4, we discuss two widely used discrete probability distributions in biology: the binomial distribution and the Poisson distribution. The normal distribution, the most widely used continuous probability distribution, is covered in detail in Section 3.5, along with an introduction to inferential statistics. In Section 3.6, we digress a little and discuss an important topic – how errors or variability associated with measured

Box 3.1 Physical activity of children aged 9–15

Obesity in youth is associated with lack of sufficient physical activity. The US Department of Health and Human Services has proposed that children require a minimum of 60 minutes of moderate to vigorous physical activity on a daily basis. A study was conducted during the period 1991–2007 to assess the activity levels of children of ages ranging from 9 to 15 on weekdays and weekends (Nader *et al.*, 2008). The purpose of this study was to observe the patterns of physical activity that the youth engaged in. Their findings were significant, in that the amount of moderate to vigorous physical activity of these children showed a steep decline with age.

Participants were recruited from ten different cities from across the USA. The children were required to wear an accelerometer around their waist throughout the day for seven days. The accelerometer measured the level of physical activity, which was then converted to energy expended in performing activities of varying intensities using established empirical equations. Figure 3.1 shows histograms or frequency distribution curves depicting the amount of moderate to vigorous activity that children of ages 9, 11, 12, and 15 engaged in; n indicates the number of participants in the sample. Note that n decreased as the children increased in age, either due to some families opting out of the study or older children experiencing discomfort wearing the accelerometer (an obtrusive and bulky gadget) during school hours.

A **histogram** is a frequency distribution plot, in which the number of occurrences of an observation (here the duration of moderate to physical activity) is plotted. A histogram is commonly used to provide a pictorial description of data and to reveal patterns that lend greater insight into the nature of the phenomenon or process generating the data. Easy comparison is enabled by setting the axis limits equal in each subplot.

Figure 3.1

Histograms of the amount of time that children of ages 9–15 spent in performing moderate to vigorous physical activity during the weekday (Nader *et al.*, 2008).

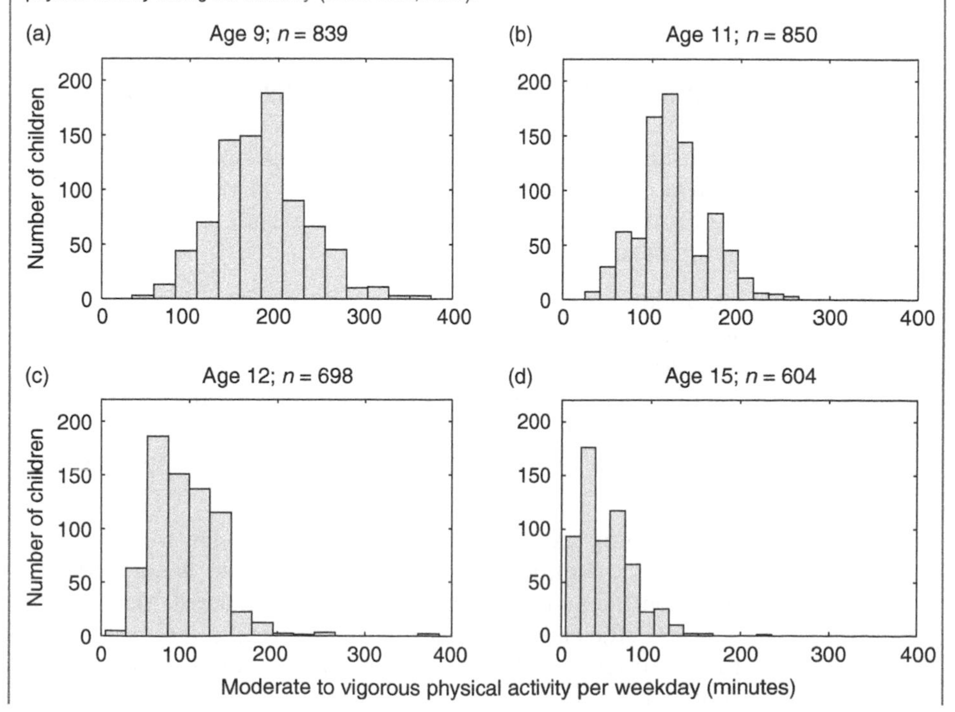

Table 3.1. *Average daily physical activity for the four age groups studied*

Age group	Mean moderate to vigorous physical activity per weekday (minutes)
9	181.8
11	124.1
12	95.6
15	49.2

The mean value for the duration of moderate-to-vigorous physical activity is calculated for each age group, and given in Table 3.1.

The mean value shifts to the left, i.e. reduces with increase in age. Why do you think the activity levels reduce with age? A drawback of this study is the **bias** inherent in the study. The activity levels of children were underpredicted by the accelerometer, since it was not worn during periods of the day when children participated in swimming and contact sports.

The spread of the value about the mean can also be calculated using various statistics described in later sections. Just by visually inspecting the histograms, you can observe that the amount of the spread about the mean value reduces with age. Study Figure 3.1. What does the amount of spread and its variation with age tell you?

quantities propagate in arithmetic operations that combine these quantities. The concepts developed in Section 3.6 require you to be familiar with material covered in Sections 3.2 and 3.5. In the last section of the chapter (Section 3.7) we revisit the linear least-squares regression method of parameter estimation that was discussed in Chapter 2. Here, we show how to calculate the error associated with predictions of the linear least-squares regression technique.

3.2 Characterizing a population: descriptive statistics

Laboratory experiments/testing are often performed to generate quantifiable results, such as oxygen consumption of yeast cells in a bioreactor, mobility of fibroblasts, or concentration of drug in urine. In most cases, each experimental run will produce different results due to

(1) variability in the individuals that make up the population of interest,
(2) slight discrepancies in experimental set-up such as quantity of media or amount of drug administered,
(3) fluctuations in measuring devices such as weighing balance, pH meter, or spectrophotometer, and
(4) differences in environmental conditions such as temperature or amount of sunlight exposure.

Publishing experimental results in a tabular format containing long lists of measured quantities provides little insight and is cumbersome and inefficient for the scientific and engineering community. Instead, experimental results or sample data are efficiently condensed into numerical quantities called **descriptive statistics** that characterize the population variable of interest. An important distinction should be noted. Any numerical measure derived from sample data is called a

statistic. Statistics are used to infer characteristics of the population from which the sample is obtained. A statistic is interpreted as an approximate measure of the population characteristic being studied. However, if the sample includes every individual from the population, the statistic is an accurate representation of the population behavior with respect to the property of interest, and is called a **population parameter**.

In order to define some useful and common descriptive statistics, we define the following:

- n is the sample size (number of individuals in the sample) or the number of experiments,
- x_i are the individual observations from the experiments or from the sample; $1 \leq i \leq n$.

3.2.1 Measures of central tendency

The primary goal in analyzing any data set is to provide a single measure that effectively represents all values in the set. This is provided by a number whose value is the center point of the data.

The most common measure of central tendency is the **mean** or average. The mean of a data set x_i is denoted by the symbol $\bar{x}$, and is calculated as

$$\bar{x} = \frac{\sum_{i=1}^{n} x_i}{n}. \tag{3.1}$$

While the mean of a sample is written as $\bar{x}$, the mean of a population is represented by the Greek letter μ to distinguish it as a population parameter as opposed to a sample statistic. If the population size is N, the population mean is given by

$$\mu = \frac{\sum_{i=1}^{N} x_i}{N}.$$

Equation (3.1) calculates the **simple mean** of the sample. Other types of means include the weighted mean, geometric mean, and harmonic mean. While the mean of the sample will lie somewhere in the middle of the sample, it will usually not coincide with the mean of the population from which the sample is derived. Sometimes the population mean may even lie outside the sample's range of values.

The simple mean is a useful statistic when it coincides with a point of convergence or a peak value near which the majority of the data points lie. If the data points tend to lie closely adjacent to or at the mean value, and this can be verified by plotting a histogram (see Figure 3.1), then the probability of selecting a data point from the sample with a value close to the mean is large. This is what one intuitively expects when presented with an average value. If Figure 3.1(a) instead looked like Figure 3.2, then using the mean to characterize the data would be misleading. This distribution shows two peaks, rather than one peak, and is described as **bimodal** (i.e. having two modes; see the definition of "mode" in this section).

Another useful measure of the central tendency of a data set is the **median**. If the data points are arranged in ascending order, then the value of the number that is the middle data point of the ordered set is the median. Half the numbers in the data set are larger than the median and the other half of the data points are smaller. If n is odd, the median is the $((n+1)/2)$th number in the set of data points arranged in increasing order. If n is even, then the median is the average of the $(n/2)$th and the $(n/2+1)$th numbers in the set of data points arranged in increasing order. The

Figure 3.2

Bimodal distribution.

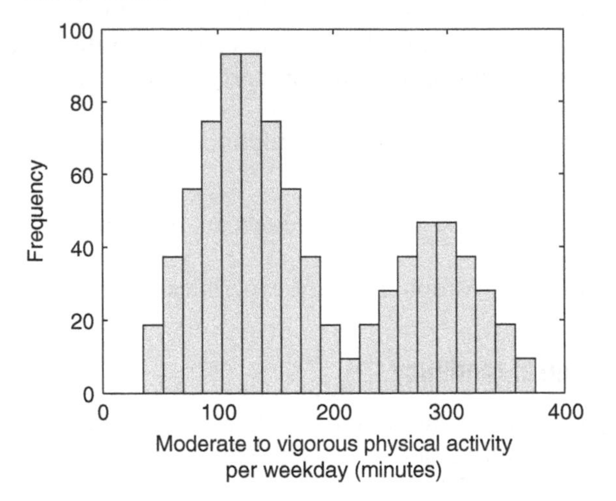

median can be a particularly useful statistic when data points far removed from the median (outliers) are present in a sample, which tend to shift the mean away from its point of central tendency.

The **mode** is the number that appears with greatest frequency and is another method of characterizing the central tendency of data.

Using MATLAB

MATLAB offers several functions to calculate measures of central tendency. The MATLAB functions `mean`, `median`, and `mode` are among the six different options that MATLAB provides to measure the central tendency of data. One syntax for `mean` is

```
mu = mean(x)
```

where $\boldsymbol{x}$ can be a vector or a matrix. If $\boldsymbol{x}$ is a matrix, then MATLAB[1] returns a row vector containing the means of each column of $\boldsymbol{x}$. If $\boldsymbol{x}$ is a vector, MATLAB returns a scalar value.

3.2.2 Measures of dispersion

Dispersion refers to the spread of data points about the mean. Three common measures of dispersion in a data set are range, variance, and standard deviation. The **range** is the difference between the largest and smallest values in the data set or sample. While the range is very easy to calculate, it only carries information regarding two data points and does not give a comprehensive picture of the spread of the other $(n-2)$ data points about the mean.

Variance is the average of the squared deviations of each data point from the mean value. For a *population* of size N, the variance, σ^2, is calculated as

$$\sigma^2 = \frac{\sum_{i=1}^{N}(x_i - \mu)^2}{N}, \tag{3.2}$$

[1] When MATLAB encounters a matrix during any statistical operation, by default it treats the matrix rows as individual observations and views each column as a different variable.

where $(x_i - \mu)$ is the **deviation** of each data point from the mean. By squaring the deviations, more weight is placed on points that lie further away from the mean. The variance, σ^2, is a population parameter and quantifies the dispersion in the entire population (size N). The variance for a sample is denoted by s^2 and is calculated differently, as shown below:

$$s^2 = \frac{\sum_{i=1}^{n}(x_i - \bar{x})^2}{n - 1}. \tag{3.3}$$

It is important to recognize that the sample variance is often used as an estimate of the variance of the population. We wish to obtain the most unbiased estimate possible. Because the sample contains only a subset of the population values, its variance will tend to be smaller than the population variance. The $n - 1$ term in the denominator corrects for the extreme values that are usually excluded from smaller samples. It can be proven that Equation (3.3) is the unbiased estimator for the population variance, and that replacing $n - 1$ with n would produce a smaller and biased estimate (see Section 3.5.3 for the proof).

The definition of sample variance also involves the concept of **degrees of freedom**. The mean $\bar{x}$ of a sample is a fixed value calculated from the n data points of the sample. Out of these n x_i numbers, $n - 1$ can be varied freely about the mean, while the nth number is constrained to some value to fix the sample mean. Thus, $n - 1$ represents the degrees of freedom available for calculating the variance.

The **standard deviation** of a sample is the square root of the variance and is calculated as follows:

$$s = \sqrt{s^2} = \sqrt{\frac{\sum_{i=1}^{n}(x_i - \bar{x})^2}{(n - 1)}}. \tag{3.4}$$

Accordingly, $\sigma = \sqrt{\sigma^2}$ is the population standard deviation.

Using MATLAB

The measures of dispersion or spread of data discussed in this section are all available as functions in MATLAB. The function `range` calculates the difference between the maximum and minimum values in a data set. The function `var` calculates the variance in the data set according to Equation (3.3) (unbiased estimator) and the function `std` calculates the square root of the variance in the data set. The input parameter for these three functions is either a vector or a matrix. If a matrix is the input parameter for any of these functions, MATLAB outputs a row vector containing the respective measure of dispersion calculated for each column of the matrix. For additional information about the MATLAB functions `range`, `var`, and `std`, see MATLAB help or type `help range`, `help var`, or `help std` at the command line.

3.3 Concepts from probability

Probability theory originated in the seventeenth century when an avid gambler, puzzled by a series of failed attempts at winning a game of dice, sought the advice of a mathematician. The gambler was a French nobleman, Antoine Gombauld, also known as Chevalier de Méré, who indulged in a popular gambling pastime, betting on dice. In accordance with his purported winning strategy, de Méré repeatedly bet even money that at least one 12 (two sixes) would occur in 24 rolls of a pair of dice. He conjectured that the probability of obtaining a 12 in one roll of two dice is 1/36, and therefore that the chances of getting at least one 12 in 24 rolls must be 24/36 or 2/3. However, this method of calculation is flawed. The true odds of getting at least one 12 in 24 rolls of a pair of dice is actually a little less than 0.5, and de Méré's betting strategy failed him. He contacted Blaise Pascal (1623–1662), a proficient mathematician and philosopher at the time, for help in solving the mystery behind his gambling fallout. Pascal calculated that the chance of winning such a bet is 49%, and this of course explained de Méré's unfortunate losses.

Soon after, Pascal collaborated with another famous mathematician, Pierre de Fermat, and together they developed a foundation for probability theory. The realm of probability theory initially focused only on games of chance to address the queries of gamblers ignorant in the rigors of mathematics. Jacob (also known as James or Jacques) Bernoulli (1654–1705) and Abraham de Moivré (1667–1754), both prolific mathematicians in the seventeenth and eighteenth centuries, made significant contributions to this field. Due to the pioneering work of Pierre-Simon Laplace (1749–1827), probability theory grew to address a variety of scientific problems in mathematical statistics, genetics, economics, and engineering, with great advances also brought about by Kolmogorov, Markov, and Chebyshev.

Getting back to our topic of statistics, we may ask, "How is the theory of probability related to statistical evaluations of a population parameter?" As discussed in Section 3.1, statistical values are *numerical estimates*. Statistical analyses are accompanied by uncertainty in the results. Why is this so? It is because statistics are derived from samples that do not usually contain every individual in the population under study. Because a sample cannot perfectly represent the population's properties, it has an inherent deficiency. The purpose of deriving any statistic is mainly to extrapolate our findings to a much larger population of interest compared to the sample size under study. Our certainty in the value of a statistic is often conveyed in terms of **confidence limits**. In other words, an upper and lower bound is specified that demarcates a range of values within which the statistic is known to lie with some probability, say 95% probability or *95% confidence*. A statistical value implies a probable range of values. Therefore no statistic is known with complete certainty unless the entire population of individuals is used to calculate the statistic, in which case the statistical number is no longer an estimate but an exact description of the population.[2]

Probability theory applied to statistics allows us to draw inferences on population characteristics. Probabilities are very useful for comparing the properties of two different populations such as a control (or normal) group and a treated group. A treated group could consist of living beings, cells, tissues, or proteins that have been treated with some drug, while a control group refers to the untreated group. Following treatment procedures, the experimentalist or clinical researcher wishes to compare the differences in the statistical parameters (indicators), such as the mean value of some observed characteristic of the population, that the treatment is believed to alter. By comparing the properties of the treated group with that of the untreated, one can use statistical methods to discern whether a treatment really had an effect. One uses "probabilities" to convey the confidence in the observed differences. For example, we may say that we are 99.9% confident that drug XYZ causes a reduction in the blood pressure of adults (or that there is a 0.1% chance that the

[2] Barring numerical error in the calculation of the statistic as discussed in Chapter 1.

difference observed between the control and treated study group could have arisen from simple random variation).

You will notice that probabilities are usually conveyed in terms of percentages, such as "the chance of survival of a patient is 50%" or "the chance of a particular medical treatment being successful is 70%." **Probability** is a branch of mathematics that prescribes the methods used to calculate the chance that a particular event will occur. Any process or experiment can have one or more possible outcomes. A single outcome or a collective set of outcomes from a process is called an **event**. If a process yields only one outcome, then this outcome is certain to occur and is assigned a probability of 1. If an event cannot possibly occur, it is called an **impossible event** and is given the probability of 0. The probability P of any event E can take values $0 \leq P(E) \leq 1$.

If two events cannot occur simultaneously, they are called **mutually exclusive** events. For example, a light bulb can either be on or off, or a room can either be occupied or empty, but not both at the same time. A card picked from a deck of 52 is either a spade, a diamond, a club, or a heart – four mutually exclusive events. No playing card can belong to more than one suit. For the case of a well-shuffled deck of cards, the probability of any randomly drawn card belonging to any one of the four suits is equal and is ¼. The probability that the drawn card will be a heart or a spade is ¼ + ¼ = ½. Note that we have simply added the individual probabilities of two mutually exclusive events to get the total probability of either event occurring. The probability that either a spade, heart, diamond, or club will turn up is ¼ + ¼ + ¼ + ¼ = 1. In other words, it is certain that one of these four events will occur.

If a process yields n mutually exclusive events $E_1, E_2, \ldots, E_n$, and if the probability of the event E_i is $P(E_i)$ for $1 \leq i \leq n$, then the probability that any of n mutually exclusive events will occur is

$$P(E_1 \cup E_2 \cup \cdots \cup E_n) = \sum_{i=1}^{n} P(E_i). \tag{3.5}$$

The sign $\cup$ denotes "or" or "union of events."

The probabilities of mutually exclusive events are simply added together. If the n mutually exclusive events are **exhaustive**, i.e. they account for all the possible outcomes arising from a process, then

$$P(E_1 \cup E_2 \cup \cdots \cup E_n) = \sum_{i=1}^{n} P(E_i) = 1.$$

If the probability of event E is $P(E)$, then the probability that E will not occur is $1 - P(E)$. This is called the **complementary event**. Events E and the complementary event "not E" are mutually exclusive. Thus, if an event is defined as E: "the card is a spade" and the probability that the drawn card is a spade is $P(E) = 1/4$, then the probability that the card is not a spade is $P(E') = 1 - 1/4 = 3/4$.

3.3.1 Random sampling and probability

Sampling is the method of choosing a sample or a group of individuals from a large population. It is important to obtain a **random sample** in order to faithfully mimic the population without bias. If every individual of the population has an equal opportunity in being recruited to the sample, then the resulting sample is said to be random. A sample obtained by choosing individuals from the population in a completely random manner, and which reflects the variability inherent in the population, is called a

Table 3.2. *Age distribution of healthy adult participants*

Age range	Number of healthy human subjects
21–30	16
31–40	10
41–50	10
51–60	4

representative sample. If a sample does not adequately reflect the population from which it is constructed, then it is said to be **biased**. A sample that is formed by haphazard selection of individual units will not form a representative sample, but will most likely be biased. When the sample is too homogeneous, the variability of the population is poorly represented, and the sample represents the behavior of only a specific segment of the population. Extrapolation of biased results to the larger population from which the sample was drawn produces misleading predictions.

The importance of *random* sampling can be appreciated when we apply the concepts of probability to statistics. Probability theory assumes that sampling is random.

> If an individual is *randomly* selected from a group, the probability that the individual possesses a particular trait is equal to the fraction (or proportion) of the population that displays that trait.

If our sampling method is biased (i.e. certain individuals of the population are preferred over others when being selected), then the probability that the selected individual will have a particular trait is either less than or greater than the population proportion displaying that trait, depending on the direction of the bias.

Example 3.1

We are interested in creating a random sample of a group of 20 healthy adults selected from 40 healthy human subjects in a new vaccine study. The randomly sampled group of 20 will serve as the treated population and will receive the vaccination. The other 20 subjects will serve as the control. Out of the 40 subjects who have consented to participate in the trial, half are men and the other half are women.[3] The distribution of ages is tabulated in Table 3.2.

The ratio of the number of males to females in each age range is 1:1.

(1) What is the probability that the first person selected from the group of 40 subjects is female?
(2) What is the probability that the first person selected from the group of 40 subjects is female and is between the ages of 31 and 50?

We define the following events:

E_1: person is female,
E_2: person belongs to the 31–50 age group.

On the first draw, the probability that the first person randomly selected is female is equal to the proportion of females in the group, and is given by

[3] When ascertaining the effects of a drug, outcome on exposure to a toxin, or cause of a disease, two study groups are employed: one is called the treated group and the other is the control or untreated group. Often other factors like age, sex, diet, and smoking also influence the result and can thereby confound or distort the result. Matching is often used in case-control (retrospective) studies and cohort (prospective) studies to ensure that the control group is similar to the treated group with respect to the confounding factor. They are called matched case-control studies and matched cohort studies, respectively.

$$P(E_1) = \frac{20 \text{ females}}{40 \text{ subjects}} = \frac{1}{2}.$$

The four age groups are mutually exclusive. A person cannot be in two age ranges at the same time. Therefore, the probability that a randomly selected person is in the age group 31–40 or 41–50 can be simply added. If

E_3: Person belongs to the 31–40 age group,
E_4: Person belongs to the 41–50 age group,

then $P(E_2) = P(E_3 \cup E_4) = P(E_3) + P(E_4)$.
Since $P(E_3) = 10/40 = 1/4$; and $P(E_4) = 10/40 = 1/4$,

$$P(E_2) = \frac{1}{4} + \frac{1}{4} = \frac{1}{2}.$$

We wish to determine $P(E_1 \cap E_2)$, where $\cap$ denotes "and." Now, since the ratio of male to female is 1:1 in each age category, $P(E_2)$ = ½ applies regardless of whether the first chosen candidate is male or female. The probability of E_2 is not influenced by the fact that E_1 has occurred. Events E_1 and E_2 are independent of each other.

Events whose probabilities are not influenced by the fact that the other has occurred are called **independent events**.

A **joint probability** is the probability of two different events occurring simultaneously. When two independent events occur simultaneously, the joint probability is calculated by multiplying the individual probabilities of the events of interest.
If E_1 and E_2 are two independent events, then

$$P(E_1 \cap E_2) = P(E_1)P(E_2). \tag{3.6}$$

The sign $\cap$ denotes "and" or "intersection of events."

Here, we wish to determine the joint probability that the first person selected is both E_1: female and E_2: within the 31–50 age group:

$$P(E_1 \cap E_2) = \frac{1}{2} \times \frac{1}{2} = \frac{1}{4}.$$

Example 3.2

We reproduce Pascal's probability calculations that brought to light de Méré's true odds of winning (or losing!) at dice.

In a single roll of one die, the probability of a six turning up is 1/6.

In a single roll of two dice, the probability of a six turning up on one die is independent of a six turning up on the second die. The joint probability of two sixes turning up on both dice is calculated using Equation (3.6):

$$P(\text{two sixes on two dice}) = \frac{1}{6} \times \frac{1}{6} = \frac{1}{36}.$$

The probability of two sixes on a pair of dice is equivalent to the probability of a 12 on a pair of dice, since 6 + 6 is the only two-dice combination that gives 12. The probability of 12 not turning up on a roll of dice = $1 - 1/36 = 35/36$:

$$\begin{aligned} &P(\text{at least one 12 on a pair of dice in 24 rolls of dice}) \\ &\quad = 1 - P(\text{12 not observed on a pair of dice in 24 rolls}). \end{aligned}$$

The outcome of one roll of the dice is independent of the outcome of the previous rolls of the dice.
Using Equation (3.6), we calculate the joint probability of

$$P(\text{12 does not turn up on a pair of dice in 24 rolls})$$

$$= \frac{35}{36} \times \frac{35}{36} \times \cdots \qquad \text{24 times}$$

$$= \left(\frac{35}{36}\right)^{24}.$$

Then,

$$P(\text{at least one 12 on a pair dice in 24 rolls of dice})$$

$$= 1 - \left(\frac{35}{36}\right)^{24} = 1 - 0.5086 = 0.4914.$$

Thus, the probability of getting at least one 12 in 24 rolls of the dice is 49.14%.

The joint probability of two mutually exclusive events arising from the same experiment is always 0. Thus, mutually exclusive events are not independent events. If E_1 and E_2 are two mutually exclusive events that result from a single process, then the knowledge of the occurrence of E_1 indicates that E_2 did not occur. Thus, if E_1 has occurred, then the probability of E_2 once we know that E_1 has occurred is 0.

If the probability of any event E_i changes when event E_j takes place, then the event E_i is said to be dependent on E_j. For example, if it is already known that the first person selected from a study group of 40 people has an age lying between 31 and 50, i.e. E_2 has occurred, then the probability of E_3: person is in the age group 31–40, is no longer ¼. Instead

$$P(E_3|E_2) = \frac{\text{10 people in group 31–40}}{\text{20 people in group 31–50}} = \frac{1}{2}.$$

$P(E_3|E_2)$ denotes the *conditional probability* of the event E_3, given that E_2 has occurred. The *unconditional probability* of selecting a person from the study group with age between 31 and 40 is ¼.

> The **conditional probability** is the probability of the event when it is known that some other event has occurred.

In the example above,

$$P(E_3|E_2) \neq P(E_3),$$

which signifies that event E_3 is not independent from E_2. Once it is known that E_2 has occurred, the probability of E_3 changes.

> The **joint probability of two dependent events** E_1 and E_2 is calculated by multiplying the probability of E_1 with the conditional probability of E_2 given that E_1 has occurred. This is written as
>
> $$P(E_1 \cap E_2) = P(E_1)P(E_2|E_1). \tag{3.7}$$
>
> On the other hand, for two *independent* events,
>
> $$P(E_2|E_1) = P(E_2). \tag{3.8}$$
>
> The knowledge that E_1 has occurred does *not* alter the probability of E_2.

Substituting Equation (3.8) into Equation (3.7) returns Equation (3.6).

Table 3.3. *Age and gender distribution of healthy adult participants*

Age range	Number of healthy human subjects	Male to female ratio
21–30	16	5:3 (10:6)
31–40	10	2:3 (4:6)
41–50	10	2:3 (4:6)
51–60	4	1:1 (2:2)
Total	40 participants	20 men : 20 women

Example 3.3

In Example 3.1, it was stated that out of the 40 subjects who have consented to participate in the trial, half are men and the other half are women, and that the ratio of males to females in each age group was the same. Assume now that the ratio of males to females varies in each age group as shown in Table 3.3.

(1) What is the probability that the first person selected from the group of 40 subjects is female?
(2) What is the probability that the first person selected from the group of 40 subjects is female *and* is between the ages of 31 and 50?

Again we define

E_1: person is female,
E_2: person belongs to age range 31 to 50.

The answer to the first question remains the same. Without prior knowledge of the age of the first selected candidate, $P(E_1) = 20/40$ or ½.

The answer to the second question changes. The joint probability of E_1 and E_2 can be calculated from either Equation (3.7),

$$P(E_1 \cap E_2) = P(E_1)P(E_2|E_1),$$

or

$$P(E_1 \cap E_2) = P(E_2)P(E_1|E_2), \tag{3.9}$$

since E_1 and E_2 are not independent events. As shown in Example 3.1,

$$P(E_1) = \frac{1}{2}; \qquad P(E_2) = \frac{1}{2}.$$

Now, $P(E_2|E_1)$ is the probability that the randomly selected person belongs to the age range 31–50, if we already know that the person is female. Since there are 12 females in the age range 31–50, and there are 20 females in total, the probability that the female selected belongs to the 31–50 age group is given by

$$P(E_2|E_1) = \frac{\text{females in age range 31–50}}{\text{total no. of females}} = \frac{12}{20} = \frac{3}{5}.$$

Substituting into Equation (3.7), we have $P(E_1 \cap E_2) = 3/10$.

We can also calculate $P(E_1 \cap E_2)$ using Equation (3.9):

$$P(E_1|E_2) = \frac{\text{females in age range 31–50}}{\text{total no. in age range 31–50}} = \frac{12}{20} = \frac{3}{5}.$$

Substituting into Equation (3.9), we have $P(E_1 \cap E_2) = 3/10$.

Example 3.4

If two people are randomly selected from the group of 40 people described in Example 3.3, what is the probability that both are male?

The random selection of more than one person from a group of people can be modeled as a sequential process. If n people are to be selected from a group containing $N \geq n$ people, then the sample size is n. Each individual that constitutes the sample is randomly chosen from the N-sized group one at a time until n people have been selected. Each person that is selected is not returned to the group from which the next selection is to be made. The size of the group decreases as the selection process continues. This method of sampling is called **sampling without replacement**. Because the properties of the group change as each individual is selected and removed from the group, the probability of an outcome at the ith selection is dependent on the outcomes of the previous $(i-1)$th selections. To obtain a truly random (unbiased) sample, it is necessary that each selection made while constructing a sample be *independent* of any or all of the previous selections. In other words, selection of any individual from a population should not influence the probability of selection of other individuals from the population. For example, if the individual units in a sample share several common characteristics that are not present in the other individuals within the population represented by the sample, then the individual sample units are *not independent*, and the sample is biased (not random).

In our example, the probability of the second selected person being male is dependent on the outcome of the first selection. The probability that the first person selected is male is 20/40 or ½. The probability that the second person selected is male depends on whether the first individual selected is male. If this is the case, the number of men left in the group is 19 out of a total of 39. If

E_1: first person selected is male,
E_2: second person selected is male,

then,

$$P(E_1 \cap E_2) = P(E_1)P(E_2|E_1),$$

$$P(E_2|E_1) = \frac{19}{39},$$

$$P(E_1 \cap E_2) = \frac{1}{2} \times \frac{19}{39} = \frac{19}{78}.$$

The calculations performed in Example 3.4 are straightforward and easy because we are only considering the selection of two people. How would we approach this situation if we had to determine outcomes pertaining to the selection of 20 people? We would need to create a highly elaborate tree structure detailing every possible outcome at every step of the selection process – a cumbersome and laborious affair. This problem is solved by theories that address methods of counting – permutations and combinations.

3.3.2 Combinatorics: permutations and combinations

A *fundamental principle of counting* that forms the basis of the theory of combinatorics is as follows. If

- event A can happen in exactly m ways, and
- event B can happen in exactly n ways, and
- the occurrence of event B follows event A,

then, these two events can happen together in exactly $m \times n$ ways.

If we have a group of ten people numbered 1 to 10, in how many ways can we line them up? The first person to be placed in line can be chosen in ten ways. Now that the first person in line is chosen, there remain only nine people in the group who can still be assigned positions in the line. The second person can be chosen in nine ways. When one person is remaining, he/she can be chosen in only one way. Thus, the number of ways in which we can order ten people is $10 \times 9 \times 8 \times 7 \times 6 \times 5 \times 4 \times 3 \times 2 \times 1 = 10!$ (read as ten factorial; $n! = n(n-1)(n-2) \cdots 3 \times 2 \times 1; 0! = 1$). What if we wish to select three people out of ten and determine the total number of ordered sets of three possible? In that case, the total number of ways in which we can select three ordered sets of individuals, or the total number of **permutations** of three individuals selected out of ten, is $10 \times 9 \times 8$. This can also be written as $\frac{10!}{7!}$.

> If r objects are to be selected out of n objects, the total number of **permutations** possible is given by
>
> $$P_r^n = \frac{n!}{(n-r)!}. \tag{3.10}$$

What if we are only interested in obtaining sets of three individuals out of ten, and the ordering of the individuals within each set is not important? In this case, we are interested in knowing the number of **combinations** we can make from randomly choosing any three out of ten. We are not interested in permuting or arranging the objects within each set. Now, three objects in a set of three can be ordered in

$$P_3^3 = \frac{3!}{0!} = 3!$$

ways. And r objects can be ordered or permuted in

$$P_r^r = \frac{r!}{0!} = r!$$

ways. Thus, every set containing r objects can undergo $r!$ permutations or have $r!$ arrangements.

We can choose r out of n objects in a total of $n \times (n-1) \times (n-2) \times \cdots \times (n-r+1)$ ways. Every possible set of r individuals formed is counted $r!$ times (but with different ordering).

> Since we wish to disregard the ways in which we can permute the individual objects in each set of r objects, we instead calculate the number of **combinations** we can make, or ways that we can select sets of r objects out of n objects as follows:
>
> $$C_r^n = \frac{n \times (n-1) \times (n-2) \times \cdots \times (n-r+1)}{r!} = \frac{n!}{r!(n-r)!}. \tag{3.11}$$

We use Equation (3.11) to determine the number of sets containing three people that we can make by selecting from a group of ten people:

$$C_3^{10} = \frac{10!}{3!(10-3)!} = 120.$$

Example 3.5

We wish to select 20 people from a group of 40 such that exactly 10 are men and exactly 10 are women. The group contains 20 men and 20 women. In how many ways can this be done?

We wish to select 10 out of 20 men. The order in which they are selected is irrelevant. We want to determine the total number of combinations of 10 out of 20 men possible, which is

$$C_{10}^{20} = \frac{20!}{10!(20-10)!} = 184\ 756.$$

We wish to select 10 out of 20 women. The total number of possible combinations of 10 out of 20 women is also

$$C_{10}^{20} = \frac{20!}{10!(20-10)!} = 184\ 756.$$

Since both the selection of 10 out of 20 men and 10 out of 20 women can happen in 184 756 ways, then the total possible number of combinations that we can make of 10 men and 10 women is $184\,576 \times 184\,576 = 3.41 \times 10^{10}$ (see the fundamental principle of counting). Remember that the ordering of the 20 individuals within each of the 3.41×10^{10} sets is irrelevant in this example.

Box 3.2 Medical testing

Diagnostic devices are often used in clinical practice or in the privacy of one's home to identify a condition or disease, e.g. kits for HIV testing, Hepatitis C testing, drug testing, pregnancy, and genetic testing. For example, a routine screening test performed during early pregnancy detects the levels of alphetoprotein (AFP) in the mother's blood. AFP is a protein secreted by the fetal liver as an excretory product into the mother's blood. Low levels of AFP indicate Down's syndrome, neural tube defects, and other problems associated with the fetus. This test method can accurately detect Down's syndrome in 60% of affected babies. The remaining 40% of babies with Down's syndrome go undetected in such tests. When a test incorrectly concludes the result as negative, the conclusion is called a **false negative**. The AFP test is somewhat controversial because it is found to produce positive results even when the baby is unaffected. If a test produces a positive result when the disease is not present, the result is called a **false positive**.

Assume that a diagnostic test for a particular human disease can detect the presence of the disease 90% of the time. The test can accurately determine if a person does not have the disease with 0.8 probability. If 5% of the population has this disease, what is the probability that a randomly selected person will test positive? What is the conditional probability that, if tested positive, the person is actually afflicted with the disease?

Let H represent a healthy person and let D represent a person afflicted with the disease.

If Y indicates a positive result, then

$$P(Y|D) = 0.9 \text{ and } P(Y|H) = 0.2.$$

If a person is randomly chosen from the population, the probability that the person has the disease is $P(D) = 0.05$ and the probability that the person does not have the disease is $P(H) = 0.95$. Also,

probability that the person has the disease and tests positive $= P(D \cap Y)$
$= P(D)P(Y|D) = 0.05 \times 0.9 = 0.045$;

probability that the person does not have the disease and tests positive (false positive)
$= P(H \cap Y) = P(H)P(Y|H) = 0.95 \times 0.2 = 0.19$;

probability that a random person will test positive $= P(D \cap Y) + P(H \cap Y)$
$= 0.045 + 0.19 = 0.235$;

conditional probability that person actually has the disease if tested positive

$$= \frac{P(D \cap Y)}{P(D \cap Y) + P(H \cap Y)} = \frac{0.045}{0.235} = 0.191.$$

So, 19% of the people that test positive actually have the disease. The test appears to be much less reliable than one would intuitively expect.

3.4 Discrete probability distributions

A widely used technique to display sample data graphically is to prepare a **histogram** (etymological roots – Greek: *histos* – web, or something spun together; *gram* – drawing). A histogram is a frequency distribution plot containing a series of rectangles or bars with widths equal to the class intervals and with heights equal to the frequency of observations for that interval. The area of each bar is proportional to the observed frequency for the class interval or range that it represents. A histogram provides a pictorial view of the distribution of frequencies with which a certain characteristic is observed in a population. Four histograms are plotted in Figure 3.1 (see Box 3.1). The observed variable in these histograms, which is plotted along the x-axis, is the moderate to vigorous physical activity of children per weekday.

The observed variable in Figure 3.1 is a **continuous variable**. The duration of physical activity measured in minutes can take on any value including fractional numbers. Continuous variables can be contrasted with **discrete variables** that take on discontinuous values. For example, the number of cars in a parking lot on any given day can only be whole numbers. The number of kittens in a litter cannot take on fractional or negative values. A measured or observed variable in any experiment for which we can produce a frequency distribution or probability distribution plot is called a random variable. A **random variable** is a measurement whose value depends on the outcome of an experiment. A probability distribution plot specifies the probability of occurrence of all possible values of the random variable.

Using MATLAB

The function `hist` in MATLAB generates a histogram from the supplied data – a vector $\boldsymbol{x}$ containing all the experimentally obtained values. MATLAB automatically determines the bin (class interval) size and number of bins (default $n = 10$) for the plot. The syntax for the `hist` function is

```
hist(x) or hist(x, n)
```

where n is the user-specified number of bins.

Another syntax is

```
[f, y] = hist(x, n)
```

where $\boldsymbol{f}$ is a vector containing the frequency for each bin, and $\boldsymbol{y}$ is a returned vector containing the bin centers. One may also manually specify bin locations by feeding a vector of values as an input argument.

The `bar` function in MATLAB draws a bar graph. The `bar` function plots the random variable vector $\boldsymbol{x}$ along the x-axis and draws bars at each x value with a height that corresponds to their frequency of observations, stored in variable $\boldsymbol{y}$. The syntax for the bar function is

```
bar(x, y)
```

where $\boldsymbol{x}$ and $\boldsymbol{y}$ are both vectors.

The frequency of observations of a population variable is an indicator of the probability of its occurrence within the population. One can easily compute these probabilities by determining the relative frequency of each observation. For

Table 3.4. *Tabulated probability distribution of the duration of moderate to vigorous physical activity per weekday for children of age 9 in the USA (Nader* et al., *2008)*

x: moderate to vigorous physical activity for age 9 children (minutes)	$P(x)$
35.0–59.3	0.0036
59.3–83.6	0.0155
83.6–107.9	0.0524
107.9–132.1	0.0834
132.1–156.4	0.1728
156.4–180.7	0.1764
180.7–205.0	0.2241
205.0–229.3	0.1073
229.3–253.6	0.0787
253.6–277.9	0.0536
277.9–302.1	0.0119
302.1–326.4	0.0131
326.4–350.7	0.0036
350.7–375.0	0.0036

Figure 3.3

Relative frequency distribution or probability distribution of the duration of moderate to vigorous physical activity per weekday for children of age 9 in the USA (Nader *et al.*, 2008).

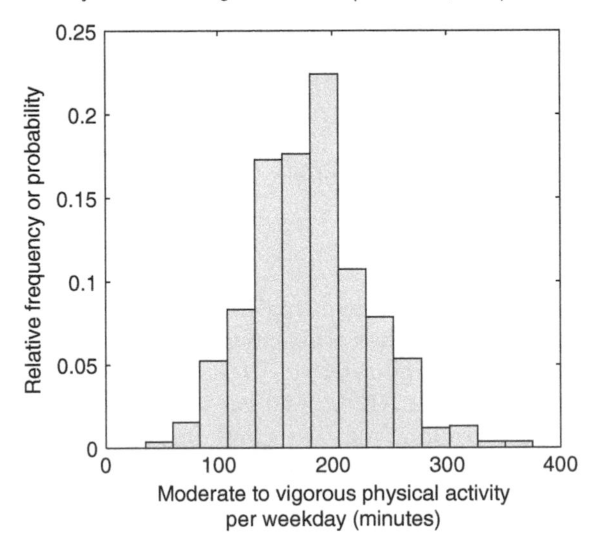

example, the histogram (age 9) in Figure 3.1(a) is replotted in Figure 3.3. The relative frequency distribution is obtained by dividing all frequency bars by $n = 839$. We now have a probability distribution of the duration of physical activity that children of age 9 in the USA engaged in during the time period 1991 to 2007. In order to generate this plot, `[f, y] = hist(x, n)` was first used to determine $\boldsymbol{f}$ and $\boldsymbol{y}$. The relative frequency distribution was plotted using the `bar` function after dividing $\boldsymbol{f}$

by 839 (i.e. `bar(y, f/839)`). The information provided in Figure 3.3 is also conveyed in Table 3.4. Note that the assumption of random sampling, perfect heterogeneity of the sample, and no bias in the sampling is implicit when a relative frequency diagram is used to approximate a probability distribution. To calculate the probability or relative frequency of occurrence of the population variable within one or more intervals, determine the fraction of the histogram's total area that lies between the subset of values that define the interval(s). This method is very useful when the population variable is continuous and the probability distribution is approximated by a smooth curve.

The shape of the histogram conveys the nature of the distribution of the observed variable within the sample. For example, in Figure 3.1 the histogram for age 9 children has a single peak, and the distribution falls off symmetrically about the peak in either direction. The three histograms for ages 11, 12, and 15 are progressively skewed to the right and do not show a symmetrical distribution about the peak. What does this tell us? If we are randomly to choose one child from the age 9 population, there is a greater probability that his/her moderate to vigorous activity level will be close to the peak value. Also, it is with equal probability that we would find a child with greater activity levels or with lower activity levels than the peak value. These expectations do not hold for the 11-, 12-, and 15-year-old population based on the trends shown in the histograms plotted in Figure 3.1. In a skewed distribution, the peak value is usually not equal to the mean value.

3.4.1 Binomial distribution

Some experiments have only two mutually exclusive outcomes, often defined as "success" or "failure," such as, live or dead, male or female, child or adult, HIV+ or HIV−. The outcome that is of interest to us is defined as a "successful event." For example, if we randomly choose a person from a population to ascertain if the person is exposed to secondhand smoke on a daily basis, then we can classify two outcomes: either the person selected from the population is routinely exposed to secondhand smoke, or the person is not routinely exposed to secondhand smoke. If the selection process yields an individual who is exposed to secondhand smoke daily, then the event is termed as a success. If three people are selected from a population, and all are routinely exposed to secondhand smoke, then this amounts to three "successes" obtained in a row.

An experiment that produces only two mutually exclusive and exhaustive outcomes is called a **Bernoulli trial**, named after Jacques (James) Bernoulli (1654–1705), a Swiss mathematician who made significant contributions to the development of the binomial distribution. For example, a coin toss can only result in heads or tails. For an unbiased coin and a fair flip, the chance of getting a head or a tail is 50 : 50 or equally likely. If obtaining a head is defined as the "successful event," then the probability of success on flipping a coin is 0.5. If several identical Bernoulli trials are performed in succession, and each Bernoulli trial is independent of the other, then the experiment is called a **Bernoulli process**.

A Bernoulli process is defined by the following three characteristics:

(1) each trial yields only two events, which are mutually exclusive;
(2) each trial is independent;
(3) the probability of success is the same in each trial.

Example 3.6

Out of a large population of children from which we make five selections, the proportion of boys to girls in the population is 50 : 50. We wish to determine the probability of choosing exactly three girls and two boys when five individual selections are made. Once a selection is made, any child that is selected is not returned back to the population. Since the population is large, removing several individuals from the population does not change the proportion of boys to girls.

The probability of selecting a girl is 0.5. The probability of selecting a boy is also 0.5. Each selection event is independent from the previous selection event. The probability of selecting exactly three girls and two boys with the arrangement GGGBB is $0.5^3 \times 0.5^2 = 0.5^5 = 1/2^5$. The probability of selecting exactly three girls and two boys with the arrangement GBBGG is also $0.5^3 \times 0.5^2 = 0.5^5 = 1/2^5$. In fact, the total number of arrangements that can be made with three girls and two boys is

$$C_3^5 = \frac{5!}{3!(5-3)!} = 10.$$

An easy way to visualize the selection process is to consider five empty boxes: □□□□□. We want to mark exactly three of them with "G." Once three boxes each have been marked with a "G," the remaining boxes will be marked with a "B." The total number of combinations that exists for choosing any three of the five boxes and marking with "G" is equal to C_3^5. Thus, we are concerned with the arrangement of the "G"s with respect to the five boxes. The total number of arrangements in which we can mark three out of five boxes with "G" is equal to the total number of combinations available to choose three girls and two boys when five selections are made. The order in which the three girls are selected is irrelevant.

The probability of selecting exactly three girls and two boys is given by

$$\text{number of combinations possible} \times P(\text{girl}) \times P(\text{girl}) \times P(\text{girl}) \times P(\text{boy}) \times P(\text{boy}) = \frac{C_3^5}{2^5}.$$

Thus, the probability of choosing exactly three girls in five children selected from a population where the probability of choosing a girl or a boy is equally likely is

$$\frac{C_3^5}{2^5} = \frac{10}{32}.$$

What if our population of interest has a girl to boy ratio of 70 : 30? Since the proportion of girls is now greater to that of boys in the population, it is more likely for the selection group of five children to contain more girls than boys. Therefore selecting five boys in succession is less probable than selecting five girls in succession. Since every selection event is a Bernoulli trial and is independent of every other selection event, the probability of selecting exactly three girls in five selection attempts from the population is equal to $C_3^5 \times (0.7)^3(0.3)^2$.

What is the probability of selecting four girls and one boy, or two girls and three boys, or one girl and four boys? If the selection of a girl is termed as a success, then the probability of obtaining 0, 1, 2, 3, 4, or 5 successes in a Bernoulli process that consists of five Bernoulli trials is given by the *binomial distribution*.

> If a Bernoulli trial is repeated n times, and each Bernoulli trial is independent of the other, then the probability of achieving k out of n successes, where $0 \le k \le n$, is given by the **binomial distribution**. If p is the probability of success in a single trial, then the probability (or relative frequency) of k successes in n independent trials is given by
>
> $$P(k \text{ successes}) = C_k^n p^k (1-p)^{n-k}. \tag{3.12}$$

A binomial distribution is an idealized probability distribution characterized by only two parameters, n – the number of independent trials and p – the probability of success in a single Bernoulli trial. Many experiments in manufacturing, biology, and public health studies have outcomes that closely follow the binomial distribution.

Figure 3.4

Binomial distribution in which the event of selecting a girl from a given population of children has probability $p = 0.7$ and is denoted as a "success." Five independent trials are performed in succession.

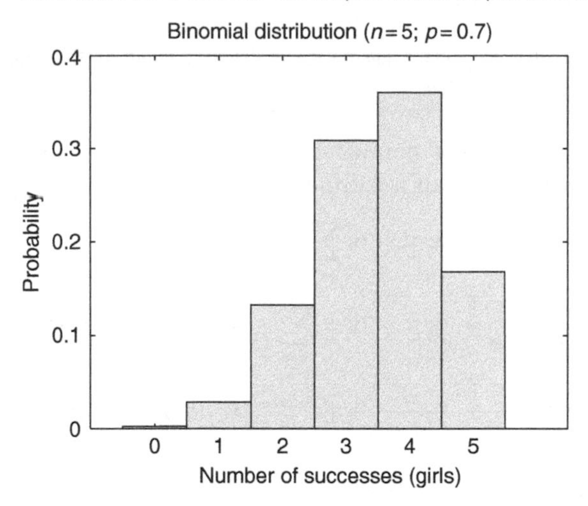

Using the formula given by Equation (3.12), we can generate a binomial distribution for Example 3.6. If $n = 5$ and $p = 0.7$, Figure 3.4 is a plot of the binomial distribution that characterizes the probabilities of all possible outcomes of the experiment.

There are six possible outcomes of a Bernoulli experiment consisting of five independent trials: 0, 1, 2, 3, 4, and 5 successes.

If p is the probability of success in a Bernoulli process, and q is the probability of failure, then the sum of the probabilities of all outcomes in a binomial distribution is given by

$$\sum_{k=0}^{n} C_k^n p^k q^{n-k} = q^n + npq^{n-1} + \frac{n(n-1)}{2} p^2 q^{n-2} + \cdots + np^{n-1}q + p^n$$
$$= (p+q)^n.$$

In other words, the sum of the binomial distribution probabilities is equal to the binomial expansion of $(p+q)^n$. Since $q = 1 - p$,

$$\sum_{k=0}^{n} C_k^n p^k q^{n-k} = 1^n = 1. \tag{3.13}$$

A binomial distribution is strictly valid when sampling from a population is done with replacement since this ensures that the probability p of success or failure does not change with each trial when the population is finite. When sampling is performed without replacement, which is usually the case, the binomial distribution is still applicable if the population size N is large compared to the sample size n (or, $n \ll N$) such that p changes minimally throughout the sampling process.

The "measured variable" or random variable in a binomial experiment is the number of successes x achieved, and is called a **binomial variable**. Let's calculate the mean value and the variance of the binomial probability distribution of the number of successes x. The mean value is the number of successes, on average, that one would expect if the Bernoulli experiment were repeated many, many times. The mean and variance serve as numerical descriptive parameters of a probability distribution.

For any discrete probability distribution $P(x)$, the **expected value** or **expectation** of the measured variable x is equal to the mean μ of the distribution and is defined as

$$E(x) = \mu = \sum_x xP(x). \tag{3.14}$$

The mean is a weighted average of the different values that the random variable x can assume, weighted by their probabilities of occurrence.

The expected value for the number of successes x in any binomial experiment consisting of n Bernoulli trials is calculated as follows:

$$\begin{aligned} E(x) &= \sum_x xP(x \text{ successes}) = \sum_{x=0}^{n} xC_x^n p^x q^{n-x} \\ &= \sum_{x=0}^{n} x \frac{n!}{x!(n-x)!} p^x q^{n-x} = \sum_{x=1}^{n} \frac{n!}{(x-1)!(n-x)!} p^x q^{n-x} \\ &= np \sum_{x=1}^{n} \frac{(n-1)!}{(x-1)!(n-x)!} p^{x-1} q^{n-x} = np \sum_{x=1}^{n} C_{x-1}^{n-1} p^{x-1} q^{n-x}, \end{aligned}$$

where $\sum_{x=1}^{x=n} C_{x-1}^{n-1} p^{x-1} q^{n-x}$ is the binomial probability distribution for $n - 1$ independent trials (substitute $y = x - 1$ to verify this).

From Equation (3.13),

$$E(x) = \mu = np \cdot 1 = np. \tag{3.15}$$

For a binomial distribution, the average number of successes is equal to np.

Variance was defined in Section 3.2 as the average of the squared deviations of the data points from the mean value μ of the data set. The variance of the random variable x that follows a discrete probability distribution $P(x)$ is equal to the expectation of $(x - \mu)^2$.

For any discrete probability distribution $P(x)$, the **variance** of x is defined as

$$\sigma^2 = E((x-\mu)^2) = \sum_x (x-\mu)^2 P(x). \tag{3.16}$$

Expanding Equation (3.16), we obtain

$$\begin{aligned} \sigma^2 &= E\left((x-\mu)^2\right) = E\left(x^2 - 2\mu x + \mu^2\right) \\ &= E\left(x^2\right) - 2\mu E(x) + \mu^2 \\ &= E\left(x^2\right) - 2\mu^2 + \mu^2 \end{aligned}$$

$$\sigma^2 = E\left(x^2\right) - \mu^2 = \sum_x x^2 P(x) - \mu^2. \tag{3.17}$$

Substituting Equation (3.12) into Equation (3.17), we obtain

$$\sigma^2 = \sum_{x=0}^{n} x^2 \frac{n!}{x!(n-x)!} p^x q^{n-x} - \mu^2.$$

Note that x^2 cannot cancel with the terms in $x!$, but $x(x-1)$ can. Therefore, we add and subtract $E(x)$ as shown below:

$$\sigma^2 = \sum_{x=0}^{n}(x^2 - x)\frac{n!}{x!(n-x)!}p^x q^{n-x} + \sum_{x=0}^{n} x\frac{n!}{x!(n-x)!}p^x q^{n-x} - \mu^2.$$

Substituting results from Equation (3.15), we obtain

$$\sigma^2 = \sum_{x=2}^{x=n} x(x-1)\frac{n!}{x!(n-x)!}p^x q^{n-x} + np - n^2p^2;$$

$$\sigma^2 = n(n-1)p^2\sum_{x=2}^{x=n}\frac{(n-2)!}{(x-2)!(n-x)!}p^{x-2}q^{n-x} + np - n^2p^2.$$

The summation term is the binomial distribution for $n - 2$ independent Bernoulli trials, and sums to 1.

The above equation simplifies as follows:

$$\begin{aligned}\sigma^2 &= n^2p^2 - np^2 + np - n^2p^2,\\ \sigma^2 &= np(1-p).\end{aligned} \tag{3.18}$$

The mean and the variance of the probability distribution shown in Figure 3.4 is calculated using Equations (3.15) and (3.18), and are 3.5 and 1.05, respectively. The standard deviation is simply the square root of the variance and is 1.025.

3.4.2 Poisson distribution

The Poisson distribution is a limiting form of the binomial distribution and has enjoyed much success in describing a wide range of phenomena in the biological sciences. Poisson distributions are used to describe the probability of rare events occurring in a finite amount of time or space (or distance). For example, water-pollution monitoring involves sampling the number of bacteria per unit volume of the water body. Increased run-off of phosphate-rich fertilizers is believed to be a cause of "cyanobacteria blooms" – a rapid growth of this very common bacteria phylum. Cyanobacteria produce cyanotoxins, which if ingested can cause gastro-intestinal problems, liver disorders, paralysis, and other neurological problems. The concentration of bacteria in a water sample can be modeled by the Poisson distribution. Given the average number of bacteria per milliliter of water obtained from the water body or aqueous effluent, the Poisson formula can determine the probability that the concentration of bacteria will not exceed a specified amount.

Another example of the application of the Poisson distribution in biology is in predicting the number of adhesive receptor–ligand bonds formed between two cells. Cells display a wide variety of receptors on their surface. These receptors specifically bind to either ligands (proteins/polysaccharides) in the extracellular space or counter-receptors located on the surface of other cells (see Figure 3.5). When two cells are brought into close contact by any physical mechanism (e.g. blood cells contacting the vessel wall by virtue of the local hemodynamic flow field), the physical proximity of adhesion receptors and counter-receptors promotes attachment of the two cells via bond formation between the complementary receptor molecules. Leukocytes have counter-receptors or ligands on their surface that allow them to bind to activated endothelial cells of the vessel wall in a region of inflammation. The activated endothelial cells express selectins that bind to their respective ligand presented by the leukocyte surface. Selectin–ligand bonds are short-lived, yet their role in the inflammation pathway is critical since they serve to capture and slow

Figure 3.5

Leukocyte transiently attached to the vascular endothelium via selectin bonds.

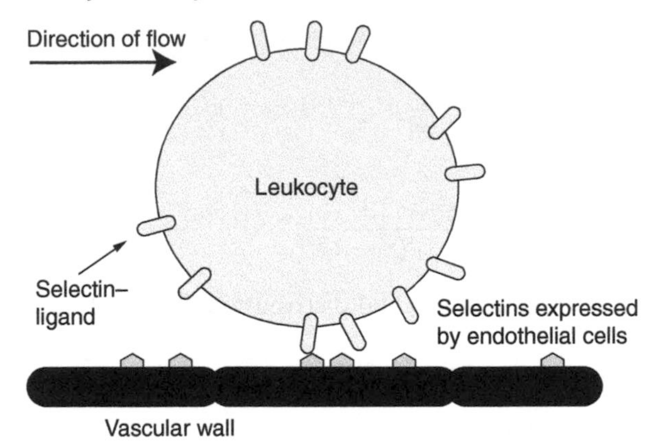

down the leukocytes so that the downstream events of firm adhesion, transmigration through the vessel, and entry into the inflamed interstitial tissue can occur. The attachment of a leukocyte to the vessel wall is termed an adhesion event. The fraction of cell–cell adhesion events that result from single-bond formation, double-bond formation, or more can be described by the Poisson distribution.

We are interested in events that take place within a certain time interval or specified volume. The time period or volume can be partitioned into many smaller subintervals so narrow that only one event can occur (e.g. death of a patient, failure of medical equipment, presence of a single microscopic organism) in that subinterval. If p is the probability that the event (or "success") occurs in the subinterval, and there are n subintervals in the specified time period or volume, then, as long as p does not vary from subinterval to subinterval, i.e. the events in each subinterval are independent, the experimental outcome will be governed by the binomial probability distribution.

If $n \to \infty$, and $\gamma = np$ (mean of the binomial distribution), then the binomial probability for x successes out of n independent trials

$$P(x) = \frac{n!}{x!(n-x)!} p^x (1-p)^{n-x}$$

becomes

$$P(x) = \lim_{n\to\infty} \frac{n!}{x!(n-x)!} \left(\frac{\gamma}{n}\right)^x \left(1 - \frac{\gamma}{n}\right)^{n-x}$$

$$= \frac{\gamma^x}{x!} \lim_{n\to\infty} \frac{(n-1)(n-2)\cdots(n-x+1)}{n^{x-1}} \left(1-\frac{\gamma}{n}\right)^n \left(1-\frac{\gamma}{n}\right)^{-x}$$

$$= \frac{\gamma^x}{x!} \lim_{n\to\infty} \left(1-\frac{1}{n}\right)\left(1-\frac{2}{n}\right)\cdots\left(1-\frac{x-1}{n}\right) \lim_{n\to\infty}\left(1-\frac{\gamma}{n}\right)^n \lim_{n\to\infty}\left(1-\frac{\gamma}{n}\right)^{-x}.$$

The $\lim_{n\to\infty}(1-\gamma/n)^n$ when expanded using the binomial theorem is the Taylor series expansion (see Section 1.6 for an explanation on the Taylor series) of $e^{-\gamma}$. Then

$$P(x) = \frac{\gamma^x}{x!} \cdot 1 \cdot e^{-\gamma} \cdot 1$$

$$= \frac{\gamma^x e^{-\gamma}}{x!}. \tag{3.19}$$

The summation of the discrete Poisson probabilities obtained from a Poisson probability distribution for all $x = 0, 1, 2, \ldots, \infty$ must equal 1. This is easily shown to be true:

$$\sum_{x=0}^{\infty} \frac{\gamma^x e^{-\gamma}}{x!} = e^{-\gamma} \sum_{x=0}^{\infty} \frac{\gamma^x}{x!} = e^{-\gamma}\left(1 + \frac{\gamma}{1!} + \frac{\gamma^2}{2!} + \frac{\gamma^3}{3!} + \cdots\right) = e^{-\gamma} e^{\gamma} = 1.$$

The mean and variance of the Poisson probability distribution are given by

$$\mu = \gamma; \qquad \sigma^2 = \gamma. \tag{3.20}$$

The proof of this derivation is left as an exercise (see Problem 3.6).

Box 3.3 Poisson distribution of the number of bonds formed during transient leukocyte–vessel wall attachments

Bioadhesive phenomena involve the attachment of cells to other cells, surfaces, or extracellular matrix and are important in angiogenesis, blood clotting, metastasis of cancer cells, and inflammation. Several experimental techniques have been devised to characterize and quantify the adhesive properties of biological macromolecules. Of particular interest to bioadhesion researchers are the effects of varying intensities of pulling force on bond strength. Force application on receptor–ligand bonds mimics the hydrodynamic forces that act upon blood cells or cancer cells that stick to surfaces or other cells under flow.

One experimental set-up used to study leukocyte ligand binding to selectins uses a glass microcantilever coated with E-selectin molecules (Tees *et al.*, 2001). A microsphere functionalized with the selectin ligand is pressed against the cantilever to ensure sufficient contact between the biomolecules (see Figure 3.6). The microsphere is then retracted away from the cantilever, which undergoes visually measurable deflection.

Figure 3.6

An adhesive microcantilever reversibly attaches to a retracting microsphere.

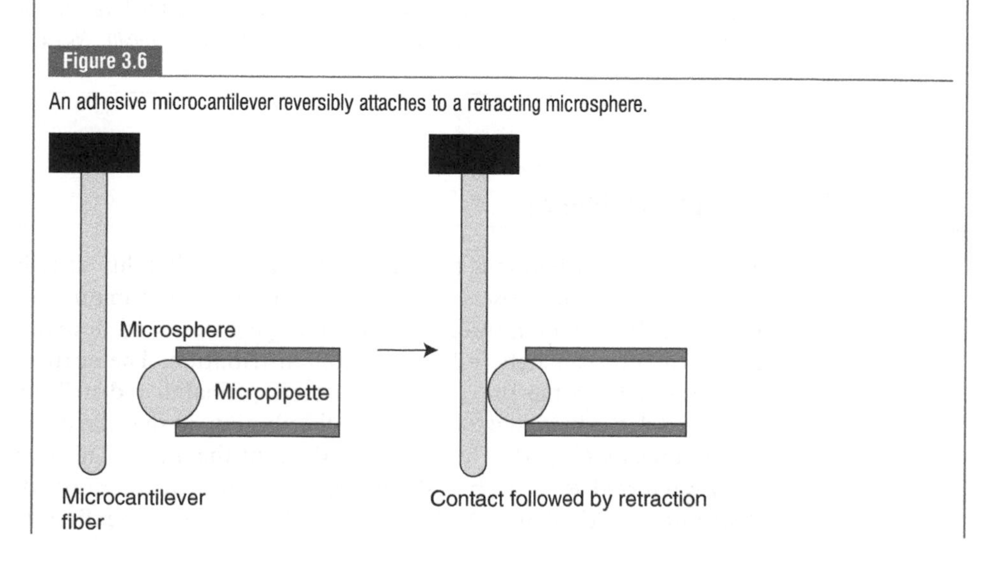

The bead contacts the cantilever surface for a fixed duration; this time interval can be subdivided into many smaller intervals over which the probability of a single bond forming is some small number p and the probability of multiple bonds forming over the very small time interval is practically 0. It was determined that 30% of the contact events between the microsphere and the microcantilever resulted in adhesion events. An adhesion event is defined to occur when the microsphere attaches to the microcantilever and causes a measurable deflection when retraction begins. Mathematically, we write

$$P(N_b > 0) = 0.3 \quad \text{or} \quad P(N_b = 0) = 0.7.$$

The distribution of the number of bonds N_b that form between the two opposing surfaces follows the Poisson formula:

$$P(N_b) = \frac{\gamma^{N_b} e^{-\gamma}}{N_b!},$$

where γ is the mean number of bonds, $\langle N_b \rangle$, that form during contact. Note that $\langle N_b \rangle$ is calculated by using the known result for $P(N_b = 0)$:

$$P(N_b = 0) = 0.7 = e^{-\langle N_b \rangle},$$

$$\langle N_b \rangle = -\ln 0.7 = 0.357.$$

Now we are in a position to calculate $P(N_b > 0)$:

$$P(N_b = 1) = \frac{0.357^1 e^{-0.357}}{1!} = 0.250,$$

$$P(N_b = 2) = \frac{0.357^2 e^{-0.357}}{2!} = 0.045,$$

$$P(N_b = 3) = \frac{0.357^3 e^{-0.357}}{3!} = 0.005.$$

The fraction of adhesion events that involve a single bond, double bond, or triple bond $P(N_b | N_b > 0)$ are 0.833, 0.15, and 0.017, respectively. How did we calculate this? Note that while the fractional frequency of formation of one, two, and three bonds sum to 1 ($0.833 + 0.15 + 0.017 = 1$), this is an artifact of rounding, and the probability of formation of greater number of bonds (>3) is small but not quite zero.

The mean number of bonds $\langle N_b \rangle$ calculated here includes the outcome of contact events that result in no binding, $N_b = 0$. What is the mean number of bonds that form if we are only to include contact events that lead to adhesion? Hint: Start out with the definition of *mean* $\langle N_b \rangle$ (see Equation (3.14)). Answer: 1.19.

3.5 Normal distribution

The normal distribution is the most widely used probability distribution in the field of statistics. Many observations in nature such as human height, adult blood pressure, Brownian motion of colloidal particles, and human intelligence are found to approximately follow a normal distribution. The normal distribution sets itself apart from the previous two discrete probability distributions discussed in Section 3.4, the binomial and Poisson distribution, in that the normal distribution is continuous *and* is perfectly symmetrical about the mean. This distribution is commonly referred to as a "bell-shaped curve" and is also described as the Gaussian distribution. Abraham De Moivre (1667–1754) was the first mathematician to propose the normal distribution, which is discussed in his seminal book on probability, *The Doctrine of Chances*. This distribution is, however, named after Carl Friedrich Gauss (1777–1855), who put forth its mathematical definition, and used the assumption of a normal distribution of errors to rigorously prove the linear least-squares method for estimation of model parameters. Before we discuss the normal distribution any further, we must first introduce the concept of the *probability density function*.

3.5.1 Continuous probability distributions

As discussed earlier, histograms pictorially convey the frequency distribution of a population variable or a random variable. When the random variable is a continuous variable, the class intervals can be constructed of any width. As the number of observations and the number of class intervals within the entire interval of study are simultaneously increased, the resulting histogram begins to take on the shape of a well-defined curve. (What happens to the shape of a histogram if we have only a few observations and we repeatedly decrease the size of the class intervals?) Figure 3.7 shows the relative frequency distribution or probability distribution after halving the size of the class intervals in Figure 3.3. Note that this has effectively reduced the individual class interval probabilities by a factor of 2. A smooth curve can theoretically be obtained if the class intervals become infinitely small and the number of class intervals rises to infinity. The total number of observations plotted thereby proceeds to infinity. Note that if the measured variable is of discrete nature, you should not attempt to draw a curve connecting the midpoints of the top edges of the bars, since continuous interpolation for a discrete variable is not defined.

When the relative frequencies displayed on the y-axis are converted to a density scale such that the entire area under the curve is 1, then the smooth curve is termed as a **probability density curve** or simply a **density curve** (see Figure 3.8). The area under the curve between any two values of the measured variable (random variable) x, say x_1 and x_2, is equal to the probability of obtaining a value of x within the interval $x_1 \leq x \leq x_2$.

Figure 3.7

Halving the size of the class intervals adds definition to the shape of the relative frequency histogram (area = 12.1 minutes).

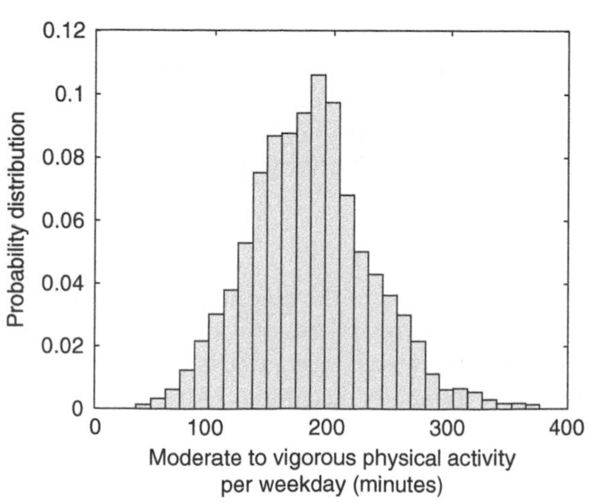

Figure 3.8

Density curve.

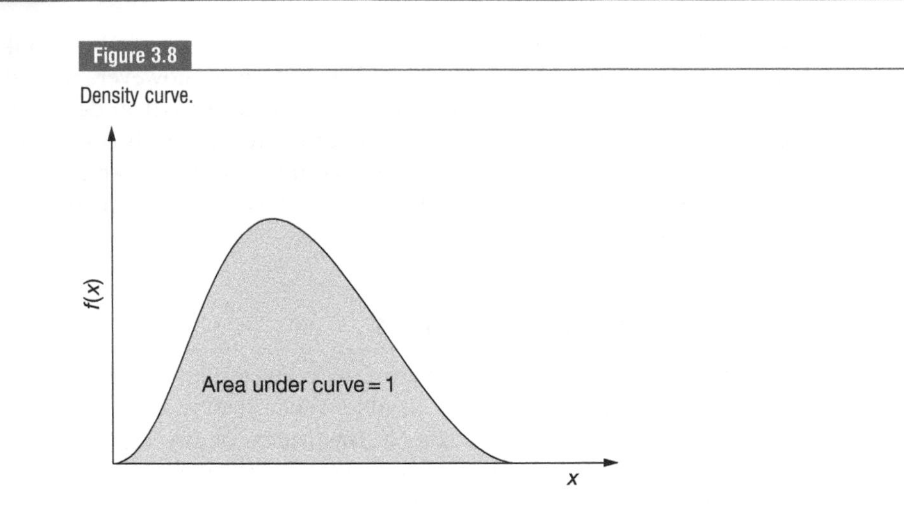

The mathematical formula that describes the shape of the density curve is called the **density function** or simply the **density**, $f(x)$. Thus, the probability density function satisfies the following requirement: the area under the density curve is given by

$$\int_{-\infty}^{\infty} f(x)dx = 1. \tag{3.21}$$

The probability $P(x)$ that the measurement x will lie in the interval $[x_1, x_2]$ is given by

$$P(x) = \int_{x_1}^{x_2} f(x)dx. \tag{3.22}$$

Equation (3.22) implies that the probability at any point x along the curve is 0. This may seem puzzling, since the density curve has non-zero density for a range of x values. Note that a line erected at any value x along the curve is considered to be infinitesimally thin and therefore possesses no area. Therefore, even if the density $f(x)$ is non-zero at x, the probability or area under the curve at x is exactly 0, and one must specify a range of x values to obtain a probability value other than 0.

Using MATLAB

Statistics Toolbox offers the function `pdf` that uses a specific probability density function (pdf) to generate a probability distribution. For example, the MATLAB function `binopdf` computes the discrete binomial probability distribution (Equation (3.12)), i.e. the probability of obtaining k successes in n trials of a binomial process characterized by a probability of success p in each trial. The syntax for `binopdf` is

```
y = binopdf(x, n, p)
```

where n and p are the parameters of the binomial distribution and $\boldsymbol{x}$ is the vector of values or outcomes whose probabilities are desired. Note that $\boldsymbol{x}$ must contain positive integral values, otherwise a zero value is returned.

To plot the binomial probability distribution for the range of values specified in $\boldsymbol{x}$, use the bar function,

```
bar(x, y)
```

where x is the vector of outcomes and y contains the respective probabilities (obtained from the `pdf` function).

The `poisspdf` function generates the Poisson probability distribution using the probability mass function specified by Equation (3.19). Note that a function that generates probabilities for the entire range of outcomes of a discrete random variable is called a **probability mass function**. A density function is defined only for continuous random variables. Unlike a mass function, a density function must be integrated over a range of values of the random variable to obtain the corresponding probability.

The `normpdf` function calculates the normal density (see Equation (3.23)). The syntax for `normpdf` is

```
y = normpdf(x, mu, sigma)
```

where `mu` and `sigma` are parameters of the normal probability distribution and x is the vector of values at which the normal density is evaluated.

3.5.2 Normal probability density

The **normal probability density** is a continuous unimodal function symmetric about μ, the mean value of the distribution, and is mathematically defined as

$$f(x) = \frac{1}{\sqrt{2\pi}\sigma} e^{-(x-\mu)^2/2\sigma^2}. \tag{3.23}$$

The mean μ and variance σ^2 are the only numerical descriptive parameters of this probability distribution. A normal curve is denoted by the symbol $N(\mu, \sigma)$. The location of the normal distribution along the x-axis is determined by the parameter μ, the mean of the distribution. The shape of the distribution – height and width – is fixed by σ^2, the variance of the distribution. The normal distribution is plotted in Figure 3.9 for three different sets of parameter values. Note how the shape of the curve flattens with increasing values of σ. The normal density curve extends from $-\infty$ to ∞.

Figure 3.9

Three normal density curves, which include the standard normal probability distribution ($\mu = 0$, $\sigma = 1$).

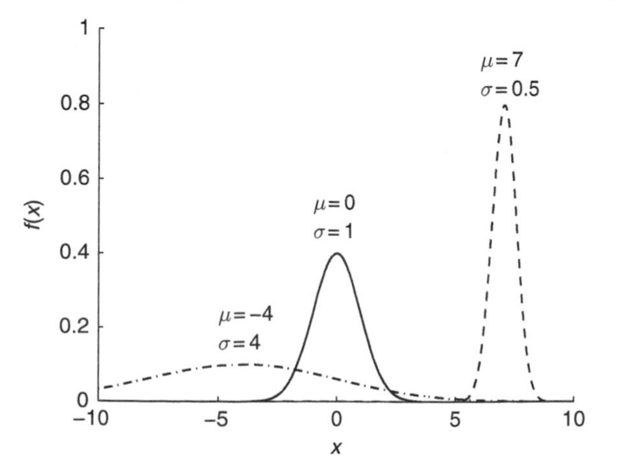

We now show that the mean and variance of the normal distribution are indeed the parameters μ and σ in Equation (3.23). We first need to define the expectation of a continuous random variable x. An **expectation** or an **expected value**, $E(x)$, is the average value obtained from performing many measurements or experiments.

The **expected value** or **expectation** of a random variable x that has a continuous probability distribution $f(x)$ is equal to the mean μ of the distribution and is defined as

$$E(x) = \mu = \int_{-\infty}^{\infty} x f(x) dx. \tag{3.24}$$

The expected value of a random variable x that follows a normal distribution $N(\mu, \sigma)$ is calculated by substituting Equation (3.23) into Equation (3.24):

$$\begin{aligned} E(x) &= \int_{-\infty}^{\infty} x \frac{1}{\sqrt{2\pi}\sigma} e^{-(x-\mu)^2/2\sigma^2} dx \\ &= \int_{-\infty}^{\infty} \frac{x - \mu + \mu}{\sqrt{2\pi}\sigma} e^{-(x-\mu)^2/2\sigma^2} dx \\ &= \int_{-\infty}^{\infty} \frac{(x-\mu)}{\sqrt{2\pi}\sigma} e^{-(x-\mu)^2/2\sigma^2} dx + \mu \int_{-\infty}^{\infty} \frac{1}{\sqrt{2\pi}\sigma} e^{-(x-\mu)^2/2\sigma^2} dx. \end{aligned}$$

Substituting $z = x - \mu$ and $\int_{-\infty}^{\infty} f(x) dx = 1$ (Equation (3.21)), we get

$$E(x) = \int_{-\infty}^{\infty} \frac{z}{\sqrt{2\pi}\sigma} e^{-z^2/2\sigma^2} dz + \mu \cdot 1.$$

The first integral can be solved using geometric arguments. It consists of an odd function $(z/\sqrt{2\pi}\sigma)$ multiplied by an even function $(e^{-z^2/2\sigma^2})$, which yields an odd function. The integral of an odd function over an even interval vanishes. Therefore, for a normally distributed variable x,

$$E(x) = \mu.$$

Thus, μ in Equation (3.23) represents the mean of the normal distribution.

For any continuous probability distribution $f(x)$, the **variance** is the expected value of $(x - \mu)^2$ and is defined as

$$\sigma^2 = E((x-\mu)^2) = \int_{-\infty}^{\infty} (x-\mu)^2 f(x) dx. \tag{3.25}$$

The variance of a normal distribution $N(\mu, \sigma)$ is calculated by substituting Equation (3.23) into Equation (3.25):

$$\sigma^2 = E((x-\mu)^2) = \int_{-\infty}^{\infty} (x-\mu)^2 \frac{1}{\sqrt{2\pi}\sigma} e^{-(x-\mu)^2/2\sigma^2} dx.$$

Let $z = (x - \mu)$. Then

$$\sigma^2 = \frac{1}{\sqrt{2\pi}\sigma} \int_{-\infty}^{\infty} z^2 \cdot e^{-z^2/2\sigma^2} dz$$

The function $z^2 e^{-z^2/2\sigma^2}$ is an even function. Therefore the interval of integration can be halved:

$$\frac{2}{\sqrt{2\pi}\sigma}\int_0^{\infty} z^2 \cdot e^{-z^2/2\sigma^2}\, dz.$$

Let $y = z^2/2$. Then $dy = z\, dz$. The integral becomes

$$\frac{2}{\sqrt{2\pi}\sigma}\int_0^{\infty} \sqrt{2y} \cdot e^{-y/\sigma^2}\, dy.$$

Integrating by parts $\left(\int uv\, dy = u\int v\, dy - \int\left[u'\int v\, dy\right]dy\right)$, we obtain

$$E((x-\mu)^2) = \frac{2}{\sqrt{2\pi}\sigma}\left[\sqrt{2y}\cdot e^{-y/\sigma^2}\cdot\left(-\sigma^2\right)\Big|_0^{\infty} - \int_0^{\infty}\frac{1}{\sqrt{2y}}\cdot\left(-\sigma^2\right)\cdot e^{-y/\sigma^2}\, dy\right].$$

The first term on the right-hand side is zero. In the second term we substitute z back into the equation to get

$$E\left((x-\mu)^2\right) = 0 + 2\sigma^2\int_0^{\infty}\frac{1}{\sqrt{2\pi}\sigma}e^{-z^2/2\sigma^2}\, dz = 2\sigma^2\cdot\frac{1}{2} = \sigma^2.$$

Thus, the variance of a normal distribution is σ^2, and the standard deviation is σ.

3.5.3 Expectations of sample-derived statistics

Now that you are familiar with some commonly used statistical measures such as mean and variance, and the concept of *expected value*, we proceed to derive the expected values of sample-derived statistics. Because the sample is a subset of a population, the statistical values derived from the sample can only approximate the true population parameters. Here, we derive the relation between the sample statistical predictions and the *true* population characteristics. The relationships for the expected values of statistics obtained from a sample *do not* assume that the random variable obeys any particular distribution within the population.

By definition, the expectation $E(x)$ of a measured variable x obtained from a single experiment is equal to the average of measurements obtained from many, many experiments, i.e. $E(x_i) = \mu$. If we obtain n observations from n independently performed experiments or n different individuals, then our sample size is n. We can easily determine the mean $\bar{x}$ of the n measurements, which we use as an estimate for the population mean. If we obtained many samples each of size n, we could measure the mean of each of the many samples. Each mean of a particular sample would be slightly different from the means of other samples. If we were to plot the distribution of the many sample means, we would obtain a frequency plot that would have a mean value $\mu_{\bar{x}}$ and a variance $s^2_{\bar{x}}$. The expectation of the sample mean $E(\bar{x})$ is the average of many, many such sample means, i.e. $\mu_{\bar{x}}$; so

$$E(\bar{x}) = E\left(\frac{1}{n}\sum_{x=1}^{n} x_i\right) = \frac{1}{n}\sum_{x=1}^{n} E(x_i) = \frac{1}{n}\sum_{x=1}^{n}\mu = \mu. \qquad (3.26)$$

Note that because $E(\cdot)$ is a linear operator, we can interchange the positions of $\sum(\cdot)$ and $E(\cdot)$. Note also that $E(\bar{x})$ or $\mu_{\bar{x}}$ (the mean of all sample means) is equal to the population mean μ.

> In statistical terminology, $\bar{x}$ is an *unbiased estimate* for μ. A sample-derived statistic r is said to be an **unbiased estimate** of the population parameter ρ if $E(r) = \rho$.

What is the variance of the distribution of the sample means about μ? If x is normally distributed, the sum of i normal variables, $\sum_i x_i$, and the mean of i normal variables,

Figure 3.10

Normal distribution $N(\mu, \sigma = 2)$ of variable x and normal distribution $N(\mu_{\bar{x}} = \mu, s_{\bar{x}} = \sigma/\sqrt{10})$ of the means of samples of size $n = 10$.

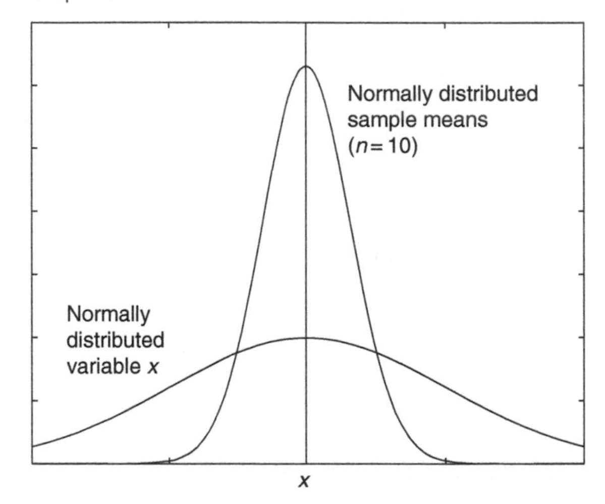

$\bar{x} = \sum_i x_i/n$, also follow a normal distribution. Therefore, the distribution of the means of samples of size n will also be normally distributed. Figure 3.10 shows the normal distribution of x as well as the distribution of the means of samples of size n. The standard deviation of the distribution of sample means is always smaller than the standard deviation of the distribution of the random variable, unless the sample size n is 1.

The variance of the distribution of sample means is defined as

$$s_{\bar{x}}^2 = E\left((\bar{x} - \mu)^2\right).$$

Now,

$$E\left((\bar{x} - \mu)^2\right) = E\left(\left(\frac{1}{n}\sum_{i=1}^{n} x_i - \mu\right)^2\right) = \frac{1}{n^2}E\left(\left(\sum_{i=1}^{n} x_i - n\mu\right)^2\right)$$
$$= \frac{1}{n^2}E\left(\left(\sum_{i=1}^{n}(x_i - \mu)\right)^2\right).$$

The squared term can be expanded using the following rule:

$$(a + b + c)^2 = a^2 + b^2 + c^2 + 2ab + 2ac + 2bc.$$

So

$$E\left((\bar{x} - \mu)^2\right) = \frac{1}{n^2}E\left(\sum_{i=1}^{n}(x_i - \mu)^2 + \sum_{i=1}^{n}\sum_{j=1, j\neq i}^{n}(x_i - \mu)(x_j - \mu)\right).$$

The double summation in the second term will count each $(x_i - \mu)(x_j - \mu)$ term twice. The order in which the expectations and the summations are carried out is now switched:

$$E\left((\bar{x} - \mu)^2\right) = \frac{1}{n^2}\sum_{i=1}^{n}E\left((x_i - \mu)^2\right) + \frac{1}{n^2}\sum_{i=1}^{n}\sum_{j=1, j\neq i}^{n}E((x_i - \mu)(x_j - \mu)).$$

Let's investigate the double summation term:

$$E\left((x_i - \mu)(x_j - \mu)\right) = \int_{-\infty}^{\infty}\int_{-\infty}^{\infty}(x_i - \mu)(x_j - \mu)f(x_i)f(x_j)dx_i\, dx_j.$$

If x_i and x_j are independent measurements (i.e. the chance of observing x_i is not influenced by the observation of x_j), then

$$E\left((x_i - \mu)(x_j - \mu)\right) = \int_{-\infty}^{\infty}(x_i - \mu)f(x_i)dx_i\int_{-\infty}^{\infty}(x_j - \mu)f(x_j)dx_j = 0 \text{ for } i \neq j.$$

Therefore,

$$E\left((\bar{x} - \mu)^2\right) = \frac{1}{n^2}\left(\sum_{i=1}^{n} E\left((x_i - \mu)^2\right)\right) + 0.$$

By definition, $E\left((x_i - \mu)^2\right) = \sigma^2$, and

$$s_{\bar{x}}^2 = E\left((\bar{x} - \mu)^2\right) = \frac{\sigma^2}{n}. \tag{3.27}$$

Equation (3.27) states that the variance of the distribution of sample means is equal to the variance of the population divided by the sample size n. Since our goal is always to minimize the variance of the distribution of the sample means and therefore obtain a precise estimate of the population mean, we should use as large a sample size as practical.

> The standard deviation of the distribution of sample means is
>
> $$s_{\bar{x}} = \frac{\sigma}{\sqrt{n}} \tag{3.28}$$
>
> and is called the **standard error of the mean** or **SEM**.

The SEM is a very important statistic since it quantifies how precise our estimate of $\bar{x}$ is for μ, or, in other words, the confidence we have in our estimate for μ. Many researchers report the error in their experimental results in terms of the SEM. We will repeatedly use SEM to estimate confidence intervals and perform tests of significance in Chapter 4. It is important to note that Equation (3.28) is valid only if the observations are truly independent of each other. Equation (3.28) holds if the population from which the sampling is done is much larger than the sample, so that during the random (unbiased) sampling process, the population from which each observation or individual is chosen has the same properties.

Let's briefly review what we have covered in this section so far. The mean of a sample depends on the observations that comprise the sample and hence will differ for different samples of the same size. The sample mean $\bar{x}$ is the best (unbiased) estimate of the population mean μ. Also, μ is the mean of the distribution of the sample means. The variance of the distribution of sample means $s_{\bar{x}}^2$ is σ^2/n. The standard deviation of the distribution of sample means, or the *standard error of the mean*, is $\sigma/\sqrt{n}$. If the sample means are normally distributed, then the distribution of sample means is defined as $N(\mu, \sigma/\sqrt{n})$.

We are also interested in determining the expected variance $E(s^2)$ of the sample (not to be confused with the expected variance of the sample mean, $s_{\bar{x}}^2$, which we determined above). The sample variance s^2 is defined by Equation (3.3). We also make use of the following identity:

$$\sum_{i=1}^{n}(x_i-\mu)^2=\sum_{i=1}^{n}(x_i-\bar{x})^2+n(\bar{x}-\mu)^2. \tag{3.29}$$

Equation 3.29 can be easily proven by expanding the terms. For example,

$$\begin{aligned}\sum_{i=1}^{n}(x_i-\bar{x})^2 &= \sum_{i=1}^{n}(x_i^2-2x_i\bar{x}+\bar{x}^2)\\ &= \sum_{i=1}^{n}x_i^2-2\bar{x}\sum_{i=1}^{n}x_i+n\bar{x}^2\\ &= \sum_{i=1}^{n}x_i^2-n\bar{x}^2.\end{aligned}$$

Therefore

$$\sum_{i=1}^{n}(x_i-\bar{x})^2=\sum_{i=1}^{n}(x_i^2-\bar{x}^2). \tag{3.30}$$

You should expand the other two terms and prove the equality of Equation (3.29).

Substituting Equation (3.29) into Equation (3.3), we get

$$s^2=\frac{\sum_{i=1}^{n}(x_i-\bar{x})^2}{n-1}=\frac{\sum_{i=1}^{n}(x_i-\mu)^2-n(\bar{x}-\mu)^2}{n-1}.$$

Then, the expected variance of the sample is given by

$$E(s^2)=\frac{\sum_{i=1}^{n}E\big((x_i-\mu)^2\big)-nE\big((\bar{x}-\mu)^2\big)}{n-1}.$$

From Equations (3.25) and (3.27),

$$E(s^2)=\frac{\sum_{i=1}^{n}\sigma^2-n\frac{\sigma^2}{n}}{n-1}=\frac{n\sigma^2-\sigma^2}{n-1}=\sigma^2.$$

Therefore, the expectation of s^2 is equal to the population variance σ^2. Note that s^2 as defined by Equation (3.3) is an *unbiased estimate* for σ^2. As described in Section 3.2, the term $n-1$ in the denominator of Equation (3.3) comes from using the same sample to calculate both the mean $\bar{x}$ and the variance s^2. Actually, only $n-1$ observations provide information about the value of s^2. One degree of freedom is lost when the data are used to estimate a population parameter. The sample variance is expected to be less than the population variance since the sample does not contain all the observations that make up the population. Hence the $n-1$ term in the denominator adjusts for the inherent underestimation of the variance that results when a sample is used to approximate the population.

Keep in mind that the following relationships, derived in this section, for expected values:

$$E(\bar{x})=\mu, \qquad E(s^2)=\sigma^2, \qquad E\big((\bar{x}-\mu)^2\big)=\frac{\sigma^2}{n},$$

are valid for any form of distribution of the random variable.

3.5.4 Standard normal distribution and the *z* statistic

The population density curve is a fundamental tool for calculating the proportion of a population that exhibits a characteristic of interest. As discussed in Section 3.5.1, the area under a density curve is exactly equal to 1. If the area under a population density curve is computed as a function of the observed variable x, then determining the area under the curve that lies between two values of x will yield the percentage of the population that exhibits the characteristic within that range. The area under a normal curve from $-\infty$ to x is the cumulative probability, i.e.

$$P(x) = \int_{-\infty}^{x} \frac{1}{\sqrt{2\pi}\sigma} e^{-(x-\mu)^2/2\sigma^2}. \tag{3.31}$$

If an observation x, such as blood pressure, follows a normal distribution, then the area under the normal curve defining the population distribution from $-\infty$ to x estimates the fraction of the population that has blood pressure less than x. Integration of Equation (3.23), the normal equation, does not lend itself to an analytical solution. Numerical solutions to Equation (3.31) have been computed and tabulated for the purpose of calculating probabilities. Generating numerical solutions and constructing tables for every normal curve is impractical. Instead, we simply convert any normal curve to a standardized normal curve.

A normal curve $N(0, 1)$ is called the **standard normal distribution** and has a probability density given by

$$f(x) = \frac{1}{\sqrt{2\pi}} e^{-z^2/2}, \tag{3.32}$$

where z is a normally distributed variable with $\mu = 0$ and $\sigma = 1$.

Any variable x that follows a normal distribution can be rescaled as follows to produce a standard normal distribution using Equation (3.33):

$$z = \frac{x-\mu}{\sigma}; \tag{3.33}$$

this is known as the ***z* statistic** and is a measure of the number of standard deviations between x and the mean μ. Note that $z = \pm 1$ represents a distance from the mean equal to one standard deviation, $z = \pm 2$ represents a distance from the mean equal to two standard deviations, and so on. The area enclosed by the standard normal curve between $z = -1$ and $z = +1$ is 0.6827, as demonstrated in Figure 3.11. This means that 68.27% of all observations are within ± 1 standard deviation from the mean. Similarly, the area enclosed by the standard normal curve between $z = -2$ and $z = +2$ is 0.9544, or 95.44% of all observations are within ± 2 standard deviations from the mean. Note that 99.74% of the enclosed area under the curve lies between $z = \pm 3$.

Using MATLAB

The `normpdf` function can be used in conjunction with the `quad` function to determine the probability of obtaining z within a specific range of values. The `quad` function performs numerical integration and has the syntax

```
i = quad(func, a, b)
```

where func is a function handle, and a and b are the limits of the integration. Numerical integration is the topic of focus in Chapter 6.

First, we create a function handle to normpdf:

```
>> f_z = @(x)normpdf(x, 0 , 1);
```

The normal distribution parameters are $\mu = 0$, $\sigma = 1$, and $\boldsymbol{x}$ is any vector variable.

The probability of z assuming any value between -1 and $+1$ is obtained by typing in the MATLAB Command Window the following:

```
>> prob = quad(f_z, -1, 1)
prob =
    0.6827
```

Similarly, the probability of z assuming any value between -2 and $+2$ is

```
>> prob = quad(f_z, -2, 2)
prob =
    0.9545
```

A very useful MATLAB function offered by Statistical Toolbox is cdf, which stands for cumulative distribution function. For a continuous distribution, the **cumulative distribution function** Φ (cdf) is defined as

$$\Phi(y) = \int_{-\infty}^{y} f(x)dx = P(x \leq y), \tag{3.34}$$

where $f(x)$ is the probability density function (pdf) that specifies the distribution for the random variable x. We are particularly interested in the function normcdf, which calculates the cumulative probability distribution for a normally distributed random variable (Equation (3.31)). The syntax is

```
p = normcdf (x, mu, sigma)
```

where mu and sigma are the parameters of the normal distribution and $\boldsymbol{x}$ is the vector of values at which the cumulative probabilities are desired.

To determine the probability of z lying between -1 and $+1$, we compute

```
>> p = normcdf(1, 0, 1) - normcdf(-1, 0, 1)
p =
   0.6827
```

What is the probability of observing $z \geq 2.5$? The answer is given by

```
>> p = 1 - normcdf(2.5, 0, 1);
```

We can also ask the question in reverse – what value of z represents the 95th percentile? In other words, what is the value of z such that 0.95 of the total area under curve lies between $-\infty$ and z? The value of z that separates the top 5% of the population distribution from the remaining 95% is 1.645. A similar question is, "For an enclosed area of 0.95 symmetrical positioned about the mean, what are the corresponding values of z?" The answer: the area enclosed by the standardized normal curve lying between $z = \pm 1.96$ contains 95% of the entire population. If the population parameters are known, we can use Equation (3.33) to calculate the corresponding value of x, i.e. $x = \mu + z\sigma$.

Figure 3.11

Areas under the standard normal curve.

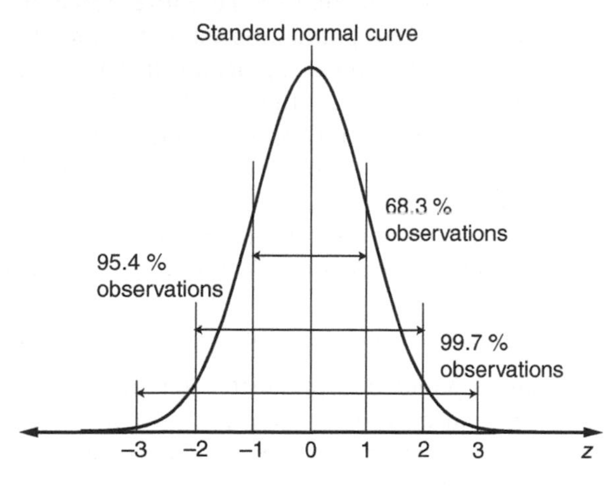

Using MATLAB

Statistics Toolbox provides an inverse cumulative distribution function, `inv`, for a variety of probability distributions. The `inv` function accepts cumulative probability values P and returns the corresponding values of the random variable x with respect to a specific probability distribution. Of particular interest to us is the `norminv` function. The syntax for `norminv` is

```
x = norminv(P, mu, sigma)
```

where `mu` and `sigma` are parameters of the normal distribution and P is the vector of probabilities at which the corresponding value(s) of the random variable x is evaluated.

If we want to know the value of the z statistic that denotes the 95th percentile, we compute

```
>> x = norminv(0.95, 0, 1)
```

MATLAB output:

```
x =
  1.6449
```

On the other hand, if we want to determine the z values that bracket 0.95 of the enclosed area symmetrically about μ, we type

```
>> x = norminv([0.025 0.975], 0, 1])
```

MATLAB output:

```
x =
  -1.9600  1.9600
```

3.5.5 Confidence intervals using the z statistic and the t statistic

Since μ and σ of a population are usually not known in any experiment, we must use data from a sample to calculate $\bar{x}$, s^2, and s, which serve as our estimates of the population statistics. The primary goal in an experiment is to estimate the mean of the population under consideration. Because discrepancies exist between the statistical properties of the sample and that of the entire population, uncertainty is associated with our sample-derived population estimates, and we proceed to make **statistical inferences**. Based on the data available, we *infer* the statistical properties of the population. The statistical estimates that are derived from the data or sample should be accompanied by a probability and a confidence interval, which convey, respectively, the level of accuracy and precision with which we have obtained our estimate. **Inferential statistics** is the branch of statistics that deals with obtaining and using sample-derived estimates of population characteristics to draw appropriate conclusions about the reference population. We now shift our focus to the topic of statistical inference.

Statistical inference has two broad purposes as detailed below.

(1) **Estimation**. To estimate a population parameter (e.g. mean, median, or variance) from the data contained in a random sample drawn from the population. The statistic calculated from the sample data can be of two types:
  (a) **point estimate**, or a single numeric value that estimates the population parameter;
  (b) **interval estimate**, or a set of two values that specify a range called a confidence interval that is believed to contain the population parameter. A probability is specified along with the interval to indicate the likelihood that the interval contains the parameter being estimated.

(2) **Hypothesis testing**. To draw a conclusion concerning one or more populations based on data obtained from samples taken from these populations. This topic is discussed in Chapter 4.

In this and the following sections, we learn how to estimate a population parameter by constructing a confidence interval. A confidence interval is an **interval estimate** of a population parameter, as opposed to a **point estimate** (single number), which you are more familiar with. When working with samples to derive population estimates, a point estimate of a parameter, e.g. $\bar{x}$ or s^2, is not sufficient. Why? Even if the measurements associated with our data are very accurate or free of measurement error, sample statistics derived from the measurements are still prone to **sampling error**. It is very likely that point estimates of a parameter based on our samples will not be correct since the sample is an incomplete representation of the entire population.

The distribution of sample means $\bar{x}$ is a frequency distribution plot of all possible values of $\bar{x}$ obtained when all possible samples of the same size are drawn from a reference population, and is called the **sampling distribution of the mean**. When samples are drawn from a normal population (i.e. x is distributed as $N(\mu, \sigma)$ within the population), the sampling distribution of the mean will also be normal, with a mean equal to μ and standard deviation equal to the standard error of the mean (SEM) $\sigma/\sqrt{n}$. The sample mean can theoretically lie anywhere along the normal curve that represents the sampling distribution of the mean. The sampling error or uncertainty in our estimate of μ is conveyed by the SEM $= \sigma/\sqrt{n}$, which measures the spread of distribution of the sample means.

For any normally distributed population, 95% of the measurements lie within ± 1.96 standard deviations ($z = 1.96$) from the mean. Specifically, in the case of the distribution of sample means, 95% of the sample means lie within a distance of $\pm 1.96\sigma/\sqrt{n}$ from μ. Of course, we do not know the actual population statistics.

Let's assume for the time being that we have knowledge of σ. We cannot determine the interval $\mu \pm 1.96\sigma/\sqrt{n}$ that will contain 95% of the sampling means, since we do not know μ. But we can determine the interval $\bar{x} \pm 1.96\sigma/\sqrt{n}$ *and* we can make the following statement.

> If the interval $\mu \pm 1.96\ \sigma/\sqrt{n}$ contains 95% of the sample means, then 95% of the intervals $\bar{x} \pm 1.96\ \sigma/\sqrt{n}$ must contain the population mean.

The importance of this statement cannot be overemphasized, so let's discuss this further. Of all possible samples of size n that can be drawn from a population of mean μ and variance σ^2, 95% of the samples will possess a mean $\bar{x}$ that lies within $1.96\ \sigma/\sqrt{n}$ standard deviations from the population mean μ. Mathematically, we write this as

$$P\left(\mu - 1.96\frac{\sigma}{\sqrt{n}} < \bar{x} < \mu + 1.96\frac{\sigma}{\sqrt{n}}\right) = 0.95.$$

We subtract $\bar{x}$ and μ from all sides of the inequality. Then, taking the negative of each side, we obtain

$$P\left(\bar{x} - 1.96\frac{\sigma}{\sqrt{n}} < \mu < \bar{x} + 1.96\frac{\sigma}{\sqrt{n}}\right) = 0.95.$$

Therefore, 95% of all samples obtained by repeated sampling from the population under study will have a mean such that the interval $\bar{x} \pm 1.96\sigma/\sqrt{n}$ includes the population mean μ. Figure 3.12 illustrates this. On the other hand, 5% of the sample means $\bar{x}$ lie further than $1.96\sigma/\sqrt{n}$ units from the population mean. Then, 5% of the intervals, $\bar{x} \pm 1.96\sigma/\sqrt{n}$, generated by all possible sample means, will not contain the population mean μ. Therefore, if we conclude that the population mean lies within 1.96 standard deviations from our current sample mean estimate, we will be in error 5% of the time.

Figure 3.12

The sampling distribution of all sample means $\bar{x}$ derived from a population that exhibits an $N(\mu, \sigma)$ distribution. The areas under the curve that lie more than 1.96 standard deviations on either side from the population mean μ, i.e. the extreme 5% of the population, are shaded in gray. A 95% confidence interval for μ is drawn for the sample mean $\bar{x}_1$.

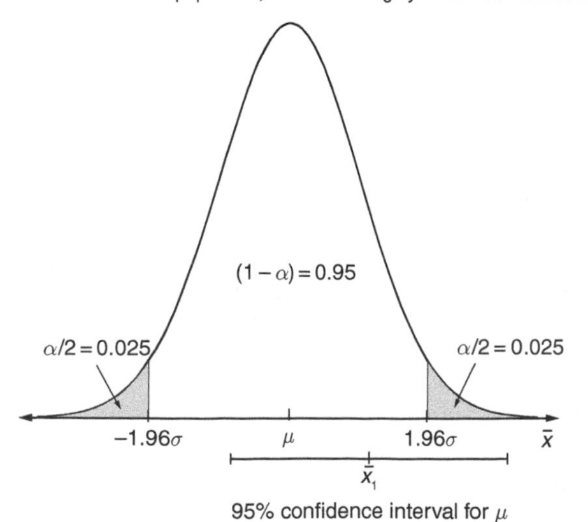

In most cases, we have only one sample mean obtained from our experiment (our sample). We infer from our data with 95% confidence that the interval $\bar{x} \pm 1.96\sigma/\sqrt{n}$ contains the true population mean.

$\bar{x} \pm 1.96\sigma/\sqrt{n}$ is called the **95% confidence interval** for the population mean μ, and the endpoints of this interval are called the **95% confidence limits**.

When we construct a 95% confidence interval, it is inferred that any mean value lying outside the confidence interval belongs to a population that is **significantly different** from the population defined by the 95% confidence interval for μ. Since 5% of the time the confidence interval constructed from the sample mean $\bar{x}$ will not contain its true population mean μ, we will wrongly assert a significant difference between the sample and the population from which the sample is derived 5% of the time. The accuracy, or level of confidence, expressed in the terms of probabilities such as 90%, 95%, or 99% confidence, with which a confidence interval is constructed, depends on the chosen **significance level** α.

Let's have a second look at the sampling distribution for the sample mean $\bar{x}$ shown in Figure 3.12. The interval $\mu \pm 1.96\sigma/\sqrt{n}$ excludes 5% of the area under the curve located within the two tails. This area (shaded in gray) corresponds to a 0.05 probability that a sample mean obtained from a population with mean μ will be excluded from this interval. In other words, the significance level for this interval is set to $\alpha = 0.05$. The excluded area within both tails is equally distributed and is $\alpha/2$ at either end, i.e. 2.5% of the sample means are located to the right of the interval and the other 2.5% are situated on the left. The critical z values that satisfy the cumulative probability equations $\Phi(z) = \alpha/2; \Phi(z) = 1 - \alpha/2$ are $z_{\alpha/2} = -1.96$ and $z_{(1-\alpha/2)} = 1.96$. The area under the curve within the interval $\mu \pm 1.96\sigma/\sqrt{n}$ is $1 - \alpha$. As the significance level α is reduced, the interval about μ for means of samples drawn from the same population widens and more sample means are included within the interval $\mu \pm z_{1-\alpha/2}(\sigma/\sqrt{n})$. If $\alpha = 0.01$, then 99% of all samples means are included within the interval centered at μ. When we create a $100(1 - \alpha)\%$ confidence interval for μ using a sample mean $\bar{x}$, we must use $z_{\alpha/2}$ and $z_{(1-\alpha/2)}$ to define the width of the interval.

Although it is a common choice among scientists and engineers, it is not necessary to restrict our calculations to 95% confidence intervals. We could choose to define 90% or 99% confidence intervals, depending on how small we wish the interval to be (precision) and the required accuracy (probability associated with the confidence interval). Table 3.5 lists the z values associated with different levels of significance α, when the confidence interval is symmetrically distributed about the sample mean (two-tailed distribution).

Table 3.5. *The z values associated with different levels of significance for a two-sided distribution*

Level of significance, α	Type of confidence interval	$z_{\alpha/2}, z_{(1-\alpha/2)}$
0.10 ($\alpha/2 = 0.05$)	90%	±1.645
0.05 ($\alpha/2 = 0.25$)	95%	±1.96
0.01 ($\alpha/2 = 0.005$)	99%	±2.575
0.005 ($\alpha/2 = 0.0025$)	99.5%	±2.81
0.001 ($\alpha/2 = 0.0005$)	99.9%	±3.27

The $100(1-\alpha)\%$ confidence interval for μ for a two-sided distribution is

$$\bar{x} \pm z_{1-\alpha/2}\frac{\sigma}{\sqrt{n}}.$$

Box 3.4 Bilirubin level in infants

Red blood cells have a normal lifespan of 3–4 months. Bilirubin is a component generated by the breakdown of hemoglobin and is further processed in the liver. Bilirubin contains a yellow pigment. Build up of bilirubin in the body due to liver malfunction gives skin a yellowish hue. Bilirubin is removed from fetal blood through placental exchanges. Once the child is born, the immature newborn liver must begin processing bilirubin. The liver is initially unable to keep up with the load of dying red blood cells. Bilirubin levels increase between days 2 and 4 after birth. Bilirubin concentrations usually fall substantially by the end of the first week. Many other factors can aggravate bilirubin concentration in blood during the first week of life. High concentrations of bilirubin can have severely adverse effects, such as permanent hearing loss, mental retardation, and other forms of brain damage.

The average measured bilirubin concentration in 20 newborns on day 5 was 6.6 mg/dl. If the bilirubin levels in 5-day-old infants are normally distributed with a standard deviation of 4.3 mg/dl, estimate μ by determining the 95% and 99% confidence intervals for the population mean bilirubin levels.

The 95% confidence interval for the mean bilirubin level in 5-day-old infants is

$$\bar{x} \pm 1.96\frac{\sigma}{\sqrt{n}},$$

i.e.

$$6.6 \pm 1.96\frac{4.3}{\sqrt{20}} = 4.7, 8.5.$$

We are 95% confident that the mean bilirubin level of 5-day-old infants lies between 4.7 and 8.5 mg/dl.

The 99% confidence interval for the mean bilirubin level in 5-day-old infants is

$$\bar{x} \pm 2.575\frac{\sigma}{\sqrt{n}},$$

i.e.

$$6.6 \pm 2.575\frac{4.3}{\sqrt{20}} = 4.1, 9.1.$$

We are 99% confident that the mean bilirubin level of 5-day-old infants lies between 4.1 and 9.1 mg/dl.

When we do not know the population standard deviation σ, we make use of the sample-derived standard deviation to obtain the best estimate of the confidence interval for μ, i.e. $\bar{x} \pm z_{1-\alpha/2}(s/\sqrt{n})$. Large sample sizes will usually yield s as a reliable estimate of σ. However, small sample sizes are prone to sampling error. Estimates of SEM that are less than their true values will underestimate the width of the confidence interval. William Sealy Gosset (1876–1937), an employee of Guinness brewery, addressed this problem. Gosset constructed a family of t curves, as modifications of the normal curve. The shape of each t curve is a function of the degrees of freedom f associated with the calculation of the standard deviation, or alternatively a function of the sample size. Recall that for a sample of size n, the degrees of freedom or independent data points available to calculate the standard deviation is $n-1$. (Note that if there is only one data point available, we have no information on the dispersion within the data, i.e. there are

Figure 3.13

Distributions of the z and t statistics.

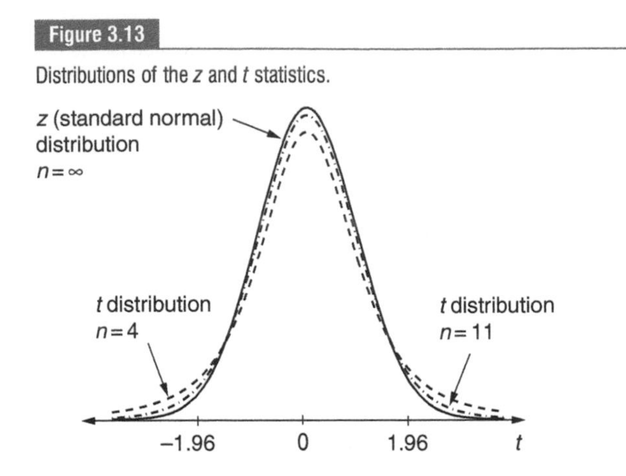

no degrees of freedom.) Gosset published his work under the pseudonym "Student" to hide his true identity since there had been a publication freeze within the Guinness company. For this reason, the t distribution is widely associated with the name "Student." If the population distribution is estimated using sample statistics, $N(\bar{x}, s)$, then the confidence interval for the population μ for any desired significance level can be calculated using the ***t* statistic**. The t statistic is given by

$$t = \frac{\bar{x} - \mu}{s/\sqrt{n}}. \tag{3.35}$$

The t statistic follows the t distribution, which features the characteristic symmetric bell-shape of the normal distribution except with bulkier tails and a shorter central region. The exact shape of the t distribution is dependent on the sample size n (or degrees of freedom, $f = n - 1$) and approaches the normal distribution as $n \to \infty$. Figure 3.13 illustrates the shape dependence of the t distribution on sample size.

The desired confidence interval for μ is calculated as

$$\bar{x} \pm t_{1-\alpha/2, f} \frac{s}{\sqrt{n}},$$

where $t_{1-\alpha/2,f}$ is the t statistic for a sample size n, degrees of freedom f, and a significance level equal to α for a **two-tailed (two-sided) distribution**. In a two-tailed distribution, the areas of significance are split symmetrically on both sides of the probability distribution curve at both ends (tails). The critical value of the t statistic, i.e. $t_{1-\alpha/2,f}$, is always greater than the corresponding (for the same sample size) critical z value $z_{1-\alpha/2}$ for the same significance level and approaches $z_{1-\alpha/2}$ for $n \to \infty$. Because Gossett adopted the pseudonym of "Student" under which he published his work, the t curves are commonly known as "Student's t distributions." While the use of t distributions is strictly valid when sampling is done from a normal population, many researchers apply t distributions for assessing non-normal populations that display at least a unimodal (single-peaked) distribution.

Using MATLAB

The MATLAB Statistics Toolbox contains several functions that allow the user to perform statistical analysis using Student's t distribution. Two useful functions are `tcdf` and `tinv`. Their respective syntax are given by

Box 3.5 Shoulder ligament injuries

An orthopedic research group studied injuries to the acromioclavicular (AC) joint and coracoclavicular (CC) joint in the shoulder (Mazzocca *et al.*, 2008). In the study, AC and CC ligaments were isolated from fresh-frozen cadaver shoulders with an average age of 73 (±13) years. Ten out of 40 specimens were kept intact (no disruption of the ligament) and loaded to failure to assess normal failure patterns. The mean load to failure of the CC ligaments was 474 N, with a standard deviation of 174 N. We assume that these ten specimens constitute a random sample from the study population. If it can be assumed that the population data are approximately normally distributed, we wish to determine the 95% confidence interval for the mean load to failure.

The sample statistics are: $\bar{x} = 474, s = 174, n = 10, f = n - 1 = 9$.

We estimate the SEM using statistics derived from the sample:

$$\text{SEM} = \frac{s}{\sqrt{n}} = \frac{174}{\sqrt{10}} = 55.0.$$

We use the `tinv` function to calculate the critical value of the t statistic, which defines the width of the confidence interval:

```
t = tinv(0.975, 9);
```

Then, $-t_{0.025} = t_{0.975} = 2.262$ for nine degrees of freedom.

The desired 95% confidence interval for mean load to failure of the intact CC ligament is thus given by

$474 \pm 2.262(55.0)$ or 349 N, 598 N.

```
p = tcdf(t, f)
t = tinv(p, f)
```

where f is the degrees of freedom, and $\boldsymbol{t}$ and $\boldsymbol{p}$ have their usual meanings (see the discussion on `normcdf` and `norminv`).

The function `tcdf` calculates the cumulative probability (area under the t curve; see Equation (3.34)) for values of $\boldsymbol{t}$, a random variable that follows Student's t distribution, and `tinv` calculates the t statistic corresponding to the cumulative probability specified by the input parameter p.

3.5.6 Non-normal samples and the central-limit theorem

In Sections 3.5.4 and 3.5.5 we assumed that the population from which our samples are drawn is normally distributed. Often, you will come across a histogram or frequency distribution plot that shows a non-normal distribution of the data. In such cases, are the mathematical concepts developed to analyze statistical information obtained from normally distributed populations still applicable? The good news is that sometimes we can use the normal distribution and its accompanying simplifying characteristics to estimate the mean of populations that are not normally distributed.

The **central-limit theorem** states that if a sample is drawn from a non-normally distributed population with mean μ and variance σ^2, and the sample size n is large ($n > 30$), the distribution of the sample means will be approximately normal with mean μ and variance σ^2/n.

Therefore, we can apply statistical tools such as the z statistic and the Student's t distribution, discussed in the previous section, to analyze and interpret non-normally distributed data. How large must the sample size be for the central-limit theorem to apply? This depends on the shape of the population distribution; if it is exactly normal, then n can be of any value. As the distribution curve deviates from normal, larger values of n are required to ensure that the sampling distribution of the means is approximately normal. Remember that random sampling requires that $n \ll N$ (population size). Since N is often very large ($N > 10^5$) in biological studies, a large sample size say of $n = 100$ or 500 will not approach N (e.g. the number of bacteria in the human gut, number of coffee drinkers in Spain, or number of patients with type II diabetes).

If the data are grossly non-normal, and the applicability of the central-limit theorem to a limited sample size is doubtful, then we can often use mathematical techniques to transform our data into an approximately normal distribution. Some

Box 3.6 Brownian motion of platelets in stagnant flow near a wall

Platelets, or thrombocytes, the smallest of the three main blood cell types, are discoid particles approximately 2 μm in diameter. At the time of vascular injury, platelets are recruited from the bloodstream to the exposed subendothelial layer at the lesion, where they promptly aggregate to seal the wound. These microparticles fall within the realm of Brownian particles, primarily due to their submicron thickness (~0.25 μm). Brownian motion is an unbiased random walk with normal distribution. The mean displacement of a Brownian particle in the absence of bulk flow is 0. Mody and King (2007) performed a computational study to investigate the Brownian motion of a platelet in order to quantify its impact on platelet flow behavior and platelet adhesive behavior when bound to a surface or to another platelet. The platelet was placed at various heights H from the surface in the absence of an external flow-field. The duration of time required for the platelet to contact the surface by virtue of its translational and rotational Brownian motion was determined (see Figure 3.14).

Two hundred surface-contact observations were sampled for each initial platelet centroid height. Figure 3.15 shows a histogram of the time taken T_c for a platelet to contact the surface by virtue of

Figure 3.14

Flow geometry in which a single platelet (oblate spheroid of aspect ratio = 0.25) is translating and rotating in stagnant fluid by virtue of translational and rotational Brownian motion near an infinite planar surface. At the instant shown in the figure, the platelet is oriented with its major axis parallel to the surface and its centroid located at a distance H from the surface.

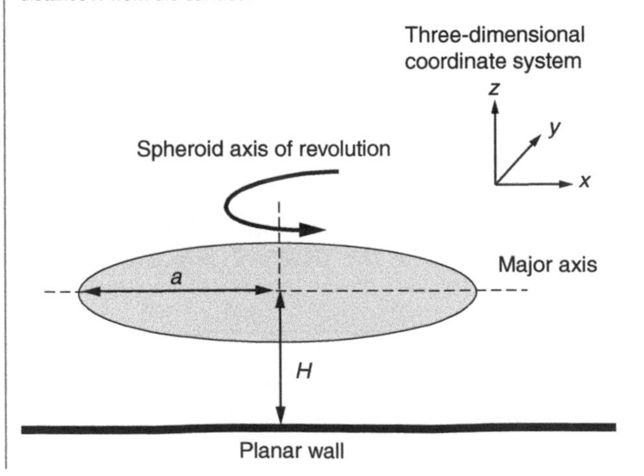

Figure 3.15

Histogram of the time taken T_c for a platelet to contact the surface when undergoing translational and rotational Brownian motion in a quiescent fluid for initial height $H = 0.625$.

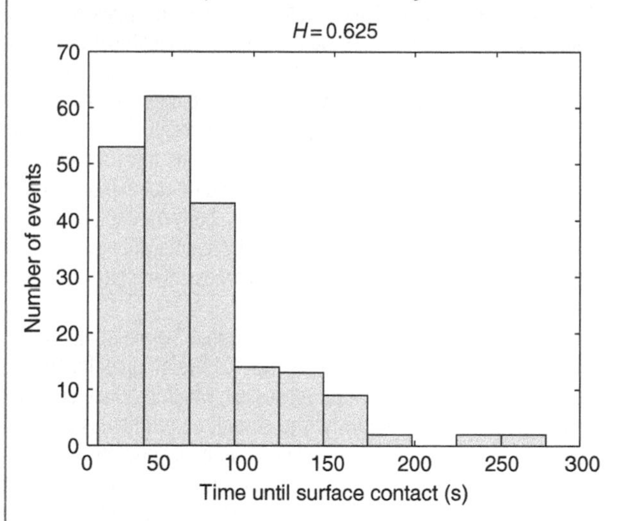

Figure 3.16

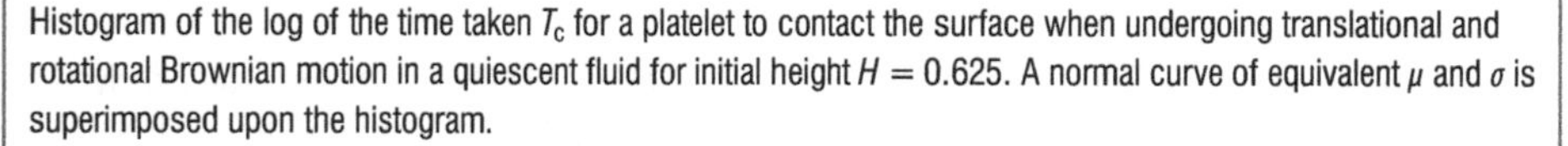
Histogram of the log of the time taken T_c for a platelet to contact the surface when undergoing translational and rotational Brownian motion in a quiescent fluid for initial height $H = 0.625$. A normal curve of equivalent μ and σ is superimposed upon the histogram.

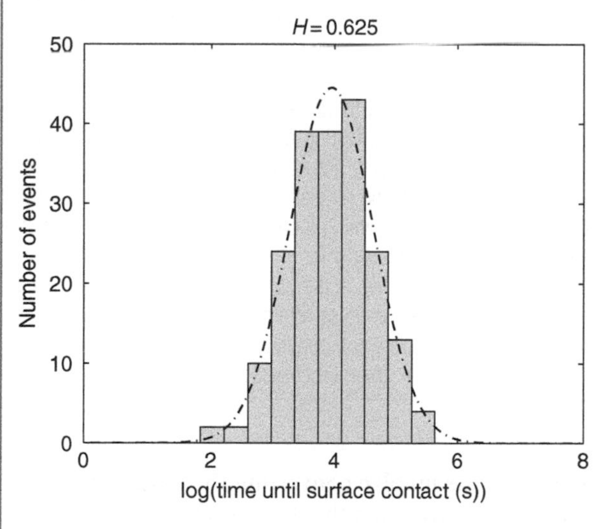

Brownian motion only, for an initial centroid height $H = 0.625$. The mean and variance of the data are 65 and 2169. The distribution curve has an early peak and a long tail.

Since we want to center the distribution peak and equalize the tails on both sides of the peak, we perform a natural logarithm transformation of the data. Figure 3.16 shows that the transformed data have an approximately **log normal** distribution. The mean and variance of the transformed distribution is 3.95 and 0.47. A normal curve with these properties and area equal to that enclosed by the histogram is superimposed upon the histogram plotted in Figure 3.16. The transformation has indeed converted the heavily skewed distribution into an approximately normal distribution.

Box 3.7 Nonlinear (non-normal) behavior of physiological variables

Physiological functioning of the body, such as heart rate, breathing rate, gait intervals, gastric contractions, and electrical activity of the brain during various phases of sleep, are measured to ascertain the state of health of the body. These mechanical or electrical bodily functions fluctuate with time and form chaotic patterns when graphed by e.g. an electrocardiograph or an electroencephalograph. It has been long believed that a healthy body strives to maintain these physiological functions at certain "normal" values according to the principle of homeostasis. The fluctuations that are observed in these rates are the result of impacts on the body from various environmental factors. Fluctuations in heart rate or stride rate that occur are primarily random and do not exhibit any pattern over a long time. Along this line of reasoning, it is thought that every human body may establish a slightly different normal rate based on their genetic make-up. However, all healthy people will have their bodily functioning rates that lie within a normal range of values. Thus, any rate process in the body is quantified as a mean value, with fluctuations viewed as random errors or noise.

Recently, studies have uncovered evidence showing that the variability in these physiological rate processes is in fact correlated over multiple time scales. Physiological functions seem to be governed by complex processes or nonlinear dynamical processes, which produce chaotic patterns in heart rate, walking rate, breathing rate, neuronal activity, and several other processes. Chaotic processes, which are complex phenomena, are inherently associated with uncertainty in their outcomes. Complex systems are sensitive to initial conditions and slight random events that influence the path of the process. "The butterfly effect" describes an extreme sensitivity of a process to the initial conditions such that even a small disturbance or change upfront can lead to a completely different sequence of events as the process unfolds. This phenomenon can result from processes governed by nonlinear dynamics and was initially discovered by Edward Lorenz, a meteorologist who was instrumental in developing modern chaos theory.

The variability in our physiological functioning is neither completely random (noise) nor is it completely correlated or regular. In fact, a state of good health is believed to require processes that are correlated to an optimal degree such that both randomness and memory retained over numerous time scales play synergistic roles to ensure versatility and robustness of our body functions (West, 2006). Disease states are represented by a loss of variability or complexity. For example, heart failure results from either a poor oxygen supply to the heart or defective or weak muscles and/or valves due to previous disease or congenital defects. During heart failure, the heart is incapable of pumping blood with sufficient force to reach the organs. Heart rate during heart failure is characterized by loss of variability, or, in other words, increasing regularity and loss of randomness (Goldberger *et al.*, 1985). Aging also appears to be synonymous with the onset of increasing predictability in rate processes and loss of complexity or robustness necessary for survival (Goldberger *et al.*, 2002). For certain physiological processes, much information lies in the observable chaotic patterns, which convey the state of health and should not be discarded as noise. Such variability cannot be described by simple normal distributions, and instead can be more appropriately represented by scale-free or fractal distributions. Normal distributions are suitable for linear processes that are additive and not multiplicative in nature.

common transformations involve converting the data to log scale ($\ln x$), calculating the square root ($\sqrt{x}$), or inverting the data ($1/x$).

3.6 Propagation of error

All experimental measurements and physical observations are associated with some uncertainty in their values due to inherent variability associated with natural phenomena, sampling error, and instrument error. The error or variability in the measurements of a random variable is often reported in terms of the standard deviation. When arithmetically combining measurements in a mathematical calculation, the combined error must also be calculated. The end result of a mathematical calculation that combines one or more random variables should be reported in terms of the mean value of the calculated result and the standard deviation associated with it. In this section, we derive the mathematical rules for combining measurements associated with error.

Let x_i and y_i be two random variables or measurements with mean values of $\bar{x}$ and $\bar{y}$ and associated variances s_x^2 and s_y^2, respectively, where

$$s_x^2 = \frac{1}{n-1}\sum_{i=1}^{n}(x_i - \bar{x})^2 \quad \text{and} \quad s_y^2 = \frac{1}{n-1}\sum_{i=1}^{n}(y_i - \bar{y})^2,$$

$1 \leq i \leq n$. Also let C_1 and C_2 be two known constants.

3.6.1 Addition/subtraction of random variables

Consider the general linear combination $z_i = C_1x_i + C_2y_i$. We would like to derive the equations that relate $\bar{z}$ and s_z^2 to our statistical estimates for x and y.

We begin by determining the mean value of z:

$$\bar{z} = \frac{1}{n}\sum_{i=1}^{n} z_i.$$

Substituting the functional dependence of z on x and y:

$$\bar{z} = \frac{C_1}{n}\sum_{i=1}^{n} x_i + \frac{C_2}{n}\sum_{i=1}^{n} y_i,$$

$$\bar{z} = C_1\bar{x} + C_2\bar{y}. \tag{3.36}$$

Next, we determine the variance associated with z:

$$s_z^2 = \frac{1}{n-1}\sum_{i=1}^{n}(z_i - \bar{z})^2 = \frac{1}{n-1}\sum_{i=1}^{n}(z_i^2 - \bar{z}^2)$$

(from Equation (3.30)). Squaring z_i,

$$z_i^2 = C_1^2x_i^2 + C_2^2y_i^2 + 2C_1C_2x_iy_i.$$

Squaring Equation (3.36),

$$\bar{z}^2 = C_1^2\bar{x}^2 + C_2^2\bar{y}^2 + 2C_1C_2\bar{x}\bar{y}.$$

Substituting the expressions for z_i^2 and $\bar{z}^2$ into that for s_z^2, we obtain

$$s_z^2 = \frac{1}{n-1}\sum_{i=1}^{n}\left(C_1^2(x_i^2 - \bar{x}^2) + C_2^2(y_i^2 - \bar{y}^2) + 2C_1C_2(x_iy_i - \bar{x}\bar{y})\right)$$

$$= C_1^2s_x^2 + C_2^2s_y^2 + \frac{2C_1C_2}{n-1}\sum_{i=1}^{n}(x_iy_i - \bar{x}\bar{y}).$$

The last term in the above equation,

$$s_{xy}^2 = \frac{1}{n-1}\sum_{i=1}^{n}(x_i - \bar{x})(y_i - \bar{y}) = \frac{1}{n-1}\sum_{i=1}^{n}(x_iy_i - \bar{x}\bar{y}),$$

is called the covariance of x and y. The **covariance** of x and y is defined as $\sigma_{xy}^2 = E((x_i - \mu_x)(y_i - \mu_y))$. If y increases as x increases, then the two variables are positively correlated and have a covariance greater than 0. If y decreases as x increases, then the two variables are negatively correlated and have a covariance less than 0.

When adding two random variables x and y such that $z_i = C_1x_i + C_2y_i$, the *associated variance* is calculated as

$$s_z^2 = C_1^2s_x^2 + C_2^2s_y^2 + 2C_1C_2s_{xy}^2. \tag{3.37}$$

If x and y are independent random variables, then the values of x have no bearing on the observed values of y, and it can be easily shown that $E(s_{xy}^2) = 0$ (see the derivation of Equation (3.27)). In that case,

$$s_z^2 = C_1^2s_x^2 + C_2^2s_y^2. \tag{3.38}$$

Often the covariance between two variables is not reported, and one is forced to assume it is zero when using measured values reported in the literature. Nevertheless, you should remain aware of the limitations of this assumption.

The mean and variance of a result that involves a subtraction operation between two random variables can be derived in a similar fashion. If $z_i = C_1x_i - C_2y_i$, it can be shown that

$$\bar{z} = C_1\bar{x} - C_2\bar{y} \tag{3.39}$$

and

$$s_z^2 = C_1^2s_x^2 + C_2^2s_y^2 - 2C_1C_2s_{xy}^2. \tag{3.40}$$

This is left to the reader as an exercise.

Estimates of the mean (expected value) and variance of a linear combination of variables, which are calculated using Equations (3.36) – (3.40), do not require the assumption of a normal distribution of the variables. However, if the random variables are all normally distributed *and* independent, then it can be shown that their linear combination is also normally distributed with the mean given by an extension of Equation (3.36) and the variance given by an extension of Equation (3.38). In other words, if all x_i for $i = 1, \ldots, n$ are normally distributed, and

$$z = \sum_{i=1}^{n} C_ix_i,$$

then z is normally distributed with estimated mean and variance, respectively,

$$\bar{z} = \sum_{i=1}^{n} C_i\bar{x}_i \quad \text{and} \quad s_z^2 = \sum_{i=1}^{n} C_i^2s_i^2.$$

3.6.2 Multiplication/division of random variables

In this section, we calculate the mean and variance associated with the result of a multiplication or division operation performed on two random variables. Let $z_i = x_i/y_i$. We do not include C_1 and C_2 since they can be factored out as a single term and then multiplied back for calculating the final result. We write x_i and y_i as

$$x_i = \bar{x} + x_i', \qquad y_i = \bar{y} + y_i',$$

where x'_i, y'_i are the deviations of x and y from their means, respectively. For either of the variables, the summation of all deviations equals 0 by definition, i.e. $\sum_{i=1}^{n} x'_i = 0, \sum_{i=1}^{n} y'_i = 0$. We assume that x_i and y_i are normally distributed and that the relative errors $S_x/\bar{x}, S_y/\bar{y} \ll 1$. This implies that x'_i, y'_i are both small quantities. We have

$$z_i = \frac{x_i}{y_i} = \frac{\bar{x} + x'_i}{\bar{y} + y'_i} = \frac{\bar{x}\left(1 + x'_i/\bar{x}\right)}{\bar{y}\left(1 + y'_i/\bar{y}\right)}.$$

Expanding $\left(1 + y'_i/\bar{y}\right)^{-1}$ using the binomial expansion,

$$z_i = \frac{\bar{x}}{\bar{y}}\left(1 + \frac{x'_i}{\bar{x}}\right)\left(1 - \frac{y'_i}{\bar{y}} + \left(\frac{y'_i}{\bar{y}}\right)^2 - \left(\frac{y'_i}{\bar{y}}\right)^3 + \cdots\right)$$

$$= \frac{\bar{x}}{\bar{y}}\left(1 + \frac{x'_i}{\bar{x}} - \frac{y'_i}{\bar{y}} + O\left(\frac{x'_i y'_i}{\bar{x}\bar{y}}, \left(\frac{y'_i}{\bar{y}}\right)^2\right)\right),$$

and ignoring the second- and higher-order terms in the expansion, we get

$$z_i \approx \frac{\bar{x}}{\bar{y}}\left(1 + \frac{x'_i}{\bar{x}} - \frac{y'_i}{\bar{y}}\right). \tag{3.41}$$

Equation (3.41) expresses z_i as an arithmetic sum of $x'_i/\bar{x}$ and $y'_i/\bar{y}$, and is used to derive expressions for $\bar{z}$ and s_z^2:

$$\bar{z} = \frac{1}{n}\sum_{i=1}^{n} z_i \approx \frac{1}{n}\sum_{i=1}^{n} \frac{\bar{x}}{\bar{y}}\left(1 + \frac{x'_i}{\bar{x}} - \frac{y'_i}{\bar{y}}\right).$$

The last two terms in the above equation sum to 0, therefore

$$\bar{z} \approx \frac{\bar{x}}{\bar{y}}. \tag{3.42}$$

Now, calculating the associated variance, we obtain

$$s_z^2 = \frac{1}{n-1}\sum_{i=1}^{n}\left(z_i^2 - \bar{z}^2\right) = \frac{1}{n-1}\sum_{i=1}^{n}\left[\left(\frac{\bar{x}}{\bar{y}}\left(1 + \frac{x'_i}{\bar{x}} - \frac{y'_i}{\bar{y}}\right)\right)^2 - \left(\frac{\bar{x}}{\bar{y}}\right)^2\right]$$

$$= \left(\frac{\bar{x}}{\bar{y}}\right)^2 \frac{1}{n-1}\sum_{i=1}^{n}\left[\left(\frac{x'_i}{\bar{x}}\right)^2 + \left(\frac{y'_i}{\bar{y}}\right)^2 + 2\frac{x'_i}{\bar{x}} - 2\frac{y'_i}{\bar{y}} - 2\frac{x'_i y'_i}{\bar{x}\bar{y}}\right],$$

$$s_z^2 = \left(\frac{\bar{x}}{\bar{y}}\right)^2\left[\frac{1}{\bar{x}^2}\cdot\frac{1}{n-1}\sum_{i=1}^{n}\left(x'_i\right)^2 + \frac{1}{\bar{y}^2}\cdot\frac{1}{n-1}\sum_{i=1}^{n}\left(y'_i\right)^2 - \frac{2}{\bar{x}\bar{y}}\cdot\frac{1}{n-1}\sum_{i=1}^{n} x'_i y'_i\right].$$

Now,

$$s_x^2 = \frac{1}{n-1}\sum_{i=1}^{n}\left(x'_i\right)^2, \quad s_y^2 = \frac{1}{n-1}\sum_{i=1}^{n}\left(y'_i\right)^2, \quad \text{and} \quad s_{xy}^2 = \frac{1}{n-1}\sum_{i=1}^{n} x'_i y'_i.$$

When dividing two random variables x and y such that $z = x/y$, the variance associated with z is given by

$$s_z^2 = \left(\frac{\bar{x}}{\bar{y}}\right)^2\left[\frac{s_x^2}{\bar{x}^2} + \frac{s_y^2}{\bar{y}^2} - 2\frac{s_{xy}^2}{\bar{x}\bar{y}}\right]. \tag{3.43}$$

On similar lines one can derive the expressions for $\bar{z}$ and s_z^2 when two random variables are multiplied together.

If $z_i = x_i y_i$ and $S_x/\bar{x}, S_y/\bar{y} \ll 1$, then

$$\bar{z} \approx \bar{x}\bar{y} \tag{3.44}$$

and

$$\frac{s_z^2}{\bar{z}^2} = \left[\frac{s_x^2}{\bar{x}^2} + \frac{s_y^2}{\bar{y}^2} + 2\frac{s_{xy}^2}{\bar{x}\bar{y}}\right]. \tag{3.45}$$

The derivation of Equations (3.44) and (3.45) is left to the reader as an exercise.

3.6.3 General functional relationship between two random variables

Let's now consider the case where $z = f(x, y)$, where f can be any potentially messy function of x and y. As before, we express x and y in terms of the mean value and a deviation, $x_i = \bar{x} + x_i'$, $y_i = \bar{y} + y_i'$, and $f(x, y)$ is expanded using the two-dimensional Taylor series:

$$f(x_i, y_i) = f(\bar{x}, \bar{y}) + x_i' \left.\frac{df}{dx}\right|_{\bar{x},\bar{y}} + y_i' \left.\frac{df}{dy}\right|_{\bar{x},\bar{y}} + O\left(x_i'^2, y_i'^2, x_i'y_i'\right).$$

Again, we assume that $S_x/\bar{x}, S_y/\bar{y} \ll 1$, so that we can ignore the second- and higher-order terms since they are much smaller than the first-order terms. The mean function value is calculated as follows:

$$\bar{z} = \frac{1}{n}\sum_{i=1}^{n} f(x_i, y_i) \approx \frac{1}{n}\sum_{i=1}^{n} f(\bar{x}, \bar{y}) + \left.\frac{df}{dx}\right|_{\bar{x},\bar{y}} \frac{1}{n}\sum_{i=1}^{n} x_i' + \left.\frac{df}{dy}\right|_{\bar{x},\bar{y}} \sum_{i=1}^{n} y_i' \approx f(\bar{x}, \bar{y}). \tag{3.46}$$

The variance associated with $f(x, y)$ is given by

$$s_z^2 = \frac{1}{n-1}\sum_{i=1}^{n}\left(z_i^2 - \bar{z}^2\right)$$

$$\approx \frac{1}{n-1}\sum_{i=1}^{n}\left[\left(f(\bar{x}, \bar{y}) + x_i' \left.\frac{df}{dx}\right|_{\bar{x},\bar{y}} + y_i' \left.\frac{df}{dy}\right|_{\bar{x},\bar{y}}\right)^2 - \left(f(\bar{x}, \bar{y})\right)^2\right]$$

$$\approx \frac{1}{n-1}\sum_{i=1}^{n}\left[x_i'^2\left(\left.\frac{df}{dx}\right|_{\bar{x},\bar{y}}\right)^2 + y_i'^2\left(\left.\frac{df}{dy}\right|_{\bar{x},\bar{y}}\right)^2 + 2x_i'y_i'\left(\left.\frac{df}{dx}\right|_{\bar{x},\bar{y}}\right)\left(\left.\frac{df}{dy}\right|_{\bar{x},\bar{y}}\right)\right.$$

$$\left. + 2f(\bar{x}, \bar{y})\left(x_i' \left.\frac{df}{dx}\right|_{\bar{x},\bar{y}} + y_i' \left.\frac{df}{dy}\right|_{\bar{x},\bar{y}}\right)\right].$$

On opening the brackets and summing each term individually, the last term is found to equal 0, so we have

$$s_z^2 \approx s_x^2\left(\left.\frac{df}{dx}\right|_{\bar{x},\bar{y}}\right)^2 + s_y^2\left(\left.\frac{df}{dy}\right|_{\bar{x},\bar{y}}\right)^2 + 2s_{xy}^2\left(\left.\frac{df}{dx}\right|_{\bar{x},\bar{y}}\right)\left(\left.\frac{df}{dy}\right|_{\bar{x},\bar{y}}\right). \tag{3.47}$$

Equation (3.47) is incredibly powerful in obtaining error estimates when combining multiple random variables and should be committed to memory.

Box 3.8 Sphericity of red blood cells

Red blood cells vary considerably in size even within a single individual. The area (A) and volume (V) of human red cells have been measured to be 130 ± 15.8 μm^2 and 96 ± 16.1 μm^3, respectively (Waugh *et al.*, 1992). Cells can also be characterized by a sphericity S, defined as

$$S = \frac{4\pi}{(4\pi/3)^{2/3}} \cdot \frac{V^{2/3}}{A}.$$

Using micropipettes, the sphericity of red cells was measured to be 0.78 ± 0.02, after correcting for minor changes to the area and volume due to temperature changes (Waugh *et al.*, 1992). First, estimate the variance in sphericity based on the mean values in area and volume, the reported standard deviations in area and volume, and by neglecting the covariance s_{VA}^2 between A and V. Next, solve for s_{VA} by keeping the covariance term as an unknown variable and setting your estimate for s_S equal to the reported standard deviation of 0.02.

We make use of Equation (3.47) and neglect the covariance term. Calculating the first two terms on the right-hand side of Equation (3.47), we obtain

$$\left(\frac{4\pi}{(4\pi/3)^{2/3}}\right)^2 \left[s_V^2\left(\frac{df}{dV}\bigg|_{\bar{V},\bar{A}}\right)^2 + s_A^2\left(\frac{df}{dA}\bigg|_{\bar{V},\bar{A}}\right)^2\right]$$

$$= 23.387 \times \left[16.1^2\left(\frac{2}{3A}V^{-1/3}\bigg|_{\bar{V},\bar{A}}\right)^2 + 15.8^2\left(-V^{2/3}A^{-2}|_{\bar{V},\bar{A}}\right)^2\right]$$

$$= 0.008 + 0.009 = 0.017,$$

where $f = V^{2/3}/A$. The estimated variance in the sphericity of red cells s_S^2 neglecting the covariance between A and V is 0.017. The covariance term has the value

$$2s_{VA}^2\left(\frac{df}{dV}\bigg|_{\bar{V},\bar{A}}\right)\left(\frac{df}{dA}\bigg|_{\bar{V},\bar{A}}\right) = 0.02^2 - 0.017 = -0.013.$$

Evaluating the left-hand side, we obtain

$$2s_{VA}^2\left(\frac{4\pi}{(4\pi/3)^{2/3}}\right)^2\left(\frac{2}{3A}V^{-1/3}\bigg|_{\bar{V},\bar{A}}\right)\left(-V^{2/3}A^{-2}|_{\bar{V},\bar{A}}\right) = -0.013$$

or

$$2s_{VA}^2 \times 23.387 \times 0.00112 \times -0.00124 = -0.013,$$

from which we obtain $s_{VA}^2 = 200.1$ μm^5. (In general, one should not take the square root of the covariance, since it is not restricted to positive values.)

In this section we have derived methods to combine errors in mathematical expressions containing two random variables. Generalizations for multidimensional or n-variable systems are discussed in more advanced texts.

3.7 Linear regression error

Hans Reichenbach (1891–1953), a leading philosopher of science and a prominent educator, stated in *The Rise of Scientific Philosophy* (Reichenbach, 1951), that

A mere report of relations observed in the past cannot be called knowledge; if knowledge is to reveal objective relations of physical objects, it must include reliable predictions.

The purpose of any regression method is to extract, from a mass of data, information in the form of a predictive mathematical equation. The sources of scientific and engineering data include experimental observations obtained in a laboratory or out in the field, instrument output such as temperature, concentration, pH, voltage, weight, and tensile strength, or quality control measurements such as purity, yield, shine, and hardness. Engineering/scientific data sets contain many paired values, i.e. a dependent variable paired with one or more independent variables. The topic of linear least-squares regression as a curve-fitting method is discussed extensively in Chapter 2. In Sections 2.8–2.11, we derived the normal equations, which upon solution yield estimates of the parameters of the *linear model*, and introduced the residual, SSR (sum of the squared residuals), and coefficient of determination R^2. Note that for any mathematical model amenable to linear least-squares regression, linearity of the equation is with respect to the undetermined parameters only. The equation may have any form of dependency on the dependent and independent variables. The method of linear regression involves setting up a mathematical model that is linear in the model parameters to explain the trends observed in the data. If y depends on the value of x, the independent variable, then the linear regression equation is set up as follows:

$$y = \beta_1 f_1(x) + \beta_2 f_2(x) + \cdots + \beta_n f_n(x) + \epsilon,$$

where f can be any function of x, and $\beta_1, \beta_2, \ldots, \beta_n$ are the model parameters. The individual functions $f_i(x)$ $(1 \leq i \leq n)$ may depend nonlinearly on x. The dependent variable y is a random variable that exhibits some variance σ^2 about a mean value $\mu_y = \beta_1 f_1(x) + \beta_2 f_2(x) + \cdots + \beta_n f_n(x)$. The term ϵ represents the random error or variability associated with y. When the number of data pairs in the data set (x, y) exceeds the number of unknown parameters in the model, the resulting linear system of equations is overdetermined. As discussed in Chapter 2, a linear least-squares curve does not pass through every (x_i, y_i) data point.

The least-squares method is an estimation procedure that uses the limited data at hand to quantify the parameters of the proposed mathematical model and consequently yield the following predictive model:

$$\hat{y} = c_1 f_1(x) + c_2 f_2(x) + \cdots + c_n f_n(x),$$

where $\hat{y}$ are the model predictions of y. The hat symbol ($\wedge$) indicates that the y values are generated using model parameters derived from a data set. Our data are subject to sampling error since the values of the regression-derived model coefficients are dependent on the data sample. Quantifications of the regression equation coefficients $c_1, c_2, \ldots, c_n$ serve as estimates of the true parameters $\beta_1, \beta_2, \ldots, \beta_n$ of the model. We use confidence intervals to quantify the uncertainty in using $c_1, c_2, \ldots, c_n$ as estimates for the actual model parameters $\beta_1, \beta_2, \ldots, \beta_n$. Sometimes, one or more of the c_i parameters have a physical significance beyond the predictive model, in which case the confidence interval has an additional importance when reporting its best-fit value. Before we can proceed further, we must draw your attention to the basic assumptions made by the linear regression model.

(1) The measured values of the independent variable x are known with perfect precision and do not contain any error.

(2) Every y_i value is normally distributed about its mean μ_{y_i} with an unknown variance σ^2. The variance of y is independent of x, and is thus the same for all y_i.

(3) The means of the dependent variable y_i, for the range of x_i values of interest, obey the proposed model $\mu_{y_i} = \beta_1 f_1(x_i) + \beta_2 f_2(x_i) + \cdots + \beta_n f_n(x_i)$.
(4) A y value at any observed x is randomly chosen from the population distribution of y values for that value of x, so that the random sampling model holds true.
(5) The y measurements observed at different x values are independent of each other so that their covariance $\sigma^2_{y_i y_j} = E((y_i - \mu_{y_i})(y_j - \mu_{y_j})) = 0$.

The departure of each y measurement from its corresponding point on the regression curve is called the **residual,** $r_i = (y_i - \hat{y}_i)$. For the least-squares approximation, it can be easily shown that the sum of the residuals $\sum_{i=1}^{m} r_i = \sum_{i=1}^{m}(y_i - \hat{y}_i) = 0$, where m is the number of data points in the data set.

> The **residual variance** measures the variability of the data points about the regression curve and is defined as follows:
>
> $$s_y^2 = \frac{1}{m-n}\sum_{i=1}^{m}(y_i - \hat{y}_i)^2 = \frac{\| r \|_2^2}{m-n}, \tag{3.48}$$
>
> where m is the number of data pairs and n is the number of undetermined model parameters. Note that $\| r \|_2^2$ is the sum of the squared residuals (SSR) (the subscript indicates $p = 2$ norm).

The degrees of freedom available to estimate the residual variance is given by $(m-n)$. Since n data pairs are required to fix the n parameter values that define the regression curve, the number of degrees of freedom is reduced by n when estimating the variance. If there are exactly n data points, then all n data points are exhausted in determining the parameter values, and zero degrees of freedom remain to estimate the variance. When the **residual standard deviation** s_y is small, the data points are closely situated near the fitted curve. A large s_y indicates considerable scatter about the curve; s_y serves as the estimate for the true standard deviation σ of y. Since the data points are normally distributed vertically about the curve, we expect approximately 68% of the points to be located within ± 1SD (standard deviation) from the curve and about 95% of the points to be situated within ± 2SD.

3.7.1 Error in model parameters

Earlier, we presented five assumptions required for making statistical predictions using the linear least-squares regression model. We also introduced Equation (3.48), which can be used to estimate the standard deviation σ of the normally distributed variable y. With these developments in hand, we now proceed to estimate the uncertainty in the calculated model parameter values.

The equation that is fitted to the data,

$$y = \beta_1 f_1(x) + \beta_2 f_2(x) + \cdots + \beta_n f_n(x),$$

can be rewritten in compact matrix form $\boldsymbol{Ac} = \boldsymbol{y}$, for m data points, where

$$\boldsymbol{A} = \begin{bmatrix} f_1(x_1) & f_2(x_1) & \cdots & f_n(x_1) \\ f_1(x_2) & f_2(x_2) & \cdots & f_n(x_2) \\ & & \vdots & \\ f_1(x_m) & f_2(x_m) & \cdots & f_n(x_m) \end{bmatrix}, \quad \boldsymbol{c} = \begin{bmatrix} \beta_1 \\ \beta_2 \\ \vdots \\ \beta_n \end{bmatrix}, \quad \boldsymbol{y} = \begin{bmatrix} y_1 \\ y_2 \\ \vdots \\ y_m \end{bmatrix}.$$

Box 3.9A Kidney functioning in human leptin metabolism

Typically, enzyme reactions

$$E + S \rightarrow E + P$$

exhibit Michaelis–Menten ("saturation") kinetics, as described by the following equation:

$$R = -\frac{dS}{dt} = \frac{R_{max} S}{K_m + S}.$$

Here, R is the reaction rate in concentration/time, S is the substrate concentration, R_{max} is the maximum reaction rate for a particular enzyme concentration, and K_m is the Michaelis constant; K_m can be thought of as the reactant concentration at which the reaction rate is half of the maximum R_{max}. This kinetic equation can be rearranged into linear form:

$$\frac{1}{R} = \frac{K_m}{R_{max}} \frac{1}{S} + \frac{1}{R_{max}}.$$

Thus, a plot of $1/R$ vs. $1/S$ yields a line of slope K_m/R_{max} and intercept $1/R_{max}$. This is referred to as the Lineweaver–Burk analysis.

The role of the kidney in human leptin metabolism was studied by Meyer *et al.* (1997). The data listed in Table 3.6 were collected for renal leptin uptake in 16 normal post-operative volunteers with varying degrees of obesity.

In this case the arterial plasma leptin is taken as the substrate concentration S (ng/ml), and the renal leptin uptake represents the reaction rate R (nmol/min).

We first perform a linear regression on these data using the normal equations to determine the Michaelis constant and R_{max}. The matrices of the equation $\mathbf{Ac} = \mathbf{y}$ are defined as follows:

Table 3.6.

Subject	Arterial plasma leptin	Renal leptin uptake
1	0.75	0.11
2	1.18	0.204
3	1.47	0.22
4	1.61	0.143
5	1.64	0.35
6	5.26	0.48
7	5.88	0.37
8	6.25	0.48
9	8.33	0.83
10	10.0	1.25
11	11.11	0.56
12	20.0	3.33
13	21.74	2.5
14	25.0	2.0
15	27.77	1.81
16	35.71	1.67

$$\mathbf{A} = \begin{bmatrix} \frac{1}{S_1} & 1 \\ \frac{1}{S_2} & 1 \\ \cdot & \\ \cdot & \\ \cdot & \\ \frac{1}{S_{16}} & 1 \end{bmatrix}, \qquad \mathbf{y} = \begin{bmatrix} \frac{1}{R_1} \\ \frac{1}{R_2} \\ \cdot \\ \cdot \\ \cdot \\ \frac{1}{R_{16}} \end{bmatrix}.$$

The coefficient vector

$$\mathbf{c} = \begin{bmatrix} \frac{K_m}{R_{max}} \\ \frac{1}{R_{max}} \end{bmatrix} = \left(\mathbf{A}^T\mathbf{A}\right)^{-1}\mathbf{A}^T\mathbf{y}.$$

Performing this calculation in MATLAB (verify this yourself), we obtain

$$c_1 = \frac{K_m}{R_{max}} = 6.279, \qquad c_2 = \frac{1}{R_{max}} = 0.5774.$$

Also

$$\mathrm{SSR} = \|\mathbf{r}\|_2^2 = \sum_{i=1}^{16}(y_i - \hat{y}_i)^2 = 12.6551,$$

where $\hat{\mathbf{y}} = 1/\hat{\mathbf{R}} = \mathbf{Ac}$. The coefficient of determination $R^2 = 0.8731$, thus the fit is moderately good. Since $m = 16$ and $n = 2$,

$$s_y = \sqrt{\frac{12.6551}{16 - 2}} = 0.9507.$$

We assume that the $y_i = 1/R_i$ values are normally distributed about their mean $\hat{y}_i$ with a standard deviation estimated to be 0.9507. Next, we plot in Figure 3.17 the transformed data, the best-fit model, and curves that represent the 95% confidence interval for the variability in the y data. Since the residual standard deviation is estimated with $16 - 2 = 14$ degrees of freedom, the 95% confidence interval is given by

$$\hat{y} \pm t_{0.975, f=14} s_y.$$

Using the MATLAB `tinv` function, we find that $t_{0.975, f=14} = 2.145$.

We expect the 95% confidence interval to contain 95% of all observations of $1/R$ for the range of substrate concentrations considered in this experiment. Out of 16, 15 of the observations are located within the two dashed lines, or $15/16 \sim 94\%$ of the observations are contained within the 95% confidence region.

Figure 3.17

Plot of the transformed data, the linear least-squares fitted curve (solid line), and the 95% confidence interval for y (situated within the two dashed lines).

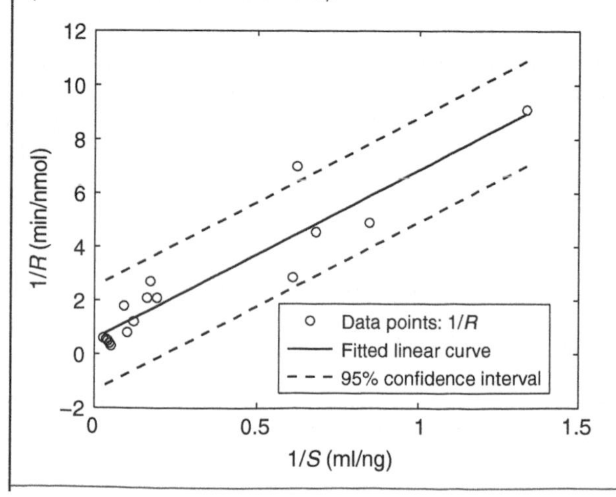

The best-fit least-squares solution $c_1, c_2, \ldots, c_n$ for the undetermined parameters $\beta_1, \beta_2, \ldots, \beta_n$ is given by the normal equations (see Section 2.9) $\boldsymbol{c} = \left(\boldsymbol{A}^{\mathrm{T}}\boldsymbol{A}\right)^{-1}\boldsymbol{A}^{\mathrm{T}}\boldsymbol{y}$. Since $\boldsymbol{A}$ is an $m \times n$ matrix, $\boldsymbol{A}^{\mathrm{T}}$ is therefore an $n \times m$ matrix, and $\boldsymbol{A}^{\mathrm{T}}\boldsymbol{A}$ and $\left(\boldsymbol{A}^{\mathrm{T}}\boldsymbol{A}\right)^{-1}$ are $n \times n$ matrices. The matrix product $\left(\boldsymbol{A}^{\mathrm{T}}\boldsymbol{A}\right)^{-1}\boldsymbol{A}^{\mathrm{T}}$ produces an $n \times m$ matrix which undergoes matrix-vector multiplication with $\boldsymbol{y}$, an $m \times 1$ vector. Let $\boldsymbol{K} = \left(\boldsymbol{A}^{\mathrm{T}}\boldsymbol{A}\right)^{-1}\boldsymbol{A}^{\mathrm{T}}$. We can then write

$$\boldsymbol{c} = \boldsymbol{K}\boldsymbol{y} \text{ or } c_i = \sum_{j=1}^{m} K_{ij}y_j, \text{ for } 1 \leq i \leq n.$$

Each coefficient c_i of the model is equal to a linear combination of m normally distributed random variables. It follows that c_i will be normally distributed. Equation (3.37) guides us on how to combine the variances $s_{y_j}^2$ associated with each random variable y_j. The second assumption in our list states that the variance σ^2 is the same for all y. Now, s_y^2 is our estimate for σ^2, so

$$s_{c_i}^2 = \sum_{j=1}^{m} K_{ij}^2 s_{y_j}^2 = \sum_{j=1}^{m} K_{ij}^2 s_y^2. \tag{3.49}$$

Equation (3.49) is used to estimate the variance or uncertainty associated with the calculated model coefficients.

3.7.2 Error in model predictions

The utility of regression models lies in their predictive nature. Mathematical models fitted to data can be used to predict the mean value of y for any given x_0 that lies within the range of x values for which the model is defined. As noted earlier, an actual observed value of y varies about its true mean with a variance of σ^2. A predicted value of y at x_0 will have a mean

$$\mu_{y_0} = E(y|x_0) = \beta_1 f_1(x_0) + \beta_2 f_2(x_0) + \cdots + \beta_n f_n(x_0).$$

The best prediction we have of the mean of y at x_0, when no actual observation of y at x_0 has been made, is

$$\hat{y}_0 = c_1 f_1(x_0) + c_2 f_2(x_0) + \cdots + c_n f_n(x_0).$$

The model prediction of the mean $\hat{y}_0$ is not perfect because the estimated regression model parameters are themselves associated with a level of uncertainty quantified by their respective confidence intervals. To convey our confidence in the prediction of $\hat{y}_0$, we must specify confidence intervals that bracket a range in which we expect the actual mean of y_0 to lie. Although y at any x is associated with a variance of σ^2, $\hat{y}_0$, our prediction of the mean at $x = x_0$, will be associated with an error that will depend on the uncertainty in the model coefficients. Let's now calculate the variance in the predicted value $\hat{y}_0$, which serves as our estimate of the true mean μ_{y_0}.

Suppose the linear regression model is $\hat{y} = c_1 f_1(x) + c_2 f_2(x)$. Using Equation (3.37), we calculate the variance associated with $\hat{y}_0$ as

$$s_{\hat{y}_0}^2 = (f_1(x_0))^2 s_{c_1}^2 + (f_2(x_0))^2 s_{c_2}^2 + 2 f_1(x_0) f_2(x_0) s_{c_1c_2}^2.$$

The covariance term cannot be dropped because $s_{c_1c_2}^2 \neq 0$. We must determine the covariance in order to proceed. We have

Box 3.9B Kidney functioning in human leptin metabolism

In Box 3.9A we calculated $s_y^2 = 0.9039$. In this example, $\boldsymbol{K} = \left(\boldsymbol{A}^{\mathrm{T}}\boldsymbol{A}\right)^{-1}\boldsymbol{A}^{\mathrm{T}}$ is a 2×16 matrix. Squaring each of the elements in $\boldsymbol{K}$ (in MATLAB you can perform element-wise operations on a matrix such as squaring by using the dot operator) we get

$$\mathtt{K.\hat{}2} = \begin{bmatrix} .2106 & .0570 & .0266 & .0186 & .0172 & .0035 & .0046 & .0053 & .0082 & .0099 & .0109 & .0150 & .0154 & .0161 & .0166 & .0175 \\ .0071 & .0002 & .0001 & .0004 & .0004 & .0066 & .0071 & .0073 & .0084 & .0089 & .0092 & .0103 & .0105 & .0106 & .0108 & .0110 \end{bmatrix};$$

$$s_{c_1}^2 = \sum_{j=1}^{m} K_{1j}^2 s_y^2 = s_y^2 \sum_{j=1}^{m} K_{1j}^2 = 0.9039(0.4529) = 0.4094;$$

$$s_{c_2}^2 = s_y^2 \sum_{j=1}^{m} K_{2j}^2 = 0.9039(0.1089) = 0.0985.$$

Although we have extracted the variance associated with the coefficients, we are not done yet. We wish to determine the variances for K_{m} and R_{max}, which are different from that of $c_1 = K_{\mathrm{m}}/R_{\mathrm{max}}$ and $c_2 = 1/R_{\mathrm{max}}$.

First we calculate the mean values of the model parameters of interest. If we divide c_1 by c_2 we recover K_{m}. Obtaining R_{max} involves a simple inversion of c_2. So,

$$K_{\mathrm{m}} = 10.87, \qquad R_{\mathrm{max}} = 1.732.$$

We use Equation (3.43) to estimate the variances associated with the two parameters above. We assume that the model parameters also exhibit normality in their distribution curves.

Let's find the variance for R_{max} first. Using Equation (3.43) and $\bar{z} = R_{\mathrm{max}} = 1/c_2$, we obtain

$$\frac{s_{R_{\mathrm{max}}}^2}{R_{\mathrm{max}}^2} = \frac{s_{c_2}^2}{c_2^2} \qquad \text{or} \qquad s_{R_{\mathrm{max}}}^2 = 1.732^2 \frac{0.0985}{0.5774^2} = 0.8863.$$

The 95% confidence interval is given by

$$R_{\mathrm{max}} \pm t_{0.975,f=14} \times s_{R_{\mathrm{max}}}.$$

Since $t_{0.025,f=14} = 2.145$,

$$1.732 \pm 2.145(0.941) \text{ or } -0.286, 3.751$$

is the 95% confidence interval for R_{max}. Since negative values of R_{max} are not possible, we replace the lower limit with zero. We can say with 95% confidence that the true value of R_{max} lies within this range.

Next, we determine the variance associated with K_{m} using Equation (3.43). Note that Equation (3.43) has a covariance term for which we do not yet have an estimate. Once we investigate the method (discussed next) to calculate the covariance between coefficients of a model, we will revisit this problem (see Box 3.9C).

$$\sigma_{c_1c_2}^2 = E((c_1 - \beta_1)(c_2 - \beta_2)).$$

Since $c_i = \sum_{j=1}^{m} K_{ij}y_j$ and $\beta_i = \sum_{j=1}^{m} K_{ij}\mu_{y_j}$, where $\boldsymbol{K} = \left(\boldsymbol{A}^{\mathrm{T}}\boldsymbol{A}\right)^{-1}\boldsymbol{A}^{\mathrm{T}}$ and $\mu_{y_i} = E(y_i)$,

$$\sigma_{c_1c_2}^2 = E\left(\left(\sum_{j=1}^{m} K_{1j}\left(y_j - \mu_{y_j}\right)\right)\left(\sum_{k=1}^{m} K_{2k}\left(y_k - \mu_{y_k}\right)\right)\right)$$

$$= E\left(\sum_{j=1}^{m} K_{1j}K_{2j}\left(y_j - \mu_{y_j}\right)^2 + \sum_{j=1}^{m}\sum_{k=1,k\neq j}^{m} K_{1j}K_{2k}\left(y_j - \mu_{y_j}\right)\left(y_k - \mu_{y_k}\right)\right).$$

Since

$$E\left(\sum_{j=1}^{m}\sum_{k=1,k\neq j}^{m} K_{1j}K_{2k}\left(y_j-\mu_{y_j}\right)\left(y_k-\mu_{y_k}\right)\right)$$

$$=\sum_{j=1}^{m}\sum_{k=1,k\neq j}^{m} K_{1j}K_{2k}E\left(\left(y_j-\mu_{y_j}\right)\left(y_k-\mu_{y_k}\right)\right)=0,$$

we have

$$\sigma^2_{c_1c_2}=E\left(\sum_{j=1}^{m}K_{1j}K_{2j}\left(y_j-\mu_{y_j}\right)^2\right)=\sum_{j=1}^{m}K_{1j}K_{2j}E\left(\left(y_j-\mu_{y_j}\right)^2\right),$$

$$\sigma^2_{c_1c_2}=\sum_{j=1}^{m}K_{1j}K_{2j}\sigma^2_y. \tag{3.50}$$

Using Equation (3.50) to proceed with the calculation of $s^2_{\hat{y}_0}$, we obtain

$$s^2_{\hat{y}_0}=(f_1(x_0))^2 s^2_{c_1}+(f_2(x_0))^2 s^2_{c_2}+2f_1(x_0)f_2(x_0)\sum_{j=1}^{m}K_{1j}K_{2j}s^2_y. \tag{3.51}$$

Equation (3.51) calculates the variance of the predicted mean value of y at x_0.

The foregoing developments in regression analysis can be generalized as shown below using concepts from linear algebra. We begin with m observations of y_i associated with different values of x_i $(1 \leq i \leq m)$. We wish to describe the dependency of y on x with a predictive model that consists of n parameters and n modeling functions:

$$y_i \approx \sum_{j=1}^{n} c_j f_j(x_i),$$

where c_j are the parameters and f_j are the modeling functions. As described earlier, the above fitting function can be rewritten as a matrix equation, $\boldsymbol{y} = \boldsymbol{Ac}$, where $A_{ij} = f_j(x_i)$. The solution to this linear system is given by the normal equations, i.e. $\boldsymbol{c} = \left(\boldsymbol{A}^{\mathrm{T}}\boldsymbol{A}\right)^{-1}\boldsymbol{A}^{\mathrm{T}}\boldsymbol{y}$, or $\boldsymbol{c} = \boldsymbol{Ky}$, where $\boldsymbol{K} = \left(\boldsymbol{A}^{\mathrm{T}}\boldsymbol{A}\right)^{-1}\boldsymbol{A}^{\mathrm{T}}$. The normal equations minimize $\| \boldsymbol{r} \|_2^2$, where $\boldsymbol{r}$ is defined as the residual: $\boldsymbol{r} = \boldsymbol{y} - \boldsymbol{Ac}$.

Recall that the covariance of two variables y and z is defined as $\sigma^2_{yz} = E((y-\mu_y)(z-\mu_z))$, and the variance of variable y is defined as $\sigma^2_y = E((y-\mu_y)^2)$. Accordingly, the matrix of covariance of the regression parameters is obtained as

$$\boldsymbol{\Sigma}^2_c \equiv E\left((\boldsymbol{c}-\boldsymbol{\mu}_c)(\boldsymbol{c}-\boldsymbol{\mu}_c)^{\mathrm{T}}\right)$$

where $\boldsymbol{\mu}_c = E(\boldsymbol{c}) = \boldsymbol{\beta}$, and $\boldsymbol{\Sigma}^2_c$ is called the **covariance matrix of the model parameters**. Note that the vector–vector multiplication $\boldsymbol{vv}^{\mathrm{T}}$ produces a matrix (a dyad), while the vector–vector multiplication $\boldsymbol{v}^{\mathrm{T}}\boldsymbol{v}$ is called the dot product and produces a scalar. See Section 2.1 for a review on important concepts in linear algebra. Substituting into the above equation $\boldsymbol{c} = \boldsymbol{Ky}$ and $\boldsymbol{\mu}_c = \boldsymbol{K\mu}_y$, we get

$$\boldsymbol{\Sigma}^2_c = \boldsymbol{K}E\left(\left(\boldsymbol{y}-\boldsymbol{\mu}_y\right)\left(\boldsymbol{y}-\boldsymbol{\mu}_y\right)^{\mathrm{T}}\right)\boldsymbol{K}^{\mathrm{T}}.$$

Note that $(\boldsymbol{K}(\boldsymbol{y}-\boldsymbol{\mu}_y))^{\mathrm{T}} = (\boldsymbol{y}-\boldsymbol{\mu}_y)^{\mathrm{T}}\boldsymbol{K}^{\mathrm{T}}$ (since $(\boldsymbol{AB})^{\mathrm{T}} = \boldsymbol{B}^{\mathrm{T}}\boldsymbol{A}^{\mathrm{T}}$).

Now, $E((\boldsymbol{y}-\boldsymbol{\mu}_y)(\boldsymbol{y}-\boldsymbol{\mu}_y)^{\mathrm{T}})$ is the covariance matrix of $\boldsymbol{y}$, $\boldsymbol{\Sigma}_y^2$. Since the individual y_i observations are independent from each other, the covariances are all identically zero, and $\boldsymbol{\Sigma}_y^2$ is a diagonal matrix (all elements except those along the diagonal are zero). The elements along the diagonal are the variances associated with each y_i observation. If all of the observations $\boldsymbol{y}$ have the same variance, then $\boldsymbol{\Sigma}_y^2 = \boldsymbol{I}\sigma_y^2$, where σ_y^2 can be estimated from Equation (3.48). Therefore,

$$\boldsymbol{\Sigma}_c^2 = \sigma_y^2\boldsymbol{K}\boldsymbol{K}^{\mathrm{T}}. \tag{3.52}$$

We can use the fitted model to predict y for k different values of x. For this we create a vector $\boldsymbol{x}^{(m)}$ that contains the values at which we wish to estimate $\boldsymbol{y}$. The model predictions $\hat{\boldsymbol{y}}^{(m)}$ are given by

$$\hat{\boldsymbol{y}}^{(m)} = \boldsymbol{A}^{(m)}\boldsymbol{c},$$

where $A_{ij}^{(m)} = f_j\left(x_i^{(m)}\right)$ and $\boldsymbol{A}^{(m)}$ is a $k \times n$ matrix. We want to determine the error in the prediction of $\hat{\boldsymbol{y}}^{(m)}$. We construct the matrix of covariance for $\hat{\boldsymbol{y}}^{(m)}$ as follows:

$$\boldsymbol{\Sigma}_{\hat{\boldsymbol{y}}^{(m)}}^2 = E\left(\left(\hat{\boldsymbol{y}}^{(m)} - \boldsymbol{\mu}_y^{(m)}\right)\left(\hat{\boldsymbol{y}}^{(m)} - \boldsymbol{\mu}_y^{(m)}\right)^{\mathrm{T}}\right).$$

Substituting $\hat{\boldsymbol{y}}^{(m)} = \boldsymbol{A}^{(m)}\boldsymbol{c}$, we obtain

$$\boldsymbol{\Sigma}_{\hat{\boldsymbol{y}}^{(m)}}^2 = \boldsymbol{A}^{(m)}E\left((\boldsymbol{c}-\boldsymbol{\mu}_c)(\boldsymbol{c}-\boldsymbol{\mu}_c)^{\mathrm{T}}\right)\boldsymbol{A}^{(m)\mathrm{T}}$$

$$= \boldsymbol{A}^{(m)}\boldsymbol{\Sigma}_c^2\boldsymbol{A}^{(m)\mathrm{T}}.$$

Therefore

$$\boldsymbol{\Sigma}_{\hat{\boldsymbol{y}}^{(m)}}^2 = \sigma_y^2\boldsymbol{A}^{(m)}\boldsymbol{K}\boldsymbol{K}^{\mathrm{T}}\boldsymbol{A}^{(m)\mathrm{T}}. \tag{3.53}$$

The diagonal elements of $\boldsymbol{\Sigma}_{\hat{\boldsymbol{y}}^{(m)}}^2$ are what we seek, since these elements provide the uncertainty (variance) in the model predictions. Note the distinction between σ_y^2 and the diagonal elements of $\boldsymbol{\Sigma}_{\hat{\boldsymbol{y}}^{(m)}}^2$. The former is a measure of the observed variability of $\boldsymbol{y}$ about its true mean $\boldsymbol{\mu}_y$, while the latter measures the error in the prediction of $\boldsymbol{\mu}_y$ yielded by the fitted model and will, in general, be a function of the independent variable.

The $\boldsymbol{x}^{(m)}$ vector should not contain any x values that correspond to the data set that was used to determine the model parameters. Can you explain why this is so? Replace $\boldsymbol{A}^{(m)}$ with $\boldsymbol{A}$ in Equation (3.53) and simplify the right-hand side. What do you get?

3.8 End of Chapter 3: key points to consider

(1) Statistical information can be broadly classified into two types: **descriptive statistics**, such as mean, median, and standard deviation, and **inferential statistics**, such as confidence intervals. In order to draw valid statistical inferences from the data, it is critical that the sample is chosen using methods of random sampling.

Box 3.9C Kidney functioning in human leptin metabolism

Let's complete the task we began in Box 3.9B of determining the variance of $K_m = \bar{z} = c_1/c_2$.
We have

$$s_y^2 = 0.9039, c_1 = 6.279, c_2 = 0.5774, s_{c_1}^2 = 0.4094, s_{c_2}^2 = 0.0985,$$

$$s_{c_1c_2}^2 = s_y^2 \sum_{j=1}^{m} K_{1j}K_{2j} = 0.9039(-0.1450) = -0.1311.$$

From Equation (3.43),

$$\frac{s_z^2}{\bar{z}^2} = \left[\frac{s_{c_1}^2}{c_1^2} + \frac{s_{c_2}^2}{c_2^2} - 2\frac{s_{c_1c_2}^2}{c_1c_2}\right] = \left[\frac{0.4094}{6.279^2} + \frac{0.0985}{0.5774^2} - 2\frac{-0.1311}{6.279(0.5774)}\right] = 0.3782.$$

So,

$$s_z^2 = 0.3782(10.87^2) = 44.68.$$

The 95% confidence interval is given by

$$K_m \pm t_{0.975, f=14} \times s_{K_m}.$$

Since $t_{0.975, f=14} = 2.145$,

$$10.87 \pm 2.145(6.684) \qquad or \qquad -3.467, 25.208$$

is the 95% confidence interval for K_m. However, K_m cannot be less than zero. We modify the confidence interval as (0, 25.2). The interval is large due to the large uncertainty (s_{K_m}) in the value of K_m.

Next, we predict kidney leptin uptake at a plasma leptin concentration S of 15 ng/ml. Our mathematical model is $1/\hat{R} = 6.279(1/S) + 0.5774$, from which we obtain $1/\hat{R} = 0.996$. Here, $f_1(S) = 1/S$ and $f_2(S) = 1$. The variance associated with this estimate of the mean value at $1/S = 1/15$ is calculated using Equation (3.51):

$$s_{\hat{y}_0}^2 = \left(\frac{1}{15}\right)^2 0.4094 + 0.0985 + \frac{2}{15}(-0.1311) = 0.0828.$$

Note that the variance $s_{\hat{y}_0}^2$ associated with the prediction of the mean $\hat{y}_0$ is much less than the variance s_y^2 associated with an individual observation of y_i about its mean value $\hat{y}_i$. Why is this so?

We also wish to estimate the error in the model predictions over the interval $S \in [0.75, 35.71]$ or $1/S \in [0.028\ 1.333]$. Let

$$x = \frac{1}{S} \text{ and } \boldsymbol{x}^{(m)} = \begin{bmatrix} 0.03 \\ 0.1 \\ 0.25 \\ 0.5 \\ 0.75 \\ 1.0 \\ 1.3 \end{bmatrix}.$$

Then,

$$\boldsymbol{A}^{(m)} = \begin{bmatrix} x_1^{(m)} & 1 \\ x_2^{(m)} & 1 \\ \cdot & \\ \cdot & \\ \cdot & \\ x_7^{(m)} & 1 \end{bmatrix}.$$

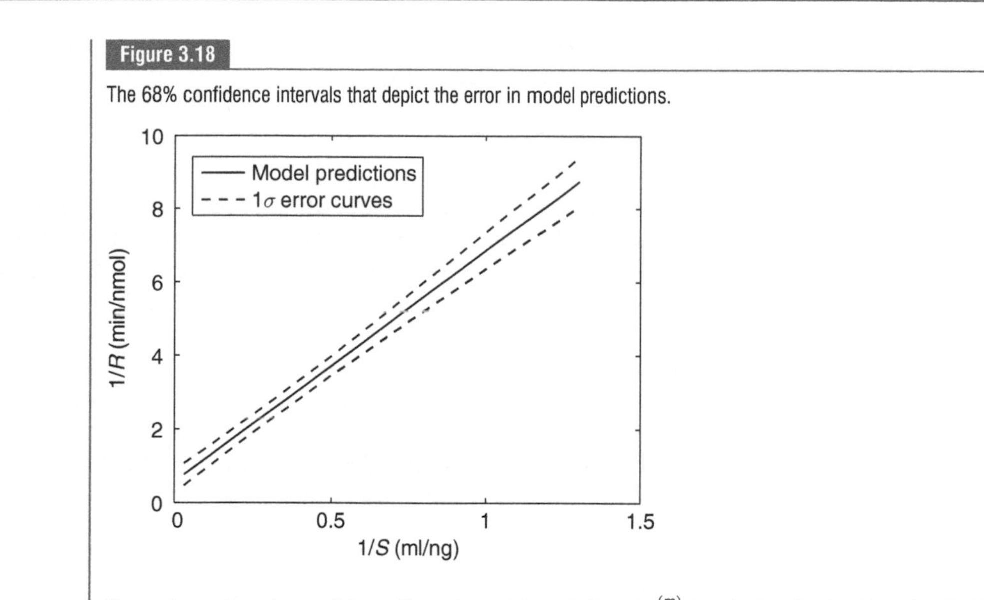

Figure 3.18

The 68% confidence intervals that depict the error in model predictions.

The variance (error) associated with each model prediction at $x_i^{(m)}$ is calculated using Equation (3.53). The diagonal of $\boldsymbol{\Sigma}^2_{\hat{\boldsymbol{y}}^{(m)}}$ gives us the variance vector $\boldsymbol{s}^2_{\hat{y}^{(m)}}$, i.e. the variance (error) associated with each prediction contained in the vector $\hat{\boldsymbol{y}}^{(m)} = \boldsymbol{A}^{(m)}\boldsymbol{c}$. Upon taking the square root of each element of $s^2_{\hat{\boldsymbol{y}}^{(m)}}$, we obtain the standard deviations associated with the model predictions $\hat{\boldsymbol{y}}^{(m)}$. Figure 3.18 shows a plot of the predicted $\hat{\boldsymbol{y}}^{(m)}$ values and the curves depicting the 68% (1σ) confidence intervals (magnitude of error or uncertainty) for the model predictions.

Upon careful examination you will note that the 1σ curves are not parallel to the regression curve. The error curves lie closest to the regression curve at the mean value of $\boldsymbol{x} = 1/\boldsymbol{S}$ of the original data set. In general, the uncertainty increases in the model's predictions as one moves away from the mean and towards the limits of the measured range.

(2) The mean

$$\bar{x} = \frac{\sum_{i=1}^{n} x_i}{n}$$

of a sample drawn by random sampling is an unbiased estimator of the population mean μ. The variance

$$s^2 = \frac{\sum_{i=1}^{n} (x_i - \bar{x})^2}{n - 1}$$

of a sample drawn by random sampling is an unbiased estimator of the population variance σ^2.

(3) A **Bernoulli trial** can have only two outcomes – success or failure – which are mutually exclusive. A binomial process involves several independent Bernoulli trials, where the probability of success remains the same in each trial. x successes out of n trials is a binomial variable. The **binomial probability distribution** is given by $C_x^n p^x (1-p)^{n-x}$ for $0 \le x \le n$, where x is an integer.

(4) The **Poisson distribution** describes the probability of events occurring in a finite amount of time or space. If x is a discrete random variable that takes on positive integral values and γ is the mean or expected value of x, then the Poisson distribution is given by $\gamma^x e^{-\gamma}/x!$.

(5) A **density curve** is a probability distribution plot for a continuous random variable x that has an area under the curve equal to 1.

(6) The **normal density function** describes a unimodal bell-shaped curve symmetric about the peak or mean value and is given by

$$f(x) = \frac{1}{\sqrt{2\pi}\sigma} e^{-(x-\mu)^2/2\sigma^2},$$

where μ is the mean and σ is the standard deviation of the normal distribution $N(\mu, \sigma)$.

(7) The **standard error of the mean** or **SEM** is the standard deviation of the **sampling distribution of a sample mean**, i.e. the probability distribution of all possible values of sample means obtained from a single population, where each sample is of size n; $\text{SEM} = s_{\bar{x}} = \sigma/\sqrt{n}$.

(8) The **z statistic** follows the **standard normal distribution** $N(0, 1)$, where $z = (x - \mu)/\sigma$, and x is a normally distributed random variable. Note that 95% of all observations of x are located within 1.96 standard deviations from the mean μ of the distribution.

(9) $\bar{x} \pm 1.96\,\sigma/\sqrt{n}$ is called the **95% confidence interval** for the population mean μ because it estimates, with 95% confidence, the range of values which contain the population mean.

(10) The **Student's t distribution** is used to estimate the width of the confidence interval when the population standard deviation is not known and the sample size n is small. The shape of the t curve is a function of the degrees of freedom available in calculating the sample standard deviation.

(11) The **central-limit theorem** states that if a sample is drawn from a non-normally distributed population with mean μ and variance σ^2, and the sample size n is large, the distribution of the sample means will be approximately normal with mean μ and variance σ^2/n.

(12) When combining measurements associated with error, one must also combine the errors (variances) in accordance with prescribed methods.

(13) It is not enough merely to fit a linear least-squares curve to a set of data thereby estimating the linear model parameters. One should also report the confidence one has in the values of the model coefficients and model predictions in terms of standard deviation (error) or confidence intervals.

3.9 Problems

Problems that require advanced critical analysis are labeled **.

3.1. A study on trends in cigarette smoking among US adolescents has a sample size of 16 000 students from grades 8, 10, and 12. It was found that in the year 2005, 25% of 12th-grade boys and 22% of 12th-grade girls were current smokers, i.e. they had smoked at least one cigarette in the past 30 days (Nelson *et al.*, 2008).

(a) If a class contains seven girls and eight boys, what is the probability that all students in the class are smokers?

(b) If a class contains seven girls and eight boys, what is the probability that *none* of the students in the class are smokers?

3.2. A cone and plate viscometer is used to shear a solution of suspended platelets. If the concentration of platelets in the solution is 250 000 platelets/μl, how many unique two-body collisions can occur per microliter?

3.3. Obesity is a rapidly growing epidemic in the USA. On average, an adult needs a daily intake of 2000 calories. However, there has been an increase in daily calorie intake of 200–300 calories over the past 40 years. A fast-food meal usually has high energy content, and habitual intake of fast-food can lead to weight gain and obesity. In fact,

Table 3.7.

Fast-food chain	Purchases <1000 calories (%)	Purchases 1000–1250 calories (%)	Purchases ≥1250 calories (%)
All[a] except Subway	66.5	19	14.5
Subway	78.7	11.9	9.4

[a] Includes Burger King, McDonald's, Wendy's, Kentucky Fried Chicken, Popeye's, Domino's, Papa John's, Pizza Hut, Au Bon Pain, and Taco Bell.

Table 3.8.

Subway customers	Purchases <1000 calories (%)	Purchases 1000–1250 calories (%)	Purchases ≥1250 calories (%)
Customers did not see calorie info.	77.0	12.7	10.3
Customers saw calorie info.	82.6	10.0	7.4
Calorie info. had effect on purchase	88.0	8.0	4.0
Calorie info. had no effect on purchase	79.8	11.1	9.1

in 2007, 74% of all food purchased from a US restaurant was fast-food. In 2007, most fast-food chains did not provide calorie or nutritional information, with the exception of Subway, which posted calorie information at the point of purchase. A study (Bassett *et al.*, 2008) surveyed the eating habits of numerous customers of New York City fast-food chains to assess the benefits of providing calorie information to patrons. A summary of their results is provided in Tables 3.7 and 3.8.

From Tables 3.7 and 3.8, the following is evident.

- On average, the calorie consumption of Subway customers who saw the calorie information reduced.
- The calorie purchasing habits of Subway customers who did not view the calorie information and of those who viewed the calorie information but disregarded it are similar. This indicates that it is not only the calorie-conscious population that view the calorie information.

Determine the following.

(a) What is the probability that three people walking into a fast-food restaurant categorized as "All except Subway" will *all* purchase a very high calorie content meal (≥1250 calories)? Calculate this probability for Subway customers for comparison.

(b) What is the probability that exactly five out of ten people walking into a fast-food restaurant categorized as "All except Subway" will purchase a normal calorific meal of <1000 calories? What is this probability for Subway customers?

(c) Using the `bar` function in MATLAB, plot a binomial distribution of the probabilities that i out of ten people ($0 \leq i \leq 10$) entering a fast-food chain categorized as "All except Subway" will purchase a normal calorific meal of < 1000 calories. Repeat for Subway customers.

(d) What is the probability that three Subway patrons do not see calorie information and yet all three purchase a very high calorie content meal ($\geq$1250 calories)?

(e) What is the probability that three Subway patrons see calorie information and yet all three purchase a very high calorie content meal ($\geq$1250 calories)?

(f) **If there are ten people who enter a Subway chain, and five individuals see the calorie information and five do not, what is the probability that exactly five people will purchase a normal calorific meal of < 1000 calories? Compare your answer with that available from the binomial distribution plotted in (c). Why is the probability calculated in (f) less than that in (c)?

(g) **If all ten people who purchase their meal from a Subway restaurant admit to viewing the calorie information before making their purchase, and if three out of the ten people report that the information did not influence their purchase, then determine the probability that exactly three of the ten customers purchased a meal with calorie content greater than 1000. Compare your answer with that available from the binomial distribution plotted in (c). How is your answer different, and why is that so?

3.4. An experiment W yields n outcomes. If S is the sample space then $S = (\xi_1, \xi_2, \xi_3, \ldots, \xi_n)$, where ξ_i is any of the n outcomes. An event is a subset of any of the n outcomes produced when W is performed, e.g. E_1: (ξ_1, ξ_2) or E_2: (ξ_1, ξ_2, ξ_3). An event is said to occur when any of the outcomes defined within the set occurs. For example, if a six-sided die is thrown and the event E is defined as (1, 2, 3), then E occurs if either a 1, 2, or 3 shows up. An event can contain none of the outcomes (the impossible event), can contain all of the outcomes (the certain event), or be any combination of the n outcomes. Using the binomial theorem,

$$(x + a)^n = \sum_{k=0}^{n} C_k^n x^k a^{n-k}$$

and your knowledge of combinatorics, determine the total number of events (subsets) that can be generated from W.

3.5. Reproduce Figure 3.4 by simulating the binomial experiment using a computer. Use the MATLAB uniform random number generator `rand` to generate a random number x between 0 and 1. If $x \leq p$, then the outcome of a Bernoulli trial is a success. If $x > p$, then the outcome is a failure. Repeating the Bernoulli trial n times generates a Bernoulli process characterized by n and p. Perform the Bernoulli process 100, 500, and 1000 times and keep track of the number of successes obtained for each run. For each case

(a) plot the relative frequency or probability distribution using the `bar` function;

(b) calculate the mean and variance and compare your result with that provided by Equations (3.15) and (3.18).

3.6. Show that the mean and variance for a Poisson probability distribution are both equal to the Poisson parameter γ.

3.7. Dizziness is a side-effect of a common type II diabetic drug experienced by 1.7 people in a sample of 100 on average. Using a Poisson model for the probability distribution, determine the probability that

(a) no person out of 100 patients treated with this drug will experience dizziness;

(b) exactly one person out of 100 patients treated with this drug will experience dizziness;

(c) exactly two people out of 100 patients treated with this drug will experience dizziness;

(d) at least two people out of 100 patients treated with this drug will experience dizziness.

3.8. Prolonged exposure to silica dust generated during crushing, blasting, drilling, and grinding of silica-containing materials such as granite, sandstone, coal, and some metallic ores causes silicosis. This disease is caused by the deposition of silica particles in the alveoli sacs and ducts of the lungs, which are unable to remove the silica particles by coughing or expelling of mucous. The deposited silica particles obstruct gaseous exchange between the lung and the blood stream. This sparks an inflammatory response which recruits large numbers of leukocytes (neutrophils) to the affected region. Roursgaard *et al.* (2008) studied quartz-induced lung inflammation in mice. The increase in the number of neutrophils is associated with higher levels of MIP-2 (anti-macrophage inflammatory protein 2). Bronchoalveolar fluid in the lungs of 23 mice exposed to 50 μg quartz was found to have a mean MIP-2 expression level of 11.4 pg/ml of bronchoalveolar fluid with a standard deviation of 5.75.

(a) What would you predict is the standard error of the mean of the MIP-2 expression levels?

(b) The expression levels may not be normally distributed, but since $n = 23$ we assume that the central-limit theorem holds. Construct a 95% confidence interval for the population mean.

3.9. Clinical studies and animal model research have demonstrated a link between consumption of black tea and a reduced risk for cancer. Theaflavins are polyphenols – biologically active components – abundantly present in black tea that have both anti-oxidative and pro-oxidative properties. Their pro-oxidant behavior is responsible for generating cytotoxic ROS (reactive oxygen species) within the cells. Normal cells have counteracting mechanisms to scavenge or reduce ROS with anti-oxidant enzymes. Cancer cells, on the other hand, are more sensitive to oxidative stress in their cells. Theaflavins were recently shown to be more toxic to tongue carcinoma cells than to normal gingival human fibroblasts, resulting in pronounced growth inhibition and apoptosis in cancer cells. These results suggest the useful chemopreventive properties of black tea for oral carcinoma. Exposure of human fibroblasts to 400 mM TF-2A (theaflavin 3-gallate) for 24 hours resulted in a mean viability of 82% with a standard error of the mean 3.33 ($n = 4$). Construct the 95% confidence interval for the population mean viability of normal fibroblasts. Exposure of carcinoma cells to 300 mM TF-2A (theaflavin 3-gallate) for 24 hours resulted in a mean viability of 62% with a standard error of the mean of 2.67 ($n = 4$). Construct the 95% confidence interval for the population mean viability of carcinoma cells.

3.10. A 20.00 ml sample of a 0.1250 molarity $CuSO_4$ solution is diluted to 500.0 ml. If the error in measurement of the molarity is ± 0.0002, that of the 20.00 ml pipet is ± 0.03 ml, and that of the volume of the 500 ml flask is ± 0.15 ml, determine how the molarity of the resulting solution should be reported. That is, what is the molarity of the resulting solution, and what error is there in this measurement? Recall that the molarity M of a solution is given by $M = n/V$, where n is the number of moles of solute and V is the volume of the solution in liters. Thus, in our case, $M_{\text{init}}V_{\text{init}} = M_{\text{final}}V_{\text{final}}$, or $M_{\text{final}} = M_{\text{init}}V_{\text{init}}/V_{\text{final}}$.

3.11. The pH of a solution is found to be 4.50 ± 0.02. What is the error in the concentration of H^+ ions?

3.12. In quantitative models of human thermal regulation, the convective heat transfer coefficient (h) combines with the conductive heat transfer coefficient through clothing (h_c) to yield an overall heat transfer coefficient H in the following way:

$$H = \frac{1}{1/h + 1/h_c}.$$

If the convective heat transfer coefficient is measured as $h = 43.0 \pm 5.2\,\mathrm{W/(m^2\,K)}$, and the conductive heat transfer coefficient through clothing is measured as $h_c = 26.1 \pm 1.8\,\mathrm{W/(m^2\,K)}$, calculate the overall heat transfer coefficient H and report the error in that estimate.

3.13. Many collections of biomolecular fibers are modeled as semi-flexible polymer solutions. An important length scale in such systems is the mesh size ξ, which represents the typical spacing between polymers in solution. The following equation relates the mesh size to the fiber radius a and the fiber volume fraction ϕ:

$$\xi = \frac{a}{\sqrt{\phi}}.$$

In the adventitia of the thoracic artery of the rat, the collagen fibrils have a diameter of 55 ± 10(SD) nm and a volume fraction of 0.0777 ± 0.0141(SD). Using the error propagation formulas, calculate the mean mesh size in the adventitia of rat thoracic artery, along with its 95% confidence interval.

3.14. A sample of acute myeloid leukemia (AML) cells are run through a Coulter Z2 Cell Counter to determine the distribution of cell sizes. This instrument determines the volume of each cell in the sample quite accurately, and reports a mean cell volume of 623.6 ± 289.1(SD) $\mu\mathrm{m}^3$. Recognizing that the spherical cell diameter (d) is simply related to the volume (V) by the relation

$$d = \left(\frac{6V}{\pi}\right)^{1/3},$$

use the error propagation formulas to calculate the mean cell diameter and standard deviation in diameter for this sample of AML cells.

3.15. A linear regression model is $\hat{y} = c_1 f_1(x) + c_2 f_2(x)$, where $f_1(x) = x$, $f_2(x) = 1$, and you have two data points (x_1, y_1) and (x_2, y_2). The predicted mean value of y at x_0 is $\hat{y}_0 = c_1 x_0 + c_2$, where c_1 and c_2 have already been determined. The variance associated with y is σ^2. You know from Equation (3.51) that the variance associated with $\hat{y}_0$ is given by

$$s_{\hat{y}_0}^2 = (f_1(x_0))^2 s_{c_1}^2 + (f_2(x_0))^2 s_{c_2}^2 + 2 f_1(x_0) f_2(x_0) \sum_{j=1}^{m} K_{1j} K_{2j} s_y^2,$$

$$K = (A'A)^{-1} A', \text{ where } A = \begin{bmatrix} x_1 & 1 \\ x_2 & 1 \end{bmatrix}.$$

(a) Find K in terms of the given constants and data points.

(b) Using Equation (3.49) and your result from (a), show that $s_{c_1}^2 = \sigma^2 / \sum (x_i - \bar{x})^2$.

(c) Using Equation (3.49) and your result from (a), show that $s_{c_2}^2 = \sigma^2 \sum x_i^2 / m \sum (x_i - \bar{x})^2$.

(d) Using Equation (3.50) and your result from (a), show that $s_{c_1 c_2}^2 = -\sigma^2 \bar{x} / \sum (x_i - \bar{x})^2$.

Note that the summation term indicates summation of all x from $i = 1$ to m, where $m = 2$ here.

Table 3.9.

	Reaction rate (nmol/(cells min))	
DOG conc. (mM)	With progesterone	Control case
0	0	0
5	3.25	2.4
10	2.4	3.75
20	5.1	–
25	6.0	4.05
30	7.2	5.4
40	7.7	6.5
50	9.1	–
60	9.05	7.8

Table 3.10.

G0 (mM)	G10 (mM)
10	5
8	3.1
6	2
4	1
2	0.5

Now prove that, for a linear equation in one independent variable, i.e. $\hat{y} = c_1 x + c_2$, the variance associated with the prediction of the mean, $\hat{y}$, at x_0 is

$$s_{\hat{y}_0}^2 = \sigma^2 \left[\frac{1}{m} + \frac{(x_0 - \bar{x})^2}{\sum (x_i - \bar{x})^2} \right].$$

What can you say about the effect of the magnitude of x_0 on the magnitude of the variance, $s_{\hat{y}_0}^2$?

3.16. Medina *et al.* (2004) studied the effect that estrogen replacement therapy has on deoxyglucose (DOG) transport in endometrial cancer cells. The data are presented in Table 3.9, showing the transport (i.e. "reaction") rate as a function of DOG concentration, either with or without 10 nM progesterone treatment.

Note that the control data set contains two fewer points. Perform a linear regression on these data using the normal equations to determine the Michaelis constant and R_{max}. Report each of your results with a 68% (1σ) confidence interval, and the units of measure. Plot the transformed data, your best-fit model, and curves that represent $\pm 1\sigma$ of your model prediction on the same set of axes. Use a legend to identify the different curves.

Are the Michaelis constants, with and without progesterone treatment, significantly different? In other words, do the 95% confidence intervals of the two parameters overlap? (If there is overlap of the two intervals, than we cannot conclude that they are significantly different.)

3.17. A certain enzyme acts on glucose. In a laboratory reaction (37 ºC, pH 7), measurements were made of the glucose concentration at time zero ($G0$) and after ten minutes ($G10$), and are given in Table 3.10.

First, convert the concentration measurements to reaction rates by subtracting $G10$ from $G0$ and dividing by 10 min. Estimate K_m and R_{max} by performing a linear regression of the Michaelis–Menten equation using the normal equations. Report the 68% confidence intervals and units. Plot the transformed data, your best-fit model, and curves that represent $\pm 1\sigma$ of your model prediction. Use a legend to identify the different curves.

References

Goldberger, A. L., Bhargava, V., West, B. J., and Mandell, A. J. (1985) On a Mechanism of Cardiac Electrical Stability. The Fractal Hypothesis. *Biophys. J.*, **48**, 525–8.

Goldberger, A. L., Amaral, L. A., Hausdorff, J. M., Ivanov, P., Peng, C. K., and Stanley, H. E. (2002) Fractal Dynamics in Physiology: Alterations with Disease and Aging. *Proc. Natl. Acad. Sci. USA*, **99** (Suppl 1), 2466–72.

Lee, J. S., FitzGibbon, E., Butman, J. A., Dufresne, C. R., Kushner, H., Wientroub, S., Robey, P. G. and Collins, M. T. (2002) Normal Vision Despite Narrowing of the Optic Canal in Fibrous Dysplasia. *N. Engl. J. Med.*, **347**, 1670–6.

Mazzocca, A. D., Spang, J. T., Rodriguez, R. R., Rios, C. G., Shea, K. P., Romeo, A. A., and Arciero, R. A. (2008) Biomechanical and Radiographic Analysis of Partial Coracoclavicular Ligament Injuries. *Am. J. Sports. Med.*, **36**, 1397–402.

Medina, R. A., Meneses, A. M., Vera, J. C. (2004) Differential Regulation of Glucose Transporter Expression by Estrogen and Progesterone in Ishikawa Endometrial Cancer Cells. *J. Endocrinol.*, **182**, 467–78.

Meyer, C., Robson, D., Rackovsky, N., Nadkarni, V., and Gerich, J. (1997) Role of the Kidney in Human Leptin Metabolism. *Am. J. Physiol.*, **273**, E903–7.

Mody, N. A. and King, M. R. (2007) Influence of Brownian Motion on Blood Platelet Flow Behavior and Adhesive Dynamics near a Plane Wall. *Langmuir*, **23**, 6321–8.

Nader, P. R., Bradley, R. H., Houts, R. M., McRitchie, S. L., and O'Brien, M. (2008) Moderate-to-Vigorous Physical Activity from Ages 9 to 15 Years. *Jama*, **300**, 295–305.

Nelson, D. E., Mowery, P., Asman, K., Pederson, L. L., O'Malley, P. M., Malarcher, A., Maibach, E. W., and Pechacek, T. F. (2008) Long-Term Trends in Adolescent and Young Adult Smoking in the United States: Metapatterns and Implications. *Am. J. Public Health*, **98**, 905–15.

Reichenbach, H. (1951) *The Rise of Scientific Philosophy* (Berkeley, CA: University of California Press).

Roursgaard, M., Poulsen, S. S., Kepley, C. L., Hammer, M., Nielsen, G. D., and Larsen, S. T. (2008) Polyhydroxylated C(60) Fullerene (Fullerenol) Attenuates Neutrophilic Lung Inflammation in Mice. *Basic Clin. Pharmacol. Toxicol.*, **103**, 386–8.

Tees, D. F., Waugh, R. E., and Hammer, D. A. (2001) A Microcantilever Device to Assess the Effect of Force on the Lifetime of Selectin-Carbohydrate Bonds. *Biophys. J.*, **80**, 668–82.

Waugh, R. E., Narla, M., Jackson, C. W., Mueller, T. J., Suzuki, T., and Dale, G. L. (1992) Rheologic Properties of Senescent Erythrocytes: Loss of Surface Area and Volume with Red Blood Cell Age. *Blood*, **79**, 1351–8.

West, B. J. (2006) *Where Medicine Went Wrong: Rediscovering the Path to Complexity* (Singapore: World Scientific Publishing Co.).

4 Hypothesis testing

4.1 Introduction

Often, a researcher, engineer, or clinician will formulate a statement or a belief about a certain method, treatment, or behavior with respect to one or more populations of people, animals, or objects. For example, a bioengineer might be interested in determining if a new batch of drug-coated stents has a drug coat thickness that is different from a previous batch of stents. The engineer makes several measurements on a small number of stents chosen at random. A substantial deviation in drug coating thickness from the standard range of acceptable values indicates that the process conditions have unexpectedly changed. The stents from the new batch may have either a thicker or a thinner drug coating compared to those of the previous batches, depending on the direction of deviation in coat thickness. Or, it is possible that the drug coat thickness widely fluctuates in the new batches of stents beyond the range specified by design criteria. The engineer formulates a hypothesis based on her preliminary survey. She will need to perform statistical tests to confirm her hypothesis.

Suppose a biomaterials researcher is testing a group of new promising materials that enhance the durability or lifetime of a contact lens. The researcher poses a number of hypotheses regarding the performance qualities of different lens materials. She must use statistical techniques to test her hypotheses and draw appropriate conclusions. As another example, a clinician may test a number of variations of a therapeutic procedure to determine which one is most effective in treating patients. He randomly assigns patients to the treatment groups and then records patient outcomes with respect to time. A comparison is drawn among the treatment groups using statistical tests, which yield a probability value that indicates whether the clinical evidence is in favor (or not in favor) of the clinician's hypothesis.

Hypothesis testing is a critical part of the method of scientific investigation. The investigator develops a hypothesis on the basis of previous experience, data, or simply intuitive reasoning. Experiments are performed or surveys conducted to generate data that characterize the process or phenomenon for which the hypothesis is made. The data are inspected for consistency with the hypothesis. If the data are consistent with the hypothesis, they offer evidence in support of the hypothesis. Because such studies are conducted using samples from their respective populations, and different samples from the same population exhibit variability in their characteristics, sample data suffer from sampling error. If we were to repeat the study, the new data set obtained would not be exactly the same as the data set from the previous study. If sampling variability in our data is unavoidable, how do we know if any discrepancies between our study findings and our hypothesis are simply due to sampling error or arise because the hypothesis is incorrect? Here, inferential statistics plays an important role. Hypothesis tests are statistical and probabilistic methods that help the researcher or engineer make a decision regarding the hypothesis based on objectivity and consistency rather than subjective opinions.

In Chapter 3, one application of statistical inference was demonstrated: estimation of population parameters. Probabilistic methods were used to generate $100(1-\alpha)\%$ confidence intervals, where α is the level of significance. Confidence intervals characterize the precision with which one can estimate a population parameter such as the population mean. In this chapter, we show how inferential statistics can be used to perform an even more powerful role: one of guiding decision-making. Using the concepts of statistics and probability in Chapter 3, such as probability distributions, error propagation formulas, the standard normal distribution, and the central-limit theorem, we develop statistical tests of significance to perform hypothesis testing, i.e. to retain or reject a hypothesis.

We begin this chapter by considering how one should formulate a hypothesis (Section 4.2). Once a hypothesis is made and worded in such a way that a statistical test can be applied to test it, how would one choose the most appropriate statistical test to use from the bewildering array of test procedures? Each statistical test is appropriate only for certain situations, such as when the observed variable is distributed normally within the population, or when multiple populations being compared have equal variances. A plethora of statistical tests have been designed by statisticians but their suitability for use depends on the nature of the sample data (e.g. cardinal, ordinal, or categorical), the nature of distribution of the random variable within the population (e.g. normal, binomial, Poisson, or non-normal), the number of populations to be compared, equality of variances among the populations being compared, and the sample sizes. Section 4.3 demonstrates how a statistical test is performed, how the probability value that is derived from the statistical test should be interpreted, and the likelihood of making an error when using a statistical test. The variety of statistical tests can be broadly categorized into parametric and non-parametric tests. Methods to assess whether the population distribution is normal are discussed in Section 4.4. Sections 4.5–4.7 cover parametric tests for hypotheses concerning a single population or comparison between two populations. A parametric test called ANOVA for comparing multiple samples is discussed in Section 4.8. Several important non-parametric tests are considered in Sections 4.9 and 4.10.

4.2 Formulating a hypothesis

The scientific exercise of formulating and testing hypotheses is the backbone of the scientific method of conducting and interpreting experimental work. Contrary to popular notion, hypothesis testing is not restricted to the scientific research community. Engineers, doctors, nurses, public health officials, environmentalists, meteorologists, economists, and statisticians often need to make an intuitive guess or prediction related to a treatment, process, or situation. The best course of action is determined after analyzing and interpreting available data. The data can either support the hypothesis or provide evidence against it. First, let's familiarize ourselves with what a hypothesis is. A **hypothesis** is a formal scientific statement that a person makes based on a conjecture or educated guess. For example, one may hypothesize from prior observation,

"Drinking tea reduces the incidence of cancer."

There may be no concrete evidence to back the above hypothetical statement. Proof must be sought to confirm or disprove the statement. One must perform a thorough scientific investigation to decide whether to reject (disprove) or retain a hypothesis. The investigation is scientific because it follows the scientific principles of objective exploration. A hypothesis (or a set of hypotheses) that is consistently retained time and again and has received overwhelming evidence to support it becomes a theory.

4.2.1 Designing a scientific study

The scientific study that is performed to uncover evidence that either supports or negates the hypothesis can be of two types: an experimental study or an observational study.

Experimental study

The investigator of an experimental study actively manipulates one or more conditions (or variables) keeping all other variables as constant as possible. The results of an experimental study can be used to infer a cause–effect relationship between the **experimental variables** (manipulated variables) and the **response variable** (observed outcome of interest).

In any experimental study that is performed to investigate the validity of a hypothesis, it is important to create a **control group** that mimics the treated group in every way except in receiving the treatment itself. If any marked deviation is observed in a patient's health status after treatment, the cause of change may not necessarily be the result of the treatment or drug itself, but is possibly due to several other factors involved in the treatment procedure. More than one control group may have to be created to test each of these factors. Often a control group may be included that does not experience any form of treatment but is simply observed for changes in the response variable with time. Several studies have uncovered the powerful psychological effect that a placebo has on treating a patient's ailments. It has been shown in several cases that a medication offered to a group of people worked just as well as a placebo, as long as the individuals were not aware of which type of treatment was being offered. The only effect of the treatment was to give the individual a psychological boost, convincing him or her that "help is on the way." If the control group had not been included in the experiment, the "placebo effect" would not have been discovered.

Bias in an experimental result is reduced by minimizing the differences between the treatment groups (which includes the control group) prior to administering treatments. A single sample of experimental subjects should be chosen initially that represents the population. Subjects from this sample should then be randomly (not haphazardly) allocated to all study groups. In this way, any inter-individual differences will be uniformly dispersed among all study groups, and the study groups will appear identical with respect to all other variables except the treatment conditions. This method of experimental design is called **completely randomized design**.

Observational study

The investigator of an observational study simply observes an existing situation to determine an association or pattern between a risk factor and an outcome. The risk factor(s) whose effect is studied is called the **exposure variable**(s). Examples of risk

Box 4.1A Use of cyclosporin to reduce reperfusion injury during acute myocardial infarction

Blockage of blood flow in coronary arteries due to thrombosis or lodging of emboli severely reduces oxygen supply to the cardiac muscle tissue. Deprived of oxygen, myocardial tissue will die and form an infarct. This condition is called acute myocardial infarction. Interventions such as stenting allow reperfusion of the affected tissue by opening the passageways in the occluded artery. Ironically, reperfusion of tissue can cause further myocardial injury. The sudden availability of oxygen to the cardiac tissue produces an overload of reactive oxygen species (ROS) within the cells. The restoration of blood flow brings an abundance of white blood cells, which, upon contact with the damaged tissue, release pro-inflammatory factors and free radicals. Sudden excessive amounts of ROS invokes trauma in the ischemic cells and may trigger the apoptotic cascade, possibly through a route involving permeabilization of mitochondria within the myocardial cells.

Cyclosporin is a well-known and widely used immunosuppressant for allogeneic organ transplantation. It is a fungal peptide, which also blocks the opening of mitochondrial permeability-transition pores. The formation of these pores (or holes) in the mitochondrial membrane is responsible for causing mitchondrial collapse and signaling end-stage apoptosis.

An experimental single-blinded pilot study (Piot *et al.*, 2008) was performed in France to determine whether administration of cyclosporin along with stenting, when treating a patient suffering from acute myocardial infarction, would limit infarct size and thereby reduce reperfusion injury. Two study groups were created. In the **treatment group**, the patients were given a single intravenous bolus injection of cyclosporin dissolved in saline, while the **control group** of patients received an equivalent amount of saline alone. Patients suffering from chest pain, who underwent coronary angiography and were deemed eligible for the stenting procedure, were randomly assigned to either the treatment group or to the control group. Each patient was given a sealed envelope that contained a number revealing the study group that he or she was assigned to. The experimental trial was **single-blinded** because the patients were aware of the study group they were assigned to, but the researchers who analyzed the data were unaware of the assignments. If the trial had been **double-blinded**, then the patients themselves would not have known which study group they were being assigned to. The two study groups were found to be similar with respect to control variables such as age, gender ratio, body-mass index, physical health conditions such as diabetes, hypertension, and smoking, area of tissue at risk for infarction, and other baseline characteristics.

Creatine kinase and troponin I concentrations in blood were measured repeatedly over a period of 72 hours following the stenting procedure to assess the size of the infarct. The size of the infarct was measured using MRI 5 days following reperfusion. A comparison of the creatine kinase concentrations as well as the infarct size in the two study groups led to the conclusion that administering cyclosporin at the time of reperfusion resulted in reduced infarct size compared to that observed with placebo. Later in this chapter, we show how this conclusion was obtained using the Wilcoxon rank-sum test.

factors are substance abuse, exposure to UV rays, chemicals, x-rays, diet, body weight, and physical activity. The outcome is measured in terms of the outcome variable or response variable. Examples of an outcome variable are change in physical or mental health status, mortality, birth rate, birth weight, preterm labor, and obesity. Often, researchers are interested in finding a link between some type of exposure (risk factor) and disease (response variable). In an observational study, no attempt is made to manipulate the variables of interest or the subjects of the study. The disadvantage of this type of study is that a positive or negative correlation between two observed variables does not necessarily indicate a cause–effect relationship between the two variables. The observed correlation may be due to an extraneous variable not considered in the study that is associated with the

Box 4.2A Treatment of skeletal muscle injury using angiotensin II receptor blocker

Skeletal muscle injuries are particularly common among athletes. Recovery from muscle strains often require a lengthy rehabilitation period and time away from being on the field or courts. Importantly, injury to skeletal muscle tissue predisposes orthopedic patients to recurring injuries because of weakening of the tissue. Following injury, regeneration of muscle tissue is compromised by simultaneous formation of fibrous (connective) tissue within the regenerated tissues. Patients who undergo surgical procedures that involve incisions made in skeletal muscle tissue also face the same problem of compromised healing and longer recovery times due to fibrosis occurring in the healing tissue.

Angiotensin II is a vasoconstricting peptide. Several antagonists have been developed as drugs (e.g. Cozaar (losartan) by Merck & Co.) to block angiotensin II receptors as a means to relieve hypertension. Losartan inhibits angiotensin II binding to its receptor and prevents the activation of precise signaling pathways that induce vasoconstriction. It has recently been found that angiotensin II upregulates the expression of transforming growth factor-beta1 (TGF–β1) in fibroblasts; TGF–β1 induces fibroblast proliferation and formation of fibrotic (collagenous) tissue.

Bedair and co-workers from the University of Pittsburgh (Bedair *et al.*, 2008) hypothesized that treating animals suffering from skeletal muscle injury with an angiotensin receptor blocker (ARB) may reduce fibrosis in the regenerating tissue. In order to test their hypothesis, they formed a study group of 40 mice on which lacerations to the gastrocnemius muscle were made to simulate muscle strain injury. The mice were randomly assigned to either

(1) a control group that received tap water for drinking over a period of 3 weeks and 5 weeks;
(2) a low-dose ARB treatment group that received tap water containing dissolved losartan at a concentration of 0.05 g/l, or
(3) a high-dose ARB treatment group that received tap water containing dissolved losartan at a concentration of 0.5 g/l.

The mice were allowed to drink water or ARB solutions at amounts and frequencies of their own discretion. The mice were sacrificed at the end of 3 weeks or 5 weeks and the number of regenerating muscle fibers and extent of fibrous tissue development were determined by histological analysis.

In later sections of this chapter we will perform *t* tests to identify significant differences, if any, among the extents of muscle fiber regeneration and fibrosis occurring in each of the three study groups.

exposure variable and may be causally related to the outcome variable. Or, the extraneous variable may have a cause–effect relationship with both the exposure and outcome variables.

While examining results obtained from a scientific study, one must be aware of the possibility that extraneous variables can influence the outcome. When searching for the nature and extent of association between the exposure variable and the response variable, these extraneous variables confound or obscure the true relationship sought. Response of individuals to exercise, diet, and prescription drugs can depend on age, gender, body weight, smoking status, daily alcohol consumption, race, and other factors. Suppose we are interested in determining the type of association that exists between smoking and body weight. Does heavy smoking have a tendency to increase body weight? Or do overweight people have a higher tendency to smoke? If the smoking population on average has a higher alcohol consumption, or relatively unhealthy eating habits, then the amount of alcohol consumption or type of diet may influence body weight more than smoking. Alcohol consumption and diet are called **confounding variables**, and may be linked either to smoking status, to body weight status, or to both **categorical variables**.

Box 4.3 Fluoridation of water supply linked to cancer

A controversy spanning several decades continues over the health effects of fluoridation of the US water supply. It is well established that fluoride helps to prevent tooth decay caused by plaque – a sticky bacterial film that forms on teeth. When breaking down sugars in food particles, bacteria secrete acids that damage the hard enamel surface of teeth and expose the delicate inner layers of the teeth (dentin, pulp) and thereby cause cavities. Fluoride interferes with the action of bacterial acids. It not only prevents acid demineralization or dissolution of the enamel, but can reverse low levels of tooth decay by promoting repair or remineralization of teeth. Moreover, in children under the age of six, fluoride that is ingested is incorporated into developing permanent teeth, thereby increasing resistance of teeth to demineralization. Accordingly, fluoride was added to the public water supply to aid in the prevention of tooth decay.

Much skepticism has arisen recently over the benefit of adding fluoride to the water supply. It is believed that the protective action of fluoride results primarily from topical application as opposed to ingestion. More importantly, animal and insect studies have shown that low levels of fluoride have mutagenic effects on chromosomal DNA and may therefore act as cancer-causing agents. Several epidemiological studies have been conducted to assess the useful/harmful effects of fluoridation of the water supply with respect to

(1) improving dental health (reducing tooth decay), and
(2) carcinogenic effects of fluoride.

The first of such epidemiological (observational) studies to pinpoint a relationship between fluoridation of water supply and death due to cancer was carried out by Yiamouyiannis and Burk (1977). They chose to compare the cancer incidence rates observed in ten fluoridated cities (treatment group) and ten non-fluoridated cities (control group) across the USA. Prior to commencement of fluoridating the water supply in any of these cities, cancer death rates (obtained from the National Center for Health Statistics) showed similar trends in both groups – fluoridated cities and non-fluoridated cities – between 1940 and 1950. Fluoridation of the water supply was initiated in the treatment group from 1952 to 1956. In 1969, the cancer death rates in the fluoridated cities were significantly higher than those observed in the non-fluoridated cities. Cancer death rate (CDR) is a function of many variables and strongly depends on age, ethnicity, and exposure to carcinogens (toxins, pollutants, chemicals, sunlight). Yiamouyiannis and Burk (1977) divided the population in each of these cities into four age groups: 0–24, 25–44, 45–64, and 65+. The changing trends of cancer death rates with time from 1950 to 1970 were studied in each of these four age groups. It was found that cancer death rate decreased significantly in the first two age groups from 1950 to 1970, while the cancer death rate increased significantly in the third age group in both fluoridated and non-fluoridated cities. An increase in cancer death rate for the 65+ population was only found in fluoridated cities and not in the control group. Their most important finding was a significantly greater rise in cancer death rate in the two age groups 45–64 and 65+ in the fluoridated cities compared to the control group, while there was no difference in the CDR within the 0–44 population of the treatment and control groups. Table 4.1 illustrates the method by which CDRs in the treatment and control groups were compared.

No differences in CDR of fluoridated and non-fluoridated cities were found on the basis of ethnicity differences within the two groups.

While Yiamouyiannis and Burk (1977) controlled for age and ethnicity in their study, they did not recognize or draw attention to any other confounding variables that could be responsible for the CDR differences observed between the treated and control groups. Thus, a major drawback of this study is that the authors assumed that the cities were all similar in every way except for fluoridation status, age, and ethnicity. If the cities were truly similar, then how does one explain the observed differences in CDR of the fluoridated and non-fluoridated cities in 1952 (prior to fluoridation), especially for the 45–64 and 65+ age groups?

The authors directly compared the values of CDRs from both groups. However, can cancer death rates from different cities be compared directly to assess for carcinogenicity of a chemical? Many other

Table 4.1. *Difference in cancer death rates (CDR) of fluoridated and non-fluoridated cities in the USA*

Age group	Difference in CDR of fluoridated cities and non-fluoridated cities 1952	1969	Change in difference in CDR of fluoridated and non-fluoridated cities from 1952 to 1969	Calculated p value
0–24	−0.1929	+0.2157	+0.41	not significant
25–44	+2.212	+2.370	+0.16	not significant
45–64	+12.98	+28.13	+15.2	<0.02
65+	+57.36	+92.75	+35.4	<0.05

extraneous variables exist that can confound the result. It is possible that the fluoridated cities had something else in common, e.g. exposure to pollutants. Perhaps the treated cities were more industrialized, because of which the public came in contact with pollutants to a greater degree than inhabitants of cities of the control group. In the absence of identification of and control for the additional extraneous factors, it is difficult to interpret the results.

Doll and Kinlen (1977) repeated this study and found that the average death rate (actually the ratio of observed death rate to expected death rate), adjusted for age, gender, and ethnicity, slightly declined in the fluoridated cities from 1950 to 1970, while they remained the same in the non-fluoridated cities over this duration. In this study, the decreasing or increasing trend in cancer death rates of each group of cities with time was studied without having calculating differences between dissimilar groups of cities in time. A study designed in this fashion, in which confounders are better controlled, yields a more convincing result.

Doll and Kinlen's findings do not rule out the possibility that fluoridation of our water supply may be harmful. Perhaps the amount of data available to them was insufficient to prove that fluoridation has negative effects. However, it is very important when analyzing data and drawing statistical inferences to take into account as many confounding variables as possible and to interpret results responsibly. It is critical that one does not try to extract more information out of one's result than what is scientifically justifiable.

An observational study can be of three types:

(1) **Prospective study or cohort study** In this study, two or more groups of disease-free individuals that differ based on exposure to certain risk factors are selected using random sampling procedures. These individuals are followed up until some point in the future. During this time development of the disease (response variable) in the individuals is noted. The individuals under study are collectively referred to as a **cohort**.

(2) **Retrospective study or case-control study** Two or more groups of individuals are identified from the population who either have one or more conditions (diseases) or who do not. A study is performed to determine which of these individuals were exposed to one or more risk factors in the past. The "case" denotes the existence of the condition and is represented by the group of individuals that have developed that condition. The "control" group does not show signs of disease. The disadvantages of a case-control study are that (i) the past must be reconstructed from memory

Box 4.4 Smoking and cognitive decline

A **prospective cohort study** (1995–2007) was performed in Doetinchem, Netherlands, to determine the effect of smoking on decline of cognitive functions in middle-aged men and women over a 5-year period (Nooyens *et al.*, 2008). Reduced cognitive functioning occurs naturally with aging and is also a symptomatic precursor to neurodegenerative disorders such as dementia and Alzheimer's disease. If smoking causes accelerated cognitive decline in the elderly, then intervention programs to prevent or end the smoking habit must be implemented early on. In this prospective study, participants between the ages of 43 and 70 years underwent cognitive testing, once at **baseline**, i.e. when they signed up for participation, and once at **follow-up**, i.e. at the end of the 5-year period after entering the study. The study group ($n = 1904$) was divided into four categories that described their smoking status:

(1) those who had never smoked (34.9%),
(2) ex-smokers (those who quit smoking before baseline) (43.8%),
(3) recent quitters (those who quit smoking in between baseline and follow-up) (6.3%), and
(4) consistent smokers (15.1%).

The four areas of cognitive testing were

(1) memory,
(2) speed of cognitive processes,
(3) cognitive flexibility, and
(4) global cognitive function.

The distributions of the scores from different cognitive tests were unimodal and skewed to the right. The results of the tests were transformed to a normal distribution and then converted to standardized z scores using the means and standard deviations of the respective test scores at baseline. The standard deviations of the test score distributions at follow-up were the same as those at baseline testing.

Cognitive decline (or reduction in cognitive test scores) was observed in all four areas of testing over the 5-year period irrespective of the smoking status due to the aging-related decline among middle-aged people. It was found that smokers were significantly worse off at baseline than those who had never smoked in the areas of cognitive speed, flexibility, and global function. At follow-up, the decline in all areas of cognitive functioning over 5 years was found to be more rapid among smokers than among those who had never smoked. The ex-smokers and recent quitters exhibited cognitive declines intermediate to that of those who had never smoked and consistent smokers. Variables recognized as confounders were age, gender, education level, body-mass index, hypertension, cardiovascular disease, HDL cholesterol levels, physical activity, alcohol consumption, energy intake, fat intake, and antioxidant intake. These variables exhibited an association either with smoking status or with cognitive function decline. A multivariable linear regression analysis was performed to adjust (standardize) the test results based on differences within the study groups with respect to the confounding variables.

It was concluded that smoking impacts cognitive decline, and that the areas of cognitive functioning may be impacted differently by the negative effects of smoking in a time-dependent, order-specific fashion.

(authentic baseline data are not available) and (ii) the case group is often a sample of convenience and therefore not necessarily a random sample.

(3) **Cross-sectional study** A study group is identified and the variables of interest are measured at the same point in time to determine if the variables are associated. The study group is not followed in time into the future or the past. The purpose of this study is to ascertain prevalence of disease as opposed to the incidence of disease; the latter is the goal of prospective and retrospective studies.

Because sociodemographic factors such as age, education, income, race/ethnicity, marital status, and employment status influence people in many ways, it is important to control for these variables in any broad scientific study. The experiment must be carefully designed to prevent potential confounders such as sociodemographic factors, gender, diet, and substance abuse from influencing the result. One can use simple standardization procedures to adjust the population parameter estimate to correspond to the same distribution of the confounding variable(s) in all study samples. Alternatively, special experimental designs are employed called **restricted randomization techniques. Blocking** is one method of restricted randomization in which the study group is divided into blocks. Each block contains the same number of experimental subjects. Randomized allotment of subjects to each treatment group is performed within each block. A key feature of the blocking method is that, within each block, variation among subjects due to effects of extraneous variables except for the treatment differences is minimized. Extraneous variability is allowed between the blocks. **Stratification** is another method of restricted randomization in which the treatment or observed groups are divided into a number of categories (strata) that are defined by the control variable. Individuals are then randomly allocated to the treatment groups within each strata.

Randomization is part and parcel of every experimental design. It minimizes biases in the result, many of which are unknown to the experimentalist. Blocking methods or stratification designs help to improve the precision of the experiment. These methods are different from the **completely randomized design** discussed earlier, in which the study group is not divided into strata or blocks. This is the only experimental design considered in this book. Other experimental designs require more sophisticated techniques of analysis which are beyond the scope of this book. There are many options available for designing an experiment, but the choice of the best design method depends upon the nature of the data and the experimental procedures involved. The **design of experiments** is a separate subject in its own right, and is closely linked to the field of statistics and probability.

In this section, we have distinguished between two broadly different methods of scientific study – experimental and observational – used to test a hypothesis. Henceforth, we refer to all scientific or engineering studies that are performed to generate data for hypothesis testing as "experiments." We will qualify an "experiment" as an experimental or observational study, as necessary.

4.2.2 Null and alternate hypotheses

Before conducting a scientific study (or engineering analysis that involves hypothesis testing), the original hypothesis should be reconstructed as two complementary hypotheses. A hypothesis cannot be proven or deemed as "true," no matter how compelling the evidence is. There is no absolute guarantee that one will not find evidence in the future that contradicts the validity of the hypothesis. Therefore, instead of attempting to prove a hypothesis, we use experimental data as evidence in hand to disprove a hypothesis. If the data provide evidence contrary to the statement made by the hypothesis, then we say that "the data are not compatible with the hypothesis."

Let's assume we wish to test the following hypothesis:

"People with type-2 diabetes have a blood pressure that is different from that of healthy people."

In order to test our hypothesis, a sample must be selected from the adult population diagnosed with type-2 diabetes, and blood pressure readings from each individual in the group must be obtained. The mean blood pressure values of this sample are compared with the population mean blood pressure of the non-diabetic fraction of the adult population. The exact method of comparison will depend upon the statistical test chosen.

We also rewrite our original hypothesis in terms of two contradicting statements. The first statement is called the **null hypothesis** and is denoted by H_0:

"People with type-2 diabetes have the same blood pressure as healthy people."

This statement is the opposite of what we suspect. It is *this statement* that is tested and will be either retained or rejected based on the evidence obtained from the study. Due to inherent sampling error, we do not expect the differences between the two populations to be exactly zero even if H_0 is true. Some difference in the mean blood pressures of the samples derived from both populations will be observed irrespective of whether H_0 is true or not. We want to know whether the observed dissimilarity between the two samples is due to natural sampling variation or due to an existing difference in the blood pressure characteristics of the two populations. Hypothesis testing searches for this answer. A statistical test assumes that the null hypothesis is true. The magnitude of deviation of the sample mean (mean sample blood pressure of type-2 diabetics) from the expected value under the null hypothesis (mean blood pressure of healthy individuals) is calculated. The extent of deviation provides clues as to whether the null hypothesis should be retained or rejected.

We also formulate a second statement:

"People with type-2 diabetes have a blood pressure that is different from that of healthy people."

This statement is the same as the original research hypothesis, and is called the **alternate hypothesis**. It is designated by H_A. Note that H_0 and H_A are mutually exclusive events and together exhaust all possible outcomes of the conclusion.

If the blood pressure readings are similar in both populations, then the data will not appear incompatible with H_0. We say that "we fail to reject the null hypothesis" or that "there is insufficient evidence to reject H_0." *This does not mean that we accept the null hypothesis.* We do not say that "the evidence supports H_0." It is possible that the data may support the alternate hypothesis. Perhaps the evidence is insufficient, thereby yielding a false negative, and availability of more data may establish a significant difference between the two samples and thereby a rejection of H_0. Thus, we do not know for sure which hypothesis the evidence supports. Remember, while our data cannot prove a hypothesis, the data can contradict a hypothesis.

If the data are not compatible with H_0 as inferred from the statistical analysis, then we reject H_0. The criterion for rejection of H_0 is determined before the study is performed and is discussed in Section 4.3. Rejection of H_0 automatically means that the data are viewed in favor of H_A since the two hypotheses are complementary. Note that we do not accept H_A, i.e. we cannot say that H_A is true. There is still a possibility that H_0 is true and that we have obtained a false positive. However, a "significant result," i.e. one that supports H_A, will warrant further investigation for purposes of confirmation. If H_A appears to be true, subsequent action may need to be taken based on the scientific significance (as opposed to the statistical significance) of the result.[1] On the other hand, the statement posed by the null hypothesis is

[1] A statistically significant result implies that the populations are different with respect to a population parameter. But the magnitude of the difference may not be **biologically (i.e. scientifically) significant**. For example, the statistical test may provide an interval estimate of the difference in blood pressure of

one that usually indicates no change, or a situation of status quo. If the null hypothesis is not rejected, in most cases (except when the null hypothesis does not represent status quo; see Section 4.6.1 for the paired sample *t* test) no action needs to be taken.

Why do we test the null hypothesis rather than the alternate hypothesis? Or, why do we assume that the null hypothesis is true and then proceed to determine how incompatible the data is with H_0? Why not test the alternate hypothesis? If instead, we assume that H_0 is false, then H_A must be true. Now, H_0 always includes the sign of equality between two or more populations with respect to some property or population variable. Accordingly, H_A does not represent equality between two population parameters. If we assume H_A to be true, then we do not know a priori the magnitude of difference between the two populations with respect to a particular property. In that case, it is not possible to determine how incompatible the data are with H_A. Therefore, we must begin with the equality assumption.

4.3 Testing a hypothesis

The procedure of testing a hypothesis involves the calculation of a **test statistic**. A test statistic is calculated after a hypothesis has been formulated, an experiment has been designed to test the null hypothesis, and data from the experiment have been collected. In this chapter, we are primarily concerned with formulating and testing hypotheses concerning the population mean or median. The method chosen to calculate the test statistic depends on the choice of the hypothesis test. The value of the test statistic is determined by the extent of deviation of the data (e.g. sample mean) from expected values (e.g. expected population mean) under the null hypothesis. For illustrative purposes, we discuss the **one-sample *z* test**.

Assume that we have a random sample of $n = 100$ men of ages 30 to 55 that have disease X. This sample is representative of the population of all men between ages 30 and 55 with disease X. The heart rate within population X is normally distributed. The sample statistics include mean $\bar{x}$ and sample variance s^2. Because the sample is fairly large, we assume that the population variance σ^2 is equal to the sample variance.

We wish to know if population X has a heart rate similar to that of the population that does not have disease X. We call the latter population Y. The heart rate at rest in the adult male Y population is normally distributed with mean μ_0 and variance σ^2, which is the same as the variance of population X.

The non-directional null hypothesis H_0 is

"The heart rate of the X population is the same as that of the Y population."

If the mean heart rate of population X is μ, then H_0 can be stated mathematically as H_0: $\mu = \mu_0$.

The non-directional alternate hypothesis H_A is

"The heart rate of the X population is different from that of the Y population,"

or H_A: $\mu \neq \mu_0$.

type-2 diabetics from that of the healthy population as 0.5–1 mm Hg. The magnitude of this difference may not be important in terms of medical treatment and may have little, if any, biological importance. A statistically significant result simply tells us that the difference observed is attributable to an actual difference and not caused by sampling error.

4.3.1 The *p* value and assessing statistical significance

We want to know if population X is equivalent to population Y in terms of heart rate. In Chapter 3, we discussed that the means $\overline{x}$ of samples of size n taken from a normal population (e.g. population Y) will be distributed normally $N(\mu_0, \sigma/\sqrt{n})$ with population mean μ_0 and variance equal to σ^2/n. If the null hypothesis is true, then we expect the means of samples ($n = 100$) taken from population X to be normally distributed with the population mean μ_0 and variance σ^2/n.

We can construct an interval from the population Y mean

$$\mu_0 \pm z_{1-\alpha/2}\frac{\sigma}{\sqrt{n}}$$

using the population parameters such that $100(1-\alpha)\%$ of the samples of size n drawn from the reference population will have a mean that falls within this range. If the single sample mean we have at our disposal lies within this range, we conclude that the mean of the sample drawn from population X is not different from the population Y mean. We do not reject the null hypothesis. If the sample mean lies outside the confidence interval constructed using population Y parameters, then the mean of the population X from which the sample is derived is considered to be different from the mean of population Y. This result favors the alternate hypothesis.

Alternatively, we can calculate the ***z* statistic**,

$$z = \frac{\overline{x} - \mu_0}{\sigma/\sqrt{n}}, \tag{4.1}$$

which is the standardized deviation of the sample mean from the population mean if H_0 is true. The z statistic is an example of a test statistic. The magnitude of the z statistic depends on the difference between the sample mean and the expected value under the null hypothesis. This difference is normalized by the standard error of the mean. If the difference between the observed and expected values of the sample mean is large compared to the variability of the sample mean, then the likelihood that this difference can be explained by sampling error is small. In this case, the data do not appear compatible with H_0. On the other hand, if the difference is small, i.e. the value of z is close to zero, then the data appear compatible with H_0.

Every test statistic exhibits a specific probability distribution under the null hypothesis. When H_0 is true, the z statistic is distributed as $N(0, 1)$. For any value of a test statistic, we can calculate a probability that tells us how common or rare this value is if the null hypothesis is true. This probability value is called the ***p* value**.

> The ***p* value** is the probability of obtaining a test statistic value at least as extreme as the observed value, under the assumption that the null hypothesis is true. The p value describes the extent of compatibility of the data with the null hypothesis. A p value of 1 implies perfect compatibility of the data with H_0.

In the example above, if H_0 is true, we expect to see a small difference between the sample mean and its expected value μ_0 most of the time. A small deviation of the sample mean $\overline{x}$ from the expected population mean μ_0 yields a small z statistic, which is associated with a large p value. Small deviations of the sample mean from the expected mean can easily be explained by sampling differences. A large p value demonstrates compatibility of the data with H_0. On the other hand, a very small p value is associated with a large difference between the mean value of the sample data and the value expected under the null hypothesis. A small p value implies that when the null hypothesis is true, the chance of observing such a large difference is rare. Large discrepancies between the data and the expected values under the null hypothesis are more easily explained by a real difference between the two populations than by sampling error, and therefore suggest incompatibility of the data with H_0.

A cut-off p value must be chosen that clearly delineates test results compatible with H_0 from those not compatible with H_0.

The criterion for rejection of the null hypothesis is a threshold p value called α, the **significance level**. The null hypothesis is rejected when $p \leq \alpha$.

For the z test, α corresponds to the area under the standard normal curve that is associated with the most extreme values of the z statistic. For other hypothesis tests, the test statistic may not have a standard normal distribution, but may assume some other probability distribution.

The significance level α is the area under the probability distribution curve of the test statistic that constitutes the **rejection region**. If a p value less than α is encountered, then the discrepancy between the test statistic and its expected value under the null hypothesis is interpreted to be large enough to signify a true difference between the populations. The test result is said to be "**statistically significant**" and the populations are said to be "significantly different at the α level of significance."

A statistically significant result leads to rejection of H_0. The $(1 - \alpha)$ area under the distribution curve represents the region that does not lead to rejection of H_0. If $p > \alpha$, then the test result is said to be "statistically insignificant." The data do not provide sufficient evidence to reject H_0. This explains why hypothesis tests are also called **tests of significance**. For purposes of objectivity in concluding the statistical significance of a test, the significance level is set at the time of experimental design before performing the experiment.

So far we have alluded to **non-directional** or **two-sided hypotheses**, in which a specific direction of the difference is not investigated in the alternate hypothesis. A non-directional hypothesis is formulated either when a difference is not expected to occur in any particular direction, or when the direction of difference is immaterial to the experimenter. A non-directional hypothesis can be contrasted with a directional hypothesis, discussed later in this section. In a non-directional hypothesis test, only the magnitude and not the sign of the test statistic is important for assessing data compatibility with H_0. For a non-directional z test, if the significance level is α, then the rejection region is equally distributed at both ends (tails) of the standard normal curve as shown in Figure 4.1. Hence, a non-directional z test is also called a **two-tailed test**. Each tail that contributes to the rejection region has an area of $\alpha/2$ (see Figure 4.1). If $\alpha = 0.05$, then 2.5% of the leftmost area and 2.5% of the rightmost area under the curve (tail regions) together form the rejection region. The critical values of z that demarcate the rejection regions are $z_{\alpha/2}$ and $z_{1-\alpha/2}$ such that

$$\Phi\left(z_{\alpha/2}\right) = \alpha/2 \quad \text{and} \quad \Phi\left(z_{1-\alpha/2}\right) = 1 - \alpha/2,$$

where Φ is the cumulative distribution function for the normal distribution (see Section 3.5.4, Equation (3.34)). If the calculated z value falls in either tail region, i.e. $z < z_{\alpha/2}$ or $z > z_{1-\alpha/2}$, then the result is statistically significant and H_0 is rejected (see Figure 4.1).

Figure 4.1

Two-sided z test; z follows the $N(0,1)$ distribution curve when H_0 is true. The shaded areas are the rejection regions.

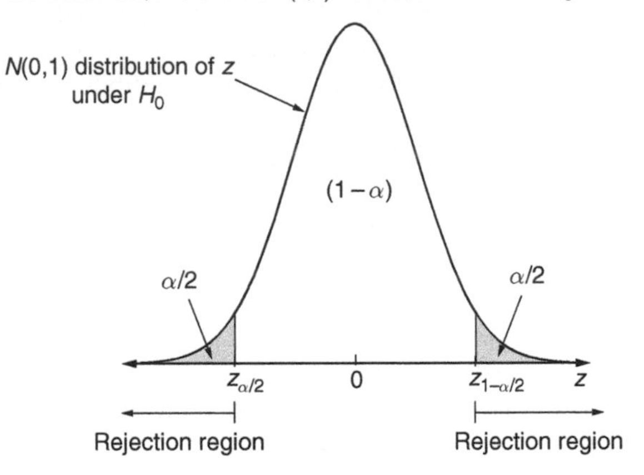

How is the p value calculated for a two-sided z test? If $z < 0$, then the probability of obtaining a value of the test statistic at least as extreme as what is observed is $p = \Phi(z)$, when considering only negative values of z. For a non-directional hypothesis test, large positive or negative values of z can lead to rejection of H_0. It is the absolute value of z that is important. When calculating the p value, we must also take into account the probability of observing positive z values that are at least as extreme as the observed absolute value of z. When $z < 0$, the p value is calculated as

$$p = \Phi(z) + 1 - \Phi(|z|).$$

Since the z distribution is perfectly symmetric about 0, $\Phi(z) = 1 - \Phi(|z|)$ for $z < 0$. Then,

$$p = 2\Phi(z).$$

If $z > 0$, then

$$p = 2(1 - \Phi(z)).$$

We can condense the above equations into a single equation by using the absolute value of z.

For a two-sided z test, p is calculated as

$$p = 2(1 - \Phi(|z|)). \quad (4.2)$$

There are two routes one can take to determine if the test result is statistically significant.

(1) Using the p value. The p value associated with the calculated test statistic is compared with the significance level of the test, α.
 (a) If $p < \alpha$, H_0 is rejected.
 (b) If $p > \alpha$, H_0 is retained.
(2) Using the test statistic. The calculated test statistic is compared with the critical value. In the z test, z is compared with the threshold value of z, which is $z_{1-\alpha/2}$.
 (a) If $|z| > z_{1-\alpha/2}$, H_0 is rejected.
 (b) If $|z| < z_{1-\alpha/2}$, H_0 is retained.

The p value method is preferred for the reason that the p value precisely conveys how strong the evidence is against H_0. A very small p value < 0.001 usually indicates a highly significant result, while $p \sim 0.05$ may indicate a result that is barely significant. Reporting the p value in addition to stating whether your results are significant at the α level is good practice. Other professionals in the same or related fields who use and interpret your data will have the freedom to set their own criteria for the significance level and decide for themselves the significance of the result.

The alternate hypothesis can also be **directional** or **one-sided**. Sometimes, a difference between two populations is expected to occur in only one direction. For example, 15-year-old children are expected to be taller than 10-year-old children. Administering an analgesic is expected to lower the pain and not increase it. In these cases, it is natural to formulate the alternate hypothesis in a single direction:

H_A: "On average, a 15-year-old girl is taller than a 10-year-old girl."
H_A: "Treatment with analgesic A will alleviate pain."

If analgesic A does happen to increase the pain, the result is insignificant to us. The drug will be worthless if it does not reduce the pain. In this case, we are interested in a difference in a single direction, even if a difference may occur in either direction.

When the alternate hypothesis is one-sided, H_0 must include either a greater than or less than sign (depending on the direction of H_A) in addition to the equality. The null hypotheses that complement the alternate hypotheses stated above are:

H_0: "On average, a 15-year-old girl is the same height or shorter than a 10-year-old girl."
H_0: "Treatment with analgesic A will either not change the amount of pain or increase it."

When the hypothesis is one-sided, we look for statistical significance in only one direction – the direction that coincides with the alternate hypothesis. In a one-sided test, the $100\alpha\%$ of the most extreme outcomes that qualify rejection of H_0 are confined to one side of the probability distribution curve of the test statistic. In other words, the rejection region is located either on the left or the right side of the null distribution, but not on both. On the other hand, test values that fall on the opposite side of the rejection region reinforce the null hypothesis. The procedure to evaluate a one-sided hypothesis test is demonstrated for the z test. The two-sided hypothesis on heart rate in populations X and Y presented earlier is converted to a single-sided hypothesis. The one-sided alternate hypothesis H_A is:

"The heart rate of population X is greater than the heart rate of population Y."

H_A is mathematically written as H_A: $\mu > \mu_0$.

The null hypothesis H_0 is:

"The heart rate of population X is either less than or equal to the heart rate of population Y."

H_0 can be stated mathematically as H_0: $\mu \leq \mu_0$.

Note that while the null hypothesis statement contains the condition of equality, it also includes infinite other possibilities of $\mu < \mu_0$. However, rejection of H_0: $\mu = \mu_0$ spawns automatic rejection of H_0: $\mu < \mu_0$. You can demonstrate this yourself. We only need to check for incompatibility of the data with H_0: $\mu = \mu_0$. For this reason,

Figure 4.2

One-sided z test for an alternate hypothesis with positive directionality. The shaded area located in the right tail is the rejection region.

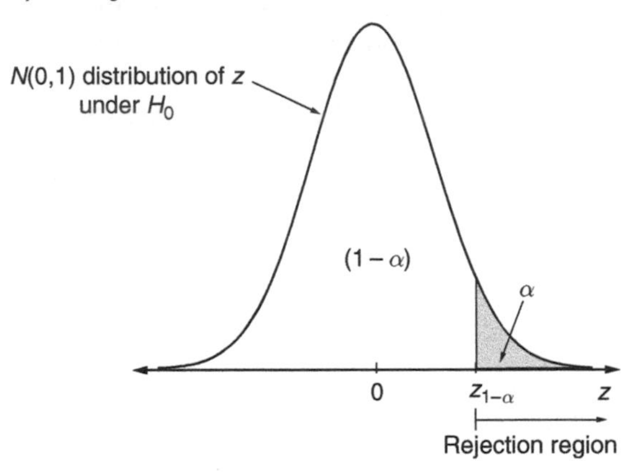

the z statistic for a one-sided hypothesis is calculated in the same way as for a non-directional hypothesis:

$$z = \frac{\bar{x} - \mu_0}{\sigma/n}.$$

Large positive values of z will lead to rejection of the null hypothesis. On the other hand, small or large negative values of z will not cause rejection. In fact, for $z \leq 0$, the null hypothesis is automatically retained, because the direction of the difference is the opposite of that stated in the alternate hypothesis and $p > 0.5$. Only when $z > 0$ does it become necessary to determine if the value of the test statistic is significant. The p value associated with the z statistic is the probability that if H_0 is true, one will obtain a value of z equal to or greater than the observed value. If α is the significance level, the rightmost area under the curve equal to α forms the rejection region, as shown in Figure 4.2.

The critical value of z that defines the lower bound of the rejection region, i.e. $z_{1-\alpha}$, satisfies the criterion

$$1 - \Phi(z_{1-\alpha}) = \alpha.$$

If $\alpha = 0.05$, then $z_{1-\alpha} = 1.645$.

A one-sided alternate hypothesis may express negative directionality, i.e.

$$H_A : \mu < \mu_0.$$

Then, the null hypothesis is formulated as H_0: $\mu \geq \mu_0$.

For this case, large negative values of z will lead to rejection of the null hypothesis, and small or large positive values of z will not cause rejection. If $z \geq 0$, the null hypothesis is automatically retained. One needs to look for statistical significance only if $z < 0$. The p value associated with the z statistic is the probability that if H_0 is true, one will obtain a value of z equal to or less than the observed value. If α is the significance level, the leftmost area under the curve equal to α forms the rejection region, as shown in Figure 4.3. The definition of the p value can be generalized for any one-sided hypothesis test.

Figure 4.3

One-sided z test for an alternate hypothesis with negative directionality. The shaded area located in the left tail is the rejection region.

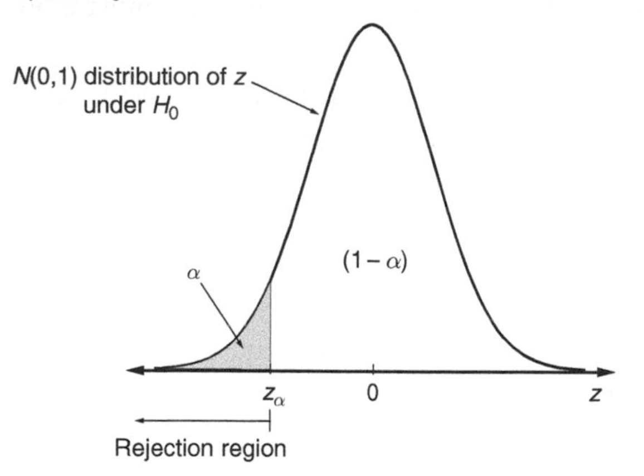

The p value of a directional hypothesis test is the probability, if H_0 is true, of obtaining a test statistic value at least as extreme as the observed value, measured in the direction of H_A.

The critical value of z that defines the upper bound of the rejection region, i.e. z_α, satisfies the criterion

$$\Phi(z_\alpha) = \alpha.$$

If $\alpha = 0.05$, then $z_\alpha = -1.645$.

Irrespective of the direction of the alternate hypothesis, the p value for a one-sided z test can be calculated using Equation (4.3) presented below.

For a one-sided z test, p is calculated as

$$p = (1 - \Phi(|z|)). \tag{4.3}$$

Upon comparing Equations (4.2) and (4.3), you will notice that for the same value of z, the p value differs by a factor of 2 depending on whether the hypothesis is directional. This has serious implications. The p value of a directional hypothesis will be less (half for symmetrically distributed null distributions) than that for a non-directional statement. Some researchers convert a non-significant result to a significant result simply by switching the nature of the hypothesis from one of no directionality to a directional one. This constitutes bad science. The choice between a directional and non-directional hypothesis should be made before collecting and inspecting the data. If a directional hypothesis is chosen, the direction should be guided by logical and scientific implications. If we formulate a directional hypothesis after surveying the data, we might be tempted to choose a direction that facilitates rejection of H_0. This is dangerous, since the probability of making a type I error (false positive) doubles when we perform an unwarranted directional analysis. If, in reality, the null hypothesis is correct, subsequent studies are likely to yield contradictory results, undermining your conclusions. When in doubt, you should employ a two-sided test, which, although more conservative than a one-sided test, cannot be called into question.

As discussed earlier for the two-sided test, you can use either of the following methods to assess statistical significance when performing a one-sided hypothesis test.

(1) Using the p value. The p value associated with the calculated test statistic is compared with the significance level of the test, α.
 (a) If $p < \alpha$, H_0 is rejected.
 (b) If $p > \alpha$, H_0 is retained.

(2) Using the test statistic. The calculated test statistic is compared with the critical value. In the z test, z is compared with the threshold value of z that defines the rejection region.
 (a) When testing for positive directionality
  (i) if $z > z_{1-\alpha}$, H_0 is rejected;
  (ii) if $z < z_{1-\alpha}$, H_0 is retained.
 (b) When testing for negative directionality
  (i) if $z < z_\alpha$, H_0 is rejected;
  (ii) if $z > z_\alpha$, H_0 is retained.

The p values that correspond to various possible values of a test statistic are published in tabular format in many standard statistics textbooks, and also on the Internet, for a variety of statistical tests (e.g. z test, t test, F test, and χ^2 test). A p value can also be calculated using statistical functions offered by software that perform statistical analysis of data, such as Minitab, SAS, and Statistics Toolbox in MATLAB. In this chapter, you will be introduced to a variety of Statistics Toolbox functions that perform tests of significance.

4.3.2 Type I and type II errors

If the value of the test statistic lies in the rejection region as defined by α, we naturally conclude the test result to be significant and reject the null hypothesis. However, it may be possible that the observed differences in the populations are due to sampling variability and no actual difference exists. Our choice of α (made prior to the calculation of the test statistic) is the probability that if the null hypothesis is true, we will still reject it. In other words, if we were to repeat the study many times, and if the null hypothesis were true, we would end up rejecting the null hypothesis (and thereby make an error) $100\alpha\%$ of the time. Thus, when we choose a significance level α, we are, in fact, deciding the probability that, when the null hypothesis is true, we will commit a false positive error.

> Rejection of the null hypothesis H_0, when H_0 is actually true, constitutes a **type I error** or a **false positive**. The probability of making a type I error, when H_0 is true, is the significance level α.

When we reject the null hypothesis, there is always the possibility that we are making a type I error. We want to minimize the risk of incorrectly rejecting H_0, and therefore we choose a small value for α. A significance level of 5%, or $\alpha = 0.05$, is widely used in most studies since it usually affords acceptable security against a type I error. In certain situations, one must choose a more conservative value of α, such as 0.01 or 0.001.

Table 4.2. *Conditions under which type I and type II errors are made*

	Test result	
	Test result is insignificant → H_0 retained	Test result is significant → H_0 rejected
H_0 is True	correct	type I error (false positive)
H_A is True	type II error (false negative)	correct

If the test result is not significant, we fail to reject the null hypothesis. There is a chance that a difference between the populations truly exists, but this may not be evident from the data. Or perhaps the chosen significance level is so small that even large differences between the samples yield an insignificant test statistic value. If the null hypothesis is actually false, but our test result is not statistically significant, we have a false negative result. Table 4.2 identifies all possible test result scenarios.

When we do not reject a false null hypothesis, we make a **type II error**. If the null hypothesis is false, the probability of making a type II error is β. In other words, β is the probability of obtaining a statistically insignificant result, even when the null hypothesis is incorrect.

If β is the probability of failing to reject a false null hypothesis, then $(1-\beta)$ is the probability of rejecting a false null hypothesis, thereby taking the correct action when the alternate hypothesis is true.

The **power** of the hypothesis test is given by $(1-\beta)$. It is the probability of rejecting a false null hypothesis.

When rejecting the null hypothesis, one only has to worry about making a type I error. Similarly, when failing to reject the null hypothesis, one only needs to consider the risk of making a type II error. Of course, you will not know if your conclusion is correct or wrong. You cannot calculate the probability of having made the right decision. For example, the statement "There is an 80% chance that the null hypothesis is true" is illogical. Such calculations are not possible. The null hypothesis is either 100% true or 100% false. Thus, when you reject, or fail to reject, the null hypothesis, you are either right or wrong. But, provided you know α and β, you will at least be cognizant of the chances of committing an error when the null hypothesis is true, or when it is false.

Since α is a parameter that you fix before you conduct the hypothesis test, you have the ability to exert control over the risk of making a type I error. However, you will not know β at the outset unless you estimate it. The power β depends on many factors, such as sample size, variance of the null distribution[2], and α. If the null hypothesis is *not* rejected, and the risk of making a type II error is unknown, the decision to retain H_0 will not hold much weight. For example, if $\beta = 0.7$, and H_0 is false, there is a 30% chance that we will obtain a test result that correctly identifies

[2] The null distribution is the probability (or sampling) distribution of the test statistic under the assumption that the null hypothesis is true.

this and leads us to reject H_0. In such situations, we will end up failing to reject the null hypothesis most of the time regardless of what is true. Since the value of α strongly impacts the power of a test, α must be chosen carefully. If α is very small, say $\alpha = 0.001$, then the probability that we will not reject the null hypothesis even if it is false is greatly increased. As we increase α (make the test less conservative), the chance of rejecting the null hypothesis in favor of the alternate hypothesis increases. An optimal value of α should be appropriately chosen to control both type I and type II errors. In many cases, a significance level of 0.05 affords sufficiently small probability of a type I error, but isn't so small that the power of the test is unnecessarily compromised.

Because β is often large and uncontrolled, there is substantial risk of making a type II error when the null hypothesis is not rejected. This explains why we avoid the terminology "accepting the null hypothesis" and instead use "fail to reject," even when we have compelling evidence at our disposal supporting the null hypothesis. Undoubtedly then, estimation of β is very useful, and it substantiates the credibility of our assertion that the data favor the null hypothesis. For many tests, α is chosen as 0.05 (or less) and $\beta > \alpha$. Often, one finds a β value of 0.2 or a power of 80% acceptable. In Section 4.5, we show how to calculate the power of a z test. The decision of choosing suitable values for α and β rests on the consequences of making a type I and a type II error, respectively. For example, if much is at stake when a type II error is made (e.g. testing for dangerous side-effects of a new drug), it is critical to design the experiment so that β is as small as possible.

Remember that null hypotheses are often statements of status quo, and, if true, no action needs to be taken. The alternate hypotheses are the action-packed statements that suggest a change be made. Any action that is suggested must be backed up with compelling proof, i.e. a hypothesis test that yields a convincing result. Therefore, the risk of a type I error should be kept small so that we are confident of our decision when we reject H_0.

Example 4.1

Intake of a food type X in a person's diet is believed to reduce premature hair loss for those who are predisposed to this condition. A researcher hypothesizes the following:

H_0: "X does not reduce premature hair loss."
H_A: "X reduces premature hair loss."

If the data reveal that X decelerates premature hair loss, when actually there is no effect of X, then the researcher has made a type I error. If this result is published, unnecessary hype and hope will be generated with the public. People afflicted with this condition, or even others who do not have this problem, may alter their diets to incorporate X. Depending on the nutritive value of X, moderate or excessive amounts of intake of X may be unhealthy or may predispose those consuming X to other conditions, such as metabolic disorders, cardiovascular disease, or kidney disease. On the other hand, if the data do not show any effect of X, when in fact X does alleviate premature hair loss, a type II error has been committed. The usefulness of X will be dismissed and may be reconsidered only when additional tests are performed to demonstrate H_A. Note that the outcome of making a type I error is much different than a type II error. Which type of error do you think is more dangerous in this case?

4.3.3 Types of variables

It is important to recognize the characteristics of the variables whose values are collected in an experiment. For example, arithmetic manipulations can be done on variables that represent speed, weight, age, energy units, length, pressure, temperature, and altitude. We are subject to greater restrictions if we wish to process numerically a variable that represents hair color, eye color, gender, or geographic location. The type of hypothesis test that can be applied to make statistical inferences depends on the nature of the random variable(s) measured in the experiment.

Variables can be broadly classified into three categories based on the type of measurement scale used.

(1) **Cardinal variables** Cardinal variables are numeric measurements that are of either continuous or discrete nature. These variables follow one of two measurement scales.
   (a) **Ratio scale** According to this scale, the number 0 means absence of the quantity being measured. Examples include age, weight, length, pressure, charge, current, speed, and viscosity. The zero point of a ratio scale is immutable. A ratio of two numbers on this scale indicates the proportionality between the two quantities. If box A weighs 40 kg, box B weighs 20 kg and box C weighs 10 kg, then box A is twice as heavy as box B, and box A is four times heavier than box C. Naturally then, box B is twice as heavy as box C. Even if we convert the measurement scale from kilograms to pounds, we would find the proportionality to be exactly the same. Measurements of intervals and ratios on this scale provide meaningful information.
   (b) **Interval scale** The zero position on this scale does not mean absence of the quantity being measured. As a result, ratios of numbers have little meaning on this scale. Examples include the Fahrenheit scale and Celsius scale for measuring temperature, in which the zero point is arbitrary and has different meaning for each scale.

(2) **Ordinal variables** Variables of this type are numbers specifying categories that can be ranked in a particular order. For example, a child's mastery of a skill may be ranked as: not developed, in progress, or accomplished. Each skill category is given a number such that a higher number indicates greater progress. More examples of ordinal variables include movie ratings, ranking of items according to preference, ranking of children in a class according to height or test scores, ranking of pain on a scale of 1 to 10, and Apgar scores ranging from 0 to 10 assigned to indicate the health of a baby at the time of birth. There is no uniformity in the differences between any two rankings. The ranks are based on subjective criteria. Therefore, limited arithmetic can be performed on ordinal variables, and mean and standard deviation of the distribution of ordinal data is not defined. Median and mode are relevant descriptive statistics for this variable type.

(3) **Nominal or categorical variables** These variables are not numbers that can be manipulated or ranked in any way but are observations according to which individuals or objects in a sample can be classified. The non-numeric values that categorical variables take on are mutually exclusive traits, attributes, classes, or categories that all together are exhaustive. Examples of categorical variables include gender: male or female, bacteria type: gram positive or gram negative, smoking status: smoker or non-smoker, natural hair color: black, brown, red or blonde, and marital status: single, married, divorced, or widowed. Based on the distribution of a categorical variable within a sample, numbers can be assigned to each value of the categorical variable, indicating the proportion of that category or trait within the sample.

In any experiment, measurements are made on people, animals, and objects such as machines, solutions contained in beakers, or flowing particles. Every individual or experimental run that provides one measurement for a particular outcome variable is called an **experimental unit**. If only a single variable is measured per experimental unit, then the data set is said to be **univariate**. **Bivariate** data result from measuring two variables per experimental unit. When three or more variables are measured on each experimental unit, the data are **multivariate**.

4.3.4 Choosing a hypothesis test

A number of factors must be considered when choosing a hypothesis test such as (i) measurement scale of the random variable and (ii) characteristics of the population distribution of the variable. Based on these two criteria alone, hypothesis tests are broadly classified into three types.

(1) **Parametric tests**: These tests assume that the measured variable exhibits a known form (or shape) of distribution within the population. Parametric tests are geared towards testing a hypothesis made about a population parameter such as mean or variance of the population distribution. To make a statistical inference regarding the value of a population parameter (e.g. mean), we must use the **sampling distribution**[3] of the corresponding sample-derived statistic under H_0. The sampling distribution of, say, the sample mean can only be determined if the probability distribution of the random variable is well-defined. (The central-limit theorem relaxes this constraint in many instances.) For example, the probability (sampling) distribution of the mean of a sample (size n) derived from a normal population is also normal. The variance of a sample (size n) drawn from a normal population follows the χ^2 **distribution**.
(2) **Distribution-free tests** These tests do not require knowledge of the nature of the distribution of the measured variable within the population. Some distribution-free methods, such as the sign test, are designed to test hypotheses on the population median.
(3) **Non-parametric tests** Non-parametric methods, by definition, do not make any inference concerning population parameters such as means, medians, or variances. Examples of non-parametric tests are the χ^2 goodness of fit test and the χ^2 test for independence. However, these tests are not necessarily distribution-free. The χ^2 tests assume that the sample sizes are large enough for the central-limit theorem to apply.

Distribution-free tests and non-parametric tests are usually grouped under a single heading – **non-parametric statistics**. Non-parametric statistical analysis is performed on all data that cannot be subjected to parametric tests. When the form of variable distribution within a population is known, parametric tests are much superior to non-parametric tests for hypothesis testing. Parametric tests have the advantage of giving the most "bang for the buck" since they utilize information contained in the data to the maximum potential to generate statistical inferences regarding the concerned populations. Parametric tests have more statistical power (less risk of making a type II error) than non-parametric tests and should always be preferred if the data qualify for parametric testing.

[3] Samples of the same size, when randomly drawn from the same population, will yield somewhat different values of the same descriptive statistic, e.g. mean, median, or variance. A **sampling distribution** is the probability distribution of all possible values of that statistic when all possible samples of some pre-specified size are drawn from the reference population.

When can you use a parametric test?

- If the probability distribution of the measured variable in the population can be described by a known, well-defined functional form, then the sampling distribution of the relevant sample-derived statistic (e.g. the sample mean) can be calculated. It is then appropriate to use a **parametric test** to answer questions regarding population characteristics such as the mean and variance.
- If the population variable is cardinal, i.e. follows a ratio scale or interval scale, then estimation of population parameters, such as mean and variance, is useful.

When should you use a non-parametric test?

- When the variable type is ordinal or nominal.
- When the underlying distribution of the variable within the population is not known and the sample size is not large enough for the central-limit theorem to apply.
- When gross inequality of sample variances exist which preclude use of a *t* test or ANOVA.
- When transformation of a highly non-normal population variable does not produce normality, or transformation does not resolve heteroscedasticity (inequality of variances) in multi-sample data.

To summarize, we list below some important factors that you should consider when choosing a hypothesis test. Specifics on how to choose a method of testing will become clearer as we discuss each test separately.

Factors that govern the choice of the hypothesis test

(1) The type of variable: cardinal, ordinal, or nominal.
(2) The shape of the distribution curve of the variable: normal or non-normal.
(3) Equality or inequality of population variances when comparing two or more samples.
(4) Sample size, for applicability of the central-limit theorem.
(5) Information that we seek from a test.

4.4 Parametric tests and assessing normality

Parametric tests are powerful because they allow you to define confidence intervals for population parameters and test hypotheses with greater precision. Many parametric procedures require that the underlying distribution of the variables be normal,[4] e.g. the z test, t test, and ANOVA (ANalysis Of VAriance). Use of the ANOVA method hinges on a normal distribution of the population variable. For the z test and t test, the normality requirement is not as strict when working with large samples. Fortunately, the central-limit theorem relaxes the normality condition for large samples that have a non-normally distributed variable. The mean of a large sample taken from a non-normal population will have an approximately normal sampling distribution that follows $N(\mu, \sigma/\sqrt{n})$, where μ is the population mean, σ is the population standard deviation, and n is the sample size. In this chapter, we only consider parametric tests that are designed to analyze normal population distributions.

[4] Other forms of continuous probability distributions include the uniform, exponential, lognormal, and Weibull distributions.

Before you can use a parametric test, you should ensure that your data meet the normality criteria. The data do not need to be perfectly normal. Parametric tests such as the z test, t test, and ANOVA are sufficiently robust that modest deviations of the data from normality do not adversely affect the reliability of the test result. Many distributions found in nature are close to normal with some modifications. Some distributions are symmetric, but in some cases may have a narrower central region and longer tails than the normal distribution, or alternatively a bulky central region and shorter tails. Many times distributions are found to be approximately normal but with asymmetric tails. Such distributions that have one tail longer than the other are called **skewed distributions**. To assess whether the distribution is acceptably normal, you need to test your data using one or more of several techniques that check for normality.

A simple method to check for normality of data is to plot a histogram of the data. You can visually inspect the histogram for symmetry and unimodal distribution. It is often difficult to tell if the distribution conforms to a normal shape, especially if the number of data points is small. To check if the histogram shape is Gaussian, you should calculate the mean μ and variance σ^2 of the data and superimpose the normal curve $N(\mu, \sigma)$ on the histogram, after rescaling the area of the histogram to 1. An overlay of both plots may yield more clues regarding the normality of the data. However, when data points are few, the shape of the histogram is not obvious and it may not be practical to use this method. Moreover, concluding whether a plot is bell-shaped can be a subjective exercise.

Another method to check for normality is to generate a **normal probability plot**. A probability plot graphs the data points arranged in ascending order against the corresponding z (standard normal variable) values expected if the data were perfectly normal. If the data follow a normal distribution, then the plot will yield an approximately straight line. Before we proceed with the description of this method, we must introduce the concept of a *quantile*.

The cumulative distribution function $\Phi(y) = \int_{-\infty}^{y} f(x)dx$ of a random variable y, which was defined in Chapter 3, calculates the cumulative probability of $P(x \leq y)$ (f is the probability density function).

> If n values of y, where y is defined by the cumulative distribution function $\Phi(y) = \int_{-\infty}^{y} f(x)dx$; $y \in (-\infty, \infty)$, $\Phi \in [0, 1]$, are chosen such they mark out $(n + 1)$ equally spaced intervals on the Φ scale within the entire interval of [0, 1], then each y point is called a **quantile**. Quantiles corresponding to $\Phi(y) = 0$ or $\Phi(y) = 1$ are not defined. If $n = 3$, each quantile is called **quartile** (4-quantile) since the cumulative distribution has been divided into four equally sized intervals, and if $n = 99$ each quantile is called a **percentile** (100-quantile).

For instance, if the cumulative distribution of a continuous random variable x is divided into ten equally spaced intervals by nine 10-quantiles and the fifth 10-quantile is m, then $\Phi(m) = P(x \leq m) = 5/10 = 0.5$, which is the median of the distribution. For the kth $(n + 1)$-quantile y, $\Phi(y) = P(x \leq y) = k/(n + 1)$. So far in this discussion we have assumed that the probability density function used to calculate the cumulative distribution is a continuous and well-defined mathematical function. If the data set is finite and the underlying probability distribution is unknown, the quantiles must be calculated from the data points. For example, a data set given below has ten values arranged in ascending order:

1.0, 1.4, 1.8, 2.0, 2.1, 2.15, 2.2, 2.3, 2.5, 3.0.

Table 4.3 *Data points and their corresponding z quantiles*

Serial number, i	Data points, x_i	kth quantile, $\Phi(z) = \frac{(i-0.5)}{n}$	Standard normal quantile, z
1	1.0	0.05	−1.6449
2	1.4	0.15	−1.0364
3	1.8	0.25	−0.6745
4	2.0	0.35	−0.3853
5	2.1	0.45	−0.1257
6	2.15	0.55	0.1257
7	2.2	0.65	0.3853
8	2.3	0.75	0.6745
9	2.5	0.85	1.0364
10	3.0	0.95	1.6449

Each data point x_i is randomly drawn and independent from each other, and contributes equally to the cumulative probability. The median of this data set, which is 2.125, divides the data into two subsets of equal intervals on the cumulative probability scale, i.e. $P(x \leq 2.125) = 1/2 = 0.5$; 50% of the data lie below the median. The median of the first subset is 1.8 and the median of the second subset is 2.3. If the data are divided into quartiles, then 25% of the data are less than the first quartile, 50% of the data lie below the second quartile, and 75% of the data fall below the third quartile. The first, second, and third quartiles are 1.8, 2.215, and 2.3, respectively. Thus, the kth $(n+1)$-quantile is the point below which $k/(n+1) \times$ 100% of the data lie.

Now let's create a normal probability plot to see if the data set given above is normal. The data are already arranged in ascending order. We already have ten quantiles but we need to calculate which kth quantile each data point represents. There is no one method universally employed to calculate the value of a quantile, and several methods are in use. For a data set containing n data points, the ith data point is associated with the $(i - 0.5)/n$th quantile, i.e. the data points in ascending order are the $(0.5/n)$th, $(1.5/n)$th, $(2.5/n)$th, $\ldots$, $((n - 0.5)/n)$th quantiles. The data points along with the kth quantiles that they represent are listed in Table 4.3.

The z values that correspond to the $((i - 0.5)/n)$th quantiles of the standard normal distribution are determined using the `norminv` function provided by Statistics Toolbox. In other words, for each value of $\Phi(z)$ we want to determine z. The corresponding z values are listed in Table 4.3.

Now, if the data points are truly normal, then the relationship between x and z is given by $x = \sigma z + \mu$. If an approximately straight line results, we may conclude that the data points follow a normal distribution. In Figure 4.4 x is plotted against z. Keep in mind that even a set of perfectly normal data points will not produce a straight line because of inherent sampling variability. If the probability plot takes on a distinctly curved shape, such as a "C" or an "S" shape, then the distribution of the data is non-normal. In Figure 4.4, the plot has several bends and kinks, indicating that the distribution of the data is probably non-normal.

Figure 4.4

Normal probability plot.

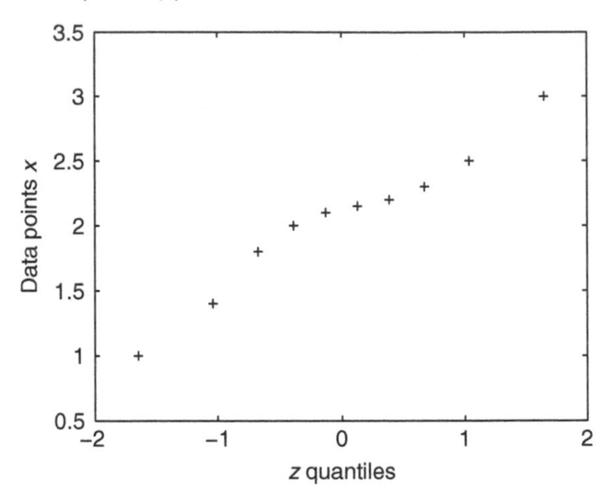

Using MATLAB

The function `normplot` offered by Statistics Toolbox creates a normal probability plot. The syntax for this function is

```
normplot(x)
```

where $\boldsymbol{x}$ is the data set to be assessed for normality. The function `normplot` plots the data points along the x-axis and the cumulative probability (in place of the z quantiles) along the y-axis. On this plot, the 25th and 75th percentiles are joined by a solid red line. This line extends in either direction as a dashed line. Try using the `normplot` function to obtain a probability plot for the data set above.

Statistical tests that quantitatively measure the fit of the observed sample distribution to the expected normal distribution are called goodness-of-fit tests. These tests are more sophisticated and also more conclusive than using normal probability plots. The χ^2 goodness-of-fit test is one example of a goodness-of-fit test that can be used to calculate the deviations of data from normality. This test is discussed in Section 4.9.1. Other popular statistical tests for assessing normality include the Kolmogorov–Smirnov one-sample test and the Shapiro–Wilks test. The latter method uses a normal probability plot for calculating the Shapiro–Wilk statistic. These methods are not discussed here. Statistics Toolbox supplies functions that perform the χ^2 goodness-of-fit test and the Kolmogorov–Smirnov one-sample test.

If the data are found to be non-normal, you can either choose to evaluate the data using a non-parametric test or use a transformation to convert the data to a normal distribution. Not all data can be transformed to a bell-shaped distribution. Another option is to use a parametric test based on another probability distribution. You will need to refer to advanced statistical texts for these statistical methods. Finally, you can proceed with a parametric test anyway, despite the non-normality of the data. The test may produce inaccurate results and you should be cautious of how you interpret the p value. You may want to use a more conservative significance level, such as $\alpha = 0.01$, to improve the reliability of your conclusion.

4.5 The z test

You are already familiar with several aspects of the **z test** from the discussion on hypothesis testing in Section 4.3. The z test is a parametric test used to compare the means of two samples or compare a sample mean to an expected value. In order to apply the z test, the variance of the populations from which the samples are drawn must be known. The z test is used in the following two ways.

(A) The **one-sample z test** is used to compare a sample mean with a population mean. Here, you want to determine if the sample was obtained from a certain population with already established parameters, i.e. the mean and variance of the population is known. This scenario was discussed in Section 4.3.

(B) The **two-sample z test** is used to compare the means of two random independent samples such as a control sample and a treated sample. Here you are trying to determine whether there is any effect of the treatment, for instance.

The conditions for using the z test are as follows.

(1) The population must be normal if the sample size is small. This ensures that the sampling distribution of the sample mean is also normally distributed. However, if the sample is large enough so that the central-limit theorem applies, then the sampling distribution of the mean will be normal regardless of the population distribution of the random variable.

(2) The variance of the populations being compared must be known. This may appear to be a serious limitation, since, most often, the purpose of the experiment is to estimate unknown population parameters and therefore we do not know the population variance beforehand. However, if the sample size is large (usually $n > 30$) the sample variance often reliably estimates the population variance.

Note the dual advantage of a large sample size – the central-limit theorem ensures approximate normality of the sampling distribution of the mean *and* we obtain an accurate estimate of the population variance. A large sample drawn from a non-normal population whose variance is unknown will still satisfy both requirements of the z test. A large sample size has several additional advantages, such as greater precision of an interval estimate of the population mean and improved power of a test.

If the sample size is small, the sample variance will not be an accurate estimate of the population variance. In this case, you can use the t test provided that the variable is normally distributed within the reference population. The t test is discussed in Section 4.6. For small samples drawn from non-normal populations, you can choose to either transform the data to normalize the distribution or use a non-parametric test (see Section 4.10).

With this knowledge in hand, we redefine situations for which the z test can be used.

(A) To determine if a sample is drawn from a population with known parameters. The population must be normal if the sample size is small. The normality requirement is relaxed if the sample size is large.

(B) To compare the means of two independent *large* random samples.

4.5.1 One-sample z test

A one-sample z test is used to determine if a sample belongs to a population whose statistics – namely mean and variance – are established. The method of using this test was illustrated in detail in Section 4.3. A one-sample z test can also be carried out in MATLAB.

Using MATLAB

Statistics Toolbox contains the function `ztest`, which performs a one-sample z test. The syntax for `ztest` is

```
[h, p] = ztest(x, mu, sigma, alpha, tail)
```

`ztest` has a variety of syntax options of which only one is listed above. The function arguments are

`x`: the vector of sample values,
`mu`: mean of the population from which the sample is derived under the null hypothesis,
`sigma`: standard deviation of the population from which the sample has come from under the null hypothesis,
`alpha`: the significance level, and
`tail`: specifies the directionality of the hypothesis and takes on the values '`both`', '`right`', or '`left`'.

The `ztest` function outputs the following:

`h`: the hypothesis test result – 1 if null hypothesis is rejected at the alpha significance level and 0 if null hypothesis is not rejected at the alpha significance level.
`p`: p value.

Type `help ztest` for additional features of this function.

In this section we explore how to determine the power of a one-sample z test. The power of a test $(1-\beta)$ is its ability to recognize a true difference between the populations under study. Let's set up a non-directional hypothesis:

$$H_0: \mu = \mu_0, \qquad H_A: \mu \neq \mu_0,$$

where μ is the mean of the population from which our sample has been taken and μ_0 is the expected value under the null hypothesis. We want to know if the population that our sample represents has the same mean as that of a certain population characterized by the statistics (μ_0, σ^2). The sample mean is calculated to be $\bar{x}$ from n individual units. The z test requires that either n is large or that the population variable x is distributed normally. The variance σ^2 is known and assumed to be the same for either population.

If the null hypothesis is true, then the sample mean $\bar{x}$ follows the normal distribution $N(\mu_0, \sigma/\sqrt{n})$.

If the alternate hypothesis is true, then $\mu = \mu_1$, where $\mu_1 \neq \mu_0$. In that case, $\bar{x}$ follows the normal distribution $N(\mu_1, \sigma/\sqrt{n})$. The mean or midpoint of the distribution of sample means under the null hypothesis and under the alternate hypothesis will be separated by a distance equal to $\mu_1 - \mu_0$ (see Figure 4.5). The two distributions predicted by the two complementary hypotheses may overlap each other. The extent of overlap of the two distributions depends on the width of each distribution, which

Figure 4.5

The power of a non-directional one-sample z test. The shaded region is the area under the H_A distribution that contributes to the power of the z test. We have $\sigma = 4$ and $\alpha/2 = 0.025$. (a) $\mu_1 - \mu_0 = 1.5$; (b) $\mu_1 - \mu_0 = 4.0$. The power increases as μ_1 moves further away from μ_0.

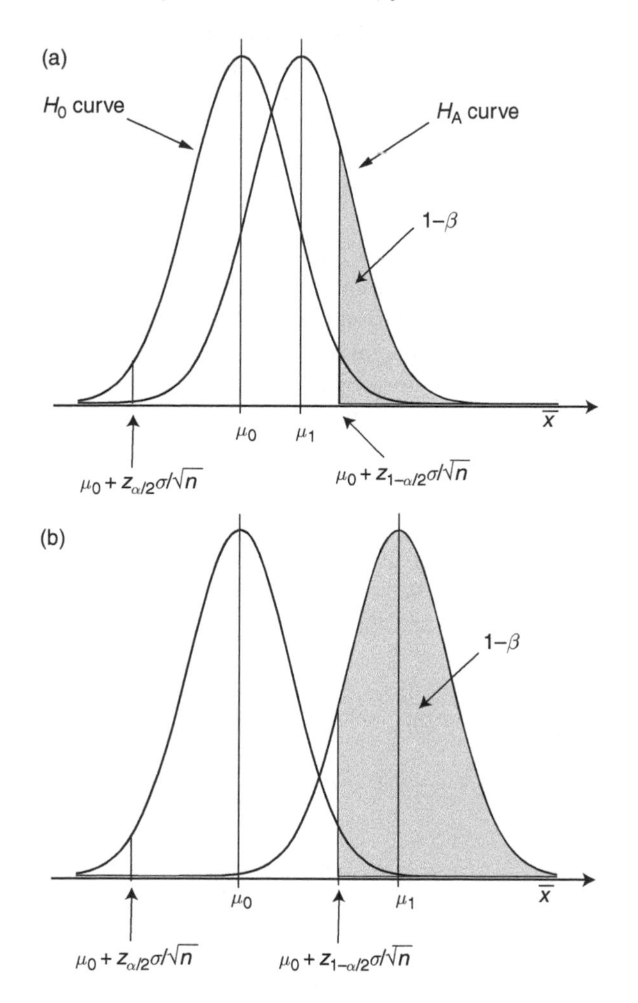

is determined by $\sigma/\sqrt{n}$ and on the difference between the population means, $\mu_1 - \mu_0$. The area under the curves that are free from overlap determines the ability to discern or "resolve" a difference between the two population means. If a large amount of area under the $N(\mu_1, \sigma/\sqrt{n})$ distribution curve lies within the $N(\mu_0, \sigma/\sqrt{n})$ distribution curve, the calculated z statistic will most likely yield a non-significant result, even if H_A is true. Under these circumstances, the power of the test is low and the risk of making a type II error is high. On the other hand, minimal overlap of the null distribution with the distribution corresponding to the alternate hypothesis affords large power to the z test.

We want to have as large a power as economically feasible so that if H_A is true, we will have maximized our chances to obtain a significant difference. Let's assume that the alternate hypothesis is *true*, and additionally that $\mu_1 > \mu_0$. In Figure 4.5, both distributions representative of the null and the alternate hypotheses are shown. The significance level that defines the null hypothesis rejection criteria is fixed at α. If the sample mean $\bar{x}$ lies in the interval

$$\mu_0 + z_{\alpha/2}\frac{\sigma}{\sqrt{n}} < \bar{x} < \mu_0 + z_{1-\alpha/2}\frac{\sigma}{\sqrt{n}},$$

then we retain the null hypothesis. This interval represents $100(1-\alpha)$ % of the area under the null distribution that forms the non-rejection region for the z test. When H_A is true and the sample mean falls within this interval, a type II error (false negative) results. On the other hand, if

$$\bar{x} < \mu_0 + z_{\alpha/2}\frac{\sigma}{\sqrt{n}} \quad \text{or} \quad \bar{x} > \mu_0 + z_{1-\alpha/2}\frac{\sigma}{\sqrt{n}},$$

then $\bar{x}$ lies in the rejection region (see Figure 4.1) and we reject H_0. Since $\bar{x}$ follows the $N(\mu_1, \sigma/\sqrt{n})$ distribution, the area under this distribution that overlaps with the non-rejection region of the null distribution defines the probability β that we will make a type II error. Accordingly, the area under the alternate distribution of sample means that does not overlap with the non-rejection region of the null distribution defines the power $(1-\beta)$ of the z test (see Figure 4.5). The power is calculated as follows:

$$1-\beta = P\left[\bar{x} < \mu_0 + z_{\alpha/2}\frac{\sigma}{\sqrt{n}}\right] + P\left[\bar{x} > \mu_0 + z_{1-\alpha/2}\frac{\sigma}{\sqrt{n}}\right].$$

The probability is calculated using the cumulative distribution function of the standard normal distribution. We convert the $\bar{x}$ scale shown in Figure 4.5 to the z scale. The power is calculated as follows:

$$\begin{aligned} 1-\beta &= P\left[\frac{\bar{x}-\mu_1}{\sigma/\sqrt{n}} < \frac{\mu_0-\mu_1}{\sigma/\sqrt{n}} + z_{\alpha/2}\right] + P\left[\frac{\bar{x}-\mu_1}{\sigma/\sqrt{n}} > \frac{\mu_0-\mu_1}{\sigma/\sqrt{n}} + z_{1-\alpha/2}\right] \\ &= P\left[z < \frac{\mu_0-\mu_1}{\sigma/\sqrt{n}} + z_{\alpha/2}\right] + P\left[z > \frac{\mu_0-\mu_1}{\sigma/\sqrt{n}} + z_{1-\alpha/2}\right] \\ &= \Phi\left(\frac{\mu_0-\mu_1}{\sigma/\sqrt{n}} + z_{\alpha/2}\right) + 1 - \Phi\left(\frac{\mu_0-\mu_1}{\sigma/\sqrt{n}} + z_{1-\alpha/2}\right). \end{aligned}$$

Although we have developed the above expression for the power of a non-directional one-sample z test assuming that $\mu_1 > \mu_0$, this expression also applies for the case where $\mu_1 < \mu_0$. You should verify this for yourself.

The power of a non-directional one-sample z test is given by

$$1-\beta = \Phi\left(\frac{\mu_0-\mu_1}{\sigma/\sqrt{n}} + z_{\alpha/2}\right) + 1 - \Phi\left(\frac{\mu_0-\mu_1}{\sigma/\sqrt{n}} + z_{1-\alpha/2}\right). \tag{4.4}$$

When H_A is true, the probability of making a type II error is

$$\beta = \Phi\left(\frac{\mu_0-\mu_1}{\sigma/\sqrt{n}} + z_{1-\alpha/2}\right) - \Phi\left(\frac{\mu_0-\mu_1}{\sigma/\sqrt{n}} + z_{\alpha/2}\right). \tag{4.5}$$

Just by looking at Figures 4.5(a) and (b), we can conclude that the area under the H_A curve to the left of the first critical value $z_{\alpha/2}$, which is equal to $\Phi\left(\frac{\mu_0-\mu_1}{\sigma/\sqrt{n}} + z_{\alpha/2}\right)$, is negligible compared to the area under the H_A curve to the right of the second critical value $z_{1-\alpha/2}$. The power of a non-directional one-sample z test when $\mu_1 > \mu_0$ is approximately

$$1-\beta \sim 1-\Phi\left(\frac{\mu_0-\mu_1}{\sigma/\sqrt{n}}+z_{1-\alpha/2}\right).$$

What if the hypothesis is one-sided? Let's revise the alternate hypothesis to H_A: $\mu > \mu_0$. For this one-sided test, the rejection region is confined to the right tail of the null distribution.

The power of the *directional* one-sample z test for H_A: $\mu > \mu_0$ is calculated as

$$1-\beta = 1-\Phi\left(\frac{\mu_0-\mu_1}{\sigma/\sqrt{n}}+z_{1-\alpha}\right). \tag{4.6}$$

Similarly, the power of the directional one-sample z test for H_A: $\mu < \mu_0$ is calculated as

$$1-\beta = \Phi\left(\frac{\mu_0-\mu_1}{\sigma/\sqrt{n}}+z_{\alpha}\right). \tag{4.7}$$

The derivation of Equations (4.6) and (4.7) is left to the reader.

To estimate the power of a test using Equations (4.4), (4.6), and (4.7), you will need to assume a value for μ_1. An estimate of the population variance is also required. A pilot experimental run can be performed to obtain preliminary data for this purpose. Frequently, these formulas are used to calculate the sample size n in order to attain the desired power $1-\beta$. Calculating sample size is an important part of designing experiments; however, a detailed discussion of this topic is beyond the scope of this book.

Based on the equations developed in this section for calculating the power of a one-sample z test, we can list the factors that influence statistical power.

(1) $\mu_1 - \mu_0$ The difference between the true mean and the expected value under the null hypothesis determines the separation distance between the two distributions. Figure 4.5(a) demonstrates the low power that results when $\mu_1 - \mu_0$ is small, and Figure 4.5(b) shows the increase in power associated with a greater $\mu_1 - \mu_0$ difference. This factor cannot be controlled; it is what we are trying to determine.

(2) σ The variance determines the spread of each distribution and therefore affects the degree of overlap. This factor is a property of the system and cannot be controlled.

(3) n The sample size affects the spread of the sampling distribution and is the most important variable that can be increased to improve power. Figure 4.6 illustrates the effect of n on power.

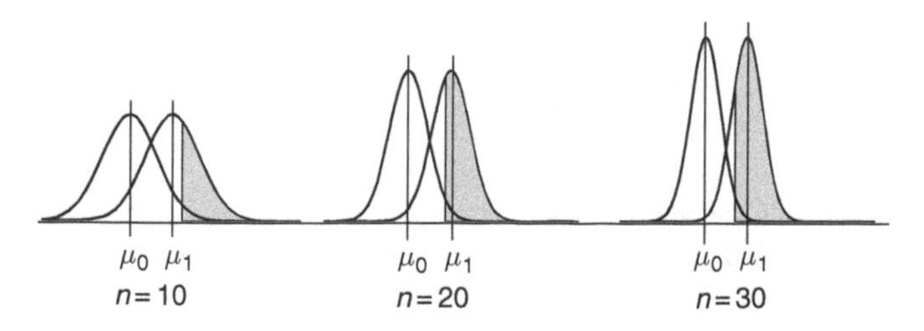

Figure 4.6

The power of the one-sample z test improves as sample size n is increased. The shaded area under the H_A distribution is equal to the power of the z test. We have $\sigma = 4$, $\alpha/2 = 0.025$, and $\mu_1 - \mu_0 = 2.0$.

(4) z_α Critical values of z that specify the null hypothesis rejection criteria reduce power when α is made very small. Suitable values of z_α are chosen in order to balance the risks of type I and type II errors.

Example 4.2

A vaccine is produced with a mean viral concentration of 6000 plaque-forming units (PFU) per dose and a standard deviation of 2000 PFU per dose. A new batch of vaccine has a mean viral concentration of 4500 PFU per dose based on a sample size of 15. The efficacy of a vaccine decreases with a decrease in viral count. Is the mean viral concentration of the new batch of vaccines significantly different from the desired amount? In order to determine this we perform a one-sample z test.

We state our hypotheses as follows:

H_0: "The new batch of vaccines has a mean viral concentration equal to the specified value, $\mu = \mu_0$."
H_A: "The new batch of vaccines has a mean viral concentration different from the specified value, $\mu \neq \mu_0$."

Our sample mean is $\bar{x} = 4500$. The mean of the population from which the sample is taken under the null hypothesis is $\mu_0 = 6000$. The standard deviation of the population is $\sigma = 2000$.

The z statistic is calculated as

$$z = \frac{4500 - 6000}{\sqrt{2000^2/15}} = -2.905.$$

The p value associated with this test statistic value is calculated using Equation (4.2) for a two-sided hypothesis. Statistics Toolbox's `normcdf` function can be used to obtain the p value. Typing the following statement in MATLAB

```
>> p = 2*(1-normcdf(2.905, 0, 1))
```

MATLAB outputs

```
p =
   0.0037
```

Thus, the probability of obtaining a value of the test statistic as far removed or further from zero as 2.905 is 0.0037. Since $0.0037 < 0.05$ (by an order of magnitude), this result is highly significant and we conclude that there is sufficient evidence to reject the null hypothesis. The new batch of vaccines has a mean viral concentration that is different (less) than the desired concentration.

Alternatively, we can perform the one-sample z test using the MATLAB hypothesis testing function `ztest`. The significance level is set at $\alpha = 0.05$.

In the Command Window, we type

```
>> [h, p] = ztest(4500, 6000, sqrt(2000^2/15), 0.05, 'both')
```

Note that because we do not have a vector of values, but instead are providing the sample mean, MATLAB does not have an estimate of the sample size. Therefore, we supply in place of the standard deviation of the population, the SEM. MATLAB outputs

```
h =
   1
p =
   0.0037
```

Next, we calculate the power of the one-sample z test to resolve a population mean viral concentration of $\mu = 4500$ PFU per dose from the population specified by quality control. The power $(1 - \beta)$ is the probability that we will obtain a z statistic that will lead us to reject the null hypothesis when the sample has been drawn from a population with mean viral concentration 4500 PFU per dose and the hypothesized mean under H_0 is 6000 PFU per dose. Using Equation (4.4), we estimate the power of the test. Since $\mu < \mu_0$,

$$1-\beta \sim \Phi\left(\frac{\mu_0 - \mu_1}{\sigma/\sqrt{n}} + z_{\alpha/2}\right).$$

Typing the following statement

```
>> power = normcdf((6000-4500)/(2000/sqrt(15)) - 1.96)
```

we obtain

```
power =
      0.8276
```

The probability that we will correctly reject the null hypothesis is 0.8276. In other words, if this experiment is repeated indefinitely (many, many independent random samples of size $n = 15$ are drawn from a population described by $N(4500, 2000)$), then 82.8% of all experiments will yield a p value ≤ 0.05. Thus, there is a 17.24% chance of making a type II error.

4.5.2 Two-sample *z* test

There are many situations that require us to compare two populations for anticipated differences. We may want to compare the product yield of two distinct processes, the lifetime of material goods supplied by two different vendors, the concentration of drug in a solution produced by two different batches, or the effect of temperature or humidity on cellular processes. The two-sample z test, which is used to determine if a statistical difference exists between the populations, assumes that the random samples derived from their respective populations are independent. For example, if two measurements are derived from every individual participating in the study, and the first measurement is allotted to sample 1 while the second measurement is allotted to sample 2, then the samples are related or dependent. In other words, the data in sample 1 influence the data in sample 2 and they will be correlated. The two-sample z test is inappropriate for dependent samples. Another method called the paired t test should be used to test hypotheses when the experiment produces dependent samples.

Assume that two random independent samples are drawn from two normal populations or two large samples are drawn from populations that are not necessarily normal. The two complementary non-directional hypotheses concerning these two populations are:

H_0: "The population means are equal, $\mu_1 = \mu_2$."
H_A: "The population means are unequal, $\mu_1 \neq \mu_2$."

If the null hypothesis of no difference is true, then both populations will have the same mean. We expect the respective sample means $\bar{x}_1$ and $\bar{x}_2$ to be equal or similar in value. If we subtract one sample mean from the other, $\bar{x}_1 - \bar{x}_2$, then the difference of the means will usually be close to zero. Since the sampling distribution of each sample mean is normal, the distribution of the difference between two sample means is also normal. The mean of the sampling distribution of the difference between the two sample means, $\bar{x}_1 - \bar{x}_2$, when H_0 is true, is

$$E(\bar{x}_1 - \bar{x}_2) = E(\bar{x}_1) - E(\bar{x}_2) = \mu_1 - \mu_2 = 0.$$

What is the variance of the sampling distribution of the difference of two sample means? In Chapter 3, you learned that when we combine two random variables arithmetically, the variance of the result is calculated based on certain prescribed rules. When one random variable is subtracted from the other,

$$z = x - y,$$

the variance of the result is

$$\sigma_z^2 = \sigma_x^2 + \sigma_y^2 - 2\sigma_{xy}^2. \tag{3.40}$$

When the random variable measurements are independent of each other, the covariance $\sigma_{xy}^2 = 0$.

Let $z = \bar{x}_1 - \bar{x}_2$. Suppose σ_1^2 and n_1 are the variance and size of sample 1, and σ_2^2 and n_2 are the variance and size of sample 2, respectively. The standard deviation of the sampling distribution of the difference of two sample means is called the **standard error** and is given by

$$\sigma_z = \sqrt{\frac{\sigma_1^2}{n_1} + \frac{\sigma_2^2}{n_2}}.$$

When the null hypothesis (H_0: $\mu_1 = \mu_2$) is true, then the statistic calculated below follows the standard normal distribution:

$$z = \frac{(\bar{x}_1 - \bar{x}_2) - (\mu_1 - \mu_2)}{\sqrt{\frac{\sigma_1^2}{n_1} + \frac{\sigma_2^2}{n_2}}}. \tag{4.8}$$

Equation (4.8) is the z statistic used to test for equality of two population means. Since $\mu_1 - \mu_2 = 0$, we simplify Equation (4.8) to

$$z = \frac{\bar{x}_1 - \bar{x}_2}{\sqrt{\frac{\sigma_1^2}{n_1} + \frac{\sigma_2^2}{n_2}}}. \tag{4.9}$$

Example 4.3 Properties of biodegradable and metallic stents

Stents are used to prop open the interior of lumenal structures, such as arteries, gastrointestinal tracts, such as the esophagus, duodenum, and colon, and the urinary tract. Such interventions are required when diseased conditions such as atherosclerosis, advanced-stage tumors, or kidney stones restrict the flow of fluid or interfere with the passage of material through the conduit. Because of the wide range of applications that stents serve, these prosthetic devices have a multitude of design options, such as size, material, mechanical properties, and structure. Metallic stents were developed initially, but certain drawbacks with using permanent metallic stents have spurred the increased interest in designing biodegradable polymeric stents. A major challenge associated with polymer stents lies in overcoming difficulties associated with deployment to the intended site. Specifically, the goal is to design polymeric stents with suitable mechanical/structural properties that enable quick and safe delivery to the application site, rapid expansion, and minimal recoil.

A new biodegradable polymeric stent has been designed with a mean elastic recoil of 4.0 ± 2.5%. The recoil is measured after expansion of the stent by a balloon catheter and subsequent deflation of the balloon. The sample size used to estimate recoil characteristics is 37. Metallic stents ($n = 42$) with comparable expansion properties have a mean recoil value of 2.7 ± 1.6%. We want to know if the recoil property is different for these two populations of stents.

The complementary hypotheses are:

H_0: "The recoil is the same for both metallic stents and polymeric stents considered in the study, $\mu_1 = \mu_2$."

H_A: "The recoil is different for the two types of stents, $\mu_1 \neq \mu_2$."

We set the significance level $\alpha = 0.05$. Since n_1 and $n_2 > 30$, we can justify the use of the two-sample z test. The critical values of z for the two-sided test are -1.96 and 1.96.

The test statistic is calculated as

$$z = \frac{4.0 - 2.7}{\sqrt{\frac{2.5^2}{37} + \frac{1.6^2}{42}}} = 2.708.$$

The p value corresponding to this test statistic is $2(1 - \Phi(z))$.

At the MATLAB command prompt, we type

```
>> z=2.708;
>> p=2*(1-normcdf(z,0,1))
```

MATLAB outputs

```
p =
    0.0068
```

Since $p < 0.05$, we reject H_0. We conclude that the mean recoil properties of the two types of stents are not equal. We can also use the MATLAB function `ztest` to perform the hypothesis test. Since $\bar{x}_1 - \bar{x}_2 = 1.3$ and the standard error

$$\mathrm{SE} = \sqrt{\frac{\sigma_1^2}{n_1} + \frac{\sigma_2^2}{n_2}} = 0.48,$$

```
>> [h, p] = ztest(1.3, 0, 0.48, 0.05, 'both')
h =
    1
p =
    0.0068
```

The power of a two-sample z test is calculated in a similar fashion as demonstrated for a one-sample z test. For a one-sample z test, we wish to maximize power by increasing the resolvability between the sampling distribution of the *sample mean* under the null hypothesis and that under the alternate hypothesis. In a two-sample z test, we calculate the degree of non-overlap between the sampling distributions of the *difference between the two sample means* under the null hypothesis and the alternate hypothesis. We assume a priori that the true difference between the two population means under H_A: $\mu_1 \neq \mu_2$ is some specific value $\mu_1 - \mu_2 = \Delta\mu \neq 0$, and then proceed to calculate the power of the test for resolving this difference.

The power of a non-directional two-sample z test, where z is calculated using Equation (4.9), is

$$1 - \beta = \Phi\left(\frac{-\Delta\mu}{\sqrt{\frac{\sigma_1^2}{n_1} + \frac{\sigma_2^2}{n_2}}} + z_{\alpha/2}\right) + 1 - \Phi\left(\frac{-\Delta\mu}{\sqrt{\frac{\sigma_1^2}{n_1} + \frac{\sigma_2^2}{n_2}}} + z_{1-\alpha/2}\right), \tag{4.10}$$

where $\Delta\mu$ is the assumed difference between the two population means under H_A.

Using the method illustrated in Section 4.5.1, you may derive Equation (4.10) on your own.

4.6 The *t* test

As discussed in Section 4.5, use of the z test is limited since the variance of the population distribution is usually unknown and the sample sizes are often small. For normal population(s), when the only available estimate for the population variance is the sample variance, the relevant statistic for constructing confidence intervals and performing hypothesis tests is the t statistic given by Equation (3.35),

$$t = \frac{\bar{x} - \mu}{s/\sqrt{n}}. \tag{3.35}$$

The Student's t distribution, which was discussed in Chapter 3, is the sampling distribution of the t statistic and is a function of the degrees of freedom f available for calculating the sample standard deviation s. In this section, we discuss how the t statistic can be used to test hypotheses pertaining to the population mean. The method of conducting a t test is very similar to the use of the z tests discussed in Section 4.5. The p value that corresponds to the calculated value of the test statistic t depends on whether the formulated hypothesis is directional.

For a two-sided t test, p is calculated as

$$p = 2(1 - \Phi_f(|t|)), \tag{4.11}$$

whereas for a one-sided t test

$$p = (1 - \Phi_f(|t|)), \tag{4.12}$$

where Φ_f is the cumulative distribution function for the Student's t distribution and f is the degrees of freedom associated with t.

4.6.1 One-sample and paired sample *t* tests

The **one-sample *t* test** is used when you wish to test if the mean $\bar{x}$ of a single sample is equal to some expected value μ_0 and the sample standard deviation is your only estimate of the population standard deviation. The variable of interest must be normally distributed in the population from which the sample is taken. The t test is found to be accurate for even moderate departures of the population from normality. Therefore, the t test is suitable for use even if the population distribution is slightly non-normal. As with the z test, the sample must be random and the sample units must be independent of each other to ensure a meaningful t test result.

Using MATLAB

Statistics Toolbox contains the function `ttest`, which performs a one-sample t test. The syntax for `ttest` is

```
[h, p] = ttest(x, mu, alpha, tail)
```

This is one of many syntax options available. The function parameters are:

`x`: the vector of sample values,
`mu`: mean of the population from which the sample is derived under the null hypothesis,

`alpha`: the significance level, and
`tail`: specifies the directionality of the hypothesis and takes on values: `'both'`, `'right'`, or `'left'`.

MATLAB computes the sample standard deviation, sample size, and the degrees of freedom from the data values stored in $\boldsymbol{x}$.
The `ttest` function outputs the following:

h: the hypothesis test result: 1 if null hypothesis is rejected at the alpha significance level and 0 if null hypothesis is not rejected at the alpha significance level,
p: p value.

Type `help ttest` for viewing additional features of this function.

The method of conducting a one-sample t test is illustrated in Example 4.4.

Example 4.4

The concentration of a hemoglobin solution is measured using a visible light monochromatic spectrophotometer. The mean absorbance of green light of a random sample consisting of ten batches of hemoglobin solutions is 0.7 and the standard deviation is 0.146. The individual absorbance values are 0.71, 0.53, 0.63, 0.79, 0.75, 0.65, 0.74, 0.93, 0.83, and 0.43. The absorbance values of the hemoglobin solutions are approximately normally distributed. We want to determine if the mean absorbance of the hemoglobin solutions is 0.85.
The null hypothesis is

$$H_0: \mu = 0.85.$$

The alternate hypothesis is

$$H_A: \mu \neq 0.85.$$

Calculating the t statistic, we obtain

$$t = \frac{0.7 - 0.85}{0.146/\sqrt{10}} = -3.249.$$

The degrees of freedom associated with t is $f = n - 1 = 9$.
The p value associated with $t(f = 9) = -3.249$ is calculated using the `tcdf` function in MATLAB:

```
>> p=2*(1-tcdf (3.249,9))
p =
   0.0100
```

Since $p < 0.05$, we reject H_0.
The test procedure above can be performed using the MATLAB `ttest` function. We store the ten absorbance values in a vector $\boldsymbol{x}$:

```
>> x=[0.71, 0.53, 0.63, 0.79, 0.75, 0.65, 0.74, 0.93, 0.83, 0.43];
>> [h, p] = ttest(x, 0.85, 0.05, 'both')
h =
   1
p =
   0.0096
```

Alternatively, we can test the null hypothesis by constructing a confidence interval for the sample mean (see Section 3.5.5 for a discussion on defining confidence intervals using the t statistic).

The 95% confidence interval for the mean population absorbance is given by

$$0.7 \pm t_{0.975,9}\frac{0.146}{\sqrt{10}}.$$

Using the MATLAB `tinv` function we get $t_{0.975,9} = 2.262$.

The 95% confidence interval for the population mean absorbance is 0.596, 0.804. Since the interval does not include 0.85, it is concluded that the hypothesized mean is significantly different from the population mean from which the sample was drawn.

Note that the confidence interval gives us a *quantitative measure* of how far the hypothesized mean lies from the confidence interval. The p value provides a *probabilistic measure* of observing such a difference, if the null hypothesis is true.

Many experimental designs generate data in pairs. Paired data are produced when a measurement is taken either from the same experimental unit twice or from a pair of similar or identical units. For example, the effects of two competing drug candidates or personal care products (e.g. lotions, sunscreens, acne creams, and pain relief creams) can be compared by testing both products on the same individual and monitoring various health indicators to measure the performance of each product. If two drugs are being compared, then the experimental design protocol usually involves administering one drug for a certain period of time, followed by a wash-out period to remove the remnant effects of the first drug before the second drug is introduced. Sometimes, the potency of a drug candidate may be contrasted with the effect of a placebo using a paired experimental design. In this way, the individuals or experimental units also serve as their own control. In cross-sectional observational studies, repeated measurements on the same individual are not possible. Pairing in such studies usually involves finding matched controls.

Paired designs have a unique advantage over using two independent samples to test a hypothesis regarding a treatment effect. When two measurements x_1 and x_2 are made on the same individual under two different conditions (e.g. administration of two different medications that are meant to achieve the same purpose, measurement of breathing rate at rest and during heavy exercise) or at two different times (i.e. progress in wound healing under new or old treatment methods), one measurement can be subtracted from the other to obtain a difference that measures the effect of the treatment. Many extraneous factors that arise due to differences among individuals, which can potentially confound the effect of a treatment on the measured variable, are eliminated when the difference is taken. This is because, for each individual, the same set of extraneous variations equally affects the two measurements x_1 and x_2. Ideally, the difference $d = x_1 - x_2$ exclusively measures the real effect of the treatment on the individual. The observed variations in d from individual to individual are due to biological (genetic and environmental) differences and various other extraneous factors such as diet, age, and exercise habits.

Many extraneous factors that contribute to variations observed in the population are uncontrollable and often unknown (e.g. genetic influences). When pairing data, it is the contributions of these entrenched extraneous factors that cancel each other and are thereby negated. Since each d value is a true measure of the treatment effect and is minimally influenced by confounding factors, the **paired design** is more efficient and lends greater precision to the experimental result than the **independent**

sample design. In an independent sample design, the effects of the uncontrolled confounding factors are mixed with the treatment effects and potentially corrupt the test result.

Even though a paired experimental design generates two data sets, the data points in each set pair up in a regular way, and therefore the data sets are not independent. A hypothesis test meant for testing two independent samples, such as the two-sample z test or two-sample t test, should not be used to analyze paired data. Instead, special hypothesis tests that cater to paired data sets should be used. If the measured variable is cardinal and the difference in measurements, d, is normally distributed, you can use a one-sample t test, called a **paired sample *t* test**, to assess statistical significance of the difference in treatment(s). Note that the individual d's must be independent of each other.

In a paired sample t test, if x_{1i} and x_{2i} are two paired data points from a data set containing n pairs, then the difference between the two paired data points is calculated as $d_i = x_{1i} - x_{2i}$. The sample mean difference is given by

$$\bar{d} = \frac{1}{n}\sum_{i=1}^{n}(x_{1i} - x_{2i}) = \bar{x}_1 - \bar{x}_2.$$

The expected value of the sample mean difference is

$$E(\bar{d}) = \mu_d = E(\bar{x}_1 - \bar{x}_2) = \mu_1 - \mu_2.$$

Thus, the population mean difference between the paired data points, μ_d, is equal to the difference of the two population means, $\mu_1 - \mu_2$. If the two population means are the same, then $\mu_d = 0$. If $\mu_1 > \mu_2$, then $\mu_d > 0$. The only estimate available of the standard deviation of the population distribution of differences between the data pairs is s_d, the standard deviation of the sample differences.

The relevant hypotheses for the paired sample t test are:

H_0: "The treatment has no effect, $\mu_d = 0$."
H_A: "The treatment has some effect, $\mu_d \neq 0$."

We have assumed in the hypotheses stated above that the null hypothesis is a statement of no effect. This does not have to be the case. We can also write the null hypothesis as

$$H_0\text{: } \mu_d = \mu_{d_0},$$

where μ_{d_0} is the expected population mean difference under the null hypothesis. The alternate hypothesis becomes

$$H_\text{A}\text{: } \bar{d} \neq \mu_{d_0}.$$

This is an example in which the null hypothesis does not necessarily represent the status quo. A non-zero mean difference μ_{d_0} is expected under the null hypothesis. If d is normally distributed within the population, and if the null hypothesis is true, then the sampling distribution of $(\bar{d} - \mu_{d_0})/(s_d/\sqrt{n})$ is the t distribution with $n - 1$ degrees of freedom. Since we wish to consider the null hypothesis of status quo, we set $\mu_{d_0} = 0$.

The test statistic for the paired sample t test is

$$t = \frac{\bar{d}}{s_d/\sqrt{n}}. \tag{4.13}$$

Box 4.5 Narrowing of the optic canal in fibrous dysplasia

Fibrous dysplasia of the bone is a rare condition that results from replacement of normal bone tissue with fibrous tissue. The fibrous growth causes expansion and weakening of the bone(s). This disease is caused by a somatic cell mutation of a particular gene and is therefore acquired after birth. This condition may be present in only one bone (monostotic) or in a large number of bones (polyostotic). In the latter case, expansion of the bone within the anterior base of the cranium is often observed, which leads to compression of the optic nerve. A study (Lee *et al.*, 2002) was performed to determine if compression of the optic nerve is symptomatic of fibrous dysplasia, and if it correlates with vision loss.

The areas of the right and left optic canals of individuals belonging to two groups, (1) patients with fibrous dysplasia and (2) age- and gender-matched controls, were measured using computed tomographic (CT) imaging. The results are presented in Table 4.4.

The non-directional null hypothesis to be tested is

$$H_0\colon \mu_d(\text{right}) = 0; \qquad H_0\colon \mu_d(\text{left}) = 0.$$

The mean difference in the areas of the right and left optic canals between group 1 and matched controls in group 2 were found to be

$$\text{right canal: } \bar{d} = -2.37 \pm 4.8, \qquad n = 32, \qquad f = 31,$$

$$\text{left canal: } \bar{d} = -1.83 \pm 4.9, \qquad n = 35, \qquad f = 34.$$

We test the hypothesis for the right optic canal. The t statistic is

$$t = \frac{-2.37}{4.8/\sqrt{32}} = -2.793.$$

The corresponding p value is obtained using the MATLAB `tcdf` function. The statement

```
p = 2*(1-tcdf(2.793,31))
```

gives us $p = 0.009$. Since $0.009 < 0.05$, we reject H_0 for the right eye. Thus, the areas of the right optic nerve canals of patients with craniofacial fibrous dysplasia are narrower than the dimensions observed in healthy individuals.

Now we test the hypothesis for the left optic canal. The t statistic is

$$t = \frac{-1.83}{4.9/\sqrt{34}} = 2.178.$$

The corresponding p value is 0.036. Since $0.036 < 0.05$, we reject H_0 for the left eye and draw the same conclusion as that for the right eye.

What if we had treated the two groups as independent samples and performed a two-sample z test[5] (or two-sample t test) to evaluate the hypothesis? What p value would you get? Calculate this. You will

Table 4.4. *Measurements of left and right optic canal areas in the study population (Lee* et al. *2002)*

		Group 1	Group 2
Sample size, n		32	35
Area (mm^2)	right	9.6 ± 3.8	12.0 ± 2.9
	left	9.9 ± 3.6	11.9 ± 2.7

[5] The t distribution approximates the z distribution when $n > 30$.

find that the two-sample test yields a similar result. Why do you think this is so? Compare the SE for both the paired test and the unpaired test. You will notice that there is little difference between the two. This means that precision was not gained by pairing. Matching on the basis of age and gender may not be important, as possibly these factors may not be biasing the measurements. Perhaps height, weight, and head circumference have more influence on the variability in optic canal cross-sectional area. Pairing of data improves the power of the test when the SE of paired data is smaller than the SE of unpaired data.

By treating paired data as independent samples, essentially we unpair the data, and thereby lose the precision that is gained when related measurements are paired. In a paired design the variation is minimized between the paired observations since the paired measurements have certain key identical features. Sometimes, a well-designed paired analysis analyzed using an independent two-sample test will yield a non-significant result even when a paired sample test demonstrates otherwise.

One shortcoming of the paired design is that it assumes that a treatment or condition will have the same effect on all individuals or experimental units in the population. If the treatment has different effects (e.g. large and positive on some, large and negative on others) on different segments of the population, a paired analysis will not reveal this and may in fact show negligible or small treatment effects when actually the opposite may be true. This limitation exists in independent-sample designs as well. Therefore all data points should be carefully inspected for such trends.

Using MATLAB

The `ttest` function can be used to perform a paired sample analysis. The syntax for the `ttest` for this purpose is

```
[h, p] = ttest(x, y, alpha, tail)
```

where x and y are the vectors storing data values from two matched samples. Each data point of x is paired with y, and therefore the two vectors must have equal lengths.

4.6.2 Independent two-sample *t* test

When we wish to make inferences regarding the difference of two population means, $\mu_1 - \mu_2$, and the random samples that are drawn from their respective populations are independent or not related to one another, we use the **independent or unpaired samples *t* test**. When the sample sizes are small, it is imperative that the populations in question have a normal distribution in order to use the t test. If the sample sizes are large, the requirement for normality is relaxed. For large n, the sampling distribution of $\bar{x}_1 - \bar{x}_2$ becomes increasingly normal. The shape of the t curve approaches the standard normal distribution and the z test may be used instead.

As discussed in Section 4.5.2, when the normality conditions mentioned above are met, the difference of two sample means $\bar{x}_1 - \bar{x}_2$ is distributed as $N\left(\mu_1 - \mu_2, \sqrt{\sigma_1^2/n_1 + \sigma_2^2/n_2}\right)$. The best estimate we have for the standard error uses the sample standard deviation, i.e.

$$s^2_{\bar{x}_1 - \bar{x}_2} = \frac{s_1^2}{n_1} + \frac{s_2^2}{n_2}.$$

The method used to calculate the test statistic hinges on whether the populations are homoscedastic (population variances are equal). The following two situations are possible.

Population variances are equal

The best estimate of the population variance is obtained by pooling the two estimates of the variance. We calculate the **pooled sample variance**, which is a weighted average of the sample variances. The weights are the degrees of freedom. This means that if the samples are of unequal size, then the larger sample provides a more accurate estimate of the sample variance and will be given more weight over the other sample variance based on a smaller sample size. The pooled sample variance is given by

$$s^2_{\text{pooled}} = \frac{(n_1 - 1)s_1^2 + (n_2 - 1)s_2^2}{n_1 + n_2 - 2}; \tag{4.14}$$

s^2_{pooled} is associated with $n_1 + n_2 - 2$ degrees of freedom, which is the sum of the degrees of freedom for sample 1, $n_1 - 1$, and sample 2, $n_2 - 1$. The larger number of degrees of freedom implies greater precision in the statistical analysis. The standard error (SE) based on the pooled sample variance is given by

$$\text{SE} = s_{\text{pooled}}\sqrt{\frac{1}{n_1} + \frac{1}{n_2}}.$$

The sample estimate of the difference of two population means is $\bar{x}_1 - \bar{x}_2$, which when standardized follows the t distribution with $n_1 + n_2 - 2$ degrees of freedom. The test statistic for the unpaired sample t test for equal population variances is

$$t = \frac{(\bar{x}_1 - \bar{x}_2) - (\mu_1 - \mu_2)}{s_{\text{pooled}}\sqrt{\frac{1}{n_1} + \frac{1}{n_2}}}.$$

If the null hypothesis is H_0: $\mu_1 - \mu_2 = 0$, then

$$t = \frac{(\bar{x}_1 - \bar{x}_2)}{s_{\text{pooled}}\sqrt{\frac{1}{n_1} + \frac{1}{n_2}}} \tag{4.15}$$

is distributed under the null hypothesis according to the t distribution with $n_1 + n_2 - 2$ degrees of freedom.

The population variances are not equal

If the sample variances differ by a factor of 4, or there is reason to believe that the population variances are not equal, then the best estimate of the standard error of the difference of two sample means is

$$\text{SE} = \sqrt{\frac{s_1^2}{n_1} + \frac{s_2^2}{n_2}}.$$

The t statistic is calculated as

$$t = \frac{(\bar{x}_1 - \bar{x}_2)}{\sqrt{\frac{s_1^2}{n_1} + \frac{s_2^2}{n_2}}}, \tag{4.16}$$

but does not exactly follow a t distribution when H_0 is true. The best approximate t distribution followed by the t statistic calculated in Equation (4.16) corresponds to f degrees of freedom calculated using Satterthwaite's formula:

$$f = \frac{\left(\frac{s_1^2}{n_1} + \frac{s_2^2}{n_2}\right)^2}{\frac{1}{(n_1 - 1)}\left(\frac{s_1^2}{n_1}\right)^2 + \frac{1}{(n_2 - 1)}\left(\frac{s_2^2}{n_2}\right)^2}. \tag{4.17}$$

If both sample sizes are the same, then the t statistic calculation is identical for both the equal variance and unequal variance case. Show that when $n_1 = n_2$, $\mathrm{SE}_{\sigma_1^2 \neq \sigma_2^2} = \mathrm{SE}_{\sigma_1^2 = \sigma_2^2}$ for the two-sample t test.

Using MATLAB

The two-sample t test can be performed for both equal and unequal sample variances using the Statistics Toolbox's function `ttest2`. One form of syntax for `ttest2` is

```
[h, p] = ttest2(x, y, alpha, tail, vartype)
```

where x and y are vectors containing the data values of the two independent samples and do not need to have equal lengths; `alpha` and `tail` have their usual meanings; `vartype` specifies whether the variances are equal, and can be assigned one of two values, either "equal" or "unequal." If "equal" is specified, then MATLAB calculates the pooled sample variance and uses Equation (4.15) to calculate the t statistic. If "unequal" is specified, MATLAB calculates the t statistic using Equation (4.16). The degrees of freedom for unequal variances are calculated using Satterthwaite's approximation (Equation (4.17)); h and p have their usual meanings.

4.7 Hypothesis testing for population proportions

Sometimes, an engineer/scientist needs to estimate the fraction or **proportion** of a population that exhibits a particular characteristic. On the basis of a single characteristic, the entire population can be divided into two parts:

(1) one that has the characteristic of interest, and
(2) the other that is devoid of that characteristic.

Any member chosen from the population will either possess the characteristic ("successful outcome") or not possess it ("failed outcome"). The event – the presence of absence of a characteristic – constitutes a random variable that can assume only two non-numeric values. A nominal variable of this type is called a **dichotomous variable** because it divides the population into two groups or categories. We are now in a position to link the concept of **population proportion** to the theory of the Bernoulli trial and the binomial distribution that was developed in Section 3.4.1.

Box 4.6 Flavone-8-acetic acid (FAA) reduces thrombosis

Flavone-8-acetic acid (FAA) is a non-cytotoxic drug that has antitumor activity in mice. It has also been found that FAA inhibits human platelet aggregation by blocking platelet surface receptor GPIbα binding to von Willebrand factor, a large plasma protein. An *in vivo* study using a porcine animal model was conducted to determine the effect of FAA on platelet aggregation (thrombosis) (Mruk *et al.*, 2000). Pigs were randomly assigned either to a treatment group that received an FAA bolus followed by continuous infusion or to a control group that was administered placebo (saline solution). Balloon angioplasty was performed for the carotid artery of all 20 pigs. Platelet deposition occurred in deeply injured arterial regions that underwent dilatation. The mean deposited platelet concentration for the treated group was $(13 \pm 3) \times 10^6$ cells/cm^2, and for the control group was $(165 \pm 51) \times 10^6$ cells/cm^2. The individual values are listed in the following:

treated group: 5.5, 6.5, 8, 8, 9.5, 11.5, 14.5, 15, 15, and 33×10^6 /cm^2, $n_1 = 10$,
control group: 10, 17, 60, 85, 90, 170, 240, 250, and 450×10^6 /cm^2, $n_2 = 9$.

To reduce the large variability in platelet deposition amounts, the data are transformed by taking the natural logarithm. We want to determine if FAA has an effect on platelet deposition. The variances of the log-transformed data of both groups differ by a factor of 6. We perform the *t* test for unequal variances using the `ttest2` function:

```
>> x1 = [5.5 6.5 8 8 9.5 11.5 14.5 15 15 33]*1e6;
>> x2 = [10 17 60 85 90 170 240 250 450]*1e6;
>> [h, p] = ttest2(log(x1), log(x2), 0.05, 'both', 'unequal')
h =
  1
p =
  8.5285e-004
```

Thus, the log of the amount of platelet deposition (thrombosis) in the porcine carotid artery during FAA administration is significantly different from that observed in the absence of FAA dosing.

Fibrinogen deposition for both groups was also measured, and was $(40 \pm 8) \times 10^{12}$ molecules/cm^2 for the treated group and $(140 \pm 69) \times 10^{12}$ molecules/cm^2 for the control group.

The individual values are listed in the following:

treated group: 14, 16, 19, 22, 37, 39, 41, 50, 65, and 96×10^{12} /cm^2, $n_1 = 10$,
control group: 19, 20, 37, 38, 70, 120, 125, 200, and 700×10^{12} /cm^2, $n_2 = 9$.

To reduce the large variability in fibrinogen deposition amounts, the data are transformed by taking the natural logarithm:

```
>> x1 = [14 16 19 22 37 39 41 50 65 95]*1e12;
>> x2 = [19 20 37 38 70 120 125 200 700]*1e12;
>> [h, p] = ttest2(log(x1), log(x2), 0.05, 'both', 'unequal')
h =
  0
p =
  0.0883
```

Since $p > 0.05$, we conclude that the log of the amount of fibrinogen deposition in the porcine carotid artery during FAA administration is not significantly different from that observed in the absence of FAA dosing. These results indicate that FAA interferes with platelet binding to the injured vessel wall but does not affect coagulation.

Box 4.2B Treatment of skeletal muscle injury using angiotensin II receptor blocker

The mean areas of fibrosis within the injured zone observed in the three groups after 3 weeks were

control group: 34% ± 17%, $n_1 = 4$;
low-dose ARB group: 16% ± 9%, $n_2 = 8$;
high-dose ARB group: 7% ± 4%, $n_3 = 8$.

We wish to compare the extent of fibrous tissue development in the control group with that observed in both the low-dose group and the high-dose group. We use the two-sample t test for dissimilar variances to test the null hypothesis of no difference.

We first compare the control group with the low-dose group. The t statistic is calculated as

$$t = \frac{(16 - 34)}{\sqrt{\frac{9^2}{8} + \frac{17^2}{4}}} = -1.983.$$

The degrees of freedom associated with this test statistic are

$$f = \frac{\left(\frac{9^2}{8} + \frac{17^2}{4}\right)^2}{\frac{1}{(8-1)}\left(\frac{9^2}{8}\right)^2 + \frac{1}{(4-1)}\left(\frac{17^2}{4}\right)^2} = 3.9.$$

The p value corresponding to this test statistic is 0.1202. If the null hypothesis is true, then we will observe a difference of this magnitude between the control group and the low-dose treated group about 12.02% of the time. Since $p > 0.05$, we fail to reject the null hypothesis. The data do not provide evidence that fibrosis is reduced by administering low doses of ARB for 3 weeks. Next, we compare the control group with the high-dose group. The t statistic is calculated as

$$t = \frac{(7 - 34)}{\sqrt{\frac{4^2}{8} + \frac{17^2}{4}}} = -3.133.$$

The degrees of freedom associated with this test statistic are

$$f = \frac{\left(\frac{4^2}{8} + \frac{17^2}{4}\right)^2}{\frac{1}{(8-1)}\left(\frac{4^2}{8}\right)^2 + \frac{1}{(4-1)}\left(\frac{17^2}{4}\right)^2} = 3.167.$$

The p value corresponding to $t_{f=3.167} = -3.133$ is 0.0483. Since $p < 0.05$, the evidence reveals a significant difference between the extent of fibrosis occurring in the untreated group and that in the high-dose treated group.

In this section, we take the concept of a Bernoulli trial further. The "experiment" – referred to in Chapter 3 – that yields one of two mutually exclusive and exhaustive outcomes is defined here as the process of sampling from a large population. In a Bernoulli trial, the members of a population are solely described by a dichotomous nominal variable, which is called a Bernoulli random variable x_i (i refers to the Bernoulli trial number). A dichotomous variable can have only two values that represent two mutually exclusive and exhaustive outcomes such as

(1) age > 20 or age ≤ 20,
(2) sleeping or awake, and
(3) pH of solution < 7 or pH ≥ 7.

When a member is selected from a population, it is classified under only one of the two possible values of the Bernoulli variable. One of the two mutually exclusive and exhaustive outcomes is termed as a "success" and x_i takes on the value of 1 when a success occurs. The probability with which a success occurs in each Bernoulli trial is equal to p. If the selected member does not possess the characteristic of interest, then the outcome is a "failure" and the Bernoulli variable takes on the value of 0. A failure occurs with probability $1 - p$.

The expected value of the Bernoulli variable x_i is the mean of all values of the variable that turn up when the trial is repeated infinitely. The probability of obtaining a success (or failure) depends on the proportion of success (or failure) within the population, and this determines the frequency with which a success (or failure) occurs when the experiment is performed many times. The expected value of x_i is calculated using Equation (3.14):

$$E(x_i) = \mu_i = \sum_{x_i} x_i P(x_i) = 1 \cdot p + 0 \cdot (1 - p) = p. \tag{4.18}$$

The variance associated with the Bernoulli variable x_i is calculated using Equations (3.16) and (3.17) to yield

$$\begin{aligned} E\left((x_i - \mu_i)^2\right) = \sigma_i^2 = E(x_i^2) - E(\mu_i^2) = \sum_{x_i} x_i^2 P(x_i) - p^2 \\ = 1^2 \cdot p + 0^2 \cdot (1 - p) - p^2 = p(1 - p). \end{aligned} \tag{4.19}$$

As discussed in Chapter 3, the binomial distribution is a discrete probability distribution of the number of successes obtained when a Bernoulli trial is repeated n times. A detailed description of the properties of this distribution is presented in Section 3.4.1. The binomial variable x is a discrete random variable that assumes integral values from 0 to n and represents the number of successes obtained in a binomial process. The binomial variable x is simply the sum of n Bernoulli variables produced during the binomial experiment,

$$x = \sum_{i=1}^{n} x_i \tag{4.20}$$

The expected values of the mean and variance of the binomial variable x are calculated below using Equations (3.14), (3.16), and (4.20):

$$E(x) = \mu = E\left(\sum_{i=1}^{n} x_i\right) = \sum_{i=1}^{n} E(x_i) = np \tag{4.21}$$

and

$$\begin{aligned} E\left((x - \mu)^2\right) = \sigma^2 &= E\left(\left(\sum_{i=1}^{n}(x_i) - np\right)^2\right) = E\left(\left(\sum_{i=1}^{n}(x_i - \mu_i)\right)^2\right) \\ &= E\left(\sum_{i=1}^{n}(x_i - \mu_i)^2 + \sum_{i=1}^{n}\sum_{j=1, i \neq j}^{n}(x_i - \mu_i)(x_j - \mu_j)\right) \\ &= \sum_{i=1}^{n} E\left((x_i - \mu_i)^2\right) + \sum_{i=1}^{n}\sum_{j=1, i \neq j}^{n} E((x_i - \mu_i)(x_j - \mu_j)). \end{aligned}$$

The expectation for $(x_i - \mu_i)(x_j - \mu_j)$ can be easily calculated for a binomial process, since the individual trials are independent of each other:

$$E((x_i - \mu_i)(x_j - \mu_j)) = \sum_i \sum_j (x_i - \mu_i)(x_j - \mu_j)P(x_i)P(x_j) \tag{4.22}$$

The double summation in Equation (4.22) occurs over all possible values of x_i and x_j. Since x_i and x_j both take on only two values, the double summation consists of only four terms. Substituting the values of the variables and their respective probabilities, we can show that $E((x_i - \mu_i)(x_j - \mu_j)) = 0$. Try to obtain this result yourself.

Substituting Equation (4.19) into the above equation, we get

$$E\left((x-\mu)^2\right) = \sigma^2 = \sum_{i=1}^{n} p(1-p) = np(1-p). \tag{4.23}$$

Equations (4.21) and (4.23) are identical to the derived mean and variance of a binomial distribution given by Equations (3.15) and (3.18).

A sample derived from a large population can be used to estimate the population proportion of a characteristic. A sample of size n is obtained through a binomial sampling process that consists of n Bernoulli trials. The sample proportion $\hat{p}$ is calculated as

$$\hat{p} = \frac{1}{n}\sum_{i=1}^{n} x_i = \frac{x}{n},$$

where x_i is a Bernoulli variable and x is a binomial variable, and the hat above p signifies a statistic calculated from observed data; $\hat{p}$ can be defined as a sample mean of the Bernoulli variable x_i. Different samples of the same size will yield different values of $\hat{p}$. The sampling distribution of $\hat{p}$ is the relative frequency distribution of all possible values of $\hat{p}$ obtained from samples of size n, and is discrete in nature. Each sample of size n can be viewed as a single binomial experiment consisting of n Bernoulli trials. Since the sample proportion is simply the binomial variable divided by the sample size n, the sampling distribution of a sample proportion is given by the binomial probability distribution, where the x-axis binomial variable of the binomial distribution curve is divided by n.

The expected value of the sample proportion is the mean of the sampling distribution of $\hat{p}$, and is equal to the population proportion p:

$$E(\hat{p}) = E\left(\frac{1}{n}\sum_{i=1}^{n} x_i\right) = \frac{E(x)}{n} = p. \tag{4.24}$$

The variance associated with the sampling distribution of $\hat{p}$ is given by

$$E\left((\hat{p}-p)^2\right) = \sigma^2 = E\left(\left(\frac{x}{n} - p\right)^2\right) = E\left(\frac{1}{n^2}(x-np)^2\right).$$

Substituting Equation (4.23) into the above expression, we get

$$E\left((\hat{p}-p)^2\right) = \sigma^2 = \frac{p(1-p)}{n}. \tag{4.25}$$

Figure 4.7 demonstrates the effect of sample size n (the number of independent Bernoulli trials) on the shape of the sampling distribution of $\hat{p}$. In the figure, the population proportion p is equal to 0.7. If n is small, then the sampling distribution

Figure 4.7

The discrete sampling distribution of the sample proportion contracts into an approximately normal distribution as sample size n increases ($p = 0.7$).

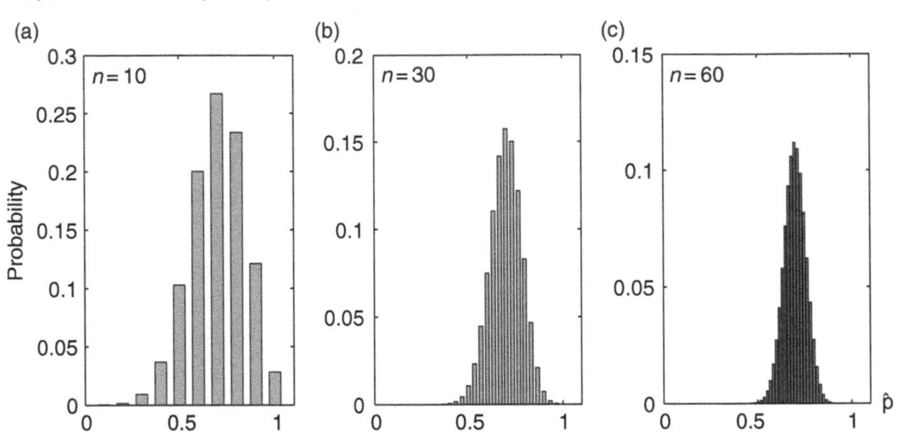

of the sample proportion is considerably discrete. The sample will usually not provide an accurate estimate of the true population proportion, and the sampling distribution of $\hat{p}$ will be relatively spread out over a larger range of $\hat{p}$ values. When n is large, the sample proportion is a better approximation of the true population proportion and the sampling distribution of $\hat{p}$ is more compact and spread over a narrower range. In fact, if n is large, such that $n\hat{p} > 5$ and $n(1 - \hat{p}) > 5$, $\hat{p}$ will have a sampling distribution that can be approximated by a continuous distribution – the normal curve. This is a consequence of the central-limit theorem, which states that the sampling distribution of the sample mean when n is large tends to a normal distribution. This theorem applies even if the random variable is not a continuous variable. Since $\hat{p}$ is essentially the sample mean of a set of 1's and 0's, i.e. Bernoulli variables, its sampling distribution at large n is dictated by the central-limit theorem. This theorem also applies to the sum $\sum_{i=1}^{n} x_i$ of all sample values x_i. When n is large, the sampling distribution of the sum of n values of a sample will be normal. Thus, the binomial distribution of x ($x = n\hat{p}$), which is the sampling distribution of the sum of n Bernoulli variables, will be approximately normal for large n.

When the conditions of $n\hat{p} > 5$ and $n(1 - \hat{p}) > 5$ are satisfied, $\hat{p}$ will be distributed approximately as

$$\hat{p} \sim N\left(p, \sqrt{\frac{p(1-p)}{n}}\right).$$

When the above conditions are satisfied, a confidence interval for p can be obtained using the normal probability distribution. We can also perform hypothesis testing on population proportions using methods similar to those discussed in Sections 4.5 and 4.6.

4.7.1 Hypothesis testing for a single population proportion

When n is large, a hypothesis for a single population proportion can be tested using the one-sample z test. A non-directional null hypothesis assumes that the population proportion $p = p_0$. According to the null hypothesis, the sample proportion $\hat{p}$ is distributed as

$$\hat{p} \sim N\left(p_0, \sqrt{\frac{p_0(1-p_0)}{n}}\right).$$

The sampling distribution of $\hat{p}$ under the null hypothesis can be converted to the standard normal distribution, and hence the z statistic serves as the relevant test statistic. The z statistic is calculated as

$$z = \frac{\hat{p} - p_0}{\sqrt{\frac{p_0(1-p_0)}{n}}}. \tag{4.26}$$

If the p value associated with z is less than the significance level α, the null hypothesis is rejected; otherwise, it is retained. A one-sided hypothesis test can also be constructed as described in Section 4.3.

4.7.2 Hypothesis testing for two population proportions

When two large samples from two populations are drawn and we want to know if the proportion within each population is different, we test whether the difference between the two population proportions $p_1 - p_2$ equals zero using the two-sample z test. The null hypothesis states that both population proportions are equal to p. Under H_0, the sample proportion $\hat{p}_1$ is distributed normally with mean p and variance $p(1-p)/n_1$, and the sample proportion $\hat{p}_2$ is distributed normally with mean p and variance $p(1-p)/n_2$. According to the null hypothesis, the difference between the two sample proportions, $\hat{p}_1 - \hat{p}_2$, is distributed normally with mean zero and variance

$$\sigma^2_{\hat{p}_1-\hat{p}_2} = \sigma^2_{\hat{p}_1} + \sigma^2_{\hat{p}_2} = \frac{p(1-p)}{n_1} + \frac{p(1-p)}{n_2}.$$

Since we do not know the true population proportion p specified under the null hypothesis, the best estimate for the population proportion p, when H_0 is true, is $\bar{p}$, which is the proportion of ones in the two samples combined:

$$\bar{p} = \frac{n_1\hat{p}_1 + n_2\hat{p}_2}{n_1 + n_2}.$$

The variance of the sampling distribution of the difference between two population proportions is

$$\sigma^2_{\hat{p}_1-\hat{p}_2} = \bar{p}(1-\bar{p})\left(\frac{1}{n_1} + \frac{1}{n_2}\right).$$

Thus, the difference between two sample proportions is distributed under H_0 as

$$\hat{p}_1 - \hat{p}_2 \sim N\left(0, \sqrt{\bar{p}(1-\bar{p})\left(\frac{1}{n_1} + \frac{1}{n_2}\right)}\right).$$

Box 4.7 In vitro fertilization (IVF) treatment

The IVF procedure is used to overcome the limitations that infertility poses on pregnancy. A single IVF cycle involves in vitro fertilization of one or more retrieved oocytes followed by implantation of the resulting embryos into the patient's body. Several embyros may be implanted in a single cycle. If the first round of IVF is unsuccessful, i.e. a live birth does not result, the patient has the choice to undergo a second cycle. Malizia and co-workers (Malizia *et al.*, 2009) performed a longitudinal study in which they tracked live-birth outcomes for a cohort (6164 patients) over six IVF cycles. Their aim was to provide an estimate of the cumulative live-birth rate resulting from IVF over the entire course of treatment. Traditionally, the success rate of IVF is conveyed as the probability of achieving pregnancy in a single IVF cycle. With a thorough estimate of the outcome of the IVF treatment course involving multiple cycles available, doctors may be in a position to better inform the patient regarding the likely duration of the treatment course and expected outcomes at each cycle.

In the study, the researchers followed the couples who registered for the treatment course until either a live birth resulted or the couple discontinued treatment. Indeed, some of the couples did not return for a subsequent cycle after experiencing unfavorable outcomes with one or more cycles of IVF. Those couples that discontinued treatment were found to have poorer chances of success versus the group that continued through all cycles until a baby was born or the sixth cycle was completed. Since the outcomes of the discontinued patients were unknown, two assumptions were possible:

(1) the group that discontinued treatment had the exact same chances of achieving live birth as the group that returned for subsequent cycles, or
(2) the group that discontinued treatment had zero chances of a pregnancy that produced a live birth.

The first assumption leads to an **optimistic analysis**, while the second assumption leads to a **conservative analysis**. The optimistic cumulative probabilities for live births observed at the end of each IVF cycle are displayed in Table 4.5. The cohort is divided into four groups based on maternal age (this procedure is called **stratification**, which is done here to reveal the effect of age on IVF outcomes since age is a potential confounding factor).

The cohort of 6164 women is categorized according to age (see Table 4.6).

We wish to determine at the first and sixth IVF cycle whether the cumulative birth rates for the age groups of less than 35 years of age and more than 40 years of age are significantly different. Let's designate the age group <35 years as group 1 and the age group >40 years as group 2. In the first IVF cycle,

$$p_1 = 0.33, \qquad n_1 = 2678;$$

$$p_2 = 0.09, \qquad n_2 = 1290.$$

Table 4.5. *Optimistic probability of live birth for one to six IVF cycles*

	Maternal age (years)			
IVF cycle	<35	35 to <38	38 to <40	≥40
1	0.33	0.28	0.21	0.09
2	0.52	0.47	0.35	0.16
3	0.67	0.58	0.47	0.24
4	0.76	0.67	0.55	0.32
5	0.82	0.74	0.62	0.37
6	0.86	0.78	0.67	0.42

Table 4.6. *Classification of study population according to age*

Age	Number of individuals
<35	2678
35 to <38	1360
38 to <40	836
≥40	1290

So we have

$$\bar{p} = \frac{2678 \cdot 0.33 + 1290 \cdot 0.09}{2678 + 1290} = 0.252;$$

$$\sigma^2_{\hat{p}_1-\hat{p}_2} = 0.252(1 - 0.252)\left(\frac{1}{2678} + \frac{1}{1290}\right) = 2.165 \times 10^{-4};$$

$$z = \frac{0.33 - 0.09}{0.0147} = 16.33.$$

The p value associated with this non-directional two-sample z test is $< 10^{-16}$. Thus, the probability that we will observe this magnitude of difference in the proportion of live births for these two age groups following IVF treatment is negligible if the null hypothesis of no difference is true. The population proportion (probability) of live births resulting from a single IVF cycle is significantly different in the two age populations of women.

For the 6th IVF cycle,

$$p_1 = 0.86, \qquad n_1 = 2678;$$

$$p_2 = 0.42, \qquad n_2 = 1290.$$

So we have

$$\bar{p} = \frac{2678 \cdot 0.86 + 1290 \cdot 0.42}{2678 + 1290} = 0.717;$$

$$\sigma^2_{\hat{p}_1-\hat{p}_2} = 0.717(1 - 0.717)\left(\frac{1}{2678} + \frac{1}{1290}\right) = 2.33 \times 10^{-4};$$

$$z = \frac{0.86 - 0.42}{0.0153} = 28.76.$$

The p value associated with this non-directional two-sample z test is $<10^{-16}$. The result is highly significant and the population proportion of live births resulting over six IVF cycles in the two age populations are unequal.

We may conclude that the decline in fertility associated with maternal aging influences IVF treatment outcomes in different age populations, i.e. older women have less chances of success with IVF compared to younger women.

The appropriate test statistic to use for testing a difference between two population proportions is the z statistic, which is distributed as $N(0, 1)$ when the null hypothesis is true, and is calculated as

$$z = \frac{\hat{p}_1 - \hat{p}_2}{\sqrt{\bar{p}(1 - \bar{p})\left(\dfrac{1}{n_1} + \dfrac{1}{n_2}\right)}}. \tag{4.27}$$

Use of Equation (4.27) for hypothesis testing of two population proportions is justified when $n_1\hat{p}_1 \geq 5, n_1(1-\hat{p}_1) \geq 5, n_2\hat{p}_2 \geq 5, n_2(1-\hat{p}_2) \geq 5$. This ensures that each sample proportion has a normal sampling distribution, and therefore the difference of the sample proportions is also normally distributed. The criterion for rejection of the null hypothesis is discussed in Section 4.3.

4.8 One-way ANOVA

So far we have considered statistical tests (namely, the z test and the t test) that compare only two samples at a time. The tests discussed in the earlier sections can be used to draw inferences regarding the differences in the behavior of two populations with respect to each other. However, situations can arise in which we need to compare multiple samples simultaneously. Often experiments involve testing of several conditions or treatments, e.g. different drug or chemical concentrations, temperature conditions, materials of construction, levels of activity or exercise, and so on. Sometimes only one factor or condition is varied in the experiment, at other times more than one factor can be varied together, e.g. temperature and concentration may be varied simultaneously. Each factor is called a **treatment variable**, and the specific values that the factor is assigned are called **levels of the treatment**. Our goal is to find out whether the treatment variable(s) has an effect on the **response variable**, which is a measured variable whose value is believed to be influenced by one or more treatment levels.

For example, suppose an experiment is conducted to study the effect of sleep on memory retention. Based on the varied sleep habits of students of XYZ School, the experimentalist identifies eight different sleep patterns among the student population. A total of 200 students chosen in a randomized fashion from the student pool are asked to follow any one out of eight sleep patterns over a time period. All students are periodically subjected to a battery of memory retention tests. The composite score on the memory tests constitutes the response variable and the sleep pattern is the treatment variable.

How should we compare the test scores of all eight samples that ideally differ only in terms of the students' sleeping habits over the study period? One may suggest choosing all possible pairs of samples from the eight samples and performing a t test on each pair. The conclusions from all t tests are then pooled together to infer which populations have the same mean test performance at the end of the study period. The number of unique sample pairs that can be obtained from a total of eight samples is calculated using the method of combinations discussed in Chapter 3. The total number of unique sets of two samples that we can select out of a set of eight different samples is $C_2^8 = 28$. If the significance level is set at $\alpha = 0.05$, and if the null hypothesis of no difference is globally true, i.e. all eight populations have the same mean value, then the probability of a type I error for any t test is 0.05. The probability of correctly concluding the null hypothesis is true in any t test is 0.95. The probability that we will correctly retain the null hypothesis in all 28 t tests is $(0.95)^{28}$. The probability that we will make at least one type I error in the process is $1-(0.95)^{28} = 0.762$. Thus, there is a 76% chance that we will make one or more type I errors! Although the type I error is constrained at 5% for a single t test, the risk is much greater when multiple comparisons are made. The increased likelihood of committing errors is a serious drawback of using t tests to perform the repeated multiple comparisons. Also, this method is tedious and cumbersome, since many t tests must be performed to determine if there is a difference among the populations. Instead, we wish to use a method that

(1) can compare the means of three or more samples simultaneously, and
(2) maintains the error risk at an acceptably low level.

R. A. Fisher (1890–1962), a renowned statistician and geneticist, devised a technique called **ANalysis Of VAriance (ANOVA)** that draws inferences regarding the means of several populations simultaneously by comparing variations within each sample to the variations among the samples. In a mathematical sense, ANOVA is an extension of the independent samples t test to three or more samples. The null hypothesis of an ANOVA test is one of no difference among the sample means, i.e.

$$H_0: \mu_1 = \mu_2 = \mu_3 = \cdots = \mu_k,$$

where k is either the number of treatment levels or the number of combinations of treatments when two or more factors are involved. Numerous designs have been devised for the ANOVA method, and the most appropriate one depends on the number of factors involved and details of the experimental design. In this book, we concern ourselves with experiments in which only a single factor is varied and the experimental design is completely randomized. For experiments that produce univariate data and have three or more treatment levels, a **one-way ANOVA** is suitable for testing the null hypothesis, and is the simplest of all ANOVA designs. When two factors are varied simultaneously, interaction effects may be present between the two factors. **Two-way ANOVA** tests are used to handle these situations. Experiments that involve stratification or blocking can be analyzed using two-way ANOVA. An introduction to two-way ANOVA is beyond the scope of this book, but can be found in standard biostatistics texts.

The ANOVA is a parametric test and produces reliable results when the following conditions hold.

(1) The random samples are drawn from normally distributed populations.
(2) All data points are independent of each other.
(3) Each of the populations has the same variance σ^2. Populations that have equal variances are called **homoscedastic**. Populations whose variances are not equal are called **heteroscedastic**.

Henceforth, our discussion of ANOVA pertains to the one-way analysis. We make use of the specific notation scheme defined below to identify individual data points and means of samples within the multi-sample data set.

Notation scheme

k is the total number of independent, random samples in the data set.
x_{ij} is the jth data point of the ith sample or treatment.
n_i is the number of data points in the ith sample.
$\bar{x}_{i\cdot} = \frac{\sum_{j=1}^{n_i} x_{ij}}{n_i}$ is the mean of the ith sample. The dot indicates that the mean is obtained by summation over all j data points in sample i.
$N = \sum_{i=1}^{k} n_i$ is the total number of observations in the multi-sample data set.
$\bar{x}_{\cdot\cdot} = \frac{\sum_{i=1}^{k}\sum_{j=1}^{n_i} x_{ij}}{N}$ is the **grand mean**, i.e. the mean of all data points in the multi-sample data set. The double dots indicate summation over all data points $j = 1, \ldots, n_i$ within a sample and over all samples $i = 1, \ldots, k$.

Each treatment process yields one sample, which is designated henceforth as the **treatment group** or simply as the **group**. We can use the multi-sample data set to estimate the common variance of the k populations in two independent ways.

(1) **Within-groups mean square estimate of population variance** Since the population variances are equal, the k sample variances can be pooled together to obtain an improved estimate of the population variance. Pooling of sample variances was first introduced in Section 4.6.2. Equation (4.14) shows how two sample variances can be pooled together. The quality of the estimate of the population variance is conveyed by the associated degrees of freedom. Since the multi-sample data set represents populations that share the same variance, the best estimate of the population variance is obtained by combining the variations observed within all k groups. The variance within the ith group is

$$s_i^2 = \frac{\sum_{j=1}^{n_i}\left(x_{ij} - \bar{x}_{i\cdot}\right)^2}{n_i - 1}.$$

The pooled sample variance s_p^2 is calculated as

$$s_p^2 = \frac{(n_1 - 1)s_1^2 + (n_2 - 1)s_2^2 + \cdots + (n_k - 1)s_k^2}{(n_1 - 1) + (n_2 - 1) + \cdots + (n_k - 1)}.$$

We obtain

$$s_p^2 = \text{MS(within)} = \frac{\sum_{i=1}^{k}\sum_{j=1}^{n_i}\left(x_{ij} - \bar{x}_{i\cdot}\right)^2}{N - k}. \tag{4.28}$$

The numerator in Equation (4.28) is called the **within-groups sum of squares** and is denoted as SS(within). It is called "within-groups" because the variations are measured within each sample, i.e. we are calculating the spread of data within the samples. The degrees of freedom associated with the pooled sample variance is $N - k$. When the within-groups sum of squares is divided by the degrees of freedom associated with the estimate, we obtain the **within-groups mean square**, denoted as MS(within), which is the same as the pooled sample variance. The MS(within) is our first estimate of the population variance and is an unbiased estimate regardless of whether H_0 is true. However, it is necessary that the equality of population variances assumption hold true, otherwise the estimate given by Equation (4.28) is not meaningful.

(2) **Among-groups mean square estimate of population variance** If all samples are derived from populations with the same mean, i.e. H_0 is true, then the multi-sample data set can be viewed as containing k samples obtained by sampling from the same population k times. Therefore, we have k estimates of the population mean, which are distributed according to the sampling distribution for sample means with mean μ and variance equal to

$$\sigma_{\bar{x}}^2 = \frac{\sigma^2}{n},$$

where n is the sample size, assuming all samples have the same number of data points. Alternatively, we can express the population variance σ^2 in terms of $\sigma_{\bar{x}}^2$, i.e.

$$\sigma^2 = n\sigma_{\bar{x}}^2.$$

We can use the k sample means to calculate an estimate for $\sigma_{\bar{x}}^2$. For this, we need to know the population mean μ. The grand mean $\bar{\bar{x}}$ is the best estimate of the

Figure 4.8

Three different scenarios for data variability in a multi-sample data set. The circles represent data points and the thick horizontal bars represent the individual group means.

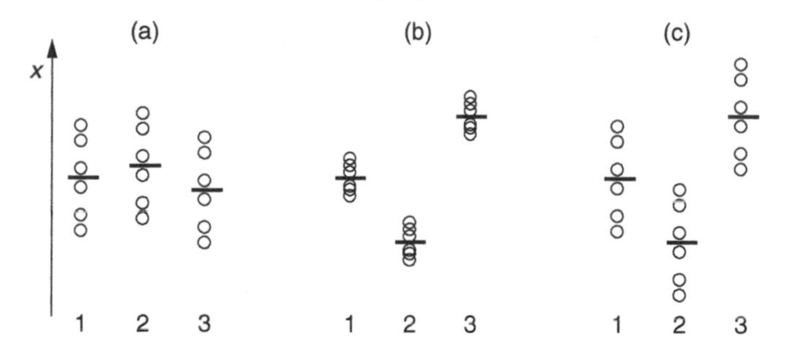

population mean μ, if H_0 is true. An unbiased estimate of the variance of the sampling distribution of the sample means is given by

$$s_{\bar{x}}^2 = \frac{\sum_{i=1}^{k}(\bar{x}_{i\cdot} - \bar{x}_{\cdot\cdot})^2}{k-1}.$$

If the sample sizes are all equal to n, we can calculate an estimate of the population variance called the **among-groups mean square** as

$$\mathrm{MS(among)} = \frac{n\sum_{i=1}^{k}(\bar{x}_{i\cdot} - \bar{x}_{\cdot\cdot})^2}{k-1}.$$

More likely than not, the sample sizes will vary. In that case the mean square among groups is calculated as

$$\mathrm{MS(among)} = \frac{\sum_{i=1}^{k}n_i(\bar{x}_{i\cdot} - \bar{x}_{\cdot\cdot})^2}{k-1}. \quad (4.29)$$

The numerator of Equation (4.29) is called the **among-groups sum of squares** or SS(among) since the term calculates the variability among the groups. The degrees of freedom associated with SS(among) is $k-1$, where k is the number of treatment groups. We subtract 1 from k since the grand mean is calculated from the sample means and we thereby lose one degree of freedom. By dividing SS(among) by the corresponding degrees of freedom, we obtain the **among-groups mean square** or MS(among). Equation (4.29) is an unbiased estimate of the common population variance only if H_0 is true and the individual groups are derived from populations with the same variance.

Figure 4.8 depicts three different scenarios of variability in a multi-sample data set ($k = 3$). In Figure 4.8(a), data variability within each of thc three groups seems to explain the variation among the group means. We expect MS(within) to be similar in value to MS(among) for this case. Variation among the groups is much larger than the variance observed within each group for the data shown in Figure 4.8(b). For this case, we expect the group means to be significantly different from each other. The data set in Figure 4.8(c) shows a large scatter of data points within each group. The group means are far apart from each other. It is not obvious from visual inspection of the data whether the spread of data within each group (sampling error) can account for the large differences observed among the group means. A statistical test must be

Figure 4.9

Three F distributions for different numerator degrees of freedom and the same denominator degrees of freedom. The `fpdf` function in MATLAB calculates the F probability density.

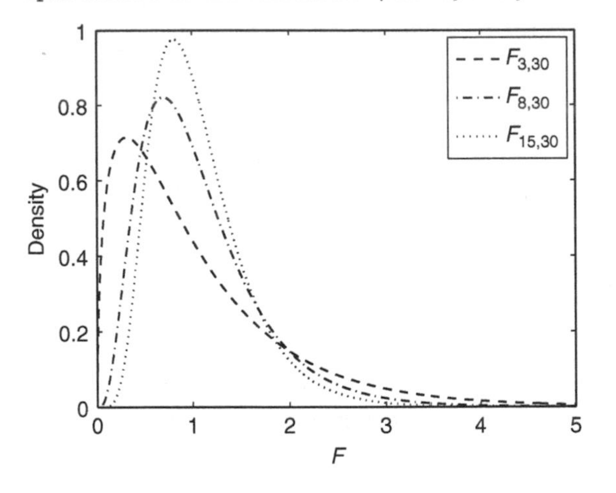

performed to determine if the large population variance can account for the differences observed among the group means.

Note that the mean squares calculated in Equations (4.28) and (4.29) are both independent and unbiased estimates of σ^2 when H_0 is true. We can calculate the ratio of these two estimates of the population variance:

$$F = \frac{\text{MS(among)}}{\text{MS(within)}}. \tag{4.30}$$

The ratio calculated in Equation (4.30) is called the ***F* statistic** or the ***F* ratio** because, under H_0, the ratio of the two estimates of the population variance follows the F probability distribution (named after the statistician R. A. Fisher who developed this distribution). The F distribution is a family of distributions (just as the Student's t distribution is a family of distributions) of the statistic $(s_1^2/\sigma_1^2)/(s_2^2/\sigma_2^2)$, where the subscripts 1 and 2 denote different populations. For the ANOVA test, both MS(among) or s_1^2 and MS(within) or s_2^2 estimate the same population variance when the null hypothesis is true, i.e. $\sigma_1^2 = \sigma_2^2$. The F distribution reduces to the sampling distribution of the ratio s_1^2/s_2^2. The shape of the individual F distribution is specified by two variables: the degrees of freedom associated with MS(among), i.e. the numerator of the F ratio, and the degrees of freedom associated with MS(within), i.e. the denominator of the F ratio. Thus, the calculated F statistic has an F distribution with $k-1$ numerator degrees of freedom and $N-k$ denominator degrees of freedom. The relevant F distribution for an ANOVA test is identified by attaching the degrees of freedom as subscripts, $F_{k-1,N-k}$. A specific F distribution defines the probability distribution of all possible values that the ratio of two sample variances s_1^2/s_2^2 can assume based on the degrees of freedom or accuracy with which the numerator and denominator are calculated. Three F distributions for three different numerator degrees of freedom and denominator degrees of freedom equal to 30 are plotted in Figure 4.9. Usage of the F distribution in statistics is on a much broader scale than that described here; F distributions are routinely used to test the hypothesis of equality of population variances. For example, before performing an ANOVA, one may wish to test if the samples are drawn from populations that have the same variance. Although not discussed here, you can find a discussion on performing hypothesis tests on population variances in numerous statistical textbooks.

If H_0 is true, then all populations have mean equal to μ. The "within-groups" and "among-groups" estimates of the population variance will usually be close in value. The group means will fluctuate about the grand (or estimated population) mean according to a SEM of $\sigma/\sqrt{n}$, and within each group the data points will vary about their respective group mean according to a standard deviation of σ. Accordingly, the F ratio will have a value close to unity. We do not expect the value of the test statistic to be exactly one even if H_0 is true since unavoidable sampling error compromises the precision of our estimates of the population variance σ^2.

If H_0 is not true, then at least one population mean differs from the others. In such situations, the MS(among) estimate of the population variance will be greater than the true population variance. This is because at least one group mean will be located much further from the grand mean than the other means. It may be possible that all of the population means are different and, accordingly, that the group means vary significantly from one another. When the null hypothesis is false, the F ratio will be much larger than unity. By assigning a significance level to the test, one can define a rejection region of the F curve. Note that the rejection region lies on the right-hand side of the F curve. If the calculated F statistic is greater than the critical value of F that corresponds to the chosen significance level, then the null hypothesis is rejected and we conclude that the population means are not all equal. When we reject H_0, there is a probability equal to the significance level that we are making a type I error. In other words, there is a possibility that the large deviation of the group means from the grand mean occurred by chance, even though the population means are all equal. However, because the probability of such large deviations occurring is so small, we conclude that the differences observed are real, and accordingly we reject H_0.

A limitation of the one-way ANOVA is now apparent. If the test result is significant, all we are able to conclude is that not all population means are equal. An ANOVA cannot pinpoint populations that are different from the rest. If we wish to classify further the populations into different groups, such that all populations within one group have the same mean, we must perform a special set of tests called **multiple comparison procedures**. In these procedures, all unique pairs of sample means are compared with each other to test for significant differences. The samples are then grouped together based on whether their means are equal. The advantage of these procedures over the t test is that if the null hypothesis is true, then the overall probability of finding a significant difference between any pair of means and thereby making a type I error is capped at a predefined significance level α. The risk of a type I error is well managed in these tests, but the individual significance tests for each pair are rather conservative and thus lower in power than a single t test performed at significance level α. Therefore, it is important to clarify when use of multiple comparison procedures is advisable.

If the decision to make specific comparisons between certain samples is made prior to data collection and subsequent analysis, then the hypotheses formed are specific to those pairs of samples only, and such comparisons are said to be "planned." In that case, it is correct to use the t test for assessing significance. The conclusion involves only those two populations whose samples are compared.

For a t test that is performed subsequent to ANOVA calculations, the estimate of the common population variance s^2_{pooled} can be obtained from the MS(within) estimate calculated by the ANOVA procedure. The advantages of using MS(within) to estimate the population variance instead of pooling the sample variances only from the two samples being compared are: (1) MS(within) is a superior estimate of the population variance since all samples are used to obtain the estimate; and (2), since it is associated with more degrees of freedom ($N - k$), the corresponding t distribution is relatively narrower, which improves the power of the test.

You should use multiple comparison procedures to find populations that are significantly different when the ANOVA yields a significant result and you need to perform rigorous testing to determine differences among all pairs of samples. Typically, the decision to perform a test of this sort is made only after a significant ANOVA test result is obtained. These tests are therefore called "unplanned comparisons." Numerous pairwise tests are performed and significant differences are tested for with respect to all populations. In other words, the final conclusion of such procedures is global and applies to all groups in the data set. Therefore you must use special tests such as the Tukey test, Bonferroni test, or Student–Newman–Keuls designed expressly for multiple comparisons of samples analyzed by the ANOVA procedure. These tests factor into the calculations the number of comparisons being made and lower the type I error risk accordingly.

The **Bonferroni adjustment** is one of several multiple comparison procedures and is very similar to the t test. If k samples are compared using a one-way ANOVA, and the null hypothesis is rejected, then we proceed to group the populations based on their means using a multiple comparisons procedure that involves comparing C_2^k pairs of samples. According to the Bonferroni method, the t statistic is calculated for each pair:

$$t = \frac{\bar{x}_1 - \bar{x}_2}{\text{MS(within)}\sqrt{\frac{1}{n_1} + \frac{1}{n_2}}}.$$

If the overall significance level for the multiple comparisons procedure is set at α, so that α is the overall risk of making a type I error if H_0 is true, then the significance level for the individual t tests is set to $\alpha^* = \alpha / C_2^k$. It can be easily shown (see below) that this choice of α^* leads to an overall significance level capped at α.

The overall probability of a type I error if H_0 is true is

$$1 - (1 - \alpha^*)^{C_2^k}; \tag{4.31}$$

$\left(1 - \frac{\alpha}{C_2^k}\right)^{C_2^k}$ can be expanded using the binomial theorem as follows:

$$\left(1 - \frac{\alpha}{C_2^k}\right)^{C_2^k} = 1 - C_2^k \frac{\alpha}{C_2^k} + \frac{C_2^k(C_2^k - 1)}{2}\left(\frac{\alpha}{C_2^k}\right)^2 + \cdots + (-1)^{C_2^k}\left(\frac{\alpha}{C_2^k}\right)^{C_2^k}.$$

As the number of samples increases, C_2^k becomes large. Since α is a small value, the quantity α / C_2^k is even smaller, and the higher-order terms in the expansion are close to zero. We can approximate the binomial expansion using the first two terms only:

$$\left(1 - \frac{\alpha}{C_2^k}\right)^{C_2^k} \sim 1 - \alpha.$$

Thus, the overall probability of making a type I error when using the Bonferroni adjustment is obtained by substituting the above into Equation (4.31), from which we recover α.

The null hypothesis for each pairwise t test modified with the Bonferroni adjustment is H_0: $\mu_1 = \mu_2$. The null hypothesis is rejected when the p value of the test statistic is less than α/C_2^k. If $p > \alpha/C_2^k$, then the null hypothesis is retained.

Using MATLAB

Statistics Toolbox lends MATLAB the capability of performing ANOVA calculations. The MATLAB function `anova1` performs a one-way ANOVA. The syntax is

```
[p, table, stats] =anova1(X, group)
```

The multi-sample data set is passed to the `anova1` function as either a matrix or vector $\boldsymbol{X}$. Supplying a matrix of values, where each column contains values of a single sample (or treatment), is recommended when the sample sizes are equal. An optional character array `group` contains the names of each treatment group. When the sample sizes are unequal, all sample data can be combined into a single vector $\boldsymbol{x}$. The `group` variable is then supplied as a vector of equal size to $\boldsymbol{x}$. Each entry of `group` must contain the name of the group that the corresponding element of $\boldsymbol{x}$ belongs to. The elements of $\boldsymbol{x}$ are arranged into groups based on their group names.

The following is output by the `anova1` function:

`p`: the p value of the F statistic;
`table`: a cell array containing the results of the one-way ANOVA test;
`stats`: a structure that serves as the input to the MATLAB `multcompare` function that performs multiple comparison procedures.

The output includes a figure that contains an ANOVA table, which is a tabulation of the sum of squares, degrees of freedom, mean squares, the F ratio, and the corresponding p value. The sum of the squares are displayed for the following.

(1) Within-groups, i.e. $\sum_{i=1}^{k}\sum_{j=1}^{n_i}\left(x_{ij} - x_{i\cdot}\right)^2$. This category is denoted "error" by MATLAB and some statisticians because it represents the degree of uncertainty or noise within each group.
(2) Among-groups, i.e. $\sum_{i=1}^{k} n_i(\bar{x}_{i\cdot} - \bar{x}_{\cdot\cdot})^2$. This category is named "columns" by MATLAB since it represents the SS differences among the columns of the input matrix.
(3) Total error, i.e. $\sum_{i=1}^{k}\sum_{j=1}^{n_i}\left(x_{ij} - \bar{x}_{\cdot\cdot}\right)^2$. The SS(total) is actually equal to SS(within) + SS(among). This is illustrated below. We write $x_{ij} - \bar{x}_{\cdot\cdot} = \left(x_{ij} - \bar{x}_{i\cdot}\right) + \left(\bar{x}_{i\cdot} - \bar{x}_{\cdot\cdot}\right)$. When we square both sides and sum over all observations, we obtain, as a final result,

$$\sum_{i=1}^{k}\sum_{j=1}^{n_i}\left(x_{ij} - \bar{x}_{\cdot\cdot}\right)^2 = \sum_{i=1}^{k}\sum_{j=1}^{n_i}\left(x_{ij} - \bar{x}_{i\cdot}\right)^2 + \sum_{i=1}^{k}\sum_{j=1}^{n_i}\left(\bar{x}_{i\cdot} - \bar{x}_{\cdot\cdot}\right)^2.$$

Another figure, called a **box plot**,[6] is output with this function. A box plot plots the median of each group along with the interquartile ranges (see Section 4.4 for a discussion on quartiles and quantiles) for each group. The box plot provides a visual display of trends in the raw data and is not a result produced by ANOVA calculations.

The `multcompare` function performs the multiple comparisons procedure to group populations together whose means are the same. The syntax for this function is

```
table = multcompare(stats, 'alpha', alpha, 'ctype', ctype)
```

The first input argument of the `multcompare` function is the `stats` variable that is computed by the `anova1` function. The second variable `alpha` specifies the overall

[6] Statistics Toolbox contains the `boxplot` function for creating box plots. See MATLAB help for more information.

significance level of the procedure. The default is the 5% significance level. The third variable `ctype` specifies the method to determine the critical value of the test statistic, the options for which are `'tukey-kramer'`, `'dunn-sidak'`, `'bonferroni'`, and `'scheffe'`. The `tukey-kramer` method is the default.

The output includes a table that contains five columns. The first two columns specify the treatment pair being compared. The fourth column is the estimate of the difference in sample means, while the third and fifth columns are the confidence limits.

An interactive graphical output is also produced by the `multcompare` function that displays the treatment means and their respective confidence intervals. The important feature of this plot is that it color codes samples derived from populations with either the same or different means to distinguish alike populations from statistically dissimilar populations.

Before you can use the ANOVA method, it is necessary to ensure that the data fulfil the conditions of normality, homoscedasticity, and independence. The quality of the data, i.e. randomness and independence of samples, is contingent upon good experimental design. It is important to inspect the methods of data generation and the data sources to ensure that the sample data are independent. For example, if different levels of treatment are applied to the same individuals, the data sets are *not* independent from one another. The condition of normality can be checked by plotting histograms and/or normal probability plots as discussed in Section 4.4. A histogram and/or normal probability plot can be drawn for each sample. Alternatively, the deviations of data points from their respective means can be combined from all data sets and plotted on a single normal probability plot, as done in Box 4.8.

The equality of population variances is a stringent requirement when sample sizes are small, but can be relaxed somewhat when sample sizes are large, since the ANOVA method is found to produce reliable results when large samples are drawn from populations of unequal variances. It will usually suffice to use ANOVA when the ratio of the largest sample variance to the smallest sample variance is less than 4. One can create side-by-side dot plots of the deviations $\left(x_{ij} - \bar{x}_{i.}\right)$ observed within samples to compare variability trends in each sample. Hypothesis testing procedures are available to determine if population variances are equal, but they are not covered in this book. If the sample variances differ from each other considerably, mathematical transformation of the data may help equalize the variances of the transformed samples. If the standard deviation is observed to increase linearly with the sample mean, i.e. $\sigma \propto \mu$, then either logarithmic transformation of the data or square root transformation of the data is recommended. Note that transformation can also be used to equalize variances across all treatment levels when performing linear regression of the data. You may also need to transform data that exhibit considerable skew so that the population distribution is more or less normal. Data transformations must be accompanied by modifications of the null and alternate hypotheses. For example, a square root transformation of data will change the null hypothesis

H_0: "The eight sleep patterns considered in this study have no effect on memory retention test results."

to

H_0: "The eight sleep patterns considered in this study have no effect on the square root of memory retention test results."

Box 4.8 Mountain climbing and oxygen deprivation at high altitudes

The atmospheric pressure, which is 760 mm Hg or 1 bar at sea level, progressively decreases with ascent from sea level. Mountain dwellers, mountain climbers, and tourists in mountainous regions are exposed to hypobaric conditions at high altitudes. The partial pressure of oxygen (~160 mm Hg at sea level) drops to precipitously low levels at very high altitudes. Decreased availability of atmospheric oxygen compromises body function due to inadequate synthesis of ATP in the cells. The human body has some ability to acclimatize to oxygen deprivation (hypoxia). Acclimatization is a natural adaptive mechanism of the human body that occurs when exposed to different weather or climate conditions (pressure, temperature, and humidity) for a period of time. Medical scientists are interested in understanding the limits of the body to handle adverse climate conditions. For example, researchers in environmental medicine are interested in estimating lower limits of the partial pressure of oxygen in arterial blood tolerable by man that still permit critical body functions such as locomotion and cognitive abilities (Grocott *et al.*, 2009). At normal atmospheric conditions, hemoglobin in arterial blood is 97% saturated with O_2. This level of hemoglobin saturation occurs due to a relatively large arterial blood partial pressure of 95 mm Hg, under healthy conditions. It has been observed that the body acclimatizes to hypobaric conditions by elevating the hemoglobin content in the blood to improve the efficiency of O_2 capture at the blood–lung interface and thereby maintain arterial oxygen content at optimum levels.

In a fictitious study, the arterial oxygen content of seven mountain climbers was measured at four different altitudes during ascent of the climbers to the summit of Mt. Tun. The climbers ascended gradually to allow the body to acclimatize to the changing atmospheric pressure conditions. The data are presented in Table 4.7.

We want to determine if the mean arterial oxygen content is the same at all altitudes or if it changes with altitude. We have four treatment groups, where the "treatments" are the differing heights above sea level. The ratio of the largest group variance to smallest group variance is 3.9. We cautiously assume that the population variances are the same for all groups. A normal probability plot (see Figure 4.10) of the deviations in all four groups combined is fairly linear. This supports the assumption that all four populations of arterial oxygen content are normally distributed.

To obtain the F ratio we must first calculate the sum of squares and mean squares within the groups and among the groups.

$$SS(\text{within}) = \sum_{i=1}^{k}\sum_{j=1}^{n_i}(x_{ij} - \bar{x}_{i\cdot})^2$$

$$= (215 - 203.1)^2 + (241 - 203.1)^2 + \cdots + (155 - 166)^2 = 14\,446;$$

Table 4.7. *Arterial oxygen content (ml/l)*

	Altitude			
Individuals	Group 1 2000 m	Group 2 4000 m	Group 3 6000 m	Group 4 8000 m
1	215	221	250	141
2	241	196	223	170
3	159	235	196	184
4	222	188	249	167
5	211	190	181	175
6	178	196	212	170
7	196	147	208	155
Mean $\bar{x}_i$	203.1	196.1	217.0	166
Variance σ_i^2	770.5	774.5	665.3	197.3

Figure 4.10

Normal probability plot of the deviations of arterial oxygen content measurements from their respective means at all four altitudes.

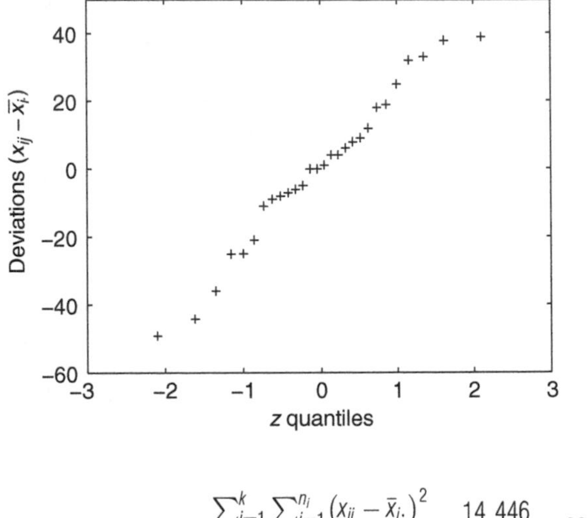

$$\text{MS(within)} = \frac{\sum_{i=1}^{k}\sum_{j=1}^{n_i}(x_{ij} - \bar{x}_{i\cdot})^2}{N - k} = \frac{14\,446}{24} = 601.9.$$

The grand mean is calculated to be

$$\bar{x}_{\cdot\cdot} = \frac{\bar{x}_{1\cdot} + \bar{x}_{2\cdot} + \bar{x}_{3\cdot} + \bar{x}_{4\cdot}}{4} = 195.6.$$

We can calculate $\bar{x}_{\cdot\cdot}$ this way only because the sample sizes are equal. Otherwise, we would need to weight each sample mean according to its respective sample size, and then divide the sum by the total number of observations.

Now,

$$\begin{aligned}\text{SS(among)} &= n\sum_{i=1}^{k}(\bar{x}_{i\cdot} - \bar{x}_{\cdot\cdot})^2\\ &= 7\left((203.1 - 195.6)^2 + (196.1 - 195.6)^2 + (217 - 195.6)^2 + (166 - 195.6)^2\right)\\ &= 9739.1\end{aligned}$$

and

$$\text{MS(among)} = \frac{n\sum_{i=1}^{k}(\bar{x}_{i\cdot} - \bar{x}_{\cdot\cdot})^2}{k - 1} = \frac{9739.1}{3} = 3246.4.$$

A MATLAB program can be written to perform the above calculations. Then,

$$F = \frac{\text{MS(among)}}{\text{MS(within)}} = \frac{3246.4}{601.9} = 5.4.$$

The p value associated with the calculated F ratio is obtained using the cumulative distribution function for F provided by MATLAB, or from a standard statistics table.

The syntax for the `fcdf` function is

```
y = fcdf(x, fnum, fdenom)
```

where ***x*** is the F ratio value, `fnum` is the degrees of freedom of the numerator, and `fdenom` is the degrees of freedom of the denominator.

For this problem we type into MATLAB

Figure 4.11

The ANOVA table generated by MATLAB using the `anova1` function.

ANOVA Table

Source	SS	df	MS	F	Prob>F
Columns	9739.1	3	3246.38	5.39	0.0056
Error	14445.7	24	601.9		
Total	24184.9	27			

```
>> p = 1- fcdf(5.4,3,24)
```

MATLAB outputs

```
p =
    0.0055
```

If our significance level is set at 0.05, then $p < \alpha$, and the result is significant. We reject the null hypothesis. The mean levels of arterial oxygen are not same within the body at all four altitudes.

We can also use the MATLAB `anova1` function to calculate the F statistic. Since the sample sizes are equal, we create a matrix $\boldsymbol{A}$ containing all of the observations. The first column of the matrix represents the first group or measurements taken at a height of 2000 m; the second column contains measurements taken at a height of 4000 m, and so on.

We type the following statement into MATLAB:

```
>> [p, table, stats] = anova1(A)
```

MATLAB outputs Figure 4.11.

The slight discrepancy in the p value obtained by the `anova1` function and our step-by-step calculations is due to rounding of our calculated values.

We would like to determine which of the four groups have significantly different means. The `multcompare` function can be used for this purpose. The output variable `stats` from the `anova1` function is the input into the `multcompare` function.

```
table = multcompare(stats)
table =
1.0000   2.0000   -29.1760     7.0000   43.1760
1.0000   3.0000   -50.0331   -13.8571   22.3188
1.0000   4.0000     0.9669    37.1429   73.3188
2.0000   3.0000   -57.0331   -20.8571   15.3188
2.0000   4.0000    -6.0331    30.1429   66.3188
3.0000   4.0000    14.8240    51.0000   87.1760
```

The first two columns list the pair of treatments that were compared. The fourth column is the calculated difference in the means of the two groups under comparison, while the third and fifth columns are the lower and upper bounds, respectively, of the confidence interval for the difference in the population means of the two treatments. For example, the first row tells us that the difference in the means of treatment group 1 and treatment group 2 is 7.0, and the confidence interval for the difference between the two population means contains 0.0. When a confidence interval for the difference between two population means does not contain zero, we may conclude that the observed difference between the two group means is statistically significant. This is the case for groups 1 vs. 4 and 3 vs. 4.

MATLAB creates an interactive figure (Figure 4.12) that allows you to check visually the differences between the groups. In the figure, the confidence interval for group 4 is selected.

Figure 4.12

Plot generated by the MATLAB `multcompare` function. The figure created by MATLAB highlights all groups that are significantly different from the selected group (blue) in red and the non-significantly different groups in light gray. Groups that have overlapping confidence intervals are not significantly different, while groups that have gaps between their confidence intervals have significantly different means.

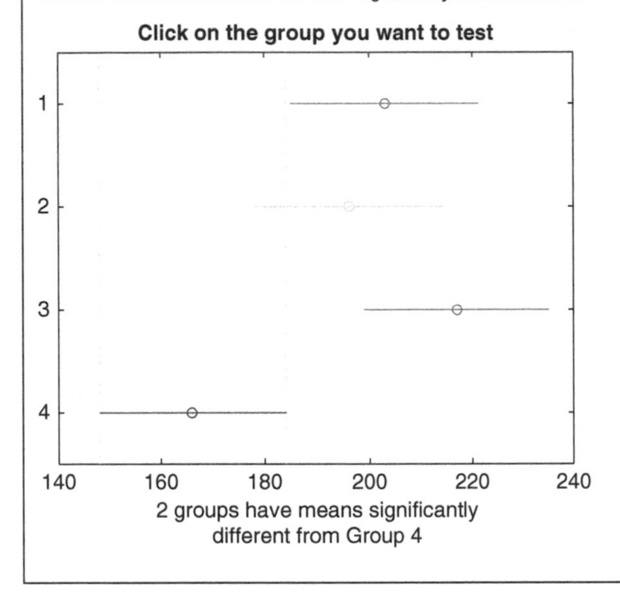

Box 4.9 Error bars in figures

Most often, experimental results are presented in a condensed graphical form and are displayed as figures in journal publications for the purpose of disseminating scientific discovery or engineering advances. For example, the result of the experiment discussed in Box 4.8 can be plotted as a bar graph of the mean arterial oxygen content observed at four different altitudes. However, simply plotting the mean values does not convey adequate information about the physiological behavior studied. The extent of variation in the variables is also a key piece of information that must also be plotted along with the mean values. If our purpose is to convey the variability in the data, it is advisable to plot error bars along with the mean, with the total length of each bar equal to the standard deviation. Such error bars are called **descriptive error bars**. If the purpose of the figure is to compare samples and infer differences between populations based on statistical analysis of the experimental data, then it is advisable to plot error bars depicting the standard error of the mean or SEM along with the mean values. Such error bars are called **inferential error bars**. The length of an SEM bar communicates the amount of uncertainty with which the mean is known. The lines of each SEM error bar stretch out from the mean to a distance equal to the SEM in both directions. The error bars are a graphical depiction of the 66.3% confidence interval of the corresponding population means (when the sampling distribution of the sample mean is normal). Using SEM error bars, one can compare the means and error bar lengths and visually assess if statistically significant differences exist between the samples or if differences in the sample means are simply fluctuations due to sampling error. Figure 4.13 is a plot of the mean arterial oxygen content for all four altitudes, with error bars that associate each mean value with its SEM.

Using MATLAB

The MATLAB function `errorbar` is a handy tool that plots data points along with error bars of specified lengths. The syntax for `errorbar` is

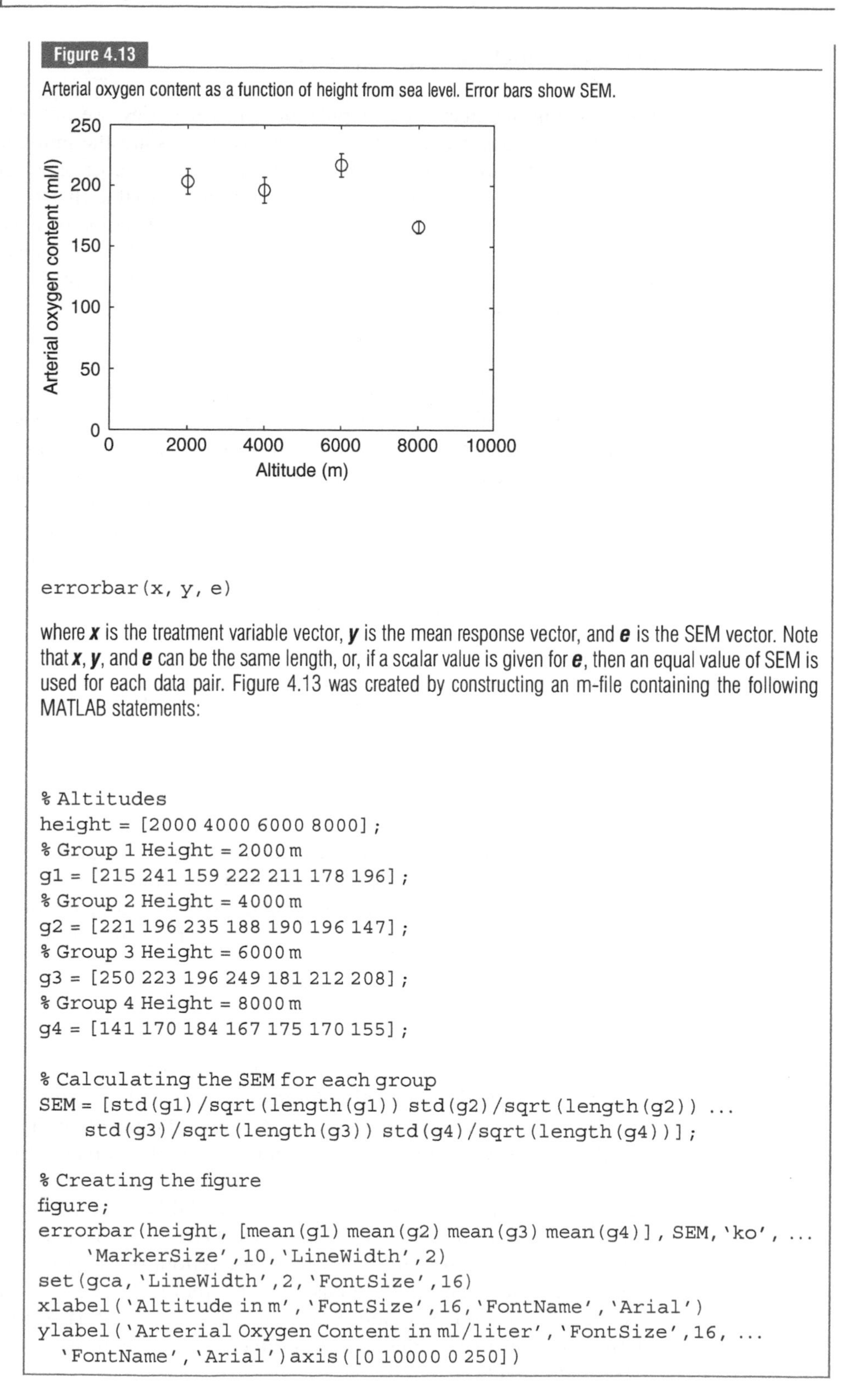

Figure 4.13

Arterial oxygen content as a function of height from sea level. Error bars show SEM.

```
errorbar(x, y, e)
```

where ***x*** is the treatment variable vector, ***y*** is the mean response vector, and ***e*** is the SEM vector. Note that ***x***, ***y***, and ***e*** can be the same length, or, if a scalar value is given for ***e***, then an equal value of SEM is used for each data pair. Figure 4.13 was created by constructing an m-file containing the following MATLAB statements:

```
% Altitudes
height = [2000 4000 6000 8000];
% Group 1 Height = 2000 m
g1 = [215 241 159 222 211 178 196];
% Group 2 Height = 4000 m
g2 = [221 196 235 188 190 196 147];
% Group 3 Height = 6000 m
g3 = [250 223 196 249 181 212 208];
% Group 4 Height = 8000 m
g4 = [141 170 184 167 175 170 155];

% Calculating the SEM for each group
SEM = [std(g1)/sqrt(length(g1)) std(g2)/sqrt(length(g2)) ...
    std(g3)/sqrt(length(g3)) std(g4)/sqrt(length(g4))];

% Creating the figure
figure;
errorbar(height, [mean(g1) mean(g2) mean(g3) mean(g4)], SEM, 'ko', ...
    'MarkerSize',10,'LineWidth',2)
set(gca,'LineWidth',2,'FontSize',16)
xlabel('Altitude in m','FontSize',16,'FontName','Arial')
ylabel('Arterial Oxygen Content in ml/liter','FontSize',16, ...
  'FontName','Arial')axis([0 10000 0 250])
```

4.9 Chi-square tests for nominal scale data

When sorting members of a population into categories, we first specify a nominal variable, i.e. a non-numeric data type, whose values are the mutually exclusive and exhaustive categories. The nominal variable has $n \geq 2$ number of categories, where n is an integer. For example, we can create a nominal variable that specifies a mutually exclusive physical attribute such as the inherited hair color of a person. The four categories of hair color could be defined as "blonde," "red," "brown," and "black." To estimate the proportion of each category within the population, a random sample is chosen from the population, and each individual is classified under one of n possible categories. The proportion of the population that belongs to each category is estimated from frequency counts in each category. In Section 4.7, we looked at the relationship between population proportion and the binomial probability distribution, where a dichotomous nominal variable was used for classification purposes. We used the frequencies observed in the "success" category to calculate the sample proportion. If the number of frequencies is large enough, the sample proportion is a good estimate of the population proportion and the central-limit theorem applies. Under these conditions, the sampling distribution of the population proportion is approximately normally distributed, and the z test can be used to test hypotheses for a single population proportion, or for the equality of proportions within two populations.

In this section, we introduce a group of methods called χ^2 **tests** (chi-square tests). We apply these procedures to test hypotheses when data consist of frequency counts under headings specified by one or more nominal variables. In any χ^2 test, the χ^2 test statistic must be calculated. The χ^2 statistic is related to the standard normal random variable z as follows:

$$\chi_k^2 = \sum_{i=1}^{k} z_i^2 = \sum_{i=1}^{k} \frac{(x_i - \mu_i)^2}{\sigma_i^2}, \tag{4.32}$$

where k is the degrees of freedom associated with the value of χ^2 and the x_i's are k normally distributed independent random variables, i.e. $x_i \sim N(\mu_i, \sigma_i)$. The χ^2 statistic with k degrees of freedom is calculated by adding k independent squared random variables z_i^2 that all have a $N(0, 1)$ distribution. For example, if $k = 1$, then

$$\chi_1^2 = z^2,$$

and χ^2 has only one degree of freedom. The distribution that the χ^2 statistic follows is called the χ^2 **(chi-square) distribution**. This probability distribution is defined for only positive values of χ^2, since χ^2 cannot be negative. You already know that the sum of two or more normal random variables is also normally distributed. Similarly, every z^2 random variable follows a χ^2 distribution of one degree of freedom, and addition of k independent χ_1^2 variables produces another χ^2 statistic that follows the χ^2 distribution of k degrees of freedom. The χ^2 distribution is thus a family of distributions, and the degrees of freedom specify the unique χ^2 distribution curve. The χ_k^2 distribution has a mean equal to k, i.e. the degrees of freedom associated with the χ^2 statistic. You should try showing this yourself. (Hint: Look at the method used to derive the variance of a normal distribution, which was presented in Section 3.5.2.)

Broadly, there are three types of χ^2 tests:

(1) χ^2 goodness-of-fit test;
(2) χ^2 test for independence of two nominal variables;
(3) χ^2 test for homogeneity of two or more populations.

Example 4.5

A random sample of 100 people is chosen and each individual in the sample is classified under any one of four categories that specify natural hair color. The observations are as follows:

blonde, 17,
red, 3,
brown, 45,
black, 35.

The numbers listed above are the observed frequencies o_i for each category.

The null hypothesis for the distribution of the nominal variable "natural hair color" within the population is

H_0: "The distribution of natural hair color is uniform among the individuals of the population."

Based on the null hypothesis, the expected frequencies e_i in the four categories are:

blonde, 25,
red, 25,
brown, 25,
black, 25.

The χ^2 statistic is calculated from frequency count data as follows:

$$\chi^2 = \sum_{i=1}^{k} \frac{(o_i - e_i)^2}{e_i}, \tag{4.33}$$

where o_i is the observed frequency in the ith category, e_i is the expected frequency in the ith category, and k is the number of categories. The **observed frequencies** are the experimental observations of the frequencies or number of sample units within each category of the nominal variable(s). (Each member of the sample is a sample unit, and one unit can be assigned to only one category.) The **expected frequencies** are calculated based on the null hypothesis that is tested. It is important to note that the frequencies used in the calculations are the actual frequency counts and not relative frequencies. The degrees of freedom associated with χ^2 are usually less than k, and the reduction in the degrees of freedom depends on the number of terms that are constrained to a particular value. As each χ^2 test is individually discussed, the method used to calculate degrees of freedom will be explained. The calculated χ^2 statistic will approximately follow the χ^2 distribution if all observations are independent, and the random samples are large enough so that the frequency in any category is not too small.

If the observed frequencies are close in value to the expected frequencies, then the value of the χ^2 statistic will be small, and there will be little reason to reject the null hypothesis. If there are large differences between the values of any of the observed and expected frequencies, then the test statistic will be large in magnitude. We can identify a rejection region of the χ^2 curve by specifying a significance level and determining the corresponding χ^2 critical value that demarcates the rejection region. For χ^2 tests, the rejection region will always be located on one side of the χ^2 distribution, even if the hypothesis is two-sided.

Using MATLAB

Methods to calculate the χ^2 probability density function and the χ^2 cumulative distribution function are available in Statistics Toolbox. The function `chi2pdf` has the syntax

```
y = chi2pdf(x, f)
```

where $\boldsymbol{x}$ is a vector containing χ^2 values at which the probability density values are desired, f is the degrees of freedom, and $\boldsymbol{y}$ is the output vector containing the probability density values.

The function `chi2cdf` has the syntax

```
p = chi2cdf(x, f)
```

where $\boldsymbol{x}$ is a vector containing χ^2 values at which the cumulative probabilities are desired, f is the degrees of freedom, and $\boldsymbol{y}$ is the output vector containing the cumulative probabilities.

4.9.1 Goodness-of-fit test

A goodness-of-fit test allows us to test the hypothesis that an observed frequency distribution follows some known function or distribution. The purpose of the goodness-of-fit test is to compare the distribution of a characteristic within the sample to a hypothesized or expected distribution. To perform a goodness-of-fit test, we require a large random sample drawn from the population of interest. The χ^2 **goodness-of-fit test** can be used to compare the frequency of sample observations in all categories specified by the nominal variable to the expected distribution within the population. In addition to testing the distribution of a nominal variable, the χ^2 goodness-of-fit test can be used to make inferences regarding the distribution of a cardinal variable within a population; i.e. you can use this test to determine whether the distribution of a characteristic within a population is normal, binomial, Poisson, lognormal, exponential, or any other theoretical distribution. Thus, the characteristic of interest can be a cardinal variable and can have a discrete or a continuous distribution within the population,[7] or can be a nominal variable, in which case the distribution among the variable's categories is described in terms of proportions (or frequencies).

Example 4.6

A process for manufacturing lip gloss has a scrap rate of 15%. Scrap rate is defined as the fraction of the finished product that must be discarded due to failure issues and thus cannot be sold on the market. A scrap rate of more than 15% is deemed economically unviable for this production process.
A slight change has been made in the process conditions, specifically the filling process. Out of 9200 lip gloss tubes produced, 780 were scrapped. The engineer needs to determine if the scrap rate has changed.

The problem describes a binomial process. The expected probability of a single tube of lip gloss going to market is 0.85, and is defined as a success. The hypotheses are set up as follows:

H_0: $p = 0.85$,
H_A: $p \neq 0.85$.

[7] The Kolmogorov–Smirnov one-sample test is more suited to test the goodness-of-fit of continuous distributions. Statistics Toolbox contains the `kstest` function that performs this test.

To test the null hypothesis, we perform the χ^2 goodness-of-fit test. The hypothesized population has a population proportion of successes equal to 0.85. We want to see if our sample is drawn from a population that has the same distribution of successes and failures as the hypothesized distribution. We first calculate the expected sample frequency in each category. If the null hypothesis is true, then, out of 9200 tubes, 7820 are expected to make it to market and 1380 are expected to be defective. We can compare the observed and expected frequencies using Equation (4.33) to generate the χ^2 test statistic:

$$\chi^2 = \frac{(8420 - 7820)^2}{7820} + \frac{(780 - 1380)^2}{1380} = 306.9.$$

There are two categories – success and failure, i.e. $k = 2$. However, the observed and expected frequencies of scrapped (failed) material are dependent on the observed and expected frequencies in the success category. This is because the total number of tubes is constrained to 9200. In other words, since $p + q = 1$, then $q = 1 - p$. Once we know p, we will always know q. So, if we know the number of successes and the total number of outcomes, then we are bound to know the number of failures. In other words, the frequency of only one category can be varied freely, while the frequency in other category is fixed by the first. Thus, the frequency count in the failure category, i.e. the last category, provides no new information. The degrees of freedom associated with this problem are $f = k - m - 1$, where m is the number of population parameters such as mean and variance calculated from the data. Since no parameters are calculated for this problem, $m = 0$, and $f = k - 0 - 1 = 1$.

To clarify the point made above regarding the number of degrees of freedom, we recalculate the χ^2 test statistic using the definition of χ^2 provided by Equation (4.32) as follows:

$$\chi^2 = \frac{(x - np)^2}{npq},$$

where x is the frequency of successes observed, np is the mean, npq is the variance of the binomial distribution, and $z = (x - np)/\sqrt{npq}$. This statistic is associated with one degree of freedom since it has only one χ^2 term. Alternatively, we can use Equation (4.33) to calculate the test statistic. This equation has two terms, which we write in terms of binomial variables as follows:

$$\chi^2 = \frac{(o_1 - np)^2}{np} + \frac{(o_2 - nq)^2}{nq}.$$

Since $o_1 + o_2 = np + nq = n$,

$$\begin{aligned}\chi^2 &= \frac{(o_1 - np)^2}{np} + \frac{(n - o_1 - (n - np))^2}{nq} \\ &= \frac{(o_1 - np)^2}{np} + \frac{(o_1 - np)^2}{nq} = \frac{(p + q)(o_1 - np)^2}{npq} \\ &= \frac{(x - np)^2}{npq}.\end{aligned}$$

Here, we have shown that Equations (4.32) and (4.33) are equivalent and that the test statistic is associated with only one degree of freedom.

Now let's calculate the p value associated with the calculated test statistic. We set the significance level at 0.01, since we want to be certain whether a change in scrap rate has occurred:

```
>> p = 1- chi2cdf(306.9, 1)
p =
    0
```

Since $p \ll 0.01$, we are very confident that the scrap rate has been lowered from 15%. In other words, the success rate has increased. Note that, even though our hypothesis is non-directional, the direction of the change is known because we can compare the observed and expected failure frequencies. Because there are only two categories, it is obvious that if the frequencies in one category significantly increase, then the frequencies in the second category have concomitantly decreased significantly. If there were more than two categories, then the χ^2 test cannot tell us which category is significantly different from the others.

We could have also performed the hypothesis test for a single population proportion (using Equation (4.26)) to test the hypothesis of $p = 0.85$. Try this yourself. What z statistic and p value do you get? The p value should be exactly the same since the two tests are essentially equivalent. Since $\chi^2 = z^2$, a z value of $\sqrt{306.9}$ is obtained.

We have assumed that use of the χ^2 statistic was valid in Example 4.5. Recall that the z test was permitted for testing population proportions only if $n\hat{p} > 5$ and $n(1-\hat{p}) > 5$. Example 4.5 fulfils this condition. The use of the χ^2 test procedure is valid only if the sampling distribution of the binomial variable x can be approximated by a normal distribution. Now we discuss the condition of validity for use of the χ^2 goodness-of-fit test in general. There is some difference in opinion among statisticians over the minimum allowable expected frequency count in any category. One recommendation that has been gaining increasing acceptance is that if n is not large, the minimum expected frequency in any category should not be less than 5, otherwise the minimum expected frequency in any category should not be less than 1. If the expected frequency in any category is found to be less than 1, the category should be combined with an adjacent category so that the combined expected frequency is greater than 1 (or 5).

Box 4.10 Distribution of birthweights in two different racial groups in England and Wales

A statistical analysis of birthweights according to racial group for all recorded human births occurring in England and Wales in the year 2005 was performed (Moser *et al.*, 2008). Table 4.8 presents the birthweight distributions observed in two specific racial groups that reside in England and Wales.

We want to find out if the birthweights exhibit a normal distribution within each racial population in England and Wales. The null hypothesis is

H_0: The human birthweight distribution within any racial subpopulation of England and Wales is normal.

To calculate the expected or hypothesized frequencies in each birthweight category, we need first to convert the birthweight ranges that define each category into deviations on the standard normal scale. We then calculate the area under the standard normal curve defined by the z ranges for each category. The population parameters μ and σ are estimated from the human birthweight data for each racial group. Because the observed frequency data are used to obtain two parameters of the hypothesized frequency distribution, the degrees of freedom associated with the test statistic reduce by 2. We perform the χ^2 goodness-of-fit test to assess normality of birthweight distribution within the White British racial group. The standardized birthweight variable z is calculated using the equation $z = (x - \mu)/\sigma$.

The first birthweight category is $x < 1.0$ kg. The z range for the first birthweight category is from

$$z = -\infty \qquad \text{to} \qquad z = \frac{1.0 - 3.393}{0.488} = -4.904.$$

The probability of observing birthweights in this category, under H_0, is equal to the area under the z curve from $z = -\infty$ to $z = -4.904$. Using the MATLAB `normcdf` function, we can calculate this probability. For this function, the default values for μ and σ are zero and one, respectively. Multiplying the total observed births with the probability gives us the expected frequency in that category:

```
>> p = normcdf(-4.904)
p =
4.6952e-007
```

Table 4.8. *Frequency distribution of birthweights of live singletons (singly born babies) for the White British and Asian (Indian) racial groups in England and Wales*

Birthweight, x (kg)	White British	Asian or Asian British (Indian)
<1	1209	91
1–1.5	2015	122
1.5–2.0	4432	261
2.0–2.5	14506	1159
2.5–3.0	60036	4580
3.0–3.5	143447	6189
3.5–4.0	125718	2552
4.0–4.5	43518	478
>4.5	8059	45
Total live single births	402942	15477
Mean birthweight, $\bar{x}$	3.393	3.082
Standard deviation, s	0.488	0.542

The expected frequency in category 1 is $402942 \times 4.6952 \times 10^{-7} = 0.1892$. Since the expected frequency in this category is less than unity, we will need to combine this category with the second birthweight category.

The second category is $1.0\,\text{kg} \leq x < 1.5\,\text{kg}$. The z range of the second category is from

$$z = -4.904 \text{ to } z = \frac{1.5 - 3.393}{0.488} = -3.879.$$

The area under the z curve from $z = -4.904$ to $z = -3.879$ gives us the expected relative frequency (probability) and is obtained as follows:

```
>> p = normcdf(-3.879) - normcdf(-4.904)
p =
5.1974e-005
```

The expected frequency in category 2 is $402942 \times 5.1974 \times 10^{-5} = 20.94$.

In a similar fashion we can calculate the expected frequencies for the remaining seven categories.

Table 4.9 shows the method used to calculate the χ^2 test statistic. The degrees of freedom associated with the test statistic are equal to the (number of categories) − 1 − (number of parameters estimated from the data). Therefore,

$$f = k - m - 1 = 9 - 2 - 1 = 6.$$

The p value for the calculated test statistic $\chi^2 = 506\,926$ is

```
>> p = 1 - chi2cdf(506926, 6)
p =
0
```

On inspecting the data, we notice that the observations appear to conform somewhat to a normal distribution for the middle categories but are grossly non-normal in the first two categories. If we plot the data, we will see that the observed distribution of birthweights has heavier tails and is skewed more to the left. Note that the number of observations, $n = 402\,942$, is very large, and in such cases any small

Table 4.9. *Calculation of the x^2 test statistic based on the observed and expected frequencies in the nine birthweight categories for the White British racial group*

Category	Birthweight, x (kg)	z range for White British	Relative frequency	Expected frequency, e_i	Observed frequency, o_i	$\frac{(o_i - e_i)^2}{e_i}$
1	<1	<−4.904	4.695×10^{-7}	0.2 }	1209 }	
2	1–1.5	−4.904 to −3.879	5.197×10^{-5}	20.9 }	2015 }	486
				21.1	3224	188
3	1.5–2.0	−3.879 to −2.854	0.0021	848.7	4432	15 129
4	2.0–2.5	−2.854 to −1.830	0.0315	12 679.1	14 506	263
5	2.5–3.0	−1.830 to −0.805	0.1768	71 234.1	60 036	1760
6	3.0–3.5	−0.805 to 0.219	0.3763	151 613.0	143 447	440
7	3.5–4.0	0.219 to 1.244	0.3066	123 532.0	125 718	39
8	4.0–4.5	1.244 to 2.268	0.0951	38 313.8	43 518	707
9	>4.5	>2.268	0.0116	4700.2	8059	2400
Sum total			~1.00	402 942	402 942	$\chi^2 = 506\,926$

deviations from normality will be sharply magnified. The large number of observations in the low-birthweight categories may correspond to pre-term births. The percentage of pre-term births (under 37 weeks gestational age) of live singletons among the White British category was found to be 6.1%. Full term is considered to be a gestational age of 40 weeks. The mean gestational age for this group was 39.31 weeks.

A similar exercise can be performed to determine if the birthweight distribution within the British Indian racial group is normal (see Problem 4.9).

We also want to know if the mean birthweights in these two racial groups of England and Wales are different. We can use a z test to compare the mean birthweights of the two samples. Since the sample sizes are large, we may assume that the standard deviation of the birthweights is an accurate estimate of the population standard deviation. The test statistic is calculated according to Equation (4.9) as follows:

$$z = \frac{\bar{x}_1 - \bar{x}_2}{\sqrt{\frac{\sigma_1^2}{n_1} + \frac{\sigma_2^2}{n_2}}} = \frac{3.393 - 3.082}{\sqrt{\frac{0.488^2}{402942} + \frac{0.542^2}{15477}}} = 70.3.$$

The p value associated with this test statistic is zero. The difference between the mean birthweights of the two racial groups is highly significant. The 95% confidence interval for the difference in the mean birthweights of the two racial populations is 311 ± 9 g (10.95 ± 0.32 oz). We may conclude that, on average, babies of the White British population are heavier than those of the British Indian population.

Using MATLAB

Statistics Toolbox supplies the function `chi2gof` that allows users to perform the χ^2 goodness-of-fit test to determine if the observed frequency distribution conforms to a hypothesized discrete or continuous probability distribution. This function automatically groups the observed data into bins or categories. The default number of bins is ten. By default, `chi2gof` tests the fit of the data to the normal distribution. If the frequency in any bin is less than 5, then adjacent bins are grouped together to bring the frequency count to greater than 5. The simplest syntax for this function is

```
[h, p] = chi2gof(x)
```

where h is the result of the test, 0 conveys no significant difference between the observed and expected distributions, 1 conveys rejection of the null hypothesis, p is the p value of the χ^2 test statistic, and $\mathbf{x}$ is a vector containing the data points. MATLAB automatically creates ten bins, estimates the population mean and variance from $\mathbf{x}$, and calculates the expected frequencies based on the normal distribution specified by the estimated population parameters.

More details of the test can be specified using the following syntax:

```
[h, p] = chi2gof(x, 'parametername1', parametervalue1,
'parametername2', parametervalue2, ...)
```

where `'parametername'` is an optional argument and can be any one or more of the following:

`'nbins'`: number of bins to use,
`'alpha'`: the significance level,
`'emin'`: the least number of counts in any category,
`'cdf'`: the expected cumulative distribution function,
`'expected'`: the expected counts in each bin.

Either the values of `'cdf'` or `'expected'` can be specified as arguments; the two parameters should not be specified together.

For more information on usage and other features of this function type `help chi2gof` at the MATLAB command prompt.

4.9.2 Test of independence

The χ^2 test is frequently used to determine whether an association or dependency exists between two nominal variables. For instance, a medical practitioner may wish to know if there is a link between long-term use of an allopathic medication and susceptibility to a particular disease. As another example, a public health care official may be interested in determining whether smoking habits are linked to body weight and/or mental health. To perform the χ^2 test of independence, a single sample is drawn from the reference population and each individual or object within the sample is classified based on the categories specified by the two variables. The frequency count or number of observations in the sample that belong to a category is recorded in a **contingency table**. The r number of categories of one variable is listed as r rows in the table, while the c number of categories of another variable is listed as c columns. The construction of a contingency table is illustrated below using an example.

Public health officials are interested in determining if the smoking status of a person is linked to alcohol consumption. If the sample size is large enough, the χ^2 test

Table 4.10. *A 2× 2 contingency table for smoking and drinking status*

	Current drinker	Non-drinker	Total
Current smoker	w	x	$w+x$
Non-smoker	y	z	$y+z$
Total	$w+y$	$x+z$	$w+x+y+z=n$

can be used to look for a dependence of smoking behavior on alcohol consumption. In this example, the first nominal variable "smoking status" contains two categories: "current smoker" and "non-smoker," and the second nominal variable "drinking status" has two categories: "current drinker" and "non-drinker." We can construct a contingency table (see Table 4.10) that has two rows based on the two categories for "smoking status" and two columns based on the categories for "drinking status." Note that "smoking status" can also be listed column-wise and "drinking status" row-wise, i.e. we can transpose the rows and columns of the table since there is no particular reason to place one variable on the rows and the other on the columns. Since this contingency table has two rows and two columns, it is called a **2 × 2 contingency table**. Upon combining the categories of both nominal variables we obtain four mutually exclusive and exhaustive categories:

(1) current smoker and drinker,
(2) current smoker but non-drinker,
(3) non-smoker but current drinker, and
(4) non-smoker and non-drinker.

Each of the four categories comprises one **cell** in the contingency table. The row totals and the column totals are called **marginal totals**. The total of all column totals is equal to the total of all row totals, and is called the **grand total**. The grand total is equal to the sample size n.

To test for an association between the two variables, we draw a random sample of size n from the population. The individuals within the sample are classified into any one of the four cells and the frequency counts are listed in Table 4.10. The frequencies observed for each cell are w, x, y, and z. Since the cells represent categories that are mutually exclusive and exhaustive, $w+x+y+z=n$.

The null hypothesis of the χ^2 statistical test of independence states that the two classification variables are independent of each other. Note that the null hypothesis does not contain any references to a population parameter, and thus the χ^2 statistical test of independence is a true non-parametric test. If there is no association between smoking tobacco and drinking alcoholic beverages then the smoking population will have the same distribution of drinkers vs. non-drinkers as the non-smoking population. The two variables of classification "smoking status" and "drinking status" are said to be *independent* when any change in the population distribution of one variable does not influence the population distribution of the other variable. For example, if 30% of the population consists of smokers, and this proportion drops to 20%, the proportion of drinkers in the population will not change if the two variables are independent. In the language of probability, we say that when two events are independent of each other, the probability of one event occurring is not influenced by the knowledge of whether the second event has occurred. If "smoking status" and "drinking status" are independent outcomes, and the following outcomes are defined:

S: person is a smoker,
NS: person is a non-smoker,
D: person is a drinker,
ND: person is a non-drinker,

then, if a person is randomly chosen from the population, the conditional probability that the person drinks if he or she is a smoker is given by

$$P(D|S) = P(D),$$

and the conditional probability that the person drinks if he or she is a non-smoker is given by

$$P(D|NS) = P(D).$$

Therefore, the joint probability of selecting a person that smokes and drinks is

$$P(S \cap D) = P(S)P(D|S) = P(S)P(D).$$

The joint probability that the person smokes but does not drink is

$$P(S \cap ND) = P(S)P(ND|S) = P(S)P(ND).$$

The theory of joint probability of independent events is used to calculate the expected frequencies under the null hypothesis for each cell of the contingency table. We use the frequency counts from the sample to estimate the joint probabilities. The probability that a person in the population is a current smoker is calculated as $(w + x)/n$. The probability that a person in the population is a current drinker is estimated as $(w + y)/n$. If the null hypothesis is true, the joint probability of encountering a person who smokes and drinks is simply the product of the individual probabilities, or

$$P(S \cap D) = P(S)P(D) = \left(\frac{w + x}{n}\right)\left(\frac{w + y}{n}\right).$$

To obtain the expected frequency of observing individuals who smoke and drink for the sample size n, we need to multiply the probability $P(S \cap D)$ by n. The expected frequency for the event $S \cap D$ is calculated as

$$\frac{(w + x)(w + y)}{n}.$$

To obtain the expected frequency for the cell in the (1,1) position, we simply multiply the first column total with the first row total and divide the product by the grand total. Similarly, the expected frequency for the cell in the (2,1) position is estimated as

$$\frac{(y + z)(w + y)}{n},$$

which is the product of the second row total and the first column total divided by the grand total. Once the expected frequencies are calculated for all $r \times c$ cells, the χ^2 test statistic is calculated using the formula given by Equation (4.33), where k, the number of categories, is equal to $r \times c$ and the degrees of freedom associated with the test statistic is $(r-1) \times (c-1)$. Remember that the observed or expected frequency in the last (rth or cth) category is dependent upon the distribution of frequencies in the other ($r-1$ or $c-1$, respectively) categories. The last category supplies redundant information and is therefore excluded as a degree of freedom. Let's illustrate the χ^2 test of independence with an example.

Example 4.7

In the IVF outcomes study described in Box 4.7, one large sample of 6164 patients was drawn from the population defined as female residents of the state of Massachusetts who sought IVF treatment. The subjects of the sample were classified on the basis of two nominal variables: maternal age group and outcome of IVF treatment at each cycle. The sample is believed to represent adequately the age distribution of women who seek IVF treatment and the likelihood of success of IVF treatment within the population. The authors of the study wished to find out if the probability of success, i.e. live-birth outcome, following IVF treatment is dependent on or associated with maternal age. In this example, we will be concerned with outcomes after the first IVF cycle, but prior to subsequent IVF cycles. Subjects were classified into any one of four categories based on maternal age, which is the first variable. The other classification variable is the outcome of one cycle of IVF treatment, i.e. whether a live-birth event occurred. The second variable is a dichotomous variable that can have only two values, "yes" or "no." At the end of the first IVF cycle, two outcomes are possible for each patient age group. We want to find out if the probability of a live-birth event is independent of maternal age. We construct a 2×4 contingency table as shown in Table 4.11. The age groups are listed as columns and the treatment outcomes are listed as rows. Alternatively, we could create a 4×2 contingency table with the age groups listed as rows and the IVF live-birth outcomes listed column-wise.

The complementary statistical hypotheses for this problem are as follows:

H_0: "Live birth outcome as a result of IVF is independent of maternal age."
H_A: "Live birth outcome as a result of IVF is dependent on maternal age."

The expected frequencies for each cell of the 2×4 contingency table are calculated and listed in Table 4.12.

The test statistic is given by

$$\chi^2 = \sum_{i=1}^{2\times4} \frac{(o_i - e_i)^2}{e_i} = \frac{(866 - 656.5)^2}{656.5} + \frac{(370 - 333.4)^2}{333.4} + \cdots + \frac{(1183 - 973.8)^2}{973.8} = 286.0.$$

The associated degrees of freedom is given by $f = (4-1) \times (2-1) = 3$. The p value is calculated using the `chi2cdf` function of MATLAB and is found to be $<10^{-16}$. If we choose $\alpha = 0.05$, $p \ll \alpha$ and the

Table 4.11. *A 2× 4 contingency table of live births resulting from IVF treatment in four maternal age groups*

	Maternal age (years)				
	<35	35 to <38	38 to <40	≥40	Total
Live birth	866	370	168	107	1511
No live birth	1812	990	668	1183	4653
Total	2678	1360	836	1290	6164

Table 4.12. *Expected frequencies of live births resulting from IVF treatment in four maternal age groups*

	Maternal age (years)				
	<35	35 to <38	38 to <40	≥40	Total
Live birth	656.5	333.4	204.9	316.2	1511
No live birth	2021.5	1026.6	631.1	973.8	4653
Total	2678	1360	836	1290	6164

test result is highly significant. We reject the null hypothesis and conclude that the response to IVF treatment is linked with maternal age. However, we do not know if our conclusion applies to all maternal age groups considered in the study, or to only some of them. When the number of categories *k* within either of the two nominal variables exceeds 2, rejection of the null hypothesis simply tells us that the variables are associated. We cannot conclude whether the alternate hypothesis applies to all categories of the variable (s) or only to some of them. This limitation is similar to that encountered with the one-factor ANOVA test.

Since the χ^2 test assumes that the sample proportion within each category has a normal sampling distribution, the sample size must be large enough to justify use of this test. For contingency tables with one degree of freedom, all cell counts should be greater than 5 in order to use the χ^2 test of independence. If there is more than one degree of freedom, the minimum cell count allowed may decrease to as low as 1, as long as the frequency count of at least 80% of the cells is greater than or equal to 5.

4.9.3 Test of homogeneity

The χ^2 test of homogeneity is used when independent random samples are drawn from two or more different populations, and we want to know if the distribution of the categories of a nominal variable is similar in those populations. In other words, we want to know if the populations are *homogeneous* with respect to the nominal variable of interest. The χ^2 test of homogeneity tests the null hypothesis that all populations are similar or homogenous with respect to the distribution of a nominal variable. The χ^2 test of homogeneity is performed in a very similar fashion to the χ^2 test of independence. The homogeneity test also requires the construction of a **contingency table**, which is a tabular description of the frequency of observations within each category for each sample. The names of the different samples can be listed either row-wise or column-wise in the table. In this book, each sample is listed as a column in the table. If there are k samples, the contingency table has k columns. The different categories of the nominal variable are listed as rows. If there are r categories and k independent samples drawn from k populations, then we have an $\boldsymbol{r \times k}$ **contingency table**.

We reconsider the illustration provided in Section 4.9.2 of a study that seeks a relationship between smoking status and alcohol status. We modify the test conditions so that we begin with two populations: (1) individuals that consume alcoholic beverages regularly, and (2) individuals that do not consume alcoholic beverages. A random sample of size n_i is drawn from each population, where $i = 1$ or 2. We want to determine whether smoking is more prevalent in the drinking population. Thus, we wish to compare the frequency distribution of smoking status within both

Table 4.13. *A 2 × 2 contingency table for smoking distribution within the drinking and non-drinking population*

	Drinking population	Non-drinking population	Total
Current smoker	w	x	$w+x$
Non-smoker	n_1-w	n_2-x	n_1+n_2-w-x
Total	n_1	n_2	$n_1+n_2=n$

samples to determine if the two populations are homogeneous with respect to the variable "smoking status." The null hypothesis states that

H_0: The proportion of smokers is the same in both the drinking and non-drinking population.

The alternate hypothesis is

H_A: The proportion of smokers in the drinking population is different from that in the non-drinking population.

The 2 × 2 contingency table is shown in Table 4.13. Note that the column totals are fixed by the experimental investigator and do not indicate the proportion of drinkers in the population. The grand total is equal to the sum of the individual sample sizes.

Under the null hypothesis, the samples are derived from populations that are equivalent with respect to distribution of smoking status. The best estimate of the true proportion of smokers in either population can be obtained by pooling the sample data. If H_0 is true, then an improved estimate of the proportion of smokers within the drinking and non-drinking population is $(w+x)/n$. Then, the expected frequency of smokers for the sample derived from the drinking population is equal to the proportion of smokers within the population multiplied by the sample size, or

$$n_1 \cdot \frac{w+x}{n}.$$

Thus, the expected frequency for cell (1, 1) is equal to the product of the first row total and the first column total divided by the grand total n. Similarly, the expected frequency for cell (1, 2) is equal to the product of the first row total and the second column total divided by the grand total n. Note that the formula for calculating the expected frequencies is the same for both χ^2 tests of independence and homogeneity, even though the nature of the null hypothesis and the underlying theory from which the formula is derived are different. Limitations of the use of the χ^2 test of homogeneity when small frequencies are encountered are the same as that discussed for the χ^2 test of independence.

For a χ^2 test of homogeneity involving a 2 × 2 contingency table, the χ^2 method is essentially equivalent to the z test for equality of two population proportions. The p value that is obtained by either method should be exactly the same for the same set of data, and you can use either statistical method for analyzing sample data. Both test methods assume a normal sampling distribution for the sample proportion, and therefore the same conditions must be met for their use.

Example 4.8

Employees who work night-shifts or varying shifts may suffer from sleepiness at night on the job because the body's circadian rhythm routinely falls out of sync with the changing daily sleeping and waking patterns. For some people, chronic misalignment of the circadian cycle with the sleep cycle can adversely affect productivity (efficiency) at work and personal health. A fraction of shift workers are unable to cope well with disrupted sleep cycles and are diagnosed with chronically excessive sleepiness. Patients with shift-work sleep disorders are more prone to accidents, depression, and ulcers. A clinical study (Czeisler *et al.*, 2005) was performed to examine if administering the drug modafinil benefited patients with this disorder. Modafinil is a stimulant drug used to treat daytime sleepiness, a key symptom of narcolepsy and obstructive sleep apnea. The researchers wished to evaluate the efficacy of the drug for treating night-time sleepiness suffered by shift-work sleep disorder patients.

The experimental study consisted of a double-blinded trial conducted over a 3 month period. The treatment group consisted of 96 patients, who received 200 mg of modafinil tablets to be ingested before commencing night-shift work. The placebo group had 108 patients who received identical-looking tablets. The severity of sleepiness at baseline (prior to administering drug or placebo) in individuals in both groups was determined by measuring sleepiness level on the Karolinska Sleepiness scale. Table 4.14 presents sleepiness severity data for both groups.

A χ^2 test is used to ensure that the null hypothesis is true at baseline, i.e. that the distribution of severity of sleepiness is the same in both treatment and control groups. The expected frequencies under the null hypothesis are tabulated in Table 4.15.

The test statistic is calculated as

$$\chi^2 = \sum_{i=1}^{4\times 2} \frac{(o_i - e_i)^2}{e_i} = \frac{(49-48)^2}{48} + \frac{(53-54)^2}{54} + \cdots + \frac{(4-3.2)^2}{3.2} = 0.529.$$

Table 4.14. *A 4 × 2 contingency table for sleepiness severity at baseline for modafinil and placebo groups*

	Modafinil	Placebo	Total
Moderately sleepy	49	53	102
Markedly sleepy	29	34	63
Severely sleepy	16	17	33
Among the most extremely sleepy	2	4	6
Total	96	108	204

Table 4.15 *Expected frequencies for sleepiness severity at baseline for modafinil and placebo groups*

	Modafinil	Placebo	Total
Moderately sleepy	48.0	54.0	102
Markedly sleepy	29.7	33.3	63
Severely sleepy	15.5	17.5	33
Among the most extremely sleepy	2.8	3.2	6
Total	96	108	204

Table 4.16 *Distinct characteristics of the χ^2 test of independence and the χ^2 test of homogeneity*

Property	χ^2 test of independence	χ^2 test of homogeneity
Sample data	single random sample from a population	k independent random samples from k populations
Classification	the sample is categorized on the basis of two nominal variables	the k samples are categorized on the basis of one nominal variable
Null hypothesis	the two nominal variables are independent with respect to each other	the k populations are homogeneous with respect to a nominal variable
Expected frequencies	calculation is based on probability of independent events	calculation is based on combining sample data to obtain a global estimate of population proportions of the different categories of the nominal variable

The associated degrees of freedom is given by $f = (4 - 1) \times (2 - 1) = 3$. The p value is calculated using the `chi2cdf` function and is found to be 0.91. The test result is insignificant and close to unity. This tells us that if H_0 is true, the chances of observing differences of this nature randomly within the two samples are very high. We conclude that at baseline the two groups have an equal distribution of the nominal variable "severity of sleepiness."

The conditions for use of the χ^2 test of homogeneity can be contrasted with that for the χ^2 test of independence. For the test of independence, we require only one random sample from a population that is subjected to classification based on two criteria. The only known or fixed quantity within the contingency table is the grand total. The column and row totals are dependent on the distribution of the two variables within the population and are thus a property of the system being investigated. For the test of homogeneity, k independent random samples are chosen from k populations that differ from each other in at least one aspect. Thus, the k column totals will be equal to the k sample sizes, and are fixed by the experimenter. Table 4.16 contrasts the two χ^2 tests.

When the sample size is too small to meet the validity criteria of the χ^2 tests, you can still calculate the probabilities corresponding to the test statistic using the theory of combinatorics. Fisher's Exact test rigorously calculates all possible outcomes of the samples and the total number of ways in which one obtains outcomes as extreme as or more extreme than the observed outcome, when H_0 is true. The exact p value can be calculated using this method. The method can be found in most statistics textbooks (see, for instance, *Statistics for the Life Sciences* by Samuels and Witmer (2003)).

4.10 More on non-parametric (distribution-free) tests

Non-parametric (distribution-free) tests are routinely used for drawing statistical inferences when the population distribution of one or more random variables is either unknown or deviates greatly from normality. Also, non-parametric procedures must be used to test hypotheses that are based on data of ordinal nature, i.e. the variable of interest is measured in terms of rankings. Use of non-parametric tests is especially common when the sample size is small and the central-limit theorem does not apply.

Some non-parametric tests make no assumption of the form or nature of distribution of the random variable within the population and are called **distribution-free tests**. In these tests, the observations within one or more samples are ranked, and a test statistic is calculated from the sum total of all ranks allotted to each sample. In this way, cardinal data are converted to ordinal type resulting in loss of information. Since distribution-free tests do not use all the information available in the data (e.g. shape of population distribution of the random variable), they are less powerful than the equivalent parametric tests. When data are suitable for analysis using a parametric procedure, those methods are thus preferred over a non-parametric analysis.

Non-parametric tests such as the χ^2 goodness-of-fit test and the χ^2 test of independence do not test hypotheses concerning population parameters. Such tests use inferential statistics to address questions of a different nature such as "Do two nominal variables demonstrate an association within a population?". Examples of such tests were discussed in the previous section. In this section, we introduce three distribution-free statistical tests: the sign test and the Wilcoxon signed-rank test for analyzing paired data, and the Wilcoxon rank-sum test for determining if two independent random samples belong to the same population.

4.10.1 Sign test

The sign test is a non-parametric procedure used to analyze either a single sample or paired data. The appeal of the **sign test** lies in its simplicity and ease of use. A hypothesis tested by a sign test concerns either (1) the *median* of a single population, or (2) the *median* of the differences between paired observations taken from two different populations. No assumptions are made regarding the shape of the population distribution. For this reason we cannot draw inferences regarding the population mean, unless the distribution is symmetric about the median, in which case the median and mean will be equal regardless of the shape of the distribution. A population median is defined as that value that is numerically greater than exactly 50% of all observed values in the population. If we were to choose only one observation from the population, the probability that its value is less than the population median is 0.5. Equivalently, the probability that the observed value is greater than the population median is 0.5.

The null hypothesis of a sign test asserts that the median of a population m is equal to some value, i.e.

$$H_0: \quad m = m_0.$$

If the population consists of differences $d_i = x_i - y_i$ calculated from paired data, then under H_0 the hypothesized median of the differences m_d is zero, i.e.

$$H_0: \quad m_d = 0.$$

The sign test does not use the magnitude of the differences to generate the test statistic. Therefore, the sign test is appropriate for use in studies where the difference between two observations is not accurately known, the data points are ordinal, or the experiment/data collection is incomplete. The only information needed is the direction of the differences, which is recorded either as a "+" sign or a "–" sign. If the null hypothesis is true, we would expect 50% of the differences to be positive and 50% of them to be negative.

When performing the sign test, the deviation d of each observation in the sample from the hypothesized median is tabulated as either (+) or (–). When paired data are being analyzed, the difference d between paired values is either greater than zero or less than zero and is marked by a (+) or (–) sign accordingly. If a data pair yields a difference of exactly zero, or if the deviation of an observation from the hypothesized median is equal to zero, the data point is ignored in the analysis and n is reduced by 1. The number of (+) signs and (–) signs are calculated. The test statistic for the sign test is the number of (+) signs, S_+, or the number of (–) signs, S_-. The sidedness of the hypothesis dictates whether S_+ or S_- is used as the test statistic.

A non-directional null hypothesis for the sign test is expressed in mathematical terms as

$$H_0:\quad P(d > 0) = P(d < 0) = 0.5.$$

The complementary alternate hypothesis is

$$H_A:\quad P(d > 0) \neq P(d < 0).$$

When the nature of the hypothesis is non-directional, the test statistic is chosen as either S_+ or S_-, whichever is least. If there are n non-zero d's, and if H_0 is true, the expected value for both S_+ and S_- is $n/2$. If H_0 is true, the probability that S_+ and S_- will be close in value is high. On the other hand, if a majority of the observations are greater (or less) than the hypothesized median, we are less inclined to believe the null hypothesis since the outcome is less easily explained by sampling variation than by the alternate hypothesis. The objective criterion for rejection of H_0 is the same as that for all other statistical tests: the p value of the test statistic should be less than the chosen significance level α.

The p value for the test is calculated using the binomial theorem. The null hypothesis states that an observed difference d has an equal probability of being positive or negative. For a sample size n, we have n independent outcomes, each of which can assume two values, either (+) or (–) with a probability of 0.5, when H_0 is true. Thus, we have a binomial process with parameters n and p, where $p = 0.5$. Let S be either S_+ or S_-, whichever is less. Therefore, S is a binomial random variable and is either the number of (+) signs or the number of (–) signs. The probability that one will observe S successes (or failures) out of n trials is given by

$$C^n_S 0.5^S 0.5^{n-S}.$$

The probability that one will observe S successes (or failures) *or fewer* out of n trials is given by

$$P = \sum_{x=0}^{S} C^n_x 0.5^x 0.5^{n-x}. \tag{4.34}$$

Equation (4.34) is the probability of observing S or fewer plus or minus signs, whichever sign is in the minority. The p value for a non-directional hypothesis is twice the probability calculated using Equation (4.34).

For a directional alternate hypothesis such as

H_A: $P(d > 0) > 0.5$,

the corresponding null hypothesis is

H_0: $P(d > 0) \leq P(d < 0)$ or $P(d > 0) \leq 0.5$.

If H_A is true, the observed differences should contain many more positive differences than negative differences. If H_0 is true, then we expect the number of (−) signs to be either roughly equal to or larger than the number of (+) signs. The appropriate test statistic S is equal to the number of (−) signs, S_-. The null distribution of S_- is a binomial distribution with $\mu = n/2$. If $S_- > n/2$, the null hypothesis is automatically retained since the corresponding p value is > 0.5. If $S_- < n/2$, then the p value associated with S_- is calculated using Equation (4.34) and compared with the significance level α. The rejection region is located only in the left tail of the distribution.

Similarly, for the directional alternate hypothesis

H_A: $P(d < 0) > 0.5$,

the test statistic is S_+. The null distribution of S_+ is a binomial distribution with $\mu = n/2$. The p value associated with S_+ is calculated. The rejection region is located only in the left tail of the distribution. If p is less than the significance level α, the null hypothesis H_0: $P(d < 0) \leq 0.5$ is rejected in favor of H_A.

Example 4.9

Marfan's syndrome is an inherited autosomal dominant disease that often ends in premature death of the patient. A genetic defect in the formation of fibrillin-1 causes abnormalities in the structure of the extracellular matrix. As a result, these patients tend to have abnormally long limbs, eyesight problems, and numerous heart conditions. Fibrillin-1 is known to suppress transforming growth factor β (TGF-β) signaling activity by binding these molecules, thereby limiting their activity in the body. Defective fibrillin-1 is associated with enhanced signaling of TGF-β, and the latter is found to cause progressive dilation of the aortic root in Marfan's syndrome patients.

In Box 4.2A, a study was presented that tested angiotensin II receptor blockers (ARB) for controlling or preventing upregulation of TGF-β (Bedair *et al.*, 2008). A small clinical trial was performed in recent years (Brooke *et al.*, 2008) to evaluate the effect of ARB on aortic root enlargement.

In a fictitious study, the annual rate of aortic root enlargement is measured in 12 Marfan syndrome patients prior to administration of losartan, an ARB drug. At one year following commencement of ARB therapy, the annual rate of aortic root enlargement is measured. It is found that the annual rate of aortic root enlargement decreased for 11 patients and increased for only one patient. Does losartan have a beneficial effect on aortic root enlargement?

The alternate hypothesis is

H_A: "Treatment with ARB is effective, i.e. ARB acts to reduce the rate of aortic root expansion." Mathematically, this is written as $P(d < 0) > 0.5$.

The null hypothesis is

H_0: "Treatment with ARB is not effective," i.e. $P(d < 0) \leq 0.5$.

Note that we have posed a directional hypothesis in this case. The experiment yields 12 data pairs. Each pair contains two values of the annual rate of aortic root enlargement, one measured at baseline and one at follow-up. The difference d for each pair is already calculated and the direction of the difference is supplied in the problem statement. We have $S_- = 11$ and $S_+ = 1$. Since the direction of the alternate hypothesis is negative, the test statistic is S_+. A small number of (+) signs will lead to rejection of the null hypothesis. The data appear to support the alternate hypothesis, but we can confirm this only after we calculate the p value.

According to Equation (4.34),

$$P = C_0^n 0.5^0 0.5^{n-0} + C_1^n 0.5^1 0.5^{n-1} = 0.0032.$$

If the null hypothesis were non-directional, an equally likely extreme outcome under H_0 would be $S_- = 1$ and $S_+ = 11$, and the probability calculated from Equation (4.34) must be doubled to obtain the p value. Here, because the hypothesis is one-sided, the probability calculated above is the p value itself and no doubling is required. Note that for the directional hypothesis considered here, the reverse outcome of $S_+ = 11$ and $S_- = 1$ is not an extreme outcome when H_0 is true, but is, in fact, a highly likely observation. If $\alpha = 0.05$, then $p = 0.0032 < \alpha$, and we conclude that the result supports the alternate hypothesis, i.e. ARB produces the desired effect.

If n is large, calculating the p value using Equation (4.34) can become cumbersome. Instead, the normal approximation of the binomial distribution of S for large n (see Section 4.7) can be used to calculate the p value. The test statistic S is normalized to obtain the z statistic and is calculated as follows:

$$z = \frac{S - n/2}{\sqrt{n \times 0.5 \times 0.5}}, \tag{4.35}$$

where z is the number of standard deviations of the binomial variable S from its mean, when H_0 is true and n is large. The p value corresponding to z is obtained using the cumulative distribution function for the normal probability curve.

Using MATLAB

You can perform a sign test in MATLAB using Statistics Toolbox function `signtest`. There are several variations in the syntax as shown below.

The following statement carries out a sign test for the non-directional null hypothesis, H_0: $m = m_0$:

```
p = signtest(x, m0)
```

where $\boldsymbol{x}$ is the vector of values and m_0 is the hypothesized median.

The following statement performs a sign test for paired data to test the null hypothesis, H_0: $m_d = 0$:

```
p = signtest(x, y)
```

where $\boldsymbol{x}$ and $\boldsymbol{y}$ are vectors that represent two samples with matched data pairs, and whose sizes must be equal. You can also specify the method to perform the sign test, i.e. the exact method using combinatorics or the approximate method using the normal approximation. See MATLAB help for more details.

4.10.2 Wilcoxon signed-rank test

The **Wilcoxon signed-rank test** is a non-parametric test also used for analyzing paired data, and is essentially an extension of the sign test. This test procedure is the method

Box 4.11 Neurostimulation as a treatment option for Parkinson's disease

Parkinson's disease is a chronic and progressive neurodegenerative disorder of impaired motor functioning. This disease is characterized by involuntary tremors, muscle rigidity or stiffness, slowness of voluntary movements, and poor posture and balance. Currently, there is no cure for this disease; however, treatments to alleviate symptoms are available. Prescription medications such as levodopa are accompanied by side-effects such as dyskinesia (or involuntary movements) and hallucinations. Recently, deep-brain stimulation was proposed as an alternate treatment method for patients suffering from drug-related complications. This is a surgical procedure in which a pacemaker-type device is implanted within the body to stimulate a particular region of the brain continuously with electricity.

An unblinded clinical study (Deuschl *et al.*, 2006) was conducted in Germany and Austria to determine if deep-brain stimulation was more effective in attenuating the motor problems of Parkinson's disease patients compared to medication alone. Eligible patients for the study were paired up, and one member of a pair was randomly assigned to electrostimulation treatment and the other was provided with conventional medical treatment. The efficacy of the treatment was measured in terms of the change in quality of life over 6 months from baseline to follow-up as assessed using the Parkinson's Disease Questionnaire summary index. Where follow-up values were missing, the worst possible outcome score was used for the patients allotted to neurostimulation and the best possible outcome score was used for patients allotted to medical treatment to yield a more conservative test.

For 78 pairs of patients, the index favored neurostimulation in 50 pairs and medical treatment in 28 pairs. There were no ties in the scores within any pair. Based on the study data, we wish to determine whether neurostimulation is effective as a treatment method for Parkinson's disease.

Since $n = 78$ is large, we use the normal approximation given by Equation (4.35) to calculate the test statistic. The test is two-sided, so S is chosen to be the smaller of either of the two totals ($S_+ = 50$, $S_- = 28$). We obtain

$$z = \frac{28 - 78/2}{\sqrt{78 \times 0.5 \times 0.5}} = -2.49.$$

The p value associated with the test statistic value is 0.0128 (verify this yourself). We conclude that deep-brain stimulation is more effective than standard medical treatment in treating the symptoms of Parkinson's disease.

of choice when experimental data are on an ordinal scale, or when the data values are cardinal but the population distribution is considerably non-normal. The Wilcoxon signed-rank test is a slightly more involved procedure than the sign test, and it is more powerful because it not only uses information about the sign of the difference between the paired data values, but also considers the magnitude of the absolute difference. The sign test is preferred over the Wilcoxon signed-rank test only when a study produces incomplete data and ranking of observations is not possible, or when requirements of the signed-rank test such as symmetric distribution of the random variable about the median are not met.

In this method, the difference between each of the paired values is calculated. Observations that yield differences exactly equal to zero are removed from the analysis, and the number of pairs n is reduced accordingly. The non-zero differences are *ranked* in ascending order of their absolute magnitude. The ranks are prefixed with the positive or negative signs of the differences to which the ranks correspond. All positive ranks are summed to obtain a positive rank sum R_+. Similarly, the negative ranks are summed to obtain a negative rank sum R_-. The appropriate test statistic R_+ or R_- is chosen based on the directionality of the hypothesis.

Table 4.17. *Calculation of the test statistic for the Wilcoxon signed-rank method*

Patient	Statin A (mg/dl)	Statin B (mg/dl)	Difference (mg/dl)	Rank	Signed rank
1	35	23	12	8	+ 8
2	41	35	6	5.5	+ 5.5
3	29	22	7	7	+ 7
4	38	41	−3	2	− 2
5	32	36	−4	3.5	− 3.5
6	40	34	6	5.5	+ 5.5
7	37	33	4	3.5	+ 3.5
8	33	31	2	1	+ 1

$R_+ = 30.5$, $R- = 5.5$.

Example 4.10

A crossover trial is performed to determine the effectiveness of two statin drugs in lowering LDL cholesterol levels in patients with high cardiovascular risk. Eight patients receive both drugs one at a time in randomized order with a wash-out period between the two treatments. The reduction in LDL cholesterol level from baseline is measured in each patient following treatment with each drug. The results are presented in Table 4.17. We wish to determine if one drug is more effective than the other.

The reduction of LDL cholesterol levels quantifies the effect of each drug on the body. The difference between the effects of Statin A and Statin B is calculated for each individual. The differences are ranked based on their absolute values. If some of the differences are tied, the rank assigned to each of the tied differences is equal to the average of the ranks that would have been assigned had the difference values been slightly unequal. For example, for patients 5 and 7, the absolute difference in reduced LDL cholesterol values is 4. Since the next two ranks that can be allotted are 3 and 4, the average rank is calculated as 3.5. The signs are added to their respective ranks, and the positive and negative rank totals, R_+ and R_-, respectively, are calculated. For this problem, the hypothesis is non-directional and we choose the lesser value of either R_+ or R_- as the test statistic.

The Wilcoxon signed-rank test requires that the following conditions are met.

(1) The distribution of d_i's about zero (or the median) is symmetrical. When this is true, the population mean difference and population median difference are one and the same. Note that we make no assumption about the shape of the distribution of the differences d_i. A symmetrical distribution about the median is needed to use the Wilcoxon signed-rank method to test a hypothesis concerning the population mean difference, μ_d. This method is distribution-free as we do not use the actual data values but convert the data to an ordinal scale before we calculate the magnitude of the test statistic.

(2) The data should be on a continuous scale. In other words, no ties should occur between the d_i's. If ties between difference values d_i occur, a correction term must be included in the formula; otherwise the calculated p value is approximate.

The null hypothesis tested by this non-parametric procedure is

H_0: $\mu_d = 0$.

When H_0 is true, the population mean difference μ_d and population median difference m_d is 0, and we expect the positive rank total R_+ to be close in value to the negative rank total R_-. The distribution of the positive rank total R_+ or negative rank total R_- under the null hypothesis is calculated using combinatorics. If $R_+ \sim R_-$, H_0 is retained. If either R_+ or R_- is very small, then it is unlikely that the mean (or median) of the differences is zero.

The non-directional alternate hypothesis is

$$H_A: \quad \mu_d \neq 0.$$

Here, we illustrate how the p value is calculated for this test, assuming that no ties among ranks occur. Suppose the number of pairs $n = 8$, then we have a set of eight ranks,

1 2 3 4 5 6 7 8.

If H_0 is true, each rank can be assigned a (+) sign or a (–) sign with 0.5 probability, since the population median difference is zero and the distribution of differences about the median is symmetrical. The probability of obtaining the combination

$$\begin{array}{cccccccc} + & + & - & - & + & + & + & + \\ 1 & 2 & 3 & 4 & 5 & 6 & 7 & 8 \end{array}$$

is 0.5^8. For this combination, $R_+ = 29$ and $R- = 7$. Other combinations that yield the same values for R_+ and R_- are

$$\begin{array}{cccccccc} + & + & + & + & + & + & - & + \\ 1 & 2 & 3 & 4 & 5 & 6 & 7 & 8 \end{array} \qquad \begin{array}{cccccccc} - & + & + & + & + & - & + & + \\ 1 & 2 & 3 & 4 & 5 & 6 & 7 & 8 \end{array}$$

$$\begin{array}{cccccccc} + & - & + & + & - & + & + & + \\ 1 & 2 & 3 & 4 & 5 & 6 & 7 & 8 \end{array} \qquad \begin{array}{cccccccc} - & - & + & - & + & + & + & + \\ 1 & 2 & 3 & 4 & 5 & 6 & 7 & 8 \end{array}$$

Thus, the probability of obtaining the result $R_+ = 29$ and $R_- = 7$ is 5×0.5^8.

The p value is equal to the probability of obtaining the result $R_+ = 29$ and $R_- = 7$, and even more extreme results. Try to determine the probability of obtaining the following outcomes:

$R_+ = 30$ and $R_- = 6$,
$R_+ = 31$ and $R_- = 5$,
$R_+ = 32$ and $R_- = 4$,
$R_+ = 33$ and $R_- = 3$,
$R_+ = 34$ and $R_- = 2$,
$R_+ = 35$ and $R_- = 1$,
$R_+ = 36$ and $R_- = 0$.

When you sum all the probabilities you get a one-sided p value of

$$(5 + 4 + 3 + 2 + 2 + 1 + 1 + 1) \times 0.5^8 = 0.0742.$$

The two-sided probability or two-sided p value is $0.0742 \times 2 = 0.1484$.

Since this process of calculating the p value is cumbersome and we may encounter ties in our difference values (such as in Example 4.10), it is best to use the computer to determine the p value.

Using MATLAB

A two-sided Wilcoxon signed-rank test can be performed using the `signrank` function of Statistics Toolbox. The syntax for using `signrank` to analyze paired data is

```
p = signrank(x,y)
```

where $\boldsymbol{x}$ and $\boldsymbol{y}$ are two vectors of equal length. It is assumed that the differences $\boldsymbol{x}-\boldsymbol{y}$ represent a random sample derived from a continuous distribution symmetric about the median. This function corrects for tied ranks.

Example 4.10 (continued)

Let's now calculate the p value associated with the calculated test statistic. We can do this in one of two ways. We first round the test statistic values to $R_+ = 30$ and $R_- = 6$. Using the method illustrated above, the two-sided p value associated with this outcome is
$2 \times (4+3+2+2+1+1+1) \times 0.5^8 = 0.1094$.

Now we round the test statistic values to $R_+ = 31$ and $R- = 5$. The p value associated with this outcome is $2 \times (3+2+2+1+1+1) \times 0.5^8 = 0.0781$.

Using the MATLAB function `signrank` to calculate the p value, we obtain $p = 0.0938$, a value intermediate to the two values calculated above.

When n is large, the sampling distribution of the test statistic R_+ or R_- assumes a normal distribution and the z statistic can be calculated for this test. The Wilcoxon signed-rank test can also be used to test a hypothesis regarding the mean (or median) of a single sample. When applying the signed-rank method to test the null hypothesis H_0: $\mu = \mu_0$, μ_0 is subtracted from each data value and ranks are allotted to the differences. The test then proceeds as usual.

The parametric equivalent of the Wilcoxon signed-rank test is the paired t test. When data are normally distributed, or when the central-limit theorem applies, the paired t test should be preferred over the signed-rank test.

4.10.3 Wilcoxon rank-sum test

The **Wilcoxon rank-sum test** is a non-parametric procedure used to test the hypothesis of the equality of two population medians. Its parametric equivalent is the independent two sample t test. This procedure does not require that the variable of interest have any particular distribution within the two populations. The conditions that must be met in order to use this test procedure are

(1) the random variable is continuous, and
(2) the shape of the variable distribution is the same in both populations.

To perform the rank-sum test we start with two independent representative samples drawn from two populations. If the median of a population is represented by m, the (non-directional) null hypothesis is

$$H_0: \quad m_1 = m_2,$$

and the corresponding alternate hypothesis is

$$H_A: \quad m_1 \neq m_2.$$

The alternate hypothesis can also be directional such as H_A: $m_1 > m_2$ or H_A: $m_1 < m_2$, in which case the respective null hypotheses are H_0: $m_1 \leq m_2$ and H_0: $m_1 \geq m_2$. The test procedure is illustrated in Box 4.1B.

How is the p value calculated for the Wilcoxon rank-sum test? Again the method of calculation is based on the theory of combinatorics. If there are no ties in the ranks, i.e. the distribution of the variable is continuous, then the calculated p value is exact. If some of the ranks are tied, the calculated p value is only approximate. If

Box 4.1B Use of cyclosporin to reduce reperfusion injury during acute myocardial infarction

The size of the infarct was estimated using MRI images from 11 patients in the control group and in 16 patients in the treated group 5 days after performing percutaneous coronary intervention (PCI). The baseline characteristics (prior to PCI) of both treated and control groups with respect to the size of the area at risk were similar. The MRI estimates of the infarct sizes are presented below (Piot *et al.*, 2008):

Control Group (g) $n_1 = 11$:
20, 30, 44, 45, 46, 56, 57, 60, 65, 68, 115
Treated Group (g) $n_2 = 16$:
14, 16, 19.5, 20, 22, 23, 24, 35, 38, 38, 40, 48, 53, 64, 64, 72

It is assumed that the requirements for the Wilcoxon rank-sum test are met. The total number of observations is $n = n_1 + n_2 = 27$.

All 27 observations from the two samples are combined and ranked in ascending order (see Table 4.18).

Table 4.18. *Calculation of the Wilcoxon rank sum test statistic*

Control Group (1)	Rank	Treated Group (2)	Rank
20	4.5	14	1
30	9	16	2
44	14	19.5	3
45	15	20	4.5
46	16	22	6
56	19	23	7
57	20	24	8
60	21	35	10
65	24	38	11.5
68	25	38	11.5
115	27	40	13
		48	17
		53	18
		64	22.5
		64	22.5
		72	26
Rank Sums:	$R_1 = 194.5$		$R_2 = 183.5$

According to the non-directional null hypothesis, the median infarct size of the two groups is the same. If H_0 is true, the numbers of low and high ranks should be distributed roughly equally between both groups. If H_0 is not true, then the medians of the two groups are not equal. If the median of group 1 is larger than that of group 2, we expect more values in group 1 to be greater than the observations in group 2. The rank sum for group 1 will be significantly larger than the rank sum for group 2.

Since the hypothesis is two-sided, either R_1 or R_2 can function as the test statistic.

Table 4.19. *Wilcoxon rank sum test*

Ranks	Rank sum
Group 1: 1, 3, 4	$R_1 = 8$
Group 2: 2, 5, 6, 7	$R_2 = 20$

many ties are present, a correction term should be included in the formula used to calculate the p value. Let's consider a simple problem in which there are two groups of sizes $n_1 = 3$ and $n_2 = 4$. We allot seven ranks in total, one to each of the seven observations in both groups. If the null hypothesis is true, i.e. the population distribution is the same for both groups, each rank number has equal probability of being assigned to either of the two groups. The allotment of three rank numbers to groups 1 and 4 rank numbers to group 2 can be viewed as a selection process in which any three ranks are chosen without replacement from a set of seven ranks. The total number of distinct combinations of two sets of ranks is $C_3^7 = 35$.

Suppose we have the outcome shown in Table 4.19 from an experiment. There are three other possible rank combinations that produce a result atleast as extreme as this, i.e. three sets of ranks (1, 2, 3), (1, 2, 4), and (1, 2, 5) produce a value of $R_1 \leq 8$.

The probability of obtaining any of these four rank combinations is $4/35 = 0.1142$. A two-sided hypothesis requires that this probability be doubled since extreme observations observed in the reverse direction also lead to rejection of the null hypothesis. For this example, the four extreme observations for group 1 in the reverse direction are (5, 6, 7), (4, 6, 7), (3, 6, 7), and (4, 5, 7), which produce $R_1 \geq 16$. The p value corresponding to the test statistic $R_1 = 8$ is $8/35 = 0.2286$.

For the problem illustrated in Box 4.1, the total number of rank combinations possible in both groups is C_{11}^{27}. Calculation of all possible rank combinations that will produce an R_1 test statistic value ≥ 194.5 is too tedious to perform by hand. Instead we make use of statistical software to calculate the p value associated with the calculated rank sums.

Using MATLAB

The `ranksum` function available in Statistics Toolbox performs the Wilcoxon rank-sum test. The syntax is

```
p = ranksum(x, y)
```

where $\boldsymbol{x}$ and $\boldsymbol{y}$ are two vectors, not necessarily of equal length, that store the observations of two independent samples. The p value is generated for a two-sided hypothesis of equal population medians.

The Wilcoxon rank-sum test is mathematically equivalent to another non-parametric statistical procedure called the **Mann Whitney U test**. The test procedures for either test are slightly different but the underlying probability calculations are equivalent and both tests yield the same p value for a given data set. The rank-sum test does not use all the information contained in the data, i.e. the absolute magnitude of the observations are not factored into the calculation of the test statistic. Thus, this method is not

Box 4.1C Use of cyclosporin to reduce reperfusion injury during acute myocardial infarction

The p value is calculated using the MATLAB `ranksum` function:

```
>> x = [4.5 9 14 15 16 19 20 21 24 25 27];
>> y = [1 2 3 4.5 6 7 8 10 11.5 11.5 13 17 18 22.5 22.5 26];
p = ranksum(x, y)
p =
   0.0483
```

The test statistic reaches significance for a significance level set at $\alpha = 0.05$. We conclude that the median infarct sizes for the treated and control groups are different.

as powerful as the t test, yet it is very useful when the data distribution is far from normal or when the data points are few in number.

Non-parametric procedures have been developed for comparing three or more independent samples. The **Kruskal–Wallis test** is an extension of the Wilcoxon rank-sum test. The parametric equivalent of the Kruskal–Wallis test is the one-factor ANOVA. Discussion of non-parametric tests for more than two samples is beyond the scope of this book.

4.11 End of Chapter 4: key points to consider

A summary of the hypothesis tests discussed in this chapter is provided as two handy flowcharts (Figures 4.14 and 4.15) outlining the criteria used to select a hypothesis test for (1) cardinal data and (2) nominal data.

4.12 Problems

4.1. **Statistical tests.** Match up the following studies with the most appropriate statistical test.
- Are three sample means different from one another? H_0: $\mu_1 = \mu_2 = \mu_3$.
- Compare the mean rolling velocity of 2500 leukocytes at a low fluid shear rate, with the mean rolling velocity of 3000 leukocytes at a high fluid shear rate.
- Compare the red blood cell counts of 20 patients before and after treatment with erythropoietin.
- When do fibroblasts stop spreading on fibronectin? Compare the average area of 12 cells at $t = 30$ min with the average area of 10 cells at $t = 120$ min.
- Mortality rates of different groups of leukemic mice are presented in a 4×2 contingency table. Test for differences among the populations.

(A) Two-sample z test.
(B) Paired t test.
(C) Two-sample t test.
(D) χ^2 test.
(E) Analysis of variance.

4.2. **Treatment of Kawasaki disease using pulsed corticosteroid therapy** Kawasaki disease is most common in young children and is characterized by inflammation of the arteries, skin, and mucous membranes. This disease is responsible for causing

Figure 4.14

Hypothesis test selection criteria for cardinal data.

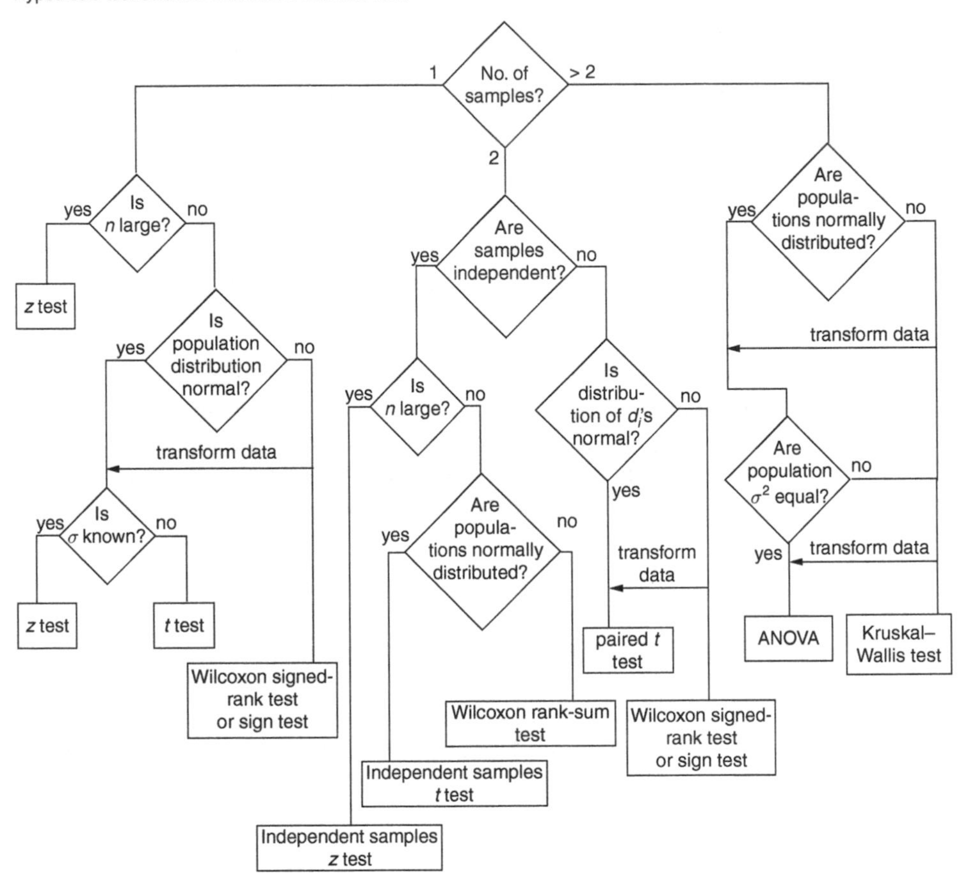

coronary artery aneurysms. Standard treatment of Kawasaki disease consists of administering intravenous immunoglobulin and aspirin. Despite treatment, a small percentage of the children still develop aneurysms. A double-blinded, placebo-controlled clinical trial was conducted in eight centers in North America to study the benefits of including corticosteroid therapy (intravenous methylprednisolone) with conventional treatment for preventing the development of aneurysms (Newburger *et al.*, 2007). Two-dimensional echocardiography was used to measure the internal diameters of several coronary arteries. The lumen diameters were standardized using body surface area and converted to z scores. The mean maximum z score after 5 weeks of administering methylprednisolone was found to be 1.31 ± 1.55 for 95 patients in the treatment group and 1.39 ± 2.03 for 95 patients in the placebo group. Use the two-sample t test to determine if methylprednisolone is effective in reducing the arterial diameter. First, construct the null hypothesis. Next, calculate the standard error of the null distribution of the test statistic. Calculate the test statistic and draw your conclusion based on the corresponding p value.

4.3. **Treatment of skeletal muscle injury using angiotensin II receptor blocker** In the study of the effects of angiotensin receptor blocker (ARB) on the healing of skeletal muscle (Box 4.2), the extent of muscle fibrosis was measured 5 weeks after injury.

Figure 4.15

Hypothesis test selection criteria for nominal data.

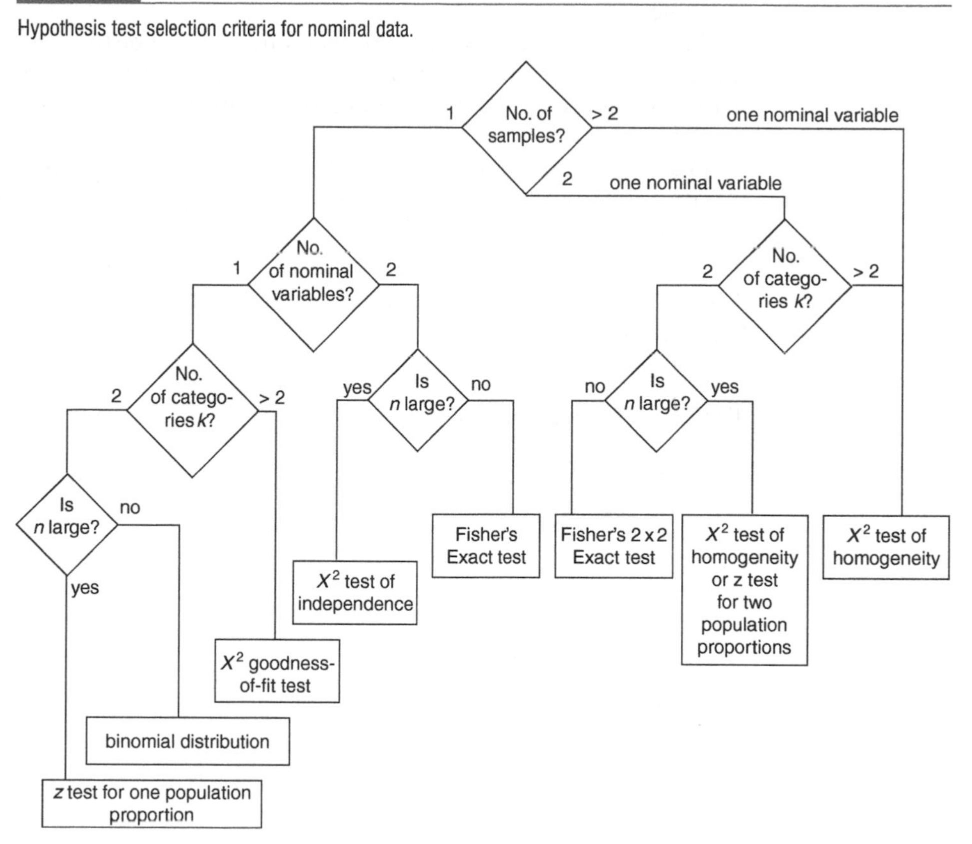

The mean areas of fibrosis within the injured zone observed in the three groups after 5 weeks are

control group: 45% ± 22%, $n_1 = 5$;
low-dose ARB group: 19% ± 9%, $n_2 = 8$;
high-dose ARB group: 14% ± 11%, $n_3 = 7$.

You are asked to determine if administration of low-dose and high-dose ARB drug (losartan) reduces fibrosis of skeletal muscle in mice. Using the two-sample (non-directional) *t* test for unequal variances, compare the mean area of fibrosis within the injured skeletal muscle:

(1) between the control group and the low-dose ARB group and
(2) between the control group and the high-dose ARB group.

4.4. ***In vitro* fertilization (IVF) treatment** Box 4.7 highlights the study performed by Malizia *et al.* (2009) to determine the cumulative live-birth rates resulting from IVF treatment over a course of six cycles. In Box 4.7, a comparison of population proportion of live births for two different age groups was performed when an optimistic approach was adopted to account for the outcomes of non-returnees in the cohort. Here, we look for differences in population proportions of live births within the age groups when a conservative approach is used, i.e. all non-returnees to subsequent IVF cycles (i.e. cycles 2 to 6) are assumed to have failed outcomes.

The conservative cumulative probabilities for live births observed at the end of each IVF cycle are displayed in Table P4.1. The cohort is divided into four groups based on maternal age. The cohort of 6164 women is categorized according to age as in Table P4.2.

Determine at the end of the sixth IVF cycle whether the probabilities of live births for the age groups of *less than 35 years of age* and *equal to or more than 35 years of age* are significantly different.

4.5. The independent-samples t test for equal variances is related to the F test (one-way ANOVA) by the following simple relationship: the square of the t statistic is equal to the F statistic, i.e.

$$t^2 = F.$$

The t statistic is calculated as

$$t = \frac{\bar{x}_1 - \bar{x}_2}{\sqrt{s_{\text{pooled}}^2\left(\frac{1}{n_1} + \frac{1}{n_2}\right)}},$$

where

$$s_{\text{pooled}}^2 = \frac{(n_1 - 1)s_1^2 + (n_2 - 1)s_2^2}{n_1 + n_2 - 2}.$$

If the two sample sizes are equal to n, then

$$t = \frac{\bar{x}_1 - \bar{x}_2}{s_{\text{pooled}}\sqrt{2/n}} \quad \text{and} \quad s_{\text{pooled}}^2 = \frac{s_1^2 + s_2^2}{2}.$$

Table P4.1. *(Conservative) probability of live birth*

	Maternal age (years)			
IVF cycle	<35	35 to <38	38 to <40	≥40
1	0.33	0.28	0.21	0.09
2	0.49	0.44	0.32	0.15
3	0.59	0.51	0.40	0.19
4	0.63	0.55	0.43	0.21
5	0.65	0.56	0.45	0.22
6	0.65	0.57	0.46	0.23

Table P4.2.

Age	Number of individuals
<35	2678
35 to <38	1360
38 to <40	836
<40	1290

Accordingly,

$$t^2 = \frac{n(\bar{x}_1 - \bar{x}_2)^2}{2s_{\text{pooled}}^2}.$$

Show that the F statistic calculated for two independent samples drawn from normally distributed populations with equal variances reduces to the equation above for t^2.

4.6. Suppose that you want to test for differences in the deformability of red blood cells among four different age groups of people. Thus, the null hypothesis is:

$$H_0: \quad \mu_1 = \mu_2 = \mu_3 = \mu_4.$$

How many pairwise t tests would be necessary to test all possible differences? If we want to set the overall probability for making a type I error at 0.10 (10%), what should the level of significance be for each individual t test using the Bonferroni adjustment?

4.7. In Box 4.8 we used the MATLAB `multcompare` function to determine which groups' means are significantly different from each other. The default multiple comparison procedure `'tukey-kramer'` was used. Perform the comparison again using the other three comparison procedures `'dunn-sidak'`, `'bonferroni'`, and `'scheffe'` available in MATLAB. Do any of these three procedures find additional significant differences between group means that the Tukey–Kramer method did not? Also, do other procedures also find groups 1 and 4 and groups 3 and 4 significantly different? What can you say about these three multiple comparison procedures? Are they each more or less conservative than the Tukey–Kramer method?

4.8. **Weight gain from type 2 diabetic treatments** Type 2 diabetes mellitus is also termed as the "insulin resistance" disease because the problem lies not with the production of insulin, but with the signal transduction process within cells that respond to insulin. Insulin binds to receptors on cell surfaces to promote uptake of glucose for various purposes, such as production of glycogen and fat (in adipocytes and hepatocytes), and conversion of glucose into energy (in myocytes). Decreased sensitivity of these cells to insulin will result in reduced transport of glucose from blood into the cells resulting in hyperglycemia. This disease is managed with anti-diabetic drugs that work in a variety of ways, such as by suppressing the production of glucose in the liver, by increasing insulin sensitivity, and/or by promoting insulin release from the pancreas. When blood sugar cannot be controlled by using anti-diabetic drugs and strict diet and exercise, insulin may be prescribed. Insulin intake is accompanied by side-effects such as hypoglycemia and weight gain. Three different insulin + anti-diabetic oral drug therapies are compared with respect to a number of outcomes including weight gain. At baseline, the adult patients recruited to the study were not treated with insulin previously. The weight gain observed after one year of therapy is presented in Table P4.3. Determine whether the diabetic treatments produce differential weight gain in the patients. Formulate the hypothesis and test it using the one-way ANOVA. Calculate the F ratio and use the `fcdf` function to calculate the p value. Use the `anova1` function to check your result. What are your conclusions?

4.9. **Inheritance of Huntington's disease** Huntington's disease is a neurodegenerative hereditary disorder characterized by loss of limb and muscle coordination and decline in mental abilities. The disease is caused by a mutation in the Huntingtin

Table P4.3.

	Weight gain (kg)		
Patient	Treatment 1	Treatment 2	Treatment 3
1	13.5	–0.7	0.1
2	6.9	16.8	9.0
3	4.2	7.6	13.2
4	9.6	4.4	7.9
5	–7.6	7.6	10.5
6	5.5	3.9	9.0
7	3.2	4.2	4.3
8	–3.6	10.7	7.1
9	0.1	10.3	
10	3.3	10.4	
11		7.4	

gene that encodes for an abnormal protein, which interferes with brain functioning. This autosomal (not linked to sex genes) disease is inherited by children from parents in a dominant fashion. Thus, if a person is heterozygous with one copy of the mutated gene, he or she will present the disease. A person having one copy of this gene will pass on the disease with 50% probability to each child that he or she has with an unaffected person. In other words, the offspring have a 50 : 50 chance of inheriting the gene and therefore the disease. Table P4.4 explains the Mendelian inheritance of the genes, where H is the normal allele of the gene and H′ is the mutated allele of the gene that produces the disease.

A person with the HH′ pair is heterozygous with one copy of each allele and will develop the disease.

A study is conducted to confirm that the disease is indeed passed from generation to generation according to Mendelian rules of complete dominance. A total of 120 couples are selected such that only one partner is heterozygous with a single copy of each allele, while the other has only the normal alleles. Out of 217 children of the 120 couples, 122 developed the disease. Use the χ^2 goodness-of-fit method to determine whether the observations support the null hypothesis of complete dominance of the mutated Huntingtin gene.

4.10. **Distribution of birthweights in two different racial groups in England and Wales** In Box 4.10 we used the χ^2 goodness-of-fit test to determine if the birthweight distribution among the White British population in England and Wales is normally distributed. We found that the birthweight distribution was negatively skewed. You are asked to use the χ^2 goodness-of-fit test to determine if the birthweight distribution among the British Indian population is normal. Prepare a table similar to Table 4.9 and report the relevant test statistic and corresponding p value. What are your conclusions?

4.11. **Administering modafinil to alleviate night-time sleepiness in patients with shift-work sleep disorder** In Example 4.8, a study (Czeisler *et al.*, 2005) that examines the efficacy of using modafinil to treat patients suffering from chronic sleepiness at night

Table P4.4. *Punnett square*

		Parent 1	
		H	H’
Parent 2	H	HH	HH’
	H	HH	HH’

A Punnett square is a diagram that lists all the possible combinations of two alleles, one of which is obtained from the mother and the other from the father. One of the listed combinations is inherited by each offspring. Using a Punnett square, one can calculate the probability of observing a particular genotype (and phenotype) in the offspring.

Table P4.5.

	Modafinil-treated group, $n_1 = 96$	Placebo group, $n_2 = 108$
Number of patients that experienced an accident or accident-like situation while commuting from work to home	28	58

during night-shifts at work was discussed. Following treatment with modafinil, the number of patients in both control and treatment groups that reported accidents or near-accidents during their travel from work to home after a night-shift was recorded and is given in Table P4.5.

Prepare a 2 × 2 contingency table for the frequency of patients that underwent an accident or close-encounters when returning home from work for both the treatment and control groups. Is the proportion of accident-prone patients different in the treated group versus the control group? Did the use of modafinil make a difference?

4.12. **Wearing of Hylamer polyethylene hip-joint components** Hip-joint implants were synthesized from Hylamer polyethylene in the 1990s. Hylamer is a modified version of ultra-high molecular weight polyethylene (UHMWPE) manufactured by DuPont. Components made from this material are prone to increased wear and tear if sterilization using gamma rays is performed in the presence of air. Wear of prosthetic implants generates debris causing loosening of the implant components, a phenomenon called periprosthetic osteolysis. A 2005 orthopedic study in Bologna, Italy (Visentin *et al.*, 2005) characterized the polyethylene particulate matter generated from both (1) Hylamer components and (2) traditional UHMWPE synthesized components, gamma sterilized with O_2 present. The morphology of the particles was studied using a scanning electron microscope. The distribution of granular particles (aspect ratio ≤ 2) and fibrillar particles (aspect ratio > 2) in both groups is presented in Table P4.6.

Is the particulate nature of the wear particles for Hylamer PE different from that of traditional UHMWPE? Formulate a null hypothesis. Which test will you use to test your hypothesis? What is the p value associated with your calculated test statistic? What are your conclusions?

4.13. **Rise of multidrug-resistant streptococcus pneumoniae in the USA** The Active Bacterial Core Surveillance program of the Centers for Disease Control and Prevention published an epidemiological study (Whitney *et al.*, 2000) on the trends in antibiotic resistance of S. pneumoniae strains that cause invasive pneumococcal infections (e.g. meningitis and bacteremia). A pneumococcal disease is considered invasive if the bacteria enters and colonizes in otherwise sterile body regions such as blood, joints, and cerebrospinal fluid. A large sample of patients was selected from various regions across the USA (Portland, Oregon; San Francisco County, California; greater Minneapolis and St. Paul; greater Baltimore; greater Atlanta; Tennessee; greater Rochester, New York; and Connecticut).

The incidence of penicillin-resistant pneumococcal infections among various age groups is given in Table P4.7.

Determine whether penicillin-resistant S. pneumoniae infections are equally prevalent among all invasive pneumococcal infected age groups.

4.14. **Neurostimulation as a treatment option for Parkinson's disease** In Box 4.11, a study on using deep-brain stimulation to symptomatically treat Parkinson's disease was outlined. In this study (Deuschl *et al.*, 2006), the severity of motor symptoms, measured using the Unified Parkinson's Disease Rating Scale, was compared pair-wise between patients receiving neurostimulation and patients when they were not on any medication. Among 78 pairs of patients, 55 pairs of neurostimulated patients fared better, and 21 pairs demonstrated that non-medicated patients had a better outcome. The outcome was the same for two pairs. Test the hypothesis that the

Table P4.6.

	Hylamer PE inserts	UHMWPE inserts
Granules	318	123
Fibrils	509	149

Table P4.7.

	Age (years)			
	<5	5–17	18–64	≥65
Penicillin-resistant infections	295	22	282	240
Penicillin susceptible infections	620	83	1179	753

severity of motor symptoms is different among patients receiving neurostimulation versus patients without medication.

4.15. **Efficacy of antivenom for severe cases of neurotoxicity caused by scorpion stings** Some varieties of scorpions produce neurotoxic venom that causes muscle hyperactivity, respiration difficulties, and visual problems when injected into the human body. A scorpion sting often produces severe illness among children but milder symptoms in adults. Treatment of severe cases requires heavy sedation to control violent involuntary movements, and assisted ventilation. A double-blinded study was conducted to test the efficacy of a scorpion antivenom (Boyer *et al.*, 2009). Eight children were randomly assigned to the treatment group that received the antivenom dug, while seven children were randomly placed in the placebo group. The placebo group received a higher sedative dose than the treatment group. At baseline, the mean venom concentrations in blood were comparable in both groups. The plasma levels of venom were measured at 1 hour from treatment time, (values are close to, but are not exactly, the true measurements). These are presented in Table P4.8. At 1 hour from treatment, are the median plasma venom levels of the two groups significantly different? Construct the null hypothesis, choose the appropriate hypothesis test, and state your conclusions based on the value of the calculated test statistic.

4.16. **Treatment of familial adenomatous polyposis** Formation of adenomatous polyps in the colon is often followed by development of colon cancer. Non-steroidal anti-inflammatory drugs (NSAIDs), such as aspirin, are cyclooxygenase (cox) inhibitors and have been linked in multiple studies to regression of colorectal adenomas and a lowering of colon cancer incidence. Mutation of the adenomatous polyposis coli (APC) gene predisposes a person to colon adenoma. Those who inherit the mutated APC gene develop familial adenomatous polyps and are at greater risk for developing colon cancer.

In a randomized, double-blinded study (Steinbach *et al.*, 2000), a cox-2 inhibitor called celecoxib was administered to patients who had developed familial adenomatous polyps. Placebo was given to another group of patients as a control case. The percent changes from baseline after 6 months of treatment in the number of colorectal polyps in patients treated with celecoxib twice per day and with placebo are listed below.

Table P4.8.

Plasma venom levels (ng/ml)	
Antivenom treatment	Placebo treatment
0.007	1.212
0.001	0.026
0.022	8.676
0.032	0.334
0.003	3.634
0.041	0.885
0.009	2.419
	1.795

Placebo ($n = 15$)
20, 17, 14, 0, 0, 0, 0, 0, −2, −8, −10, −13, −17, −20, −45

400 mg of celecoxib administered 2 × daily ($n = 30$)
20, 4, 0, 0, 0, 0, 0, −10, −14, −16, −16.5, −18, −19, −19.5, −28.5, −30, −30.5, −38, −38, −40, −40, −43, −45, −47, −50, −52, −55, −58, −70, −80.

Using a non-parametric method, test the hypothesis that celecoxib reduces the occurrence of adenomatous polyps in patients with familial adenomatous polyposis. What assumptions are necessary to apply the appropriate hypothesis test?

References

Bassett, M. T., Dumanovsky, T., Huang, C., Silver, L. D., Young, C., Nonas, C., Matte, T. D., Chideya, S. and Frieden, T. R. (2008). Purchasing Behavior and Calorie Information at Fast-Food Chains in New York City, 2007. *Am J Public Health*, **98**, 1457–9.

Bedair, H. S., Karthikeyan, T., Quintero, A., Li, Y., and Huard, J. (2008) Angiotensin II Receptor Blockade Administered after Injury Improves Muscle Regeneration and Decreases Fibrosis in Normal Skeletal Muscle. *Am. J. Sports Med.*, **36**, 1548–54.

Boyer, L. V., Theodorou, A. A., Berg, R. A., Mallie, J., Chavez-Mendez, A., Garcia-Ubbelohde, W., Hardiman, S., and Alagon, A. (2009) Antivenom for Critically Ill Children with Neurotoxicity from Scorpion Stings. *N. Engl. J. Med.*, **360**, 2090–8.

Brooke, B. S., Habashi, J. P., Judge, D. P., Patel, N., Loeys, B., and Dietz, H. C., 3rd. (2008) Angiotensin Ii Blockade and Aortic-Root Dilation in Marfan's Syndrome. *N. Engl. J. Med.*, **358**, 2787–95.

Czeisler, C. A., Walsh, J. K., Roth, T. *et al.* (2005) Modafinil for Excessive Sleepiness Associated with Shift-Work Sleep Disorder. *N. Engl. J. Med.*, **353**, 476–86.

Deuschl, G., Schade-Brittinger, C., Krack, P. *et al.* (2006) A Randomized Trial of Deep-Brain Stimulation for Parkinson's Disease. *N. Engl. J. Med.*, **355**, 896–908.

Doll, R. and Kinlen, L. (1977) Fluoridation of Water and Cancer Mortality in the U.S.A. *Lancet*, **1**, 1300–2.

Grocott, M. P., Martin, D. S., Levett, D. Z., McMorrow, R., Windsor, J., and Montgomery, H. E. (2009) Arterial Blood Gases and Oxygen Content in Climbers on Mount Everest. *N. Engl. J. Med.*, **360**, 140–9.

Malizia, B. A., Hacker, M. R., and Penzias, A. S. (2009) Cumulative Live-Birth Rates After In Vitro Fertilization. *N. Engl. J. Med.*, **360**, 236–43.

Moser, K., Stanfield, K. M., and Leon, D. A. (2008) Birthweight and Gestational Age by Ethnic Group, England and Wales 2005: Introducing New Data on Births. *Health Stat. Q.*, issue no. 39, 22–31, 34–55.

Mruk, J. S., Webster, M. W., Heras, M., Reid, J. M., Grill, D. E., and Chesebro, J. H. (2000) Flavone-8-Acetic Acid (Flavonoid) Profoundly Reduces Platelet-Dependent Thrombosis and Vasoconstriction after Deep Arterial Injury in Vivo. *Circulation*, **101**, 324–8.

Newburger, J. W., Sleeper, L. A., McCrindle, B. W. *et al.* (2007) Randomized Trial of Pulsed Corticosteroid Therapy for Primary Treatment of Kawasaki Disease. *N. Engl. J. Med.*, **356**, 663–75.

Nooyens, A. C., van Gelder, B. M., and Verschuren, W. M. (2008) Smoking and Cognitive Decline among Middle-Aged Men and Women: The Doetinchem Cohort Study. *Am. J. Public Health*, **98**, 2244–50.

Piot, C., Croisille, P., Staat, P., *et al.* (2008) Effect of Cyclosporine on Reperfusion Injury in Acute Myocardial Infarction. *N. Engl. J. Med.*, **359**, 473–81.

Samuels, M. L. and Witmer, J. A. (2003) *Statistics for the Life Sciences* (Upper Saddle River; NJ: Pearson Education, Inc.).

Steinbach, G., Lynch, P. M., Phillips, R. K. *et al.* (2000) The Effect of Celecoxib, a Cyclooxygenase-2 Inhibitor, in Familial Adenomatous Polyposis. *N. Engl. J. Med.*, **342**, 1946–52.

Visentin, M., Stea, S., Squarzoni, S., Reggiani, M., Fagnano, C., Antonietti, B., and Toni, A. (2005) Isolation and Characterization of Wear Debris Generated in Patients Wearing Polyethylene Hylamer Inserts, Gamma Irradiated in Air. *J. Biomater. Appl.*, **20**, 103–21.

Whitney, C. G., Farley, M. M., Hadler, J. *et al.* (2000) Increasing Prevalence of Multidrug-Resistant Streptococcus Pneumoniae in the United States. *New Engl. J. Med.*, **343**, 1917–24.

Yiamouyiannis, J. and Burk, D. (1977) Fluoridation and Cancer: Age-Dependence of Cancer Mortality Related to Artificial Fluoridation. *Fluoride*, **10**, 102–125.

5 Root-finding techniques for nonlinear equations

5.1 Introduction

As previously discussed in Chapter 2, a system of *linear equations* can be expressed as $\boldsymbol{Ax} = \boldsymbol{b}$, where $\boldsymbol{A}$ and $\boldsymbol{b}$ are not functions of $\boldsymbol{x}$. On the other hand, a system of equations written as $\boldsymbol{Ax} = \boldsymbol{b}$, in which $\boldsymbol{A}$ and/or $\boldsymbol{b}$ *is* a function of $\boldsymbol{x}$, contains one or more *nonlinear equations*. A polynomial in x of order ≥ 2 has nonlinear dependency on the variable x and is an example of a nonlinear equation. Equations involving trigonometric and/or transcedental functions such as $\sin(x)$, $\cos(x)$, $\log(x)$, e^x, and $\coth(x)$ are also nonlinear in x. The solution x of a nonlinear equation $f(x) = 0$ is termed a **root** of the equation or a **zero** of the function $f(x)$. Application of numerical techniques to obtain the roots of nonlinear equations is the focus of this chapter.

Nonlinear equations that are not amenable to an analytical solution may be solved using iterative numerical schemes. A system of consistent linear equations will always yield a solution. In fact, linear equations that are consistent but not underdetermined will always have a unique solution, regardless of the technique used to solve the system of linear equations. On the other hand, a numerical method used to solve a nonlinear equation may not necessarily converge to a solution. Often, certain conditions must be met before a particular numerical technique can successfully produce a solution. For example, convergence of a method may be contingent on the magnitude of the derivative of a function lying within bounded limits, or may depend on how close the initial guess value is to the root. In many cases, nonlinear equations do not have unique solutions, i.e. an equation may admit several roots. The root that is obtained will usually depend on the numerical method chosen as well as the starting guess value.

Numerical schemes for solving nonlinear equations in one variable can be broadly classified into two categories: **bracketing methods** and **open interval methods**. Sections 5.2 and 5.3 discuss the bisection method and the regula-falsi method, which are numerical root-finding methods of the first category. Bracketing methods of solution require that a closed interval on the number line containing the desired root be specified at the beginning of the first iteration. With each iteration, the interval that brackets the root shrinks, allowing the location of the root to be known with increasing accuracy. When employing bracketing methods, convergence is usually guaranteed unless the initially specified bracket does not contain the root, or contains a singularity point[1] instead of a zero. The latter can be avoided by plotting the function graphically to study the behavior of the function and location of the roots. Bracketing methods of solution are reliable but converge slowly. Faster methods of convergence are usually preferred. This explains the popularity of open

[1] A singularity point of a function $f(x)$ is the x value at which the magnitude of the function blows up, i.e. tends towards plus and/or minus infinity.

Box 5.1A Rheological properties of blood

Blood is responsible for the convective transport of various vital compounds in the body such as oxygen, nutrients, and proteins. Blood consists of plasma and formed elements, i.e. red blood cells (RBCs) or erythrocytes, white blood cells (WBCs) or leukocytes, and platelets. The distinct size, shape, and deformability of each of the three blood cell types – erythrocytes, leukocytes, and platelets strongly influence the flow characteristics of these cells within the vasculature.

The volume fraction of RBCs in blood is called the **hematocrit** and typically ranges from 40–50% in humans. RBCs have a biconcave disc shape with a diameter of 6–8 μm and thickness of 2 μm, and are highly deformable. The distinctive shape of RBCs facilitates aggregation of red blood cells during flow. Aggregates of RBCs flow along the centerline of blood vessels, pushing other blood cells (leukocytes and platelets) to the wall periphery. Thus, flowing blood segregates into two layers: (1) an **RBC-rich core** surrounded by (2) a concentric **cell-depleted plasma layer** near the wall. Figure 5.1 illustrates this. The overall hematocrit of blood is less than the hematocrit exclusively observed in the central flow (core) region where the RBC concentration is highest.

The tendency of RBCs to flow in the center of a vessel gives rise to a flow phenomenon, known as the **Fahraeus effect**, in which the apparent hematocrit of blood that discharges from a vessel is greater than the actual hematocrit of the flowing blood. This results from the greater velocity of the core region of the flowing blood and therefore a greater concentration of RBCs in the collected discharge from the vessel.

The viscosity[2] of blood in the core region is much greater than that of the concentric plasma layer at the periphery.

The viscosity of plasma is given by

$$\mu_{\text{plasma}} = 1.2 \text{ cP}$$

(cP = centiPoise, where 1 Poise (g/(cm s)) = 100 cP). The viscosity of the core region can be calculated as follows:

$$\mu_{\text{core}} = \mu_{\text{plasma}} \frac{1}{1 - aH_{\text{core}}}, \tag{5.1}$$

where H_{core} is the core region hematocrit,

$$a = 0.070 \exp\left(2.49H_{\text{core}} + \frac{1107}{T} \exp(-1.69H_{\text{core}})\right),$$

Figure 5.1

Cross-sectional and longitudinal schematic view of blood flow in a blood vessel.

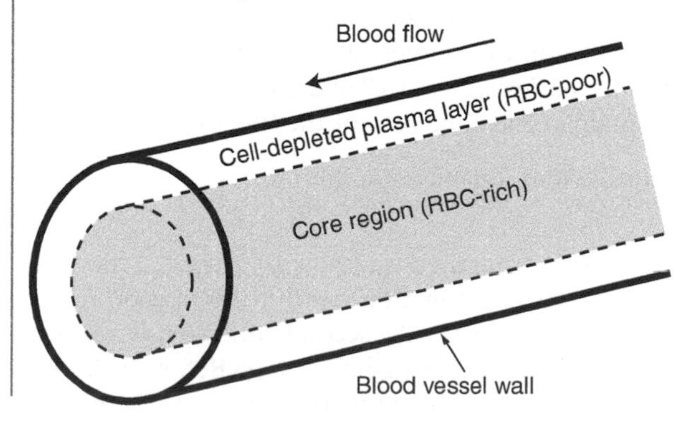

[2] Viscosity is a physical property of a fluid that describes the ease with which a fluid can be sheared. A larger viscosity means that more energy is required to induce and maintain flow.

and T is the temperature in Kelvin.

As an investigator of blood rheology, you want to calculate the hematocrit of the core region in a blood vessel. You have obtained the following expression for the core hematocrit (Fournier, 2007):

$$\frac{H_{\text{core}}}{H_{\text{inlet}}} = 1 + \frac{(1-\sigma^2)^2}{\sigma^2\left[2(1-\sigma^2) + \sigma^2 \frac{\mu_{\text{plasma}}}{\mu_{\text{core}}}\right]} \quad \text{for} \quad H_{\text{core}} < 0.6, \tag{5.2}$$

where

$$\sigma = 1 - \frac{\delta}{R},$$

δ is the thickness of the plasma layer, and R is the radius of the blood vessel.

If H_{inlet} (hematocrit at inlet of blood vessel) $= 0.44$, $\delta = 2.92\ \mu\text{m}$, $R = 15\ \mu\text{m}$, and $T = 310$ K are the estimated values, then σ is calculated as 0.8053. On substituting Equation (5.1) into (5.2), we obtain

$$H_{\text{core}} = H_{\text{inlet}}\left[1 + \frac{(1-\sigma^2)^2}{\sigma^2[2(1-\sigma^2) + \sigma^2(1 - \alpha H_{\text{core}})]}\right]. \tag{5.3}$$

Nonlinear root-finding techniques are required to solve Equation (5.3).

interval methods such as Newton's method and the secant method discussed in Sections 5.5 and 5.6, which have higher orders of convergence but do not always guarantee a solution.

Often, a biomedical engineer will face the challenge of solving engineering problems consisting of a system of two or more simultaneous nonlinear equations. Newton's method for solution of multiple nonlinear equations is discussed in Section 5.7. MATLAB provides a built-in function called `fzero` that uses a hybrid numerical method to determine the zeros of any nonlinear function. The MATLAB `fzero` function to solve for the roots of an equation is discussed in Section 5.8.

5.2 Bisection method

The bisection method is perhaps the simplest method devised for nonlinear root finding. The first step in this numerical procedure involves determining an interval $[a, b]$ that contains the desired root x^* of a nonlinear function $f(x)$. The interval $[a, b]$ is certain to contain at least one zero of the function if

(1) $f(x)$ is continuous within the interval, and
(2) $f(a) \cdot f(b) < 0$.

If the first condition is true, then we know that the function has no discontinuities (e.g. a singularity or a step function) within the bracketing interval and thus the function must cross the x-axis if the second condition is also true.

How do you choose a bracketing interval(s)? Before tackling any nonlinear root-finding problem, regardless of the numerical method of solution chosen, you should first plot the function $f(x)$. The graphical plot supplies a host of information, such as the number of roots, approximate location of each root, presence of any discontinuities or singularities, behavior of the function in proximity of each root, and

Figure 5.2

Plot of a function that contains a double root. At the double root, the function does not cross the x-axis, but is tangent to the axis.

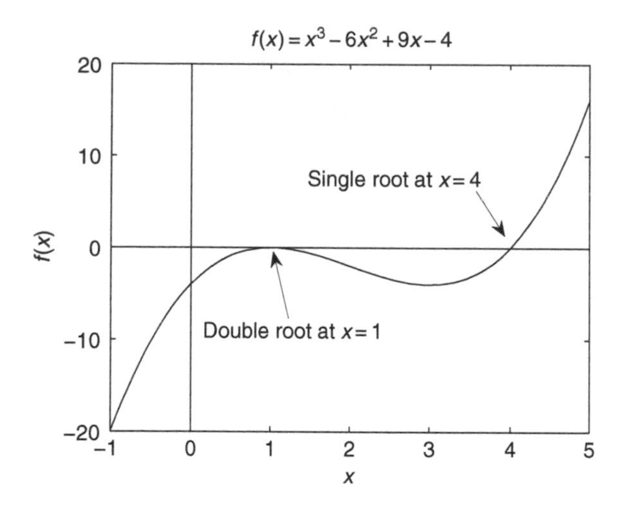

existence of multiple roots. A root that occurs more than once in the solution is called a **multiple root**. For example, the equation

$$f(x) = x^3 - 6x^2 + 9x - 4 = 0 \tag{5.4}$$

can be factored as

$$(x - 1)(x - 1)(x - 4) = 0.$$

This equation has a double root at $x = 1$.

Once a plot of $f(x)$ has revealed the approximate location of the desired root, we can easily obtain the interval $[a, b]$ encapsulating the root. Next, we determine the value of the function at the endpoints of the interval. If the interval contains one root, then $f(a) \cdot f(b)$ must be less than zero, i.e. $f(x)$ evaluated at a and b will lie on opposite sides of the x-axis. This may not be the case if the interval contains more than one root. For example, on both sides of a double root, the value of the function remains on the same side of the x-axis. Equation (5.4) is plotted in Figure 5.2 to illustrate the behavior of a function at a double root.

The behavior of a function at a multiple root is discussed in greater detail in Section 5.5. The bisection method will fail to uncover a root if the function does not cross the x-axis at the root, and thus cannot be used to solve for a double root.

The bisection method searches for the root by depleting the size of the bracketing interval *by half* so that the location of the root is known with an accuracy that doubles with every iteration. If our initial interval is $[a, b]$, then we proceed as follows (Figure 5.3 illustrates these steps).

(1) Bisect the interval such that the midpoint of the interval is $x_1 = (a + b)/2$. Unless x_1 is precisely the root, i.e. unless $f(x_1) = 0$, the root must lie either in the interval $[a, x_1]$ or in $[x_1, b]$.
(2) To determine which interval contains the root, $f(x_1)$ is calculated and then compared with $f(a)$ and $f(b)$.

Figure 5.3

Illustration of the bisection method.

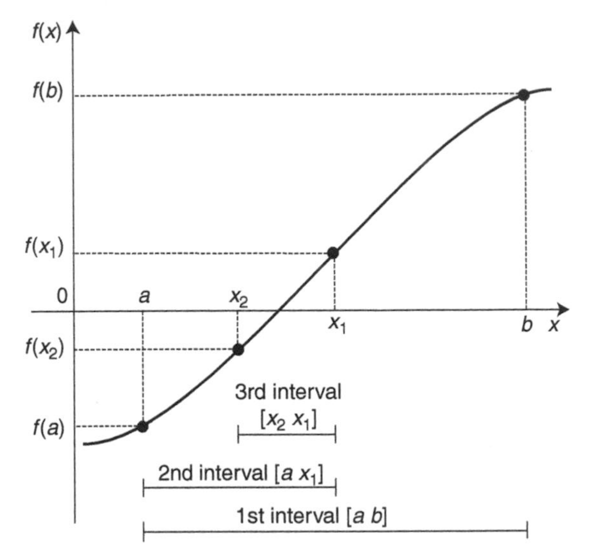

(a) If $f(a) \cdot f(x_1) < 0$, then the root is located in the interval $[a, x_1]$, and the current value of b is replaced by x_1, i.e. $b = x_1$. The new interval is half the size of the previous starting interval.

(b) If $f(a) \cdot f(x_1) > 0$, the root lies within $[x_1, b]$, and a is set equal to x_1.

(3) The bisection step is repeated to find the midpoint, x_2, of the new bracketing interval, which is either $[a, x_1]$ or $[x_1, b]$. Then, the next smaller interval that encloses the root is selected. In Figure 5.3, the interval $[x_2, x_1]$ is found to bracket the root. This new interval is one-quarter the size of the original interval $[a, b]$.

(4) The iterative process is stopped when the algorithm has converged upon a solution. The criteria for attaining convergence are discussed below.

We must specify certain criteria that establish whether we have arrived at a solution. After every iteration, the solution should be tested for convergence to determine whether to continue the bisection method or terminate the process. Tests for convergence estimate the error in the numerical solution and check if this is within the specified tolerance.

One method to test for convergence calculates the difference between the numerically calculated root and the previous root estimate. If the numerical procedure is convergent such that $|x_{n+1} - x_n| < |x_n - x_{n-1}|$ for all n, then the difference between successive estimates of the root indicates whether the solution has converged. The convergence criterion may be specified as follows:

$$|x_{n+1} - x_n| < \mathrm{Tol}(x) \quad \text{(absolute difference)} \tag{5.5}$$

or

$$\frac{|x_{n+1} - x_n|}{|x_{n+1}|} < \mathrm{Tol}(x) \quad \text{(relative difference).} \tag{5.6}$$

Here, Tol(•) is the maximum allowable error tolerance for the root or the function whose zeros we seek. For the bisection method, there is an alternative way of calculating the error in the root estimation. The bracketing interval specifies the maximum error with which the actual root is known. Within the bracketing interval $[a, b]$ the actual root lies within a distance of $|b - a|/2$ from the midpoint of the interval. At the first iteration $i = 1$, the midpoint of the interval $(a + b)/2$ is our approximated root, and the maximum possible error in root estimation is $|b - a|/2$, in which case the root lies at either of the interval endpoints. Thus, the magnitude of the error $\epsilon_1 = |x_1 - x^*|$ is between 0 and $|b - a|/2$.

At the second iteration, the interval size $|b - a|/2$ is again halved. For

$i = 2$ the error bounds in root estimation is given by $\dfrac{|b-a|}{2^2}$,

and for

$i = 3$ the error bounds in root estimation is given by $\dfrac{|b-a|}{2^3}$,

.

.

.

$i = n$ the error bounds in root estimation is given by $\dfrac{|b-a|}{2^n}$.

At every iteration, the error in root estimation reduces linearly by half.

If a tolerance for the error in estimation of the root is specified, we can define a convergence criterion

$$\frac{|b-a|}{2^n} \leq \mathrm{Tol}(x). \tag{5.7}$$

The number of iterations n that will result in a bracketing interval small enough to permit tolerable error is calculated by taking the log of both sides of Equation (5.7), so that

$$n \geq \frac{\log(|b-a|/\mathrm{Tol}(x))}{\log 2}. \tag{5.8}$$

The rate of convergence determines how quickly a numerical procedure approaches the solution or root of the equation, or, in other words, how many iterations it takes to converge upon a solution. The rate of convergence differs from one numerical root-finding method to another.

If the function has a gradual slope at the location of the root, then Equations (5.5), (5.6), or (5.8) are usually sufficient to establish a stop criterion. However, if the function has a steep slope when it crosses the x-axis, then even small deviations from the actual root can lead to substantial departures of the function value from zero. In such cases, another important criterion for convergence specifies the tolerance for $f(x)$, i.e.

$$|f(x)| < \mathrm{Tol}(f(x)); \tag{5.9}$$

$\mathrm{Tol}(f(x))$ may be specified as a fraction of the function values at the endpoints of the initial interval $[a, b]$.

The **rate of convergence** or **order of convergence *r*** for a numerical technique is defined as

$$\lim_{i \to \infty} \frac{|\epsilon_{i+1}|}{|\epsilon_i|^r} = C, \tag{5.10}$$

where

ϵ_{i+1} is the error in root estimation for the $(i + 1)$th iteration,
ϵ_i is the error in root estimation for the ith iteration; $\epsilon_i = x_i - x^*$, where x^* is the root.

The efficiency of a root-finding method depends on the values of C and r. It is desirable to have a large value of r and a small value for C. For the bisection method, $C = 0.5$ and $r = 1$.

Box 5.1B Rheological properties of blood

As described in Box 5.1A, we would like to calculate the core region hematocrit (the volume fraction of red blood cells within the core region). Equation (5.3) is rearranged to yield

$$f(H_{\text{core}}) = 1 + \frac{(1-\sigma^2)^2}{\sigma^2[2(1-\sigma^2) + \sigma^2(1 - aH_{\text{core}})]} - \frac{H_{\text{core}}}{H_{\text{inlet}}},$$

where a is given by Equation (5.1). Let's denote H_{core} by x. On combining the above equation with Equation (5.1) and substituting the known constants into the resulting equation, we obtain

$$f(x) = 1 + \frac{(1-\sigma^2)^2}{\sigma^2\left[2(1-\sigma^2) + \sigma^2\left(1 - 0.070\exp\left(2.49x + \frac{1107}{T}\exp(-1.69x)\right)x\right)\right]} - \frac{x}{0.44}, \tag{5.11}$$

and $f(x)$ is plotted in Figure 5.4 to determine the behavior of the function, the number of roots, and the location of the desired root. We seek the value of x at $f(x) = 0$.

The plot in Figure 5.4 shows that the function is well behaved near the root. Equation (5.3) is only valid for $H_{\text{core}} \leq 0.6$, and we do not expect $H_{\text{core}} < H_{\text{inlet}}$. Hence we choose the plotting interval $0.4 < x < 0.6$. On inspection of Figure 5.4 we choose the bracketing interval [0.5, 0.55].

Figure 5.4

Plot of the function specified by Equation (5.11).

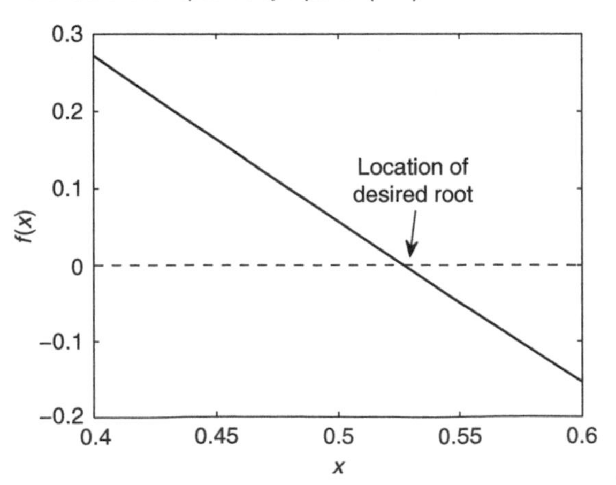

A MATLAB program is written using the bisection algorithm to solve Equation (5.11). First a function called `hematocrit` is created in a separate function m-file called hematocrit.m to evaluate $f(x)$ (Equation (5.11)). Another function is created in MATLAB called `bisectionmethod` which performs the algorithm of the bisection method. The MATLAB function `feval` is used in `bisectionmethod` to evaluate the `hematocrit` function at the specified value of x. As part of the algorithm, it is necessary to define the maximum number of iterations that the program will allow to prevent indefinite looping in situations where convergence is elusive. The program code is listed in the following. Note that the plotting commands have not been included in order to keep the code simple and straightforward. Appendix A discusses the MATLAB `plot` function.

MATLAB program 5.1

```
function f = hematocrit(x)
% Constants
delta = 2.92;  % microns : plasma layer thickness
R = 15;        % microns : radius of blood vessel
hinlet = 0.44; % hematocrit
temp = 310;    % Kelvin : temperature of system

% Calculating variables
sigma = 1 - delta/R;
sigma2 = sigma^2;

alpha = 0.070.*exp(2.49.*x + (1107/temp).*exp(-1.69.*x));
% Equation 5.3
numerator = (1-sigma2)^2;
denominator = sigma2*(2*(1-sigma2)+sigma2*(1-alpha.*x));
  % Equation 5.11
f = 1+numerator./denominator -x./hinlet;
```

MATLAB program 5.2

```
function bisectionmethod(func,ab, tolx, tolfx)
% Bisection algorithm used to solve a nonlinear equation in x

% Input variables
% func   : nonlinear function
% ab     : bracketing interval [a, b]
% tolx   : tolerance for error in estimating root
% tolfx  : tolerance for error in function value at solution

% Other variables
maxloops = 50; % maximum number of iterations allowed

% Root-containing interval [a b]
a = ab(1);
b = ab(2);
fa = feval(func, a); % feval evaluates func at point a
fb = feval(func, b); % feval evaluates func at point b
% Min number of iterations required to meet criterion of Tolx
minloops = ceil(log(abs(b-a)/tolx)/log(2));
          % ceil rounds towards +Inf

fprintf('Min iterations for reaching convergence = %2d \n',minloops)
```

```
fprintf(' i    x    f(x)   \n'); % \n is carriage return

% Iterative solution scheme
for i = 1:maxloops
    x = (a+b)/2; % mid-point of interval
    fx = feval(func,x);
    fprintf('%3d %5.4f %5.4f \n',i,x,fx);
    if (i>= minloops && abs(fx) < tolfx)
        break % Jump out of the for loop
    end
    if (fx*fa < 0) % [a x] contains root
        fb = fx; b = x;
    else           % [x b] contains root
        fa = fx; a = x;
    end
end
```

We specify the maximum tolerance to be 0.002 for both the root and the function value at the estimated root. In the Command Window, we call the function `bisectionmethod` and include the appropriate input variables as follows:

```
>> bisectionmethod('hematocrit',[0.5 0.55],0.002,0.002)
Min iterations for reaching convergence = 5
i       x          f(x)
1     0.5250       0.0035
2     0.5375      -0.0230
3     0.5313      -0.0097
4     0.5281      -0.0031
5     0.5266       0.0002
```

Figure 5.5 graphically shows the location of the midpoints calculated at every iteration.

Note from the program output that the first approximation of the root is rather good and the next two approximations are actually worse off than the first one. Such inefficient convergence is typically observed with the bisection method.

Figure 5.5.

Graphical illustration of the iterative solution of the bisection method.

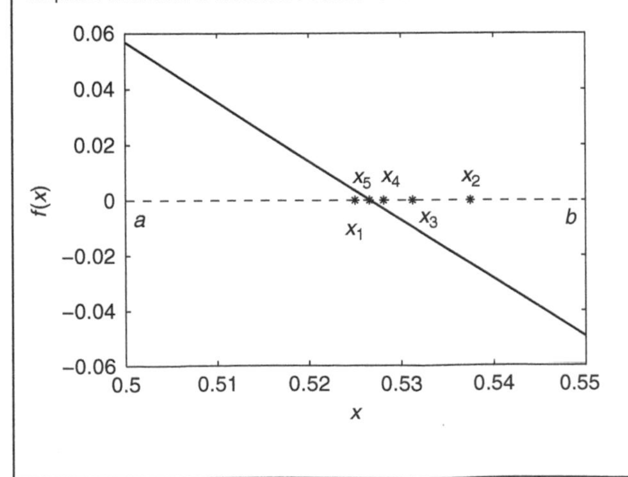

5.3 Regula-falsi method

This numerical root-finding technique is also known as "the method of false position" or the "method of linear interpolation," and is much like the bisection method, in that a root is always sought within a bracketing interval. The name "regula-falsi" derives from the fact that this iterative method produces an estimate of the root that is "false," or only an approximation, yet as each iteration is carried out according to the "rule," the method converges to the root. The methodology for determining the root closely follows the steps enumerated in Section 5.2 for the bisection algorithm. An interval $[x_0, x_1]$, where x_0 and x_1 are two initial guesses that contain the root, is chosen in the same fashion as discussed in Section 5.2. Note that the function $f(x)$ whose root is being sought must be continuous within the chosen interval (see Section 5.2). In the bisection method, an interval is bisected at every iteration so as to halve the bounded error. However, in the regula-falsi method, a line called a **secant line** is drawn connecting the two interval endpoints that lie on either side of the x-axis. The secant line intersects the x-axis at a point x_2 that is closer to the true root than either of the two interval endpoints (see Figure 5.6). The point x_2 divides the initial interval into two subintervals $[x_0, x_2]$ and $[x_2, x_1]$, only one of which contains the root. If $f(x_0) \cdot f(x_2) < 0$ then the points x_0 and x_2 on $f(x)$ necessarily straddle the root. Otherwise the root is contained in the interval $[x_2, x_1]$.

We can write an equation for the secant line by equating the slopes of

(1) the line joining points $(x_0, f(x_0))$ and $(x_1, f(x_1))$ and
(2) the line joining points $(x_1, f(x_1))$ and $(x_2, 0)$.

Thus,

$$\frac{f(x_1) - f(x_0)}{x_1 - x_0} = \frac{0 - f(x_1)}{x_2 - x_1}.$$

Figure 5.6

Illustration of the regula-falsi method.

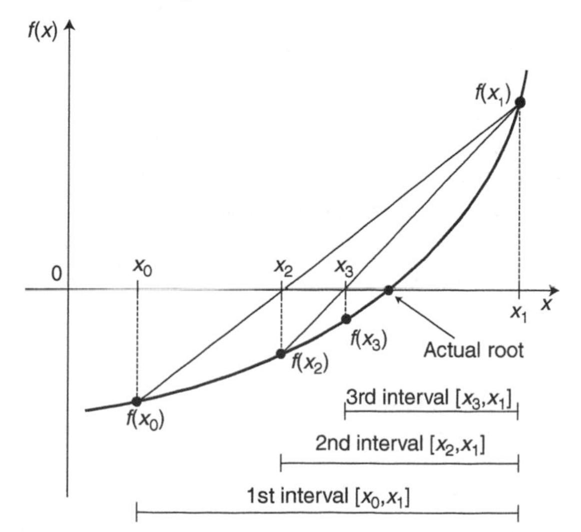

Regula-falsi method:

$$x_2 = x_1 - \frac{f(x_1)(x_1 - x_0)}{f(x_1) - f(x_0)}. \tag{5.12}$$

The algorithm for the regula-falsi root-finding process is as follows.

(1) Select an interval $[x_0, x_1]$ that brackets the root, i.e. $f(x_0) \cdot f(x_1) < 0$.
(2) Calculate the next estimate x_2 of the root within the interval using Equation (5.12). This estimate is the point of intersection of the secant line joining the endpoints of the interval with the x-axis. Now, unless x_2 is precisely the root, i.e. unless $f(x_2) = 0$, the root must lie either in the interval $[x_0, x_2]$ or in $[x_2, x_1]$.
(3) To determine which interval contains the root, $f(x_2)$ is calculated and then compared with $f(x_0)$ and $f(x_1)$.
   (a) If $f(x_0) \cdot f(x_2) < 0$, then the root is located in the interval $[x_0, x_2]$, and x_1 is replaced with the value of x_2, i.e. $x_1 = x_2$.
   (b) If $f(x_0) \cdot f(x_2) > 0$, the root lies within $[x_2, x_1]$, and x_0 is set equal to x_2.
(4) Steps (2)–(3) are repeated until the algorithm has converged upon a solution. The criteria for attaining convergence may be either or both
   (a) $|x_{\text{new}} - x_{\text{old}}| < \text{Tol}(x)$;
   (b) $|f(x_{\text{new}})| < \text{Tol}(f(x))$.

This method usually converges to a solution faster than the bisection method. However, in situations where the function has significant curvature with concavity near the root, one of the endpoints stays fixed throughout the numerical procedure, while the other endpoint inches towards the root. The shifting of only one endpoint while the other remains stagnant slows convergence and many iterations may be required to reach a solution within tolerable error. However, the solution is guaranteed to converge as subsequent root estimates move closer and closer to the actual root.

5.4 Fixed-point iteration

The fixed-point iterative method of root-finding involves the least amount of calculation steps and variable overhead per iteration in comparison to other methods, and is hence the easiest to program. It is an open interval method and requires only one initial guess value in proximity of the root to begin the iterative process. If $f(x)$ is the function whose zeros we seek, then this method can be applied to solve for one or more roots of the function, if $f(x)$ is continuous in proximity of the root(s). We express $f(x)$ as $f(x) = g(x) - x$, where $g(x)$ is another function of x that is continuous within the interval between the root x^* and the initial guess value of the root supplied to the iterative procedure. The equation we wish to solve is $f(x) = 0$, and this is equivalent to

$$x = g(x). \tag{5.13}$$

That value of x for which Equation (5.13) holds is a root x^* of $f(x)$ and is called a **fixed point** of the function $g(x)$. In geometric terms, a fixed point is located at the point of intersection of the curve $y = g(x)$ with the straight line $y = x$.

To begin the iterative method for locating a fixed point of $g(x)$, simply substitute the initial root estimate x_0 into the right-hand side of Equation (5.13). Evaluating

Box 5.1C Rheological properties of blood

We re-solve Equation (5.11) using the regula-falsi method with an initial bracketing interval of [0.5 0.55]:

$$f(x) = 1 + \frac{(1-\sigma^2)^2}{\sigma^2\left[2(1-\sigma^2) + \sigma^2\left(1 - 0.070\exp\left(2.49x + \frac{1107}{T}\exp(-1.69x)\right)x\right)\right]} - \frac{x}{0.44}. \tag{5.11}$$

From Figure 5.3 it is evident that the curvature of the function is minimal near the root and therefore we should expect quick and efficient convergence.

A MATLAB function `regulafalsimethod` is written, which uses the regula-falsi method for root-finding to solve Equation (5.11):

```
>> regulafalsimethod('hematocrit',[0.5 0.55],0.002, 0.002)
i       x           f(x)
1     0.526741    -0.000173
2     0.526660    -0.000001
```

The regula-falsi method converges upon a solution in the second iteration itself. For functions of minimal (inward or outward) concavity at the root, this root-finding method is very efficient.

MATLAB program 5.3

```
function regulafalsimethod(func, x0x1, tolx, tolfx)
% Regula-Falsi method used to solve a nonlinear equation in x

% Input variables
% func : nonlinear function
% x0x1 : bracketing interval [x0, x1]
% tolx : tolerance for error in estimating root
% tolfx: tolerance for error in function value at solution

% Other variables
maxloops = 50;
% Root-containing interval [x0, x1]
x0 = x0x1(1); x1 = x0x1(2);
fx0 = feval(func, x0); fx1 = feval(func, x1);
fprintf(' i    x     f(x)     \n');
xold = x1;
% Iterative solution scheme
for i = 1:maxloops
    % intersection of secant line with x-axis
    x2 = x1 - fx1*(x1 - x0)/(fx1 - fx0);
    fx2 = feval(func,x2);
    fprintf('%3d %7.6f %7.6f \n',i,x2,fx2);
    if (abs(x2 - xold) <=tolx && abs(fx2) < tolfx)
        break % Jump out of the for loop
    end
    if (fx2*fx0 < 0) % [x0 x2] contains root
        fx1 = fx2; x1 = x2;
    else             % [x2 x1] contains root
        fx0 = fx2; x0 = x2;
    end
    xold = x2;
end
```

Figure 5.7

Illustration of the fixed-point iterative method.

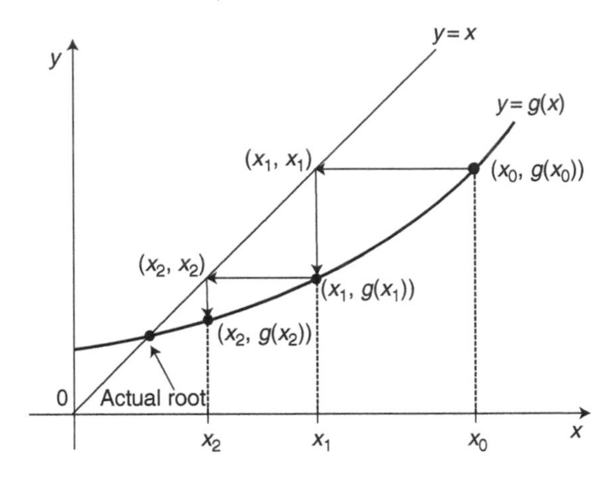

$g(x)$ at x_0 produces another value of x, x_1, which is an improved approximation of the root compared to the initial guess value x_0. Thus,

$$x_1 = g(x_0).$$

Again calculate the value of $g(x)$ at x_1 to obtain an even better estimate of the root,

$$x_2 = g(x_1).$$

This iterative procedure is continued until the approximation $x_{n+1} = g(x_n)$ obtained meets the imposed convergence criteria, e.g. $|x_{n+1} - x_n| < \mathrm{Tol}(x)$ and/or $f(x) < \mathrm{Tol}(f(x))$.

The fixed-point iterative method of solution is portrayed graphically in Figure 5.7. The pair of equations $y = g(x)$ and $y = x$ are plotted on the same axes. The point $(x_0, g(x_0))$ is located on the plot $y = g(x)$. Since $g(x_0) = x_1$, by moving horizontally from the point $(x_0, g(x_0))$ towards the $y = x$ line, one can locate the point (x_1, x_1). Next, a vertical line from the $y = x$ axis to the curve $y = g(x)$ is drawn to reveal the point $(x_1, g(x_1))$. Again, by constructing a horizontal line from this point to the $y = x$ line, the point (x_2, x_2) can be determined. Continuing in this iterative fashion, the final approximation of the root can be obtained.

Figure 5.7 depicts a converging solution. However, convergence of this iterative technique is guaranteed only under certain conditions. First, one must ascertain the existence of a fixed point in the interval of interest, and whether this point is unique or whether multiple fixed points exist. Once we know the approximate location of a root, we need to check whether the iterative process will converge upon the root or exhibit divergence. The following theorems are stated here without proof. Proofs can be found in many standard numerical analysis texts (we recommend *Numerical Analysis* by Burden and Faires (2005)).

If the function variable $x \in [a, b]$, $g(x)$ is continuous on the interval $[a, b]$, and $g'(x)$ exists on the same interval, then we have the following.

Theorem 5.1: If function $g(x)$ maps x into the same interval $[a, b]$, i.e. $g(x) \in [a, b]$, then g has at least one fixed point in the interval $[a, b]$.

Figure 5.8

Situations where divergence is encountered when using the fixed-point iterative method to determine the location of the root.

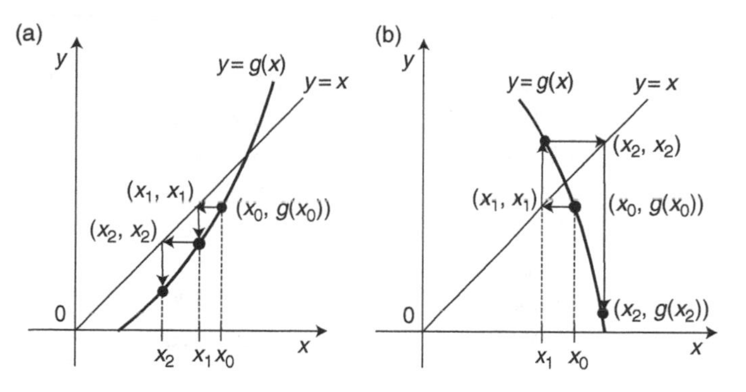

Theorem 5.2[3]: If $0 < k < 1$ and $|g'(x)| \leq k$ for all $x \in [a, b]$, then

(a) a single fixed point exists in the defined interval, i.e. the fixed point is unique;
(b) for $x_0 \in [a, b]$ the sequence $x_0, x_1, x_2, \ldots, x_n$ generated by the equation $x_n = g(x_{n-1})$ will **linearly** converge to the unique fixed point x^*.

Note the following.

(1) Theorems 5.1 and 5.2(a) are sufficient but not necessary conditions for the existence of a unique fixed point x^* in the interval $[a, b]$. Thus, a fixed point may exist in the interval $[a, b]$, even if $g(x)$ maps x partly outside the interval.
(2) If $k = 0$, then the order of convergence is not linear but of second order. See Newton's method for an example of a situation where $g'(x) = 0$, which leads to a quadratic convergence rate.
(3) For $0 < k < 1$, if C is the error constant (see Equation (5.10)), then $\lim_{n\to\infty} C = |g'(x^*)|$. Thus the smaller the value of k, the faster the rate of convergence.
(4) If $0 < g'(x) < 1$, then the convergence will occur monotonically; otherwise, if $-1 < g'(x) < 0$, the successive approximations will oscillate about the root.

Figure 5.8 illustrates two situations where the solution technique exhibits divergence. In both cases depicted in Figure 5.8, one can visually observe that $|g'(x)| > 1$ at all points in the interval under consideration. Figure 5.8(a) shows that when $g'(x) > 1$, the solution monotonically diverges away from the location of the root, while in Figure 5.8(b) the solution oscillates with increasing amplitude as it progresses away from the root.

[3] Theorem 5.2 can be easily demonstrated. Suppose $g(x)$ has a fixed point in the interval $[a, b]$. We can write

$$x_{n+1} - x^* = g(x_n) - g(x^*) = g'(\xi)(x_n - x^*),$$

where ξ is some number in the interval $[x_n, x^*]$ (according to the mean value theorem; see Section 7.2.1.1 for the definition of the mean value theorem). If $|g'(x)| \leq k$ over the entire interval, then $|e_{n+1}| = k|e_n|$ and $|e_{n+1}| = k^{n+1}|e_0|$, where $e_n = x_n - x^*$. If $k < 1$, then $\lim_{n\to\infty} e_{n+1} = 0$.

Table 5.1.

Iteration	$x = 2e^x(2 - x^2)$	$x = \sqrt{\frac{4 - xe^{-x}}{2}}$	$x = 4 - 2x^2 - xe^{-x} + x$
0	1.0	1.0	1.0
1	5.4366	1.34761281	2.6321
2	−12656.6656	1.35089036	−7.4133
3	0.0000	1.35094532	12177.2917
4	4.0000	1.35094624	−296560686.9631
5	−1528.7582	1.35094626	∞
6	0.0000	1.35094626	
7	4.0000	1.35094626	
8	−1528.7582	1.35094626	

Example 5.1

Consider the equation $f(x) = 2x^2 + xe^{-x} - 4 = 0$, which can be expressed in the form $x = g(x)$ in three different ways:

$$x = 2e^x\left(2 - x^2\right), \tag{5.14}$$

$$x = \sqrt{\frac{4 - xe^{-x}}{2}}, \tag{5.15}$$

and

$$x = 4 - 2x^2 - xe^{-x} + x. \tag{5.16}$$

Note that $f(x)$ contains a root in the interval [0, 2] and thus there exists a fixed point of g within this interval. We wish to find the root of $f(x)$ within this interval by using fixed-point iteration. We start with an initial guess value, $x_0 = 1$. The results are presented in Table 5.1.

Equations (5.14)–(5.16) perform very differently when used to determine the root of $f(x)$ even though they represent the same function $f(x)$. Thus, the choice of $g(x)$ used to determine the fixed point is critical. When the fixed-point iterative technique is applied to Equation (5.14), cyclic divergence results. Equation (5.15) results in rapid convergence. Substitution of the fixed point $x = 1.3509$ into $f(x)$ verifies that this is a zero of the function. Equation (5.16) exhibits oscillating divergence to infinity. We analyze each of these three equations to study the properties of $g(x)$ and the reasons for the outcome observed in Table 5.1.

Case A: $g(x) = 2e^x(2 - x^2)$

First we evaluate the values of $g(x)$ at the endpoints of the interval: $g(0) = 4$ and $g(2) = -29.56$. Also $g(x)$ has a maxima of $\sim$6.0887 at $x \sim 0.73$. Since for $x \in [a, b]$, $g(x) \notin [a, b]$, $g(x)$ does not exclusively map the interval into itself. While a fixed point does exist within the interval [0, 2], the important question is whether the solution technique is destined to converge or diverge.

Next we determine $g'(x) = 2e^x(2 - 2x - x^2)$ at the endpoints: $g'(0) = 4$ and $g'(2) = -88.67$. At x_0, $g'(1) = -5.44$. Although $g'(0.73) < 0.03$ near the point of the maxima, $|g'(x)| > 1$ for [1, 2] within which the root lies. Thus this method based on $g(x) = 2e^x(2 - x^2)$ is expected to fail.

Case B: $g(x) = \sqrt{\frac{4 - xe^{-x}}{2}}$

$1.34 < g(x) < 1.42$ for all $x \in [0, 2]$. For case B, g maps [1, 2] into [1, 2],

$$g'(x) = \frac{e^{-x}}{4}\frac{(x-1)}{\sqrt{\frac{4-xe^{-x}}{2}}}$$

so $g'(0) = -0.177$ and $g'(2) = 0.025$. For $x \in [0, 2]$, $|g'(x)| < 0.18$; $k = 0.18$, and this explains the rapid convergence for this choice of g.

Case C: $g(x) = 4 - 2x^2 - xe^{-x} + x$

For $x \in [0, 2]$, $-2.271 \le g(x) \le 4$; $g(x)$ does not map the interval into itself. Although a unique fixed point does exist in the interval, because $g'(x) = 1 - 4x + e^{-x}(x-1) = 0$ at $x = 0$ and is equal to 6.865 at $x = 2$, convergence is not expected in this case either.

Box 5.1D Rheological properties of blood

We re-solve Equation (5.11) using fixed-point iteration and an initial guess value of 0.5:

$$f(x) = 1 + \frac{(1-\sigma^2)^2}{\sigma^2\left[2(1-\sigma^2) + \sigma^2(1 - 0.070\exp(2.49x + \frac{1107}{T}\exp(-1.69x))x)\right]} - \frac{x}{0.44}. \quad (5.11)$$

Equation (5.11) is rearranged into the form $x = g(x)$, as follows:

$$x = 0.44\left[1 + \frac{(1-\sigma^2)^2}{\sigma^2\left[2(1-\sigma^2) + \sigma^2(1 - 0.070\exp(2.49x + \frac{1107}{T}\exp(-1.69x))x)\right]}\right].$$

A MATLAB function program called `fixedpointmethod` is written to solve for the fixed point of

$$g(x) = 0.44\left[1 + \frac{(1-\sigma^2)^2}{\sigma^2\left[2(1-\sigma^2) + \sigma^2(1 - 0.070\exp(2.49x + \frac{1107}{T}\exp(-1.69x))x)\right]}\right].$$

A simple modification is made to the function program hematocrit.m in order to evaluate $g(x)$. This modified function is given the name `hematocritgx`. Both `hematocrit` and `hematocritgx` are called by `fixedpointmethod` so that the error in both the approximation of the fixed point as well as the function value at its root can be determined and compared with the tolerances. See MATLAB programs 5.4 and 5.5.

MATLAB program 5.4

```
function g = hematocritgx(x)
% Constants
delta = 2.92;    % microns : plasma layer thickness
R = 15;          % microns : radius of blood vessel
hinlet = 0.44;   % hematocrit
temp = 310;      % Kelvin : temperature of system
% Calculating variables
```

```
sigma = 1 - delta/R;
sigma2 = sigma^2;

alpha=0.070.*exp(2.49.*x+(1107/temp).*exp(-1.69.*x)); %Equation 5.1
numerator = (1-sigma2)^2; % Equation 5.3
denominator=sigma2*(2*(1-sigma2)+sigma2*(1-alpha.*x)); %Equation 5.3
g = hinlet*(1+numerator./denominator);
```

MATLAB program 5.5

```
function fixedpointmethod(gfunc, ffunc, x0, tolx, tolfx)
% Fixed-Point Iteration used to solve a nonlinear equation in x

% Input variables
% gfunc : nonlinear function g(x) whose fixed-point we seek
% ffunc : nonlinear function f(x) whose zero we seek
% x0 : initial guess value
% tolx : tolerance for error in estmating root
% tolfx : tolerance for error in function value at solution

% Other variables
maxloops = 20;
fprintf('  i     x(i+1)      f(x(i))    \n');
% Iterative solution scheme
for i = 1:maxloops
    x1 = feval(gfunc, x0);
    fatx1 = feval(ffunc,x1);
    fprintf('%2d %9.8f %7.6f \n',i,x1,fatx1);
    if (abs(x1 - x0) <=tolx && abs(fatx1) <= tolfx)
        break % Jump out of the for loop
    end
    x0 = x1;
end
```

The function `fixedpointmethod` is run from the MATLAB Command Window with the following parameters:

function specifying g(x): `hematocritgx`,
function specifying f(x): `hematocrit`,
$x_0 = 0.5$,
tolerance for root = 0.002,
Tolerance for $f(x)$ = 0.002.

MATLAB outputs

```
>> fixedpointmethod('hematocritgx','hematocrit',0.5, 0.002, 0.002)
i      x(i+1)        f(x(i))
1    0.52498704    0.003552
2    0.52654981    0.000233
```

The solution is arrived at in only two steps, which shows rapid convergence. Note that $g(x)$ maps the interval [0.5 0.55] into itself and $|g'(x)|$ at $x = 0.5265$ is 0.1565 while $|g'(0.5)| = 0.1421$. Of course the derivative of $g(x)$ is somewhat messy and requires careful calculation.

5.5 Newton's method

Newton's method, also called the Newton–Raphson method, is an open interval root-finding method. It is a popular choice because of its second-order or **quadratic rate of convergence** ($r = 2$ in Equation (5.10)). Newton's method is applicable only when both $f(x)$ and $f'(x)$ are continuous and differentiable. Only a single initial guess x_0 is required to begin the root-finding algorithm. The x-intercept of the tangent to $f(x)$ at x_0 yields the next approximation, x_1, to the root. This process is repeated until the solution converges to within tolerable error. Unlike the brackcting methods discussed in Sections 5.2 and 5.3, an open interval method does not always converge to a solution.

The solution technique is derived from the Taylor series expansion of $f(x)$ about an initial guess point x_0 located close to the root:

$$f(x_0 + \Delta x) = f(x_0) + f'(x_0)\Delta x + \frac{f''(\zeta)\Delta x^2}{2}, \qquad \zeta \epsilon [x_0, x_0 + \Delta x]. \tag{5.17}$$

If $x = x_0 + \Delta x$ is the root of the function then $f(x_0 + \Delta x) = 0$ and the Taylor series expansion can be rewritten as follows:

$$0 = f(x_0) + f'(x_0)(x - x_0) + 0.5f''(\zeta)(x - x_0)^2. \tag{5.18}$$

If only the first two terms on the right-hand side of the Taylor expansion series in Equation (5.18) are retained, then

$$x = x_0 - \frac{f(x_0)}{f'(x_0)}. \tag{5.19}$$

The validity of Equation (5.19) is contingent on our initial guess value x_0 being close to the root enough that the error term squared $(x - x_0)^2$ is much smaller than $(x - x_0)$. Thus, the last term will be much smaller than thc first two terms in Equation (5.18), which justifies neglecting the f'' term in the series expansion. Equation (5.19) is a first-order approximation of the function as a straight line through point $(x_0, f(x_0))$ with slope equal to $f'(x_0)$. Because $f(x)$ is not linear in x, x obtained from Equation (5.19) will not be the root, but an approximation.

Equation (5.19) can be graphically interpreted as follows. If the function $f(x)$ is approximated by the tangent to the curve at an initial guess value of the root x_0, then the x-intercept of this tangent line produces the next approximation x_1 of the actual root (Figure 5.9). The slope of this tangent line is calculated as

$$f'(x_0) = \frac{f(x_0) - 0}{x_0 - x_1},$$

which, on rearrangement, yields Equation (5.19) where x, or x_1, is the first approximation to the root. Equation (5.19) can be generalized as Equation (5.20).

Newton's method

$$x_{i+1} = x_i - \frac{f(x_0)}{f'(x_0)}. \tag{5.20}$$

Equation (5.20) defines the step-wise iterative technique of Newton's method. Figure 5.9 demonstrates graphically the step-wise algorithm to approach the root.

The algorithm for calculating a zero using Newton's method is as follows.

Figure 5.9

Illustration of Newton's method.

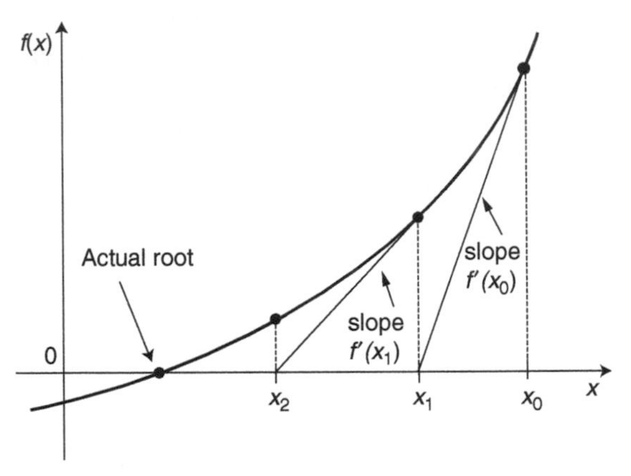

(1) Provide an initial guess x_0 of the root of $f(x) = 0$ either by plotting the function or by performing an initial bisection or regula-falsi root-approximation step.
(2) Derive the analytical function $f'(x)$.
(3) Calculate $f(x_i)$ and $f'(x_i)$ at x_i. For the first iteration, $x_i = x_0$.
(4) Evaluate $x_{i+1} = x_i - f(x_i)/f'(x_i)$.
(5) Repeat steps 3 and 4 until the method has converged upon a solution.

Example 5.2

Let's use Newton's method to solve the simple problem of determining the square root of 8, i.e.

$$f(x) = x^2 - 8 = 0.$$

Then, $f'(x) = 2x$.

The iterative formula for this problem obtained from Equation (5.20) is

$$x_{i+1} = x_i - \frac{x_i^2 - 8}{2x_i}.$$

Let's start with an initial guess value $x_0 = 3$. Then

$$x_1 = 3 - \frac{3^2 - 8}{2 \times 3} = \frac{17}{6},$$

$$x_2 = \frac{17}{6} - \frac{\frac{17^2}{6} - 8}{2 \times \frac{17}{6}} = 2.828431373,$$

$$x_3 = 2.828431373 - \frac{2.828431373^2 - 8}{2 \times 2.828431373} = 2.828427125.$$

The actual solution is $\sqrt{8} = 2.828427125$.

The error at each stage of calculation is given by

$$\varepsilon_0 = -0.1715729,$$
$$\varepsilon_1 = -0.0049062,$$
$$\varepsilon_2 = -4.24780 \times 10^{-6},$$
$$\varepsilon_3 = -3.18945 \times 10^{-12}.$$

If Newton's method is quadratically convergent then $|\varepsilon_{i+1}|/|\varepsilon_i|^r$ for $r = 2$ should yield a number C that converges to a constant value as i increases; i.e.

$$\frac{|\varepsilon_1|}{|\varepsilon_0|^2} = 0.1667,$$

$$\frac{|\varepsilon_2|}{|\varepsilon_1|^2} = 0.1765,$$

$$\frac{|\varepsilon_3|}{|\varepsilon_2|^2} = 0.1768.$$

Note the fast rate of convergence when $r = 2$.

The last term on the right-hand side of Equation (5.18) provides valuable information regarding the rate of convergence of Newton's method. Equation (5.18) may be rewritten as follows:

$$x^* = x_i - \frac{f(x_i)}{f'(x_i)} - \frac{f''(\zeta)}{2f'(x_i)}(x^* - x_i)^2. \tag{5.21}$$

After substituting Equation (5.20) into Equation (5.21) we obtain

$$(x^* - x_{i+1}) = -\frac{f''(\zeta)}{2f'(x_i)}(x^* - x_i)^2$$

or

$$\frac{|(x^* - x_{i+1})|}{|(x^* - x_i)|^2} = \frac{|\epsilon_{i+1}|}{|\epsilon_i|^2} = \frac{|f''(\zeta)|}{2|f'(x_i)|}. \tag{5.22}$$

Equation (5.22) proves that the order of convergence is $r = 2$ for Newton's method. This does not hold true if $f'(x) = 0$. Here

$$C = \frac{|f''(\zeta)|}{2|f'(x_i)|}$$

and

$$\lim_{\substack{i \to \infty \\ \text{or} \\ x_i \to x_{\text{root}}}} C = \frac{|f''(x^*)|}{2|f'(x^*)|}.$$

For Example 5.2, $x^2 - 8 = 0, f''(\sqrt{8}) = 2$, $f'(\sqrt{8}) = 5.65685$, and $C = 0.1768$.

5.5.1 Convergence issues

Convergence is not guaranteed

Newton's method converges if the initial approximation of the solution is sufficiently close to the root. How close the initial guess value must be to the root to guarantee convergence depends upon the shape of the function, i.e. the values of $f'(x)$ and $f''(x)$ at the guess values, at subsequent root approximations, and at the actual root. If the value of $f'(x)$ becomes zero during the iterative procedure, then the solution

Figure 5.10

Two situations in which Newton's method does not converge.

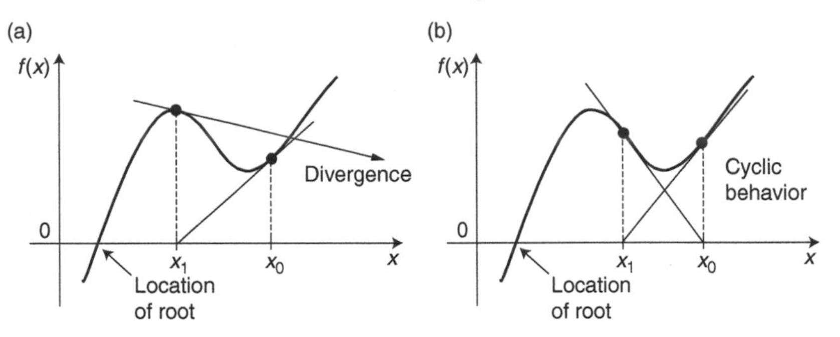

technique will diverge. In fact, a near-zero slope or inflection point anywhere between the initial guess and the root can cause persistent oscillations or divergence of the solution. Figure 5.10(a) and (b) graphically depict two situations where convergence cannot be attained.

Note that Equation (5.20) is of the form $x = g(x)$, where

$$g(x) = x - \frac{f(x)}{f'(x)}. \tag{5.23}$$

It is necessary that $|g'(x)| < 1$ near the root to establish convergence in any $x = g(x)$ fixed-point iterative scheme (see Section 5.4), where

$$g'(x) = \frac{f(x)f''(x)}{(f'(x))^2}. \tag{5.24}$$

At the root, $f(x) = 0$ and $g'(x) = 0$. Therefore, there must be a range of values of x near the root such that $|g'(x)| < 1$. Convergence of the method is certain as long as the initial guess lies within this range.

Convergence rate is not always quadratic

The convergence rate was shown to be quadratic for this method; however, this applies only at simple roots, i.e. roots of order $p = 1$.

A root x^* of $f(x)$ is said to be of **order** p if the function's derivatives are continuous and the following hold true:

$$f(x^*) = 0; \quad f'(x^*) = 0; \quad f''(x^*) = 0; \quad \cdots \quad f^{(p-1)}(x^*) = 0; \quad f^{(p)}(x^*) \neq 0.$$

For root x^* of order $p = 1$,

$$f(x^*) = 0; \quad f'(x^*) \neq 0; \quad f''(x^*) \neq 0,$$

and $f(x)$ can be written as $(x - x^*)h(x)$, where $h(x) \neq 0$ when $x = x^*$.
For root x^* of order $p = 2$,

$$f(x^*) = 0; \quad f'(x^*) = 0; \quad f''(x^*) \neq 0,$$

and $f(x)$ can be written as $(x - x^*)^2 h(x)$.

For simple roots ($p = 1$), Equation (5.24) yields $g'(x) = 0$, and from fixed-point iteration theory Newton's method will have second-order convergence. For multiple roots, i.e. $p > 1$, $g'(x)$ takes on a 0/0 form and converges to a number $0 < g'(x) < 1$, which signifies linear convergence (see Section 5.4). If $f(x) = (x - x^*)^p h(x)$, is substituted into Equation (5.24), where p is the multiplicity of the root, it can be shown that $g'(x) = (p - 1)/p$. For a double root, $p = 2$ and $g'(x) = 1/2$. Quadratic convergence can be enforced by modifying Equation (5.23) as follows:

$$g(x) = x - p\frac{f(x)}{f'(x)}, \tag{5.25}$$

where p is the order of the root x^*.

Box 5.2A Equilibrium binding of multivalent ligand in solution to cell surfaces

Macromolecules that bind to **receptors** located on the surface of cells are called **ligands**. Receptors are special proteins, like enzymes, that have specific binding properties and usually perform a function, such as transport of a ligand across a cell membrane or activation of a signal to turn on certain genes, when specific ligands are bound. When a ligand contains multiple binding sites such that it can bind to more than one receptor at any time, the ligand is said to crosslink or aggregate multiple receptors. When ligands such as antibody–antigen complexes, hormones, or other extracellular signaling molecules induce the aggregation of receptors on a cell surface, various important cellular responses are triggered, such as the expression of new protein.

A model for the equilibrium binding of multivalent ligand to cell surface receptors was proposed by Perelson (1981). In this model, the multivalent ligand is assumed to have a total of ν active sites available for binding to receptors on the surfaces of cells suspended in solution. A ligand molecule in solution can initially attach to a cell surface by forming a bond between any of its ν binding sites and a receptor on the cell surface (see Figure 5.11). However, once a single bond has formed between the ligand and a cell receptor, the ligand orientation may allow only a limited number of binding sites to interact with that cell. The total number of binding sites that a ligand may present to a cell surface for binding is f.

Figure 5.11

Binding of a multivalent ligand in solution to monovalent receptors on a cell surface.

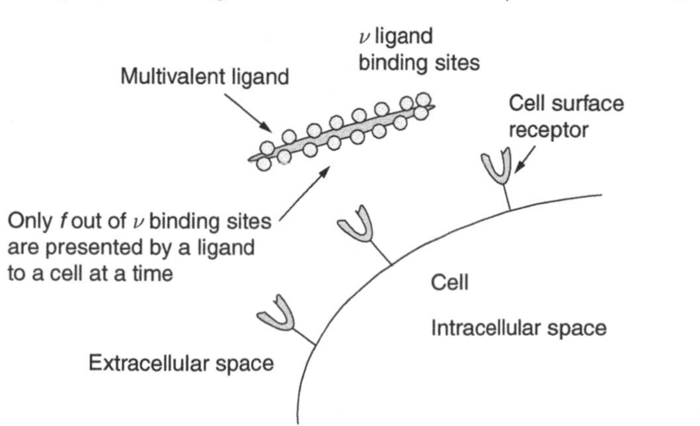

The extent of binding of a ligand containing multiple identical binding sites to receptors on a cell surface can be measured by determining the concentration of unbound receptors present on the cell surface at equilibrium, which is given by the following equation (Lauffenburger and Linderman, 1993):

$$R_T = R_{eq}\left[1 + v\left(\frac{L_0}{K_D}\right)(1 + K_x R_{eq})^{f-1}\right], \tag{5.26}$$

where

v is the total number of binding sites present on the ligand surface,
f is the total number of binding sites available for binding to a single cell,
K_x is the two-dimensional crosslinking equilibrium constant (1/(# ligands/ cell)),
L_0 is the ligand concentration in solution (M),
K_D is the three-dimensional dissociation constant for binding of ligand molecules in solution to cells (M),
R_{eq} is the equilibrium concentration of unbound receptors present on the cell surface (# receptors/ cell),
R_T is the total number of receptors present on the platelet surface.

Von Willebrand factor, a plasma protein, binds platelet cells via multiple receptor–ligand bonds. For the von Willebrand factor–platelet system, the parameters are estimated to be as follows (Mody and King, 2008):

(1) $v = 18$,
(2) $f = 9$,
(3) $K_D = 7.73 \times 10^{-5}$ M,
(4) $K_x = 5.8 \times 10^{-5}$ cell/# ligands
(5) $L_0 = 2 \times 10^{-9}$ M,
(6) $R_T = 10\,688$ # receptors/cell.

We wish to determine the equilibrium concentration R_{eq} of unbound or free receptors on a platelet. Replacing R_{eq} with x, we obtain

$$f(x) = x\left[1 + v\left(\frac{L_0}{K_D}\right)(1 + xK_x)^{f-1}\right] - R_T, \tag{5.27}$$

$$f'(x) = 1 + v\left(\frac{L_0}{K_D}\right)(1 + xK_x)^{f-1} + xK_x v(f-1)\left(\frac{L_0}{K_D}\right)(1 + xK_x)^{f-2}. \tag{5.28}$$

A function called `ReceptorLigandEqbmBinding` was written in MATLAB to calculate $f(x)$ and $f'(x)$ as given by Equations (5.27) and (5.28). A general function called `Newtonsmethod` was created to solve a nonlinear problem in a single variable x using Newton's method. When calling this function at the command prompt, we specify the function `ReceptorLigandEqbmBinding` as an input parameter to `Newtonsmethod`, which supplies the values of $f(x)$ and $f'(x)$ for successive values of x. Note that the `feval` function requires the function name to be specified in string format, and therefore the function name is included within single quotation marks when input as a parameter.

```
>> newtonsmethod('Receptor Ligand Eqbm Binding',8000,0.5,0.5)
i      x(i)          f(x(i))       f'(x(i))
1    10521.7575     55.262449     1.084879
2    10470.8187     0.036442      1.083452
3    10470.7850     0.000000      1.083451
```

Note the rapid convergence of $f(x) \to 0$. The number of receptors on the platelet surface bound to ligand when equilibrium conditions are established is $R_T - 10\,471 = 217$ receptors or 2% of the total available receptors. The number of multivalent ligand molecules (containing multiple identical binding sites) bound to receptors on a cell surface at equilibrium is given by the following equation:

$$C_{i,\mathrm{eq}} = \left[\frac{f!}{i!(f-i)!}\right] K_x^{i-1} \frac{v}{f}\left(\frac{L_0}{K_D}\right) R_{\mathrm{eq}}^i, \qquad i = 1, 2, \ldots, f,$$

where $C_{i,\mathrm{eq}}$ is the number of ligand molecules bound to the cell by i bonds. Now that you have solved Equation (5.26), determine the total number of ligand molecules $\sum_{i=1}^{f} C_i$ that bind a platelet cell at equilibrium. Use `sum(floor(Ci))` to calculate the integral number of bound ligand molecules.

MATLAB program 5.6

```
function [fx, fxderiv] = ReceptorLigandEqbmBinding (Req)
% Equilibrium binding of multivalent ligand to cell surface

% Constants
RT = 10688;      % Total number of receptors on a cell
L0 = 2.0e-9;     % Ligand concentration in medium (M)
Kd = 7.73e-5;    % Equilibrium dissociation constant (M)
v = 18;          % Total number of binding sites on ligand
f = v/2;         % Total number of binding sites that a single cell can bind
Kx = 5.8e-5;     % Crosslinking equilibrium constant (cell/#)

fx = Req*(1+v*(L0/Kd)*(1+Kx*Req)^(f-1)) - RT;
fxderiv = 1+v*(L0/Kd)*(1+Kx*Req)^(f-1) + ...
    Req*v*(L0/Kd)*(f-1)*Kx*(1+Kx*Req)^(f-2);
```

MATLAB program 5.7

```
function newtonsmethod(func, x0, tolx, tolfx)
% Newton's method used to solve a nonlinear equation in x
% Input variables
% func : nonlinear function
% x0 : initial guess value
% tolx : tolerance for error in estimating root
% tolfx: tolerance for error in function value at solution

% Other variables
maxloops = 50;

[fx, fxderiv] = feval(func,x0);
fprintf('  i     x(i)     f(x(i))     f''(x(i))   \n');
% Iterative solution scheme
for i = 1:maxloops
    x1 = x0 - fx/fxderiv;
    [fx, fxderiv] = feval(func,x1);
    fprintf('%2d %5.4f %7.6f %7.6f \n',i,x1,fx,fxderiv);
    if (abs(x1 - x0) <=tolx && abs(fx) < tolfx)
        break % Jump out of the for loop
    end
    x0 = x1;
end
```

Box 5.3 Immobilized enzyme packed bed reactor

Enzymes are proteins in the body whose main functions are to catalyze a host of life-sustaining chemical reactions, such as breaking down food into its primary components, e.g. proteins into amino acids, converting glucose to CO_2 and H_2O, synthesizing molecular components of cells, enabling transcription and translation of genes, and so on. Enzymes have tremendous specificity for the substrate acted upon and the reaction catalyzed. The rate of reaction r for an enzyme-catalyzed reaction such as

$$E + S \rightarrow E + P$$

where E is the enzyme, S is the substrate, and P is the product, is commonly described using Michaelis–Menten kinetics:

$$R = -\frac{dS}{dt} = \frac{R_{max} S}{K_m + S}, \tag{5.29}$$

where

S is the substrate concentration,
R_{max} is the maximum reaction rate for a particular enzyme concentration E,
K_m is the Michealis constant, equal to the substrate concentration at which the reaction rate is half of the maximum rate.

Equation (5.29) can be integrated to yield an analytical relationship between the substrate concentration and time as follows:

$$K_m \ln\left(\frac{S_0}{S}\right) + (S_0 - S) = R_{max} t. \tag{5.30}$$

Enzymes can be produced (using genetic engineering techniques), purified, and then used *ex vivo* in a suitably designed contacting device to catalyze similar reactions when placed in a conducive physiological environment, such as at body temperature and favorable pH. Often enzymes are immobilized on a support structure within the contacting equipment rather than kept free in solution. This is advantageous for several reasons, such as (1) ease of separation of the contact stream from the enzyme (the exiting stream must be cleared of enzyme if it is introduced into the body) and (2) reduced loss of enzymes, which is especially important when the process is expensive.

In certain diseased conditions, the inefficient breakdown of a specific accumulating molecular entity can pose a threat to the body since abnormally high concentration levels of the non-decomposed substrate may be toxic to certain body tissues. In such cases the use of an immobilized enzyme reactor to catalyze the breakdown reaction *extracorporeally* may be of therapeutic benefit. There are several options available for the type of contacting device and contacting scheme used to carry out the reaction process. A packed bed is a commonly used reactor configuration in which the catalyst particles are dispersed on a support structure throughout the reactor and an aqueous stream containing the substrate continuously passes through the reactor (McCabe *et al.*, (2004) provides an introduction to packed bed columns).

A prototype immobilized enzyme packed bed reactor has been designed whose function is to degrade a metabolite X that is normally processed by the liver except in cases of liver disease or liver malfunction. Accumulation of X in the body prevents functioning of certain cell types in the body. Enzyme AZX is a patented enzyme that is immobilized within inert pellet particles contained in the reactor (Figure 5.12) and degrades X according to Michaelis–Menten kinetics (Equation (5.29)). A simple mass balance over the reactor based on certain simplifying assumptions produces the following design equation for the reactor (Fournier, 2007):

$$K_m \ln\left(\frac{S_{feed}}{S_{out}}\right) + K(S_{feed} - S_{out}) = \frac{\eta R_{max} K A_c (1 - \varepsilon) L}{Q}, \tag{5.31}$$

Figure 5.12

Immobilized enzyme packed bed reactor.

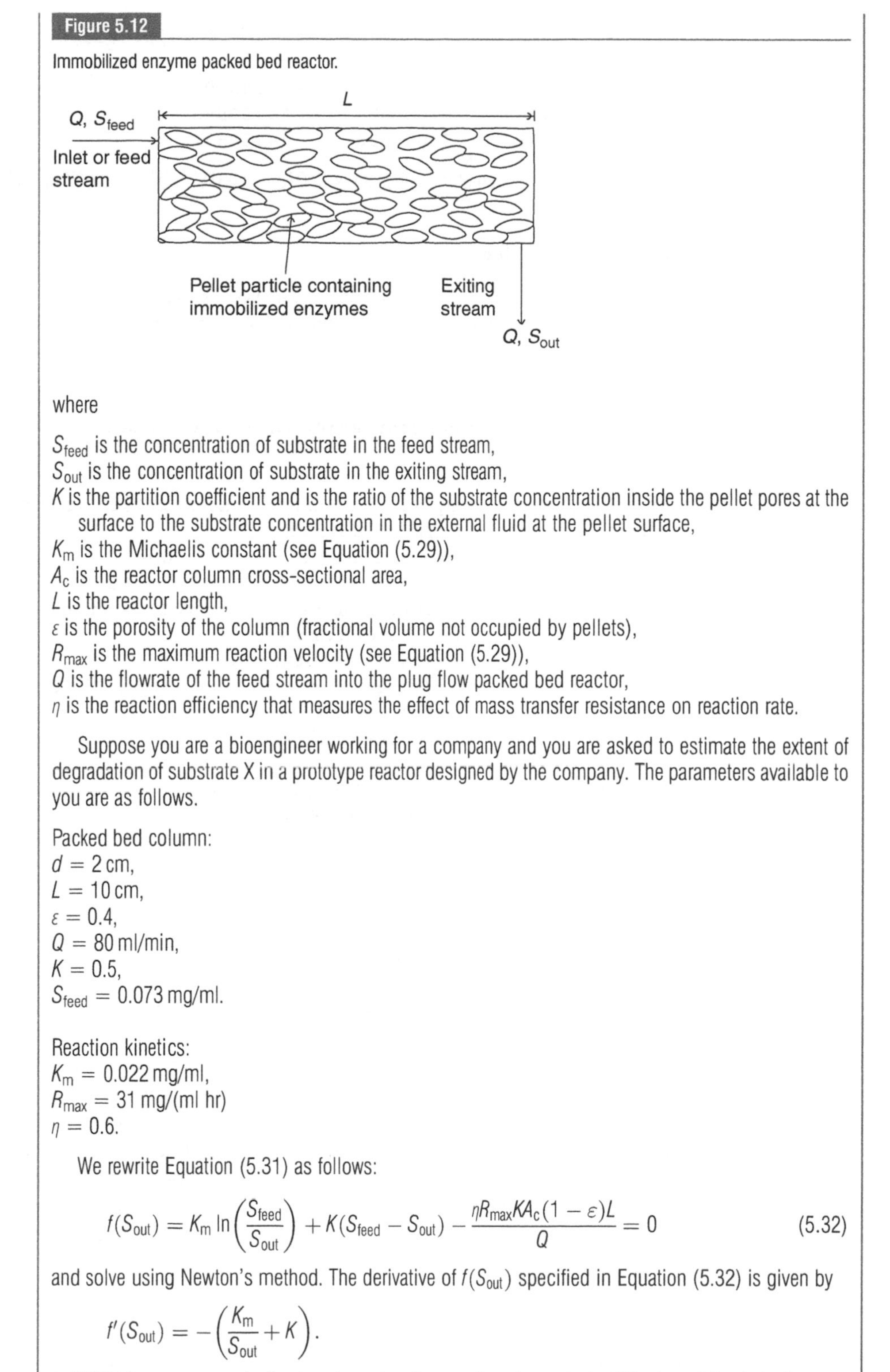

where

S_{feed} is the concentration of substrate in the feed stream,
S_{out} is the concentration of substrate in the exiting stream,
K is the partition coefficient and is the ratio of the substrate concentration inside the pellet pores at the surface to the substrate concentration in the external fluid at the pellet surface,
K_m is the Michaelis constant (see Equation (5.29)),
A_c is the reactor column cross-sectional area,
L is the reactor length,
ε is the porosity of the column (fractional volume not occupied by pellets),
R_{max} is the maximum reaction velocity (see Equation (5.29)),
Q is the flowrate of the feed stream into the plug flow packed bed reactor,
η is the reaction efficiency that measures the effect of mass transfer resistance on reaction rate.

Suppose you are a bioengineer working for a company and you are asked to estimate the extent of degradation of substrate X in a prototype reactor designed by the company. The parameters available to you are as follows.

Packed bed column:
$d = 2$ cm,
$L = 10$ cm,
$\varepsilon = 0.4$,
$Q = 80$ ml/min,
$K = 0.5$,
$S_{feed} = 0.073$ mg/ml.

Reaction kinetics:
$K_m = 0.022$ mg/ml,
$R_{max} = 31$ mg/(ml hr)
$\eta = 0.6$.

We rewrite Equation (5.31) as follows:

$$f(S_{out}) = K_m \ln\left(\frac{S_{feed}}{S_{out}}\right) + K(S_{feed} - S_{out}) - \frac{\eta R_{max} K A_c (1-\varepsilon) L}{Q} = 0 \tag{5.32}$$

and solve using Newton's method. The derivative of $f(S_{out})$ specified in Equation (5.32) is given by

$$f'(S_{out}) = -\left(\frac{K_m}{S_{out}} + K\right).$$

A MATLAB program packedbedreactor.m has been written to evaluate $f(S_{out})$ and its derivative at a supplied value of S_{out}. Here, we run our function `newtonsmethod` using relevant arguments and an initial guess of $S_{out} = 0.05$:

```
>> newtonsmethod('packedbedreactor',0.05,0.0001,0.001)
i      x(i)        f(x(i))      f'(x(i))
1    0.032239    0.001840    -1.182405
2    0.033795    0.000025    -1.150986
3    0.033816    0.000000    -1.150571
```

The extent of conversion of the substrate is (0.073 – 0.0338)/0.073 = 0.537 or 53.7%.

MATLAB program 5.8

```
function [f fderiv] = packedbedreactor(sout)
% Calculation of mass balance equation for immobilized enzyme fixed bed
% reactor

% Input Variable
% sout : substrate concentration in outlet stream

% Constants
sfeed = 0.073;     % mg/ml : substrate concentration at inlet
q = 80;            % ml/min : flowrate
dia = 2;           % cm : diameter of reactor tube
Ac = pi/4*dia^2;   % cross-sectional area of reactor
porosity = 0.4;    % fraction of unoccupied volume in reactor
l = 10;            % cm : length of reactor
K = 0.5;           % partition coefficient
vmax = 31/60;      % mg/ml/min : maximum reaction velocity
Km = 0.022;        % mg/ml : Michaelis constant
eff = 0.6;         % reaction efficiency

f = Km*log(sfeed/sout) + K*(sfeed - sout) - ...
  (1-porosity)*Ac*l*K*vmax*eff/q;
fderiv = -Km/sout - K;
```

5.6 Secant method

The secant method is a combination of two root approximation techniques – the regula-falsi method and Newton's method – and accordingly has a convergence rate that lies between linear and quadratic. You may have already noted that $f(x)$ in Equation (5.3) of Example 5.1 has a complicated and messy derivative. In many situations, the calculation of derivatives of a function may be prohibitively difficult or time-consuming. We seek an alternative to evaluating the analytical form of the derivative for such situations. In the secant method, the analytical form of the derivative of a function is substituted by an expression for the slope of the secant line connecting two points (x_0 and x_1) that lie on $f(x)$ close to the root, i.e.

$$f'(x) = \frac{f(x_1) - f(x_0)}{(x_1 - x_0)}. \tag{5.33}$$

If two initial estimates x_0 and x_1 of the root are specified, the x-intercept of the line that joins the points $(x_0, f(x_0))$ and $(x_1, f(x_1))$ provides the next improved approximation, x_2, of the root being sought. Equation (5.12), reproduced below, is

Figure 5.13

Illustration of the secant method.

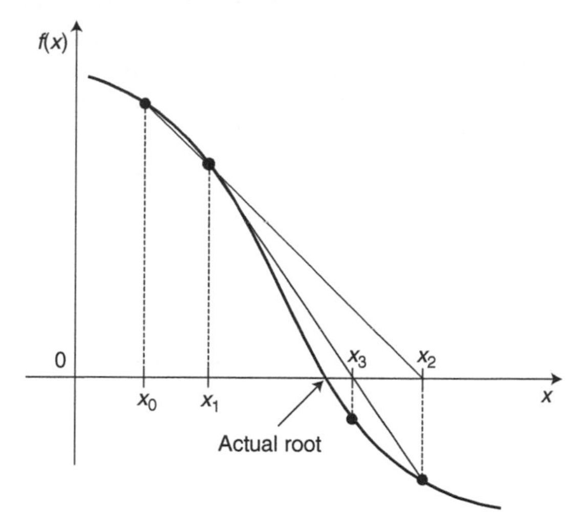

used to calculate the new estimate x_2 of the root based on the secant line connecting points x_0 and x_1:

$$x_2 = x_1 - \frac{f(x_1)(x_1 - x_0)}{f(x_1) - f(x_0)}. \tag{5.12}$$

The derivation of Equation (5.12) is described in Section 5.3. Substituting Equation (5.33) into Equation (5.19) also produces Equation (5.12). This procedure of root estimation is very similar to the regula-falsi method with the exception that the two initial estimates are not required to bracket the root. Therefore, the first two initial guesses x_0 and x_1 may be chosen such that they both lie on the same side of the x-axis with respect to the location of the root. Figure 5.13 illustrates graphically the algorithm prescribed by this method.

The next estimate x_3 of the root is obtained from the intersection of the secant line joining points $(x_2, f(x_2))$ and $(x_1, f(x_1))$ with the x-axis. Equation (5.12) can be generalized as follows:

Secant method

$$x_{i+1} = x_i - \frac{f(x_i)(x_i - x_{i-1})}{f(x_i) - f(x_{i-1})}. \tag{5.34}$$

Note that in this procedure an interval bracketing the root is not sought. At any iteration, the actual root may lie outside the interval contained by the two previous estimates that generate the secant line. Since the algorithm does not force successive approximations of the root to lie within a shrinking interval that encapsulates the root, divergence of the solution technique is possible for situations in which the calculated slope of the secant line approaches or is near 0.

A note of caution: if the function has either a large (steep) slope or a very shallow slope near the root, or if the points x_i and x_{i-1} are very close together, subtractive cancellation in the terms $(x_i - x_{i-1})$ and/or $f(x_i) - f(x_{i-1})$ can result in large round-off error due to loss of significance (see Chapter 1 for a discussion of round-off error). It is highly recommended that you monitor the values of x and $f(x)$ generated in the successive iterations to catch any inadvertent difficulties that may arise during convergence to the root. In such cases, you can choose an alternate method or a modification to the existing method to contain the errors. For example, if x_i and x_{i-1} tend to approach each other too closely, one may fix x_{i-1} and allow x_i to change to prevent significant round-off errors from corrupting the estimates.

The secant method uses the following algorithm to find a zero of $f(x)$.

(1) Determine two initial estimates x_0 and x_1 of the root of the function $f(x)$, which are not required to bracket the root.
(2) Calculate the next estimate of the root using the equation $x_{i+1} = x_i - \dfrac{f(x_i)(x_i - x_{i-1})}{f(x_i) - f(x_{i-1})}$.
(3) Repeat step 2 until the method has converged upon a solution.

The convergence rate for the secant method is given by Equation (5.35), and is superlinear for simple roots:

$$|\epsilon_{i+1}| \sim \left|\frac{f''(x^*)}{2f'(x^*)}\right|^{0.618} |\epsilon|^{1.618}. \tag{5.35}$$

Here, the order of convergence $r = \left(1 + \sqrt{5}\right)/2 = 1.618$ is the *golden ratio*, and

$$C = \left|\frac{f''(x^*)}{2f'(x^*)}\right|^{0.618}.$$

The order of convergence is linear for functions that have multiple roots.

5.7 Solving systems of nonlinear equations

Many engineering problems involve a system of nonlinear equations in two or more independent variables that must be solved simultaneously to yield a solution. Solving nonlinear systems of equations is more challenging than solving a single nonlinear equation or a system of linear equations. Iterative techniques can be employed for finding a solution to systems of nonlinear equations. As with all numerical root-finding techniques for nonlinear equations, the number of iterations to be carried out is not known a priori, convergence cannot be guaranteed unless certain pre-specified conditions are met, and multiple solutions may exist although the solution technique may not reveal all of them.

Newton's method, which was introduced in Section 5.5 to solve a single nonlinear equation, can be generalized to handle a system of two or more nonlinear equations. A system of n nonlinear equations in n independent variables can be written as follows:

$$\begin{aligned} f_1(x_1, x_2, \ldots, x_n) &= 0,\\ f_2(x_1, x_2, \ldots, x_n) &= 0,\\ &\vdots\\ f_n(x_1, x_2, \ldots, x_n) &= 0. \end{aligned}$$

If $\boldsymbol{x} = [x_1, x_2, \ldots, x_n]$, then this system of equations can be represented as follows:

Box 5.2B Equilibrium binding of multivalent ligand in solution to cell surfaces

This problem is re-solved using the secant method. A function in MATLAB was written called `secantmethod` that implements the algorithm described above to solve for the root of a nonlinear equation. To solve Equation (5.27), `secantmethod` calls the user-defined function `ReceptorLigandEqbmBinding` and begins the search for the root using two initial guesses $x_0 = 8000$ and $x_1 = 10\,400$ receptors:

```
>> secantmethod('ReceptorLigandEqbmBinding', 8000, 10400, 0.5, 0.5)
i     x(i-1)          x(i)          x(i+1)        f(x(i-1))       f(x(i))
1    8000.0000    10400.0000    10472.6067    -2609.378449    -76.622878
2   10400.0000    10472.6067    10470.7834      -76.622878      1.973693
3   10472.6067    10470.7834    10470.7850        1.973693     -0.001783
```

MATLAB program 5.9

```
function secantmethod(func, x0, x1, tolx, tolfx)
% Secant method used to solve a nonlinear equation in x

% Input variables
% func : nonlinear function
% x0 : first initial guess value
% x1 : second initial guess value
% tolx : tolerance for error in estimating root
% tolfx: tolerance for error in function value at solution

% Other variables
maxloops = 50;

fx0 = feval(func,x0);
fx1 = feval(func,x1);
fprintf(' i   x(i-1)   x(i)   x(i+1)   f(x(i-1))   f(x(i)) \n');

% Iterative solution scheme
for i = 1:maxloops
    x2 = x1 - fx1*(x1 - x0)/(fx1 - fx0);
    fx2 = feval(func,x2);
    fprintf('%2d %5.4f %5.4f %5.4f %7.6f %7.6f \n', ...
      i,x0,x1,x2, fx0,fx1);
    if (abs(x2 - x1) <=tolx && abs(fx2) < tolfx)
        break % Jump out of the for loop
    end
    x0 = x1;
    x1 = x2;
    fx0 = fx1;
    fx1 = fx2;
end
```

$$\boldsymbol{F}(\boldsymbol{x}) = \begin{bmatrix} f_1(x_1, x_2, \ldots, x_n) \\ f_2(x_1, x_2, \ldots, x_n) \\ \cdot \\ \cdot \\ \cdot \\ f_n(x_1, x_2, \ldots, x_n) \end{bmatrix} = \begin{bmatrix} 0 \\ 0 \\ \cdot \\ \cdot \\ \cdot \\ 0 \end{bmatrix}. \tag{5.36}$$

Let $\boldsymbol{x}^{(k)} = [x_1^{(k)}, x_2^{(k)}, \ldots, x_n^{(k)}]$ be the kth estimate for the root. The initial guess, which lies near the root, is $\boldsymbol{x}^{(0)} = \left[x_1^{(0)}, x_2^{(0)}, \ldots, x_n^{(0)}\right]$. Each function f can be expanded using the Taylor series about $\boldsymbol{x}^{(0)}$. If $\boldsymbol{x}^{(0)}$ is located close to the root, then the higher-order terms of $\Delta\boldsymbol{x}^{(0)}$ are much smaller than the zero- and first-order terms in the expansion and can be neglected as is done here:

$$f_1\left(x_1^{(0)} + \Delta x_1^{(0)}, \ldots, x_n^{(0)} + \Delta x_n^{(0)}\right)$$
$$\simeq f_1\left(x_1^{(0)}, \ldots, x_n^{(0)}\right) + \Delta x_1^{(0)} \left.\frac{\partial f_1}{\partial x_1}\right|_{x_1^{(0)},\ldots,x_n^{(0)}} + \Delta x_2^{(0)} \left.\frac{\partial f_1}{\partial x_2}\right|_{x_1^{(0)},\ldots,x_n^{(0)}} + \cdots$$
$$+ \Delta x_n^{(0)} \left.\frac{\partial f_1}{\partial x_n}\right|_{x_1^{(0)},\ldots,x_n^{(0)}}$$

$$\vdots$$

$$f_n\left(x_1^{(0)} + \Delta x_1^{(0)}, \ldots, x_n^{(0)} + \Delta x_n^{(0)}\right)$$
$$\simeq f_n\left(x_1^{(0)}, \ldots, x_n^{(0)}\right) + \Delta x_1^{(0)} \left.\frac{\partial f_n}{\partial x_1}\right|_{x_1^{(0)},\ldots,x_n^{(0)}} + \Delta x_2^{(0)} \left.\frac{\partial f_n}{\partial x_2}\right|_{x_1^{(0)},\ldots,x_n^{(0)}} + \cdots$$
$$+ \Delta x_n^{(0)} \left.\frac{\partial f_n}{\partial x_n}\right|_{x_1^{(0)},\ldots,x_n^{(0)}}.$$

We seek a solution $\boldsymbol{x} + \Delta\boldsymbol{x}$ to the system of equations such that $\boldsymbol{F}(\boldsymbol{x} + \Delta\boldsymbol{x}) = 0$. Therefore, if we set

$$f_1\left(x_1^{(0)} + \Delta x_1^{(0)}, \ldots, x_n^{(0)} + \Delta x_n^{(0)}\right) = 0$$

$$\vdots$$

$$f_n\left(x_1^{(0)} + \Delta x_1^{(0)}, \ldots, x_n^{(0)} + \Delta x_n^{(0)}\right) = 0,$$

we obtain the matrix equation

$$\boldsymbol{J}\left(\boldsymbol{x}^{(0)}\right)\Delta\boldsymbol{x}^{(0)} = -\boldsymbol{F}\left(\boldsymbol{x}^{(0)}\right). \tag{5.37}$$

Here,

$$\boldsymbol{J}(\boldsymbol{x}) = \begin{bmatrix} \frac{\partial f_1}{\partial x_1} & \frac{\partial f_1}{\partial x_2} & \cdots & \frac{\partial f_1}{\partial x_n} \\ & \vdots & & \\ \frac{\partial f_n}{\partial x_1} & \frac{\partial f_n}{\partial x_2} & \cdots & \frac{\partial f_n}{\partial x_n} \end{bmatrix}$$

is called the Jacobian of $\boldsymbol{F}$.

Equation (5.37) can be generalized as follows for any iteration k:

$$\boldsymbol{J}\left(\boldsymbol{x}^{(k)}\right)\Delta\boldsymbol{x}^{(k)} = -\boldsymbol{F}\left(\boldsymbol{x}^{(k)}\right), \tag{5.38}$$

where

$$\boldsymbol{x}^{(k+1)} = \boldsymbol{x}^{(k)} + \Delta\boldsymbol{x}^{(k)}. \tag{5.39}$$

Equations (5.38) and (5.39) highlight the two-step solution technique of Newton's method. Equation (5.38) forms a set of linear equations which are solved for $\Delta\boldsymbol{x}^{(k)}$. Once $\Delta\boldsymbol{x}^{(k)}$ is obtained, the next approximation of the root $\boldsymbol{x}^{(k+1)}$ is easily derived using Equation (5.39).

Convergence of Equations (5.38) and (5.39) will occur only if certain conditions similar to those discussed in Section 5.5 are met.

(1) All functions and their partial derivatives are continuous at and near the root.
(2) The Jacobian is non-singular, i.e. the determinant $|\boldsymbol{J}(\boldsymbol{x})| \neq 0$. (Refer to Section 2.2 for a discussion on singular matrices and calculation of the determinant.)
(3) The initial guess value is located close enough to the root.

The algorithm is summarized as follows.

(1) Provide an initial estimate $\boldsymbol{x}^{(0)}$ of the root of the multi-equation system $\boldsymbol{F}(\boldsymbol{x})$.
(2) Determine the analytical form of the Jacobian $\boldsymbol{J}(\boldsymbol{x})$.
(3) Calculate $\boldsymbol{F}(\boldsymbol{x})$ and $\boldsymbol{J}(\boldsymbol{x})$ at $\boldsymbol{x}^{(k)}$.
(4) Solve the equation $\boldsymbol{J}(\boldsymbol{x}^{(k)})\Delta\boldsymbol{x}^{(k)} = -\boldsymbol{F}(\boldsymbol{x}^{(k)})$ for $\Delta\boldsymbol{x}^{(k)}$.
(5) Calculate the new approximation of the root $\boldsymbol{x}^{(k+1)} = \boldsymbol{x}^{(k)} + \Delta\boldsymbol{x}^{(k)}$.
(6) Set $\boldsymbol{x}^{(k)} = \boldsymbol{x}^{(k+1)}$.
(7) Repeat (3)–(6) until the method has converged upon a solution.

Example 5.3

Let's consider an example problem in which we have two nonlinear equations in two variables:

$$f_1 = 3x_1^2 + x_2^2 - 10 = 0,$$
$$f_2 = x_1^2 + x_2 - 1 = 0.$$

This system of equations has two solutions (–1.5942, –1.5414) and (1.5942, –1.5414).[4] Note that this set of equations can easily be solved by hand to obtain the exact solution.

The Jacobian for this system of equations is

$$\boldsymbol{J} = \begin{bmatrix} \frac{\partial f_1}{\partial x_1} & \frac{\partial f_1}{\partial x_2} \\ \frac{\partial f_2}{\partial x_1} & \frac{\partial f_2}{\partial x_2} \end{bmatrix} = \begin{bmatrix} 6x_1 & 2x_2 \\ 2x_1 & 1 \end{bmatrix}.$$

Let our initial guess $\boldsymbol{x}^{(0)}$ be (1, –1), so

$$\boldsymbol{J}\left(\boldsymbol{x}^{(0)}\right) = \begin{bmatrix} 6 & -2 \\ 2 & 1 \end{bmatrix} \text{ and } \boldsymbol{F}\left(\boldsymbol{x}^{(0)}\right) = \begin{bmatrix} -6 \\ -1 \end{bmatrix}.$$

We solve the set of equations

$$\begin{bmatrix} 6 & -2 \\ 2 & 1 \end{bmatrix}\begin{bmatrix} \Delta x_1 \\ \Delta x_2 \end{bmatrix} = -\begin{bmatrix} -6 \\ -1 \end{bmatrix}$$

and obtain

[4] MATLAB graphics: One way of studying the behavior of functions in two variables involves drawing contour plots using the MATLAB `contour` function. The intersection of the contour lines for both functions at $f = 0$ provides information on the location of the root. Use of `contour` is illustrated in Chapter 8.

$$\Delta x_1^{(0)} = 0.8, \qquad \Delta x_2^{(0)} = -0.6.$$

Therefore,

$$x_1^{(1)} = x_1^{(0)} + \Delta x_1^{(0)} = 1.8,$$
$$x_2^{(1)} = x_2^{(0)} + \Delta x_2^{(0)} = -1.6.$$

With $\mathbf{x}^{(1)} = (1.8, -1.6)$, we evaluate

$$\mathbf{J}\left(\mathbf{x}^{(1)}\right) = \begin{bmatrix} 10.8 & -3.2 \\ 3.6 & 1 \end{bmatrix} \text{ and } \mathbf{F}\left(\mathbf{x}^{(1)}\right) = \begin{bmatrix} 2.28 \\ 0.64 \end{bmatrix}.$$

We solve the set of equations

$$\begin{bmatrix} 10.8 & -3.2 \\ 3.6 & 1 \end{bmatrix} \begin{bmatrix} \Delta x_1 \\ \Delta x_2 \end{bmatrix} = -\begin{bmatrix} 2.28 \\ 0.64 \end{bmatrix}$$

to obtain

$$\Delta x_1^{(1)} = -0.1939, \qquad \Delta x_2^{(1)} = 0.0581.$$

Therefore,

$$x_1^{(2)} = x_1^{(1)} + \Delta x_1^{(1)} = 1.6061,$$
$$x_2^{(2)} = x_2^{(1)} + \Delta x_2^{(1)} = -1.5419.$$

Evidently, convergence of the solution to the one of the two roots is imminent.

On the other hand, if we were to try to solve a similar set of equations,

$$f_1 = x_1^2 + x_2^2 - 4 = 0,$$
$$f_2 = x_1^2 + x_2 + 2 = 0,$$

the solution does not converge. As an exercise, verify that the Jacobian becomes singular at the root (0, –2).

A difficulty encountered with Newton's method is that convergence of the equations is contingent on the availability of good initial guesses, and this can be difficult to obtain for a system of multiple nonlinear equations. An algorithm is likely to converge if

(1) the algorithm moves closer to the solution with every iteration, i.e. $\left|\mathbf{F}\left(\mathbf{x}^{(k+1)}\right)\right| < \left|\mathbf{F}\left(\mathbf{x}^{(k)}\right)\right|$, and
(2) the correction to the estimate of the root does not become increasingly large, i.e. $\left|\mathbf{x}^{(k+1)} - \mathbf{x}^{(k)}\right| < \delta$, where δ is some upper limiting value.

An alternate approach to using linear methods to determine $\Delta\mathbf{x}^{(k)}$ is to solve a nonlinear optimization problem in which we seek $\Delta\mathbf{x}^{(k)}$ values that minimize

$$\left|\mathbf{J}\left(\mathbf{x}^{(k)}\right)\Delta\mathbf{x}^{(k)} + \mathbf{F}\left(\mathbf{x}^{(k)}\right)\right|$$

subject to $\Delta\mathbf{x}^{(k)} < \delta$. This method can be used even if $\mathbf{J}$ becomes singular. The algorithm proceeds as follows.

(1) Determine $\Delta\mathbf{x}^{(k)}$ using a nonlinear optimization technique to minimize $\left|\mathbf{J}\left(\mathbf{x}^{(k)}\right)\Delta\mathbf{x}^{(k)} + \mathbf{F}\left(\mathbf{x}^{(k)}\right)\right|$.
(2) If $\left|\mathbf{F}\left(\mathbf{x}^{(k+1)}\right)\right| > \left|\mathbf{F}\left(\mathbf{x}^{(k)}\right)\right|$, then reduce δ and return to step 1.
(3) If $\mathbf{F}\left(\mathbf{x}^{(k+1)}\right) \sim 0$, then terminate algorithm or else return to step 1 to start the next iteration.

Box 5.4 The Fahraeus–Lindquist effect

A well-known phenomenon that results from the tendency of RBCs to concentrate in the core region of a blood vessel is the decrease in apparent blood viscosity as blood vessel diameter is reduced. This phenomenon is called the **Fahraeus–Lindquist** effect and occurs due to the changing proportion of the width of the cell-free plasma layer with respect to the core RBC region. As the vessel diameter decreases, the thickness of the cell-free plasma layer δ (viscosity $\mu = 1.2$ cP) increases. Since the viscosity of this outer RBC-depleted layer is much lower than that of the RBC-rich core region, the overall tube viscosity exhibits a dependence on the vessel diameter. If μ_F is the viscosity of blood in a feed vessel (upstream vessel) of large diameter (e.g. of an artery in which $\delta \sim 0$ (Fournier, 2007)), then the viscosity ratio $(\mu_{apparent}/\mu_F) < 1.0$, where $\mu_{apparent}$ is the apparent viscosity in a blood vessel of smaller size. Equation (5.40) relates this viscosity ratio to the hematocrit (volume fraction of RBCs in blood) in the feed vessel and the core region of the vessel of interest:

$$\frac{\mu_{apparent}}{\mu_F} = \frac{1 - \alpha_F H_F}{1 - \sigma^4 \alpha_{core} H_{core}}, \tag{5.40}$$

where

H_{core} is the core region hematocrit,
H_F is the feed vessel hematocrit,
$\sigma = 1 - \delta/R$,
δ = thickness of the plasma layer,
R = radius of blood vessel,
$\alpha = 0.070 \exp\left(2.49H + \frac{1107}{T} \exp(-1.69H)\right)$, and
T is the temperature (K).

If H_F (hematocrit of feed or large blood vessel) = 0.44, $R = 15$ μm, and $T = 310$ K are the values provided to you, and experimental data (Gaehtgens, 1980; Gaehtgens *et al.*, 1980) also provide $\mu_{apparent}/\mu_F = 0.69$, use Equation (5.40) in conjunction with Equation (5.3), repeated here

$$H_{core} = H_F\left[1 + \frac{(1-\sigma^2)^2}{\sigma^2[2(1-\sigma^2) + \sigma^2(1 - \alpha_{core}H_{core})]}\right], \tag{5.3}$$

to determine the core region hematocrit H_{core} and the thickness of the plasma layer δ for this particular blood vessel diameter (Fournier, 2007).

We have two nonlinear equations in two independent variables: H_{core} and σ. First, Equations (5.3) and (5.40) must be rewritten in the form $f(H_{core}, \sigma) = 0$; i.e.

$$f_1(H_{core}, \sigma) = 1 + \frac{(1-\sigma^2)^2}{\sigma^2[2(1-\sigma^2) + \sigma^2(1 - \alpha_{core}H_{core})]} - \frac{H_{core}}{H_F} = 0 \tag{5.41}$$

and

$$f_2(H_{core}, \sigma) = \frac{1 - \alpha_F H_F}{1 - \sigma^4 \alpha_{core} H_{core}} - \frac{\mu_{apparent}}{\mu_F} = 0. \tag{5.42}$$

Our next task is to determine the Jacobian for this problem. We have

$$\frac{\partial f_1}{\partial H_{core}} = \frac{(1-\sigma^2)^2 \alpha_{core}}{[2(1-\sigma^2) + \sigma^2(1 - \alpha_{core}H_{core})]^2} \times \left[1 + H_{core}\left\{2.49 - \frac{1870.83}{T} e^{-1.69H_{core}}\right\}\right] - \frac{1}{H_F};$$

$$\frac{\partial f_1}{\partial \sigma} = \frac{4(\sigma^2 - 1)}{\sigma[2(1-\sigma^2) + \sigma^2(1 - \alpha_{\text{core}}H_{\text{core}})]} - \frac{4(1-\sigma^2)^2[1 - \sigma^2(1 - \alpha_{\text{core}}H_{\text{core}})]}{\sigma^3[2(1-\sigma^2) + \sigma^2(1 - \alpha_{\text{core}}H_{\text{core}})]^2};$$

$$\frac{\partial f_2}{\partial H_{\text{core}}} = \frac{\sigma^4(1 - \alpha_{\text{F}}H_{\text{F}})\alpha_{\text{core}}}{[1 - \sigma^4 \alpha_{\text{core}}H_{\text{core}}]^2} \cdot \left[1 + H_{\text{core}}\left\{2.49 - \frac{1870.83}{T}e^{-1.69H_{\text{core}}}\right\}\right];$$

$$\frac{\partial f_2}{\partial \sigma} = \frac{4\sigma^3(1 - \alpha_{\text{F}}H_{\text{F}})\alpha_{\text{core}}H_{\text{core}}}{[1 - \sigma^4 \alpha_{\text{core}}H_{\text{core}}]^2}.$$

In this problem the Jacobian is defined as follows:

$$\boldsymbol{J} = \begin{bmatrix} \dfrac{\partial f_1}{\partial H_{\text{core}}} & \dfrac{\partial f_1}{\partial \sigma} \\ \dfrac{\partial f_2}{\partial H_{\text{core}}} & \dfrac{\partial f_2}{\partial \sigma} \end{bmatrix}. \tag{5.43}$$

A MATLAB function program FahraeusLindquistExample.m was created to calculate the values of f_1 and f_2 (Equations (5.41) and (5.42)) as well as the Jacobian 2×2 matrix defined in Equation (5.43) for the supplied values of H_{core} and σ. Another MATLAB function program called gen_newtonmethod2.m was constructed to use the generalized Newton's method (Equations (5.38) and (5.39)) to solve a system of two nonlinear equations in two variables.

Using MATLAB

(1) At every iteration, we solve the linear system of equations $\boldsymbol{J}(\boldsymbol{x}^{(k)})\Delta\boldsymbol{x}^{(k)} = -\boldsymbol{F}(\boldsymbol{x}^{(k)})$ to determine $\Delta\boldsymbol{x}^{(k)}$. This is equivalent to solving the linear system of equations $\boldsymbol{Ax} = \boldsymbol{b}$. MATLAB includes built-in techniques to solve a system of linear equations $\boldsymbol{Ax} = \boldsymbol{b}$. The syntax: `x = A\b` (which uses the backslash operator) tells MATLAB to find the solution to the linear problem $\boldsymbol{Ax} = \boldsymbol{b}$, which reads as `b` divide by `A`. Note that MATLAB does not compute the inverse of `A` to solve a system of simultaneous linear equations (see Chapter 2 for further details).

(2) Because we use MATLAB to solve a system of simultaneous linear equations, `x` and `b` must be specified as column vectors. Note that `b` should have the same number of rows as `A` in order to perform the operation defined as `x = A\b`.

(3) Here, when using the specified tolerance values to check if the numerical algorithm has reached convergence, we enforce that the solution for all independent variables and the resulting function values meet the tolerance criteria. The relational operator $\leq$ checks whether all `dx` values and `fx` values are less than or equal to the user-defined tolerance specifications. Because we are using vectors rather than scalar values as operands, we must use the element-wise logical operator `&`, instead of the short-circuit operator `&&` (see MATLAB help for more information). When using the `&` operator, the `if` statement will pass control to the logical expression contained in the `if - end` statement *only if* all of the elements in the `if` statement argument are non-zero (i.e. true).

MATLAB program 5.10

```
function [f, J] = Fahraeus LindquistExample(x)
% Calculation of f1 and f2 and Jacobian matrix for Box 5.4

% Input Variables
hcore = x(1); % Core region hematocrit
sigma = x(2); % 1 - delta/R
```

```
% Output Variables
f = zeros(2,1); % function column vector [f1;f2] (should not be a row vector)
J = zeros(2,2); % Jacobian 2x2 matrix

% Constants
hfeed = 0.44;       % hematocrit of the feed vessel or large artery
temp = 310;         % kelvin : temperature of system
viscratio = 0.69;   % ratio of apparent viscosity in small vessel
                    % to that of feed vessel

% Calculating variables
sigma2 = sigma^2;
sigma3 = sigma^3;
sigma4 = sigma^4;
alphafeed = 0.070*exp(2.49*hfeed + (1107/temp)*exp(-1.69*hfeed));
alphacore = 0.070*exp(2.49*hcore + (1107/temp)*exp(-1.69*hcore));
f1denominator = 2*(1-sigma2)+sigma2*(1-alphacore*hcore);
f2denominator = 1 - sigma4*alphacore*hcore;

% Evaluating the function f1 and f2
% Equation 5.41
f(1) = 1+ ((1-sigma2)^2)/sigma2/f1denominator - hcore/hfeed;

% Equation 5.42
f(2) = (1 - alphafeed*hfeed)/f2denominator - viscratio;

% Evaluating the Jacobian Matrix
% J(1,1) = df1/dhcore
J(1,1) = ((1-sigma2)^2)*alphacore;
J(1,1) = J(1,1)/(f1denominator^2);
J(1,1) = J(1,1)*...
    (1 + hcore*(2.49 - 1870.83/temp*exp(-1.69*hcore))) - 1/hfeed;

% J(1,2) = df1/dsigma
J(1,2) = 4*(sigma2 - 1)/sigma/f1denominator;
J(1,2) = J(1,2)-4*(1-sigma2)^2*(1-sigma2*(1-alphacore*hcore))/...
  sigma3/f1denominator^2;

% J(2,1) = df2/dhcore
J(2,1) = sigma4*(1-alphafeed*hfeed)*alphacore/f2denominator^2;
J(2,1) = J(2,1)*(1 + hcore*(2.49 - 1870.83/temp*exp(-1.69*hcore)));

% J(2,2) = df2/dsigma
J(2,2) = 4*sigma3*(1-alphafeed*hfeed)*alphacore*hcore;
J(2,2) = J(2,2)/f2denominator^2;
```

MATLAB program 5.11

```
function gen_newtonsmethod2(func, x, tolx, tolfx)
% Generalized Newton's method to solve a system of two nonlinear
% equations in two variables
```

```
% Input variables
% func  : function that evaluates the system of nonlinear equations and
          % the Jacobian matrix
% x     : initial guess values of the independent variables
% tolx  : tolerance for error in the solution
% tolfx : tolerance for error in function value

% Other variables
maxloops = 20;

[fx, J] = feval(func,x);
fprintf('  i     x1(i+1)     x2(i+1)     f1(x(i))     f2(x(i))     \n');
% Iterative solution scheme
for i = 1:maxloops
    dx = J \ (-fx);
    x = x + dx;
    [fx, J] = feval(func,x);
    fprintf('%2d  %7.6f  %7.6f  %7.6f  %7.6f  \n', ...
      i, x(1), x(2), fx(1), fx(2));
    if (abs(dx) <=tolx & abs(fx) < tolfx)
       % not use of element-wise AND operator
       break % Jump out of the for loop
    end
end
```

We run the user-defined function `gen_newtonsmethod2` and specify the appropriate arguments to solve our nonlinear problem in two variables. We use starting guesses of $H_{\text{core}} = 0.44$ and $\delta = 3$ μm or $\sigma = 0.8$:

```
>> gen_newtonsmethod2('FahraeusLindquistExample', [0.44; 0.8],
   0.002, 0.002)
i      x1(i+1)      x2(i+1)      f1(x(i))      f2(x(i))
1      0.446066     0.876631      0.069204      0.020425
2      0.504801     0.837611     -0.006155      0.000113
3      0.500263     0.839328      0.001047     -0.000001
4      0.501049     0.839016     -0.000181     -0.000000
```

The solution obtained is $H_{\text{core}} = 0.501$ and $\sigma = 0.839$ or $\delta = 2.41$ μm.

5.8 MATLAB function `fzero`

Thus far, we solved nonlinear equations using only one technique at a time. However, if we solve for a root of a nonlinear equation by combining two or more techniques, we can, for example, achieve the coupling of the robustness of a bracketing method of root finding with the convergence speed of an open-interval method. Such combinations of methods of root finding are termed **hybrid methods** and are much more powerful than using any single method alone. The MATLAB function `fzero` is an example of using a hybrid method to find roots of a nonlinear equation. The `fzero` function combines the reliable bisection method with faster convergence methods: the secant method and inverse quadratic interpolation. The inverse quadratic interpolation method is similar to the secant method, but instead of using two endpoints to produce a secant line as an approximation of the function near the root, three data points are used to approximate the function as a quadratic curve. The point of intersection of the quadratic with the x-axis provides the next estimate of the root. When the secant method or the inverse quadratic interpolation method produces a diverging solution, the bisection method is employed. Note that, if the function has a discontinuity, `fzero` will not be able to distinguish a sign change at a discontinuity from a zero point and may report the point of discontinuity as a zero.

The basic syntax for the `fzero` function is

```
x = fzero(func, x0)
```

Here `x` is the numerically determined root, `func` is the name of the function whose zero is required, and `x0` can be one of two things: an initial guess value or a bracketing interval. If `x0` is a scalar value, then `fzero` will automatically search for the nearest bracketing interval and then proceed to find a zero within that interval. If `x0` is a vector containing two values that specify a bracketing interval, the function values evaluated at the interval endpoints must lie on opposite sides of the x-axis, otherwise an error will be generated. Note that `fzero` cannot locate points where the function touches the x-axis (no change in sign of the function) but does not cross it.

The function whose zero is sought can either be specified as an m-file function or an inline function. Examples of m-file functions are supplied throughout this chapter. Such functions are defined using the `function` keyword at the beginning of an m-file, and the m-file is saved with the same name as that of the function. An `inline` function is constructed using `inline` and providing a string expression of the function as input. For example, entering the following at the command line produces an `inline` function object `f`:

```
>> f = inline('2*x - 3')
f =
     Inline function:
     f(x) = 2*x - 3
```

`f` is now a function that can be evaluated at any point `x`. For example,

```
>> f(2)
```

produces the output

```
ans =
     1
```

Let's use `fzero` to determine a root of the equation $x^3 - 6x^2 + 9x - 4 = 0$ (Equation (5.4)). This equation has three roots, a single root: $x = 4$ and a double root at $x = 1$.

We first specify only a single value as the initial guess:

```
>> fzero(inline('x^3 - 6*x^2 + 9*x - 4'),5)
ans =
     4
```

Now we specify a bracketing interval:

```
>> fzero(inline('x^3 - 6*x^2 + 9*x - 4'), [3 5])
ans =
     4.0000
```

Let's try to find the double root:

```
>> fzero(inline('x^3 - 6*x^2 + 9*x - 4'),0)
ans =
    4
```

The root $x = 1$ could not be found.

Let's specify the interval of search as [0 2] (note that there is no sign change in this interval):

```
>> fzero(inline('x^3 - 6*x^2 + 9*x - 4'), [0 2])
??? Error using ==> fzero at 292
The function values at the interval endpoints must differ in sign.
```

On the other hand, if we specify the interval [−1 1], we obtain

```
>> fzero(inline('x^3 - 6*x^2 + 9*x - 4'), [-1 1])
ans =
    1
```

We have found the double root! Since `fzero` cannot locate double roots, can you guess how this is possible?

Other ways of calling the `fzero` function can be studied by typing in the statement `help fzero` in the Command Window.

5.9 End of Chapter 5: key points to consider

(1) Nonlinear equations of the form $f(x) = 0$ that are not amenable to analytical solution can be solved using numerical iterative procedures such as
the bisection method,
the regula-falsi method,
fixed-point iteration,
Newton's method,
the secant method.

(2) The iterative schemes listed above can be broadly stated in two categories.
  (a) **Bracketing methods**, in which the search for a root of an equation is constrained within a specified interval that encapsulates a sign change of the function, e.g. the bisection method and the regula-falsi method.
  (b) **Open interval methods**, in which one or two guesses are specified initially; however, the search for the root may proceed in any direction and any distance away from the initally supplied values, e.g. fixed-point iteration, Newton's method and the secant method.

(3) It is critical for the success of these numerical root-finding methods that $f(x)$ and/or its derivatives be continuous within the interval of interest in which the root lies.

(4) A comparison of nonlinear root-finding methods is outlined in Table 5.2.

(5) A generalized Newton's method can be applied to solve nonlinear multi-equation systems. The limitations of Newton's method for solving systems of nonlinear equations are similar to those faced when solving a nonlinear equation in one variable.

(6) **Hybrid methods** of solution combine two or more numerical schemes of root finding and are more powerful compared to individual methods of root finding. A hybrid method can marshal the strengths of two or more root-finding techniques to effectively minimize the limitations imposed by any of the individual numerical techniques.

Table 5.2. *Nonlinear root-finding methods*

Property	Bisection method	Regula-falsi method	Newton's method	Secant method	Fixed-point iteration
Type of method	bracketing	bracketing	open interval	open interval	open interval
Rate of convergence	$r = 1$ $C = 0.5$	$r = 1$	$r = 2$ for simple roots $r = 1$ for multiple roots; if Equation (4.25) is used then $r = 2$ $C = \left\lvert \frac{f''(x^*)}{2f'(x^*)} \right\rvert$	$r = 1.618$ $C = \left\lvert \frac{f''(x^*)}{2f'(x^*)} \right\rvert^{0.618}$	$r = 1$ $C = \lvert g'(x) \rvert$
Number of initial guesses required	2	2	1	2	1
Is convergence guaranteed as $i \to \infty$?	yes	yes	no	no	no
Other limitations	cannot be used to find a double root	cannot be used to find a double root	proximity of guess value to root to ensure convergence not known a priori	proximity of guess value to root to ensure convergence not known a priori	Convergence observed only if $\lvert g'(x) \rvert < 1$

(7) The MATLAB-defined function `fzero` is a hybrid method that uses the bisection method, secant method, and the inverse quadratic interpolation method to solve for a zero of a function, i.e. wherever a sign change in the function occurs (excluding singularities).

5.10 Problems

5.1. **Packed bed column** A column packed with spherical particles provides high surface area geometry that is useful in isolating specific protein(s) or other biological molecules from a cell lysate mixture. The Ergun equation relates the pressure drop through a packed bed of spheres to various fluid and geometric parameters of the bed:

$$\frac{\Delta p}{l} = 150\frac{(1-\varepsilon)^2}{\varepsilon^3}\frac{\mu u}{d_{\mathrm{p}}^2} + 1.75\frac{(1-\varepsilon)}{\varepsilon^3}\frac{\rho u^2}{d_{\mathrm{p}}},$$

where Δp is the pressure drop, l is the length of the column, ε is the porosity, μ is the fluid viscosity, u is the fluid velocity, d_{p} is the diameter of the spherical particles, and

ρ is the fluid density. For a 20 cm column, packed with 1 mm spheres and perfused with a buffer of equal viscosity and density to water ($\mu = 0.01$ P, $\rho = 1$ g/cm^3), use the *bisection method* to determine the column porosity if the pressure drop is measured to be 810.5 dyn/cm^2 for a fluid flowing with velocity $u = 0.75$ cm/s. Make sure that you use consistent units throughout your calculation.

Using a starting interval of $0.1 < \varepsilon < 0.9$, report the number of iterations necessary to determine the porosity to within 0.01 (tolerance).

5.2. **Blood rheology** Blood behaves as a non-Newtonian fluid, and can be modeled as a "Casson fluid." This model predicts that unlike simple fluids such as water, blood will flow through a tube such that the central core will move as a plug with little deformation, and most of the velocity gradient will occur near the vessel wall. The following equation is used to describe the plug flow of a Casson fluid:

$$F(\xi) = 1 - \frac{16}{7}\sqrt{\xi} + \frac{4}{3}\xi - \frac{1}{21}\xi^4,$$

where F measures the reduction in flowrate (relative to a Newtonian fluid) experienced by the Casson fluid for a given pressure gradient and ξ gives an indication of what fraction of the tube is filled with plug flow. For a value of $F = 0.40$, use the *secant method* to determine the corresponding value of ξ. Use a starting guess of ξ = 0.25 (and a second starting guess of 0.26, since the secant method requires two initial guesses). Show that the rate of convergence is superlinear and approximately equal to $r = 1.618$. Note that the function is not defined when $\xi \leq 0$. Therefore, one must choose the starting guess values carefully.

5.3. **Saturation of oxygen in water** Oxygen can dissolve in water up to a saturation concentration that is dependent upon the temperature of water. The saturation concentration of dissolved oxygen in water can be calculated using the following empirical equation:

$$\ln s_{\text{ow}} = -139.34411 + \frac{1.575701 \times 10^5}{T_\text{a}} - \frac{6.642308 \times 10^7}{T_\text{a}^2} + \frac{1.243800 \times 10^{10}}{T_\text{a}^3} - \frac{8.621949 \times 10^{11}}{T_\text{a}^4},$$

where s_{ow} = the saturation concentration of dissolved oxygen in water at 1 atm (mg/l) and T_a = absolute temperature (K). Remember that $T_\text{a} = T_\text{c} + 273.15$, where T_c = temperature (°C). According to this equation, saturation decreases with increasing temperature. For typical water samples near ambient temperature, this equation can be used to determine oxygen concentration ranges from 14.621 mg/l at 0 °C to 6.949 mg/l at 35 °C. Given a value of oxygen concentration, this formula and the *bisection method* can be used to solve for the temperature in degrees Celsius.

(a) If the initial guesses are set as 0 and 35 °C, how many bisection iterations would be required to determine temperature to an absolute error of 0.05 °C?

(b) Based on (a), develop and test a bisection m-file function to determine T as a function of a given oxygen concentration. Test your function for $s_{\text{ow}} = 8$, 10, and 14 mg/l. Check your results.

5.4. **Fluid permeability in biogels** The specific hydraulic permeability k relates the pressure gradient to the fluid velocity in a porous medium such as agarose gel or extracellular matrix. For non-cylindrical pores comprising a fiber matrix, k is given by

$$k = \frac{r_\text{f}^2 \varepsilon^3}{20(1-\varepsilon)^2},$$

where r_f is the fiber radius and $0 \leq \varepsilon \leq 1$ is the porosity (Saltzman, 2001). For a fiber radius of 10 nm and a measurement of $k = 0.4655\ \text{nm}^2$, determine the porosity using both (i) the *regula-falsi* and (ii) the *secant methods*. Calculate the first three iterations using both methods for an initial interval of $0.1 \leq \varepsilon \leq 0.5$ for regula-falsi, and initial guesses of $\varepsilon_{-1} = 0.5$ and $\varepsilon_0 = 0.55$ for the secant method. What are the relative errors in the solutions obtained at the third iteration?

5.5. **Thermal regulation of the body** Stolwijk (1980) developed a model for thermal regulation of the human body, where the body is represented as a spherical segment (head), five cylindrical segments (trunk, arms, hands, legs, and feet), and the central blood compartment. An important mechanism of heat loss is the respiratory system. To approximate the thermal energy contained in the breath, we use the following mathematical model that relates the pressure of water vapor in the expired breath p_{H_2O} (mm Hg) to temperature T, (°C) of the inhaled air:

$$p_{H_2O} = \exp\left[9.214 - \frac{1049.8}{1.985(32 + 1.8T)}\right].$$

If the water vapor in the expired breath is measured to be 0.298 mm Hg, then use *Newton's method* of nonlinear root finding to determine the inhaled gas temperature. Use a starting guess of 15 °C. Confirm your answer using the *bisection method.* Take (5 °C, 20 °C) as your initial interval.

5.6. **Osteoporosis in Chinese women** Wu *et al.* (2003) studied the variations in age-related speed of sound (SOS) at the tibia and prevalence of osteoporosis in native Chinese women. They obtained the following relationship between the SOS and the age in years, Y:

$$SOS = 3383 + 39.9Y - 0.78Y^2 + 0.0039Y^3,$$

where the SOS is expressed in units of m/s. The SOS for one research subject is measured to be 3850 m/s; use *Newton's method* of nonlinear root finding to determine her most likely age. Take $Y = 45$ years as your initial guess.

5.7. **Biomechanics of jumping** The effort exerted by a person while jumping is dependent on several factors that include the weight of the body and the time spent pushing the body off the ground. The following equation describes the force F (measured in pound-force, lb_f) exerted by a person weighing 160 lb and performing a jump from a stationary position. The time spent launching the body upwards, or the "duration of lift-off," is denoted by τ (in milliseconds):

$$F(t) = 480 \cdot \sin\left(\frac{\pi t}{\tau}\right) + 160(1 - t/\tau).$$

For $\tau = 180$ ms, use the *regula-falsi method* of nonlinear root finding to determine the time at which the force is equal to 480 lb_f. Use $1 < t < 75$ as your initial interval.

5.8. **Drug concentration profile** The following equation describes the steady state concentration profile of a drug in the tissue surrounding a spherical polymer implant:

$$\frac{C_t}{C_i} = \frac{1}{\zeta}\exp[-\phi(\zeta - 1)],$$

where $C_t(\zeta)$ is the concentration in the tissue, C_i is the concentration at the surface of the implant,

$$\phi = R\sqrt{k^*/D^*},$$

$\zeta = r/R$ is the dimensionless radial position, R is the radius of the implant, k^* is the apparent first-order elimination constant, and D^* is the apparent diffusivity of the drug (Saltzman, 2001). For parameter values of $R = 0.032$ cm, $k^* = 1.9 \times 10^{-4}$/s, and $D^* = 4 \times 10^{-7}$ cm^2/s (typical values for experimental implants used to treat brain tumors), use the *secant method* to solve for the dimensionless radial position, where the concentration of drug is equal to half that at the implant surface. Determine the first two guess values by first plotting the function. Why will the fixed-point iterative method not work here?

5.9. Using Equations (5.10) and (5.13), show that, for the fixed-point iterative method,

$$\lim_{n\to\infty} C = |g'(x^*)|.$$

5.10. **Animal-on-a-chip** Box 2.1 discussed the concept of "animal-on-a-chip" or the microscale cell culture analog (μCCA) invented by Shuler and co-workers at Cornell University as cost-effective *in vitro* drug-screening method. The transport and metabolism of naphthalene was investigated within the μCCA assuming steady state conditions. In our problem formulation discussed in Box 2.1, it was assumed that the consumption rate of naphthalene in the lung and liver followed zero-order kinetics, so that we could obtain a set of linear equations. Now we drop the assumption of zero-order kinetics. Substituting the process parameters listed in Box 2.1 into Equations (2.1) and (2.2), we have the following.

Lung compartment

$$780(1 - R) + R\left(0.5C_{\text{liver}} + 1.5C_{\text{lung}}\right) - \frac{8.75C_{\text{lung}}}{2.1 + C_{\text{lung}}}0.08 - 2C_{\text{lung}} = 0.$$

Liver compartment

$$0.5C_{\text{lung}} - \frac{118C_{\text{liver}}}{7.0 + C_{\text{liver}}}0.322 - 0.5C_{\text{liver}} = 0.$$

(a) For recycle fractions $R = 0.6$, 0.8, and 0.9, determine the steady state naphthalene concentrations in the lung and liver compartments using the *multi-equation Newton's method.* Use as the initial guess values for C_{lung} and C_{liver} the solution obtained from Problem 2.5, in which you solved for the naphthalene concentrations using linear solution methods, and zero-order kinetics was assumed for naphthalene consumption. These are provided below:

$$R = 0.6{:}\ \ C_{\text{lung}} = 360.6\,\mu\text{M},\ C_{\text{liver}} = 284.6\,\mu\text{M};$$
$$R = 0.8{:}\ \ C_{\text{lung}} = 312.25\,\mu\text{M},\ C_{\text{liver}} = 236.25\,\mu\text{M};$$
$$R = 0.9{:}\ \ C_{\text{lung}} = 215.5\,\mu\text{M},\ C_{\text{liver}} = 139.5\,\mu\text{M}.$$

(b) Determine the steady state naphthalene concentrations in the lung and liver compartments for $R = 0.95$. Can you use the solution from Problem 2.5 for $R = 0.9$ as initial guess values? Why or why not? If not, find other values to use as initial guess values. Do you get a solution at $R = 0.95$ using the multi-equation Newton's method?

(c) Compare your solutions obtained in this problem with that obtained in Problem 2.5 (listed above) for the three recycle ratios $R = 0.6$, 0.8, and 0.9. Why are the steady state naphthalene concentrations higher when zero-order naphthalene consumption rate is not assumed? Find the relative error in the Problem 2.5 (linear) solutions listed above. What is the trend in relative error with R? Can you explain this trend?

References

Burden, R. L. and Faires, J. D. (2005) *Numerical Analysis* (Belmont, CA: Thomson Brooks/Cole).

Fournier, R. L. (2007) *Basic Transport Phenomena in Biomedical Engineering* (New York: Taylor & Francis).

Gaehtgens, P. (1980) Flow of Blood through Narrow Capillaries: Rheological Mechanisms Determining Capillary Hematocrit and Apparent Viscosity. *Biorheology*, **17**, 183–9.

Gaehtgens, P., Duhrssen, C., and Albrecht, K. H. (1980) Motion, Deformation, and Interaction of Blood Cells and Plasma During Flow through Narrow Capillary Tubes. *Blood Cells*, **6**, 799–817.

Lauffenburger, D. A. and Linderman, J. J. (1993) *Receptors: Models for Binding, Trafficking and Signaling* (New York: Oxford University Press).

McCabe, W., Smith, J., and Harriott, P. (2004) *Unit Operations of Chemical Engineering* (New York: McGraw-Hill).

Mody, N. A. and King, M. R. (2008) Platelet Adhesive Dynamics. Part II: High Shear-Induced Transient Aggregation Via GPIbalpha-vWF-GPIbalpha Bridging. *Biophys. J.*, **95**, 2556–74.

Perelson, A. S. (1981) Receptor Clustering on a Cell-Surface. 3. Theory of Receptor Cross-Linking by Multivalent Ligands – Description by Ligand States. *Math. Biosci.*, **53**, 1–39.

Saltzman, W. M. (2001) *Drug Delivery: Engineering Principles for Drug Therapy* (New York: Oxford University Press).

Stolwijk, J. A. (1980) Mathematical Models of Thermal Regulation. *Ann. NY Acad. Sci.*, **335**, 98–106.

Wu, X. P., Liao, E. Y., Luo, X. H., Dai, R. C., Zhang, H., and Peng, J. (2003) Age-Related Variation in Quantitative Ultrasound at the Tibia and Prevalence of Osteoporosis in Native Chinese Women. *Br. J. Radiol.*, **76**, 605–10.

6 Numerical quadrature

6.1 Introduction

Engineering and scientific calculations often require the evaluation of integrals. To calculate the total amount of a quantity (e.g. mass, number of cells, or brightness of an image obtained from a microscope/camera) that continuously varies with respect to an independent variable such as location or time, one must perform integration. The concentration or rate of the quantity (e.g. concentration of substrate or drug, intensity of light passing through a microscope, position-dependent density of a material) is integrated with respect to the independent variable, which is called the **integration variable**. Integration is the inverse of differentiation. A derivative measures the rate of change of a quantity, e.g. the speed of travel $v(t)$ is obtained by measuring the change in distance s with respect to time elapsed, or $v(t) = ds/dt$. A definite integral I, on the other hand, is a multiplicative operation and calculates the product of a function with an infinitesimal increment of the independent variable, summed continuously over an interval, e.g. the distance traveled is obtained by integrating speed over an interval of time, or $s = \int_0^{t_f} v(t)dt$. The integration process seeks to find the function $g(x)$ when its derivative $g'(x)$ is provided. Therefore, an integral is also called an anti-derivative, and is defined below.

$$I = g(x) = \int_a^b g'(x)dx = \int_a^b f(x)dx,$$

where $f(x)$ is called the **integrand**, x is the variable of integration such that $a \leq x \leq b$, and a and b are the **limits of integration**. If the integrand is a well-behaved, continuous function, i.e. it does not become infinite or exhibit discontinuities within the bounds of integration, an analytical solution to the integral may exist. An analytical solution is the preferred method of solution because it is exact, and the analytical expression can lend insight into the process or system being investigated. The effects of modifying any of the process parameters are straightforward and intuitive.

However, not all integrals are amenable to an analytical solution. Analytical solutions to integrals do not exist for certain integrand functions or groups of functions, i.e. the anti-derivative of some functions are not defined. Or, it may be impossible to evaluate an integral analytically because the function becomes singular (blows up) at some point within the interval of integration, or the function is not continuous. Even if an analytical solution does exist, it may happen that the solution is too time-consuming to find or difficult to compute. Sometimes, the functional form of the integrand is unknown and the only available information is a discrete data set consisting of the value of the function at certain values of x, the integration variable. When any of the above situations are encountered, we use numerical methods of integration to obtain the value of the definite integral. The use of numerical methods to solve a definite integral is called **numerical quadrature**.

Figure 6.1

Graphical interpretation of the process of integration.

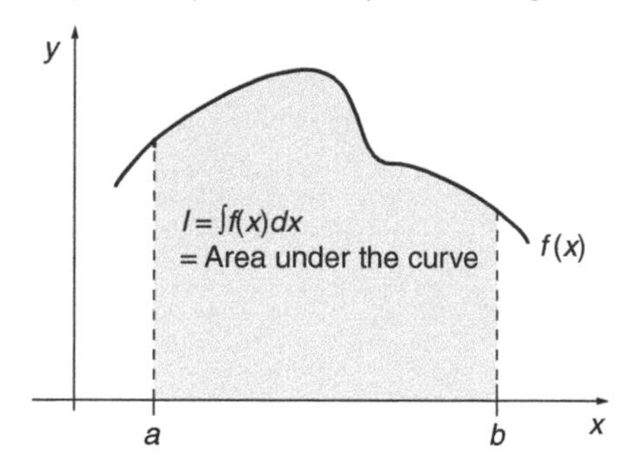

A graphical description best illustrates the process of integration. The definite integral represents an area that lies under a curve described by the function $f(x)$, and which is bounded on either side by the limits of integration. This is shown in Figure 6.1. The result of evaluating a definite integral is a number which has units equal to $f \cdot x$, and is a unique solution.

Suppose we have values of the function $f(x_i)$ $(i \in 0, 1, \ldots, n)$ specified at $n + 1$ points, i.e. at $x_0, x_1, \ldots, x_n$. Numerical quadrature proposes an approximate solution to the integral of interest according to the formula given below:

$$\int_a^b f(x)dx \approx \sum_{i=0}^{n} w_i f(x_i). \tag{6.1}$$

The known function values $f(x_i)$ are multiplied by suitable weights and are summed to yield the numerical approximation. Thc weights that are used in the formula depend on the quadrature method used.

In any numerical method of integration, the functional form of the integrand is replaced with a simpler function called an **interpolant** or **interpolating function**. The latter function is integrated in place of the original function, and the result yields the numerical solution. The interpolating function fits the data exactly at specific or known data points and interpolates the data at intermediate values. Specifically, the interpolant passes through either (1) the supplied data points if only a discrete data set is available, or (2) the function evaluated at discrete points called **nodes** within the interval, if an analytical form of the integrand is available. The interpolating function approximates the behavior of $f(x)$ as a function of x.

Suppose we are given $n + 1$ function values $f(x_i)$ at $n + 1$ equally spaced values, $x_i = a + ih$, within the interval $a \leq x \leq b$. Accordingly, $x_0 = a$ and $x_n = b$. The entire interval $[a, b]$ is now divided into n subintervals $[x_i,\ x_{i+1}]$. The width of each subinterval is $h = x_{i+1} - x_i$. Within each subinterval $x_i \leq x \leq x_{i+1}$, suppose the function $f(x)$ is approximated by a zeroth-degree polynomial, which is a constant. Let the value of the constant in any subinterval be equal to the function evaluated at the midpoint of the subinterval, $f(x_{i+0.5})$. Over any particular subinterval, the integral can be approximated as

$$\int_{x_i}^{x_{i+1}} f(x)dx = f(x_{i+0.5}) \int_{x_i}^{x_{i+1}} dx = f(x_{i+0.5})h.$$

Figure 6.2

The composite midpoint rule.

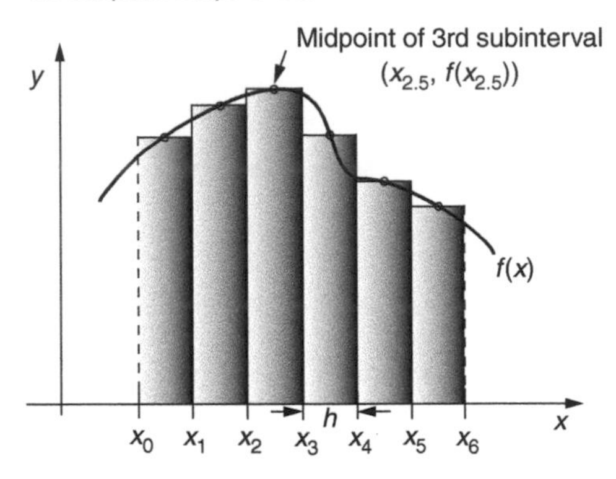

Since

$$\int_a^b f(x)dx = \int_{x_0}^{x_1} f(x)dx + \int_{x_1}^{x_2} f(x)dx + \cdots + \int_{x_{n-1}}^{x_n} f(x)dx,$$

we obtain the numerical formula for the integral

$$\int_a^b f(x)dx = h\sum_{i=0}^{n-1} f(x_{i+0.5}). \tag{6.2}$$

Equation (6.2) is called the **composite midpoint rule** and is demonstrated in Figure 6.2. The midpoint rule falls under the **Newton–Cotes rules of integration** which are discussed in Section 6.3. The term "composite" is used because the midpoint rule is evaluated multiple times over sequential segments of the interval. The composite midpoint rule substitutes $f(x)$ with a piecewise constant function. Note that as we increase the number of subintervals, the accuracy in approximating $f(x)$ will improve. As $n \to \infty$, our numerical result will become exact (limited by round-off error). However, if we were to find the function value at the midpoint $(a + b)/2$ of the interval $[a, b]$ and simply multiply that by the width of the interval $b - a$, then our numerical approximation is likely to be quite inaccurate.

In this chapter, we focus on the use of **polynomial interpolating functions** to approximate the true function or integrand. The interpolating polynomial $p(x)$ is constructed so that it coincides with the original function at the $n + 1$ node points; $p(x)$ approximates the behavior of $f(x)$ at all points between the known values. A polynomial has the distinct advantage that it can be easily integrated. It is therefore popularly used to develop numerical integration schemes. The entire interval of integration is divided into several subintervals. The data are approximated by a **piecewise polynomial interpolating function**. In piecewise interpolation, a different interpolating function is constructed for each subinterval. Each subinterval is then integrated separately, i.e. each polynomial function is integrated only over the range of x values or the subinterval on which it is defined. The numerical formula for integration, the order of the numerical method, and the magnitude of truncation error depend on the degree of the interpolating polynomial.

Function values may be evaluated at neither of the endpoints, one or both endpoints of the interval, and/or multiple points inside the interval. Sometimes, the function value at the endpoint of an interval may become infinite, or endpoint function values may not be available in the data set. Accordingly, two schemes of numerical integration have been developed: **open integration** and **closed integration**. The midpoint rule and Gaussian quadrature are examples of open integration methods, in which evaluation of both endpoint function values of the interval is not required to obtain the integral approximation. The trapezoidal rule and Simpson's rule use function values at both endpoints of the integration interval, and are examples of closed integration formulas.

Box 6.1A Solute transport through a porous membrane

Solute transport through a porous membrane separating two isolated compartments occurs in a variety of settings such as (1) diffusion of hydrophilic solutes (e.g. protein molecules, glucose, nucleic acids, and certain drug compounds) found in the extracellular matrix into a cell through aqueous pores in the cell membrane, (2) movement of solutes present in blood into the extravascular space through slit pores between adjacent endothelial cells in the capillary wall, (3) membrane separation to separate particles of different size or ultrafiltration operations to purify water, and (4) chromatographic separation processes. Solute particles travel through the pores within the porous layer by diffusion down a concentration gradient and/or by convective transport due to pressure-driven flow. The pore shape can be straight or convoluted. Pores may interconnect with each other, or have dead-ends. When the solute particle size is of the same order of magnitude as the pore radius, particles experience considerable resistance to entry into and movement within the pore. Frictional resistance, steric effects, and electrostatic repulsive forces can slow down or even preclude solute transport through the pores, depending on the properties and size of the solute particles and the pores. Since a porous membrane acts as a selective barrier to diffusion or permeation, the movement of solute through pores of comparable size is called **hindered transport**.

Mathematical modeling of the solute transport rate through a porous membrane is of great utility in (1) developing physiological models of capillary (blood vessel) filtration, (2) studying drug transport in the body, and (3) designing bioseparation and protein purification methods. The simplest model of hydrodynamic (pressure-driven) and/or diffusive (concentration gradient-driven) flow of solute particles through pores is that of a rigid sphere flowing in Stokes flow through a cylindrical pore. The hindered transport model discussed in this Box was developed by Anderson and Quinn (1974), Brenner and Gaydos (1977), and Smith and Deen (1983). A review is provided by Deen (1987).

Consider the steady state transport of a solid, rigid sphere of radius a flowing through a cylindrical pore of radius R_0 and length L. The pore walls are impermeable and non-reactive. The sphere has a time-averaged uniform velocity U in the direction of flow along length L (see Figure 6.3). The time-averaging smoothens out the fluctuations in velocity due to the Brownian motion of the particle. Each cylindrical pore in the porous membrane connects a liquid reservoir with solute concentration C_0 on one side of the membrane to another liquid reservoir of concentration C_L that lies on the opposite side of the membrane ($C_0 > C_L$). Let $C(r, z)$ be the concentration (number of sphere centers per unit volume) of the solute in the pore. Particle motion in the pore is due to convective fluid flow as well as diffusion down a concentration gradient.

The radius of the particle is made dimensionless with respect to the pore radius, $\alpha = a/R_0$. The dimensionless radial position of the particle in the pore is given by $\beta = r/R_0$.

Figure 6.3

Solute transport through a cylindrical pore.

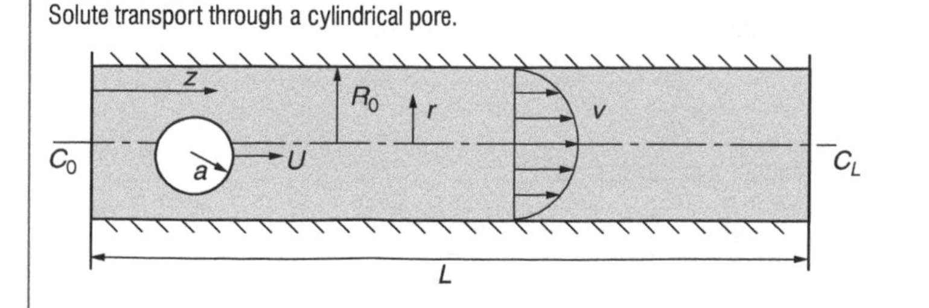

The flow within the pore is fully developed and the flow profile is parabolic, i.e. the fluid velocity is a function of radial position only:

$$v(\beta) = v_0\left(1 - \beta^2\right).$$

Fluid flow within the pores is at very low Reynolds number (Stokes flow or viscous flow). A freely flowing suspended sphere in a bounded medium has a particle flow velocity equal to Gv, where $G < 1$ and is called the "lag coefficient." Note that G characterizes the reduction in particle velocity with respect to fluid velocity due to resistance posed by the pore wall. Due to the concentration gradient (chemical potential gradient), which is the driving force for diffusion, on average, the solute particles are transported faster across the pore than the fluid itself. Since the sphere moves at an average velocity U that is larger than Gv, it is not freely suspended in the fluid, and its expedient motion relative to the fluid is retarded by wall friction. The drag force experienced by the sphere in Stokes flow is $6\pi\mu aK(U - Gv)$, where K is the "enhanced drag" (Anderson and Quinn, 1974).

At steady state, the drag forces that act to oppose motion of the sphere relative to the fluid can be equated to the chemical driving forces that induce motion of the sphere relative to the fluid. On doing so, we arrive at the equation for solute flux N (quantity transported/cross-sectional area/time).

$$N(r) = -DK^{-1}\frac{\partial C}{\partial z} + GvC.$$

At steady state, the flux is only a function of radial position and does not depend on axial position z. The average flux $\bar{N}$ is obtained by averaging $N(r)$ over the pore cross-sectional area. Solute concentration C in the pore is a function of radial and axial position:

$$C = g(z)\exp[-E(\beta)/kT], \tag{6.3}$$

where E is the potential energy of interaction between the sphere and the pore wall, k is the Boltzmann constant, and T is the temperature. An example of interaction between the sphere and the pore wall is **electrostatic repulsion/attraction**, which is observed when the surfaces have a net charge. Another example is **steric interaction**, which arises due to the finite size a of the solute particles. The annular region adjacent to the pore wall, defined as $1 - a < r \leq 1$, is devoid of the centers of moving spheres. In other words, the concentration of sphere centers in this region is zero. There is no solute flux at $r > 1 - a$.

We simplify the analysis by considering only diffusional transport of the solute, i.e. fluid velocity $v = 0$. The average flux is as follows:

$$\bar{N} = \frac{\int_0^{1-a} N(2\pi\beta)d\beta}{\int_0^1 (2\pi\beta)d\beta} = -2D\frac{dg}{dz}\int_0^{1-a} K^{-1}(\beta)\exp[-E(\beta)/kT]\beta\, d\beta. \tag{6.4}$$

The concentration $\bar{C}(z)$ averaged over the cross-sectional area is obtained by integrating Equation (6.3):

$$\frac{1}{\pi \cdot 1^2}\int_0^{1-a} C(2\pi\beta)d\beta = \bar{C} = g\int_0^{1-a}\exp[-E(\beta)/kT](2\beta)d\beta.$$

Taking the derivative with respect to z, we obtain an expression for $\partial g/\partial z$:

$$\frac{dg}{dz} = \frac{d\bar{C}/dz}{\int_0^{1-a}\exp[-E(\beta)/kT](2\beta)d\beta}. \tag{6.5}$$

Substituting this relationship into Equation (6.4), we get

$$\bar{N} = -DK_{\mathrm{d}}\frac{d\bar{C}}{dz},$$

where (Brenner and Gaydos, 1977)

$$K_{\mathrm{d}} = \overline{K^{-1}} = \frac{\int_0^{1-a} K^{-1}\exp[-E(\beta)/kT]\beta\, d\beta}{\int_0^{1-a}\exp[-E(\beta)/kT]\beta\, d\beta}. \tag{6.6}$$

During steady state diffusion, the cross-sectional average concentration $\bar{C}$ is a linear function of the distance traveled along the pore,

$$\bar{C} = \bar{C}_0 - \frac{\bar{C}_0 - \bar{C}_L}{L} \cdot z,$$

and

$$\frac{d\bar{C}}{dz} = -\frac{\bar{C}_0 - \bar{C}_L}{L}.$$

The bulk solute concentration immediately outside the pore at $z = 0, L$ is not equal to the average solute concentration $\bar{C}(z)$ just inside the pore. This is because steric and electrostatic interactions influence the distribution of solute within the pore such that the concentration within the pore may be less or more than the external concentration. If only steric interactions are present, a lower concentration of solute particles will be present in the pore at $z = 0$ than outside the pore. In this case, the concentration profile in the pore at $z = 0$ is given by

$$C(\beta)|_{z=0} = \begin{cases} C_0 & \beta \leq 1 - a \\ 0 & \beta > 1 - a. \end{cases}$$

The **partition coefficient** Φ is the ratio of the average internal pore concentration to the external concentration of solute in the bulk solution *at equilibrium*:

$$\Phi = \frac{\bar{C}_0}{C_0} = \frac{\int_0^{1-a} C_0 \cdot \exp[-E(\beta)/kT] 2\pi\beta \, d\beta}{C_0(\pi \cdot 1^2)} = 2\int_0^{1-a} \exp[-E(\beta)/kT]\beta \, d\beta. \tag{6.7}$$

The concentration gradient in the pore can be expressed in terms of the outside pore solute concentrations and the partition coefficient:

$$\frac{d\bar{C}}{dz} = -\Phi \frac{C_0 - C_L}{L}.$$

Thus, the steady-state cross-sectional average solute flux is given by

$$\bar{N} = \Phi D K_d \left(\frac{C_0 - C_L}{L}\right).$$

Smith and Deen (1983) calculated the interaction energy $E(\beta)$ between a sphere and the pore wall, both with constant surface charge density, as a function of the sphere position in the pore. The solvent was assumed to be a 1 : 1 univalent electrolyte (e.g. NaCl, KCl, NaOH, or KBr solution).

When a charged surface (e.g. negatively charged) is placed in an electrolytic solution, the counter-ions (positive ions) in the solution distribute such that they exist at a relatively higher concentration near the charged surface due to electrostatic attraction. A "double layer" of charge develops on each charged surface, and consists of the surface charge layer and the accumulated counter-charges adjacent to the surface. The thickness of the double layer depends on the concentration of the ions in the solution and the valency of each ion species in the solution. The counter-ions that assemble near the charged surface tend to screen the electrostatic effects of the charged surface. As the ionic strength of the solution decreases (concentration of salt is reduced) the counter-ion screening of the surface charges weakens and the double layer width increases.

The **Debye length** $(1/\kappa)$ can be viewed as the thickness of the electric double layer. It is strictly a property of the solvent. A large magnitude indicates that the charges on the surface are not well shielded by counter-ions present in the solution. On the other hand, a narrow double layer results when a large concentration of counter-ions adjacent to the surface charge shields the charges on the surface effectively. As the Debye length increases, the double layers of two closely situated charged surfaces interact more strongly with each other.

The interaction energy $E(\beta)$ in Equation (6.3) is equal to[1] (Smith and Deen, 1983)

$$E(\beta) = R_0 \varepsilon \varepsilon_0 \left(\frac{RT}{F}\right)^2 \Delta G, \tag{6.8}$$

where ΔG is the dimensionless change in free energy that occurs when the sphere and pore are brought from infinite separation to their interacting position.

The physical constants are as follows:

F = Faraday constant = 96485.34 C/mol,
R = gas constant = 8.314 J/(mol K)
k = Boltzmann constant = 1.38×10^{-23} J/(molecule/ K),
ε_0 = dielectric permittivity of free space = 8.8542×10^{-12} C/(V m),
ε = relative permittivity, i.e. the ratio of solvent dielectric permittivity to permittivity in a vacuum.

For constant surface charge densities q_c at the pore wall, and q_s at the sphere surface, ΔG for the system is given by (Smith and Deen, 1983)

$$\Delta G = \frac{\frac{8\pi\tau a^4 e^{\tau a}}{(1+\tau a)^2} \cdot \Lambda \sigma_s^2 + \frac{4\pi^2 a^2 I_0(\tau\beta)}{(1+\tau a) I_1(\tau)} \cdot \sigma_s \sigma_c + \left[\frac{\pi I_0(\tau\beta)}{\tau I_1(\tau)}\right]^2 \cdot \left[\frac{(e^{\tau a} - e^{-\tau a})\tau a L(\tau a)}{(1+\tau a)}\right] \cdot \sigma_c^2}{\pi\tau e^{-\tau a} - \frac{2(e^{\tau a} - e^{-\tau a})\tau a L(\tau a) \Lambda}{(1+\tau a)}}, \tag{6.9}$$

where

$\sigma_s = \frac{F R_0 q_s}{\varepsilon \varepsilon_0 RT}$ is the dimensionless sphere surface charge density,
$\sigma_c = \frac{F R_0 q_c}{\varepsilon \varepsilon_0 RT}$ is the dimensionless cylinder surface charge density,
I_ν and K_ν are the modified Bessel functions of order ν,
$L(\tau a) = \coth \tau a - (1/\tau a)$,
$\Lambda \cong \frac{\pi}{2} I_0(\tau\beta) \sum_{t=0}^{\infty} \left[\frac{\beta^t (2t)!}{2^{3t}(t!)^2} I_t(\tau\beta) \cdot [\tau K_{t+1}(2\tau) + 0.75 K_t(2\tau)]\right]$,
$\tau = R_0 \kappa = R_0 \left[\frac{F^2}{\varepsilon \varepsilon_0 RT} \sum_i z_i^2 C_{\text{electrolyte},\infty}\right]^{1/2}$, is the ratio of the pore radius R_0 to the Debye length $1/\kappa$,
$C_{\text{electrolyte},\infty}$ = bulk concentration of *electrolyte*,
z_i = valency of ion species ($i = 1$ for both the positive and negative ions that constitute the 1 : 1 univalent electrolyte).

The parameters for this system are:

$q_s = 0.01$ C/m^2,
$q_c = 0.01$ C/m^2,
$\varepsilon = 74$ for a dilute 1 : 1 univalent aqueous solution,
$T = 310$ K,
$C_{\text{electrolyte},\infty} = 0.01$ M,
$R_0 = 20$ nm,
$a = 0.25$.

You are asked to calculate the partition coefficient Φ for the solute. To perform this task, one must use numerical integration (Smith and Deen, 1983). The solution is demonstrated in Section 6.3 using the trapezoidal rule and Simpson's 1/3 rule.

[1] The model makes the following assumptions.

(1) The solute concentration is sufficiently dilute such that no solute–solute electrostatic and hydrodynamic interactions occur. Therefore, we can model the motion of one sphere independently from the motion of other spheres.

Section 6.3 discusses the Newton–Cotes rules of integration, which include popular methods such as the trapezoidal rule, Simpson's 1/3 rule, and Simpson's 3/8 rule. The composite Newton–Cotes rules approximate the functional form of the integrand with a piecewise polynomial interpolant. The interval is divided into n subintervals of equal width. The truncation errors associated with the quadrature algorithms are also derived. A convenient algorithm to improve the accuracy of the trapezoidal rule by several orders of magnitude is Romberg integration, and is considered in Section 6.4. Gauss–Legendre quadrature is a high accuracy method of quadrature for integrating functions. The nodes at which the integrand function is evaluated are the zeros of special orthogonal polynomials called Legendre polynomials. Section 6.5 covers the topic of Gaussian quadrature. Before we discuss some of the established numerical methods of integration, we start by introducing some important concepts in polynomial interpolation and specifically the Lagrange method of interpolation. The latter is used to develop the Newton–Cotes rules for numerical integration.

6.2 Polynomial interpolation

Many scientific and industrial experiments produce discrete data sets, i.e. tabulations of some measured value of a material or system property that is a function of a controllable variable (the independent variable). Steam tables, which list the thermodynamic properties of steam as a function of temperature and pressure, are an example of a discrete data set. To obtain steam properties at any temperature T and/ or pressure P intermediate to those listed in the published tables, i.e. at $T \neq T_i$, one must interpolate tabulated values that bracket the desired value. Before we can perform interpolation, we must define a function that adequately describes the behavior of the steam property data subset. In other words, to interpolate the data points, we must first draw a smooth and continuous curve that joins the data points. Often, a simple method like linear interpolation suffices when the adjacent data points that bracket the desired data point are close to each other in value.

> **Interpolation** is a technique to estimate the value of a function $f(x)$ at some value x, when the only information available is a discrete data set that tabulates $f(x_i)$ at several values of x_i, such that $a \leq x_i \leq b$, $a<x<b$, and $x \neq x_i$.

Interpolation can be contrasted with **extrapolation**. The latter describes the process of predicting the value of a function when x lies outside the interval over which the measurements have been taken. Extrapolated values are predictions made using an interpolation function and *should be regarded with caution.*

To perform interpolation, a curve, called an **interpolating function**, is fitted to the data such that *the function passes through each data point*. An interpolating function (also called an **interpolant**) is used for curve fitting of data when the data points arc accurately known. If the data are accompanied by a good deal of uncertainty, there is little merit in forcing the function to pass through each data point. When the data are error-prone, and there are more data points than undetermined equation parameters

(2) The pore length is sufficiently long to justifiably neglect entrance and exit effects on the velocity profile and to allow the particle to sample all possible radial positions within the pore.
(3) The electrical potentials are sufficiently small for the Debye–Hückel form of the Poisson–Boltzmann equation to apply (Smith and Deen, 1983).

(the unknowns in the interpolant), linear least-squares regression techniques are used to generate a function that models the trend in the data (see Chapter 2 for a discussion on linear least-squares regression).

Sometimes, a complicated functional dependency of a physical or material property on an independent variable x is already established. However, evaluation of the function $f(x)$ may be difficult because of the complexity of the functional form and/or number of terms in the equation. In these cases, we search for a simple function that exactly agrees with the true, complex function at certain points within the defined interval of x values. The techniques of interpolation can be used to obtain a simple interpolating function, in place of a complicated function, that can be used quickly and easily.

Interpolating functions can be constructed using polynomials, trigonometric functions, or exponential functions. We concentrate our efforts on polynomial interpolation functions. Polynomials are a class of functions that are commonly used to approximate a function. There are a number of advantages to using polynomial functions to create interpolants: (1) arithmetic manipulation of polynomials (addition, subtraction, multiplication, and division) is straightforward and easy to follow, and (2) polynomials are easy to differentiate and integrate, and the result of either procedure is still a polynomial. For these reasons, polynomial interpolants have been used to develop a number of popular numerical integration schemes.

A generalized polynomial in x of degree n contains $n+1$ coefficients, and is as follows:

$$p(x) = c_n x^n + c_{n-1}x^{n-1} + c_{n-2}x^{n-2} + \cdots + c_1 x + c_0. \tag{6.10}$$

To construct a polynomial of degree n, one must specify $n + 1$ points (x, y) that lie on the path of $p(x)$. The $n + 1$ points are used to determine the values of the $n + 1$ unknown coefficients. For example, to generate a line, one needs to specify only two distinct points that lie on the line. To generate a quadratic function, three different points on the curve must be specified.

> Only one polynomial of degree n exists that passes through any unique set of $n + 1$ points. In other words, an interpolating polynomial function of degree n that is fitted to $n + 1$ data points is unique.

The above statement can be proven as follows. Suppose two different polynomials $p(x)$ and $q(x)$, both of degree n, pass through the same $n + 1$ points. If $r(x) = p(x) - q(x)$, then $r(x)$ must also be a polynomial of degree either n or less. Now, the difference $p(x) - q(x)$ is exactly equal to zero at $n + 1$ points. This means that $r(x) = 0$ has $n + 1$ roots. A polynomial of degree n has exactly n roots (some or all of which can be complex). Then, $r(x)$ must be a polynomial of degree $n + 1$. This contradicts our supposition that both $p(x)$ and $q(x)$ are of degree n. Thus, two different polynomials of degree n cannot pass through the same $n + 1$ points. Only one polynomial of degree n exists that passes through all $n + 1$ points.

There are two ways to create an interpolating polynomial function.

(1) **Polynomial interpolation** Fit a polynomial of degree n to all $n + 1$ data points.
(2) **Piecewise polynomial interpolation** Fit polynomials of a lower degree to subsets of the data.[2] For example, to fit a polynomial to ten data points, a cubic polynomial

[2] The MATLAB `plot` function connects adjacent data points using piecewise linear interpolation.

Figure 6.4

Piecewise polynomial interpolation versus high-degree polynomial interpolation.

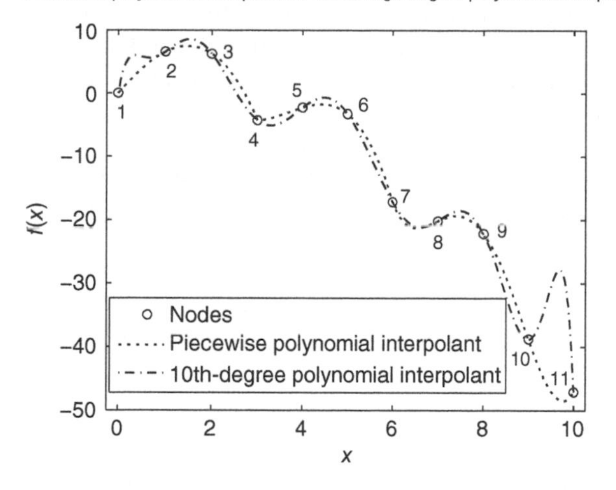

can be fitted to three sets of four consecutive points. Three cubic polynomials are constructed for the three data subsets: (x_1, x_2, x_3, x_4), (x_4, x_5, x_6, x_7), and (x_7, x_8, x_9, x_{10}). Thus, the interpolating domain $x_1 \leq x \leq x_{10}$ is divided into three subintervals, and interpolating polynomials of degree 3 are constructed for each of these subintervals.

When the number of data points in the data set is greater than five, piecewise polynomial interpolation is preferred over creating a single interpolating polynomial function that passes through each point. Fitting a polynomial of increasingly higher degrees to $n > 5$ data points is strongly discouraged. An interpolating polynomial will pass through all the points or nodes used to construct the polynomial function. However, in between the nodes, a high-degree polynomial function may exhibit pronounced oscillatory behavior. In other words, high-degree polynomial interpolants may produce intermediate values that are far removed from the true path that connects the data points. This capricious behavior of higher-degree polynomials is called **polynomial wiggle**.

Figure 6.4 shows 11 data points. A tenth-degree polynomial function is constructed using all 11 points. A piecewise polynomial function is also constructed in which the first four points (1–4) are fitted with a cubic polynomial, the next set of four points (4–7) is fitted with a cubic polynomial, while the last set of five points (7–11) is fitted with a fourth-degree polynomial. Note that the cubic functions smoothly connect each point without straying away from the direction (slope) of the line connecting two adjacent data points. The high degree polynomial function shows pronounced oscillations between data points. Between the last two data points, the function shows a large peak that is absent in the data.

If you examine Figure 6.4, you will notice that by choosing non-overlapping subsets of data points and interpolating with low-order polynomials, slope continuity of the functions may not be guaranteed at the endpoints of each subset. In this example, discontinuity of the interpolating polynomial function occurs at the fourth and seventh data points. A special polynomial interpolating function called a **cubic spline** can be used to ensure slope continuity at nodes which join two different polynomial interpolants. To learn how to construct cubic splines for data interpolation, the reader is referred to Recktenwald (2000). The `spline` function provided by the MATLAB software calculates the piecewise polynomial cubic spline interpolant of the user-supplied data.

An interpolating function approximates the function values that lie between the known data points. The error in approximation depends on (1) the true functional dependency of the property of interest on the independent variable, (2) the degree of the interpolating polynomial, and (3) the position of the $n+1$ nodes where the function values are known and used to construct the polynomial interpolant. The error does not necessarily decrease with increase in the degree of the polynomial used for interpolation due to the drawbacks of polynomial wiggle.

A straightforward method to determine the coefficients of the nth-degree polynomial is to substitute each data pair (x, y) into Equation (6.10). This produces $n+1$ linear equations. Suppose $n = 4$, and we have function values at five points: x_1, x_2, x_3, x_4, and x_5. The set of linear equations can be written in matrix notation:

$$\boldsymbol{Ac} = \boldsymbol{y},$$

where

$$\boldsymbol{A} = \begin{bmatrix} x_1^4 & x_1^3 & x_1^2 & x_1 & 1 \\ x_2^4 & x_2^3 & x_2^2 & x_2 & 1 \\ x_3^4 & x_3^3 & x_3^2 & x_3 & 1 \\ x_4^4 & x_4^3 & x_4^2 & x_4 & 1 \\ x_5^4 & x_5^3 & x_5^2 & x_5 & 1 \end{bmatrix}, \quad \boldsymbol{c} = \begin{bmatrix} c_1 \\ c_2 \\ c_3 \\ c_4 \\ c_5 \end{bmatrix}, \quad \boldsymbol{y} = \begin{bmatrix} y_1 \\ y_2 \\ y_3 \\ y_4 \\ y_5 \end{bmatrix};$$

$\boldsymbol{A}$ is called the **Vandermonde** matrix. This linear problem is not overdetermined, and we do not need to invoke least-square regression procedures. Solution of the system of linear equations using methods discussed in Chapter 2 yields the coefficients that define the interpolating polynomial. The $n+1$ values at which $\boldsymbol{y}$ is specified must be unique, otherwise $\boldsymbol{A}$ will be rank deficient and an interpolating polynomial of degree n cannot be determined.

Using MATLAB

A set of simultaneous linear equations can be solved using the MATLAB backslash operator (`c = A\y`). The use of this operator was discussed in Chapter 2. This is one method to find the coefficient matrix $\boldsymbol{c}$. Another method to determine the constants of the interpolating polynomial function is to use `polyfit`. The MATLAB `polyfit` function can be used to calculate the coefficients of a polynomial even when the problem is not overspecified. Usage of the `polyfit` function for overdetermined systems was discussed in Chapter 2. The syntax to solve a fully determinate problem (a set of linear equations that is consistent and has a unique solution) is the same and is given by

```
[c, S] = polyfit(x, y, n)
```

where n is the degree of the interpolating polynomial function and is equal to one less than the number of data points. The output vector $\boldsymbol{c}$ contains the coefficients of the polynomial of nth degree that passes exactly through the $n+1$ points.

A difficulty with this method is that the Vandermonde matrix quickly becomes ill-conditioned when either the interpolating polynomial degree ≥ 5 or the magnitude of x is large. As you may remember from Chapter 2, in an ill-conditioned system, the round-off errors are greatly magnified, producing an inaccurate solution. The condition number of the Vandermonde matrix indicates the conditioning of the linear system. When the condition number is high, the solution of the linear system is very sensitive to round-off errors.

Other methods, such as the **Lagrange interpolation formula** and **Newton's method of divided differences**, have been developed to calculate the coefficients of the interpolating polynomial. Because of the uniqueness property of an nth-degree polynomial generated by a set of $n + 1$ data points, these methods produce the same interpolating function. However, each method rearranges the polynomial differently, and accordingly the coefficients obtained by each method are different. For the remainder of this section, we focus on the Lagrange method of interpolation. An important advantage of this method is that the polynomial coefficients can be obtained without having to solve a system of linear equations. Lagrange interpolation formulas are popularly used to derive the Newton–Cotes formulas for numerical integration. For a discussion on Newton's method of interpolation, the reader is referred to Recktenwald (2000).

Suppose we wish to linearly interpolate between two data points (x_0, y_0) and (x_1, y_1). One way to express the linear interpolating function is

$$p(x) = c_1 x + c_0.$$

We can also write the linear interpolating function as

$$p(x) = c_1(x - x_1) + c_0(x - x_0). \tag{6.11}$$

When substituting $x = x_0$ in Equation (6.11), we obtain

$$c_1 = \frac{y_0}{x_0 - x_1}.$$

When $x = x_1$, we obtain

$$c_0 = \frac{y_1}{x_1 - x_0}.$$

Equation (6.11) becomes

$$p(x) = \frac{(x - x_1)}{x_0 - x_1} y_0 + \frac{(x - x_0)}{x_1 - x_0} y_1. \tag{6.12}$$

Equation (6.12) is the **first-order Lagrange interpolation formula**. We define the **first-order Lagrange polynomials** as

$$L_0(x) = \frac{(x - x_1)}{x_0 - x_1}; \quad L_1(x) = \frac{(x - x_0)}{x_1 - x_0}.$$

The interpolating polynomial is simply a linear combination of the Lagrange polynomials, i.e.

$$p(x) = L_0 y_0 + L_1 y_1,$$

so $p(x)$ is equal to the sum of two straight line equations. The first term, $\frac{(x-x_1)}{x_0-x_1} y_0$, represents a straight line that passes through two points (x_0, y_0) and $(x_1, 0)$. The second term, $\frac{(x-x_0)}{x_1-x_0} y_1$, represents a straight line that passes through the points (x_1, y_1) and $(x_0, 0)$. The three straight line equations are plotted in Figure 6.5.

Suppose we wish to interpolate between three points (x_0, y_0), (x_1, y_1), and (x_2, y_2) by constructing a quadratic function. The interpolating function is given by

$$p(x) = c_2(x - x_1)(x - x_2) + c_1(x - x_0)(x - x_2) + c_0(x - x_0)(x - x_1). \tag{6.13}$$

Figure 6.5

The interpolating polynomial $p(x)$ passes through two points (1, 1) and (2, 2). The sum of the two dashed lines produces the interpolating polynomial.

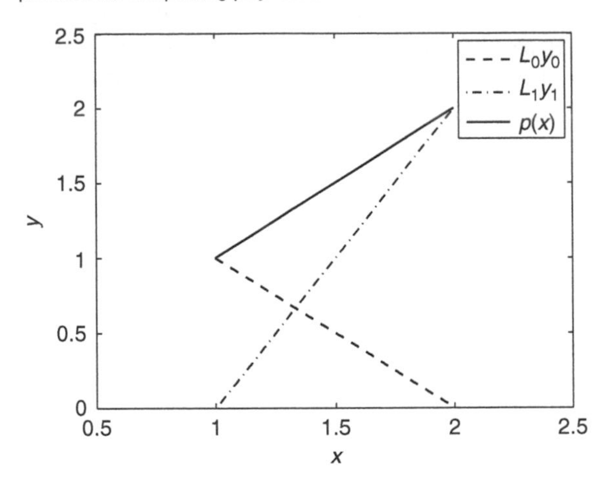

The coefficients can be derived in a similar fashion as done for the linear problem. The **second-order Lagrange interpolation formula** is given by

$$p(x) = \frac{(x - x_1)(x - x_2)}{(x_0 - x_1)(x_0 - x_2)} y_0 + \frac{(x - x_0)(x - x_2)}{(x_1 - x_0)(x_1 - x_2)} y_1 + \frac{(x - x_0)(x - x_1)}{(x_2 - x_0)(x_2 - x_1)} y_2, \tag{6.14}$$

and can be written compactly as

$$p(x) = \sum_{k=0}^{2} \left[y_k \prod_{j=0, j \neq k}^{2} \frac{(x - x_j)}{(x_k - x_j)} \right],$$

where $\prod$ stands for "the product of." There are three second-order Lagrange polynomials and each is a quadratic function as follows:

$$L_0(x) = \frac{(x - x_1)(x - x_2)}{(x_0 - x_1)(x_0 - x_2)}; \qquad L_1(x) = \frac{(x - x_0)(x - x_2)}{(x_1 - x_0)(x_1 - x_2)};$$

$$L_2(x) = \frac{(x - x_0)(x - x_1)}{(x_2 - x_0)(x_2 - x_1)}.$$

Note that $L_0(x_i)$ is equal to zero when $i = 1, 2$ and is equal to one when $i = 0$; $L_1(x_i)$ is equal to one for $i = 1$, and is zero when $i = 0$, $i = 2$. Similarly, $L_2(x_i)$ exactly equals one when $i = 2$ and is zero for the other two values of i. Because of this property of the Lagrange polynomials, we recover the exact value y_i of the function when the corresponding x_i value is plugged into the interpolation formula. Each term L_iy_i represents a quadratic polynomial. The interpolating polynomial $p(x)$ is obtained by summing the three quadratic Lagrange polynomials. Figure 6.6 demonstrates how the sum of three quadratic functions produces a second-degree interpolating polynomial.

An advantage of the Lagrange method is that round-off errors are less likely to creep into the calculation of the Lagrange polynomials.

Figure 6.6

The 2nd-degree interpolating polynomial is the sum of three quadratic Lagrange functions. The polynomial interpolates between three points (1, 1), (2, 4), and (3, 8).

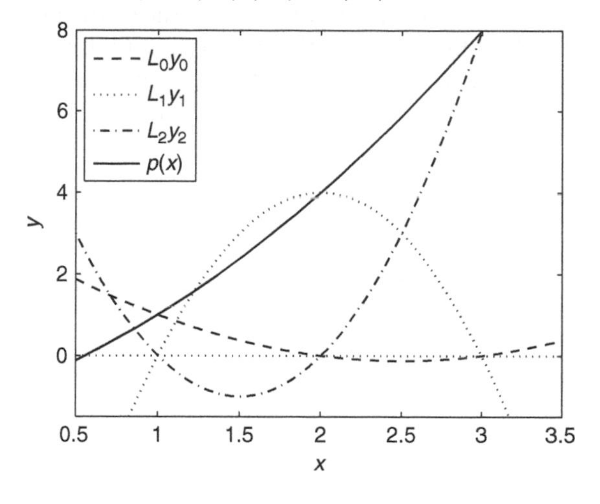

We generalize the Lagrange method for any interpolating polynomial of degree n. The kth Lagrange polynomial of order n is given by

$$L_k(x) = \frac{(x - x_0)(x - x_1)\cdots(x - x_{k-1})(x - x_{k+1})\cdots(x - x_n)}{(x_k - x_0)(x_k - x_1)\cdots(x_k - x_{k-1})(x_k - x_{k+1})\cdots(x_k - x_n)}. \tag{6.15}$$

Note that $L_k(x)$ does not contain the term $(x - x_k)$ in the numerator. When $x = x_k$, the numerator of L_k equals the denominator and $L_k = 1$. For which values of x does $L_k = 0$?

Lagrange interpolation for generating an nth-degree polynomial involves the construction of $n + 1$ Lagrange polynomials. To evaluate $p(x)$ one needs to calculate the values of the $n + 1$ nth-order Lagrange polynomials. The Lagrange interpolation formula for the nth-degree polynomial $p(x)$ is

$$p(x) = \sum_{k=0}^{n}\left[y_k \prod_{j=0,\, j\neq k}^{n} \frac{(x - x_j)}{(x_k - x_j)}\right] = \sum_{k=0}^{n} y_k L_k. \tag{6.16}$$

A drawback of the Lagrange method is that to interpolate at a new value of x it is necessary to recalculate all individual terms in the expansion for $p(x)$. Manual calculation of the Lagrange polynomials is cumbersome for interpolating polynomials of degree ≥ 3.

If the function $f(x)$ that defines the data is differentiable $n + 1$ times in the interval $[a, b]$, which contains the $n + 1$ points at which the function values are known, the error in polynomial interpolation is given by Equation (6.17):

$$E(x) = \frac{f^{n+1}(\xi(x))}{(n+1)!}(x - x_0)(x - x_1)\cdots(x - x_n) \qquad a \leq \xi \leq b. \tag{6.17}$$

The location of ξ is a function of x. Note that this term is similar to the error term of the Taylor series. The error depends on (1) the location of the x value within the interval $[a, b]$, (2) the degree of the interpolating polynomial, (3) the behavior of the actual function, and (4) the location of the $n + 1$ nodes. The derivation of Equation (6.17) requires the use of advanced calculus concepts (e.g. generalized Rolle's theorem) which are beyond the scope of this book. The interested reader is referred to Burden and Faires (2005) for the proof. Because the error cannot be evaluated, it is not possible to deduce the degree of the polynomial that will give the least error during interpolation. Increasing the degree of the interpolating polynomial does not necessarily improve the quality of the approximation.

Box 6.2 Transport of nutrients to cells suspended in cell culture media

Suspended spherical cells placed in a nutrient medium may survive (and multiply) for a period of time if critical solutes/nutrients are available. If the cell media is kept stationary (not swirled or mixed – no convective currents[3]), the nutrients in the media spread by diffusing through the liquid medium to the cell surface, where they either diffuse through the cell membrane or react with cell membrane proteins (includes receptors and channels). If uptake of a solute is rapid, the concentration of the solute at the cell surface is negligible, and the process by which the cells acquire the desired nutrient is said to be "transport-limited" or "diffusion-limited." In a spherical coordinate system, solute transport towards a spherical cell of radius a is governed by the following mass balance equation:

$$\frac{\partial C_i}{dt} = D_i \frac{1}{r^2} \frac{\partial}{\partial r}\left(r^2 \frac{\partial C_i}{\partial r}\right),$$

where D_i is the binary diffusion coefficient of diffusing species i in water. The mixture is assumed to be dilute. We are only interested in the transport of species A.

The initial and boundary conditions for this system are

$$\begin{aligned} C_A(r,t) &= C_0, && \text{for } r \geq a, t = 0; \\ C_A(r,t) &= 0, && \text{for } r = a, t > 0; \\ C_A(r,t) &= C_0, && \text{for } r \to \infty, t > 0. \end{aligned}$$

Figure 6.7 illustrates the system geometry.

Figure 6.7

Diffusion of solute through the medium to the cell surface.

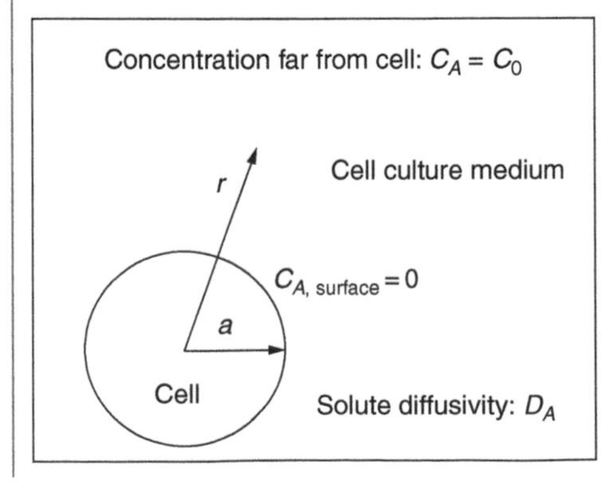

[3] While the cells will settle on the bottom surface of the flask, we will assume the wall can be neglected in analyzing the diffusion.

The mass balance equation is solved to yield the following analytical solution:

$$\frac{C_A}{C_0} = 1 - \frac{a}{r}\text{erfc}\left(\frac{r-a}{2\sqrt{D_A t}}\right). \tag{6.18}$$

Equation (6.18) describes the time-varying concentration profile of solute. It contains the complementary error function, which is defined as

$$\text{erfc}(z) = 1 - \text{erf}(z) = 1 - \frac{2}{\sqrt{\pi}}\int_0^z e^{-z^2} dz.$$

The integral $\int_0^z e^{-z^2} dz$ cannot be solved analytically. You may recognize this integral as similar in form to the cumulative normal distribution function discussed in Chapter 3. The error function has the following property:

$$\frac{2}{\sqrt{\pi}}\int_0^\infty e^{-z^2} dz = 1.$$

The integral solutions for various values of z are available in tabulated form in mathematical handbooks such as that by Abramowitz and Stegun (1965). The error function values for seven different values of z are listed in Table 6.1.

The constants in Equation (6.18) are $D_A = 10^{-7}$ cm^2/s, $a = 10$ μm. Using the values in Table 6.1, we calculate the solute concentration in the medium at six radial distances from the cell surface at $t = 10$ s. The calculations are presented in Table 6.2.

Table 6.1. *Values of error function*

z	erf(z)
0.0	0.0000000000
0.1	0.1124629160
0.2	0.2227025892
0.5	0.5204998778
1.0	0.8427007929
1.5	0.9661051465
2.0	0.9953222650

Table 6.2. *Concentration profile at t = 10 s*

r (μm)	z	erfc(z)	C_A/C_0
10	0.0	1.0000000000	0.0000000000
12	0.1	0.8875370830	0.2603857641
14	0.2	0.7772974108	0.4447875637
20	0.5	0.4795001222	0.7602499389
30	1.0	0.1572992071	0.9475669309
40	1.5	0.0338948535	0.9915262866
50	2.0	0.0046777350	0.9990644530

Figure 6.8

Piecewise cubic polynomial interpolant plotted along with the true concentration profile.

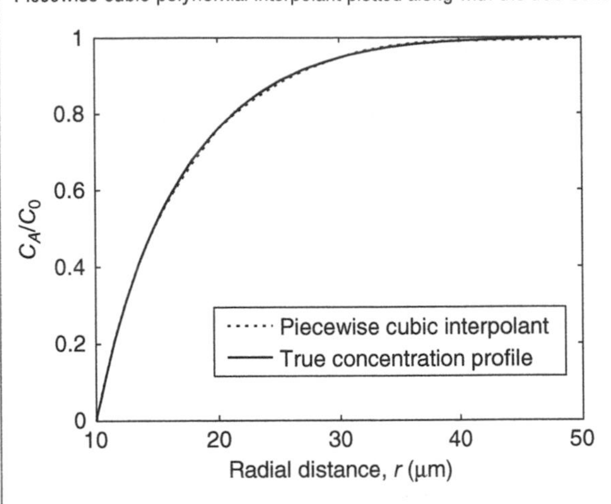

You are asked to use a third-degree piecewise polynomial interpolant to approximate the concentration profile of solution near the cell surface at time $t = 10$ s. Use the MATLAB `polyfit` function to obtain the coefficients for the two cubic interpolants.

The first cubic interpolant is created using the first four data points. The MATLAB statements

```
r = [10 12 14 20];
y = [0.0000000000 0.2603857641 0.4447875637 0.7602499389];
[p S] = polyfit(r, y, 3);
```

give us the following cubic polynomial:

$$y = 0.000454r^3 - 0.025860r^2 + 0.533675r - 3.205251 \quad \text{for} \quad 10 \le r \le 20.$$

The second cubic interpolant is given by

$$y = 0.000018r^3 - 0.002321r^2 + 0.100910r - 0.472203 \quad \text{for} \quad 20 \le r \le 50.$$

The piecewise cubic interpolating function is plotted in Figure 6.8. Using the built-in MATLAB error function `erf`, we calculate the true concentration profile (Equation (6.18)). The actual concentration profile is calculated in a MATLAB script file as follows:

```
Da = 10; % um^2/s
t = 10; % s
a = 10; % um
hold on
r = [10:0.1:50];
y = 1 - a./r.*(1 - erf((r-a)/2/sqrt(Da*t)));
```

The exact concentration profile is plotted in Figure 6.8 for comparison. The plot shows that there is good agreement between the cubic polynomial interpolant and the true function.

There is a discontinuity at $r = 20$ μm where the two piecewise interpolants meet. Calculate the slopes of both functions at $r = 20$ μm and compare them. Figure 6.9 shows a magnification of the region of discontinuity. Do your slope calculations explain the discontinuous behavior observed here?

Figure 6.9

Discontinuity at r = 20 μm.

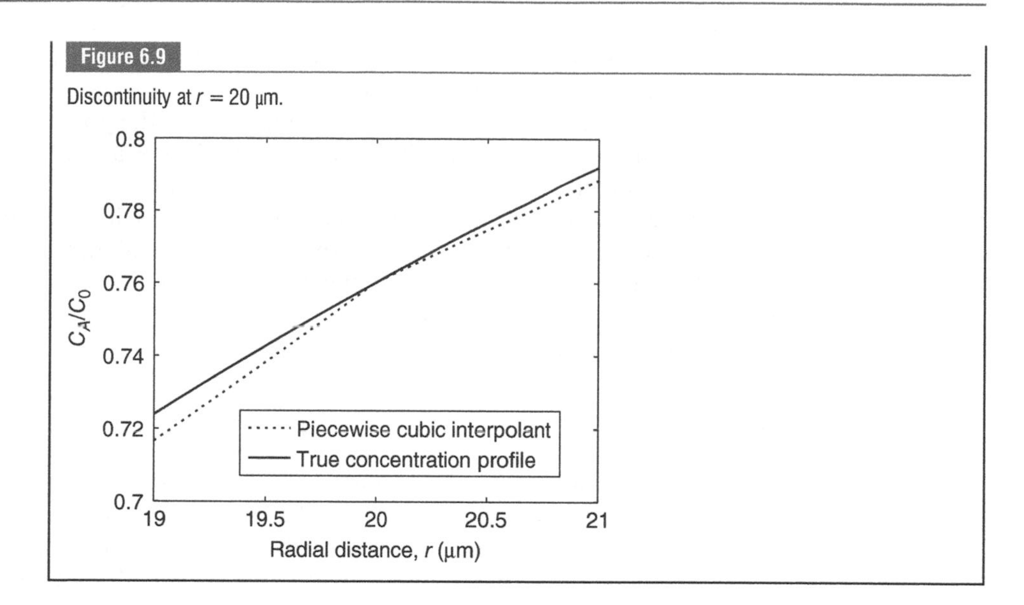

6.3 Newton–Cotes formulas

The most widely used numerical quadrature schemes are the Newton–Cotes rules. The **Newton–Cotes formulas** are obtained by replacing the integrand function with a polynomial function that interpolates between specific known values of the integrand. For example, the **midpoint rule** (which was introduced in Section 6.1) uses a zero-degree polynomial to represent the integrand function. Because the interpolating function is a constant value within the interval, the function value needs to be specified at only one node. The location of the node is at the midpoint of the interval (and not at the interval endpoints). Therefore, the midpoint rule is a **Newton–Cotes open integration rule**. To improve accuracy of the midpoint method, a piecewise constant function can be used to approximate the integrand function. The entire interval is divided into a number of subintervals and the midpoint rule, i.e. $f(x) = f(x_{i+0.5})$ for $x \in [x_i, \ x_{i+1}]$, is applied to each subinterval. This is called the **composite midpoint rule**, which was illustrated in Figure 6.2. The formula for this rule is given by Equation (6.2). The integration rules that we discuss in this section are closed integration formulas, i.e. the nodes used to construct the interpolating polynomial (nodes at which the function values are specified) include both endpoints of the interval of integration.

A first-degree polynomial (n = 1), i.e. a straight line, can be used to interpolate the two data points that lie at the ends of the interval. Since only two data points are required to define a line, and both these points are located at the ends of the interval, this scheme is a closed form of numerical integration. The straight line approximates the integrand function over the entire interval. Alternatively, the function can be represented by a series of straight lines applied piecewise over the n subintervals. Numerical integration using linear interpolation or piecewise linear interpolation of the data generates the **trapezoidal rule**.

If a second-degree polynomial (n = 2), i.e. a quadratic function, is used to interpolate three data points, two of which are located at the endpoints of the interval and a third point that bisects the first two, then the resulting integration rule is called **Simpson's 1/3 rule**. For n = 3, a cubic polynomial interpolant (that connects four data points) of the function is integrated, and the resulting Newton–Cotes formula is called **Simpson's 3/8 rule**.

A key feature of the Newton–Cotes formulas is that the polynomial function of degree n must interpolate equally spaced nodes. For example, a second-degree interpolant is constructed using function values at three nodes. Since two of the nodes are the interval endpoints, the integration interval is divided into two parts by the third node, located in the interior of the interval. The subinterval widths must be equal. In other words, if the node points are x_0, x_1, and x_2, and $x_2 > x_1 > x_0$, then $x_1 - x_0 = x_2 - x_1 = h$. In general, the $n+1$ nodal points within the closed interval $[a, b]$ must be equally spaced.

6.3.1 Trapezoidal rule

The trapezoidal rule is a Newton–Cotes quadrature method that uses a straight line to approximate the integrand function. The straight line is constructed using the function values at the two endpoints of the integration interval. The Lagrange formula for a first-degree polynomial interpolant given by Equation (6.12) is reproduced below:

$$p(x) = \frac{(x - x_1)}{x_0 - x_1} f(x_0) + \frac{(x - x_0)}{x_1 - x_0} f(x_1).$$

The function values at the integration limits, $a = x_0$ and $b = x_1$, are used to derive the straight-line polynomial. The equation above contains two first-order Lagrange polynomials, which upon integration produce the weights of the numerical integration formula of Equation (6.1). Let $h = x_1 - x_0$. The integral of the function $f(x)$ is exactly equal to the integral of the corresponding first-degree polynomial interpolant plus the integral of the truncation error associated with interpolation, i.e.

$$\int_a^b f(x)dx = \int_{x_0}^{x_1} \left[\frac{(x - x_1)}{x_0 - x_1} f(x_0) + \frac{(x - x_0)}{x_1 - x_0} f(x_1) \right] dx$$
$$+ \int_{x_0}^{x_1} \frac{f''(\xi(x))}{2} (x - x_0)(x - x_1)dx, \qquad a \le \xi \le b.$$

Performing integration of the interpolant, we obtain

$$\int_a^b f(x)dx \approx \int_{x_0}^{x_1} \left[\frac{(x - x_1)}{x_0 - x_1} f(x_0) + \frac{(x - x_0)}{x_1 - x_0} f(x_1) \right] dx$$
$$= \frac{1}{h} \left(-f(x_0) \int_{x_0}^{x_1} (x - x_1)dx + f(x_1) \int_{x_0}^{x_1} (x - x_0)dx \right)$$
$$= \frac{1}{h} \left(-f(x_0) \left(-\frac{1}{2} (x_1 - x_0)^2 \right) + f(x_1) \left(\frac{1}{2} (x_1 - x_0)^2 \right) \right);$$

$$\int_a^b f(x)dx \approx \frac{h}{2} (f(x_0) + f(x_1)). \tag{6.19}$$

Equation (6.19) is called the **trapezoidal rule**. It is the formula used to calculate the area of a trapezoid. Thus, the area under the straight-line polynomial is a trapezoid. Figure 6.10 presents the graphical interpretation of Equation (6.19).

Figure 6.10

The trapezoidal rule for numerical integration. The shaded area is a trapezoid.

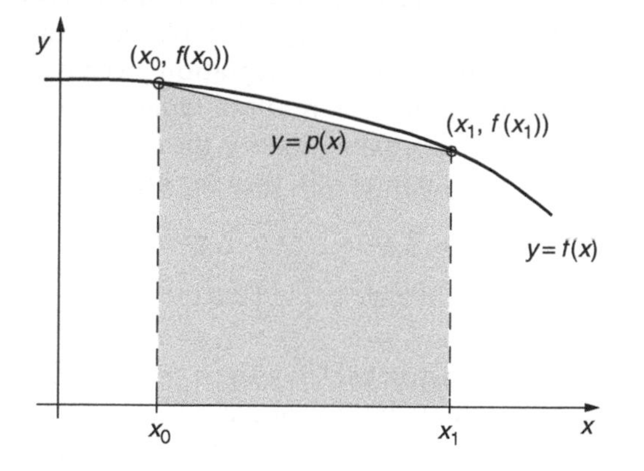

Now we calculate the error associated with using a linear interpolant to approximate the function $f(x)$. We assume that the function is twice differentiable on the interval range. For some value ξ in the interval, we have[4]

$$E = \frac{f''(\xi)}{2}\int_{x_0}^{x_1}(x - x_0)(x - x_1)dx.$$

The integral is evaluated as

$$\begin{aligned}\int_{x_0}^{x_1}(x - x_0)(x - x_1)dx &= \left[\frac{x^3}{3} - \left(\frac{x_0 + x_1}{2}\right)x^2 + x_0x_1x\right]_{x_0}^{x_1}\\ &= \frac{1}{6}\left[\left(x_0^3 - x_1^3\right) + 3x_0x_1(x_1 - x_0)\right]\\ &= \frac{(x_0 - x_1)^3}{6} = -\frac{h^3}{6}.\end{aligned}$$

Thus, the error term becomes

$$E = -\frac{h^3}{12}f''(\xi). \tag{6.20}$$

The error of the trapezoidal rule is exactly zero for any function whose second derivative is equal to zero.

> The **accuracy or precision of a quadrature rule** is defined as the largest positive integer n such that the rule integrates all polynomials of degree n and less with an error of zero.

The second derivative of any linear function is zero.

> The degree of precision of the trapezoidal rule is one. The error of the trapezoidal rule is $O(h^3)$.

Equation (6.19) is the trapezoidal rule applied over a single interval. If the interval of integration is large, a linear function may be a poor choice as an approximating

[4] We can take $f''(\xi)$ out of the interval because $(x - x_0)(x - x_1)$ does not change sign over the entire interval and the weighted mean value theorem applies. While these theorems are beyond the scope of this book, the interested reader may refer to Burden and Faires (2005), p. 8.

function over this interval, and this can lead to a large error in the result. To increase accuracy in the approximation of the integrand function, the interval is divided into a number of subintervals and the function within each subinterval is approximated by a straight line. Thus, the actual function $f(x)$ is replaced with a piecewise linear function. The trapezoidal rule for numerical integration given by Equation (6.19) is applied to each subinterval. The numerical result obtained from each subinterval is summed up to yield the integral approximation for the entire interval.

If the interval is divided into n subintervals, then the integral can be written as the sum of n integrals:

$$\int_a^b f(x)dx = \int_{x_0}^{x_1} f(x)dx + \int_{x_1}^{x_2} f(x)dx + \int_{x_2}^{x_3} f(x)dx + \cdots + \int_{x_{n-1}}^{x_n} f(x)dx.$$

Now n linear polynomials are constructed from the $n+1$ data points, i.e. values of $f(x)$ evaluated at $x_0, x_1, x_2, \ldots, x_n$ (whose values are in increasing order). The trapezoidal rule (Equation (6.19)) is applied to each subinterval. The numerical approximation of the integral is given by

$$\begin{aligned}\int_a^b f(x)dx &\approx \frac{(x_1 - x_0)}{2}(f(x_0) + f(x_1)) + \frac{(x_2 - x_1)}{2}(f(x_1) + f(x_2)) \\ &+ \frac{(x_3 - x_2)}{2}(f(x_2) + f(x_3)) + \cdots \\ &+ \frac{(x_n - x_{n-1})}{2}(f(x_{n-1}) + f(x_n)).\end{aligned} \tag{6.21}$$

Equation (6.21) is called the **composite trapezoidal rule for unequal subinterval widths**, i.e. it applies when the function is not evaluated at a uniform spacing of x. If all subintervals have the same width equal to h, then $x_i = x_0 + ih$, and Equation (6.21) simplifies to

$$\int_a^b f(x)dx \approx \frac{h}{2}\left[f(x_0) + 2\sum_{i=1}^{n-1} f(x_i) + f(x_n)\right]. \tag{6.22}$$

Equation (6.22) is the **composite trapezoidal rule for subintervals of equal width**. Comparing Equation (6.22) with Equation (6.1), you will observe that the composite trapezoidal rule for numerical integration is the summation of the function values evaluated at $n+1$ points along the interval, multiplied by their respective weights. Figure 6.11 illustrates the composite trapezoidal rule.

Figure 6.11

The composite trapezoidal rule. The six shaded areas each have the shape of a trapezoid.

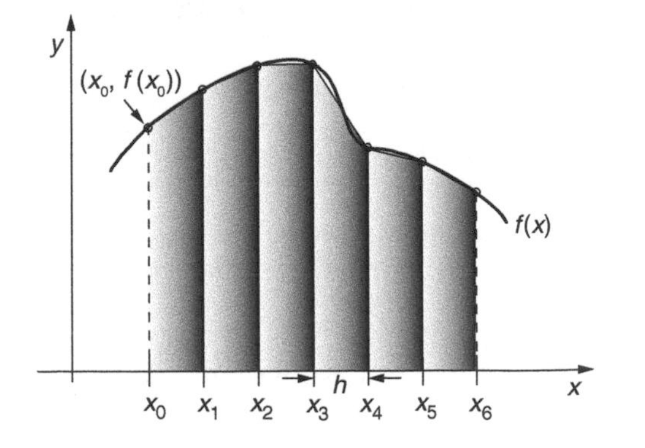

Next, we calculate the error associated with the composite trapezoidal rule when the subinterval widths are equal. The integration error of each subinterval is added to obtain the total error of integration over n subintervals as follows:

$$E = -\frac{h^3}{12}\sum_{i=1}^{n} f''(\xi_i) = -\frac{(b-a)^3}{12n^3}\sum_{i=1}^{n} f''(\xi_i),$$

where $\xi_i \in [x_{i-1}, x_i]$ is a point located within the ith subinterval. We define the average value of the second derivative over the interval of integration:

$$\overline{f''} = \frac{1}{n}\sum_{i=1}^{n} f''(\xi_i).$$

The quadrature error becomes

$$E = -\frac{(b-a)^3\overline{f''}}{12n^2} = -\frac{(b-a)h^2\overline{f''}}{12}, \tag{6.23}$$

which is $O(h^2)$. Note that the error of the composite trapezoidal rule is one order of magnitude less than the error associated with the *basic* trapezoid rule (Equation (6.19)).

Box 6.3A IV drip

A cylindrical bag of radius 10 cm and height 50 cm contains saline solution (0.9% NaCl) that is provided to a patient as a peripheral intravenous (IV) drip. The method used to infuse saline into the patient's bloodstream in this example is gravity drip. The solution flows from the bag downwards under the action of gravity through a hole of radius 1 mm at the bottom of the bag. If the only resistance offered by the flow system is the viscous resistance through the tubing, determine the time it will take for the bag to empty by 90%. The length of the tubing is 36″ (91.44 cm) and its ID (inner diameter) is 1 mm. The viscosity of the saline solution $\mu = 0.01$ Poise, and the density $\rho = 1\ \text{g/cm}^3$.

Let L be the length of the tubing, d the diameter of the tubing, and R the radius of the cylindrical bag. Then,

$L = 91.44$ cm; $d = 0.1$ cm; $R = 10$ cm.

Mechanical energy balance

The mechanical energy balance is given by the Bernoulli equation corrected for loss of mechanical energy due to fluid friction. The mechanical energy at the liquid level in the bag is equal to the mechanical energy of the fluid flowing out of the tubing minus the frictional loss. We ignore the friction loss from sudden contraction of the cross-sectional area of flow at the tube entrance. The height at which the drip fluid enters the patient is considered the datum level. The height of the fluid in the bag is z cm with respect to the bottom of the bag, and is $(z + L)$ cm with respect to the datum level (see Figure 6.12).

The pressure drop in the tubing due to wall friction is given by the Hagen–Poiseuille equation:

$$\Delta p = \frac{32Lu\mu}{d^2} = 2926.08u.$$

The Bernoulli equation for this system is as follows:

$$g(z+L) = \frac{u^2}{2} + \frac{\Delta p}{\rho} = \frac{u^2}{2} + \frac{32Lu\mu}{d^2\rho}, \tag{6.24}$$

Figure 6.12

Schematic of saline IV drip under gravity.

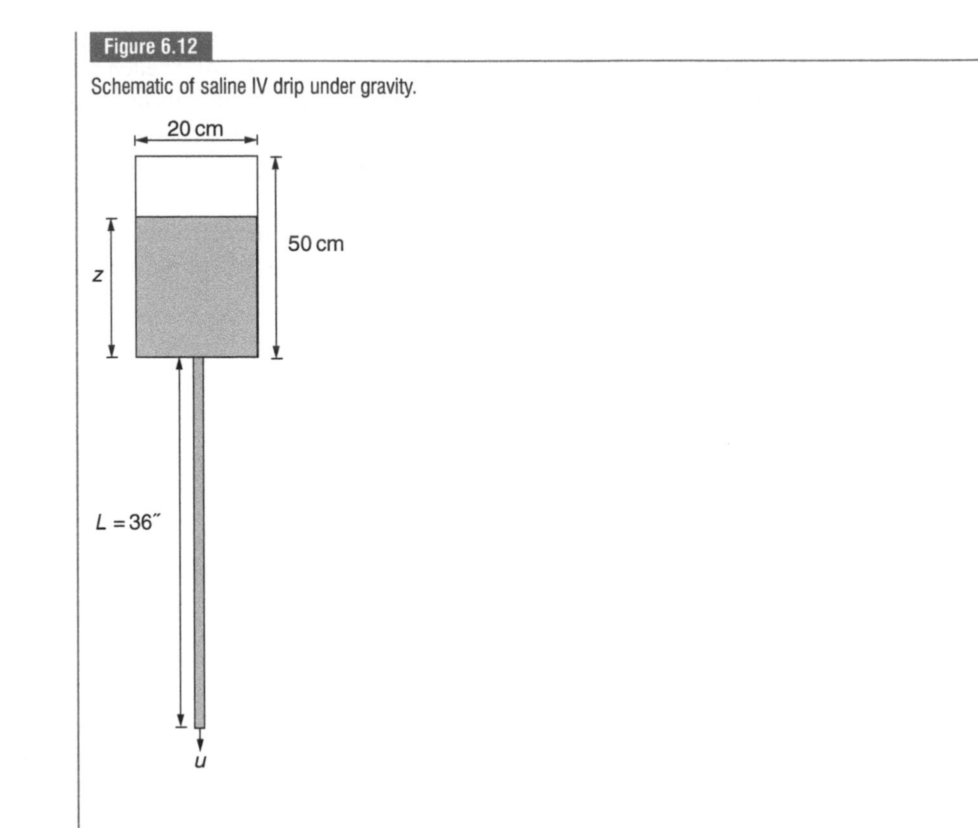

where $\Delta p/\rho$ is the term for mechanical loss due to fluid friction and has units of energy per mass.

Steady state mass balance

There is no reaction, accumulation, or depletion within the system, and we ignore the initial unsteady flow when initiating the IV drip. The mass flow rate in the bag is equal to the mass flow rate in the tubing:

$$-\rho\pi R^2 \frac{dz}{dt} = \rho\frac{\pi}{4}d^2u.$$

On rearranging, we get

$$dt = -\frac{4R^2}{d^2u}dz.$$

We can express u in terms of z by solving Equation (6.24) for u.

Letting $a = 64L\mu/d^2\rho$,

$$u = \frac{-a + \sqrt{a^2 + 8g(z+L)}}{2}.$$

Substituting the above into the steady state mass balance,

$$dt = -\frac{8R^2}{\left(-a + \sqrt{a^2 + 8g(z+L)}\right)d^2}dz.$$

Integrating the differential equation

We wish to integrate z from 50 cm to 5 cm. Integrating the left-hand side from 0 to t, we obtain

$$t = \int_{50}^{5} -\frac{8R^2}{\left(-\alpha + \sqrt{\alpha^2 + 8g(z+L)}\right)d^2}dz$$
$$= \frac{8R^2}{d^2}\int_{5}^{50} \frac{1}{\left(-\alpha + \sqrt{\alpha^2 + 8g(z+L)}\right)}dz.$$

The quadrature method chosen is the composite trapezoidal rule. Two programs were written to compute the integral. Program 6.1 is the trapezoidal integration routine and Program 6.2 calculates the integrand function for any value of z. The function routine to calculate the integrand has been vectorized to allow input of a vector variable.

Typing into the MATLAB Command Window

```
>> trapezoidal_rule('IVdrip',5, 50, 9)
```

the output is

```
0.5751
```

The time taken for 90% of the bag to empty is given by

$$t = \frac{8R^2}{d^2}(0.5751),$$

which is equal to 46 008 seconds or 12.78 hours.

MATLAB program 6.1

```
function I = trapezoidal_rule(integrandfunc, a, b, n)
% This function uses the composite trapezoidal rule to calculate an
% integral when an analytical function for the integrand
  is specified.

% Input variables
% integrandfunc : function that calculates the integrand
% a : lower limit of integration
% b : upper limit of integration
% n : number of subintervals

% Output variables
% I : value of integral

x = linspace(a, b, n + 1); % n + 1 nodes created within the interval
y = feval(integrandfunc, x);
h = (b-a)/n;
I = h/2*(y(1) + y(n+1) + 2*sum(y(2:n)));
```

MATLAB program 6.2

```
function f = IVdrip(z)
% This function calculates the integrand function of the IV drip problem.

% Parameters
```

```
L = 91.44; % cm
mu = 0.01; % Poise
d = 0.1; % cm
rho = 1; % g/cm^3
alpha = 64*L*mu/d^2/rho;
g = 981; % cm/s^2

f = 1./(-alpha + sqrt(alpha^2 + 8*g*(z + L)));
```

Box 6.1B Solute transport through a porous membrane

The integral in Equation (6.7) can be solved by numerically integrating the integrand defined by Equations (6.8) and (6.9). Calculation of ΔG is not straightforward, but also not too difficult. Note that the term Λ is defined by an infinite series. The MATLAB software contains the functions `besseli` and `besselk` to calculate the modified Bessel function of any order. Program 6.3 calculates the integrand. Program 6.1 above is used to perform the numerical integration for different numbers of subintervals, $n = 5$, 10, and 20:

```
>> alpha = 0.25;
>> trapezoidal_rule('partition_coefficient',0, 1-alpha, 5)
ans =
    0.0512
```

We double the number of subintervals to obtain

```
>> trapezoidal_rule('partition_coefficient',0, 1-alpha, 10)
```

MATLAB calculates

```
ans =
    0.0528
```

Again, on redoubling the number of subintervals we obtain

```
>> trapezoidal_rule('partition_coefficient',0, 1-alpha, 20)
```

MATLAB calculates

```
ans =
    0.0532
```

The radially averaged concentration of solute within a pore when the solute concentration is the same on both sides of the membrane (no solute gradient) is $0.0532 \times C_0$, or approximately 5% of the bulk solute concentration. The pore excludes 95% of solute particles that otherwise would have occupied the pore region if steric and electrostatic effects were absent!

If we assume that the solution for $n = 20$ is correct, the error in the first solution using five subintervals is 0.0020 and is 0.0004 when ten subintervals are used (second solution). The ratio of the errors obtained for $n = 10$ and $n = 5$ is

$$\frac{E_2}{E_1} \sim \left(\frac{h_2}{h_1}\right)^2 = \frac{1}{4}.$$

The error difference agrees with the second-order truncation error (Equation 6.23) of the composite trapezoidal rule.

MATLAB program 6.3

```
function f = partition_coefficient(beta)
% This function calculates
% (1) the dimensionless potential energy of interaction between the
% solute and pore wall, and
% (2) the integrand function that characterizes solute distribution
% within the pore.

% Input variables
% beta : dimensionless radial position of sphere center in tube

% Parameters
kB = 1.38e-23;       % J/molecule/deg K : Boltzmann constant
T = 273 + 37;        % deg K : temperature
R = 8.314;           % J/mol/deg K : gas constant
F = 96485.34;        % Coulomb/mol : Faraday constant
eps0 = 8.8542e-12; % C/V/m : dielectric permittivity of free space
releps = 74;         % solvent relative dielectric permittivity
deltabar = 0.1;      % ratio of sphere to solvent dielectric constant
R0 = 20e-9;          % m : pore radius
Cinf = 0.01*1000;  % mol/m^3 : electrolyte concentration
alpha = 0.25;        % dimensionless sphere radius
qs = 0.01;           % C/m^2 : sphere surface charge density
qc = 0.01;           % C/m^2 : cylinder surface charge density

% Additional calculations
abseps = releps*eps0; % absolute solvent dielectric permittivity
sigmas = F*R0*qs/abseps/R/T; % dimensionless surface charge density
sigmac = F*R0*qc/abseps/R/T; % dimensionless surface charge density

% Calculation of tau: ratio of pore radius to Debye length
tau = R0*sqrt(F^2/abseps/R/T*2*Cinf);

% Calculation of lambda
% Calculation of infinite series
t = 0; firstterm = 0; total = 0; lastterm = 1;
% If difference between new term and first term is less than 0.001*first
% term
while lastterm > (0.001*firstterm)
    preterm = (beta.^t)*factorial(2*t)/(2^(3*t))/((factorial(t))^2);
    lastterm = preterm.*besseli(t, tau*beta)*...
      (tau*besselk(t + 1, 2*tau) + 0.75*besselk(t, 2*tau));
    if t == 0
      firstterm = lastterm;
    end
    total = lastterm + total;
    t = t + 1;
end
lambda = pi/2*besseli(0, tau*beta).*total;

% Calculating the Langevin function
L = coth(tau*alpha)-1/tau/alpha;
```

```
% Calculating the interaction energy
delGnum1 = 8*pi*tau*alpha^4*exp(tau*alpha)/((1+tau*alpha)^2)* ...
  lambda*sigmas^2;
delGnum2 = 4*pi^2*alpha^2*besseli(0, tau*beta)/ ...
   (1 + tau*alpha)/besseli(1,tau)*sigmas*sigmac;
delGnum3 = (pi*besseli(0, tau*beta)/tau/besseli (1, tau)).^2 * ...
   (exp(tau*alpha)-exp(-tau*alpha))* ...
   tau*alpha*L/(1 + tau*alpha)*sigmac^2;
delGnum = delGnum1 + delGnum2 + delGnum3;
delGden = pi*tau*exp(-tau*alpha)-2*(exp(tau*alpha)- ...
   exp(-tau*alpha)) * ... tau*alpha*L*lambda/(1 + tau*alpha);
delG = delGnum./delGden;

% Calculating the dimensional interaction energy
E = R0*(R*T/F)^2*abseps*delG;

% Calculating the integrand
f = 2*exp(-E/kB/T).*beta;
```

6.3.2 Simpson's 1/3 rule

Simpson's 1/3 rule is a closed numerical integration scheme that approximates the integrand function with a quadratic polynomial. The second-degree polynomial is constructed using three function values: one at each endpoint of the integration interval and one at the interval midpoint. The function is evaluated at x_0, x_1, and x_2, where $x_2 > x_1 > x_0$, and x_0 and x_2 specify the endpoints of the interval. The entire integration interval is effectively divided into two subintervals of equal width: $x_1 - x_0 = x_2 - x_1 = h$. The integration interval is of width $2h$. The formula for a second-degree polynomial interpolant is given by Equation (6.14), which is reproduced below:

$$p(x) = \frac{(x - x_1)(x - x_2)}{(x_0 - x_1)(x_0 - x_2)} f(x_0) + \frac{(x - x_0)(x - x_2)}{(x_1 - x_0)(x_1 - x_2)} f(x_1) + \frac{(x - x_0)(x - x_1)}{(x_2 - x_0)(x_2 - x_1)} f(x_2).$$

The interpolating equation above contains three second-order Lagrange polynomials, which upon integration yield the weights of the numerical integration formula. The integral of the function is equal to the integral of the second-degree polynomial interpolant plus the integral of the truncation error associated with interpolation, i.e.

$$\int_{x_0}^{x_2} f(x)dx = \int_{x_0}^{x_2} \left[\frac{(x - x_1)(x - x_2)}{(x_0 - x_1)(x_0 - x_2)} f(x_0) + \frac{(x - x_0)(x - x_2)}{(x_1 - x_0)(x_1 - x_2)} f(x_1) + \frac{(x - x_0)(x - x_1)}{(x_2 - x_0)(x_2 - x_1)} f(x_2) \right] dx + \int_{x_0}^{x_2} \frac{f'''(\xi(x))}{3!} (x - x_0)(x - x_1)(x - x_2)dx, \qquad a \leq \xi \leq b.$$

Performing integration of the first term of the interpolant $p(x)$, we obtain

$$I_1 = f(x_0)\int_{x_0}^{x_2} \frac{(x - x_1)(x - x_2)}{(x_0 - x_1)(x_0 - x_2)}dx,$$

and after some algebraic manipulation we obtain

$$I_1 = \frac{(x_2 - x_0)^2(x_1/2 - (2x_0 + x_2)/6)}{(x_0 - x_1)(x_0 - x_2)}f(x_0).$$

Substituting $x_2 = x_0 + 2h$ and $x_1 = x_0 + h$ into the above, I_1 simplifies to

$$I_1 = \frac{h}{3}f(x_0).$$

Integration of the second term,

$$I_2 = f(x_1)\int_{x_0}^{x_2} \frac{(x - x_0)(x - x_2)}{(x_1 - x_0)(x_1 - x_2)}dx,$$

yields

$$I_2 = -\frac{(x_2 - x_0)^3}{6(x_1 - x_0)(x_1 - x_2)}f(x_1).$$

Again, the uniform step size allows us to simplify the integral as follows:

$$I_2 = \frac{4h}{3}f(x_1).$$

On integrating the third term,

$$I_3 = f(x_2)\int_{x_0}^{x_2} \frac{(x - x_0)(x - x_1)}{(x_2 - x_0)(x_2 - x_1)}dx,$$

we get

$$I_3 = \frac{(x_2 - x_0)^2((x_0 + 2x_2)/6 - x_1/2)}{(x_2 - x_0)(x_2 - x_1)}f(x_2).$$

Defining x_1 and x_2 in terms of x_0 and the uniform step size h, I_3 becomes

$$I_3 = \frac{h}{3}f(x_2).$$

The numerical formula for the integral is thus given by

$$\begin{aligned}\int_a^b f(x)dx &\approx I = I_1 + I_2 + I_3 \\ &= \frac{h}{3}(f(x_0) + 4f(x_1) + f(x_2)).\end{aligned} \tag{6.25}$$

Equation (6.25) is called **Simpson's 1/3 rule**. Figure 6.13 demonstrates graphically the numerical integration of a function using Simpson's 1/3 rule.

We cannot take $f'''(\xi(x))$ out of the error term integral since it is a function of the integration variable. Subsequent integration steps that retain this term in the integral require complex manipulations that are beyond the scope of this book. (The interested reader is referred to Patel (1994), where the integration of this term is

Figure 6.13

Graphical description of Simpson's 1/3 rule.

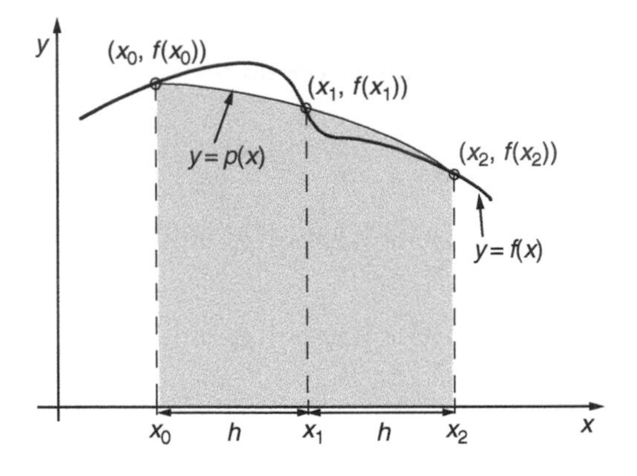

demonstrated.) To obtain the error term of Simpson's 1/3 rule, we look to the Taylor series expansion.

Using the Taylor series, we expand the integrand function $f(x)$ about the midpoint of the interval x_1:

$$f(x) = f(x_1) + (x - x_1)f'(x_1) + \frac{(x - x_1)^2}{2!}f''(x_1) + \frac{(x - x_1)^3}{3!}f'''(x_1) + \frac{(x - x_1)^4}{4!}f^{(4)}(\xi(x)),$$

where $\xi \in [x_1, x]$.

Integrating the above expansion,

$$\int_a^b f(x)dx = \int_{x_1-h}^{x_1+h}\left[f(x_1) + (x - x_1)f'(x_1) + \frac{(x - x_1)^2}{2!}f''(x_1) + \frac{(x - x_1)^3}{3!}f'''(x_1) + \frac{(x - x_1)^4}{4!}f^{(4)}(\xi(x))\right]dx$$

$$= 2hf(x_1) + 0 \cdot f'(x_1) + \frac{2(h)^3}{2.3}f''(x_1) + 0 \cdot f'''(x_1) + f^{(4)}(\xi_1)\frac{2(h)^5}{5 \cdot 4!};$$

$$\int_a^b f(x)dx = 2hf(x_1) + \frac{h^3}{3}f''(x_1) + \frac{h^5}{60}f^{(4)}(\xi_1), \tag{6.26}$$

where $\xi_1 \in [a, b]$.

Now we expand the function at the endpoints of the interval about x_1 using the Taylor series:

$$f(a) = f(x_0) = f(x_1) - hf'(x_1) + \frac{h^2}{2!}f''(x_1) - \frac{h^3}{3!}f'''(x_1) + \frac{h^4}{4!}f^{(4)}(\xi_2),$$

$$f(b) = f(x_2) = f(x_1) + hf'(x_1) + \frac{h^2}{2!}f''(x_1) + \frac{h^3}{3!}f'''(x_1) + \frac{h^4}{4!}f^{(4)}(\xi_3).$$

The above expressions for $f(x_0)$ and $f(x_2)$ are substituted into Equation (6.25) to yield

$$I = \frac{h}{3}\left[6f(x_1) + h^2 f''(x_1) + \frac{h^4}{4!}\left(f^{(4)}(\xi_2) + f^{(4)}(\xi_3)\right)\right]. \tag{6.27}$$

Let $f^{(4)}(\xi) = \max\left|\left(f^{(4)}(\xi(x))\right)\right|$ for $a \leq \xi \leq b$. Subtracting Equation (6.27) from Equation (6.26), the integration error is found to be

$$E \sim f^{(4)}(\xi)\left(\frac{h^5}{60} - \frac{h^5}{36}\right) = -\frac{h^5}{90} f^{(4)}(\xi). \tag{6.28}$$

The error is proportional to the subinterval width h raised to the fifth power. Simpson's 1/3 rule is a fifth-order method. You can compare this with the trapezoidal rule, which is a third-order method. Since the error term contains the fourth derivative of the function, this means that Simpson's rule is exact for any function whose fourth derivative over the entire interval is exactly zero. All polynomials of degree 3 and less have a fourth derivative equal to zero. It is interesting to note that, even though we interpolated the function using a quadratic polynomial, the rule allows us to integrate cubics exactly due to the symmetry of the interpolating function (cubic errors cancel out).

> The degree of precision of Simpson's 1/3 rule is 3. The error of Simpson's 1/3 rule is $O(h^5)$.

The accuracy of the numerical integration routine can be greatly improved if the integration interval is subdivided into smaller intervals and the quadrature rule is applied to each pair of subintervals. A requirement for using Simpson's 1/3 rule is that the interval must contain an even number of subintervals. Numerical integration using piecewise quadratic functions to approximate the true integrand function, with equal subinterval widths, is called the **composite Simpson's 1/3 method.**

The integration interval has n subintervals, such that $x_i = x_0 + ih$, $x_i - x_{i-1} = h$, and $i = 0, 1, 2, \ldots, n$. Since n is even, two consecutive subintervals can be paired to produce $n/2$ such pairs (such that no subinterval is repeated). Summing the integrals over the $n/2$ pairs of subintervals we recover the original integral:

$$\int_a^b f(x)dx = \sum_{i=1}^{n/2} \int_{x_{2(i-1)}}^{x_{2i}} f(x)dx.$$

Simpson's 1/3 rule (Equation (6.25)) is applied to each of the $n/2$ subintervals:

$$\int_a^b f(x)dx = \sum_{i=1}^{n/2}\left[\frac{h}{3}\left(f(x_{2i-2}) + 4f(x_{2i-1}) + f(x_{2i})\right) - \frac{h^5}{90} f^{(4)}(\xi_i)\right],$$

where $\xi_i \in [x_{2i-2}, x_{2i}]$ is a point located within the ith subinterval pair. Note that the endpoints of the paired subintervals $[x_{2i-2}, x_{2i}]$, except for the endpoints of the integration interval $[x_0, x_n]$, are each repeated twice in the summation since each endpoint belongs to both the previous subinterval pair and the next subinterval pair. Accordingly, the formula simplifies to

$$\int_a^b f(x)dx = \frac{h}{3}\left(f(x_0) + 2\sum_{i=1}^{(n/2)-1} f(x_{2i}) + 4\sum_{i=1}^{n/2} f(x_{2i-1}) + f(x_n)\right) + \sum_{i=1}^{n/2}\left[-\frac{h^5}{90} f^{(4)}(\xi_i)\right].$$

The numerical integration formula for the composite Simpson's 1/3 rule is given by

$$\int_a^b f(x)dx \approx \frac{h}{3}\left(f(x_0) + 2\sum_{i=1}^{(n/2)-1} f(x_{2i}) + 4\sum_{i=1}^{n/2} f(x_{2i-1}) + f(x_n)\right). \tag{6.29}$$

Next, we evaluate the error term of integration. The average value of the fourth derivative over the interval of integration is defined as follows:

$$\overline{f^{(4)}} = \frac{2}{n}\sum_{i=1}^{n/2} f^{(4)}(\xi_i),$$

and

$$E = -\frac{h^5}{90}\cdot\frac{n}{2}\overline{f^{(4)}}.$$

Since $n = (b-a)/h$, the error term becomes

$$E = -\frac{h^4(b-a)}{180}\overline{f^{(4)}}, \tag{6.30}$$

which is $O(h^4)$. The error of the composite Simpson's 1/3 rule has a fourth-order dependence on the distance h between two uniformly spaced nodes.

6.3.3 Simpson's 3/8 rule

Simpson's 3/8 rule is a Newton–Cotes closed integration formula that uses a cubic interpolating polynomial to approximate the integrand function. Suppose the function values $f(x)$ are known at four uniformly spaced points x_0, x_1, x_2, and x_3 within the integration interval, such that x_0 and x_3 are located at the interval endpoints, and x_1 and x_2 are located in the interior, such that they trisect the interval, i.e. $x_1 - x_0 = x_2 - x_1 = x_3 - x_2 = h$. The integration interval is of width $3h$. A cubic polynomial can be fitted to the function values at these four points to generate an interpolant that approximates the function. The formula for a third-degree polynomial interpolant can be constructed using the Lagrange polynomials:

$$p(x) = \frac{(x-x_1)(x-x_2)(x-x_3)}{(x_0-x_1)(x_0-x_2)(x_0-x_3)}f(x_0) + \frac{(x-x_0)(x-x_2)(x-x_3)}{(x_1-x_0)(x_1-x_2)(x_1-x_3)}f(x_1)$$

$$+ \frac{(x-x_0)(x-x_1)(x-x_3)}{(x_2-x_0)(x_2-x_1)(x_2-x_3)}f(x_2) + \frac{(x-x_0)(x-x_1)(x-x_2)}{(x_3-x_0)(x_3-x_1)(x_3-x_2)}f(x_3).$$

The cubic interpolant coincides exactly with the function $f(x)$ at the four equally spaced nodes. The interpolating equation above contains four third-order Lagrange polynomials, which upon integration yield the weights of the numerical integration formula. The integral of $f(x)$ is exactly equal to the integral of the corresponding third-degree polynomial interpolant plus the integral of the truncation error associated with interpolation. Here, we simply state the numerical formula that is obtained on integration of the polynomial $p(x)$ over the interval (the proof is left to the reader):

$$\int_a^b f(x)dx \approx \frac{3h}{8}(f(x_0) + 3f(x_1) + 3f(x_2) + f(x_3)). \tag{6.31}$$

Equation (6.31) is called **Simpson's 3/8 rule**. For this rule, $h = (b - a)/3$. The error associated with this numerical formula is

$$E = -\frac{3h^5}{80} f^{(4)}(\xi). \tag{6.32}$$

The error associated with Simpson's 3/8 rule is $O(h^5)$, which is of the same order as Simpson's 1/3 rule. The degree of precision of Simpson's 3/8 rule is 3.

If the integration interval is divided into $n > 3$ subintervals, Simpson's 3/8 rule can be applied to each group of three consecutive subintervals. Thus, the integrand function is approximated using a piecewise cubic polynomial interpolant. However, to apply Simpson's 3/8 rule, it is necessary for n to be divisible by 3. The **composite Simpson's 3/8 formula** is as follows:

$$\int_a^b f(x)dx \approx \frac{3h}{8}\left(f(x_0) + 3\sum_{i=1}^{n/3}\left[f(x_{3i-2}) + f(x_{3i-1})\right] + 2\sum_{i=1}^{(n/3)-1} f(x_{3i}) + f(x_n)\right). \tag{6.33}$$

The truncation error of Equation (6.33) is

$$E = -\frac{h^4(b-a)}{80}\overline{f^{(4)}}. \tag{6.34}$$

Since the truncation errors of the composite Simpson's 1/3 rule and the composite Simpson's 3/8 rule are of the same order, both methods are often combined so that restrictions need not be imposed on the number of subintervals (such as n must be even or n must be a multiple of 3). If n is even, only the composite Simpson's 1/3 rule needs to be used. However, if n is odd, Simpson's 3/8 rule can be used to integrate the first three subintervals, while the remainder of the interval can be integrated using composite Simpson's 1/3 rule. The use of the 1/3 rule is preferred over the 3/8 rule since fewer data points (or functional evaluations) are required by the 1/3 rule to achieve the same level of accuracy.

In Simpson's formulas, since the nodes are equally spaced, the accuracy of integration will vary over the interval. To achieve a certain level of accuracy, the largest subinterval width h within the most difficult region of integration that produces a result of sufficient accuracy fixes the global step size. This subinterval width must be applied over the entire interval, which unnecessarily increases the number of subintervals required to calculate the integral. A more advanced algorithm is the **adaptive quadrature method**, which selectively narrows the node spacing when the accuracy condition is not met for a particular subinterval. The numerical integration is performed twice for each subinterval, at a node spacing of h and at a node spacing of $h/2$. The difference between the two integrals obtained with different n (one twice the other) is used to estimate the truncation error. If the error is greater than the specified tolerance, the subinterval is halved recursively, and integration is performed on each half by doubling the number of subintervals. The truncation error is estimated based on the new node spacing and the decision to subdivide the subinterval further is made accordingly. This procedure continues until the desired accuracy level is met throughout the integration interval.

Box 6.1C Solute transport through a porous membrane

The integral in Equation (6.7) can be solved by numerically integrating the function defined by Equations (6.8) and (6.9) using Simpson's 1/3 and 3/8 methods used in conjunction. Program 6.4 lists the function code that performs numerical integration using combined Simpson's 1/3 rule and 3/8 rule so that any number of subintervals may be considered. The integration of Equation (6.7) is performed for $n = 5$, 10, and 20. The numerical solutions obtained are as follows:

$n = 5$: $I = 0.0546813$;

$n = 10$: $I = 0.0532860$;

$n = 20$: $I = 0.0532808$.

The radially averaged concentration of solute within a pore when the solute concentration is the same on both sides of the membrane is approximately 5% of the bulk solute concentration. With Simpson's rule, the solution converges much faster.

MATLAB program 6.4

```
function I = simpsons_rule(integrandfunc, a, b, n)
% This function uses Simpson's 3/8 rule and composite Simpson's 1/3
% rule to compute an integral when the analytical form of the
% function is provided.

% Input variables
% integrandfunc : function that calculates the integrand
% a : lower limit of integration
% b : upper limit of integration
% n : number of subintervals.

% Output variables
% I : value of integral

x = linspace(a, b, n + 1); % n + 1 nodes created within the interval
y = feval(integrandfunc, x);
h = (b-a)/n;

% Initializing
I1 = 0; I2 = 0; k = 0;

% If n is odd, use Simpson's 3/8 rule first
if mod(n,2)~=0 % n is not divisible by 2
    I1 = 3*h/8*(y(1) + 3*y(2) + 3*y(3) + y(4));
    k = 3; % Placeholder to locate start of integration using 1/3 rule
end

% Use Simpson's 1/3 rule to evaluate remainder of integral
I2 = h/3*(y(1+k) + y(n+1) + 4*sum(y(2+k:2:n)) + 2*sum(y(3+k:2:n-1)));

I = I1 + I2;
```

Using MATLAB

The MATLAB function quad performs numerical quadrature using an adaptive Simpson's formula. The formula is adaptive because the subinterval width is adjusted (reduced) while integrating to obtain a result within the specified absolute tolerance (default: 1×10^{-6}) during the integration. The syntax for the function is

```
I = quad('integrand_function', a, b)
```

or

```
I = quad('integrand_function', a, b, Tol)
```

where a and b are the limits of integration and Tol is the user-specified absolute tolerance. The function handle supplied to the quad function must be able to accept a vector input and deliver a vector output.

6.4 Richardson's extrapolation and Romberg integration

The composite trapezoidal formula for equal subinterval widths is a low-order approximation formula, since the leading order of the error term is $O(h^2)$. The truncation error of the composite trapezoidal rule follows an infinite series consisting of only even powers of h:

$$\int_a^b f(x)dx - \frac{h}{2}\left[f(x_0) + 2\sum_{i=1}^{n-1} f(x_i) + f(x_n)\right] = c_1h^2 + c_2h^4 + c_3h^6 + \cdots, \tag{6.35}$$

the proof of which is not given here but is discussed in Ralston and Rabinowitz (1978). Using **Richardson's extrapolation technique** it is possible to combine the numerical results of trapezoidal integration obtained for two different step sizes, h_1 and h_2, to obtain a numerical result, $I_3 = f(I_1, I_2)$, that is much more accurate than the two approximations I_1 and I_2. The higher-order integral approximation I_3 obtained by combining the two low-accuracy approximations of the trapezoidal rule has an error whose leading order is $O(h^4)$. Thus, using the extrapolation formula, we can reduce the error efficiently by two orders of magnitude in a single step. The goal of the extrapolation method is to combine the two integral approximations in such a way as to eliminate the lowest-order error term in the error series associated with the numerical result I_2.

Suppose the trapezoidal rule is used to solve an integral over the interval $[a, b]$. We get the numerical result $I(h_1)$ for subinterval width h_1 (n_1 subintervals), and the result $I(h_2)$ for subinterval width h_2 (n_2 subintervals). If I_E is the exact value of the integral, and the error associated with the result $I(h_1)$ is $E(h_1)$, then

$$I_E = I(h_1) + E(h_1). \tag{6.36}$$

Equation (6.23) defines the truncation error for the composite trapezoidal rule. We have

$$E(h_1) = -\frac{(b-a)h_1^2\overline{f''}}{12}.$$

If the error associated with the result $I(h_2)$ is $E(h_2)$, then

$$I_E = I(h_2) + E(h_2) \tag{6.37}$$

and

$$E(h_2) = -\frac{(b-a)h_2^2\overline{f''}}{12}.$$

If the average value of the second derivative $\overline{f''}$ over all nodes within the entire interval does not change with a decrease in step size, then we can write $E(h_1) = ch_1^2$ and $E(h_2) = ch_2^2$.

On equating Equations (6.36) and (6.37),

$$I(h_1) + ch_1^2 \approx I(h_2) + ch_2^2,$$

we obtain

$$c \approx \frac{I(h_2) - I(h_1)}{h_1^2 - h_2^2}.$$

Substituting the expression for c into Equations (6.36) or (6.37),

$$I_{\mathrm{E}} \approx \frac{(h_1^2/h_2^2)I(h_2) - I(h_1)}{(h_1^2/h_2^2) - 1}.$$

The scheme to combine two numerical integral results that have errors of the same order of magnitude such that the combined result has a much smaller error is called **Richardson's extrapolation**.

If $h_2 = h_1/2$, the above equation simplifies to

$$I_{\mathrm{E}} \approx \frac{4}{3}I(h_2) - \frac{1}{3}I(h_1). \tag{6.38}$$

In fact, Equation (6.38) is equivalent to the composite Simpson's 1/3 rule (Equation (6.29)). This is shown next. Let n be the number of subintervals used to evaluate the integral when using step size h_1; $2n$ is the number of subintervals used to evaluate the integral when using step size h_2.

According to the trapezoidal rule,

$$I(h_1) = \frac{h_1}{2}\left[f(x_0) + 2\sum_{i=1}^{n-1} f(x_{2i}) + f(x_{2n})\right]$$

and

$$I(h_2) = \frac{h_2}{2}\left[f(x_0) + 2\sum_{i=1}^{2n-1} f(x_i) + f(x_{2n})\right].$$

Combining the above two trapezoidal rule expressions according to Equation (6.38), we obtain

$$I_{\mathrm{E}} \approx \frac{4}{3}I(h_2) - \frac{1}{3}I(h_1) = \frac{2h_2}{3}\left[f(x_0) + 2\left(\sum_{i=1}^{n-1} f(x_{2i}) + \sum_{i=1}^{n} f(x_{2i-1})\right) + f(x_{2n})\right]$$
$$- \frac{h_2}{3}\left[f(x_0) + 2\sum_{i=1}^{n-1} f(x_{2i}) + f(x_{2n})\right],$$

which recovers the composite Simpson's 1/3 rule,

$$I_{\mathrm{E}} \approx \frac{h_2}{3}\left[f(x_0) + 2\sum_{i=1}^{n-1} f(x_{2i}) + 4\sum_{i=1}^{n} f(x_{2i-1}) + f(x_{2n})\right].$$

The error associated with the composite Simpson's 1/3 rule is $O(h^4)$. The error in Equation (6.38) is also $O(h^4)$. This can be shown by combining the terms in the truncation error series of the trapezoidal rule for step sizes h_1 and h_2 (Equation (6.35)) using the extrapolation rule (Equation (6.38)). The numerical integral approximations obtained using the trapezoidal rule are called the **first levels of extrapolation**. The $O(h^4)$ approximation obtained using Equation (6.38) is called the **second level of extrapolation**, where $h = h_2$.

Suppose we have two $O(h^4)$ (level 2) integral approximations obtained for two different step sizes. Based on the scheme described above, we can combine the two results to yield a more accurate value:

$$I(h_1) + ch_1^4 \approx I(h_2) + ch_2^4.$$

Eliminating c, we obtain the following expression:

$$I_E \approx \frac{(h_1/h_2)^4 I(h_2) - I(h_1)}{(h_1/h_2)^4 - 1},$$

which simplifies to

$$I_E \approx \frac{16}{15} I(h_2) - \frac{1}{15} I(h_1) \tag{6.39}$$

when $h_2 = h_1/2$. The error associated with Equation 6.39 is $O(h^6)$. Equation (6.39) gives us a **third level of extrapolation**.

If k specifies the level of extrapolation, a **generalized Richardson's extrapolation formula** to obtain a numerical integral estimate of $O(h^{2k+2})$ from two numerical estimates of $O(h^{2k})$, such that one estimate of $O(h^{2k})$ is obtained using twice the number of subintervals as that used for the other estimate of $O(h^{2k})$, is

$$I_{k+1} = \frac{4^k I_k(h/2) - I_k(h)}{4^k - 1}. \tag{6.40}$$

Note that Equation (6.40) applies only if the error term can be expressed as a series of even powers of h. If the error term is given by

$$E = c_1 h^2 + c_2 h^3 + c_3 h^4 + \cdots$$

then the extrapolation formula becomes

$$I_{k+1} = \frac{2^{k+1} I_k(h/2) - I_k(h)}{2^{k+1} - 1},$$

where level $k = 1$ corresponds to h^2 as the leading order of the truncation error.

Richardson's extrapolation can be applied to integral approximations as well as to finite difference (differentiation) approximations. When this extrapolation technique is applied to numerical integration repeatedly to improve the accuracy of each successive level of approximation, the scheme is called **Romberg integration**. The goal of Romberg integration is to achieve a remarkably accurate solution in a few steps using the averaging formula given by Equation (6.40).

To perform Romberg integration, one begins by obtaining several rough approximations to the integral whose value is sought, using the composite trapezoidal rule. The numerical formula is calculated for several different step sizes equal to $(b - a)/2^{j-1}$, where $j = 1, 2, \ldots, m$, and m is the maximum level of integration

Table 6.3. *The Romberg integration table*

	Level of integration; order of magnitude of truncation error				
	$k = 1; O(h^2)$	$k = 2; O(h^4)$	$k = 3; O(h^6)$	$k = 4; O(h^8)$	$k = 5; O(h^{10})$
Step size					
$j = 1$	$R_{1,1}$				
$j = 2$	$R_{2,1}$	$R_{2,2}$			
$j = 3$	$R_{3,1}$	$R_{3,2}$	$R_{3,3}$		
$j = 4$	$R_{4,1}$	$R_{4,2}$	$R_{4,3}$	$R_{4,4}$	
$j = 5$	$R_{5,1}$	$R_{5,2}$	$R_{5,3}$	$R_{5,4}$	$R_{5,5}$

The step size in any row is equal to $(b - a)/2^{j-1}$

(corresponding to an error $\sim O(h^{2m})$) desired; m can be any integral value ≥ 1. Let $R_{j,k}$ denote the Romberg integral approximation at level k and with step size $(b - a)/2^{j-1}$. Initially, m integral approximations $R_{j,1}$ are generated using Equation (6.22). They are entered into a table whose columns represent the level of integration (or order of magnitude of the error) (see Table 6.3). The rows represent the step size used to compute the integral. The trapezoidal rule is used to obtain all entries in the first column.

Using Equation (6.40), we can generate the entries in the second, third, . . . , and mth column. We rewrite Equation (6.40) to make it suitable for Romberg integration:

$$R_{j,k} \approx \frac{4^{k-1} R_{j,k-1} - R_{j-1,k-1}}{4^{k-1} - 1}. \tag{6.41}$$

In order to reduce round-off error caused by subtraction of two numbers of disparate magnitudes, Equation (6.41) can be rearranged as follows:

$$R_{j,k} \approx R_{j,k-1} + \frac{R_{j,k-1} - R_{j-1,k-1}}{4^{k-1} - 1}. \tag{6.42}$$

Table 6.3 illustrates the tabulation of approximations generated during Romberg integration for up to five levels of integration. Note that the function should be differentiable many times over the integral to ensure quick convergence of Romberg integration. If singularities in the derivatives arise at one or more points in the interval, convergence rate deteriorates.

A MATLAB program to perform Romberg integration is listed in Program 6.5. The function output is a matrix whose elements are the approximations $R_{j,k}$.

MATLAB program 6.5

```
function R = Romberg_integration(integrandfunc, a, b, maxlevel)
% This function uses Romberg integration to calculate an integral when
% the integrand function is available in analytical form.

% Input variables
% integrandfunc : function that calculates the integrand
% a : lower limit of integration
% b : upper limit of integration
% maxlevel : maximum level of integration
```

```
% Output variables
% I : value of integral

% Initializing variables
R(1:maxlevel,1:maxlevel) = 0;
j = [1:maxlevel];
n = 2.^(j-1); % number of subintervals vector
% Generating the first column of approximations using trapezoidal rule
for l = 1 : maxlevel
    R(l, 1) = trapezoidal_rule(integrandfunc, a, b, n(l));
end

% Generating approximations for columns 2 to maxlevel
for k = 2 : maxlevel
    for l = k : maxlevel
        R(l, k) = R(l, k - 1) + (R(l, k - 1) - R(l - 1, k - 1))/(4^(k-1) - 1);
    end
end
```

In the current coding scheme, the function values at all $2^{j-1} + 1$ nodes are calculated each time the trapezoidal rule is used to generate an approximation corresponding to step size $(b - a)/2^{j-1}$. The efficiency of the integration scheme can be improved by recycling the function values evaluated at $2^{j-1} + 1$ nodes performed for a larger step size to calculate the trapezoid integration result for a smaller step size ($2^j + 1$ nodes). To do this, the trapezoidal formula $I(h/2)$ for step size $h/2$ must be rewritten in terms of the numerical result $I(h)$for step size h. (See Problem 6.9.)

Although the maximum number of integration levels is set as the stopping criterion, one can also use a tolerance specification to decide how many levels of integration should be calculated. The tolerance criterion can be set based on the relative error of two consecutive solutions in the last row of Table 6.3 as follows:

$$\frac{R_{j,k} - R_{j,k-1}}{R_{j,k}} \leq \text{Tol.}$$

Problem 6.10 is concerned with developing a MATLAB routine to perform Romberg integration to meet a user-specified tolerance.

6.5 Gaussian quadrature

Gaussian quadrature is a powerful numerical integration scheme that, like the Newton–Cotes rules, uses a polynomial interpolant to approximate the behavior of the integrand function within the limits of integration. The distinguishing feature of the Newton–Cotes equations for integration is that the function evaluations are performed at equally spaced points in the interval. In other words, the $n + 1$ node points used to construct the polynomial interpolant of degree n are uniformly distributed in the interval. The convenience of a single step size h greatly simplifies the weighted summation formulas, especially for composite integration rules, and makes the Newton–Cotes formulas well suited to integrate tabular data. A Newton–Cotes formula derived using a polynomial interpolant of degree n has a precision equal to (1) n, if n is odd, or (2) $n + 1$, if n is even.

Box 6.3B IV drip

Romberg integration is used to integrate

$$t = \int_5^{50} \frac{1}{\left(-a + \sqrt{a^2 + 8g(z + L)}\right)} dz,$$

where

$a = 64L\mu/d^2\rho$,
length of tubing $L = 91.44$ cm,
viscosity $\mu = 0.01$ Poise,
density $\rho = 1$ g/cm^3,
diameter of tube $d = 0.1$ cm, and
radius of bag $R = 10$ cm.

We choose to perform Romberg integration for up to a maximum of four levels:

```
>> format long
>> Romberg_integration('IVdrip',5, 50, 4)
0.589010645018746                   0                   0                   0
0.578539899240569   0.575049650647843                   0                   0
0.575851321866992   0.574955129409132   0.574948827993218                   0
0.575174290383643   0.574948613222527   0.574948178810086   0.574948168505592
```

The time taken for 90% of the bag to empty is given by

$$t = \frac{8R^2}{d^2}(0.5749482),$$

which is equal to 45 996 seconds or 12.78 hours. Note the rapid convergence of the solution in the second–fourth columns in the integration table above. To obtain the same level of accuracy using the composite trapezoidal rule, one would need to use more than 2000 subintervals or steps! In Romberg integration performed here, the maximum number of subintervals used is eight.

It is possible to increase the precision of a quadrature formula beyond n or $n + 1$, even when using an nth-degree polynomial interpolant to approximate the integrand function. This can be done by optimizing the location of the $n + 1$ nodes between the interval limits. By carefully choosing the location of the nodes, it is possible to construct the "best" interpolating polynomial of degree n that approximates the integrand function with the least error. Gaussian quadrature uses a polynomial of degree n to approximate the integrand function and achieves a precision of $2n + 1$, the highest precision attainable. In other words, Gaussian quadrature produces the exact value of the integral of any polynomial function up to a degree of $2n + 1$, using a polynomial interpolant of a much smaller degree equal to n.

Like the Newton–Cotes formulas, the Gaussian quadrature formulas are weighted summations of the integrand function evaluated at $n + 1$ unique points within the interval:

$$\int_a^b f(x)dx \approx \sum_{i=0}^{n} w_i f(x_i).$$

Figure 6.14

Accuracy in numerical quadrature increases when the nodes are strategically positioned within the interval. This is shown here for a straight-line function that is used to approximate a downward concave function. (a) The two nodes are placed at the interval endpoints according to the trapezoidal rule. (b) The two nodes are placed at some specified distance from the integration limits according to the Gaussian quadrature rule.

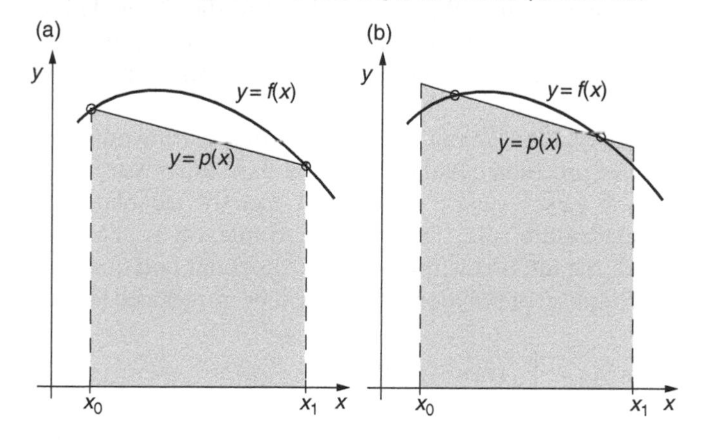

However, the nodes x_i are not evenly spaced within the interval. The weights w_i can be calculated by integrating the Lagrange polynomials once the positions of the nodes are determined. How does one determine the optimal location of the $n + 1$ nodes so that the highest degree of precision is obtained?

First, let's look at the disadvantage of using an even spacing of the nodes within the interval. Uniform node spacing can result in sizeable errors in the numerical solution. When nodes are positioned at equal intervals, the interpolating polynomial that is constructed from the function values at these nodes is more likely to be a poor representation of the true function. Consider using the trapezoidal rule to integrate the downward concave function shown in Figure 6.14. If the function's value at the midpoint of the interval is located far above its endpoint values, the straight-line function that passes through the interval endpoints will grossly underestimate the value of the integral (Figure 6.14(a)). However, if the two node points are positioned *inside* the interval at a suitable distance from the integration limits, then the line function will overestimate the integral in some regions of the interval and underestimate the integral in other regions, allowing the errors to largely cancel each other (Figure 6.14(b)). The overall error is much smaller.

Since an optimal choice of node positions is expected to improve the accuracy of the solution, intuitively we expect Gaussian quadrature to have greater precision than a Newton–Cotes formula with the same degree of the polynomial interpolant. If the step size is not predetermined (the step size is predetermined for Newton–Cotes rules to allow for equal subinterval widths), the quadrature formula of Equation (6.1) has $2n + 2$ unknown parameters: $n + 1$ weights, w_i, and $n + 1$ nodes, x_i. With the additional $n + 1$ adjustable parameters of the quadrature formula, it is possible to increase the precision by $n + 1$ degrees from n (typical of Newton–Cotes formulas) to $2n + 1$.

In theory, we can construct a set of $2n + 2$ nonlinear equations by equating the weighted summation formula to the exact value of the integral of the first $2n + 2$ positive integer power functions $(1, x, x^2, x^3, \ldots, x^{2n+1})$ with integration limits

$[-1, 1]$. On solving this set of equations, we can determine the values of the $2n + 2$ parameters of the summation formula that will produce the exact result for the integral of any positive integer power function up to a maximum degree of $2n + 1$. A $(2n + 1)$th-degree polynomial is simply a linear combination of positive integer power functions of degrees from 0 to $2n + 1$. The resulting $(n + 1)$-point quadrature formula will be exact for all polynomials up to the $(2n + 1)$th degree. We demonstrate this method to find the parameters of the two-point Gaussian quadrature formula for a straight-line approximation (first-degree polynomial interpolant) of the integrand function.

A Gaussian quadrature rule that uses an interpolating polynomial of degree $n = 1$ has a degree of precision equal to $2n + 1 = 3$. In other words, the integral of a cubic function $c_3x^3 + c_2x^2 + c_1x + c_0$ can be exactly calculated using a two-point Gaussian quadrature rule. The integration interval is chosen as $[-1, 1]$, so as to coincide with the interval within which orthogonal polynomials are defined. These are an important set of polynomials and will be introduced later in this section. Since

$$\int_{-1}^{1} f(x)dx = \int_{-1}^{1} \left(c_3x^3 + c_2x^2 + c_1x + c_0\right)dx$$
$$= c_3\int_{-1}^{1} x^3dx + c_2\int_{-1}^{1} x^2dx + c_1\int_{-1}^{1} xdx + c_0\int_{-1}^{1} dx,$$

the two-point quadrature formula must be exact for a constant function, linear function, quadratic function, and cubic function, i.e. for $f(x) = 1, x, x^2, x^3$. We simply need to equate the exact value of each integral with the weighted summation formula. There are two nodes, x_1 and x_2, that both lie in the interior of the interval. We have four equations of the form

$$\int_{-1}^{1} f(x)dx = w_1f(x_1) + w_2f(x_2),$$

where $f(x) = 1, x, x^2$, or x^3. Writing out the four equations, we have

$$\begin{aligned}
f(x) &= 1: & w_1 + w_2 &= 2;\\
f(x) &= x: & w_1x_1 + w_2x_2 &= 0;\\
f(x) &= x^2: & w_1x_1^2 + w_2x_2^2 &= 2/3;\\
f(x) &= x^3: & w_1x_1^3 + w_2x_2^3 &= 0.
\end{aligned}$$

Solving the second and fourth equations simultaneously, we obtain $x_1^2 = x_2^2$. Since the two nodes must be distinct, we obtain the relation $x_1 = -x_2$. Substituting this relationship between the two nodes back into the second equation, we get $w_1 = w_2$. The first equation yields

$$w_1 = w_2 = 1,$$

and the third equation yields

$$x_1 = -\frac{1}{\sqrt{3}}, \quad x_2 = \frac{1}{\sqrt{3}}.$$

The two-point Gaussian quadrature formula is

$$I \approx f\left(-\frac{1}{\sqrt{3}}\right) + f\left(\frac{1}{\sqrt{3}}\right), \tag{6.43}$$

which has a degree of precision equal to 3. It is a linear combination of two function values obtained at two nodes equidistant from the midpoint of the interval. However, the two nodes lie slightly closer to the endpoints of their respective sign than to the midpoint of the interval.

Equation (6.43) applies only to the integration limits -1 and 1. We will need to modify it so that it can apply to any integration interval. Let x be the variable of integration that lies between the limits of closed integration a and b. If we define a new integration variable y such that $y = x - (a+b)/2$, the integral over the interval $[a, b]$ becomes

$$\int_a^b f(x)dx = \int_{-(b-a)/2}^{(b-a)/2} f\left(\frac{a+b}{2} + y\right) dy.$$

The midpoint of the new integration interval is zero. Now divide the integration variable y by half of the interval width $(b-a)/2$ to define a new integration variable $z = 2y/(b-a) = 2x/(b-a) - (a+b)/(b-a)$. The limits of integration for the variable z are -1 and 1. A linear transformation of the new variable of integration z recovers x:

$$x = \frac{b-a}{2}z + \frac{a+b}{2}.$$

Differentiating the above equation we have $dx = (b-a)/2\ dz$. The integral can now be written with the integration limits -1 and 1:

$$I = \int_a^b f(x)dx = \frac{b-a}{2}\int_{-1}^{1} f\left(\frac{b-a}{2}z + \frac{a+b}{2}\right)dz. \tag{6.44}$$

The Gaussian quadrature rule for any integration interval $[a, b]$ is

$$\int_a^b f(x)dx = \frac{b-a}{2}\int_{-1}^{1} f\left(\frac{b-a}{2}z + \frac{a+b}{2}\right)dz \approx \frac{b-a}{2}\sum_{i=0}^{n} w_i f\left(\frac{b-a}{2}z_i + \frac{a+b}{2}\right), \tag{6.45}$$

where z_i are the $n+1$ Gaussian nodes and w_i are the $n+1$ Gaussian weights.

The two-point Gaussian quadrature rule can now be applied to any integration interval using Equation (6.45):

$$\int_a^b f(x)dx \approx \frac{b-a}{2}\left[f\left(\frac{b-a}{2}\left(-\frac{1}{\sqrt{3}}\right) + \frac{a+b}{2}\right) + f\left(\frac{b-a}{2}\left(\frac{1}{\sqrt{3}}\right) + \frac{a+b}{2}\right)\right]. \tag{6.46}$$

Gaussian quadrature formulas given by Equations (6.45) and (6.46) are straightforward and easy to evaluate once the nodes and weights are known. Conversely, calculation of the optimal node points and weights is not as straightforward. One method of finding the optimal node points and corresponding weights is by solving a system of $2n + 2$ nonlinear equations, as discussed earlier. However, this method is not practical for higher precision quadrature rules since a solution to a large set of nonlinear equations is difficult to obtain. Instead, we approximate the function $f(x)$ with a Lagrange interpolation polynomial just as we had done to derive the Newton–Cotes formulas in Section 6.3:

$$
\begin{aligned}
\int_a^b f(x)dx &\approx \int_a^b \sum_{i=0}^{n}\left[f(x_i)\prod_{j=0,j\neq i}^{n}\frac{(x-x_j)}{(x_i-x_j)}\right]dx\\
&=\sum_{i=0}^{n} f(x_i)\left[\int_a^b \prod_{j=0,j\neq i}^{n}\frac{(x-x_j)}{(x_i-x_j)}dx\right]\\
&=\frac{b-a}{2}\sum_{i=0}^{n} f(x_i)\left[\int_{-1}^{1} \prod_{j=0,j\neq i}^{n}\frac{(z-z_j)}{(z_i-z_j)}dz\right]\\
&=\frac{b-a}{2}\sum_{i=0}^{n} w_i f(x_i),
\end{aligned}
$$

where

$$
w_i = \int_{-1}^{1} \prod_{j=0,j\neq i}^{n}\frac{(z-z_j)}{(z_i-z_j)}dz \tag{6.47}
$$

and

$$
x = \frac{b-a}{2}z + \frac{a+b}{2}.
$$

To calculate the weights we must first determine the location of the nodes. The $n + 1$ node positions are calculated by enforcing that the truncation error of the quadrature formula be exactly zero when integrating a polynomial function of degree $\leq 2n + 1$. The roots of the $(n + 1)$th-degree Legendre polynomial define the $n + 1$ nodes of the Gaussian quadrature formula at which $f(x)$ evaluations are required (see Appendix B for the derivation). Legendre polynomials are a class of orthogonal polynomials. This particular quadrature method is called **Gauss–Legendre quadrature**. Other Gaussian quadrature methods have also been developed, but are not discussed in this book.

The Legendre polynomials are

$$
\Phi_0(z) = 1,
$$

$$
\Phi_1(z) = z,
$$

$$
\Phi_2(z) = \frac{1}{2}(3z^2 - 1),
$$

$$
\Phi_3(z) = \frac{1}{2}(5z^3 - 3z),
$$

$$
\Phi_n(z) = \frac{1}{2^n n!}\frac{d^n}{dz^n}\left[(z^2-1)^n\right]. \tag{6.48}
$$

Equation (6.48) is called the Rodrigues' formula.

Table 6.4. *Gauss–Legendre nodes and weights*

n	Nodes, z_i (roots of Legendre polynomial $\Phi_{n+1}(z)$)	Weights, w_i
1	−0.5773502692	1.0000000000
	0.5773502692	1.0000000000
2	−0.7745966692	0.5555555556
	0.0000000000	0.8888888889
	0.7745966692	0.5555555556
3	−0.8611363116	0.3478548451
	−0.3399810436	0.6521451549
	0.3399810436	0.6521451549
	0.8611363116	0.3478548451
4	−0.9061798459	0.2369268850
	−0.5384693101	0.4786286705
	0.0000000000	0.5688888889
	0.5384693101	0.4786286705
	0.9061798459	0.2369268850

n is the degree of the polynomial interpolant

Note the following properties of the roots of a Legendre polynomial that are presented below without proof (see Grasselli and Pelinovsky (2008) for proofs).

(1) The roots of a Legendre polynomial are simple, i.e. no multiple roots (two or more roots that are equal) are encountered.
(2) All roots of a Legendre polynomial are located symmetrically about zero within the interval $(-1, 1)$.
(3) No root is located at the endpoints of the interval. Thus, Gauss–Legendre quadrature is an open-interval method.
(4) For an odd number of nodes, zero is always a node.

Once the Legendre polynomial roots are specified, the weights of the quadrature formula can be calculated by integrating the Lagrange polynomials over the interval. The weights of Gauss–Legendre quadrature formulas are always positive. If the nodes were evenly spaced, this would not be the case. Tenth- and higher-order Newton–Cotes formulas, i.e. formulas that use an eighth- or higher-degree polynomial interpolant to approximate the integrand function, have both positive and negative weights, which can lead to subtractive cancellation of significant digits and large round-off errors (see Chapter 1), undermining the accuracy of the method. Gauss–Legendre quadrature uses positive weights at all times, and thus the danger of subtractive cancellation is precluded.

The roots of the Legendre polynomials, and the corresponding weights, calculated using Equation (6.47) or by other means, have been extensively tabulated for many values of n. Table 6.4 lists the Gauss–Legendre nodes and weights for $n = 1, 2, 3, 4$.

The nodes z_i and weights w_i are defined on the interval $[-1, 1]$. A simple transformation shown earlier converts the integral from the original interval scale $[a, b]$ to $[-1, 1]$. Equation (6.45) is used to perform quadrature once the weights are calculated.

The error associated with Gauss–Legendre quadrature is presented below without proof (Ralston and Rabinowitz, 1978):

$$E = \frac{2^{2n+3}[(n+1)!]^4}{(2n+3)[(2n+2)!]^3} f^{(2n+2)}(\xi), \qquad \xi \in [a, b].$$

Example 6.1

The nodes of the three-point Gauss–Legendre quadrature are $-\sqrt{3/5}, 0, \sqrt{3/5}$. Derive the corresponding weights by integrating the Lagrange polynomials according to Equation (6.47).

We have

$$\begin{aligned} w_0 &= \int_{-1}^{1} \prod_{j=1,2}^{n} \frac{(z - z_j)}{(z_0 - z_j)} dz \\ &= \int_{-1}^{1} \frac{(z-0)\left(z - \sqrt{\frac{3}{5}}\right)}{\left(-\sqrt{\frac{3}{5}} - 0\right)\left(-\sqrt{\frac{3}{5}} - \sqrt{\frac{3}{5}}\right)} dz \\ &= \frac{5}{6}\int_{-1}^{1} \left(z^2 - \sqrt{\frac{3}{5}} z\right) dz \\ &= \frac{5}{6}\left[\frac{z^3}{3} - \sqrt{\frac{3}{5}}\frac{z^2}{2}\right]_{-1}^{1}; \end{aligned}$$

$$w_0 = \frac{5}{9}.$$

Similarly, one can show that $w_1 = 8/9$ and $w_2 = 5/9$.

Example 6.2

Use the two-point and three-point Gauss–Legendre rule and Simpson's 1/3 rule to calculate the cumulative standard normal probability,

$$\frac{1}{\sqrt{2\pi}} \int_{-2}^{2} e^{-x^2/2} dx.$$

The integrand is the density function (Equation (3.32); see Chapter 3 for details) that describes the standard normal distribution. Note that x is a normal variable with a mean of zero and a standard deviation equal to one.

The exact value of the integral to eight decimal places is

$$\int_{-2}^{2} e^{-x^2/2} dx = 2.39257603.$$

The integration limits are $a = -2$ and $b = 2$. Let $x = 2z$, where z is defined on the interval $[-1, 1]$.[5] Using Equation (6.44),

[5] Here, z stands for a variable of integration with integration limits −1 and 1. In this example, x is the standard normal variable, not z.

$$\int_{-2}^{2} e^{-x^2/2}dx = 2\int_{-1}^{1} e^{-2z^2}\, dz.$$

According to Equation (6.45), Gauss–Legendre quadrature gives us

$$\int_{-2}^{2} e^{-x^2/2}dx \approx 2\sum_{i=0}^{n} w_i e^{-2z_i^2}.$$

For $n = 1, z_{0,1} = \pm 1/\sqrt{3}, w_{0,1} = 1.0$, and
$n = 2, z_{0,2} = \pm\sqrt{3/5}, z_1 = 0.0,\ w_{0,2} = 5/9, w_1 = 8/9.$

Two-point Gauss–Legendre quadrature yields

$$\int_{-2}^{2} e^{-x^2/2}dx \approx 2\left(2 \cdot e^{-2z_0^2}\right) = 2.05366848,$$

with an error of 14.16%.

Three-point Gauss–Legendre quadrature yields

$$\int_{-2}^{2} e^{-x^2/2}dx \approx 2\left(2\frac{5}{9}e^{-2z_0^2} + \frac{8}{9}e^{-2z_1^2}\right) = 2.44709825,$$

with an error of 2.28%.

Simpson's 1/3 rule gives us

$$\int_{-2}^{2} e^{-x^2/2}\, dx \approx \frac{2}{3}\left(e^{-(-2)^2/2} + 4e^{-0^2/2} + e^{-2^2/2}\right) = 2.84711371$$

with an error of 19.0%.

The two-point quadrature method has a slightly smaller error than Simpson's 1/3 rule. The latter requires three function evaluations, i.e. one additional function evaluation than that required by the two-point Gauss–Legendre rule. The three-point Gauss–Legendre formula has an error that is an order of magnitude smaller than the error of Simpson's 1/3 rule.

To improve accuracy further, the integration interval can be split into a number of smaller intervals, and the Gauss–Legendre quadrature rule (Equation (6.45) with nodes and weights given by Table 6.4) can be applied to each subinterval. The subintervals may be chosen to be of equal width or of differing widths. When the Gaussian quadrature rule is applied individually to each subinterval of the integration interval, the quadrature scheme is called **composite Gaussian quadrature**.

Suppose the interval $[a, b]$ is divided into N subintervals of equal width equal to d, where

$$d = \frac{b-a}{N}.$$

Let n be the degree of the polynomial interpolant so that there are $n + 1$ nodes within each subinterval and a total of $(n+1)N$ nodes in the entire interval. The Gauss–Legendre quadrature rule requires that the integration limits $[x_{i-1}, x_i]$ of any subinterval be transformed to the interval $[-1, 1]$. The weighted summation formula applied to each subinterval is

$$\int_{x_{i-1}}^{x_i} f(x)dx = \frac{d}{2}\int_{-1}^{1} f\left(\frac{d}{2}z + x_{(i-0.5)}\right)dz \approx \frac{d}{2}\sum_{k=0}^{n} w_k f\left(\frac{d}{2}z_k + x_{(i-0.5)}\right), \tag{6.49}$$

where $x_{(i-0.5)}$ is the midpoint of the ith subinterval. The midpoint can be calculated as

$$x_{(i-0.5)} = \frac{x_{i-1} + x_i}{2}$$

or

$$x_{(i-0.5)} = x_0 + d\left(i - \frac{1}{2}\right).$$

The numerical approximation of the integral over the entire interval is equal to the sum of the individual approximations:

$$\int_a^b f(x)dx = \int_{x_0}^{x_1} f(x)dx + \int_{x_1}^{x_2} f(x)dx + \cdots + \int_{x_{N-1}}^{x_N} f(x)dx;$$

$$\int_a^b f(x)dx \approx \sum_{i=1}^{i=N} \frac{d}{2} \sum_{k=0}^{n} w_k f\left(\frac{d}{2} z_k + x_{(i-0.5)}\right). \tag{6.50}$$

Equation (6.50) can be automated by writing a MATLAB program, as demonstrated in Program 6.6. While the nodes and weights listed in the program correspond to the two-point quadrature scheme, nodes and weights for higher precision $(n + 1)$-point Gauss–Legendre quadrature formulas can be also incorporated in the program (or as a separate program that serves as a look-up table). By including the value of n as an input parameter, one can select the appropriate set of nodes and weights for quadrature calculations.

MATLAB program 6.6

```
function I = composite GLrule(integrandfunc, a, b, N)
% This function uses the composite 2-point Gauss-Legendre rule to
% approximate the integral.

% Input variables
% integrandfunc : function that calculates the integrand
% a : lower limit of integration
% b : upper limit of integration
% N : number of subintervals

% Output variables
% I : value of integral

% Initializing
I = 0;

% 2-point Gauss-Legendre nodes and weights
z(1) = -0.5773502692; z(2) = 0.5773502692;
w(1) = 1.0; w(2) = 1.0;

% Width of subintervals
d = (b-a)/N;

% Mid-points of subintervals
xmid = [a + d/2:d:b - d/2];
```

```
% Quadrature calculations
for i = 1: N
    % Nodes in integration subinterval
    x = xmid(i) + z*d/2;
    % Function evaluations at node points
    y = feval(integrandfunc, x);
    % Integral approximation for subinterval i
    I = I + d/2*(w*y'); % term in bracket is a dot product
end
```

Example 6.3

Use the two-point Gauss–Legendre rule and $N = 2$, 3, and 4 subintervals to calculate the following integral:

$$\int_{-2}^{2} e^{-x^2/2} dx.$$

The exact value of the integral to eight decimal places is

$$\int_{-2}^{2} e^{-x^2/2} dz = 2.39257603.$$

The integration limits are $a = -2$ and $b = 2$. This integral is solved using Program 6.6. A MATLAB function is written to evaluate the integrand.

MATLAB program 6.7

```
function f = stdnormaldensity(x)
% This function calculates the density function for the standard normal
% distribution.
f = exp(-(x.^2)/2);
```

The results are given in Table 6.5.

The number of subintervals required by the composite trapezoidal rule to achieve an accuracy of 0.007% is approximately 65.

The Gaussian quadrature method is well-suited for the integration of functions whose analytical form is known, so that the function is free to be evaluated at any point within the interval. It is usually not the method of choice when discrete values of the function to be integrated are available, in the form of tabulated data.

Table 6.5. *Estimate of integral using composite Gaussian quadrature*

N	I	Error (%)
2	2.40556055	0.543
3	2.39319351	0.026
4	2.39274171	0.007

Using MATLAB

The MATLAB function `quadl` performs numerical quadrature using a four-point adaptive Gaussian quadrature scheme called Gauss–Lobatto. The syntax for `quadl` is the same as that for the MATLAB function `quad`.

6.6 End of Chapter 6: key points to consider

(1) **Quadrature** refers to use of numerical methods to solve an integral. The integrand function is approximated by an interpolating polynomial function. Integration of the polynomial interpolant produces the following quadrature formula:

$$\int_a^b f(x)dx \approx \sum_{i=0}^{n} w_i f(x), \qquad a \le x_i \le b.$$

(2) An **interpolating polynomial function**, or **polynomial interpolant**, can be constructed by several means, such as the Vandermonde matrix method or using Lagrange interpolation formulas.

(3) The **Lagrange interpolation formula** for the nth-degree polynomial $p(x)$ is

$$p(x) = \sum_{k=0}^{n}\left[y_k \prod_{j=0, j\neq k}^{n} \frac{(x-x_j)}{(x_k-x_j)}\right] = \sum_{k=0}^{n} y_k L_k,$$

where L_k are the Lagrange polynomials.

(4) If the function values at both endpoints of the interval are included in the quadrature formula, the quadrature method is a **closed method**; otherwise the quadrature method is an **open method**.

(5) **Newton–Cotes integration formulas** are derived by approximating the integrand function with a polynomial function of degree n that is constructed from function values obtained at equally spaced nodes.

(6) The **trapezoidal rule** is a closed Newton–Cotes integration formula that uses a first-degree polynomial (straight line) to approximate the integrand function. The trapezoidal formula is

$$\int_a^b f(x)dx \approx \frac{h}{2}(f(x_0) + f(x_1)).$$

The **composite trapezoidal rule** uses a piecewise linear polynomial to approximate the integrand function over an interval divided into n segments. The degree of precision of the trapezoidal rule is 1. The error of the composite trapezoidal rule is $O(h^2)$.

(7) **Simpson's 1/3 rule** is a closed Newton–Cotes integration formula that uses a second-degree polynomial to approximate the integrand function. Simpson's 1/3 formula is

$$\int_a^b f(x)dx \approx \frac{h}{3}(f(x_0) + 4f(x_1) + f(x_2)).$$

The **composite Simpson's 1/3 rule** uses a piecewise quadratic polynomial to approximate the integrand function over an interval divided into n segments. The degree of precision of Simpson's 1/3 rule is 3. The error of composite Simpson's 1/3 rule is $O(h^4)$.

(8) **Simpson's 3/8 rule** is a Newton–Cotes closed integration formula that uses a cubic interpolating polynomial to approximate the integrand function. The formula is

$$\int_a^b f(x)dx \approx \frac{3h}{8}(f(x_0) + 3f(x_1) + 3f(x_2) + f(x_3)).$$

The error associated with basic Simpson's 3/8 rule is $O(h^5)$, which is of the same order as basic Simpson's 1/3 rule. The degree of precision of Simpson's 3/8 rule is 3. The composite rule has a truncation error of $O(h^4)$.

(9) **Richardson's extrapolation** is a numerical scheme used to combine two numerical approximations of lower order in such a way as to obtain a result of higher-order accuracy. **Romberg integration** is an algorithm that uses Richardson's extrapolation to improve the accuracy of numerical integration. The trapezoidal rule is used to generate m $O(h^2)$ approximations of the integral using a step size that doubles with each successive approximation. The approximations are then combined pairwise to obtain $(m-1)$ $O(h^4)$ second-level approximations. The $(m-2)$ third-level $O(h^6)$ approximations are computed from the second-level approximations. This process continues until the mth level approximation is attained or a user-specified tolerance is met.

(10) **Gaussian quadrature** uses an nth-degree polynomial interpolant of the integrand function to achieve a degree of precision equal to $2n + 1$. This method optimizes the location of the $n + 1$ nodes within the interval as well as the $n + 1$ weights. The node points of the **Gauss–Legendre quadrature** method are the roots of the Legendre polynomial of degree $n + 1$. Gauss–Legendre quadrature is an open-interval method. The roots are all located symmetrically in the interior of the interval. Gaussian quadrature is well suited for numerical integration of analytical functions.

6.7 Problems

6.1. **Transport of nutrients to cells suspended in cell culture media** For the problem stated in Box 6.2, construct a piecewise cubic interpolating polynomial to approximate the solute/drug concentration profile near the cell surface for $t = 40$ s. Plot the approximate concentration profile. How well does the interpolating function approximate the exact function? Compare your plot to Figure 6.8, which shows the concentration profile for $t = 10$ s. Explain in a qualitative sense how the shape of the concentration profile changes with time.

6.2. You were introduced to the following error function in Box 6.2:

$$\text{erf}(z) = \frac{2}{\sqrt{\pi}}\int_0^z e^{-z^2}dz.$$

Use the built-in MATLAB function `quad` to calculate erf(z) and plot it for $z \in [-4, 4]$. Then calculate the error between your quadrature approximation and the built-in MATLAB function `erf`. Plot the error between your quadrature approximation and the "correct" value of erf(z). Over which range is the quadrature scheme most accurate, and why?

6.3. Use the composite Simpson's 1/3 rule to evaluate numerically the following definite integral for $n = 8, 16, 32, 64$, and 128:

$$\int_0^\pi \frac{dx}{(3+5x)(x^2+49)}.$$

Compare your results to the exact value as calculated by evaluating the analytical solution, given by

$$\int \frac{dx}{(3+5x)(7^2+x^2)} = \frac{1}{(3^2+7^2 5^2)}\left[5\cdot\log|3+5x| - \frac{5}{2}\log(7^2+x^2) + \frac{3}{7}\tan^{-1}\frac{x}{7}\right]$$

at the two integration limits.

The error scales with the inverse of n raised to some power as follows:

$$\text{error} \sim \frac{1}{n^{\alpha}}$$

If you plot the log of error versus the log of the number of nodes, N, the slope of a best-fit line will be equal to $-\alpha$:

$$\log(\text{error}) = -\alpha \log n + C.$$

Construct this plot and determine the slope from a `polyfit` regression. Determine the value of the order of the rule, α. (Hint: You can use the MATLAB plotting function `loglog` to plot the error with respect to n.)

6.4. **Pharmacodynamic analysis of morphine effects on ventilatory response to hypoxia** Researchers in the Department of Anesthesiology at the University of Rochester studied the effects of intrathecal morphine on the ventilatory response to hypoxia (Bailey *et al.*, 2000). The term "intrathecal" refers to the direct introduction of the drug into the sub-arachnoid membrane space of the spinal cord. They measured the following time response of minute ventilation (l/min) for placebo, intravenous morphine, and intrathecal morphine treatments. Results are given in Table P6.1.

One method of reporting these results is to compute the "overall effect," which is the area under the curve divided by the total time interval (i.e. the time-weighted average). Compute the overall effect for placebo, intravenous, and intrathecal morphine treatments using (i) the trapezoidal quadrature method, and (ii) Simpson's 1/3 method. Note that you should separate the integral into two different intervals, since from t = 0–2 hours the panel width is 1 hour and from 2–12 hours the panel width is 2 hours.

Bailey *et al.* (2000) obtained the values shown in Table P6.2. How do your results compare to those given in that table?

Table P6.1. *Minute ventilation (l/min)*

Hours	Placebo	Intravenous	Intrathecal
0	38	34	30
1	35	19	23.5
2	42	16	20
4	35.5	16	13.5
6	35	17	10
8	34.5	20	11
10	37.5	24	13
12	41	30	19.5

Table P6.2. *Time-weighted average of minute ventilation (l/min)*

Group	Overall effect
Placebo	36.8 ± 19.2
Intravenous morphine	20.2 ± 10.8
Intrathecal morphine	14.5 ± 6.4

Figure P6.1

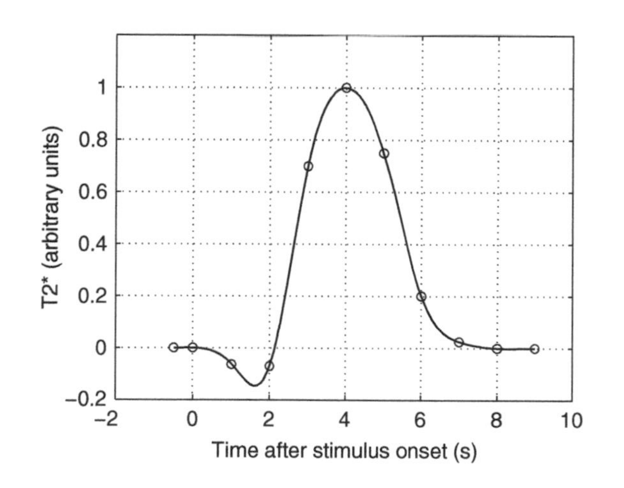

6.5. **MRI imaging in children and adults** Functional MRI imaging is a non-invasive method to image brain function. In a study by Richter and Richter (2003), the brain's response was measured following an impulse of visual stimulus. The readout of these experiments was a variable called "T2*," which represents a time constant for magnetic spin recovery. The experiments were carried out in a 3T Siemens Allegra Head Scanner, and the subject was shown a flickering checkerboard for 500 ms at a frequency of 8 Hz. The aim of the study was to look for differences in the temporal response between adults and children. Figure P6.1 resembles a typical measurement. The data upon which the plot is based are given in Table P6.3.

Such curves are characterized by calculating the area under the curve. Using Simpson's 1/3 rule, calculate, by hand, the following integral:

$$I = \int_0^8 \mathrm{T2}^* dt.$$

Since some of the area lies below the x-axis, this area will be subtracted from the positive area.

6.6. **Polymer cyclization (Jacobson–Stockmayer theory)** Polymer chains sample a large number of different orientations in space and time. The motion of a polymer chain is governed by stochastic (probabilistic) processes. Sometimes, the two ends of a linear polymer chain can approach each other within a reactive distance. If a bond forms between the two polymer ends, the reaction is termed as **cyclization.** By studying the probability with which this occurs one can estimate the rate at which

Table P6.3. *Experimental data set*

t (s)	T2*
−0.5	0
0	0
1	−0.063
2	−0.070
3	0.700
4	1.00
5	0.750
6	0.200
7	0.025
8	0
9	0

Table 6.4. *AFM data set*

Displacement (nm)	Force (nN)
10	1.2
30	1.4
50	2.1
70	5.3
90	8.6

a linear chain is converted to a circular chain. Consider a linear polymer with N links. The probability that the two ends of the chain come within a bond distance b of each other is given by the following integral:

$$\left[\frac{3}{2\pi Nb^2}\right]^{3/2} \int_0^b \exp\left(-3r^2/2Nb^2\right) 4\pi r^2 \, dr.$$

If $N = 20$ links and $b = 1$, calculate the probability the chain ends come within a distance b of each other. Use a two-segment composite trapezoidal rule and then a four-segment composite trapezoidal rule. Use these two approximations to extrapolate to an even more accurate solution by eliminating the $O(h^2)$ error term, where h is the panel width. You may perform these calculations either by hand or by writing one or more MATLAB programs to evaluate the function and carry out trapezoidal integration.

6.7. **Numerical quadrature of AFM data** An atomic force microscope is used to probe the mechanics of an endothelial cell surface. The data in Table P6.4 are collected by measuring the cantilever force during a linear impingement experiment.

Calculate the integral of these data using composite Simpson's 1/3 rule to obtain the work done on the cell surface. Express your answer in joules.

6.8. **Numerical quadrature and pharmacokinetics** In the field of pharmacokinetics, the *area under the curve* (AUC) is the area under the curve in a plot of concentration of drug in plasma against time. In real-world terms the AUC represents the total amount

Table P6.5. *Weights and nodes for the Gaussian four-point rule*

z_i	w_i
±0.8611363	0.3478548
±0.3399810	0.6521452

of drug absorbed by the body, irrespective of the rate of absorption. This is useful when trying to determine whether two formulations of the same dose (for example a capsule and a tablet) release the same dose of drug to the body. Another use is in the therapeutic monitoring of toxic drugs. For example, gentamicin is an antibiotic which displays nephro- and ototoxicities; measurement of gentamicin concentrations in a patient's plasma, and calculation of the AUC is used to guide the dosage of this drug.

(a) Suppose you want to design an experiment to calculate the AUC in a patient over a four-hour period, using a four-point Gaussian quadrature. At what times (starting the experiment at $t = 0$) should you make the plasma measurements? Show your work.

(b) Now redesign the experiment to obtain plasma measurements appropriate for using an eight-panel Simpson's rule quadrature.

Weights and nodes for the Gaussian four-point rule are given in Table P6.5.

6.9. Reformulate Equation (6.23) to show how the trapezoidal formula for step size $h/2$, $I(h/2)$, is related to the trapezoidal formula for step size h. Rewrite Program 6.5 to use your new formula to generate the first column of approximations for Romberg integration. You will not need to call the user-defined MATLAB function `trapezoidal_rule` since your program will itself perform the trapezoidal integration using the improvised formula. Verify the correctness of your program by using it to calculate the time taken for the IV drip bag to empty by 90% (see Box 6.3).

6.10. Rewrite the function `Romberg_integration` in Program 6.5 so that levels of integration are progressively increased until the specified tolerance is met according to the stopping criterion

$$\frac{R_{j,k} - R_{j,k-1}}{R_{j,k}} \leq \text{Tol}.$$

The tolerance level should be an input parameter for the function. Thus, the program will determine the maximum level required to obtain a solution with the desired accuracy. You will need to modify the program so that an additional row of integral approximations corresponding to the halved step size (including the column 1 trapezoidal rule derived approximation) is dynamically generated when the result does not meet the tolerance criterion. You may use the `trapezoidal_rule` function listed in Program 6.1 to perform the numerical integration step, or the new formula you derived in Problem 6.9 to generate the first column of approximations. Repeat the IV drip calculations discussed in Box 6.3 for tolerances Tol $= 10^{-7}$, 10^{-8}, and 10^{-9}.

6.11. Rewrite Program 6.6 to perform an $(n + 1)$-point Gauss–Legendre quadrature, where $n = 1, 2, 3$, or 4. Rework Example 6.3 for $n = 2, 3$, and 4 and N (number of segments) $= 2$. You may wish to use the `switch-case` program control statement to assign the appropriate nodes and weights to the node and weight variable based on user input. See Appendix A or MATLAB help for the appropriate syntax for `switch-case`. This control statement should be used in place of the `if-else-end` statement when several different outcomes are possible, depending on the value of the control variable.

References

Abramowitz, M. and Stegun, I. A. (1965) *Handbook of Mathematical Functions* (Mineola, NY: Dover Publications, Inc.).

Anderson, J. L. and Quinn, J. A. (1974) Restricted Transport in Small Pores. A Model for Steric Exclusion and Hindered Particle Motion. *Biophys. J.*, **14**, 130–50.

Bailey, P. L., Lu, J. K., Pace, N. L. *et al.* (2000) Effects of Intrathecal Morphine on the Ventilatory Response to Hypoxia. *N. Engl. J. Med.*, **343**, 1228–34.

Brenner, H. and Gaydos, L. J. (1977) Constrained Brownian-Movement of Spherical-Particles in Cylindrical Pores of Comparable Radius – Models of Diffusive and Convective Transport of Solute Molecules in Membranes and Porous-Media. *J. Colloid Interface Sci.*, **58**, 312–56.

Burden, R. L. and Faires, J. D. (2005) *Numerical Analysis* (Belmont, CA: Thomson Brooks/Cole).

Deen, W. M. (1987) Hindered Transport of Large Molecules in Liquid-Filled Pores. *Aiche J.*, **33**, 1409–25.

Grasselli, M. and Pelinovsky, D. (2008) *Numerical Mathematics* (Sudbury, MA: Jones and Bartlett Publishers).

Patel, V. A. (1994) *Numerical Analysis* (Fort Worth, TX: Saunders College Publishing).

Ralston, A. and Rabinowitz, P. (1978) *A First Course in Numerical Analysis* (New York: McGraw-Hill, Inc.).

Recktenwald, G. W. (2000) *Numerical Methods with Matlab Implementation and Application* (Upper Saddle River, NJ: Prentice Hall).

Richter, W. and Richter, M. (2003) The Shape of the FMRI Bold Response in Children and Adults Changes Systematically with Age. *Neuroimage*, **20**, 1122–31.

Smith, F. G. and Deen, W. M. (1983) Electrostatic Effects on the Partitioning of Spherical Colloids between Dilute Bulk Solution and Cylindrical Pores. *J. Colloid Interface Sci.*, **91**, 571–90.

7 Numerical integration of ordinary differential equations

7.1 Introduction

Modeling of dynamic processes, i.e. the behavior of physical, chemical, or biological systems that have not reached equilibrium, is done using mathematical formulations that contain one or more differential terms such as dy/dx or dC/dt. Such equations are called differential equations. The derivative in the equation indicates that a quantity is constantly evolving or changing with respect to time t or space (x, y, or z). If the quantity of interest varies with respect to a single variable only, then the equation has only one independent variable and is called an **ordinary differential equation (ODE)**. An ODE can have more than one dependent variable, but all dependent variables in the equation vary as a function of the same independent variable. In other words, all derivatives in the equation are taken with respect to one independent variable. An ODE can be contrasted with a **partial differential equation (PDE)**. A PDE contains two or more partial derivatives since the dependent variable is a function of more than one independent variable. An example of an ODE and a PDE are given below. We will not concern ourselves with PDEs in this chapter.

$$\text{ODE:} \quad \frac{dy}{dt} = 3 + 4y;$$

$$\text{PDE:} \quad \frac{\partial T}{\partial t} = 10\frac{\partial^2 T}{\partial x^2}.$$

The **order** of an ordinary differential equation is equal to the order of the highest derivative contained in the equation. A differential equation of **first order** contains only the first derivative of the dependent variable y, i.e. dy/dt. An nth-order ODE contains the nth derivative, i.e. $d^n y/dt^n$. All ODEs for which $n \geq 2$ are collectively termed **higher-order ordinary differential equations**. An analytical solution of an ODE describes the dependent variable y as a function of the independent variable t such that the functional form $y = f(t)$ satisfies the differential equation. A unique solution of the ODE is implied only if n constraints or boundary conditions are defined along with the equation, where n is equal to the order of the ODE. Why is this so? You know that the analytical solution of a first-order differential equation is obtained by performing a single integration step, which produces one integration constant. Solving an nth-order differential equation is synonymous with performing n sequential integrations or producing n constants. To determine the values of the n constants, we need to specify n conditions.

ODEs for which all n constraints apply at the beginning of the integration interval are solved as **initial-value problems**. Most initial-value ODE problems have time-dependent variables. An ODE that contains a time derivative represents a process that has not reached steady state or equilibrium.[1] Many biological and industrial processes are time-dependent. We use the t symbol to represent the independent variable of an initial-value problem (IVP). A first-order ODE is always solved as an IVP since only one boundary condition is required, which provides the initial condition, $y(0) = y_0$. An initial-value problem consisting of a higher-order ODE, such as

$$\frac{d^3y}{dt^3} = \frac{dy}{dt} + xy^2,$$

requires that, in addition to the initial condition $y(0) = y_0$, initial values of the derivatives at $t = 0$, such as $y'(0) = a; y''(0) = b$, must also be specified to obtain a solution. Sometimes boundary conditions are specified at two points of the problem, i.e. at the beginning and ending time points or spatial positions. Such problems are called **two-point boundary-value problems**. A different solution method must be used for such problems.

When the differential equation exhibits linear dependence on the derivative terms and on the dependent variable, it is called a **linear ODE**. All derivatives and terms involving the dependent variable in a linear equation have a power of 1. An example of a linear ODE is

$$e^t\frac{d^2y}{dt^2} + 3\frac{dy}{dt} + t^3 \sin t + ty = 0.$$

Note that there is no requirement for a linear ODE to be linearly dependent on the independent variable t. If some of the dependent variable terms are multiplied with each other, or are raised to multiple powers, or if the equation has transcedental functions of y, e.g. $\sin(y)$ or $\exp(y)$, the resulting ODE is **nonlinear**.

Analytical techniques are available to solve linear and nonlinear differential equations. However, not all differential equations are amenable to an analytical solution, and in practice this is more often the case. Sometimes an ODE may contain more than one dependent variable. In these situations, the ODE must be coupled with other algebraic equations or ODEs before a solution can be found. Several coupled ODEs produce a system of simultaneous ODEs. For example,

$$\frac{dy}{dt} = 3y + z + 1,$$
$$\frac{dz}{dt} = 4z + 3,$$

form a set of two simultaneous first-order linear ODEs. Finding an analytical solution to coupled differential equations can be even more challenging. Instead of

[1] A steady state process can involve a spatial gradient of a quantity, but the gradient is unchanging with time. The net driving forces are finite and maintain the gradient. There is no accumulation of depletion of the quantity of interest anywhere in the system. On the other hand, a system at equilibrium has no spatial gradient of any quantity. The net driving forces are zero.

searching for an elusive analytical solution, it can be more efficient to find the unique solution of a well-behaved (well-posed) ODE problem using numerical techniques. Any numerical method of solution for an initial-value or a boundary-value problem involves discretizing the continuous time domain (or space) into a set of finite discrete points (equally spaced or unequally spaced) or **nodes** at which we find the magnitude of $y(t)$. The value of y evolves in time according to the slope of the curve prescribed by the right-hand side of the ODE, $dy/dt = f(t, y)$. The solution at each node is obtained by marching step-wise forward in time. We begin where the solution is known, such as at $t = 0$, and use the known solutions at one or more nearby points to calculate the unknown solution at the adjacent point. If a solution is desired at an intermediate point that is not a node, it can be obtained using interpolation. The numerical solution is only approximate because it suffers from an accumulation of truncation errors and round-off errors.

In this chapter, we will discuss several classical and widely used numerical methods such as Euler methods, Runge–Kutta methods, and the predictor–corrector methods based on the Adams–Bashforth and Adams–Moulton methods for solving ordinary differential equations. For each of these methods, we will investigate the truncation error. The order of the truncation error determines the convergence rate of the error to zero and the accuracy limitations of the numerical method. Even when a well-behaved ODE has a bounded solution, a numerical solver may not necessarily find it. The conditions, or values of the step size for which a numerical method is stable, i.e. conditions under which a bounded solution can be found, are known as numerical stability criteria. These depend upon both the nature of the differential problem and the characteristics of the numerical solution method. We will consider the stability issues of each method that we discuss.

Because we are focused on introducing and applying numerical methods to solve engineering problems, we do not dwell on the rigors of the mathematics such as theorems of uniqueness and existence and their corresponding proofs. In-depth mathematical analyses of numerical methods are offered by textbooks such as *Numerical Analysis* by Burden and Faires (2005), and *Numerical Mathematics* by Grasselli and Pelinovsky (2008). In our discussion of numerical methods for ODEs, it is assumed (unless mentioned otherwise) that the dependent variable $y(t)$ is continuous and fully differentiable within the domain of interest, the derivatives of $y(t)$ of successive orders $y'(t), y''(t), \ldots$ are continuous within the defined domain, and that a unique bounded solution of the ODE or ODE system exists for the boundary (initial) conditions specified.

In this chapter, we assume that the round-off error is much smaller than the truncation error. Accordingly, we neglect the former as a component of the overall error. When numerical precision is less than double, round-off error can sizably contribute to the total error, decreasing the accuracy of the solution. Since MATLAB uses double precision by default and all our calculations will be performed using MATLAB software, we are justified in neglecting round-off error. In Sections 7.2–7.5, methods to solve initial-value problems consisting of a single equation or a set of coupled equations are discussed in detail. The numerical method chosen to solve an ODE problem will depend on the inherent properties of the equation and the limitations of the ODE solver. How one should approach stiff ODEs, which are more challenging to solve, is the topic of discussion in Section 7.6. Section 7.7 covers boundary-value problems and focuses on the shooting method, a solution technique that converts a boundary-value problem to an initial-value problem and uses iterative techniques to find the solution.

Box 7.1A HIV–1 dynamics in the blood stream

HIV–1 virus particles attack lymphocytes and hijack their genetic machinery to manufacture many virus particles within the cell. Once the viral particles have been prepared and packaged, the viral genome programs the infected host to undergo lysis. The newly assembled virions are released into the blood stream and are capable of attacking new cells and spreading the infection. The dynamics of viral infection and replication in plasma can be modeled by a set of differential equations (Perelson *et al.*, 1996). If T is the concentration of target cells, T^* is the concentration of infected cells, and V is the concentration of infectious virus particles in plasma, we can write

$$\frac{dT^*}{dt} = kVT - \delta T^*,$$

$$\frac{dV}{dt} = N\delta T^* - cV,$$

where k characterizes the rate at which the target cells T are infected by viral particles, δ is the rate of cell loss by lysis, apoptosis, or removal by the immune system, c is the rate at which viral particles are cleared from the blood stream, and N is the number of virions produced by an infected cell.

In a clinical study, an HIV–1 protease inhibitor, ritonavir, was administered to five HIV-infected individuals (Perelson *et al.*, 1996). Once the drug is absorbed into the body and into the target cells, the newly produced virus particles lose their ability to infect cells. Thus, once the drug takes effect, all newly produced virus particles are no longer infectious. The drug does not stop infectious viral particles from entering into target cells, or prevent the production of virus by existing infected cells. The dynamics of ritonavir-induced loss of infectious virus particles in plasma can be modeled with the following set of first-order ODEs:

$$\frac{dT^*}{dt} = kV_1T - \delta T^*, \tag{7.1}$$

$$\frac{dV_1}{dt} = -cV_1, \tag{7.2}$$

$$\frac{dV_X}{dt} = N\delta T^* - cV_X, \tag{7.3}$$

where V_1 is the concentration of infectious viral RNA in plasma, and V_X is the concentration of non-infectious viral particles in plasma. The experimental study yielded the following estimates of the reaction rate parameters (Perelson *et al.*, 1996):

$$\delta = 0.5/\text{day},$$

$$c = 3.0/\text{day}.$$

The initial concentrations of T, T^*, V_1, and V_X are as follows:

$V_1(t = 0) = 100/\mu l$;
$V_X(t = 0) = 0/\mu l$;
$T(t = 0) = 250$ non-infected cells/μl (Haase *et al.*, 1996);
$T^*(t = 0) = 10$ infected cells/μl (Haase *et al.*, 1996).

Based on a quasi-steady state analysis ($dT^*/dt = 0$, $dV/dt = 0$) before time $t = 0$, we calculate the following:

$k = 2 \times 10^{-4}$ μl/day/virions, and
$N = 60$ virions produced per cell.

Using these equations, we will investigate the dynamics of HIV infection following protease inhibitor therapy.

Box 7.2A Enzyme deactivation in industrial processes

Enzymes are proteins that catalyze physiological reactions, typically at body temperature, atmospheric pressure, and within a very specific pH range. The process industry has made significant advances in enzyme catalysis technology. Adoption of biological processes for industrial purposes has several key advantages over using traditional chemical methods for manufacture: (1) inherently safe operating conditions are characteristic of enzyme-catalyzed reactions, and (2) non-toxic non-polluting waste products are produced by enzymatic reactions. Applications of enzymes include manufacture of food, drugs, and detergents, carrying out intermediate steps in synthesis of chemicals, and processing of meat, hide, and alcoholic beverages.

Enzymes used in an industrial setting begin to deactivate over time, i.e. lose their activity. Enzyme denaturation may occur due to elevated process temperatures, mechanically abrasive operations that generate shearing stresses on the molecules, solution pH, and/or chemical denaturants. Sometimes trace amounts of a poison present in the broth or protein solution may bind to the active site of the enzymes and slowly but irreversibly inactivate the entire batch of enzymes.

A simple enzyme-catalyzed reaction can be modeled by the following reaction sequence:

$$E + S \rightleftharpoons ES \xrightarrow{k_2} E + P;$$

E: uncomplexed or free active enzyme,
S: substrate,
ES: enzyme–substrate complex,
P: product.

The reaction rate is given by the Michaelis–Menten equation,

$$-\frac{dS}{dt} = r = \frac{k_2 E_{\text{total}} S}{K_{\text{m}} + S}, \tag{7.4}$$

where E_{total} is the total concentration of active enzyme, i.e. $E_{\text{total}} = E + ES$.

The concentration of active enzyme progressively reduces due to gradual deactivation. Suppose that the deactivation kinetics follow a first-order process:

$$\frac{dE_{\text{total}}}{dt} = -k_d E.$$

It is assumed that only the free (unbound) enzyme can deactivate. We can write $E = E_{\text{total}} - ES$, and we have from Michaelis–Menten theory

$$ES = \frac{E_{\text{total}} S}{K_m + S}.$$

(An important assumption made by Michaelis and Menten was that $d(ES)/dt = 0$. This is a valid assumption only when the substrate concentration is much greater than the total enzyme concentration.) The above kinetic equation for dE_{total}/dt becomes

$$\frac{dE_{\text{total}}}{dt} = -\frac{k_d E_{\text{total}}}{1 + S/K_m}. \tag{7.5}$$

Equations (7.4) and (7.5) are two coupled first-order ordinary differential equations. Product formation and enzyme deactivation are functions of the substrate concentration and the total active enzyme concentration.

An enzyme catalyzes proteolysis (hydrolysis) of a substrate for which the rate parameters are

$k_2 = 21$ M/s/e.u. (e.u. $\equiv$ enzyme units);
$K_m = 0.004$ M;
$k_d = 0.03$/s.

The starting substrate and enzyme concentrations are $S_0 = 0.05$ M and $E_0 = 1 \times 10^{-6}$ units. How long will it take for the total active enzyme concentration to fall to half its initial value? Plot the change in substrate and active enzyme concentrations with time.

Box 7.3A Dynamics of an epidemic outbreak

When considering the outbreak of a communicable disease (e.g. swine flu), members of the affected population can be categorized into three groups: diseased individuals D, susceptible individuals S, and immune individuals I. In this particular epidemic model, individuals with immunity contracted the disease and recover from it, and cannot become reinfected or transmit the disease to another person. Depending on the mechanism and rate at which the disease spreads, the disease either washes out, or eventually spreads throughout the entire population, i.e. the disease reaches epidemic proportions. Let k be a rate quantity that characterizes the ease with which the infection spreads from an infected person to an uninfected person. Let the birth rate per individual within the population be β and the death rate be μ.

The change in the number of susceptible individuals in the population can be modeled by the deterministic equation

$$\frac{dS}{dt} = -kSD + \beta(S + I) - \mu S. \tag{7.6}$$

Newborns are always born into the susceptible population. When a person recovers from the disease, he or she is permanently immune. A person remains sick for a time period of γ, which is the average infectious period.

The dynamics of infection are represented by the following equations:

$$\frac{dD}{dt} = kSD - \frac{1}{\gamma}D, \tag{7.7}$$

$$\frac{dI}{dt} = \frac{1}{\gamma}D - \mu I. \tag{7.8}$$

It is assumed that the recovery period is short enough that one does not generally die of natural causes while infected. We will study the behavior of these equations for two sets of values of the defining parameters of the epidemic model.

Box 7.4A Microbial population dynamics

In ecological systems, multiple species interact with each other in a variety of ways. When animals compete for the same food, this interaction is called **competition**. A carnivorous animal may prey on a herbivore in a **predator–prey** type of interaction. Several distinct forms of interaction are observed among microbial mixed-culture populations. Two organisms may exchange essential nutrients to benefit one another mutually. When each species of the mixed population fares better in the presence of the other species as opposed to under isolated conditions, the population interaction dynamics is called **mutualism**. Sometimes, in a two-population system, only one of the two species may derive an advantage (**commensalism**). For example, a yeast may metabolize glucose to produce an end product that may serve as bacterial nourishment. The yeast however may not derive any benefit from co-existing with the bacterial species. In another form of interaction, the presence of one species may harm or inhibit the growth of another, e.g. the metabolite of one species may be toxic to the other. In this case, growth of one species in the mixed culture is adversely affected, while the growth of the second species is unaffected (**amensalism**).

Microbes play a variety of indispensable roles both in the natural environment and in industrial processes. Bacteria and fungi decompose organic matter such as cellulose and proteins into simpler carbon compounds that are released to the atmosphere in inorganic form as CO_2 and bicarbonates. They are responsible for recirculating N, S, and O through the atmosphere, for instance by fixing atmospheric nitrogen, and by breaking down proteins and nucleic acids into simpler nitrogen compounds such as NH_3, nitrites, and nitrates. Mixed cultures of microbes are used in cheese production, in fermentation broths to manufacture alcoholic beverages such as whiskey, rum, and beer, and to manufacture vitamins. Domestic (sewage) wastewater and industrial aqueous waste must be treated before being released into natural water bodies. Processes such as activated sludge, trickling biological filtration, and anaerobic digestion of sludge make use of a complex mix of sludge, sewage bacteria, and bacteria-consuming protozoa, to digest the organic and inorganic waste.

A well-known mathematical model that simulates the oscillatory behavior of predator and prey populations (e.g. plant–herbivore, insect pest–control agent, and herbivore–carnivore) in a two-species ecological system is the **Lotka–Volterra model**. Let N_1 be the number of prey in an isolated habitat, and let N_2 be the number of predators localized to the same region. We make the assumption that the prey is the only source of food for the predator. According to this model, the birth rate of prey is proportional to the size of its population, while the consumption rate of prey is proportional to the number of predators multiplied by the number of prey. The Lotka–Volterra model for predator–prey dynamics is presented below:

$$\frac{dN_1}{dt} = bN_1 - \gamma N_1 N_2, \tag{7.9}$$

$$\frac{dN_2}{dt} = \varepsilon\gamma N_1 N_2 - dN_2. \tag{7.10}$$

The parameter γ describes the ease or difficulty with which a prey can be killed and devoured by the predator, and together the term $\gamma N_1 N_2$ equals the frequency of killing events. The survival and growth of the predator population depends on the quantity of food available (N_1) and the size of the predator population (N_2); ε is a measure of how quickly the predator population reproduces for each successfully hunted prey; d is the death rate for the predator population. A drawback of this model is that it assumes an unlimited supply of food for the prey. Thus, in the absence of predators, the prey population grows exponentially with no bounds.

A modified version of the Lotka–Volterra model uses the Monod equation (see Box 1.6 for a discussion on Monod growth kinetics) to model the growth of bacterial prey as a nonlinear function of substrate concentration, and the growth of a microbial predator as a function of prey concentration (Bailey and Ollis, 1986). This model was developed by Tsuchiya *et al.* (1972) for mixed population microbial dynamics in a chemostat (a continuous well-stirred flow reactor containing a uniform suspension of microbes). A mass balance performed on the substrate over the chemostat apparatus (see Figure P7.2 in Problem 7.5), assuming constant density and equal flowrates for the feed stream and product stream, yields the following time-dependent differential equation:

$$V\frac{dS}{dt} = F(S_0 - S) - Vr_S,$$

where V is the volume of the reactor, F is the flowrate, S_0 is the substrate concentration in the feed stream, and r_S is the consumption rate of substrate. Growth of the microbial population X that feeds on the substrate S is described by the kinetic rate equation

$$r_X = \mu X,$$

where X is the cell mass, and the dependency of μ, the specific growth rate, on S is prescribed by the **Monod equation**

$$\mu = \frac{\mu_{\max} S}{K_S + S},$$

where $\mu_{\max}$ is the maximum specific growth rate for X when S is not limiting. The substrate is consumed at a rate equal to

$$r_S = \frac{1}{Y_{X|S}} \frac{\mu_{S,\max} SX}{K_S + S},$$

where $Y_{X|S}$ is the called the **yield factor** and is defined as follows:

$$Y_{X|S} = \frac{\text{increase in cell mass}}{\text{quantity of substrate consumed}}.$$

The predictive Tsuchiya equations of *E. coli* consumption by the amoeba *Dictyostelium discoidium* are listed below (Bailey and Ollis, 1986; Tsuchiya *et al.*, 1972). The limiting substrate is glucose. It is assumed that the feed stream is sterile.

$$\frac{dS}{dt} = \frac{F}{V}(S_0 - S) - \frac{1}{Y_{N_1|S}} \frac{\mu_{N_1,\max} S N_1}{K_{N_1} + S}; \tag{7.11}$$

$$\frac{dN_1}{dt} = -\frac{F}{V} N_1 + \frac{\mu_{N_1,\max} S N_1}{K_{N_1} + S} - \frac{1}{Y_{N_2|N_1}} \frac{\mu_{N_2,\max} N_1 N_2}{K_{N_2} + N_1}; \tag{7.12}$$

$$\frac{dN_2}{dt} = -\frac{F}{V} N_2 + \frac{\mu_{N_2,\max} N_1 N_2}{K_{N_2} + N_1}. \tag{7.13}$$

The kinetic parameters that define the predator–prey system are (Tsuchiya *et al.*, 1972):

$\frac{F}{V} = 0.0625$/hr;
$S_0 = 0.5$ mg/ml;
$\mu_{N_1,\max} = 0.25$/hr, $\mu_{N_2,\max} = 0.24$/hr;
$K_{N_1} = 5 \times 10^{-4}$ mg glucose/ml, $K_{N_2} = 4 \times 10^8$ bacteria/ml;
$1/Y_{N_1|S} = 3.3 \times 10^{-10}$ mg glucose/bacterium;
$1/Y_{N_2|N_1} = 1.4 \times 10^3$ bacteria/amoeba.

At $t = 0$, $N_1 = 13 \times 10^8$ bacteria/ml and $N_2 = 4 \times 10^5$ amoeba/ml. Integrate this system of equations to observe the population dynamics of this predator–prey system. What happens to the dynamics when a step change in the feed substrate concentration is introduced into the flow system?

7.2 Euler's methods

The simplest method devised to integrate a first-order ODE is the **explicit Euler method**. Other ODE integration methods may be more complex, but their underlying theory parallels the algorithm of Euler's method. Although Euler's method is seldom used, its algorithm is the most basic of all methods and therefore serves as a suitable introduction to the numerical integration of ODEs. Prior to performing numerical integration, the first-order ODE should be expressed in the following format:

$$\frac{dy}{dt} = f(t, y), \qquad y(t = 0) = y_0. \tag{7.14}$$

$f(t, y)$ is the instantaneous slope of the function $y(t)$ at the point (t, y). The slope of y is, in general, a function of t and y and varies continuously across the spectrum of values $0 < t < t_f$ and $a < y < b$. When no constraint is specified, infinite solutions exist to the first-order ODE, $dy/dt = f(t, y)$. By specifying an initial condition for y, i.e. $y(t = 0) = y_0$, one fixes the initial point on the path. From this starting point, a unique path can be traced for y starting from $t = 0$ by calculating the slope, $f(t, y)$, at each point of t for $0 < t < t_f$. The solution is said to be unique because, at any point t, the corresponding value of y cannot have two simultaneous values. There are additional mathematical constraints that must be met to guarantee a unique solution for a first-order ODE, but these theorems will not be discussed here. We will assume in this chapter that the mathematical requirements for uniqueness of a solution are satisfied.

A first-order ODE will always be solved as an **initial-value problem (IVP)** because only one constraint must be imposed to obtain a unique solution, and the constraining value of y will lie either at the initial point or the final point of the integration interval. If the constraint provided is at the final or end time t_f, i.e. $y(t_f) = y_f$, then the ODE integration proceeds backward in time from t_f to zero, rather than forward in time. However, this problem is mathematically equivalent to an IVP with the constraint provided at $t = 0$.

To perform numerical ODE integration, the interval $[0, \ t_f]$ of integration must be divided into N subintervals of equal size. The size of one subinterval is called the **step size**, and is represented by h. We will consider ODE integration with *varying* step sizes in Section 7.4. Thus, the interval is discretized into $N = t_f/h + 1$ points, and the differential equation is solved only at those points. The ODE is integrated using a forward-marching numerical scheme. The solution $y(t)$ is calculated at the endpoint of each subinterval, i.e. at $t = t_1, t_2, \ldots, t_f$, in a step-wise fashion. The numerical solution of y is therefore not continuous, but, with a small enough step size, the solution will be sufficiently descriptive and reasonably accurate.

7.2.1 Euler's forward method

In Euler's explicit method, h is chosen to be small enough such that the slope of the function $y(t)$ within any subinterval remains approximately constant. Using the first-order forward finite difference formula (see Section 1.6.3 for a discussion on finite difference formulas and numerical differentiation) to approximate a first-order derivative, we can rewrite dy/dt within the subinterval $[t_0, \ t_1]$ in terms of a numerical derivative:

$$\left.\frac{dy}{dt}\right|_{t_0=0} = f(0, y_0) \approx \frac{y_1 - y(0)}{t_1 - 0} = \frac{y_1 - y_0}{h}.$$

The above equation is rearranged to obtain

$$y_1 = y_0 + hf(0, y_0).$$

Here, we have converted a differential equation into a **difference equation** that allows us to calculate the value of y at time t_1, using the slope at the previous point t_0. Because the forward finite difference is used to derive the difference equation, this method is also commonly known as **Euler's forward method**. Note that y_1 is not the same as $y(t_1)$; y_1 is the approximate (or numerical) solution of y at t_1. The exact or true solution of the ODE at t_1 is denoted as $y(t_1)$. The difference equation is inherently associated with a truncation error (called the *local truncation error*) whose order of magnitude we will determine shortly.

Euler's forward method assumes that the slope of the y trajectory is constant within the subinterval and is equal to the initial slope. If the step size is sufficiently small and/or the slope changes slowly, then the assumption is acceptable and the error that is generated by the numerical approximation is small. If the trajectory curves sharply within the subinterval, the constancy of slope assumption is invalid and a large numerical error will be produced by this difference equation.

Because the values on the right-hand side of the above difference equation are specified by the initial condition and are therefore known to us, the solution technique is said to be **explicit**, allowing y_1 to be calculated directly. Once y_1 is calculated, we can obtain the approximate solution y_2 for y at t_2, which is calculated as

$$y_2 = y_1 + hf(t_1, y_1).$$

Figure 7.1

Numerical integration of an ODE using Euler's forward method.

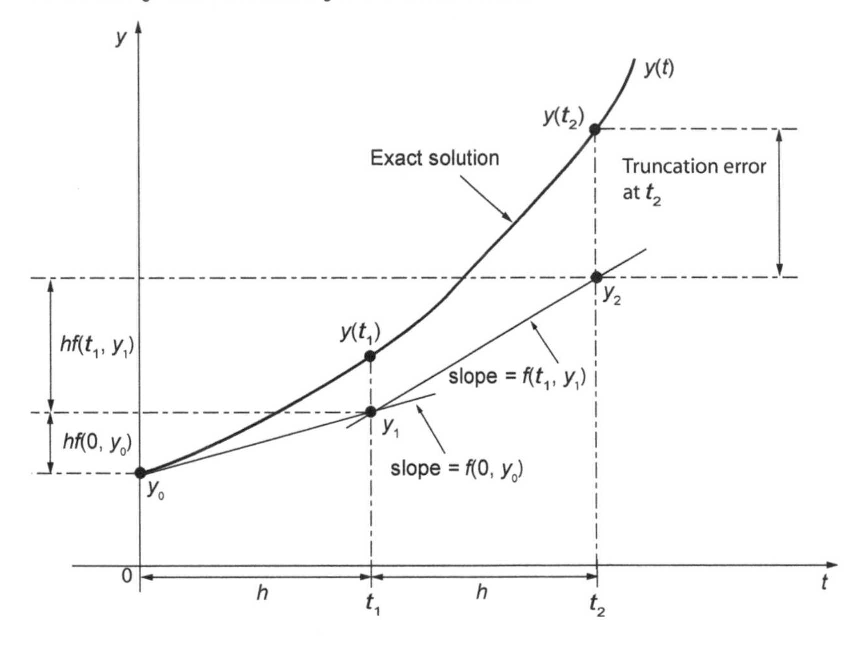

Since y_1 is the approximate solution for $y(t_1)$, errors involved in estimating $y(t_1)$ will be included in the estimate for $y(t_2)$. Thus, in addition to the inherent truncation error at the present step associated with using the numerical derivative approximation, we also observe a propagation of previous errors as we march forward from $t = 0$ to $t = t_f$. Figure 7.1 illustrates the step-wise marching technique used by Euler's method to find the approximate solution, at the first two time points of the discretized interval.

We can generalize the above equation for any subinterval $[t_k, \ t_{k+1}]$, where $k = 0, 1, 2, 3, \ldots, N-1$ and $t_{k+1} - t_k = h$:

$$y_{k+1} = y_k + hf(t_k, y_k). \tag{7.15}$$

Note that, while the true solution $y(t)$ is continuous, the numerical solution, $y_1, y_2, \ldots, y_k, y_{k+1}, \ldots, y_N$, is discrete.

Local and global truncation errors

Let us now assess the magnitude of the truncation error involved with Euler's explicit method. There are two types of errors produced by an ODE numerical integration scheme. The first type of error is the **local truncation error** that is generated at every step. The local truncation error generated in the first step is carried forward to the next step and thereby propagates through the numerical solution to the final step. As each step produces a new local truncation error, these errors accumulate as the numerical calculation progresses from one iteration to the next. The total error at any step, which is the sum of the local truncation error and the propagated errors, is called the **global truncation error**.

When we perform the first step of the numerical ODE integration, the starting value of y is known exactly. We can express the exact solution at $t = t_1$ in terms of the solution at t_0 using the Taylor series (see Section 1.6 for a discussion on Taylor series) as follows:

$$y(t_1) = y(t_0) + y'(t_0)h + \frac{y''(\xi)h^2}{2!}, \qquad \xi \in [t_0, t_1].$$

The final term on the right-hand side is the remainder term discussed in Section 1.6. Here, the remainder term is the sum total of all the terms in the infinite series except for the first two. The first two terms on the right-hand side of the equation are already known:

$$y(t_1) = y_0 + hf(0, y_0) + \frac{y''(\xi)h^2}{2!}$$

or

$$y(t_1) = y_1 + \frac{y''(\xi)h^2}{2!}.$$

Therefore,

$$y(t_1) - y_1 = \frac{y''(\xi)h^2}{2!}.$$

If $|y''(t)| \leq M$ for $0 \leq t \leq t_f$, then

$$y(t_1) - y_1 \leq \frac{Mh^2}{2!} \sim O(h^2). \tag{7.16}$$

The order of magnitude of the local truncation error associated with the numerical solution of y at t_1 is h^2. For the first step, the local truncation error is equal to the global truncation error. The local truncation error generated at each step is $O(h^2)$. For the second step, we express $y(t_2)$ in terms of the exact solution at the previous step,

$$y(t_2) = y(t_1) + hf(t_1, y(t_1)) + \frac{y''(\xi)h^2}{2!}, \qquad \xi \in [t_1, t_2].$$

Since $y(t_1)$ is unknown, we must use y_1 in its place. The truncation error associated with the numerical solution for $y(t_2)$ is equal to the sum of the local truncation error, $y''(\xi)h^2/2!$, generated at this step, the local truncation error associated with y_1, and the error in calculating the slope at t_1, i.e. $f(t_1, y(t_1)) - f(t_1, y_1)$. At the $(k+1)$th step, the true solution $y(t_{k+1})$ is expressed as the sum of terms involving $y(t_k)$:

$$y(t_{k+1}) = y(t_k) + hf(t_k, y(t_k)) + \frac{y''(\xi)h^2}{2!}, \qquad \xi \in [t_k, t_{k+1}]. \tag{7.17}$$

Subtracting Equation (7.15) from Equation (7.17), we obtain the global truncation error in the numerical solution at t_{k+1}:

$$y(t_{k+1}) - y_{k+1} = y(t_k) - y_k + h(f(t_k, y(t_k)) - f(t_k, y_k)) + \frac{y''(\xi)h^2}{2!}. \tag{7.18}$$

We use the mean value theorem to simplify the above equation. If $f(y)$ is continuous and differentiable within the interval $y \in [a, b]$, then the **mean value theorem** states that

$$\frac{f(b)-f(a)}{b-a}=\left.\frac{df(y)}{dy}\right|_{y=c},\tag{7.19}$$

where c is a number within the interval $[a, b]$. In other words, the slope of the line joining the endpoints of the interval $(a, f(a))$ and $(b, f(b))$ is equal to the derivative of $f(y)$ at the point c, which lies within the same interval.

Applying the mean value theorem (Equation (7.19)) to Equation (7.18), we obtain

$$y(t_{k+1})-y_{k+1}=y(t_k)-y_k+h\frac{\partial f(t_k,c)}{\partial y}(y(t_k)-y_k)+\frac{y''(\xi)h^2}{2!}$$

or

$$y(t_{k+1})-y_{k+1}=\left(1+h\frac{\partial f(t_k,c)}{\partial y}\right)(y(t_k)-y_k)+\frac{y''(\xi)h^2}{2!}.$$

If $|y''(t)|\leq M$ for $0\leq t\leq t_f$, $|\partial f(t,y)/\partial y|\leq C$ for $0\leq t\leq t_f$ and $a\leq y\leq b$ (also called the Lipschitz criterion), then[2]

$$y(t_{k+1})-y_{k+1}\leq(1+hC)(y(t_k)-y_k)+\frac{Mh^2}{2}.\tag{7.20}$$

Now, $y(t_k)-y_k$ is the global truncation error of the numerical solution at t_k. Since Equation (7.20) is valid for $k\geq 1$, we can write

$$y(t_k)-y_k\leq(1+hC)(y(t_{k-1})-y_{k-1})+\frac{Mh^2}{2}.$$

Substituting the above expression into Equation (7.19), we get

$$y(t_{k+1})-y_{k+1}\leq(1+hC)\left((1+hC)(y(t_{k-1})-y_{k-1})+\frac{Mh^2}{2}\right)+\frac{Mh^2}{2}$$

or

$$y(t_{k+1})-y_{k+1}\leq(1+hC)^2(y(t_{k-1})-y_{k-1})+(1+(1+hC))\frac{Mh^2}{2}.$$

Performing the substitutions for the global truncation error at $t_{k-1}, t_{k-2}, \ldots, t_2$ recursively, we obtain

$$\begin{aligned}y(t_{k+1})-y_{k+1}\leq&(1+hC)^k(y(t_1)-y_1)\\&+\left(1+(1+hC)+(1+hC)^2+\cdots+(1+hC)^{k-1}\right)\frac{Mh^2}{2};\end{aligned}$$

$y(t_1)-y_1$ is given by Equation (7.16) and only involves the local truncation error. The equation above becomes

$$y(t_{k+1})-y_{k+1}\leq\left(1+(1+hC)+(1+hC)^2+\cdots+(1+hC)^{k-1}+(1+hC)^k\right)\frac{Mh^2}{2}.$$

The sum of the geometric series $1+r+r^2+\cdots+r^n$ is $(r^{n+1}-1)/(r-1)$, and we use this result to simplify the above to

[2] C is some constant value and should not be confused with c, which is a point between a and b on the y-axis.

$$y(t_{k+1}) - y_{k+1} \leq \frac{(1+hC)^{k+1} - 1}{(1+hC) - 1} \cdot \frac{Mh^2}{2}. \tag{7.21}$$

Using the Taylor series, we can expand e^{hC} as follows:

$$e^{hC} = 1 + hC + \frac{(hC)^2}{2!} + \frac{(hC)^3}{3!} + \cdots.$$

Therefore,

$$e^{hC} > 1 + hC.$$

Substituting the above into 7.21,

$$y(t_{k+1}) - y_{k+1} < \frac{e^{h(k+1)C} - 1}{hC} \cdot \frac{Mh^2}{2}$$

If $k + 1 = N$, then $hN = t_f$, and the final result for the global truncation error at t_f is

$$y(t_N) - y_N < \left(e^{t_f C} - 1\right) \cdot \frac{Mh}{2C} \sim O(h). \tag{7.22}$$

The most important result of this derivation is that the **global truncation error** of **Euler's forward method is** $O(h)$. Euler's forward method is called a **first-order method.** Note that the order of magnitude of the global truncation error is one order less than that of the local truncation error. In fact, for any numerical ODE integration technique, the global truncation error is one order of magnitude less than the local truncation error for that method.

Example 7.1

We use Euler's method to solve the linear first-order ODE

$$\frac{dy}{dt} = t - y, \qquad y(0) = 1. \tag{7.23}$$

The exact solution of Equation (7.23) is $y = t - 1 + 2e^{-t}$, which we will use to compare with the numerical solution.

To solve this ODE, we create two m-files. The first is a general purpose function file named `eulerforwardmethod.m` that solves any first-order ODE using Euler's explicit method. The second is a function file named `simplelinearODE.m` that evaluates $f(t, y)$, i.e. the right-hand side of the ODE specified in the problem statement. The name of the second function file is supplied to the first function. The function `feval` (see Chapter 5 for usage) is used to pass the values of t and y to `simplelinearODE.m` and calculate $f(t, y)$ at the point (t, y).

MATLAB program 7.1

```
function [t, y] = eulerforwardmethod(odefunc, tf, y0, h)
% Euler's forward method is used to solve a first-order ODE

% Input variables
% odefunc : name of the function that calculates f(t, y)
% tf : final time or size of interval
% y0 : y(0)
% h : step size

% Output variables
```

```
t = [0:h:tf]; % vector of time points
y = zeros(size(t)); % dependent variable vector
y(1) = y0; % indexing of vector elements begins with 1 in MATLAB

% Euler's forward method for solving a first-order ODE
for k = 1:length(t)-1
    y(k+1) = y(k) + h*feval(odefunc, t(k), y(k));
end
```

MATLAB program 7.2

```
function f = simplelinearODE(t, y)
% Evaluate slope f(t,y) = t - y
f = t - y;
```

MATLAB program 7.1 can be called from the command line

```
>> [t, y] = eulerforwardmethod('simplelinearODE', 3, 1, 0.5)
```

Figure 7.2 shows the trajectory of y obtained by the numerical solution with step sizes $h = 0.5$, 0.2, and 0.1, and compares this with the exact solution.

We see from the figure that the accuracy of the numerical scheme increases with decreasing h. The maximum global truncation error for each trajectory is shown in Table 7.1.

Table 7.1. *Change in maximum global truncation error with step size*

Numerical solution	Step size	Maximum global truncation error
1	0.5	0.2358
2	0.2	0.0804
3	0.1	0.0384

Figure 7.2

Numerical solution of Equation (7.23) using Euler's forward method for three different step sizes.

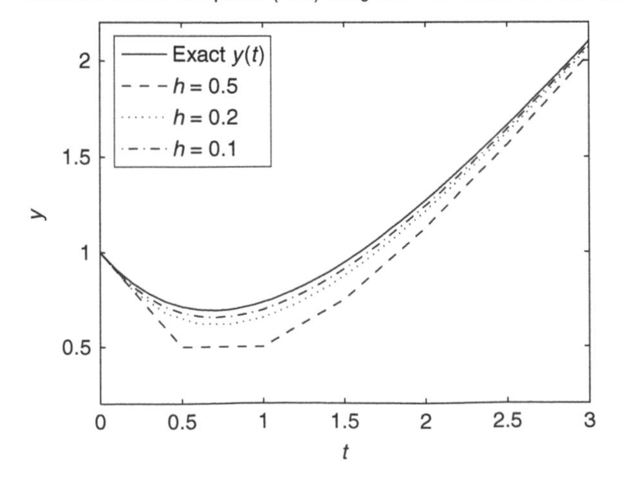

For a decrease in step size of $0.2/0.5 = 0.4$, the corresponding decrease in maximum global truncation error is $0.0804/0.2358 = 0.341$. The global error is reduced by the same order of magnitude as the step size. Similarly, a decrease in step size of $0.1/0.2 = 0.5$ produces a decrease in maximum global truncation error of $0.0384/0.0804 = 0.478$. The maximum truncation error reduces proportionately with a decrease in h. Thus, the error convergence to zero is observed to be $O(h)$, as predicted by Equation (7.22). Therefore, if h is cut by half, the total truncation error will also be reduced by approximately half.

Stability issues

Let us consider the simple first-order ODE

$$\frac{dy}{dt} = \lambda y, \qquad y(0) = \alpha. \tag{7.24}$$

The exact solution of Equation (7.24) is $y = \alpha e^{\lambda t}$. We can use the exact solution to benchmark the accuracy of Euler's forward method and characterize the limitations of this method. The time interval over which the ODE is integrated is divided into N equally spaced time points. The uniform step size is h.

At $t = t_1$, the numerical solution is

$$y_1 = y(0) + h\lambda y(0) = \alpha + h\lambda\alpha = \alpha(1 + \lambda h).$$

At $t = t_2$,

$$y_2 = y_1 + h\lambda y_1 = (1 + \lambda h)y_1 = \alpha(1 + \lambda h)^2.$$

At $t = t_k$,

$$y_{k+1} = (1 + \lambda h)y_k = \alpha(1 + \lambda h)^{k+1}.$$

And at $t = t_N$,

$$y_N = \alpha(1 + \lambda h)^N. \tag{7.25}$$

Fortunately, for this problem, we have arrived at a convenient expression for the numerical solution of y at any time t. We use this simple expression to illustrate a few basic concepts about Euler's method. If the step size $h \to 0$, the number of increments $N \to \infty$.

The expression in Equation (7.25) can be expanded using the binomial theorem as follows:

$$(1 + \lambda h)^N = 1^N + N\lambda\frac{t_f}{N} \cdot 1^{N-1} + \frac{N(N-1)}{2!}\left(\lambda\frac{t_f}{N}\right)^2 1^{N-2} + \cdots,$$

where $h = t_f/N$. When $N \to \infty$, $h \to 0$, and $\lambda h \ll 1$. The above expression becomes

$$(1 + \lambda h)^\infty \sim 1 + \lambda t_f + \frac{(\lambda t_f)^2}{2!} + \cdots = e^{\lambda t_f}.$$

Therefore,

$$y_{t_f} = \alpha e^{\lambda t_f}.$$

When $h \to 0$, the numerical solution approaches the exact solution. This example shows that for a very small step size, Euler's method will converge upon the true solution. For very small h, a large number of steps must be taken before the end of the interval is reached. A very large value of N has several disadvantages: (1) the method becomes computationally expensive in terms of the number of required operations; (2) round-off errors can dominate the total error and any improvement in accuracy is lost; (3) round-off errors will accumulate and become relatively large because now many, many steps are involved. This will corrupt the final solution at $t = t_f$.

What happens when h is not small enough? Let's first take a closer look at the exact solution. If $\lambda > 0$, the true solution blows up at large times and $y \to \infty$. We are not interested in the unstable problem. We consider only those situations in which $\lambda < 0$. For these values of λ the solution decays with time to zero. Examples of physical situations in which this equation applies are radioactive decay, drug elimination by the kidneys, and first-order (irreversible) consumption of a chemical reactant.

In a first-order decay problem,

$$\frac{dy}{dt} = -\lambda y, \quad \lambda > 0. \ y(0) > 0$$

According to the exact solution, it is always the case that $y(t_2) < y(t_1)$, where $t_2 > t_1$. The numerical solution of this ODE problem is $y_{k+1} = (1 - \lambda h) y_k$. When $|1 - \lambda h| < 1$, then we will always have $y_{k+1} < y_k$. To achieve a decaying solution, h must be constrained to the range $-1 < 1 - \lambda h < 1$ or $0 < h < 2/\lambda$. Note that h is always positive in a forward-marching scheme, so the condition that assures a decaying numerical solution is $h < 2/\lambda$. If $0 < h < 1/\lambda$, then the decay is monotonic. If $1/\lambda < h < 2/\lambda$, an oscillatory decay is observed. What happens when $h = 1/\lambda$? If $h = 2/\lambda$, then the numerical solution does not decay but instead oscillates from $-\alpha$ to α at every time step.

When $h \leq 2/\lambda$, the numerical solution is bounded and the numerical solution is said to be **stable** over this range of step size values. The accuracy of the solution will depend on the size of h. When $h > 2/\lambda$, the numerical solution diverges with time with sign reversal at each step and tends towards infinity. The solution blows up because the errors grow (as opposed to decay) exponentially and cause the solution to blow up. For this range of h values, the numerical solution is **unstable**. The magnitude of λ defines the upper bound for the step size. A large λ requires a very small h to ensure stability of the numerical technique. However, if h is too small, then the limits of numerical precision may be reached and round-off error will preclude any improvement in accuracy.

The stability of numerical integration is determined by the ODE to be solved, the step size, and the numerical integration scheme. We explore the stability limits of Euler's method by looking at how the error accumulates with each step. Equation (7.20) expresses the global truncation error at t_{k+1} in terms of the error at t_k:

$$y(t_{k+1}) - y_{k+1} \leq \left(1 + h\frac{\partial f}{\partial y}\right)(y(t_k) - y_k) + \frac{Mh^2}{2}. \tag{7.20}$$

Note that the global error at t_k, which is given by $y(t_k) - y_k$, is multiplied by an **amplification factor**, which for Euler's forward method is $1 + h(\partial f/\partial y)$. If $(\partial f/\partial y)$ is positive, then $1 + h(\partial f/\partial y) > 1$, and the global error will be amplified regardless of the step size. Even if the exact solution is bounded, the numerical solution may accumulate large errors along the path of integration. If $(\partial f/\partial y)$ is negative and h is chosen such that $-1 \leq 1 + h(\partial f/\partial y) < 1$, then the error remains bounded for all times within the time interval. However, if $(\partial f/\partial y)$ is very negative, then $|(\partial f/\partial y)|$ is large. If h is not small enough, then $1 + h(\partial f/\partial y) < -1$ or $h(\partial f/\partial y) < -2$. The errors will be magnified at each step, and the numerical method is rendered unstable.

ODEs for which $df(t, y)/dy \ll -1$ over some of the time interval are called **stiff equations**. Section 7.6 discusses this topic in detail.

Coupled ODEs

When two or more first-order ODEs are coupled, the set of differential equations are solved simultaneously as a system of first-order ODEs as follows:

$$\frac{dy_1}{dt} = f_1(t, y_1, y_2, \ldots, y_n)$$
$$\frac{dy_2}{dt} = f_2(t, y_1, y_2, \ldots, y_n)$$
$$\vdots$$
$$\frac{dy_n}{dt} = f_n(t, y_1, y_2, \ldots, y_n).$$

A system of ODEs can be represented compactly using vector notation. If $\mathbf{y} = [y_1, y_2, \ldots, y_n]$ and $\mathbf{f} = [f_1, f_2, \ldots, f_n]$, then we can write

$$\frac{d\mathbf{y}}{dt} = \mathbf{f}(t, \mathbf{y}), \qquad \mathbf{y}(0) = \mathbf{y}_0, \tag{7.26}$$

where the bold notation represents a vector (lower-case) or matrix (upper-case) (notation to distinguish vectors from scalars is discussed in Section 2.1) as opposed to a scalar.

A set of ODEs can be solved using Euler's explicit formula. For a three-variable system,

$$\frac{dy_1}{dt} = f_1(t, y_1, y_2), \qquad y_1(0) = y_{1,0},$$
$$\frac{dy_2}{dt} = f_2(t, y_1, y_2), \qquad y_2(0) = y_{2,0},$$

the numerical formula prescribed by Euler's method is

$$y_{1,k+1} = y_{1,k} + hf_1\left(t_k, y_{1,k}, y_{2,k}\right),$$
$$y_{2,k+1} = y_{2,k} + hf_2\left(t_k, y_{1,k}, y_{2,k}\right).$$

The first subscript on y identifies the dependent variable and the second subscript denotes the time point. The same forward-marching scheme is performed. At each time step, t_{k+1}, $y_{1,k+1}$, and $y_{2,k+1}$ are calculated, and the iteration proceeds to the next time step. To solve a higher-order ordinary differential equation, or a system of ODEs that contains at least one higher-order differential equation, we need to convert the problem to a mathematically equivalent system of coupled, first-order ODEs.

Example 7.2

A spherical cell or particle of radius a and uniform density ρ_s, is falling under gravity in a liquid of density ρ_f and viscosity μ. The motion of the sphere, when wall effects are minimal, can be described by the equation

$$\frac{4}{3}\pi a^3 \rho_s \frac{dv}{dt} + 6\mu\pi a v - \frac{4}{3}\pi a^3 (\rho_s - \rho_f) g = 0,$$

or, using $v = dz/dt$,

$$\frac{4}{3}\pi a^3 \rho_s \frac{d^2z}{dt^2} + 6\mu\pi a \frac{dz}{dt} - \frac{4}{3}\pi a^3 (\rho_s - \rho_f) g = 0,$$

with initial boundary conditions

$$z(0) = 0; \qquad v = \frac{dz}{dt} = 0,$$

where z is the displacement distance of the sphere in the downward direction. Downward motion is along the positive z direction. On the left-hand side of the equation, the first term describes the acceleration, the second describes viscous drag, which is proportional to the sphere velocity, and the third is the force due to gravity acting downwards from which the buoyancy force has been subtracted. Two initial conditions are included to specify the problem completely. This initial value problem can be converted from a single second-order ODE into a set of two coupled first-order ODEs.

Set $y_1 = z$ and $y_2 = dz/dt$. Then,

$$\frac{d^2y_1}{dt^2} = \frac{d}{dt}\left(\frac{dy_1}{dt}\right) = \frac{dy_2}{dt} = \frac{(\rho_s - \rho_f)}{\rho_s} g - \frac{9\mu}{2a^2\rho_s} y_2.$$

We now have two simultaneous first-order ODEs, each with a specified initial condition:

$$\frac{dy_1}{dt} = y_2, \qquad y_1(0) = 0; \tag{7.27}$$

$$\frac{dy_2}{dt} = \frac{(\rho_s - \rho_f)}{\rho_s} g - \frac{9\mu}{2a^2\rho_s} y_2, \qquad y_2(0) = 0. \tag{7.28}$$

The constants are set to

$$a = 10\,\mu\text{m}; \; \rho_s = 1.1\,\text{g/cm}^3; \; \rho_f = 1\,\text{g/cm}^3; \; g = 9.81\,\text{m/s}^2; \; \mu = 3.5\,\text{cP}.$$

We solve this system numerically using Euler's explicit method. Program 7.1 was designed to solve only one first-order ODE at a time. The function described in Program 7.1 can be generalized to solve a set of coupled ODEs. This is done by vectorizing the numerical integration scheme to handle Equation (7.26). The generalized function that operates on coupled first-order ODEs is named `eulerforwardmethodvectorized`. The variable ***y*** in the function file eulerforwardmethodvectorized.m is now a matrix with the row number specifying the dependent variable and the columns tracking the progression of the variables' values in time. The function file, settlingsphere.m, evaluates the slope vector ***f***.

MATLAB program 7.3

```
function [t, y] = eulerforwardmethodvectorized(odefunc, tf, y0, h)
% Euler's forward method is used to solve coupled first-order ODEs

% Input variables
% odefunc : name of the function that calculates f(t, y)
% tf : final time or size of interval
% y0 : vector of initial conditions y(0)
% h : step size

% Other variables
n = length(y0); % number of dependent time-varying variables

% Output variables
t = [0:h:tf]; % vector of time points
```

```
y = zeros(n, length(t)); % dependent variable vector
y(:,1) = y0; % initial condition at t = 0
             % indexing of matrix elements begins with 1 in MATLAB

% Euler's forward method for solving coupled first-order ODEs
for k = 1:length(t)-1
    y(:,k+1) = y(:,k) + h*feval(odefunc, t(k), y(:,k));
end
```

MATLAB program 7.4

```
function f = settlingsphere(t, y)
% Evaluate slopes f(t,y) of coupled equations
a = 10e-4;     % cm   : radius of sphere
rhos = 1.1;    % g/cm3  :  density of sphere
rhof = 1;      % g/cm3  :  density of medium
g = 981;       % cm/s2  :  acceleration due to gravity
mu = 3.5e-2;   % g/cm.s :  viscosity
f = [y(2); (rhos - rhof)*g/rhos - (9/2)*mu*y(2)/(a^2)/rhos];
```

Equations (7.27) and (7.28) are solved by calling the `eulerforwardmethodvectorized` function from the MATLAB command line. We choose a step size of 0.0001 s:

```
>> [t, y] = eulerforwardmethodvectorized('settlingsphere', 0.001,
        [0; 0], 0.0001);
```

Plotting the numerically calculated displacement z and the velocity dz/dt with time (Figure 7.3), we observe that the solution blows up. We know that physically this cannot occur. When the sphere falls from rest through the viscous medium, the downward velocity increases monotonically from zero and reaches a final settling velocity. At this point the acceleration reduces to zero. The numerical method is unstable for this choice of step size since the solution grows without bound even though the exact solution is bounded. You can calculate the terminal settling velocity by assuming steady state for the force balance (acceleration term is zero).

Figure 7.3

Displacement and velocity trajectories for a sphere settling under gravity in Stokes flow. Euler's explicit method exhibits numerical instabilities for the chosen step size of $h = 0.0001$ s.

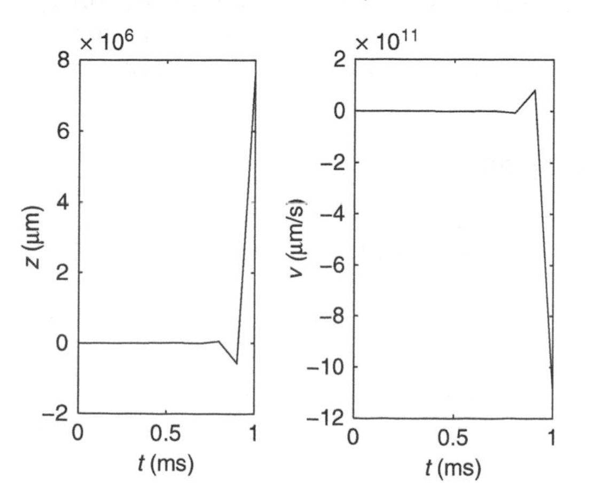

Figure 7.4

Displacement and velocity trajectories for a sphere settling under gravity in Stokes flow. Euler's explicit method is stable but the velocity trajectory is not physically meaningful for the chosen step size of $h = 10$ μs.

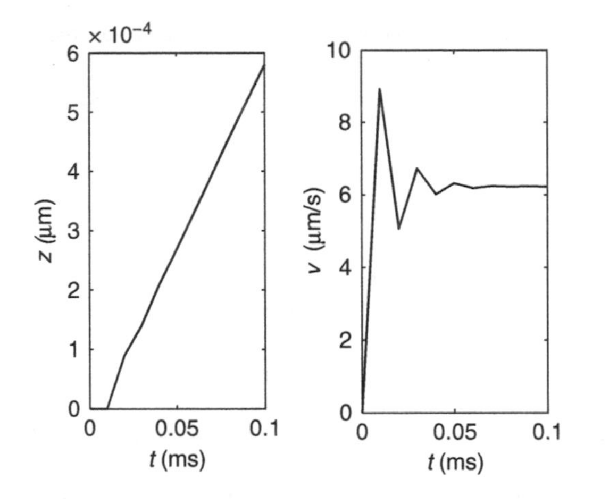

We reduce the step size by an order of magnitude to 0.00001 s:

```
>> [t, y] = eulerforwardmethodvectorized ('settlingsphere', 0.0001,
       [0; 0], 0.00001);
```

and re-plotting the time-dependent displacement and velocity (Figure 7.4), we observe that the final solution is bounded, but the velocity profile is unrealistic. The numerical method is stable for this choice of step size, yet the solution is terribly inaccurate; the steps are too far apart to predict a physically realizable situation.

Finally we choose a step size of 1 μs. Calling the function once more:

```
>> [t, y] = eulerforwardmethodvectorized ('settlingsphere', 0.0001,
       [0; 0], 0.000001);
```

we obtain a physically meaningful solution in which the settling velocity of the falling sphere monotonically increases until it reaches the terminal velocity, after which it does not increase or decrease. Figure 7.5 is a plot of the numerical solution of the coupled system using a step size of 1 μs.

Some differential equations are sensitive to step size, and require exceedingly small step sizes to guarantee stability of the solution. Such equations are termed as **stiff**. Issues regarding stability of the numerical technique and how best to tackle stiff problems are taken up in Section 7.6.

7.2.2 Euler's backward method

For deriving Euler's forward method, the forward finite difference formula was used to approximate the first derivative of y. To convert the differential problem

$$\frac{dy}{dt} = f(t, y), \qquad y(t = 0) = y_0,$$

into a difference equation, this time we replace the derivative term by the backward finite difference formula (Equation (1.23)). In the subinterval $[t_k, \ t_{k+1}]$, the derivative dy/dt at the time point t_{k+1} is approximated as

Figure 7.5

Displacement and velocity trajectories for a sphere settling under gravity in Stokes flow. Euler's explicit method is stable and physically meaningful for the chosen step size of $h = 1$ μs.

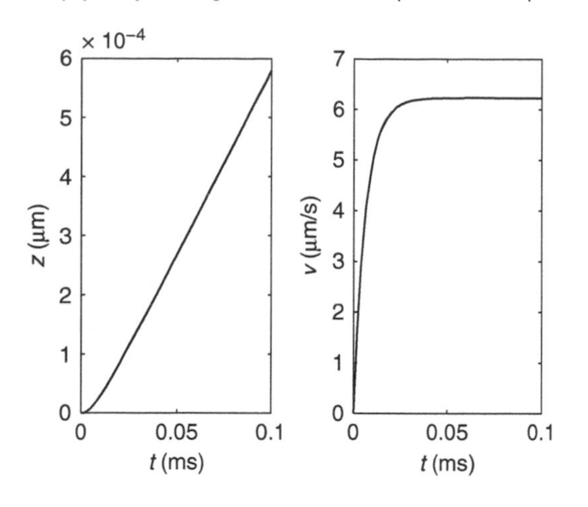

$$\left.\frac{dy}{dt}\right|_{t=t_{k+1}} = f(t_{k+1}, y_{k+1}) \approx \frac{y_{k+1} - y_k}{h},$$

where $h = t_{k+1} - t_k$. Rearranging, we obtain Euler's backward formula,

$$y_{k+1} = y_k + hf(t_{k+1}, y_{k+1}). \tag{7.29}$$

This method assumes that the slope of the function $y(t)$ in each subinterval is approximately constant. This assumption is met if the step size h is small enough. In this forward-marching scheme to solve for y_{k+1}, when y_k is known, we need to know the slope at y_{k+1}. But the value of y_{k+1} is unknown. A difference equation in which unknown terms are located on both sides of the equation is called an **implicit equation**. This can be contrasted with an explicit equation, in which the unknown term is present only on the left-hand side of the difference equation. Therefore, Euler's backward method is sometimes called **Euler's implicit method**.

To calculate y_{k+1} at each time step, we need to solve Equation (7.29) for y_{k+1}. If the equation is linear or quadratic in y_{k+1}, then the terms of the equation can be rearranged to yield an explicit equation in y_{k+1}. If the backward difference equation is nonlinear in y_{k+1}, then a nonlinear root-finding scheme must be used to obtain y_{k+1}. In the latter case, the accuracy of the numerical solution is limited by both the global truncation error of the ODE integration scheme as well as the truncation error associated with the root-finding algorithm.

The order of the local truncation error of Euler's implicit method is the same as that for Euler's explicit method, $O(h^2)$. Consider the subinterval $[t_0,\ t_1]$. The initial value of y at t_0 can be expanded in terms of the value of y at t_1 using the Taylor series:

$$y(t_1 - h) = y(t_0) = y(t_1) - y'(t_1)h + \frac{y''(\xi)h^2}{2!}, \qquad \xi \in [t_0, t_1].$$

Rearranging, we get

$$y(t_1) = y_0 + hf(t_1, y_1) - \frac{y''(\xi)h^2}{2!}.$$

Using Equation (7.29) ($k = 0$) in the above expansion, we obtain

$$y(t_1) = y_1 - \frac{y''(\xi)h^2}{2!}$$

or

$$y(t_1) - y_1 = -\frac{y''(\xi)h^2}{2!} \sim O(h^2).$$

The global truncation error for Euler's implicit scheme is one order of magnitude lower than the local truncation error. Thus, the convergence rate for global truncation error is $O(h)$, which is the same as that for the explicit scheme.

An important difference between Euler's explicit method and the implicit method is the stability characteristics of the numerical solution. Let's reconsider the first-order decay problem,

$$\frac{dy}{dt} = -\lambda y, \quad \lambda > 0.$$

The numerical integration formula for this ODE problem, using Euler's implicit method, is found to be

$$y_{k+1} = y_k - \lambda h y_{k+1}$$

or

$$y_{k+1} = \frac{y_k}{1 + \lambda h}. \tag{7.30}$$

Note that the denominator of Equation (7.30) is greater than unity for any $h > 0$. Therefore, the numerical solution is guaranteed to mimic the decay of y with time for any value of $h > 0$. This example illustrates a useful property of the Euler implicit numerical scheme.

> The Euler implicit numerical scheme is **unconditionally stable**. The numerical solution will not diverge if the exact solution is bounded, for any value of $h > 0$.

The global truncation error of the implicit Euler method at t_{k+1} is given by (see Equation (7.20) for a comparison)

$$y(t_{k+1}) - y_{k+1} \le \frac{1}{1 - h\frac{\partial f}{\partial y}}(y(t_k) - y_k) + \frac{Mh^2}{2}.$$

The amplification factor for the global truncation error of Euler's implicit method is $1/\left(1 - h\frac{\partial f}{\partial y}\right)$. If $\partial f/\partial y < 0$, i.e. the exact solution is a decaying function, the error is clearly bounded, and the method is numerically stable for all h.

Example 7.3

The second-order differential equation describing the motion of a sphere settling under gravity in a viscous medium is re-solved using Euler's implicit formula. The two simultaneous first-order ODEs are linear. We will be able to obtain an explicit formula for both y_1 and y_2:

$$\frac{dy_1}{dt} = y_2, \qquad y_1(0) = 0; \tag{7.27}$$

$$\frac{dy_2}{dt} = \frac{(\rho_s - \rho_f)}{\rho_s} g - \frac{9\mu}{2a^2\rho_s} y_2, \qquad y_2(0) = 0. \tag{7.28}$$

The numerical formulas prescribed by Euler's backward difference method are

$$y_{1,k+1} = y_{1,k} + hy_{2,k+1},$$

$$y_{2,k+1} = y_{2,k} + h\left(\frac{(\rho_s - \rho_f)}{\rho_s} g - \frac{9\mu}{2a^2\rho_s} y_{2,k+1}\right).$$

The numerical algorithm used to integrate the coupled ODEs is

$$y_{2,k+1} = \frac{y_{2,k} + \frac{(\rho_s - \rho_f)}{\rho_s} gh}{1 + \frac{9\mu}{2a^2\rho_s} h},$$

$$y_{1,k+1} = y_{1,k} + hy_{2,k+1}.$$

Because of the implicit nature of the algorithm, a general purpose program cannot be written to solve ODEs using an implicit algorithm. A MATLAB program was written with the express purpose of solving the coupled ODEs using this algorithm. (Try this yourself.)

Keeping the constants the same, setting $h = 0.0001$ s, and plotting the numerical solution for the displacement z and the velocity dz/dt with time (Figure 7.6), we observe that the solution is very well-behaved. This example demonstrates that the stability characteristics of Euler's backward method are superior to those of Euler's forward method.

7.2.3 Modified Euler's method

The main disadvantage of Euler's forward method and backward method is that obtaining an accurate solution involves considerable computational effort. The accuracy of the numerical result (assuming round-off error is negligible compared to truncation error) increases at the slow rate of $O(h)$. To reduce the error down to acceptable values, it becomes necessary to work with very small step sizes. The inefficiency in improving accuracy arises because of the crude approximation of $y'(t)$ over the entire subinterval with a single value of $f(t, y)$ calculated at either the

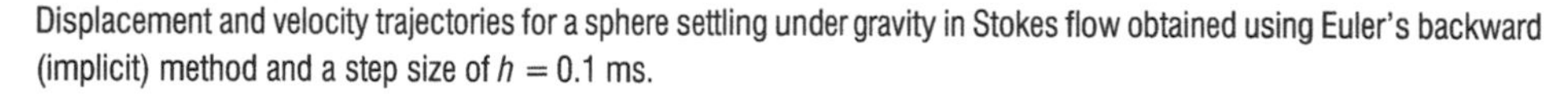

Figure 7.6

Displacement and velocity trajectories for a sphere settling under gravity in Stokes flow obtained using Euler's backward (implicit) method and a step size of $h = 0.1$ ms.

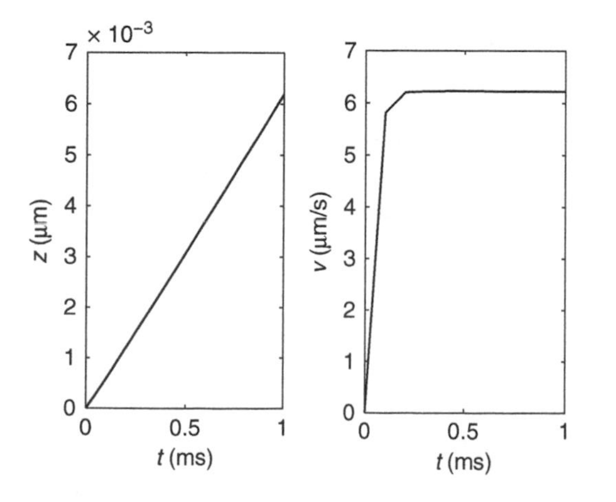

beginning or end of the subinterval. However, $y'(t)$ is expected to change over an interval of width h. What is more desirable when numerically integrating an ODE is to use a value of the slope intermediate to that at the beginning and end of the subinterval such that it captures the change in the trajectory of y over the subinterval.

The **modified Euler method** is a numerical ODE integration method that calculates the slope at the beginning and end of the subinterval and uses the average of the two slopes to increment the value of y at each time step. For the first-order ODE problem given below,

$$\frac{dy}{dt} = f(t, y), \qquad y(0) = y_0,$$

the numerical integration algorithm is implicit. At time step t_{k+1}, the numerical approximation to $y(t_{k+1})$ is y_{k+1}, which is calculated using the formula

$$y_{k+1} = y_k + h\frac{f(t_k, y_k) + f(t_{k+1}, y_{k+1})}{2}. \tag{7.31}$$

How quickly does the error diminish with a decrease in time step for the modified Euler method? Consider the Taylor series expansion of $y(t_k)$ about $y(t_{k+1})$ in the subinterval $[t_k,\ t_{k+1}]$:

$$y(t_k) = y(t_{k+1}) - y'(t_{k+1})h + \frac{y''(t_{k+1})h^2}{2!} - \frac{y'''(\xi)h^3}{3!}, \qquad \xi \in [t_k, t_{k+1}].$$

Now, $y'(t_{k+1}) = f(t_{k+1}, y(t_{k+1}))$, and $y''(t_{k+1})$ can be approximated using the backward finite difference formula:

$$y''(t_{k+1}) = \frac{df(t_{k+1}, y(t_{k+1}))}{dt} = \frac{f(t_{k+1}, y(t_{k+1})) - f(t_k, y(t_k))}{h} + O(h).$$

Substituting into the expansion series, we obtain

$$y(t_{k+1}) = y(t_k) + f(t_{k+1}, y(t_{k+1}))h - \frac{h}{2}(f(t_{k+1}, y(t_{k+1})) - f(t_k, y(t_k))) + O(h^3).$$

Rearranging, we get

$$y(t_{k+1}) = y(t_k) + \frac{h}{2}(f(t_{k+1}, y(t_{k+1})) + f(t_k, y(t_k))) + O(h^3)$$

or

$$y(t_{k+1}) - y_{k+1} \sim O(h^3).$$

The local truncation error for the modified Euler method is $O(h^3)$. The global truncation error includes the accumulated error from previous steps and is one order of magnitude smaller than the local error, and therefore is $O(h^2)$. The modified Euler method is thus called a **second-order method**.

Example 7.4

We use the modified Euler method to solve the linear first-order ODE

$$\frac{dy}{dt} = t - y, \qquad y(0) = 1. \tag{7.23}$$

The numerical integration formula given by Equation (7.31) is implicit in nature. Because Equation (7.23) is linear, we can formulate an equation explicit for y_{k+1}:

Table 7.2. *Numerical solution of $y(t)$ at $t_f = 2.8$ obtained using three different Euler schemes for solving Equation (7.23)*

h	Euler's forward method	Euler's backward method	Modified Euler method
0.4	1.8560	1.9897	1.9171
0.2	1.8880	1.9558	1.9205
0.1	1.9047	1.9387	1.9213

Table 7.3. *Maximum global truncation error observed when using three different Euler schemes for solving Equation (7.23)*

h	Euler's forward method	Euler's backward method	Modified Euler method
0.4	0.1787	0.1265	0.0098
0.2	0.0804	0.068	0.0025
0.1	0.0384	0.0353	0.000614

$$y_{k+1} = \frac{(2-h)y_k + h(t_k + t_{k+1})}{2+h}. \tag{7.32}$$

Starting from $t = 0$, we obtain approximations for $y(t_k)$ using Equation (7.32). A MATLAB program was written to perform numerical integration of Equation (7.23) using the algorithm given by Equation (7.32). Equation (7.23) is also solved using Euler's forward method and Euler's backward method. The numerical result at $t_f = 2.8$ is tabulated in Table 7.2 for the three different Euler methods and three different step sizes. The exact solution for $y(2.8)$ is 1.9216. The maximum global truncation error encountered during ODE integration is presented in Table 7.3.

Table 7.3 shows that the global error is linearly dependent on step size for both the forward method and the backward method. When h is halved, the error also reduces by half. However, global error reduction is $O(h^2)$ for the modified Euler method. The error decreases by roughly a factor of 4 when the step size is cut in half. The modified Euler method is found to converge on the exact solution more rapidly than the other two methods, and is thus more efficient.

Let us consider the stability properties of the implicit modified Euler method for the first-order decay problem:

$$\frac{dy}{dt} = -\lambda y, \qquad \lambda > 0.$$

The step-wise numerical integration formula for this ODE problem is

$$y_{k+1} = y_k + h\frac{-\lambda(y_k + y_{k+1})}{2}$$

or

$$y_{k+1} = \frac{(2-\lambda h)y_k}{2+\lambda h}. \tag{7.33}$$

Equation (7.33) guarantees decay of y with time for any $h > 0$. The numerical scheme is unconditionally stable for this problem. This is another example demonstrating the excellent stability characteristics of implicit numerical ODE integration methods.

7.3 Runge–Kutta (RK) methods

The Runge–Kutta (RK) methods are a set of explicit higher-order ODE solvers. The modified Euler method discussed in Section 7.2.3 was an implicit second-order method that requires two slope evaluations at each time step. By doubling the number of slope calculations in the interval, we were able to increase the order of accuracy by an order of magnitude from $O(h)$, characteristic of Euler's explicit method, to $O(h^2)$ for the modified Euler method. This direct correlation between the number of slope evaluations and the order of the method is exploited by the RK methods. Instead of calculating higher-order derivatives of y, RK numerical schemes calculate the slope $f(t, y)$ at specific points within each subinterval and then combine them according to a specific rule. In this manner, the RK techniques have a much faster error convergence rate than first-order schemes, yet require evaluations only of the first derivative of y. This is one reason why the RK methods are so popular.

The RK methods are easy to use because the formulas are explicit. However, their stability characteristics are only slightly better than Euler's explicit method. Therefore, a stable solution is not guaranteed for all values of h. When the RK numerical scheme is stable for the chosen value of h, subsequent reductions in h lead to rapid convergence of the numerical approximation to the exact solution. The convergence rate depends on the order of the RK method. The order of the method is equal to the number of slope evaluations combined in one time step, when the number of slopes evaluated is two, three, or four. When the slope evaluations at each time step exceed four, the correlation between the order of the method and the number of slopes combined is no longer one-to-one. For example, a fifth-order RK method requires six slope evaluations at every time step.[3] In this section, we introduce second-order (RK-2) and fourth-order (RK-4) methods.

7.3.1 Second-order RK methods

The second-order RK method is, as the name suggests, second-order accurate, i.e. the global truncation error associated with the numerical solution is $O(h^2)$. This explicit ODE integration scheme requires slope evaluations at two points within each time step. If k_1 and k_2 are the slope functions evaluated at two distinct time points within the subinterval $[t_k, \ t_{k+1}]$, then the general purpose formula of the RK-2 method is

$$
\begin{aligned}
y_{k+1} &= y_k + h(c_1 k_1 + c_2 k_2), \\
k_1 &= f(t_k, y_k), \\
k_2 &= f(t_k + p_2 h, y_k + q_{21} k_1 h).
\end{aligned}
\tag{7.34}
$$

[3] RK methods of order greater than 4 are not as computationally efficient as RK-4 methods.

The values of the constants c_1, c_2, p_2, and q_{21} are determined by mathematical constraints, or rules, that govern this second-order method. Let's investigate these rules now.

We begin with the Taylor series expansion for $y(t_{k+1})$ expanded about $y(t_k)$ using a step size h:

$$y(t_{k+1}) = y(t_k) + y'(t_k)h + \frac{y''(t_k)h^2}{2!} + O(h^3).$$

The second derivative, $y''(t)$, is the first derivative of $f(t, y)$ and can be expressed in terms of partial derivatives of $f(t, y)$:

$$y''(t) = f'(t, y) = \frac{\partial f(t, y)}{\partial t} + \frac{\partial f(t, y)}{\partial y} \cdot \frac{dy}{dt}$$

Substituting the expansion for $y''(t_k)$ into the Taylor series expansion, we obtain a series expression for the exact solution at t_{k+1}:

$$y(t_{k+1}) = y(t_k) + f(t_k, y_k)h + \frac{h^2}{2}\left(\frac{\partial f(t, y)}{\partial t} + \frac{\partial f(t, y)}{\partial y} f(t_k, y_k)\right) + O(h^3). \quad (7.35)$$

To find the values of the RK constants, we must compare the Taylor series expansion of the exact solution with the numerical approximation provided by the RK-2 method. We expand k_2 in terms of k_1 using the two-dimensional Taylor series:

$$k_2 = f(t_k, y_k) + p_2 h \frac{\partial f(t_k, y_k)}{\partial t} + q_{21} k_1 h \frac{\partial f(t_k, y_k)}{\partial y} + O(h^2).$$

Substituting the expressions for k_1 and k_2 into Equations (7.34) for y_{k+1}, we obtain

$$\begin{aligned} y_{k+1} = y_k + h\Big(&c_1 f(t_k, y_k) + c_2 f(t_k, y_k) \\ &+ c_2 p_2 h \frac{\partial f(t_k, y_k)}{\partial t} + c_2 q_{21} h f(t_k, y_k) \frac{\partial f(t_k, y_k)}{\partial y} + O(h^2)\Big) \end{aligned}$$

or

$$\begin{aligned} y_{k+1} = y_k + f(t_k, y_k)(c_1 + c_2)h + h^2\left(c_2 p_2 \frac{\partial f(t_k, y_k)}{\partial t} + c_2 q_{21} \frac{\partial f(t_k, y_k)}{\partial y} f(t_k, y_k)\right) \\ + O(h^3). \end{aligned} \quad (7.36)$$

Comparison of Equations (7.35) and (7.36) yields three rules governing the RK-2 constants:

$$c_1 + c_2 = 1,$$

$$c_2 p_2 = \frac{1}{2},$$

$$c_2 q_{21} = \frac{1}{2}.$$

For this set of three equations in four variables, an infinite number of solutions exist. Thus, an infinite number of RK–2 schemes can be devised. In practice, a family of RK-2 methods exists, and we discuss two of the most common RK-2 schemes: the explicit modified Euler method and the midpoint method.

Modified Euler method (explicit)

This numerical scheme is also called **the trapezoidal rule** and **Heun's method**. It uses the same numerical integration formula as the implicit modified Euler method discussed in Section 7.2.3. However, the algorithm used to calculate the slope of y at t_{k+1} is what sets this method apart from the one discussed earlier.

Setting $c_2 = 1/2$, we obtain the following values for the RK–2 parameters:

$$c_1 = \frac{1}{2}; \qquad p_2 = 1; \qquad q_{21} = 1.$$

Equations (7.34) simplify to

$$\begin{aligned} y_{k+1} &= y_k + \frac{h}{2}(k_1 + k_2), \\ k_1 &= f(t_k, y_k), \\ k_2 &= f(t_k + h, y_k + k_1 h). \end{aligned} \tag{7.37}$$

This RK–2 method is very similar to the implicit modified Euler method. The only difference between this integration scheme and the implicit second-order method is that here the value of y_{k+1} on the right-hand side of the equation is predicted using Euler's explicit method, i.e. $y_{k+1} = y_k + hk_1$. With replacement of the right-hand side term y_{k+1} with known y_k terms, the integration method is rendered explicit. The global error for the explicit modified Euler method is still $O(h^2)$; however, the actual error may be slightly larger in magnitude than that generated by the implicit method. Figure 7.7 illustrates the geometric interpretation of Equations (7.37).

Figure 7.7

Numerical integration of an ODE using the explicit modified Euler method.

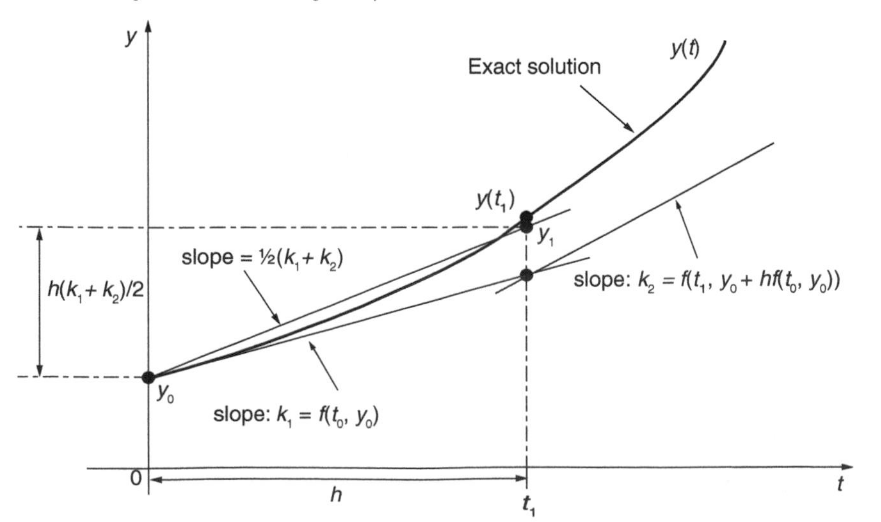

Midpoint method

The **midpoint method** uses the slope at the midpoint of the interval, i.e. $t_k + h/2$ to calculate the approximation to $y(t_{k+1})$. The value of y at the midpoint of the interval is estimated using Euler's explicit formula.

Setting $c_2 = 1$, we obtain the following values for the RK–2 parameters:

$$c_1 = 0; \qquad p_2 = \frac{1}{2}; \qquad q_{21} = \frac{1}{2}.$$

Equations (7.34) simplify to

$$\begin{aligned} y_{k+1} &= y_k + hk_2, \\ k_1 &= f(t_k, y_k), \\ k_2 &= f\left(t_k + \frac{h}{2}, y_k + \frac{1}{2}k_1 h\right). \end{aligned} \tag{7.38}$$

Numerical ODE integration via the midpoint method is illustrated in Figure 7.8.

Now consider the stability properties of the explicit modified Euler method for the first-order decay problem

$$\frac{dy}{dt} = -\lambda y, \qquad \lambda > 0.$$

The numerical integration formula for this ODE problem using the explicit modified Euler's method is

$$y_{k+1} = y_k + h\frac{-\lambda(y_k + (y_k - \lambda h y_k))}{2}$$

or

$$y_{k+1} = \left(1 - \lambda h + \frac{\lambda^2 h^2}{2}\right) y_k. \tag{7.39}$$

Figure 7.8

Numerical integration of an ODE using the midpoint method.

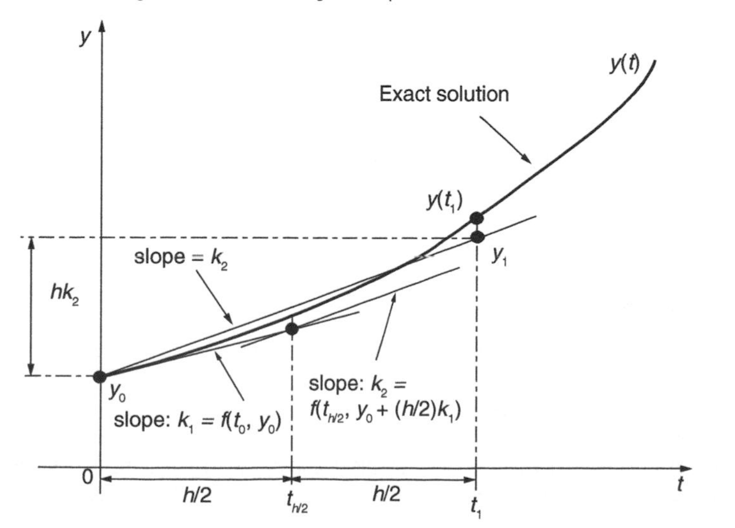

A bounded solution will be obtained if

$$-1 \leq 1 - \lambda h + \frac{\lambda^2 h^2}{2} \leq 1.$$

Let $x = \lambda h$. Values of x that simultaneously satisfy the two quadratic equations specified by the condition above ensure a stable numerical result.

The first equation, $x(x-2) \leq 0$, requires that $0 \leq x \leq 2$.
The second equation $x^2 - 2x - 4 \geq 0$ requires that $-1.236 \leq x \leq 3.236$.

Therefore, $0 \leq \lambda h \leq 2$ is the numerical stability requirement for this problem when using the explicit modified Euler method. The stability characteristics of this RK–2 method for the first-order decay problem are the same as those of the explicit Euler method. While the RK–2 method will converge faster to the exact solution compared to the forward Euler method, it requires an equally small h value to prevent instability from creeping into the numerical solution. It is left to the reader to show that the numerical stability properties of the midpoint method are the same as that of the explicit modified Euler method for the first-order decay problem.

7.3.2 Fourth-order RK methods

Fourth-order RK methods involve four slope evaluations per time step, and have a local convergence error of $O(h^5)$ and a global convergence error of $O(h^4)$. The numerical scheme for the RK–4 methods is

$$\begin{aligned}
y_{k+1} &= y_k + h(c_1 k_1 + c_2 k_2 + c_3 k_3 + c_4 k_4),\\
k_1 &= f(t_k, y_k),\\
k_2 &= f(t_k + p_2 h, y_k + q_{21} k_1 h),\\
k_3 &= f(t_k + p_3 h, y_k + q_{31} k_1 h + q_{32} k_2 h),\\
k_4 &= f(t_k + p_4 h, y_k + q_{41} k_1 h + q_{42} k_2 h + q_{43} k_3 h).
\end{aligned}$$

The derivation of the 11 equations governing the 13 constants is too tedious and complicated to be performed here. A popular fourth-order RK method in use is the classical method presented below:

$$\begin{aligned}
y_{k+1} &= y_k + \frac{h}{6}(k_1 + 2k_2 + 2k_3 + k_4)\\
k_1 &= f(t_k, y_k)\\
k_2 &= f(t_k + 0.5h, y_k + 0.5k_1 h)\\
k_3 &= f(t_k + 0.5h, y_k + 0.5k_2 h)\\
k_4 &= f(t_k + h, y_k + k_3 h)
\end{aligned} \tag{7.40}$$

Example 7.5

We use the classical RK–4 method and the midpoint method to solve the linear first-order ODE

$$\frac{dy}{dt} = t - y, \qquad y(0) = 1. \tag{7.23}$$

Functions created to perform single ODE integration by the midpoint method and the classical RK–4 method are listed below. The ODE function that evaluates the right-hand side of Equation (7.23) is listed in

Program 7.2. Another MATLAB program can be created to call these numerical ODE solvers, or the functions can be called from the MATLAB Command Window.

MATLAB program 7.5

```
function [t, y] = midpointmethod(odefunc, tf, y0, h)
% Midpoint (RK-2) method is used to solve a first-order ODE

% Input variables
% odefunc : name of the function that calculates f(t, y)
% tf : final time or size of interval
% y0 : y(0)
% h : step size

% Output variables
t = [0:h:tf]; % vector of time points
y = zeros(size(t)); % dependent variable vector
y(1) = y0; % indexing of vector elements begins with 1 in MATLAB

% Midpoint method for solving first-order ODE
for k = 1:length(t)-1
    k1 = feval(odefunc, t(k), y(k));
    k2 = feval(odefunc, t(k) + h/2, y(k) + h/2*k1);
    y(k+1) = y(k) + h*k2;
end
```

MATLAB program 7.6

```
function [t, y] = RK4method(odefunc, tf, y0, h)
% Classical RK-4 method is used to solve a first-order ODE

% Input variables
% odefunc : name of the function that calculates f(t, y)
% tf : final time or size of interval
% y0 : y(0)
% h : step size

% Output variables
t = [0:h:tf]; % vector of time points
y = zeros(size(t)); % dependent variable vector
y(1) = y0; % indexing of vector elements begins with 1 in MATLAB

% RK-4 method for solving first-order ODE
for k = 1:length(t)-1
    k1 = feval(odefunc, t(k), y(k));
    k2 = feval(odefunc, t(k)+ h/2, y(k) + h/2*k1);
    k3 = feval(odefunc, t(k)+ h/2, y(k) + h/2*k2);
    k4 = feval(odefunc, t(k)+ h, y(k) + h*k3);
    y(k+1) = y(k) + h/6*(k1 + 2*k2 + 2*k3 + k4);
end
```

The final value at $t_f = 2.8$ and maximum global error is tabulated in Table 7.4 for the two RK methods and two different step sizes. The exact solution for $y(2.8)$ is 1.9216.

From the results presented in Table 7.4, it is observed that, in the midpoint method, the maximum truncation error reduces by a factor of $4.65 \sim 2^2$ upon cutting the step size in half. In the RK–4 method,

Table 7.4. *Numerical solution of $y(t)$ at $t_f = 2.8$ and maximum global truncation error obtained using two different RK schemes to solve Equation (7.23) numerically*

h	Midpoint method		RK–4 method	
	Numerical solution of $y(2.8)$	Maximum global truncation error	Numerical solution of $y(2.8)$	Maximum global truncation error
0.4	1.9345	0.0265	1.9217	2.1558×10^{-4}
0.2	1.9243	0.0057	1.9216	1.159×10^{-5}

the maximum truncation error reduces by a factor of $18.6 \sim 2^4$ upon cutting the step size in half. For one uniform step size, the number of function calls, and therefore the computational effort, is twice as much for the RK–4 method compared to the RK–2 method. Now compare the accuracy of the two methods. The total number of slope evaluations over the entire interval performed by the RK–4 method for $h = 0.4$ is equal to that performed by the midpoint method at $h = 0.2$. However, the RK–4 method is $0.0057/2.1558 \times 10^{-4} = 27$ times more accurate at $h = 0.4$ than the midpoint method is at $h = 0.2$. Thus, the fourth-order method is much more efficient than the second-order method.

The RK–4 method can be easily adapted to solve coupled first-order ODEs. The set of equations for solving the two coupled ODEs,

$$\frac{dy_1}{dt} = f_1(t, y_1, y_2),$$
$$\frac{dy_2}{dt} = f_2(t, y_1, y_2),$$

is

$$\begin{aligned}
k_{1,1} &= f_1\left(t_k, y_{1,k}, y_{2,k}\right),\\
k_{2,1} &= f_2\left(t_k, y_{1,k}, y_{2,k}\right),\\
k_{1,2} &= f_1\left(t_k + 0.5h, y_{1,k} + 0.5k_{1,1}h, y_{2,k} + 0.5k_{2,1}h\right),\\
k_{2,2} &= f_2\left(t_k + 0.5h, y_{1,k} + 0.5k_{1,1}h, y_{2,k} + 0.5k_{2,1}h\right),\\
k_{1,3} &= f_1\left(t_k + 0.5h, y_{1,k} + 0.5k_{1,2}h, y_{2,k} + 0.5k_{2,2}h\right),\\
k_{2,3} &= f_2\left(t_k + 0.5h, y_{1,k} + 0.5k_{1,2}h, y_{2,k} + 0.5k_{2,2}h\right),\\
k_{1,4} &= f_1\left(t_k + h, y_{1,k} + k_{1,3}h, y_{2,k} + k_{2,3}h\right),\\
k_{2,4} &= f_2\left(t_k + h, y_{1,k} + k_{1,3}h, y_{2,k} + k_{2,3}h\right),\\
y_{1,k+1} &= y_{1,k} + \frac{h}{6}\left(k_{1,1} + 2k_{1,2} + 2k_{1,3} + k_{1,4}\right),\\
y_{2,k+1} &= y_{2,k} + \frac{h}{6}\left(k_{2,1} + 2k_{2,2} + 2k_{2,3} + k_{2,4}\right).
\end{aligned} \tag{7.41}$$

These equations are used to solve the coupled ODE system in Box 7.1A.

7.4 Adaptive step size methods

An important goal in developing a numerical ODE solver (or, in fact, any numerical method) is to maximize the computational efficiency. We wish to attain the desired accuracy of the solution with the least amount of computational effort. In earlier

Box 7.1B HIV–1 dynamics in the bloodstream

We make the following substitutions for the dependent variables:

$$y_1 = T^*; y_2 = V_1; y_3 = V_x.$$

The set of three first-order ODEs listed below are solved using the RK–4 method:

$$\frac{dy_1}{dl} = 0.05y_2 - 0.5y_1,$$

$$\frac{dy_2}{dt} = -3.0y_2,$$

$$\frac{dy_3}{dt} - 30y_1 - 3.0y_3.$$

The initial concentrations of T^*, V_1, and V_X are

$V_1(t = 0) = 100/\mu l$
$V_X(t = 0) = 0/\mu l$
$T^*(t = 0) = 10$ infected cells/μl. (Haase *et al.*, 1996).

The function program `HIVODE.m` is written to evaluate the right-hand side of the ODEs. The RK–4 method from Program 7.6 must be vectorized in the same way that Euler's explicit method was (see Programs 7.1 and 7.3) to solve coupled ODEs. The main program calls the ODE solver `RK4methodvectorized` and plots the time-dependent behavior of the variables.

MATLAB program 7.7

```
% Use the RK-4 method to solve the HIV dynamics problem
clear all
% Variables
y0(1) = 10;   % initial value of T*
y0(2) = 100;  % initial value of V1
y0(3) = 0;    % initial value of Vx
tf = 5;       % interval size (day)
h = 0.1;      % step size (day)

% Call ODE Solver
[t, y] = RK4methodvectorized('HIVODE', tf, y0, h);

% Plot Results
figure;
subplot(1,3,1)
plot(t,y(1,:),'k-','LineWidth',2)
set(gca,'FontSize',16,'LineWidth',2)
xlabel('{\itt} in day')
ylabel('{\itT}^* infected cells/ \muL')

subplot(1,3,2)
plot(t,y(2,:),'k-','LineWidth',2)
set(gca,'FontSize',16,'LineWidth',2)
xlabel('{\itt} in day')
ylabel('{\itV}_1 virions/ \muL')

subplot(1,3,3)
plot(t,y(3,:),'k-','LineWidth',2)
```

```
set(gca,'FontSize',16,'LineWidth',2)
xlabel('{\itt} in day')
ylabel('{\itV}_X noninfectious virus particles/ \muL')
```

MATLAB program 7.8

```
function [t, y] = RK4methodvectorized(odefunc, tf, y0, h)
% Classical RK-4 method is used to solve coupled first-order ODEs

% Input variables
% odefunc : name of the function that calculates f(t, y)
% tf : final time or size of interval
% y0 : vector of initial conditions y(0)
% h : step size

% Other variables
n = length(y0); % number of dependent time-varying variables

% Output variables
t = [0:h:tf]; % vector of time points
y = zeros(n, length(t)); % dependent variable vector

y(:,1) = y0; % initial condition at t = 0
% RK-4 method for solving coupled first-order ODEs
for k = 1:length(t)-1
    k1 = feval(odefunc, t(k), y(:,k));
    k2 = feval(odefunc, t(k)+ h/2, y(:,k) + h/2*k1);
    k3 = feval(odefunc, t(k)+ h/2, y(:,k) + h/2*k2);
    k4 = feval(odefunc, t(k)+ h, y(:,k) + h*k3);
    y(:,k+1) = y(:,k) + h/6*(k1 + 2* k2 + 2* k3 + k4);
end
```

MATLAB program 7.9

```
function f = HIVODE(t, y)
% Evaluate slope f(t,y) of coupled ODEs for HIV-1 dynamics in bloodstream
% Constants
k = 2e-4;       % uL/day/virions
N = 60;         % virions/cell
delta = 0.5;    % /day
c = 3.0;        % /day
T = 250;        % noninfected cells/uL

f = [k*T*y(2)-delta*y(1); -c*y(2); N*delta*y(1)-c*y(3)];
```

Figure 7.9 shows the rapid effect of the drug on the number of infected cells, and the number of infectious and non-infectious virus particles in the bloodstream. The number of infected cells in the bloodstream plunges by 90% after five days of ritonavir treatment.

There are a number of simplifying assumptions of the model, particularly (1) virus counts in other regions of the body such as lymph nodes and tissues are not considered in the model and (2) the number of non-infected cells is assumed to be constant over five days.

Figure 7.9

Effect of ritonavir treatment on viral particle concentration in the bloodstream.

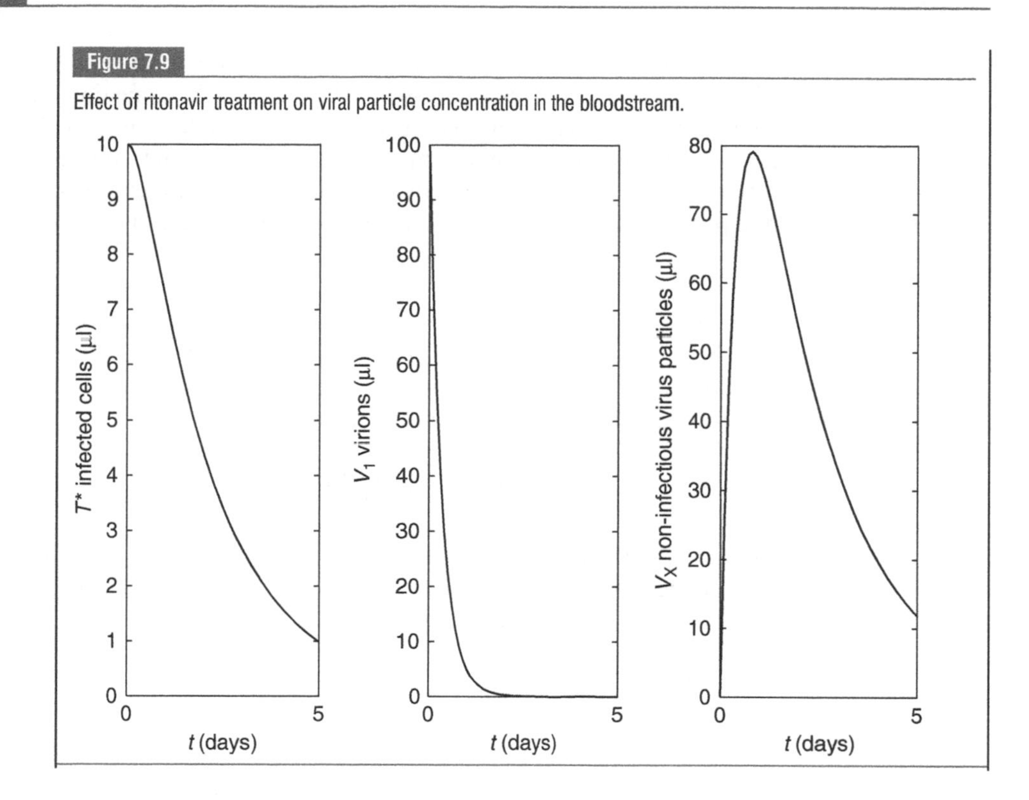

sections of this chapter, we discussed two ways to improve the accuracy of a numerical solution: (1) reduce the step size and (2) increase the order of the ODE integration routine. Our earlier discussions have focused on closing the gap between the numerical approximation and the exact solution. In practice, we seek numerical solutions that approach the true solution within a specified tolerance. The tolerance criterion sets a lower limit on the accuracy that must be met by an ODE solver. If a numerical method produces a solution whose numerical error is one or more orders of magnitude less than the tolerance requirement, then more computational effort in terms of time and resources to obtain a solution has been expended than required. We wish to minimize wasteful computations that add little, if any, value to the solution. Using a fixed time step throughout the integration process is an example of unnecessary computational work. An ODE solver that uses a fixed time step must choose a time step that resolves the most rapid change or capricious behavior of the slope function to the desired accuracy. For other regions of the time interval, where the slope evolves more slowly, the fixed step size in use is much smaller than what is required to keep the error within the tolerance limit. The integration thus proceeds more slowly than necessary in these subintervals.

To integrate more efficiently over the time interval, once can adjust the step size h at each time step. The optimal step size becomes a function of time and is chosen based on the local truncation error generated by the solution at each instant in time. ODE solvers that optimize the step size as an integration proceeds along the interval are known as **adaptive step size methods**. There are several ways to estimate the local truncation error, which measures the departure of the numerical solution from the exact solution assuming that the solution at the previous time point is known with certainty. In this section, we focus our discussion on adaptive step size algorithms for RK methods.

One technique to measure local truncation error is to solve the ODE at time t_{k+1} using a step size h, and then to re-solve the ODE at the same time point using a step size $h/2$. The difference between the two numerical solutions provides an estimate of the local truncation error at time step t_{k+1}. The exact solution at t_{k+1} can be expressed in terms of the numerical solution y_k, assuming that the solution at the time point t_k is known exactly, i.e. $y(t_k) = y_k$. For a step size h,

$$y(t_{k+1}) = y_{k+1} + Mh^{n+1}.$$

The second term on the right-hand side is the local truncation error. For a step size of $h/2$, the integration must be performed twice to reach t_{k+1}. If we assume that M is constant over the time step, and the accumulation of other errors when stepping twice to reach the end of the subinterval is negligible, then

$$y(t_{k+1}) \sim z_{k+1} + M\left(\frac{h}{2}\right)^{n+1} + M\left(\frac{h}{2}\right)^{n+1},$$

where z_{k+1} is the numerical solution obtained with step size $h/2$. Taking the difference of the two numerical solutions, $z_{k+1} - y_{k+1}$, we get an estimate of the local truncation error for a step size of h:

$$Mh^{n+1} = \frac{2^n}{2^n - 1}(z_{k+1} - y_{k+1}). \tag{7.42}$$

If ϵ is the tolerance prescribed for the local error, then ϵ can be set equal to Mh_{opt}^{n+1}, where h_{opt} is the optimum step size needed attain the desired accuracy. Substituting $M = \epsilon/h_{\text{opt}}^{n+1}$ in Equation (7.42), we get

$$\left(\frac{h_{\text{opt}}}{h}\right)^{n+1} = \frac{2^n - 1}{2^n} \cdot \frac{\epsilon}{z_{k+1} - y_{k+1}}. \tag{7.43}$$

Equation (7.43) can be used at each time step to solve for h_{opt}. However, there is a disadvantage in using two unequal step sizes at each time step to estimate the truncation error. The ODE integration must be performed over the same subinterval *three times* just to optimize the step size. Additionally, if the newly calculated h_{opt} is less than $h/2$, then the integration step must be repeated one more time with h_{opt} at the current time point to obtain the more accurate numerical solution.

Evaluating a numerical solution at any time point using two different step sizes is very useful however when assessing the stability of the numerical method. When the numerical solution obtained with two different step sizes does not differ appreciably, one can be confident that the method is numerically stable at that time point. If the two solutions differ significantly from each other, even for very small step sizes, one should check for numerical instabilities, possibly by using a more robust ODE solver to verify the numerical solution.

The **Runge–Kutta–Fehlberg (RKF) method** is an efficient strategy for evaluating the local error at each time point. This method advances from one time point to the next using one step size and a pair of RK methods, with the second method of one order higher than the first. Two numerical solutions are obtained at the current time point. For example, a fourth-order RK method can be paired with a fifth-order RK method in such a way that some of the slope evaluations for either method share the same formula. An RK–4 method requires at least four slope evaluations to achieve a local truncation error of $O(h^5)$. An RK–5 method requires at least six slope evaluations to achieve a local truncation error of $O(h^6)$. If the two methods are performed independently of each other, a total of ten slope evaluations are necessary to calculate two numerical solutions, one more accurate than the other. However, the RKF45 method that pairs a fourth-order method with a fifth-order method requires only six total slope evaluations since slopes k_1, k_3, k_4, and k_5 are shared among the two methods.

Here, we demonstrate an RKF23 method that pairs a second-order method with a third-order method, and requires only three slope evaluations. The RK–2 method in this two-ODE solver combination is the modified Euler method or trapezoidal rule (Equation (7.37)). The local truncation error associated with each time step is $O(h^3)$. We have

$$y_{k+1} = y_k + \frac{h}{2}(k_1 + k_2).$$

The RK–3 method is

$$k_1 = f(t_k, y_k),$$
$$k_2 = f(t_k + h, y_k + k_1 h),$$
$$k_3 = f(t_k + 0.5h, y_k + 0.25k_1 h + 0.25k_2 h),$$
$$y_{k+1} = y_k + \frac{h}{6}(k_1 + k_2 + 4k_3).$$

The local truncation error associated with each time step is $O(h^4)$. The RK–2 and RK–3 methods share the slopes k_1 and k_2. If we may assume that $O(h^4)$ local errors are negligible compared to the $O(h^3)$ local errors, then, on subtracting the RK–3 solution from the RK–2 numerical approximation, we obtain

$$\frac{h}{2}(k_1 + k_2) - \frac{h}{6}(k_1 + k_2 + 4k_3) = \frac{h}{3}(k_1 + k_2 - 2k_3) \sim Mh^3.$$

The tolerance for the local error is represented by ϵ. Since $\epsilon = Mh_{\text{opt}}^3$, we have

$$\left(\frac{h_{\text{opt}}}{h}\right)^3 = \frac{\epsilon}{(h/3)(k_1 + k_2 - 2k_3)}.$$

The optimal step size calculation is only approximate since global errors are neglected. For this reason, a safety factor α is usually included to adjust the prediction of the correct step size. Adaptive step size algorithms typically avoid increasing the step size more than necessary. If the current step size is larger than the calculated optimum, the algorithm must recalculate the numerical solution using the smaller step size h_{opt}. If the current step size is smaller than h_{opt}, then no recalculation is done for the current step, and the step size is adjusted at the next time step.

MATLAB offers a variety of ODE solvers that perform numerical integration of a system of ordinary differential equations. ODE solvers `ode23` and `ode45` use explicit RK adaptive step size algorithms to solve initial-value problems. The `ode45` function uses paired RK–4 and RK–5 routines to integrate ODEs numerically and is based on the algorithm of Dormand and Prince. It is recommended that this function be tried when first attempting to solve an ODE problem. The `ode23` function uses paired RK–2 and RK–3 routines (Bogacki and Shampine algorithm) to integrate ODEs numerically, and can be more efficient than `ode45` when required tolerances are higher.

Using MATLAB

All MATLAB ODE solvers (except for `ode15i`) share the same syntax. This makes it easy to switch from one function call to another. The simplest syntax for a MATLAB ODE solver is

```
[t, y] = odexxx(odefunc, tf, y0), or
[t, y] = odexxx(odefunc, [t0 tf], y0)
```

The input parameters for an ODE solver are as follows.

odefunc: The name of the user-defined function that evaluates the slope $f(t, y)$, entered within single quotation marks. For a coupled ODE problem, this ODE function should output a column vector whose elements are the first derivatives of the dependent variables, i.e. the right-hand side of the ODE equations.

[t0 tf]: The beginning and ending time points of the time interval.

y0: The value of the dependent variable(s) at time 0. If there is more than one dependent variable, y0 should be supplied as a column vector, otherwise y0 is a scalar. The order in which the initial values of the variables should be listed must correspond to the order in which their slopes are calculated in the ODE function odefunc.

The output variables are as follows.

t: A column vector of the time points at which the numerical solution has been calculated. The time points will not be uniformly spaced since the time step is optimized during the integration. If you require a solution at certain time points only, you should specify a vector containing specific time points in place of the time interval [t0, tf], e.g.

```
[t, y] = odexxx(odefunc, [0:2:10], y0)
```

The numerical solution $\boldsymbol{y}$ will be output at these six time points, although higher resolution may be used during integration.

y: For a single ODE problem, this is a column vector containing the solution at time points recorded in vector $\boldsymbol{t}$. For a coupled ODE problem, this is a matrix with each column representing a different dependent variable and each row representing a different time point that corresponds to the same row of $\boldsymbol{t}$.

The optimal step size is chosen such that the local truncation error $e(k)$ at step k, estimated by the algorithm, meets the following tolerance criteria:

```
e(k) ≤ max(RelTol * |y(k)|, AbsTol)
```

The default value for the relative tolerance RelTol is 10^{-3}. The default value for the absolute tolerance AbsTol is 1×10^{-6}. For $|y| > 10^{-3}$ the relative tolerance will set the tolerance limit. For $|y| < 10^{-3}$, the absolute tolerance will control the error. For $|y| \ll 1$, it is recommended that you change the default value of AbsTol accordingly.

Sometimes you may want to suggest an initial step size or specify the maximum step size (default is $0.1(t_f - t_0)$) allowed during integration. To change the default settings of the ODE solver you must create an options structure and pass that structure along with other parameters to the odexxx function. A **structure** is a data type that is somewhat like an array except that it can contain different types of data such as character arrays, numbers, and matrices each stored in a separate field. Each field of a structure is given a name. If the name of the structure is Options, then RelTol is one field and its contents can be accessed by typing Options.RelTol.

To create an options structure, you will need to use the function odeset. The syntax for odeset is

```
options = odeset('parameter1', 'value1', 'parameter2', 'value2', ...)
```

For example, you can set the relative tolerance by doing the following. Create an `options` structure,

```
options = odeset('RelTol', 0.0001);
```

Now pass the newly created `options` structure to the ODE solver using the syntax

```
[t, y] = ode45(odefunc, [t0 tf], y0, options)
```

There are a variety of features of the ODE solvers that you can control using the `options` structure. For more information, see MATLAB help.

Values of constants or parameters can be passed to the function that calculates the right-hand side of Equation (7.26). These parameters should be listed after the `options` argument. If you do not wish to supply an `options` structure to the ODE solver, you should use `[]` in its place. When creating the defining line of the function `odefunc`, you must pass an additional argument called flag, as shown below:

```
f = odefunc(t, y, flag, p1, p2)
```

where `p1` and `p2` are additional parameters that are passed from the calling program. See Box 7.3B for an example that uses this syntax.

Box 7.1C HIV–1 dynamics in the blood stream

We re-solve the coupled ODE problem using the MATLAB `ode45` function:

$$\frac{dy_1}{dt} = 0.05y_2 - 0.5y_1,$$
$$\frac{dy_2}{dt} = -3.0y_2,$$
$$\frac{dy_3}{dt} = 30y_1 - 3.0y_3$$

The initial concentrations of T^*, V_1, and V_X are $y_1(0) = 10; y_2(0) = 100; y_3(0) = 0$.

The `ode45` function can be called either in a script file or in the Command Window:

```
>> [t, y] = ode45('HIVODE', 5.0, [10; 100; 0]);
```

The first five and last five time points for integration are

```
>> t(1:5)                    >> t(end-4:end)
ans =                        ans =
  1.0e-006 *                       4.8440
                0                  4.8830
           0.1675                  4.9220
           0.3349                  4.9610
           0.5024                  5.0000
           0.6698
```

The time step increases by five orders of magnitude from the beginning to the end of the interval. The number of infected cells, infectious viral particles, and non-infectious viral particles at $t = 5$ days is

```
>> y(end, :)
ans -
    0.9850     0.0000     11.8201
```

Box 7.2B Enzyme deactivation in industrial processes

Defining $y_1 \equiv S$ and $y_2 \equiv E_{\text{total}}$, we rewrite the first-order differential equations as follows:

$$\frac{dy_1}{dt} = -\frac{k_2 y_2 y_1}{K_m + y_1},$$
$$\frac{dy_2}{dt} = -\frac{k_d y_2}{1 + y_1/K_m},$$

where

$k_2 = 21$ M/s/e.u. (e.u. $\equiv$ enzyme units)
$K_m = 0.004$ M
$k_d = 0.03$/s.

The initial conditions are $y_1 = 0.05$ M and $y_2 = 1 \times 10^{-6}$ units.

`enyzmedeact` is an ODE function written to evaluate the right-hand side of the coupled ODE equations. A script file is written to call `ode45` and plot the result.

MATLAB program 7.10

```
% Use ode45 to solve the enzyme deactivation problem.
clear all
% Variables
y0(1) = 0.05; % M
y0(2) = 1e-6; % units
tf = 3600; % s

% Call ODE Solver
[t, y] = ode45('enzymedeact', tf, y0);

% Plot Results
figure;
subplot(1,2,1)
plot(t,y(:,1),'k-','LineWidth',2)
set(gca,'FontSize',16,'LineWidth',2)
xlabel('{\itt} in sec')
ylabel('{\itS} in M')

subplot(1,2,2)
plot(t,y(:,2),'k-','LineWidth',2)
set(gca,'FontSize',16,'LineWidth',2)
xlabel('{\itt} in sec')
ylabel('{\itE}_t_o_t_a_l in units')
```

MATLAB program 7.11

```
function f = enzymedeact(t, y)
% Evaluate slope f(t,y) of coupled ODEs for enzyme deactivation

% Constants
k2 = 21; % M/s/e.u.
Km = 0.004; % M
kd = 0.03; % s-1
f = [-k2*y(1)*y(2)/(Km + y(1)); -kd*y(2)/(1 + y(1)/Km)];
```

Figure 7.10

Decrease in concentration of substrate and active enzyme with time.

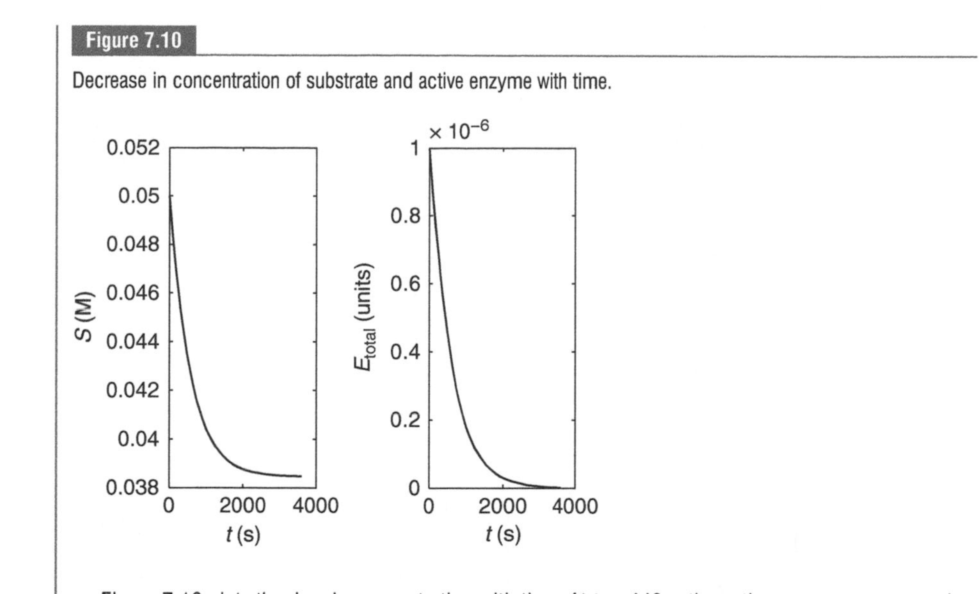

Figure 7.10 plots the drop in concentration with time. At $t = 440$ s, the active enzyme concentration reduces to half its starting value. Note the quick deactivation of the enzyme in comparison to consumption of the substrate. The enzymatic process needs to be modified to increase the life of the enzyme.

Box 7.3B Dynamics of an epidemic outbreak

Two sets of parameters are chosen to study the dynamics of the spread of disease in a population of 100 000.

Case (1) $k = 1 \times 10^{-5}$ and $\gamma = 4$. This set of values represents a disease that spreads easily and recovery is rapid, e.g. flu.

Case (2) $k = 5 \times 10^{-8}$ and $\gamma = 4000$. This set of values represents a disease that spreads with difficulty yet a person remains infectious for a longer period of time, e.g. diseases that are spread by contact of bodily fluids.

The birth and death rates are $\beta = 0.0001$; and $\mu = 0.00001$, respectively. Defining $y_1 \equiv S$, $y_2 \equiv D$, and $y_3 \equiv I$, we rewrite the first-order differential equations as

$$\frac{dy_1}{dt} = -ky_1y_2 + \beta(y_1 + y_3) - \mu y_1,$$

$$\frac{dy_2}{dt} = ky_1y_2 - \frac{1}{\gamma}y_2,$$

$$\frac{dy_3}{dt} = \frac{1}{\gamma}y_2 - \mu y_3.$$

The MATLAB ODE solver `ode23` is used to solve this problem. For case 1, we observe from Figure 7.11 that the disease becomes an epidemic and spreads through the population within a month. As large sums of people become sick, recover, and then become immune, the disease disappears because there is a lack of people left to spread the disease.

For the second case, because the recovery period is so long, we must modify the equations to account for people who die while infected. From Figure 7.12, we see that, in 7 years, most of the

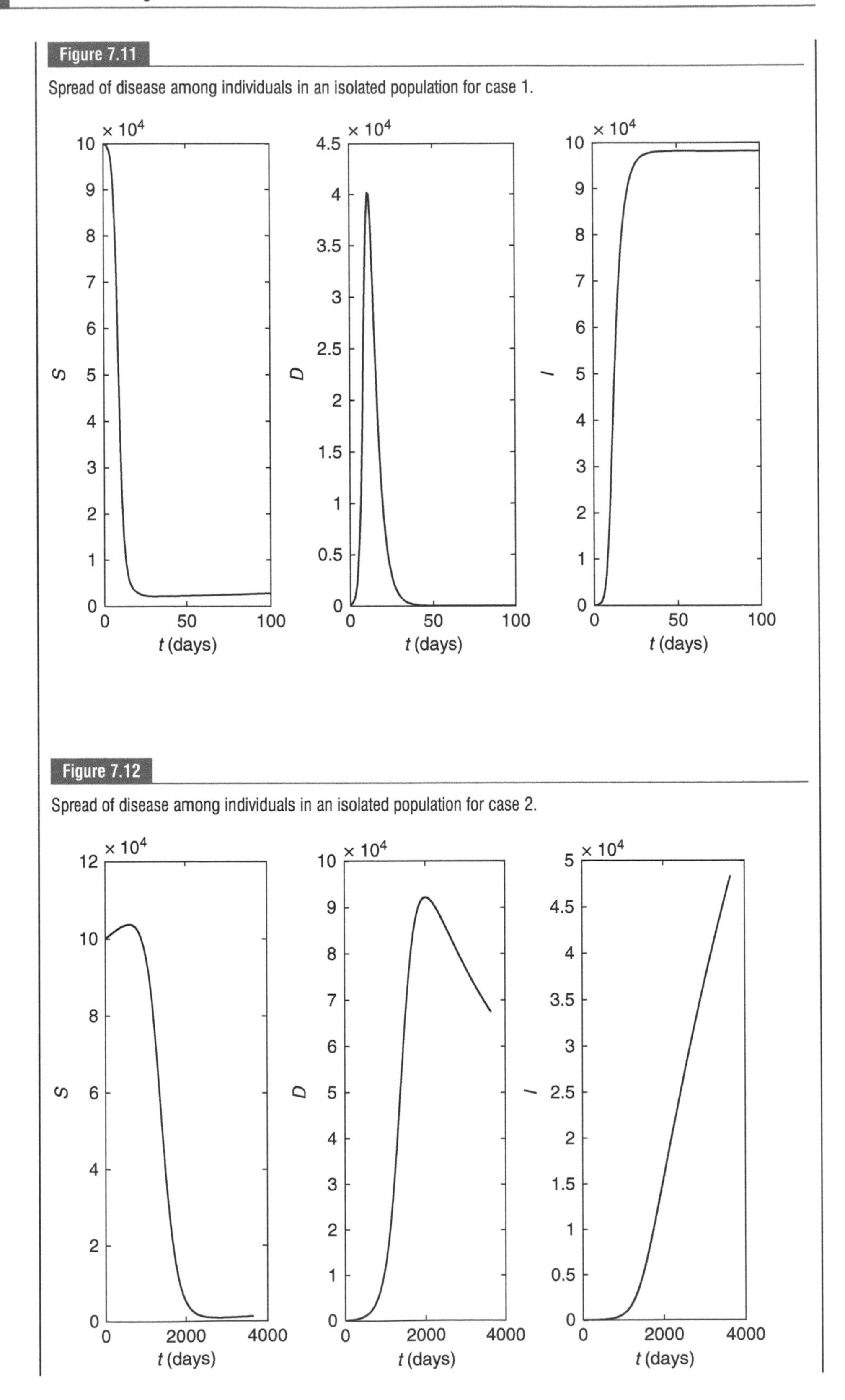

Figure 7.11

Spread of disease among individuals in an isolated population for case 1.

Figure 7.12

Spread of disease among individuals in an isolated population for case 2.

population succumbs to the disease. The disease however does not disappear, but remains embedded in the population, even after 10 years.

Note that these models are highly simplified. The equations assume that each person has an equal probability of coming into contact with another person. It has been found that some epidemic models are better explained using complex network theory that uses a power law distribution to model the number of people that an individual of a population comes into contact with. Complex network theory assigns some people the status of "hubs." Disease spreads most easily through "hubs" since they are the most well-connected (Barabasi, 2003).

MATLAB program 7.12

```
% Use ode23 to solve the epidemic outbreak problem.
clear all
% Variables
k = 1e-5; % infectiousness of disease, per person per contact per day
gamma = 4; % recovery period, days

% Initial conditions
y0(1) = 100000;
y0(2) = 100;
y0(3) = 0;

% Time interval
tf = 100; % days

% Call ODE Solver
[t, y] = ode23('epidemic', tf, y0, [], k, gamma);
```

MATLAB program 7.13

```
function f = epidemic(t, y, flag, k, gamma)
% Evaluate slope f(t,y) of coupled ODEs for dynamics of epidemic outbreak

% Constants
beta = 0.0001; % birth rate, per person per day
mu = 0.00001; % death rate, per person per day

f = [-k*y(1)*y(2) + beta*(y(1)+y(3)) - mu*y(1); ...
    k*y(1)*y(2) - 1/gamma*y(2); ...
    1/gamma*y(2) - mu*y(3)];
```

7.5 Multistep ODE solvers

The Euler and the RK methods of numerical ODE integration are called **one-step methods** because they use the solution y_k at the current time step t_k to estimate the solution at the next time step t_{k+1}. In RK methods, the formula to obtain the numerical approximation to $y(t_{k+1})$ requires evaluation of the slope at one or more time points within the subinterval $[t_k,\ t_{k+1}]$ over which integration is currently being performed. Information on the solution prior to t_k is not used to determine y_{k+1} in these methods. Thus, in these methods the integration moves forward by stepping from the current step to the next step, i.e. by a single step.

Much useful information is contained in the solution at time points earlier than t_k. The change in the values of $y_k, y_{k-1}, y_{k-2}, \ldots$ at previous time points provides a measure of the rate of change in the slope $f'(t, y)$ and higher derivatives of y. The knowledge of prior solutions over a larger time interval consisting of several time steps can be used to obtain more accurate approximations to the solution at the next time point. ODE integration methods that use not only the solution at the current step t_k to approximate the solution at the next step, but also previous solution values at $t_{k-1}, t_{k-2}, \ldots$ are called **multistep methods**.

A general-purpose numerical integration formula for a three-step method is

$$\begin{aligned} y_{k+1} =& p_1 y_k + p_2 y_{k-1} + p_3 y_{k-2} \\ &+ h[q_0 f(t_{k+1}, y_{k+1}) + q_1 f(t_k, y_k) + q_2 f(t_{k-1}, y_{k-1}) + q_3 f(t_{k-2}, y_{k-2})], \end{aligned}$$

where $p_1, p_2, p_3, q_0, q_1, q_2$, and q_3 are constants that are derived using the Taylor series.

One advantage of multistep methods for ODE integration is that only one evaluation of $f(t, y)$ is required for each step regardless of the order of the method. Function values at previous time points are already available from previous integration steps in the multistep method. However, multistep methods are not **self-starting**, because several previous solution points are required by the formula to calculate y_{k+1}. An initial condition, in most cases, consists of the value of y at one initial time point only. To begin the integration, a one-step method of order equal to that of the multistep method is used to generate the solution at the first few time points, and after that the multistep method can take over.

There are two kinds of multistep methods: explicit methods and implicit methods. Explicit multistep methods do not require an evaluation of the slope at t_{k+1} to find y_{k+1}. Implicit multistep methods include a term for the slope at t_{k+1} on the right-hand side of the integration formula. The implicit methods have better stability properties, but are harder to execute than explicit methods. Unless the slope function is linear in y, a nonlinear root-finding algorithm must be used to search for y_{k+1}. A variety of multistep algorithms have been developed. We discuss the **Adams methods** of explicit and implicit type for multistep ODE integration. MATLAB offers an ODE solver based on an Adams routine.

7.5.1 Adams methods

The **Adams–Bashforth methods** are explicit multistep ODE integration schemes in which the number of steps taken during the integration at each time step is equal to the order of the method. Keep in mind that the order of a method is equal to the order of the global truncation error produced by that method. The integration formula for an m-step Adams–Bashforth method is given by

$$y_{k+1} = y_k + h \sum_{j=1}^{m} q_j f(t_{k-j+1}, y_{k-j+1}). \tag{7.44}$$

The formula for an m-step Adams–Bashforth method can be derived using the Taylor series expansion about $y(t_k)$ with step size h, as follows:

$$y(t_{k+1}) = y(t_k) + f(t_k, y(t_k))h + \frac{f'(t_k, y(t_k))h^2}{2!} + \frac{f''(t_k, y(t_k))h^3}{3!} + O(h^4).$$

Substituting numerical differentiation formulas for derivatives of f generates a multistep difference formula. This is demonstrated for a two-step Adams–Bashforth method.

We can approximate the first derivative of f using the first-order backward difference formula (Equation (1.22)),

$$f'(t_k, y(t_k)) \sim \frac{f(t_k, y(t_k)) - f(t_{k-1}, y(t_{k-1}))}{h} + \frac{h}{2} f''(t_k, y(t_k)) + O(h^2).$$

Substituting the backward difference formula above into the Taylor expansion, we get

$$\begin{aligned} y(t_{k+1}) =& y(t_k) + f(t_k, y(t_k))h + \frac{h}{2}(f(t_k, y(t_k)) - f(t_{k-1}, y(t_{k-1}))) \\ &+ \frac{5h^3}{12} f''(t_k, y(t_k)) + O(h^4). \end{aligned}$$

This simplifies to

$$y(t_{k+1}) = y(t_k) + \frac{h}{2}(3f(t_k, y(t_k)) - f(t_{k-1}, y(t_{k-1}))) + \frac{5h^3}{12} f''(\xi, y(\xi)),$$

where $\xi \in [t_{k-1},\ t_{k+1}]$. Writing $f(t_k, y(t_k))$ as f_k, the **second-order Adams–Bashforth formula** is given by

$$y_{k+1} = y_k + \frac{h}{2}(3f_k - f_{k-1}). \tag{7.45}$$

Upon substituting a second-order backward difference approximation for the first derivative of f and a first-order backward difference formula for the second derivative of f, we obtain the **third-order Adams–Bashforth formula**:

$$y_{k+1} = y_k + \frac{h}{12}(23f_k - 16f_{k-1} + 5f_{k-2}). \tag{7.46}$$

In this way, higher-order formulas for the Adams–Bashforth ODE solver can be derived.

The **Adams–Moulton methods** are implicit multistep ODE integration schemes. For these methods, the number of steps m taken to advance to the next time point is one less than the order of the method. The integration formula for an m-step Adams–Moulton method is given by

$$y_{k+1} = y_k + h\sum_{j=0}^{m} q_j f(t_{k+1-j}, y_{k+1-j}). \tag{7.47}$$

The formula for an m-step Adams–Moulton method can be derived using a Taylor series expansion about $y(t_{k+1})$ with step size $-h$, as follows:

$$\begin{aligned} y(t_k) = y(t_{k+1}) - f(t_{k+1}, y(t_{k+1}))h + \frac{f'(t_{k+1}, y(t_{k+1}))h^2}{2!} - \frac{f''(t_{k+1}, y(t_{k+1}))h^3}{3!} \\ + O(h^4). \end{aligned}$$

Substituting numerical differentiation formulas for derivatives of f generates a multistep difference formula. Retaining only the first two terms on the right-hand side, we recover Euler's implicit method, a first-order method. Approximating the first derivative of f in the Taylor series expansion with a first-order backward difference formula (Equation (1.22)), and dropping the $O(h^3)$ terms, we obtain the numerical integration formula for the modified Euler method, a second-order method (see Section 7.2.3):

$$y_{k+1} = y_k + \frac{1}{2}(f_k + f_{k+1}). \tag{7.48}$$

The associated $O(h^3)$ error term is $-(h^3/12)f''(\xi, y(\xi))$. Try to derive the error term yourself.

Upon substituting a second-order backward difference approximation for the first derivative of f and a first-order backward difference formula for the second derivative of f, we obtain the **third-order Adams–Moulton formula**:

$$y_{k+1} = y_k + \frac{h}{12}(5f_{k+1} + 8f_k - f_{k-1}). \tag{7.49}$$

In this manner, higher-order formulas for the Adams–Moulton ODE solver can be derived.

Upon comparing the magnitude of the error term of the second-order Adams–Bashforth method and the second-order Adams–Moulton method, we notice that the local truncation error is smaller for the implicit Adams method. This is also the case for higher-order explicit and implicit Adams methods. The smaller error lends itself to smaller global truncation error and improved numerical stability. Remember that instability of a numerical solution occurs when the error generated at each step is magnified over successive time steps by the integration formula. The second-order Adams–Moulton method is stable for all step sizes when the system of ODEs is well-posed.[4] However, implicit multistep methods of order greater than 2 are not stable for all step sizes, and can produce truncation errors that grow without bound when large step sizes are used (Grasselli and Pelinovsky, 2008).

7.5.2 Predictor–corrector methods

Implicit numerical ODE integration methods have superior stability characteristics and smaller local truncation errors compared to explicit methods. However, execution of implicit methods can be cumbersome since an iterative root-finding algorithm must be employed in general to find the solution at each time step. Also, to begin the root-finding process, one must provide one or two guesses that lie close to the true solution of the implicit ODE integration formula. Rather than having to resort to iterative numerical methods for solving implicit multistep formulas, one can obtain a prediction of the solution using the explicit multistep formula. The explicit multistep ODE solver is called the **predictor**. The prediction $y_{k+1}^{(0)}$ is plugged into the right-hand side of the implicit multistep ODE solver to calculate the slope at $(t_{k+1}, y_{k+1}^{(0)})$. The implicit formula imparts a correction on the prediction of the solution made by the predictor. Therefore, the implicit ODE integration formula is called the **corrector**.

Predictor and corrector formulas of the same order are paired with each other. An example of a predictor–corrector pair is the second-order Adams–Bashforth method paired with the second-order Adams–Moulton method shown below:

[4] Well-posed ODEs have a unique solution that changes only slightly when the initial condition is perturbed. An ODE whose solution fluctuates widely with small changes in the initial condition constitutes an ill-posed problem.

$$\text{predictor:} \quad y_{k+1}^{(0)} = y_k + \tfrac{h}{2}(3f_k - f_{k-1});$$
$$\text{corrector:} \quad y_{k+1}^{(1)} = y_k + \tfrac{1}{2}(f(t_k, y_k) + f(t_{k+1}, y_{k+1}^{(0)})).$$

A numerical integration method that uses a predictor equation with error $O(h^n)$ to predict the solution and a corrector equation with error $O(h^n)$ to improve upon the solution is called a **predictor–corrector method**. In some predictor–corrector algorithms, the corrector formula is applied once per time step and $y_{k+1}^{(1)}$ is used as the final approximation of the solution at t_{k+1}. Alternatively, the approximation $y_{k+1}^{(1)}$ can be supplied back to the right-hand side of the corrector equation to obtain an improved corrector approximation of the solution $y_{k+1}^{(2)}$. This is analogous to performing fixed-point iteration (see Chapter 5). Convergence of the corrector equation is obtained when

$$\frac{\left| y_{k+1}^{(i)} - y_{k+1}^{(i-1)} \right|}{y_{k+1}^{(i)}} \leq \epsilon.$$

An advantage of the predictor–corrector algorithm is that it also provides an estimate of the local truncation error at each time step. The exact solution at t_{k+1} can be expressed in terms of the predictor solution,

$$y(t_{k+1}) = y_{k+1}^{(0)} + c_1 h^{n+1} y^{n+1}(\xi_1),$$

and the corrector solution as

$$y(t_{k+1}) = y_{k+1}^{(i)} + c_2 h^{n+1} y^{n+1}(\xi_2).$$

By assuming that $\xi_1 \sim \xi_2 = \xi$, we can subtract the two expressions above to get

$$y_{k+1}^{(i)} - y_{k+1}^{(0)} = (c_1 - c_2) h^{n+1} y^{n+1}(\xi).$$

Simple algebraic manipulations yield

$$\frac{c_2}{c_1 - c_2}\left(y_{k+1}^{(i)} - y_{k+1}^{(0)} \right) = c_2 h^{n+1} y^{n+1}(\xi). \tag{7.50}$$

Thus, the approximations yielded by the predictor and corrector formulas allow us to estimate the local truncation error at each time step. Specifically for the second-order Adams–Bashforth–Moulton method:

$$c_1 = \frac{5}{12}, \qquad c_2 = -\frac{1}{12}.$$

Thus, for the second-order predictor–corrector method based on Adams formulas, the local truncation error at t_{k+1} can be estimated with the formula

$$-\frac{1}{6}\left(y_{k+1}^{(i)} - y_{k+1}^{(0)} \right).$$

As discussed in Section 7.4, local error estimates come in handy for optimizing the time step such that the local error is always well within the tolerance limit. The local truncation error given by Equation (7.50) can be compared with the tolerance limit to determine if a change in step size is warranted. The condition imposed by the tolerance limit ϵ is given by $c_2 h_{\text{opt}}^{n+1} y^{n+1}(\xi) \leq \epsilon$. Substituting $y^{n+1}(\xi) \leq \epsilon / c_2 h_{\text{opt}}^{n+1}$ into Equation (7.50), we obtain, after some rearrangement,

Box 7.4B Microbial population dynamics

The coupled ODEs that describe predator–prey population dynamics in a mixed culture microbial system are presented below after making the following substitutions: $y_1 = S; y_2 = N_1; y_3 = N_2$.

$$\frac{dy_1}{dt} = \frac{F}{V}(y_{1,0} - y_1) - \frac{1}{Y_{N_1|S}} \frac{\mu_{N_1,\max} y_1 y_2}{K_{N_1} + y_1};$$

$$\frac{dy_2}{dt} = -\frac{F}{V} y_2 + \frac{\mu_{N_1,\max} y_1 y_2}{K_{N_1} + y_1} - \frac{1}{Y_{N_2|N_1}} \frac{\mu_{N_2,\max} y_2 y_3}{K_{N_2} + y_2};$$

$$\frac{dy_3}{dt} = -\frac{F}{V} y_3 + \frac{\mu_{N_2,\max} y_2 y_3}{K_{N_2} + y_2}.$$

Using the kinetic parameters that define the predator–prey system (Tsuchiya *et al.*, 1972), we attempt to solve this system of ODEs using `ode113`. However, the numerical algorithm is unstable and the solution produced by the ODE solver diverges with time to infinity. For the given set of parameters, this system of equations is difficult to integrate because of sudden changes in the substrate concentration and microbial density with time. Such ODEs are called **stiff differential equations**, and are the topic of Section 7.6.

$$\left(\frac{h_{\text{opt}}}{h}\right)^{n+1} \leq \frac{\epsilon(c_1 - c_2)}{c_2\left(y_{k+1}^{(i)} - y_{k+1}^{(0)}\right)}. \tag{7.51}$$

An adjustment to the step size requires recalculation of the solution at new, equally spaced time points. For example, if the solutions at t_k, t_{k-1}, and t_{k-2} are used to calculate y_{k+1}, and if the time step is halved during the interval $[t_k,\ t_{k+1}]$, then, to generate the approximation for $y_{k+1/2}$, the solutions at $t_k, t_{k-1/2}$, and t_{k-1} are required. Also, $y_{k-1/2}$ must be determined using interpolation formulas of the same order as the predictor–corrector method. Step size optimization for multistep methods entails more computational work than for one-step methods.

Using MATLAB

The MATLAB ODE solver `ode113` is based on the Adams–Bashforth–Moulton method. This solver is preferred for problems with strict tolerance limits since the multistep method on which the solver is based produces smaller local truncation errors that the one-step RK method of corresponding order.

7.6 Stability and stiff equations

In earlier sections of this chapter, we discussed the stability properties of one-step methods, specifically the Euler and explicit RK methods (implicit RK methods have been devised but are not discussed in this book.) When solving a well-posed system of ODEs, the ODE solver that yields a solution whose relative error is kept in control even after a large number of steps is said to be **numerically stable** for that particular problem. If the relative error of the solution at each time step is magnified at the next time step, the numerical solution will diverge from the exact solution. The solution produced by the numerical method will grow increasingly inaccurate with time.[5] The ODE solver is said to be **numerically unstable** for the problem at hand. An ODE solver that successfully solves one ODE problem may not necessarily produce a convergent solution for another ODE problem, when using the same step size. The numerical stability of an ODE solver is defined by a range of step sizes within which a non-divergent solution (solution does not blow up at long times) is guaranteed. This step size range is not only ODE solver specific, but also problem specific, and thus varies from problem to problem for the same solver. Some numerical integration schemes have better stability properties than others (i.e. wider step size range for producing a stable solution), and selection of the appropriate ODE solver will depend on the nature of the ODE problem.

Some ODE problems are more "difficult" to solve. Solutions that have rapidly changing slopes can cause difficulties during numerical integration. A solution that has a fast decaying component, i.e. an exponential term, $e^{-\alpha t}$, where $\alpha \gg 0$, will change rapidly over a short period. The solution is said to have a **large transient**. ODEs that have large transients are called **stiff equations**. In certain processes, often multiple phenomena with different time scales will contribute to the behavior of a system. Some of the contributing factors will rapidly decay to zero, while others will evolve slowly with time. The transient parts of the solution eventually disappear and give way to a steady state solution, which is unchanging with time. The single ODE initial-value problem given by

$$\frac{dy}{dt} = f(t, y), \qquad y(0) = y_0,$$

can be linearized to the form $dy/dt = \alpha y$, such that $\alpha = df/dy$. Note that $\alpha(t)$ varies in magnitude as a function of time. If at any point α becomes large and negative, the ODE is considered stiff.

Unless the time step is chosen to be vanishingly small throughout the entire integration interval, the decay of a transient part of the solution will not occur when using certain ODE solvers. Large errors can accumulate due to the magnification of errors at each time step when, at any time during the integration, the step size jumps outside the narrow range that guarantees a stable, convergent solution. Solvers that vary the step size solely based on local truncation error are particularly prone to this problem when solving stiff equations.

Ordinary ODE solvers must use prohibitively small step sizes to solve stiff problems. Therefore, we turn to special solution techniques that cater to this category of differential equations. Stiff ODE solvers have better stability properties and can handle stiff problems using a step size that is practical. Very small step sizes increase the solution time considerably, and may reach the limit of round-off error. In the latter case, an accurate solution cannot be found. Stiffness is more commonly observed in systems of ODEs rather than in single ODE problems.

For a system of linearized, first-order ordinary differential equations, the set of equations can be compactly represented as

$$\frac{d\mathbf{y}}{dt} = \mathbf{J}\mathbf{y}, \qquad \mathbf{y}(0) = \mathbf{y}_0,$$

[5] Although we have referred to time as the independent variable in an initial-value problem, the methods developed for solving initial-value problems apply equally well to any continuous independent variable, such as distance x, speed s, or pressure P.

where

$$J = \begin{bmatrix} \frac{\partial f_1}{\partial y_1} & \frac{\partial f_1}{\partial y_2} & \cdots & \frac{\partial f_1}{\partial y_n} \\ \frac{\partial f_2}{\partial y_1} & \frac{\partial f_2}{\partial y_2} & \cdots & \frac{\partial f_2}{\partial y_n} \\ & \cdot & & \\ & \cdot & & \\ & \cdot & & \\ \frac{\partial f_n}{\partial y_1} & \frac{\partial f_n}{\partial y_2} & \cdots & \frac{\partial f_n}{\partial y_n} \end{bmatrix}.$$

J is called the **Jacobian**, and was introduced in Chapter 5. The exact solution to this linear problem is

$$y = e^{Jt} y_0. \tag{7.52}$$

Before we can simplify Equation (7.52), we need to introduce some important linear algebra concepts. The equation $Ax = \lambda x$ comprises an **eigenvalue problem**. This linear matrix problem represents a linear transformation of x that simply changes the size of x (i.e. elongation or shrinkage) but not its direction. To obtain non-zero solutions of x, $|A - \lambda I|$ is set as equal to zero, and the values of λ for which the determinant $|A - \lambda I|$ equals zero are called the **eigenvalues** of A. The solution x for each of the n eigenvalues are called the **eigenvectors** of A.

If J (a square matrix of dimensions $n \times n$) has n distinct eigenvalues λ_i and n distinct eigenvectors x_i, then it is called a perfect matrix, and the above exponential equation can be simplified by expressing J in terms of its eigenvectors and eigenvalues. How this can be done is not shown here since it requires the development of linear algebra concepts that are beyond the scope of this book. (The proof is explained in Chapter 5 of Davis and Thomson (2000).) The above equation simplifies to

$$y = \sum_{i=1}^{n} c_i e^{\lambda_i t} x_i. \tag{7.53}$$

If an eigenvalue is complex, then the real part needs to be negative for stability. If the real part of any of the n eigenvalues λ_i is positive, the exact solution will blow up. If all λ_i are negative, the exact solution is bounded for all time t. Thus, the sign of the eigenvalues of the Jacobian reveal the stability of the true solution, i.e. whether the solution remains stable for all times, or becomes singular at some point in time. If the eigenvalues of the Jacobian are all negative and differ by several orders of magnitude, then the ODE problem is termed stiff. The smallest step size for integration is set by the eigenvalue of largest magnitude.

Using MATLAB

MATLAB offers four ODE solvers for stiff equations, `ode15s`, `ode23s`, `ode23t`, and `ode23tb`.

- `ode15s` is a multistep solver. It is suggested that `ode15s` be tried when `ode45` or `ode113` fails and the problem appears to be stiff.
- `ode23s` is a one-step solver and can solve some stiff problems for which `ode15s` fails.
- `ode23t` and `ode23tb` both use the modified Euler method, also called the trapezoidal rule or the Adams–Moulton second-order method; `ode23tb` also implements an implicit RK formula.

For stiff ODE solvers, you can supply a function that evaluates the Jacobian matrix to the solver (using `options`) to speed up the calculations. If a Jacobian matrix is not provided for the function, the components of the Jacobian are estimated by the stiff ODE solver using finite difference formulas. See MATLAB help for further details.

Box 7.4C Microbial population dynamics

We re-solve the coupled ODEs presented below that describe predator–prey population dynamics in a mixed culture microbial system:

$$\frac{dy_1}{dt} = \frac{F}{V}(y_{1,0} - y_1) - \frac{1}{Y_{N_1|S}}\frac{\mu_{N_1,\max}y_1y_2}{K_{N_1} + y_1};$$

$$\frac{dy_2}{dt} = -\frac{F}{V}y_2 + \frac{\mu_{N_1,\max}y_1y_2}{K_{N_1} + y_1} - \frac{1}{Y_{N_2|N_1}}\frac{\mu_{N_2,\max}y_2y_3}{K_{N_2} + y_2};$$

$$\frac{dy_3}{dt} = -\frac{F}{V}y_3 + \frac{\mu_{N_2,\max}y_2y_3}{K_{N_2} + y_2},$$

using the stiff ODE solver `ode15s`. For this parameter set, we get an oscillatory response of the system, shown in Figure 7.13. Changes in the initial values of the variables N_1 and N_2 do not affect the long-term oscillatory properties of the system. On the other hand, the nature of the oscillations that are characteristic of the population dynamics predicted by the Lotka–Volterra predator–prey model are affected by changes in initial predator or prey numbers.

There are a number of operating variables that can be changed, such as the feed-to-volume ratio and the substrate concentration feed. At time $t = 50$ days, we introduce a step change in the feed glucose

Figure 7.13

Evolution of number densities of *E.coli* and *Dictyostelium discoideum*, and substrate concentration with time in a chemostat for $S_0 = 0.5$ mg/ml.

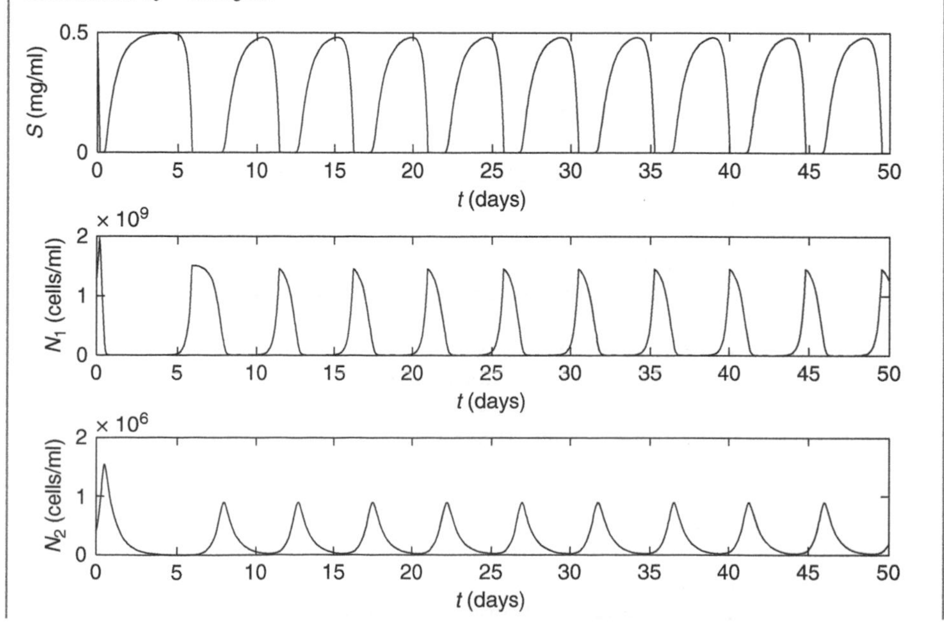

Figure 7.14

Evolution of number densities of *E.coli* and *Dictyostelium discoideum* amoebae, and substrate concentration with time, in a chemostat for $S_0 = 0.1$ mg/ml.

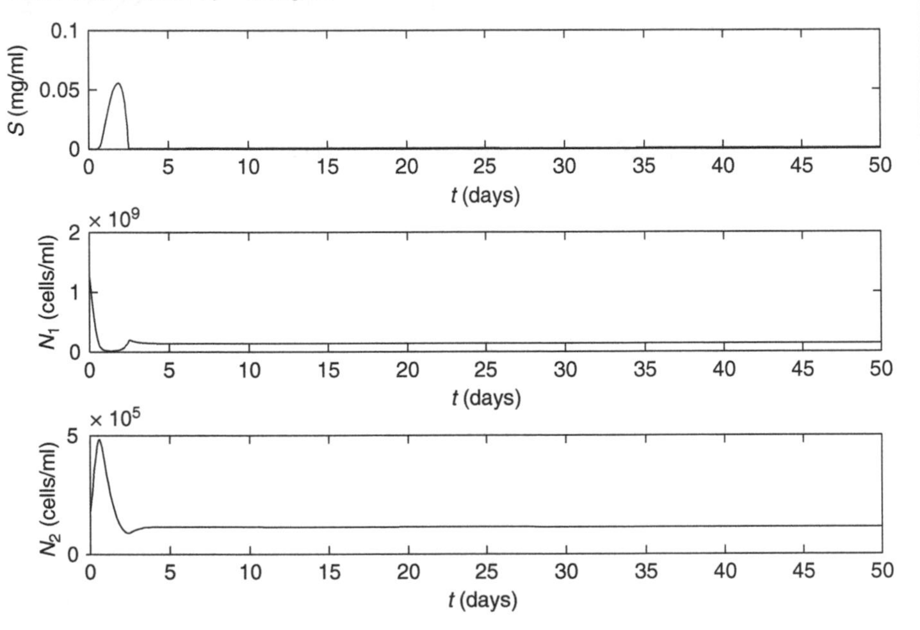

concentration from 0.5 mg/ml to 0.1 mg/ml. The behavior of the system for the next 50 days is shown in Figure 7.14.

The system quickly stabilizes to steady state values. The long-term system behavior is a function of the *F/V* ratio and the S_0 concentration. Operating at $F/V = (1/5)/\text{s}$ and $S_0 = 0.5$ mg/ml leads to washout of amoebae (predator).

In the program listing, the plotting commands have been removed for brevity.

MATLAB program 7.14

```
% Use ode15s to solve the microbial population dynamics problem.
clear all

% Variables
S0 = 0.5; % glucose concentration in feed mg/ml

% Initial conditions
y0(1) = 0.5; % mg glucose/ml
y0(2) = 13e8; % bacteria/ml
y0(3) = 4e5; % amoeba/ml

% Time interval
tf = 1200; % hr

% Call ODE Solver
[t, y] = ode15s('popdynamics', tf, y0, [], S0);
```

```
% Step change in substrate concentration in feed
S0 = 0.1; %mg glucose/ml
y0 = y(end, :); % Values of variables at last time point

% Call ODE Solver
[t, y] = ode15s('popdynamics', tf, y0, [], S0);
```

MATLAB program 7.15

```
function f = popdynamics(t, y, flag, S0)
% Evaluate slope f(t,y) of coupled ODEs for microbial population dynamics

% Constants
FbyV = 1/16;            % feed to chemostat volume ratio (/hr)
muN1max = 0.25;         % max specific growth rate for E coli (/hr)
muN2max = 0.24;         % max specific growth rate for amoeba (/hr)
KN1 = 5e-4;             % saturation constant (mg glucose/ml)
KN2 = 4e8;              % saturation constant (bacteria/ml)
invYN1S = 3.3e-10;      % reciprocal yield factor (mg glucose/bacteria)
invYN2N1 = 1.4e3;       % reciprocal yield factor (bacteria/amoeba)

f = [FbyV*(S0 - y(1)) - invYN1S*muN1max*y(1)*y(2)/(KN1 + y(1)); ...
-FbyV*y(2) + muN1max*y(1)*y(2)/(KN1 + y(1)) - ...
invYN2N1*muN2max*y(2)*y(3)/(KN2 + y(2)); ...
-FbyV*y(3) + muN2max*y(2)*y(3)/(KN2 + y(2))];
```

7.7 Shooting method for boundary-value problems

A unique solution for an ordinary differential equation of order n is possible only when n constraints are specified along with the equation. This is because, if a solution exists for the ODE, the constraints are needed to assign values to the n integration constants of the general solution, thus making the solution unique. For an initial-value problem, the n constraints are located at the initial point of integration. Such problems are often encountered when integration is performed over the time domain. When the interval of integration lies in the physical space domain, it is likely that some of the n constraints are available at the starting point of integration, i.e. at one end of the boundary, while the remaining constraints are specified at the endpoint of integration at the other end of the boundary. Constraints that are specified at two or more boundary points of the interval of integration are called **boundary conditions**. ODEs of second order and higher that are supplied with boundary conditions are called **boundary-value problems (BVPs)**.

Consider the second-order ODE

$$\frac{d^2y}{dx^2} - \frac{dy}{dx} + y = \rho.$$

To solve this non-homogeneous ODE problem, y is integrated with respect to x. The interval of integration is $a \leq x \leq b$, and ρ is some constant.

There are several types of boundary conditions that can be specified for a second-order ODE. At any boundary endpoint, three types of boundary conditions are possible.

(1) $y = \alpha$ A boundary condition that specifies the value of the dependent variable at an endpoint is called a **Dirichlet boundary condition**.
(2) $y' = \alpha$ A boundary condition that specifies the value of the derivative of the dependent variable at an endpoint is called a **Neumann boundary condition**.
(3) $c_1 y' + c_2 y = c_3$ A boundary condition that linearly combines the dependent variable and its derivative at an endpoint is called a **mixed boundary condition**.

For any second-order boundary-value ODE problem, one boundary condition is specified at $x = a$ and another condition at $x = b$. The boundary-value problem

$$y'' - y' + y = \rho, \qquad y(a) = \alpha; y(b) = \beta$$

is said to be have two Dirichlet boundary conditions.

Several numerical methods have been designed specifically to solve boundary-value problems. In one popular technique, called the **finite difference method**, the integration interval is divided into many equally spaced discrete points called nodes. The derivatives of y in the equation are substituted by finite difference approximations. This creates one algebraic equation for each node in the interval. If the ODE system is linear, the numerically equivalent system of algebraic equations is linear, and the methods discussed in Chapter 2 for solving linear equations are used to solve the problem. If the ODE system is nonlinear in y, the resulting algebraic equations are also nonlinear, and iterative methods such as Newton's method must be used.

Another method for solving boundary-value problems is the **shooting method**. This method usually provides better accuracy than finite difference methods. However, this method can only be applied to solve ODEs and not partial differential equations (PDEs). On the other hand, finite difference methods are easier to use for the higher-order ODEs and are commonly used to solve PDEs. In the shooting method, the nth-order ODE is first converted to a system of first-order ODEs. Any unspecified initial conditions for the set of first-order equations are guessed, and the ODE system is treated as an initial-value problem. The ODE integration schemes described earlier in this chapter are used to solve the problem for the set of known and assumed initial conditions. The final endpoint values of the variables obtained from the solution are compared with the actual boundary conditions supplied. The magnitude of deviation of the numerical solution at $x = b$ from the desired solution is used to alter the assumed initial conditions, and the system of ODEs is solved again. Thus, we "shoot" from $x = a$ (by assuming values for unspecified boundary conditions at the first endpoint) for a desired solution at $x = b$, our fixed target. Every time we shoot from the starting point, we refine our approximation of the unknown boundary conditions at $x = a$ by taking into account the difference between the numerically obtained and the known boundary conditions at $x = b$. The shooting method can be combined with a method of nonlinear root finding (see Chapter 5) for improving the future iterations of the initial boundary condition.

Boundary-value problems for ODEs can be classified as linear or nonlinear, based on the dependence of the differential equations on y and its derivatives. Solution techniques for linear boundary-value problems are different from the methods used for nonlinear problems. Linear ODEs are solved in a straightforward manner without iteration, while nonlinear BVPs require an iterative sequence to converge upon a solution within the tolerance provided. In this section, we discuss shooting methods for second-order ODE boundary-value problems. Higher-order boundary-value problems can be solved by extending the methods discussed here.

7.7.1 Linear ODEs

A linear second-order boundary value problem is of the form

$$y'' = p(x)y' + q(x)y + r(x), \qquad x \in [a, b], \tag{7.54a}$$

where the independent variable is x. It is assumed that a unique solution exists for the boundary conditions supplied. Suppose the ODE is subjected to the following boundary conditions:

$$y(a) = \alpha;\ y(b) = \beta. \tag{7.54b}$$

A linear problem can be reformulated as the sum of two simpler linear problems, each of which can be determined separately. The superposition of the solutions of the two simpler linear problems yields the solution to the original problem. This is the strategy sought for solving a linear boundary-value problem. We express Equation (7.54) as the sum of two linear initial-value ODE problems. The first linear problem has the same ODE as given in Equation (7.54a):

$$u'' = p(x)u' + q(x)u + r(x); \qquad u(a) = \alpha; u'(a) = 0. \tag{7.55}$$

One of the two initial conditions is taken from the actual boundary conditions specified in Equation (7.54b). The second boundary condition assumes that the first derivative is zero at the left boundary of the interval. The second linear ODE is the homogeneous form of Equation (7.54a):

$$v'' = p(x)v' + q(x)v; \qquad v(a) = 0; u'(a) = 1. \tag{7.56}$$

The initial conditions for the second linear problem are chosen so that the solution of the original ODE (Equation (7.54)) is

$$y(x) = u(x) + cv(x), \tag{7.57}$$

where c is a constant that needs to be determined. Before we proceed further, let us confirm that Equation (7.57) is indeed the solution to Equation (7.54a).

Differentiating Equation (7.57) twice, we obtain

$$y'' = u'' + cv'' = p(x)u' + q(x)u + r(x) + c(p(x)v' + q(x)v).$$

Rearranging,

$$y'' = p(u' + cv') + q(u + cv) + r.$$

We recover Equation (7.54a),

$$y'' = p(x)y' + q(x)y + r(x).$$

Thus, the solution given by Equation (7.57) satisfies Equation (7.54a).

Equation (7.57) must also satisfy the boundary conditions given by Equation (7.54b) at $x = a$:

$$u(a) + cv(a) = \alpha + c \cdot 0 = \alpha.$$

We can use the boundary condition at $x = b$ to determine c. Since

$$u(b) + cv(b) = \beta,$$

$$c = \frac{\beta - u(b)}{v(b)}.$$

If $v(b) = 0$, the solution of the homogeneous equation can be either $v = 0$ (in which case $y = u$ is the solution) or a non-trivial solution that satisfies homogeneous boundary conditions. When $v(b) = 0$, the solution to Equation (7.54) may not be unique.

For $v(b) \neq 0$, the solution to the second-order linear ODE with boundary conditions given by Equation (7.54b) is given by

$$y(x) = u(x) + \frac{\beta - u(b)}{v(b)} v(x). \tag{7.58}$$

To obtain a solution for a linear second-order boundary-value problem subject to two Dirichlet boundary conditions, we solve the initial-value problems given by Equations (7.55) and (7.56), and then combine the solutions using Equation (7.58).

If the boundary condition at $x = b$ is of Neumann type ($y'(b) = \beta$) or of mixed type, and the boundary condition at $x = a$ is of Dirichlet type, then the two initial-value problems (Equations (7.55) and (7.56)) that must be solved to obtain u and v remain the same. The method to calculate c is, for a Neumann boundary condition at $x = b$, given by

$$y'(b) = u'(b) + cv'(b) = \beta$$

or

$$c = \frac{\beta - u'(b)}{v'(b)}.$$

The initial conditions of the two initial-value ODEs in Equations (7.55) and (7.56) will depend on the type of boundary condition specified at $x = a$ for the original boundary-value problem.

7.7.2 Non-linear ODEs

Our goal is to find a solution to the nonlinear second-order boundary-value problem

$$y'' = f(x, y, y'), \qquad a \leq x \leq b, \qquad y(a) = \alpha; \; y(b) = \beta. \tag{7.59}$$

Because the ODE has a nonlinear dependence on y and/or y', we cannot simplify the problem by converting Equation (7.59) into a linear combination of simpler ODEs, i.e. two initial-value problems. Equation (7.59) must be solved iteratively by starting with a guessed value s for the unspecified initial condition, $y'(a)$, and improving our estimate of s until the error in our numerical estimate of $y(b)$ is reduced to below the tolerance limit ϵ. The deviation of $y(b)$ from its actual value β is a nonlinear function of the guessed slope s at $x = a$. We wish to minimize the error, $y(b) - \beta$, which can be written as a function of s:

$$y(b) - \beta = z(s). \tag{7.60}$$

We seek a value of s such that $z(s) = 0$. To solve Equation (7.60), we use a nonlinear root-finding algorithm. We begin by solving the initial-value problem,

$$y'' = f(x, y, y'), \qquad a \leq x \leq b, \qquad y(a) = \alpha; \; y'(a) = s_1,$$

using the methods discussed in Sections 7.2–7.6, to obtain a solution $y_1(x)$. If $z_1 = y_1(b) - \beta \le \epsilon$, the solution $y_1(x)$ is accepted.

Otherwise, the initial-value problem is solved for another guess value, s_2, for the slope at $x = a$,

$$y'' = f(x, y, y'), \qquad a \le x \le b, \qquad y(a) = \alpha;\ y'(a) = s_2,$$

to yield another solution $y_2(x)$ corresponding to the guessed value $y'(a) = s_2$. If $z_2 = y_2(b) - \beta \le \epsilon$, then the solution $y_2(x)$ is accepted. If neither guess for the slope s_1 or s_2 produce a solution that meets the boundary condition criterion at $x = b$, we must generate an improved guess s_3. The **secant method**, which approximates the function z within an interval Δs as a straight line, can be used to produce a next guess for $y'(a)$. If the two points on the function curve, $z(s)$, that define the interval, Δs, are (s_1, z_1) and (s_2, z_2), then the slope of the straight line joining these two points is

$$m = \frac{z_2(b) - z_1(b)}{s_2 - s_1}.$$

A third point lies on the straight line with slope m such that it crosses the x-axis at the point $(s_3, 0)$, where s_3 is our approximation for the actual slope that satisfies $z(s) = 0$. The straight-line equation passing through these three points is given by

$$0 - z_2(b) = m(s_3 - s_2)$$

or

$$s_3 = s_2 - \frac{(s_2 - s_1)z_2(b)}{z_2(b) - z_1(b)}.$$

We can generalize the formula above as

$$s_{i+1} = s_i - \frac{(s_i - s_{i-1})z_i(b)}{z_i(b) - z_{i-1}(b)}, \tag{7.61}$$

which is solved iteratively until $z(s_{i+1}) \le \epsilon$. A nonlinear equation usually admits several solutions. Therefore, the choice of the starting guess value is critical for finding the correct solution.

Newton's method can also be used to iteratively solve for s. This root-finding algorithm requires an analytical expression for the derivative of $y(b)$ with respect to s. Let's write the function y at $x = b$ as $y(b, s)$, since it is a function of the initial slope s. Since the analytical form of $z(s)$ is unknown, we cannot directly take the derivative of $y(b, s)$ to obtain $dy/ds(b, s)$. Instead, we create another second-order initial-value problem whose solution gives us $\partial y/\partial s(x, s)$. Construction of this companion second-order ODE requires obtaining the analytical form of the derivatives of $f(x, y, y')$ with respect to y and y'. While Newton's formula results in faster convergence, i.e. fewer iterations to obtain the desired solution, two second-order initial-value problems must be solved simultaneously for the slope s_i to obtain an improved guess value s_{i+1}. An explanation of the derivation and application of the shooting method using Newton's formula can be found in Burden and Faires (2005).

Box 7.5 Controlled-release drug delivery using biodegradable polymeric microspheres

Conventional dosage methods, such as oral delivery and injection, are not ideally suited for sustained drug delivery to diseased tissues or organs of the body. Ingesting a tablet or injecting a bolus of drug dissolved in aqueous solution causes the drug to be rapidly released into the body. Drug concentrations in blood or tissue can rise to nearly toxic levels followed by a drop to ineffective levels. The duration over which an optimum drug concentration range is maintained in the body, i.e. the period of deriving maximum therapeutic benefit, may be too short. To maintain an effective drug concentration in the blood, high drug dosages, as well as frequent administration, become necessary.

When a drug is orally administered, in order to be available to different tissues, the drug must first enter the bloodstream after absorption in the gastrointestinal tract. The rate and extent of absorption may vary greatly, depending on the physical and chemical form of the drug, presence or absence of food, posture, pH of gastrointestinal fluids, duration of time spent in the esophagus and stomach, and drug interactions. Uncontrolled rapid release of drug can cause local gastrointestinal toxicity, while slow or incomplete absorption may prevent realization of therapeutic benefit.

In recent years, more sophisticated drugs in the form of protein-based and DNA-based compounds have been introduced. However, oral delivery of proteins to the systemic circulation is particularly challenging. For the drug to be of any use, the proteins must be able to pass through the gastrointestinal tract without being enzymatically degraded. Many of these drugs have a small therapeutic concentration range, with the toxic concentration range close to the therapeutic range. Research interest is focused on the development of controlled-release drug-delivery systems that can maintain the therapeutic efficacy of such drugs. Controlling the precise level of drug in the body reduces side-effects, lowers dosage requirements and frequency, and enables a predictable and extended duration of action. Current controlled-release drug-delivery technologies include transdermal patches, implants, microencapsulation, and inhaled and injectable sustained-release peptide/protein drugs.

An example of a controlled-release drug-delivery system is the polymer microsphere. Injection of particulate suspensions of drug-loaded biodegradable polymeric spheres is a convenient method to deliver hormonal proteins or peptides in a controlled manner. The drug dissolves into the surrounding medium at a pre-determined rate governed by the diffusion of drug out of the polymer and/or by degradation of the polymer. Examples of commercially available devices for sustained drug release are Gliadel (implantable polyanhydride wafers that release drug at a constant rate as the polymer degrades) for treating brain cancer, Lupron Depot (injectable polymer microspheres) for endometriosis and prostate cancer, and Nutropin Depot for pituitary dwarfism. Lupron Depot (leuprolide) is a suspension of microspheres made of poly-lactic-glycolic acid (PLGA) polymer containing leuprolide acetate, and is administered once a month as an intramuscular injection. The drug is slowly released into the blood to maintain a steady plasma concentration of leuprolide for one month.

Since controlled-release drug-loaded microparticles are usually injected into the body, thereby bypassing intestinal digestion and first-pass liver metabolism, one must ensure that the polymeric drug carrier material is non-toxic, degrades within a reasonable time frame, and is excreted from the body without any accumulation in the tissues. Biodegradable polymers gradually dissolve in body fluids either due to enzymatic cleavage of polymer chains or hydrolytic breakdown (hydrolysis) of the polymer. PLGA is an example of a biocompatible polymer (polyester) that is completely biodegradable. As sections of the polymer chains in the outer layers of the drug–polymeric system are cleaved and undergo dissolution into the surrounding medium, drug located in the interior becomes exposed to the medium. The surrounding fluid may penetrate into the polymer mass, thereby expediting drug diffusion out of the particle. Transport of drug molecules from within the polymeric mass to the surrounding tissues or fluid is governed by several mechanisms that occur either serially or in parallel. Mathematical formulation of the diffusional processes that govern controlled release is far from trivial.

Here, we avoid dealing with the partial differential equations used to model the diffusion of drug in time and space within the particle and in the surrounding medium. We make the simplifying assumption that the transport characteristics of the drug (drug dissolution rate, diffusion of drug in three-dimensional aqueous channels within the degrading polymeric matrix, diffusion of drug in the hydrodynamic layer surrounding the particle, polymer degradation kinetics, and kinetics of cleavage of covalently attached drug from polymer) can be conveniently represented by a time-invariant mass transfer coefficient k.

A mass balance on a drug-loaded PLGA microparticle when placed in an aqueous medium is given by the first-order ODE

$$-\frac{dM}{dt} = kA(C_s - C_m),$$

where M is the mass of the drug remaining in the particle, k is the mass transfer coefficient that characterizes the flux of solute per unit surface area per unit concentration gradient, A is the instantaneous surface area of the sphere, C_s is the solubility of the drug in water, and C_m is the concentration of drug in the medium. If the particle is settling under gravity in the quiescent medium, the sedimentation rate of the particle is assumed to be slow. The concentration of drug immediately surrounding the particle is assumed to be equal to the maximum solubility of drug at body temperature.

If C_{drug} is the uniform concentration of drug within the microsphere that does not change with time during the polymer degradation process, and a is the instantaneous particle radius, we have

$$M = C_{drug}\left(\frac{4}{3}\pi a^3\right).$$

The mass balance equation can be rearranged to obtain

$$\frac{da}{dt} = -\frac{k}{C_{drug}}(C_s - C_m). \tag{7.62}$$

The size a of the microsphere shrinks with time.

For a spherical particle settling in Stokes flow under the effect of gravity, the momentum or force balance that yields a first-order ODE in velocity was introduced in Example 7.2. However, the changing particle size produces another term in the equation (adapted from Pozrikidis (2008)). The original momentum balance for a settling sphere in Stokes flow,

$$\frac{4}{3}\pi\rho_p\frac{d(a^3v)}{dt} + 6\mu\pi av - \frac{4}{3}\pi a^3(\rho_p - \rho_m)g = 0,$$

now becomes

$$a^3\frac{dv}{dt} + 3a^2v\frac{da}{dt} = a^3\frac{(\rho_p - \rho_m)}{\rho_p}g - \frac{9\mu a}{2\rho_p}v. \tag{7.63}$$

Substituting Equation 7.62 into Equation 7.63, we get

$$\frac{dv}{dt} = \frac{(\rho_p - \rho_f)}{\rho_p}g - \frac{9\mu}{2a^2\rho_p}v + \frac{3vk}{aC_{drug}}(C_s - C_m). \tag{7.64}$$

The initial condition of Equation (7.64) is $v(0) = \frac{dz}{dt}\big|_{t=0} = 0$. Equation (7.64) is nonlinear in a.

We are asked to determine the size, a_0, of the largest drug-loaded microparticle that, when added to the top of a chamber, will dissolve by the end of its travel down some vertical distance through a body of fluid. Figure 7.15 illustrates the geometry of the system. Under gravity, the drug-loaded particle sinks in low Reynolds number flow. As the particle falls through the medium, the drug is released in a controlled and continuous manner.

To calculate the size of the particle that, starting from rest, will shrink to zero radius after falling a distance L, we simultaneously integrate Equations (7.62) and (7.64). However, the time interval of integration is unknown. Integration of the system of ODEs must include the first-order equation

Figure 7.15

Schematic diagram of the process of controlled drug release from a microparticle settling in aqueous fluid.

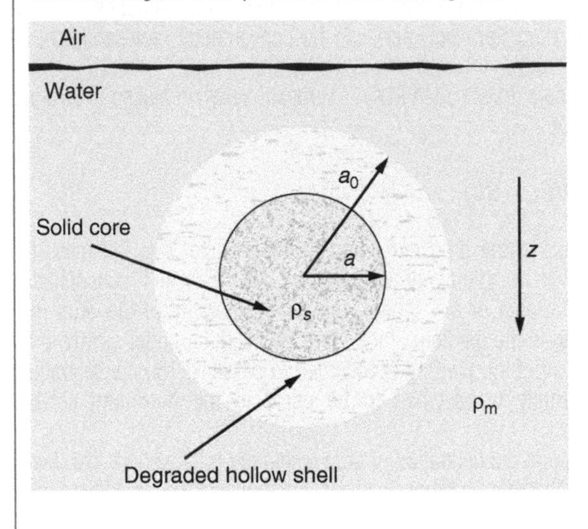

$$\frac{dz}{dt} = v$$

to calculate the distance traveled, so we can stop the integration when $z = L$. It is more convenient to integrate for a known interval length. Therefore, we change the independent variable of Equations (7.62) and (7.64) from t to z using the rule

$$\frac{df}{dz} = \frac{df}{dt}\frac{dt}{dz} = \frac{1}{v}\frac{df}{dt}.$$

Equations (7.62) and (7.64) become

$$\frac{da}{dz} = -\frac{k}{vC_{\text{drug}}}(C_s - C_m), \qquad 0 \le z \le L; \tag{7.65}$$

$$\frac{dv}{dz} = \frac{(\rho_p - \rho_m)}{\rho_p}\frac{g}{v} - \frac{9\mu}{2a^2\rho_p} + \frac{3k}{aC_{\text{drug}}}(C_s - C_m), \qquad 0 \le z \le L. \tag{7.66}$$

The boundary conditions for Equations (7.65) and (7.66) are

$$v(0) = 0; \qquad a(L) = 0.$$

We use the secant method to solve for a_0. We try different values of a_0 as suggested by Equation (7.61) to satisfy the boundary condition $a(L) = 0$.

The drug loading in a single microparticle is 20% w/w. The initial distribution of drug is uniform within the particle. The parameters are assigned the following values at $T = 37$ °C:

$C_s = 0.1$ mg/ml (Sugano, 2008),
$C_m = 0.0$ mg/ml,
$C_{\text{drug}} = 0.256$ g/cm^3,
$k = 1 \times 10^{-6}$ cm/s (Siepmann *et al.*, 2005),
$\rho_p = 1.28$ g/cm^3 (Vauthier *et al.*, 1999),
$\rho_f = 1.0$ g/cm^3,

$\mu = 0.7$ cP,
$L = 10$ cm.

The CGS system of units is used. However, the equations will be designed such that a_0 can be entered in units of μm. We make the following assumptions.

(1) As the polymer matrix-loaded drug dissolves into the aqueous surroundings, concomitantly the outer particle shell devoid of drug completely disintegrates, or is, at the least, sufficiently hollow that for all practical purposes it does not exert a hydrodynamic drag. The radius a therefore refers to the particle core that has a density of $\rho_s = 1.28$ g/cm^3.
(2) The density, ρ_p, of the particle core that contains the bulk of the drug does not change. In other words, hydrolysis of polymer and dissolution of drug occur primarily at the shell surface and penetrate inwards as a wavefront.
(3) The velocity is small enough that mass transfer is dominated by diffusional processes, and the mass transfer coefficient is independent of particle velocity (Sugano, 2008).

On making the following substitutions

$$y_1 = a; \quad y_2 = v; \quad x = z$$

into Equations (7.65) and (7.66), we obtain

$$\frac{dy_1}{dx} = -\frac{k}{y_2 C_{\text{drug}}}(C_s - C_m), \tag{7.67}$$

$$\frac{dy_2}{dx} = \frac{(\rho_p - \rho_m)}{\rho_p}\frac{g}{y_2} - \frac{9\mu}{2y_1^2\rho_p} + \frac{3k}{y_1 C_{\text{drug}}}(C_s - C_m). \tag{7.68}$$

We will not be able to solve Equations (7.67) and (7.68) using the boundary conditions stated above. Can you guess why? The solution becomes singular (blows up) at $a = 0$ and at $v = 0$. We must thus make the boundary conditions more realistic.

(1) If we want 90% of the drug contained in the microsphere to be released, i.e. 90% of the sphere volume to vanish during particle descent in the fluid, the boundary condition at $x = L$ becomes

$$y_1(x = L) = 0.464a_0.$$

In other words, when the radius of the particle drops to below 46.4% of its original size, at least 90% of the drug bound in the polymer matrix will have been released.
(2) We assume that the microparticle is introduced into the fluid volume from a height such that, when the particle enters the aqueous medium, it has attained terminal Stokes velocity in air at $T = 20$ °C (N_{RE} (in air) $\sim O(10^{-5})$). Accordingly, the initial velocity at $x = 0$ is $0.016a_0^2$ cm/s, where a_0 is measured microns.

We also want to know the time taken for drug release. Let $y_3 = t$. We add a third equation to the ODE set,

$$\frac{dy_3}{dx} = \frac{1}{v} \tag{7.69}$$

The secant equation (Equation (7.61)) is rewritten for this problem as

$$a_{0,j+1} = a_{0,j} - \frac{(a_{0,j} - a_{0,j-1})(y_{1,j}(L) - 0.464a_{0,j})}{(y_{1,j}(L) - 0.464a_{0,j}) - (y_{1,j-1}(L) - 0.464a_{0,j-1})} \tag{7.70}$$

Example 7.2 demonstrated that the problem of time-varying settling velocity is stiff. We use `ode15s` to solve the IVP defined by Equations (7.67) – (7.69) and the initial conditions

Figure 7.16

Change in microparticle size with distance traversed as the drug–polymer matrix erodes and dissolves into the medium.

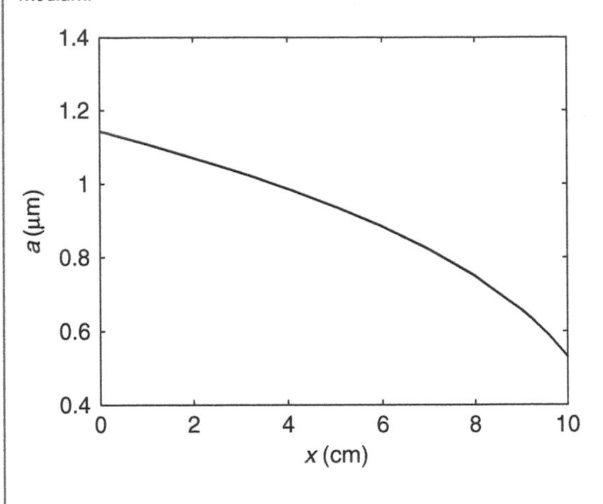

Figure 7.17

Change in microparticle settling velocity with distance traversed as the drug–polymer matrix erodes and dissolves into the medium. (a) Large retardation in velocity of the particle over the first 0.00013 μm of distance traveled. (b) Gradual velocity decrease taking place as the particle falls from 0.00013 μm to 10 μm.

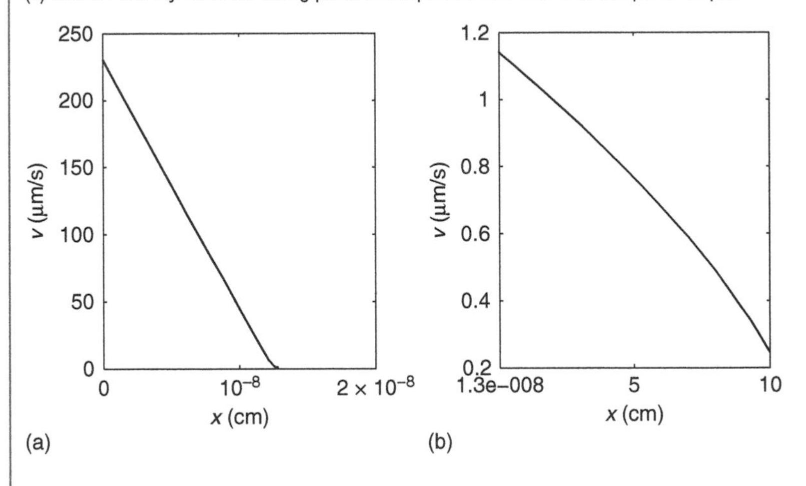

$$y_1(x=0) = a_0;\; y_2(x=0) = 0.016a_0^2;\; y_3(x=0) = 0.$$

Care must be taken to start with two guessed values for a_0 that are close to the correct initial condition that satisfies the boundary conditions at $x = L$. We choose the two guessed values for the initial particle size to be $a_{0,1} = 1.2$ μm and $a_{0,2} = 1.3$ μm. MATLAB programs 7.16 and 7.17 contain the code used to obtain the result.

After seven iterations of Equation (7.70), we obtain a solution that satisfies our tolerance condition: $a_0 = 1.14$ μm. The time taken for 90% of the particle volume to dissolve is 43.6 hours. Figures 7.16 and 7.17 depict the change in radius and velocity of the particle as it traverses downward. Note that initial velocity retardation is immediate and occurs over a very short time of $O(10^{-6})$ s. This region of integration is very stiff.

MATLAB program 7.16

```
% Use the shooting method to solve the controlled drug delivery
  problem.
clear all

% Variables
maxloops = 50; % maximum number of iterations allowed
L = 10; % length of interval (cm)
y0(3) = 0; % initial time (s)

% First guess for initial condition
a0_1 = 1.2; % radius of particle (um)
y0(1) = a0_1;
y0(2) = 0.016*a0_1^2; % velocity (cm/s)

% Call ODE Solver: First guess
[x, y] = ode15s('drugdelivery', L, y0);
y1(1) = y(end,1); % End-point value for y1

% Second guess for initial condition
a0_2 = 1.3; % radius of particle (um)
y0(1) = a0_2;
y0(2) = 0.016*a0_2^2; % velocity (cm/s)

% Call ODE Solver: Second guess
[x, y] = ode15s('drugdelivery', L, y0);
y1(2) = y(end,1); % End-point value for y1

% Use secant method to find correct initial condition
for i = 1:maxloops
    % Secant Equation
    a0_3 = a0_2 - (a0_2 - a0_1)*(y1(2)-a0_2*0.464)/ ...
          ((y1(2) - a0_2*0.464) - (y1(1) - a0_1*0.464));
    % New guess for initial condition
    y0(1) = a0_3;
    y0(2) = 0.016*a0_3^2; % velocity (cm/s)
    % Solving system of ODEs for improved value of initial condition
    [x, y] = ode15s('drugdelivery', L, y0);
    y1(3) = y(end,1);
    % Check if final value satisfies boundary condition
    if abs(y1(3) - a0_3*0.464) < a0_3*0.001 % tolerance criteria
        break
    end
    a0_1 = a0_2;
    a0_2 = a0_3;
    y1(1) = y1(2);
    y1(2) = y1(3);
end
```

MATLAB program 7.17

```
function f = drugdelivery(x, y)
% Evaluate f(t,y) for controlled-release drug delivery problem

% Constants
```

```
Cs = 0.1e-3;        % solubility of drug (g/ml)
Cm = 0.0;           % concentration of drug in medium (g/ml)
Cdrug = 0.256;      % concentration of drug in particle (g/cm^3)
rhop = 1.28;        % density of particle (g/cm^3)
rhof = 1.0;         % density of medium (g/cm^3)
mu = 0.7e-2;        % viscosity of medium (Poise)
k = 1e-6;           % mass transfer coefficient (cm/s)
g = 981;            % acceleration due to gravity (cm/s^2)

f = [-k/y(2)/Cdrug*(Cs - Cm)*1e4; ...
    (rhop-rhof)/rhop*g/y(2) - 9*mu/2/(y(1)*1e-4)^2/rhop + ...
    3*k/(y(1)*1e-4)/Cdrug*(Cs - Cm); ...
    1/y(2)];
```

7.8 End of Chapter 7: key points to consider

(1) An **ordinary differential equation (ODE)** contains derivatives with respect to only one independent variable.

(2) An ODE of order n requires n constraints or boundary values to be specified before a unique solution can be determined. An ODE is classified based on where the constraints are specified. An **initial-value problem** has all n constraints specified at the starting point of the interval and a **boundary-value problem** has some conditions specified at the beginning of the interval and the rest specified at the end of the interval.

(3) To integrate an ODE numerically, one must divide the integration interval into many subintervals. The number of subintervals depends on the step size and the size of the integration interval. The endpoints of the subintervals are called nodes or time points. The solution to the ODE or system of ODEs is determined at the nodes.

(4) A numerical integration method that determines the solution at the next time point, t_{k+1}, using slopes and/or solutions calculated at the current time point t_k, and/or prior time points (e.g. t_{k-1}, t_{k-2}) is called an **explicit method**, e.g. Euler's forward method. A numerical integration method whose difference equation requires a slope evaluation at t_{k+1} in order to find the solution at the next time point, t_{k+1}, is called an **implicit method**, e.g. Euler's backward method. The unknown terms in an implicit formula are located on both sides of the difference equation.

(5) The **local truncation error** at any time point in the integration interval is the numerical error due to the truncation of terms, if one assumes that the previous solution(s) used to calculate the solution at the next time point is exact. The **global truncation error** at any time point in the integration interval is the error between the numerical approximation and the exact solution. The global error is the sum of (i) accumulated errors from previous time steps, and (ii) the local truncation error at the current integration step.

(6) The **order** of a numerical ODE integration method is determined by the order of the global truncation error. An nth-order method has a global truncation error of $O(h^n)$.

(7) A **one-step integration method** uses the solution at the current time point to predict the solution at the next time point. A **multistep integration method** evaluates the slope function at previous time steps and collectively uses the knowledge of slope behavior at multiple time points to determine a solution at t_{k+1}.

(8) An ODE or system of ODEs is said to be stable if its solution is bounded at all time t. This is possible if the real parts of all eigenvalues of the Jacobian matrix are negative. If even one eigenvalue has a positive real part, the solution will grow in time without

Table 7.5. *MATLAB ODE functions discussed in the chapter*

ODE function	Algorithm	Type of problem
`ode45`	RKF method: RK–4 and RK–5 pair using the Dormand and Prince algorithm	non-stiff ODEs
`ode23`	RKF method: RK–2 and RK–3 pair using the Bogacki and Shampine algorithm	non-stiff ODEs when high accuracy is not required
`ode113`	Adams–Bashforth–Moulton method	non-stiff ODEs when high accuracy is required
`ode15s`	multistep method	stiff problems; use on the first try
`ode23s`	one-step method	stiff problems at low accuracy
`ode23t`	trapezoidal rule	stiff equations
`ode23tb`	implicit RK formula	stiff equations at low accuracy

bound. A numerical solution is said to be **stable** when the error remains constant or decays with time, and is **unstable** if the error grows exponentially with time. The **stability of a numerical solution** is dependent on the nature of the ODE system, the properties of the ODE solver, and the step size chosen. Implicit methods have better stability properties than explicit methods.

(9) A **stiff differential equation** is difficult to integrate since it has one or more rapidly vanishing or accelerating transients contained in the solution. To integrate a stiff ODE, one must use a numerical scheme that has superior stability properties. An integration method with poor stability properties must necessarily use vanishingly small step sizes to ensure that the numerical solution remains stable. When the step size is too small, round-off error will preclude a stable solution.

(10) The **shooting method** is used to solve boundary-value ODE problems. This method converts the boundary-value problem into an initial-value problem by guessing values for the unspecified initial conditions. The numerical estimates for the boundary endpoint values at the end of the integration are compared to the target values. One can use a nonlinear root-finding algorithm to improve iterations of the guessed initial condition(s).

(11) Table 7.5 lists the MATLAB ODE functions that are discussed in this chapter.

7.9 Problems

7.1. **Monovalent binding of ligand to cell surface receptors** Consider a monovalent ligand L binding reversibly to a monovalent receptor R to form a receptor/ligand complex C:

$$R + L \underset{k_r}{\overset{k_f}{\rightleftharpoons}} C,$$

where C represents the concentration of receptor (R) – ligand (L) complex, and k_f and k_r are the forward and reverse reaction rates, respectively. If we assume that the ligand concentration remains constant at L_0, the following differential equation governs the dynamics of C (Lauffenburger and Linderman, 1993):

$$\frac{dC}{dt} = k_f[R_T - C]L_0 - k_r C,$$

where R_T is the total number of receptor molecules on the cell surface.

(a) Determine the stability criterion of this differential equation by calculating the Jacobian. Supposing that we are interested in studying the binding of fibronectin receptor to fibronectin on fibroblast cells ($R_T = 5 \times 10^5$ sites/cell, $k_f = 7 \times 10^5$/(M min), $k_r = 0.6$/min). What is the range of ligand concentrations L_0 for which this differential equation is stable?

(b) Set up the numerical integration of the differential equation using an implicit second-order rule, i.e. where the error term scales as $\sim h^3$ and $h = 0.1$ min is the step size. Use $L_0 = 1$ μM. Perform the integration until equilibrium is reached. How long does it take to reach equilibrium (99% of the final concentration C is achieved)? (You will need to calculate the equilibrium concentration to perform this check.)

7.2. **ODE model of convection–diffusion–reaction** The following equation is a differential model for convection, diffusion, and reaction in a tubular reactor, assuming first-order, reversible reaction kinetics:

$$\frac{1}{Pe}\frac{d^2C}{dx^2} - \frac{dC}{dx} = Da \cdot C,$$

where Pe is the Peclet number, which measures the relative importance of convection to diffusion, and Da is the Damkohler number, which compares the reaction time scale with the characteristic time for convective transport.

Perform a stability analysis for this equation.

(a) Convert the second-order system of equations into a set of coupled first-order ODEs. Calculate the Jacobian $\boldsymbol{J}$ for this system.

(b) Find the eigenvalues of $\boldsymbol{J}$. (Find the values of λ such that $|\boldsymbol{A} - \lambda \boldsymbol{I}| = 0$.)

(c) Assess the stability of the ODE system.

7.3. **Chlorine loading of the stratosphere** Chlorofluorocarbons (CFCs), hydrofluorocarbons (HCFCs), and other haloalkanes are well known for their ozone-depleting characteristics. Accordingly, their widespread use in industries for refrigeration, fire extinguishing, and solvents has been gradually phased out. CFCs have a long lifespan in the atmosphere because of their relative inertness. These compounds enter the stratosphere, where UV radiation splits the molecules to produce free chlorine radicals (Cl·) that catalyze the destruction of ozone (O_3) located within the stratosphere. As a result, holes in the ozone have formed that allow harmful UV rays to reach the Earth's surface.

CFC molecules released into the atmosphere cycle every three years or so between the stratosphere and the troposphere. Halogen radicals are almost exclusively produced in the stratosphere. Once transported back to the troposphere, these radicals are usually washed out by rain and return to the soil (see Figure P7.1).

The HCFCs are not as long-lived as CFCs since they are more reactive; they break down in the troposphere and are eliminated after a few years. The concentrations of halocarbons in the stratosphere and troposphere as well as chlorine loading in the stratosphere can be modeled using a set of ordinary differential equations (Ko *et al.*, 1994). If B_S is the concentration of undissociated halocarbons in the stratosphere, B_T is the concentration of undissociated halocarbons in the troposphere, and C is the

Figure P7.1

First two atmospheric layers.

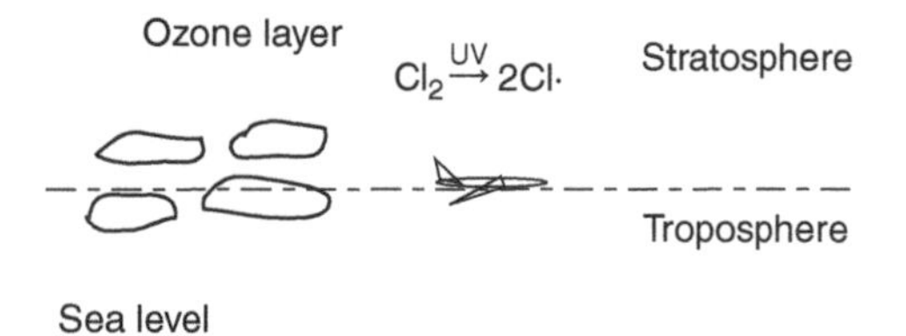

quantity of chlorine in the stratosphere, then we can formulate the following equations:

$$\frac{dB_T}{dt} = \frac{B_S - fB_T}{\tau} - \frac{B_T}{L_T},$$
$$\frac{dB_S}{dt} = \frac{fB_T - B_S}{\tau} - \frac{B_S}{L_S},$$
$$\frac{dC}{dt} = \frac{B_S}{L_S} - \frac{C}{\tau}$$

where

τ is the characteristic time for replacing stratospheric air with tropospheric air,
$f = 0.15/0.85$ is the scaled fraction of tropospheric air that enters the stratosphere (15% of atmospheric mass lies in the stratosphere),
L_T is the halocarbon lifetime in the troposphere that characterizes the time until chemical degradation,
L_S is the halocarbon lifetime in the stratosphere that characterizes the time until chemical degradation.

It is assumed that chlorine is quickly lost from the troposphere. The parameter values are set as $\tau = 3$ yr, $L_T = 1000$ yr, and $L_S = 5$ yr for the halocarbon $CFCl_3$ (Ko *et al.*, 1994).

(a) What is the Jacobian for this ODE system? Use the MATLAB function `eig` to calculate the eigenvalues for $\boldsymbol{J}$. Study the magnitude of the three eigenvalues, and decide if the system is stiff. Accordingly, choose the appropriate ODE solver.
(b) If 100 kg of CFC is released at time $t = 0$, what is the CFC loading in the troposphere and chlorine loading within the stratosphere after 10 years? And after 100 years?

7.4. **Population dynamics among hydrocarbon-consuming bacterial species** *Pseudomonas* is a class of bacteria that converts methane in the gas phase into methanol. Secreted methanol participates in a feedback inhibition control process inhibiting further growth. When grown with another bacterial species, *Hyphomicrobium*, both species benefit from an advantageous interdependent interaction (**mutualism**). *Hyphomicrobium* utilizes methanol as a substrate and thereby lowers the methanol concentration in the surrounding medium promoting growth of *Pseudomonas*. Wilkinson *et al.* (1974) investigated the dynamic behavior of bacterial growth of these two species in a chemostat. Their developments are reproduced here based on the discussion of Bailey and Ollis (1986). Let x_1 denote the *Pseudomonas* population mass and let x_2 denote the *Hyphomicrobium* population mass. Dissolved oxygen is a

Figure P7.2

Flow into and out of a chemostat.

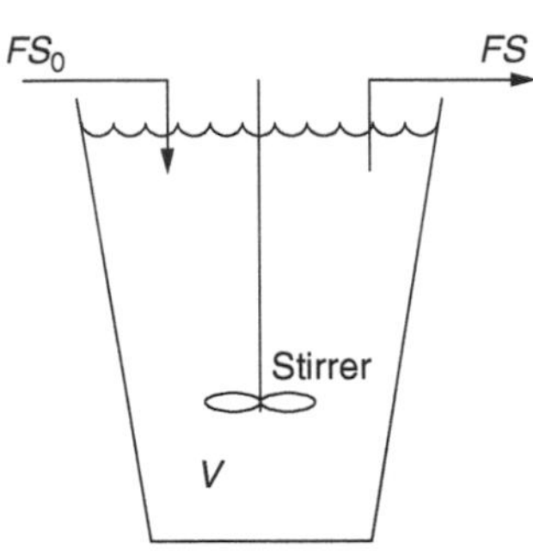

requirement for *Pseudomonas* growth, and its growth rate can be described by the following kinetic rate equation:

$$r_{x_1} = \frac{\mu_{x_1,\max} c_{O_2}}{K_{x_1} + c_{O_2}} \cdot \frac{1}{1 + S/K_i} x_1.$$

This rate equation takes into account the inhibitory action of methanol (S), where K_i is the inhibition constant. The growth of *Pseudomonas* exhibits a monod dependence on dissolved oxygen concentration, c_{O_2}, and S is the methanol concentration.

The growth of *Hyphomicrobium* is not oxygen-limited, but is substrate (methanol)-dependent, and its growth kinetics are described by the following rate equation:

$$r_{x_2} = \frac{\mu_{x_2,\max} S}{K_{x_2} + S} x_2.$$

The reactions take place in a **chemostat** (see Figure P7.2, and see Box 7.4A for an explanation of this term).

The following mass balance equations describe the time-dependent concentrations of the substrates and two bacterial populations:

$$\frac{dc_{O_2}}{dt} = k_l a (c_{O_2 s} - c_{O_2}) - \frac{1}{Y_{x_1|O_2}} r_{x_1} - \frac{F}{V} c_{O_2},$$

where $k_l a (c_{O_2 s} - c_{O_2})$ is the mass transfer rate of oxygen at the gas–liquid interface. We have

$$\frac{dS}{dt} = -\frac{F}{V} S + \frac{1}{Y_{x_1|S}} r_{x_1} - \frac{1}{Y_{x_2|S}} r_{x_2},$$

$$\frac{dx_1}{dt} = -\frac{F}{V} x_1 + r_{x_1},$$

$$\frac{dx_2}{dt} = -\frac{F}{V} x_2 + r_{x_2}.$$

The reaction parameters estimated by Wilkinson *et al.* (1974) are listed below.

$\mu_{x_1,\max} = 0.185$/hr; $\mu_{x_2,\max} = 0.185$/hr,
$K_{x_1} = 1 \times 10^{-5}$ g/l; $K_{x_2} = 5 \times 10^{-6}$ g/l,
$K_i = 1 \times 10^{-4}$ g/l,
$Y_{x_1|O_2} = 0.2$ g *pseudomonas* bacteria/g O_2 consumed,
$Y_{x_1|S} = 5.0$ g *pseudomonas* bacteria produced/g methanol produced,

$Y_{x_2|S} = 0.3$ g *hyphomicrobium* bacteria produced/g methanol consumed,
$k_l a = 42.0$/hr,
c_{O_2s} = saturated dissolved O_2 concentration = 0.008 g/l,
$\frac{F}{V} = 0.08$/hr.

The starting mass concentrations of the bacterial species are $x_1 = 0.672$ g/l and $x_2 = 0.028$ g/l, and $c_{O_2} = 1 \times 10^{-5}$ g/l. At $t = 0$, the methanol concentration in the chemostat is increased to 1.6 g/l. Evaluate the transient response of the system to the methanol shock load over the 72 hour period following the step change in methanol concentration. Is a steady state obtained after three days?

(Hint: The methanol concentration dips to zero between 16 and 17 hours of elapsed time. How does this pose a problem for a discrete integration method? Do you get a meaningful result? What corrective measures can be taken?)

7.5. **Binding kinetics for multivalent ligand in solution interacting with cell surface receptors** This problem models the dynamics of the attachment of proteins that have multiple binding sites to cell surfaces. When a single ligand molecule binds two or more cell surface receptors, the bound receptors are said to be "crosslinked" by the ligand. Viral infection involves attachment of a virus to the cell surface at multiple points. Once the virus is tightly bound to the cell surface, it injects viral genes into the host cell, hijacks the cell's gene replication machinery, and finally lyses the cell to release newly packaged viral particles within the cell. Study of the mechanism of viral docking to cell surface receptors is required to develop intervention strategies or suitable drugs that block the steps leading to viral attachment.

Perelson (1981) developed a kinetic model for multivalent ligand binding to cells (see Box 5.2A). This model makes an assumption (the "equivalent site hypothesis" assumption) that the forward and reverse crosslinking rate constants that characterize the frequency and duration, respectively, of receptors and ligand association are the same for all ligand binding sites. Note that the two-dimensional crosslinking forward rate constant is different from the three-dimensional association rate constant. The latter applies when a ligand in solution first binds to a single receptor on the cell surface. If L is the time-varying ligand concentration in solution, R is the density of unbound receptors on the surface, and C_i is the number of ligand molecules bound to a cell by i ligand binding sites, then the set of deterministic kinetic equations that govern the dynamics of binding is:

$$\frac{dL}{dt} = -\upsilon k_f L R + k_r C_1,$$

$$\frac{dC_1}{dt} = \upsilon k_f L R - k_r C_1 - (f-1) k_x C_1 R + 2k_{-x} C_2,$$

$$\frac{dC_i}{dt} = (f-i+1) k_x C_{i-1} R - (f-i) k_x C_i R - i k_{-x} C_i + (i+1) k_{-x} C_{i+1}, \quad 2 \le i \le f-1,$$

$$\frac{dC_f}{dt} = k_x C_{f-1} R - f k_{-x} C_f,$$

$$R = R_T - \sum_{i=1}^{f} C_i,$$

where k_x is the two-dimensional crosslinking forward rate constant and k_{-x} is the two-dimensional cross-linking reverse rate constant. For L, which varies with time, these $f + 2$ equations containing $f + 2$ unknowns constitute an initial-value problem that can only be solved numerically. This set of equations involves $f + 1$ first-order differential equations and one algebraic equation.

The equivalent site hypothesis model for multivalent ligand binding to monovalent receptors can be used to model the dynamic binding of von Willebrand factor (vWF), a multivalent plasma protein, to platelet cells. At abnormally high shear flows in blood, typical of regions of arterial constriction, von Willebrand factor spontaneously binds to platelet cells. This can lead to the formation of platelet aggregates called thrombi. Platelet thrombi can obstruct blood flow in narrowed regions of the arteries or microvasculature and cause ischemic organ damage that have life-threatening consequences. We use this kinetic model to study certain features of vWF binding to platelets. We adopt the following values for the following model parameters (Mody and King, 2008):

$\nu = 18,$
$f = 9,$
$k_\mathrm{f} = 3.23 \times 10^{-8}$ 1/(# molecules/cell) · 1/s
$k_\mathrm{r} = k_{-x} = 5.47$/s
$k_x = 0.0003172 \times 10^{-4}$ 1/(# molecules/cell) · 1/s

The initial conditions at time $t = 0$ are

$R = R_\mathrm{T} = 10\,700$ #receptors/cell
$L = L_0 = 4380$ #unbound molecules/cell
$C_i\ (i = 1,2, \ldots, 9) = 0$ #bound molecules/cell.

You are asked to do the following.

(a) Develop a numerical scheme based on the **fourth-order Runge–Kutta method** to determine the number of ligand molecules that bind to a platelet cell ($\sum_{i=1}^{f} C_i$) after 1 s and 5 s when the cell is exposed to vWF molecules at high shear rates that permit binding reactions. Use a time step of $\Delta t = 0.01$ s. Use `sum(floor(Ci))` after calculating C_i to calculate the integral number of bound ligand molecules.
(b) Modify the program you have written to determine the time taken for the system to reach thermodynamic equilibrium. The criterion for equilibrium is $|R(t + \Delta t) - R(t)| < 0.01\Delta t$. Also determine the number of ligand molecules bound to the cell at equilibrium.
(c) Verify the accuracy of your solution by comparing it to the equilibrium solution you obtained in Box 5.2B.
(d) In von Willebrand disease, the dissociation rate decreases six-fold due to a mutation in the ligand binding site. If k_r and k_{-x} are reduced by a factor of 6, determine
   (i) the number of bound receptors ($R_\mathrm{T} - R$) after 1 s and 5 s;
   (ii) the time required for equilibrium to be established. Also determine the number of ligand molecules bound to the cell at equilibrium.

7.6. **Settling of a microparticle in air** Determine the time it takes for a microsphere of radius $a = 30$ μm and density 1.2 g/cm^3 to attain 99% of its terminal Stokes velocity when falling from rest in air at $T = 20$ ℃; assume $\rho_\mathrm{air} = 1.206 \times 10^{-3}$ g/cm^3 (no humidity) and $\mu_\mathrm{air} = 0.0175$ cP. The governing equations (Equations (7.27) and (7.28)) are presented in Example 7.2.

References

Bailey, J. E. and Ollis, D. F. (1986) *Biochemical Engineering Fundamentals* (New York: The Mc-Graw Hill Companies).

Barabasi, A. L. (2003) *Linked: How Everything Is Connected to Everything Else. What It Means for Business, Science and Everyday Life* (New York: Penguin Group (USA)).

Burden, R. L. and Faires, J. D. (2005) *Numerical Analysis.* (Belmont, CA: Thomson Brooks/Cole).

Davis, T. H. and Thomson, K. T. (2000) *Linear Algebra and Linear Operators in Engineering* (San Diego, CA: Academic Press).

Grasselli, M. and Pelinovsky, D. (2008) *Numerical Mathematics* (Sudbury, MA: Jones and Bartlett Publishers).

Haase, A. T., Henry, K., Zupancic, M. *et al.* (1996) Quantitative Image Analysis of HIV-1 Infection in Lymphoid Tissue. *Science*, **274**, 985–9.

Ko, M. K. W., Sze, N. D., and Prather, M. J. (1994) Better Protection of the Ozone-Layer. *Nature*, **367**, 505–8.

Lauffenburger, D. A. and Linderman, J. J. (1993) *Receptors: Models for Binding, Trafficking and Signaling* (New York: Oxford University Press).

Mody, N. A. and King, M. R. (2008) Platelet Adhesive Dynamics. Part II: High Shear-Induced Transient Aggregation Via GPIbalpha-vwf-GPIbalpha Bridging. *Biophys. J.*, **95**, 2556–74.

Perelson, A. S. (1981) Receptor Clustering on a Cell-Surface. 3. Theory of Receptor Cross-Linking by Multivalent Ligands – Description by Ligand States. *Math. Biosci.*, **53**, 1–39.

Perelson, A. S., Neumann, A. U., Markowitz, M., Leonard, J. M., and Ho, D. D. (1996) Hiv-1 Dynamics in Vivo: Virion Clearance Rate, Infected Cell Life-Span, and Viral Generation Time. *Science*, **271**, 1582–6.

Pozrikidis, C. (2008) *Numerical Computation in Science and Engineering* (New York: Oxford University Press).

Siepmann, J., Elkharraz, K., Siepmann, F., and Klose, D. (2005) How Autocatalysis Accelerates Drug Release from PLGA-Based Microparticles: A Quantitative Treatment. *Biomacromolecules*, **6**, 2312–19.

Sugano, K. (2008) Theoretical Comparison of Hydrodynamic Diffusion Layer Models Used for Dissolution Simulation in Drug Discovery and Development. *Int. J. Pharm.*, **363**, 73–7.

Tsuchiya, H. M., Drake, J. F., Jost, J. L., and Fredrickson, A. G. (1972) Predator-Prey Interactions of Dictyostelium Discoideum and Escherichia Coli in Continuous Culture. *J. Bacteriol.*, **110**, 1147–53.

Vauthier, C., Schmidt, C., and Couvreur, P. (1999) Measurement of the Density of Polymeric Nanoparticulate Drug Carriers by Isopycnic Centrifugation. *J. Nanoparticle Res.*, **1**, 411–18.

Wilkinson, T. G., Topiwala, H. H., and Hamer, G. (1974) Interactions in a Mixed Bacterial Population Growing on Methane in Continuous Culture. *Biotechnol. Bioeng.*, **16**, 41–59.

8 Nonlinear model regression and optimization

8.1 Introduction

Mathematical modeling is used to analyze and quantify physical, biological, and technological processes. You may ask, "What is modeling and what purpose does a model serve in scientific research or in industrial design?" Experimental investigation generates a set of data that describes the nature of a process. The quantitative trend in the data, which describes system behavior as a function of one or more variables, can be fitted to a mathematical function that is believed to represent adequately the relationship between the observations and the independent variables. Fitting a model to data involves determining the "best-fit" values of the parameters of the model.

A mathematical model can be of two types.

(1) A model formulated from physical laws is called a **mechanistic model**. A model of this type, if formulated correctly, provides insight into the nature of the process. The model parameters can represent physical, chemical, biological, or economic properties of the process.

(2) An **empirical model**, on the other hand, is not derived from natural laws. It is constructed to match the shape or form of the data, but it does not explain the origin of the trend. Therefore, the parameters of an empirical model do not usually correspond to physical properties of the system.

Mechanistic models of scientific processes generate new insights into the mechanisms by which the processes occur and proceed in nature or in a man-made system. A mathematical model is almost always an incomplete or imperfect description of the system. If we were to describe a process completely, we would need to incorporate a prohibitively large number of parameters, which would make mathematical modeling exceedingly difficult. Often, only a few parameters influence the system in an important way, and we only need to consider those. The model analysis and development phase is accompanied by a systematic detailing of the underlying assumptions used to formulate the model.

Models serve as predictive tools that allow us to make critical decisions in our work. The parameters of the mechanistic model quantify the behavior of the system. For example, a pharmacodynamic model of the dose-response of a drug such as

$$E = E_{\max}\frac{(C/C_{50})^{\gamma}}{1 + (C/C_{50})^{\gamma}}$$

quantifies the potency of a drug, since the model parameter C_{50} is the drug concentration in plasma that produces 50% of its maximum pharmacological effect. Here E is the drug effect and γ is the shape factor. By quantifying and comparing the potency, efficacy, and lethal doses of a variety of drugs, the drug researcher or manufacturer can choose the optimal drug to consider for further development. Biochemical kinetic models that characterize the reactive properties of enzymes with regard to substrate conversion, enzyme inhibition, temperature, and pH are needed to proceed with the industrial design of a biocatalytic process. It is very important that the mathematical model serve as an adequate representation of the system. Careful attention should be paid to the model development phase of the problem. Performing nonlinear regression to fit a "poor" model to the data will likely produce unrealistic values of the system parameters, or wide confidence intervals that convey large uncertainty in the parameter estimates.

In Chapters 2 and 3, you were introduced to the method of linear least-squares regression. The "best-fit" values of a linear model's parameters were obtained by minimizing the sum of the squares of the regression error. The linear least-squares method produces the best values of the parameters when the following assumptions regarding the data are valid.

(1) The values of the independent variable x are known exactly. In other words, the error in x is much less compared to the variability in the dependent variable y.
(2) The scatter of the y points about the function curve follows a Gaussian distribution.
(3) The standard deviation that characterizes the variation in y is the same for all x values. In other words, the data are homoscedastic.
(4) All y points are independent and are randomly distributed about the curve.

The goal of least-squares regression is to minimize the sum of the squares of the error between the experimental observations y and the corresponding model predictions $\hat{y}$. We seek to minimize the quantity

$$\mathrm{SSE} = \sum_{i=1}^{m} (y_i - \hat{y}_i)^2,$$

where m is the number of independent observations. The expression above is called an **objective function**. To find the parametric values that minimize an objective function, we take the derivative of the function and equate it to zero. In Chapter 2, it was shown that, for an objective function that is linear in the unknowns, this procedure produces a matrix equation called the normal equations, $\boldsymbol{A}^{\mathrm{T}}\boldsymbol{A}\boldsymbol{c} = \boldsymbol{A}^{\mathrm{T}}\boldsymbol{y}$, the solution of which yields the best-fit model parameters $\boldsymbol{c}$.

Most mathematical models in the biological and biomedical sciences are *nonlinear* with respect to one or more of the model parameters. To obtain the best-fit values for the constants in a nonlinear model, we make use of **nonlinear regression methods**.[1] When the four assumptions stated above for the validity of the least-squares method apply – when variability of data is exclusively observed in the y direction, the observations are distributed normally about the curve in the y direction, all observations are independent, and the variance of the data is uniform along the curve – we

[1] In Chapter 2 it was shown that some forms of non-linear models can be linearized using suitable transformations. However, this may not always be a good route to pursue when the goal is to obtain a good estimate of the model parameters. The new x values of the transformed data may be a function of the dependent variable and therefore may exhibit uncertainty or variability in their values. Moreover, the transformation can change the nature of scatter about the curve converting a normal distribution to a non-normal one. As a result, the assumptions for least-squares regression may no longer be valid and the values of the parameters obtained using the linearized model may be inaccurate. Exercise caution when using non-linear transformations to convert a non-linear model into a linear one for model regression. Non-linear transformations are usually recommended when the goal is to remove heteroscedasticity in the data.

perform **nonlinear least-squares regression** to compute the optimal values of the constants.

In some cases, the variance of y is found to increase with y. When the variances are unequal, y values with wide scatter influence the shape of the curve more than y values with narrow scatter, and inaccurate values of the model parameters will result. To make the data suitable for least-squares regression, a weighting scheme must be applied to equalize the scatter. If the standard deviation is proportional to the magnitude of y, then each square $(y_i - \hat{y}_i)^2$ is weighted by $1/y_i^2$.

This is called the relative weighting scheme. In this scheme, the goal of least-squares regression is to minimize the *relative* squared error. Therefore, the objective function to be minimized is the sum of the squares of the relative error or

$$\mathrm{SSE} = \sum_{i=1}^{m} \frac{(y_i - \hat{y}_i)^2}{y_i^2}.$$

How does one calculate the minimum of an objective function? For a nonlinear problem, this almost always requires an iterative approach using a very useful mathematical tool called **optimization**. In industrial design, manufacturing, and operations research, optimization is a valuable tool used to optimize industrial processes such as daily production quantities, inventory stock, equipment sizing, and energy consumption. An **objective function** $f(x)$ that calculates some quantity to be maximized, such as profit, or to be minimized, such as cost, is formulated in terms of the process variables. The objective function is the performance criterion that is to be either minimized or maximized. The optimization problem involves finding the set of values of the process variables that yields the best possible value of $f(x)$. In this chapter, we will focus on the minimization problem since this is of primary concern when performing nonlinear least-squares regression. If $f(x)$ is to be maximized, then the problem is easily converted to the minimization of $-f(x)$.

Sometimes, limitations or constraints on the variables' values exist. Some physical variables such as mass, age, or kW of energy consumed, cannot become negative. In industrial production, availability of raw material may be limited, or demand for a certain product in the market may be capped at a certain value, and therefore production must be constrained. A variety of constraints may apply that limit the feasible space within which the solution is sought. An optimization problem that is not subject to any constraints is an **unconstrained optimization problem**. When constraints narrow the region within which the minimum of the objective function must be searched, the problem is a **constrained optimization problem**.

In mathematical terms, we define the optimization problem as the minimization of the objective function(s) $f(\boldsymbol{x})$ subject to the equality constraint(s) $\boldsymbol{h}(\boldsymbol{x}) = 0$, and the inequality constraint(s) $\boldsymbol{g}(\boldsymbol{x}) \leq 0$, where $\boldsymbol{x}$ is the variable vector to be optimized, f is the objective function, and $\boldsymbol{g}$ and $\boldsymbol{h}$ are each a set of functions.

Consider an optimization problem in two variables x_1 and x_2. Suppose the objective function is a quadratic function of x_1 and x_2 and is subject to one equality constraint,

$$h(x_1, x_2) : px_1 + qx_2 = b,$$

and three inequality constraints, of which two are

$$g_1(x_1, x_2) : x_1 \geq 0; \qquad g_2(x_1, x_2) : x_2 \geq 0.$$

In a two-variable optimization problem, the inequality constraints together define an area within which the feasible solution lies. On the other hand, an equality

Figure 8.1

Two-dimensional constrained optimization problem. The constrained minimum is the optimal solution

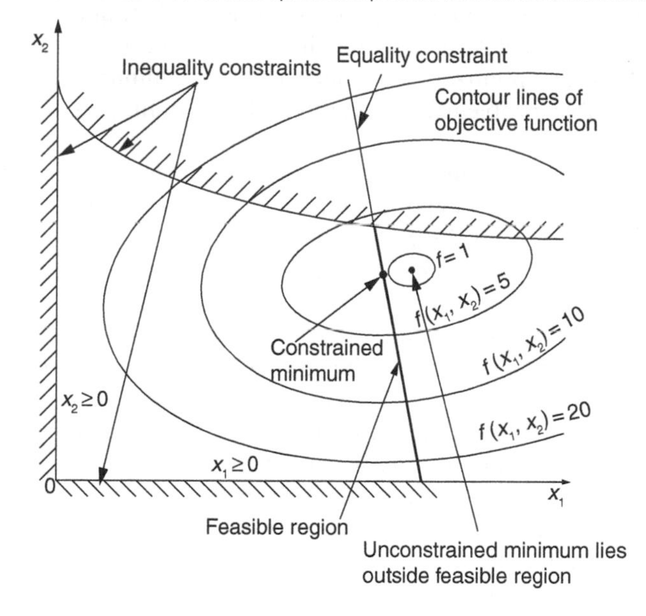

constraint defines a line on which the feasible solution lies. In this problem, if two equality constraints had been defined, then the solution of the problem would be given by the intersection of the two equality curves, and an optimization of the objective function would be unnecessary since two equality constraints reduce the feasible region to a point. Figure 8.1 illustrates the feasible region of the solution. The **feasible region** is the variable space within which the optimization solution is searched, since this space satisfies all the imposed inequality and equality constraints. Superimposed on the plot are the contour lines of the objective function in two variables. Note that the minimum of the objective function lies outside the feasible region. The minimum of the function is not a feasible solution according to the constraints of the problem.

Optimization methods can be used to solve a variety of problems as follows.

(1) The constrained minimization problem, e.g. linear programming and nonlinear programming problems.
(2) Nonlinear regression. The performance criterion of a regression problem is the minimization of the error or discrepancy between the data and the model prediction.
(3) Systems of nonlinear equations $\boldsymbol{f}(\boldsymbol{x}) = 0$, where $\boldsymbol{x}$ is a vector of the variables, and the objective function is the norm of the $\boldsymbol{f}(\boldsymbol{x})$.

Optimization techniques come in handy for solving a wide variety of nonlinear problems.

Optimization uses a systematic algorithm to select the best solution from a variety of possible solutions. Different combinations of variables (or parameters) will yield different values of the objective criterion, but usually only one set of values will satisfy the condition of the minimum value of the objective function. Sometimes, a nonlinear function has more than one minimum (or maximum). A **local minimum** $\boldsymbol{x}^*$ is defined as a point in the function space where the function values at all points $\boldsymbol{x}$ in the vicinity of $\boldsymbol{x}^*$ obey the criterion $f(\boldsymbol{x}) \geq f(\boldsymbol{x}^*)$. A local minimum may correspond to a subregion of the variable space, but may not be the smallest possible value of the function when the entire variable space is considered. A minimum point at which the function value is smallest in the domain space of the function is called a **global minimum**. While an optimization method can locate a minimum, it is usually difficult (for multivariable problems) to ascertain if the minimum found is the global minimum.

A local minimum of a function in one variable is defined in calculus as the point at which the first derivative $f'(x)$ of the function equals zero and the second derivative $f''(x)$ of the function is positive. A **global minimum** of a single variable function is defined as value of x at which the value of the function is the least in the defined interval $x \in [a, b]$. A global minimum may either be located at a local minimum, if present within the interval, or at an endpoint of a closed interval, in which case the endpoint may not be an extreme point. In Figure 8.3, the global minimum over the interval shown is at $x = 0$; $f'(x=0)$ may not necessarily equal zero at such points.

The goal of an optimization problem is to find the global minimum, which is the best solution possible. For any *linear* optimization problem, the solution, if it exists, is unique. Thus, a linear programming problem has only one minimum, which corresponds to the global minimum. However, a nonlinear function can have more than one minimum. Such functions that have multiple minima (and/or maxima) are called **multimodal functions**. In contrast, a function that has only one extreme point is called a **unimodal function**. A function $f(\boldsymbol{x})$ with a minimum at $\boldsymbol{x}^*$ is said to be **unimodal** if

$$f(\boldsymbol{x}) \geq f(\boldsymbol{x}^*) \qquad \boldsymbol{x} < \boldsymbol{x}^*;$$

$$f(\boldsymbol{x}) \geq f(\boldsymbol{x}^*) \qquad \boldsymbol{x} > \boldsymbol{x}^*,$$

where $\boldsymbol{x}$ is any point in the variable space. For a function of one or two variables, unimodality can usually be detected by plotting the function. For functions of three or more variables, assessing the modality of a function is more difficult.

Our discussion of optimization techniques in this chapter applies in most cases to objective functions that are continuous and smooth. Discontinuities in the function or in its derivatives near the optimal point can prevent certain optimization techniques from converging to the solution. Optimization problems that require a search for the optimum within a discretized function space, arising due to the discrete nature of one or more independent variables, is a more challenging topic beyond the scope of this book. Unless mentioned otherwise, it is assumed that the objective function has continuous first and second derivatives.

Box 8.1A Pharmacokinetic and toxicity studies of AZT

The thymidine nucleoside analog AZT is a drug used in conjunction with several other treatment methods to combat HIV. The importance of AZT is increased by the fact that it is found to reduce HIV transmission to the fetus during pregnancy. AZT crosses the placenta into the fetal compartment easily since its lipophilic properties support rapid diffusion of the molecule. However, AZT is found to be toxic to bone marrow, and may possibly be toxic to the placenta as well as to the developing fetus. In vitro studies have shown that AZT inhibits DNA synthesis and cell proliferation. Boal *et al.* (1997) conducted studies of the transport of nucleoside analog AZT across the dually perfused (perfused by both maternal and fetal perfusate) human term placental lobule. The purpose of this study was to characterize pharmacokinetic parameters that describe the transport and toxicity properties of AZT in placental tissues. AZT was found to transfer rapidly from the maternal perfusate (compartment) to the fetal perfusate (compartment). The concentration of AZT in the fetal perfusate never rose beyond the maternal levels of AZT, indicating that drug accumulation in the fetal compartment did not occur. Table 8.1 lists the concentration of AZT in the maternal and fetal perfusates with time.

In Chapter 2, Boxes 2.3A, B gave you an introduction to the field of pharmacokinetics and presented a mathematical model of the first-order elimination process of a drug from a single well-mixed compartment. In this example, we model the maternal compartment as a first-order drug absorption and elimination process. Figure 8.2 is a block diagram describing the transport of drug into and out of a single well-mixed compartment that follows first-order drug absorption and elimination. The following description of the model follows Fournier's (2007) pharmacokinetic analysis. Let A_0 be the total mass of drug available for absorption by the compartment. Let A be the mass of drug waiting to be absorbed at any time t. If k_a is the rate of absorption, then the depletion of unabsorbed drug is given by the following mass balance:

$$\frac{dA}{dt} = -k_a A.$$

The solution of this first-order differential equation is $A = A_0 e^{-k_a t}$.

The drug is distributed in the compartment over a volume V, and C is the concentration of drug in the compartment at time t. A stream with flowrate Q leaves the compartment and enters the elimination site where the entire drug in the stream is removed, after which the same stream, devoid of drug, returns to the compartment at the same flowrate. Note that this perfect elimination site is more of a mathematical abstraction used to define the elimination rate in terms of a flowrate Q. The concentration of drug in the compartment is governed by the first-order differential equation

$$V\frac{dC}{dt} = k_a A - QC.$$

Table 8.1. *AZT concentration in maternal and fetal perfusates following exposure of the maternal compartment to AZT at 3.8 mM*

Maternal		Fetal	
AZT (mM)	Time (min)	AZT (mM)	Time (min)
0	0	0	0
1.4	1	0.1	1
4.1	3	0.4	5
4.5	5	0.8	10
3.5	10	1.1	20
3.0	20	1.2	40
2.75	40	1.4	50
2.65	50	1.35	60
2.4	60	1.6	90
2.2	90	1.7	120
2.15	120	1.9	150
2.1	150	2.0	180
2.15	180	1.95	210
1.8	210	2.2	240
2.0	240		

Figure 8.2

Block diagram of the mass balance of drug over a single compartment in which drug absorption and elimination follow a first-order process

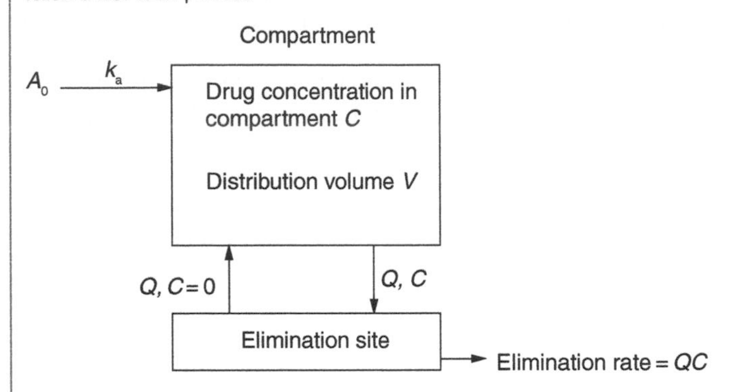

The volumetric clearance rate from the compartment per unit distribution volume is set equal to an elimination constant

$$k_e = \frac{Q}{V}.$$

The differential equation is rewritten as

$$V\frac{dC}{dt} = k_a A_0 e^{-k_a t} - k_e VC.$$

Solving the differential equation, we obtain the concentration profile of the drug in the compartment as a function of time:

$$C(t) = \frac{A_0}{V}\left(\frac{k_a}{k_a - k_e}\right)\left(e^{-k_e t} - e^{-k_a t}\right).$$

If $I_0 = A_0/V$, then

$$C(t) = I_0\left(\frac{k_a}{k_a - k_e}\right)\left(e^{-k_e t} - e^{-k_a t}\right).$$

You are asked to model the maternal and fetal compartments separately. The maternal compartment is to be modeled as a first-order drug absorption and elimination problem. You will need to perform a nonlinear regression to solve for k_e, k_a, and I_0. Check the fit by plotting the data along with your best-fit model.

The fetal compartment is to be modeled as a first-order absorption problem with no elimination ($k_e = 0$). While there is no exponential decay of the drug in the maternal perfusate (the source of drug for the fetal compartment), the first-order absorption model will mimic the reduced absorption of drug into the fetal compartment at long times when diffusion of drug across the compartmental barrier stops due to equalization of concentrations in the two compartments:

$$C_f(t) = I_0\left(1 - e^{-k_a t}\right).$$

Find suitable values of the model parameters I_0 and k_a for the fetal compartment.

The selection of the type of optimization algorithm to use depends on the nature of the objective function, the nature of the constraint functions, and the number of variables to be optimized. A variety of optimization techniques have been developed to solve the wide spectrum of nonlinear problems encountered in scientific research, industrial design, traffic control, meteorology, and economics. In this chapter, several well-established classical optimization tools are presented; however, our discussion only scratches the surface of the body of optimization literature currently available. We first address the topic of unconstrained optimization of one variable in Section 8.2. We demonstrate three classic methods used in one-dimensional minimization: (1) Newton's method (see Section 5.5 for a demonstration of its use as a nonlinear root-finding method), (2) successive parabolic interpolation method, and (3) the golden section search method. The minimization problem becomes more challenging when we need to optimize over more than one variable. Simply using a trial-and-error method for multivariable optimization problems is discouraged due to its inefficiency. Section 8.3 discusses popular methods used to perform unconstrained optimization in several dimensions. We discuss the topic of constrained optimization in Section 8.4. Finally, in Section 8.5 we demonstrate Monte Carlo techniques that are used in practice to estimate the standard error in the estimates of the model parameters.

8.2 Unconstrained single-variable optimization

The behavior of a nonlinear function in one variable and the location of its extreme points can be studied by plotting the function. Consider any smooth function f in a single variable x that is twice differentiable on the interval $[a, b]$. If $f'(x) > 0$ over any interval $x \in [a, b]$, then the single-variable function is *increasing* on that interval. Similarly, when $f'(x) < 0$ over any interval of x, then the single-variable function is *decreasing* on that interval. On the other hand, if there lies a point x^* in the interval $[a, b]$ such that $f'(x^*) = 0$, the function is neither increasing nor decreasing at that point.

In calculus courses you have learned that an extreme point of a continuous and smooth function $f(x)$ can be found by taking the derivative of $f(x)$, equating it to zero, and solving for x. One or more solutions of x that satisfy $f'(x) = 0$ may exist. The points $\boldsymbol{x}^*$ at which $f'(x)$ equals zero are called **critical points or stationary points**. At a critical point x^*, the function is parallel with the x-axis. Three possibilities arise regarding the nature of the critical point x^*:

(1) if $f'(x^*) = 0; f''(x^*) > 0$, the point is a local minimum;
(2) if $f'(x^*) = 0; f''(x^*) < 0$, the point is a local maximum;
(3) if $f'(x^*) = 0; f''(x^*) = 0$, the point may be either an **inflection point** or an extreme point. As such, this test is inconclusive. The values of higher derivatives $f'''(x), \ldots, f^{k-1}(x), f^k(x)$ are needed to define this point. If $f^k(x)$ is the first non-zero kth derivative, then if k is even, the point is an extremum (minimum if $f^k(x) > 0$; maximum if $f^k(x) < 0$), and if k is odd, the critical point is an inflection point.

The method of establishing the nature of an extreme point by calculating the value of the second derivative is called the **second derivative test**. Figure 8.3 graphically illustrates the different types of critical points that can be exhibited by a nonlinear function.

To find the local minima of a function, one can solve for the zeros of $f'(x)$, which are the critical points of $f(x)$. Using the second derivative test (or by plotting the function), one can then establish the character of the critical point. If $f'(x) = 0$ is a nonlinear equation whose solution is difficult to obtain analytically, root-finding methods such as those described in Chapter 5 can be used to solve iteratively for the

Figure 8.3

Different types of critical points

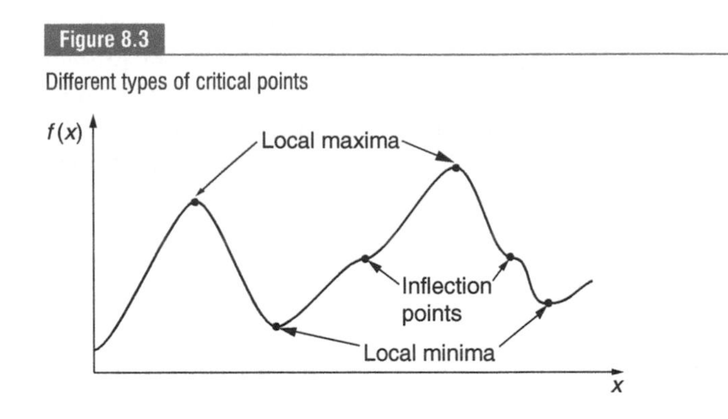

roots of the equation $f'(x) = 0$. We show how Newton's root-finding method is used to find the local minimum of a single-variable function in Section 8.2.1. In Section 8.2.2 we present another calculus-based method to find the local minimum called the polynomial approximation method. In this method the function is approximated by a quadratic or a cubic function. The derivative of the polynomial is equated to zero to obtain the value of the next guess. Section 8.2.3 discusses a non-calculus-based method that uses a bracketing technique to search for the minimum. All of these methods assume that the function is unimodal.

8.2.1 Newton's method

Newton's method of one-dimensional minimization is similar to Newton's nonlinear root-finding method discussed in Section 5.5. The unimodal function $f(x)$ is approximated by a quadratic equation, which is obtained by truncating the Taylor series after the third term. Expanding the function about a point x_0,

$$f(x) \approx f(x_0) + f'(x_0)(x - x_0) + \frac{1}{2} f''(x_0)(x - x_0)^2.$$

At the extreme point x^*, the first derivative of the function must equal zero. Taking the derivative of the expression above with respect to x, the expression reduces to

$$x = x_0 - \frac{f'(x_0)}{f''(x_0)}. \tag{8.1}$$

If you compare Equation (8.1) to Equation (5.19), you will notice that the structure of the equation (and hence the iterative algorithm) is exactly the same, except that, for the minimization problem, both the first and second derivative need to be calculated at each iteration. An advantage of this method over other one-dimensional optimization methods is its fast second-order convergence, when convergence is guaranteed. However, the initial guessed values must lie close to the minimum to avoid divergence of the solution (see Section 5.5 for a discussion on convergence issues of Newton's method). To ensure that the iterative technique is converging, it is recommended to check at each step that the condition $f(x_{i+1}) < f(x_i)$ is satisfied. If not, modify the initial guess and check again if the solution is progressing in the desired direction.

Some difficulties encountered when using this method are as follows.

(1) This method is unusable if the function is not smooth and continuous, i.e. the function is not differentiable at or near the minimum.
(2) A derivative may be too complicated or not in a form suitable for evaluation.
(3) If $f''(x)$ at the critical point is equal to zero, this method will not exhibit second-order convergence (see Section 5.5 for an explanation).

Box 8.2A Optimization of a fermentation process: maximization of profit

A first-order irreversible reaction $A \rightarrow B$ with rate constant k takes place in a well-stirred fermentor tank of volume V. The process is at steady state, such that none of the reaction variables vary with time. W_B is the amount of product B in kg produced per year, and p is the price of B per kg. The total annual sales is given by $\$pW_B$.

The mass flowrate of broth through the processing system per year is denoted by W_{in}. If Q is the flowrate (m^3/hr) of broth through the system, ρ is the broth density, and the number of hours of operation annually is 8000, then $W_{in} = 8000\rho Q$. The annualized capital (investment) cost of the reactor and the downstream recovery equipment is $\$4000W_{in}^{0.6}$.

The operating cost c is \$15.00 per kg of B produced. If the concentration of B in the exiting broth is b_{out} in molar units, then we have $W_B = 8000Qb_{out}MW_B$.

A steady state material balance incorporating first-order reaction kinetics gives us

$$b_{out} = a_{in}\left(1 - \frac{1}{1 + kV/Q}\right),$$

where V is the volume of the fermentor and a_{in} is the inlet concentration of the reactant. Note that as Q increases, the conversion will decrease.

The profit function is given by

$$(p - c)W_B - 4000(W_{in})^{0.6},$$

and in terms of the unknown variable Q, it is as follows:

$$f(Q) = 8000(p - c)MW_B Q a_{in}\left(1 - \frac{1}{1 + (kV/Q)}\right) - 4000(8000\rho Q)^{0.6}. \tag{8.2}$$

Initially, as Q increases the product output will increase and profits will increase. With subsequent increases in Q, the reduced conversion will adversely impact the product generation rate while annualized capital costs continue to increase.

You are given the following values:

$p = \$400$ per kg of product,
$\rho = 2.5\ kg/m^3$,
$a_{in} = 10\ \mu M\ (1\ M \equiv 1\ kmol/m^3)$,
$k = 2/hr$,
$V = 12\ m^3$,
$MW_B = 70\,000\ kg/kmol$.

Determine the value of Q that maximizes the profit function. This problem was inspired by a discussion on reactor optimization in Chapter 6 of Nauman (2002).

We plot the profit as a function of the hourly flowrate Q in Figure 8.4. We can visually identify an interval $Q \in [50, 70]$ that contains the maximum value of the objective function.

The derivative of the objective function is as follows:

$$f'(Q) = 8000(p - c)MW_B a_{in}\left[1 - \frac{1}{1 + (kV/Q)} - \frac{QkV}{(Q + kV)^2}\right] - 0.6 \cdot \frac{4000(8000\rho)^{0.6}}{Q^{0.4}}.$$

Figure 8.4

Profit function

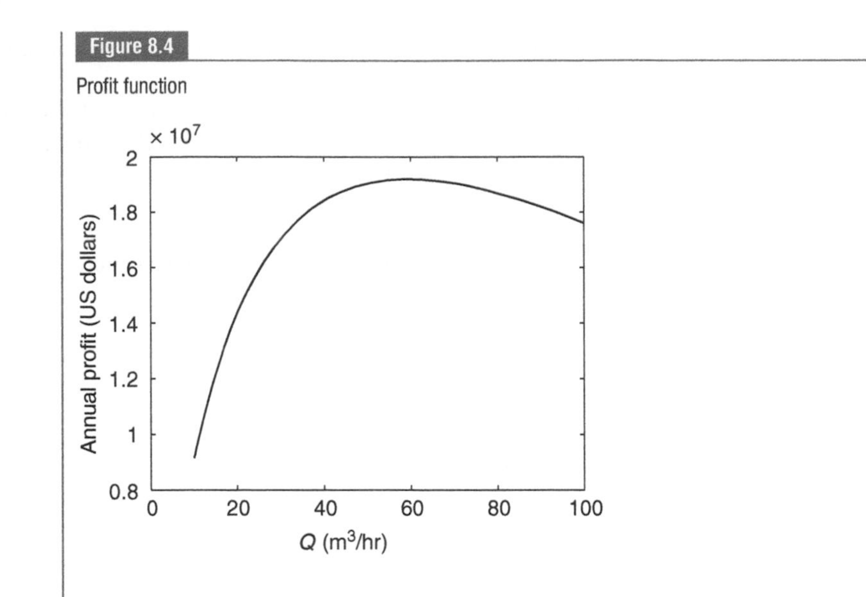

The second derivative of the objective function is as follows:

$$f''(Q) = 8000(p - c)\mathrm{MW_B}a_{\mathrm{in}}\left[-\frac{2(kV)^2}{(Q + kV)^3}\right] + 0.24 \cdot \frac{4000(8000\rho)^{0.6}}{Q^{1.4}}.$$

Two functions are written to carry out the maximization of the profit function. The first function listed in Program 8.1 performs a one-dimensional Newton optimization, and is similar to Program 5.7. Program 8.2 calculates the first and second derivative of the profit function.

MATLAB program 8.1

```
function newtons1Doptimization(func, x0, tolx)
% Newton's method is used to minimize an objective function of a single
% variable

% Input variables
% func : first and second derivative of nonlinear function
% x0 : initial guessed value
% tolx : tolerance in estimating the minimum

% Other variables
maxloops = 20;

[df, ddf] = feval(func,x0);
fprintf(' i x(i) f'' (x(i)) f'''' (x(i)) \n'); %\n is carriage return

% Minimization scheme
for i = 1:maxloops
    x1 = x0 - df/ddf;
    [df, ddf] = feval(func,x1);
    fprintf('%2d %7.6f %7.6f %7.6f \n',i,x1,df,ddf);
    if (abs(x1 - x0) <= tolx)
        break
```

```
    end
    x0 = x1;
end
```

MATLAB program 8.2

```
function [df ddf] = flowrateoptimization(Q)
% Calculates the first and second derivatives of the profit function

ain = 10e-6;    % feed concentration of A (kmol/m^3)
p = 400;        % price in US dollars per kg of product
c = 15;         % operating cost per kg of product
k = 2;          % forward rate constant (1/hr)
V = 12;         % fermentor volume (m^3)
rho = 2.5;      % density (kg/m^3)
MWB = 70000;    % molecular weight (kg/kmol)

df = 8000*(p - c)*ain*MWB*(1 - 1/(1 + k*V/Q) - Q*k*V/(Q + k*V)^2) - ...
    0.6*4000*(8000*rho)^0.6/(Q^0.4);
ddf = 8000*(p - c)*ain*MWB*(-2*(k*V)^2/(Q + k*V)^3) + ...
     0.24*4000* (8000*rho)^0.6/(Q^1.4);
```

The optimization function is executed from the command line as follows:

```
>>newtons1Doptimization('flowrateoptimization', 50, 0.01)
i       x(i)           f'(x(i))        f''(x(i))
1      57.757884      5407.076741    -3295.569935
2      59.398595      178.749716     -3080.622008
3      59.456619      0.211420       -3073.338162
4      59.456687      0.000000       -3073.329539
```

The optimum flowrate that maximizes the profit is $Q = 59.46$ m^3/hr. The second derivative is negative, which tells us that the critical point is a maximum. To use Newton's optimization method we did not need to convert the maximization problem to a minimization problem because the method searches for an extreme point and does not formally distinguish between a minimum point and a maximum point.

In this problem, we did not specify any constraints on the value of Q. In reality, the flowrate will have some upper limit that is dictated by the size of installed piping, frictional losses in the piping, pump capacity, and other constraints.

The termination criterion in Box 8.2A consisted of specifying a tolerance value for the absolute difference in the solution of two consecutive iterations. Other stopping criteria may be specified either individually or simultaneously.

(1) $|f'(x)| < \epsilon$. A cut-off value close to zero can be specified. If this criterion is met, it indicates that $f'(x)$ is close enough to zero such that an extreme point has been reached. This criterion can be used only with a minimization algorithm that calculates the derivative of the function.

(2) $\left|\frac{f(x_{i+1})-f(x_i)}{f(x_{i+1})}\right| < \epsilon$. If the function value is changing very slowly, this can signify that a critical point has been found.

(3) $\left|\frac{x_{i+1}-x_i}{x_{i+1}}\right| < \epsilon$. The relative change in the solution can be used to determine when the iterations can be stopped.

8.2.2 Successive parabolic interpolation

In the **successive parabolic interpolation method**, a unimodal function is approximated over an interval $[x_0, x_2]$ that brackets the minimum by a parabolic function $p(x) = c_2x^2 + c_1x + c_0$. Any three values of the independent variable, $x_0 < x_1 < x_2$, not necessarily equally spaced in the interval, are selected to construct the unique parabola that passes through these three points. The minimum point of the parabola is located at $p'(x) = 0$. Taking the derivative of the second-degree polynomial expression and equating it to zero, we obtain

$$x_{\min} = -\frac{c_1}{2c_2}. \tag{8.3}$$

The minimum given by Equation (8.3) will most likely not coincide with the minimum x^* of the objective function (unless the function is also parabolic); $x_{\min}$ will lie either between x_0 and x_1 or between x_1 and x_2. We calculate $f(x_{\min})$ and compare this with the function values at the three other points. The point $x_{\min}$ and two out of the three original points are chosen to form a new set of three points used to construct the next parabola. The selection is made such that the minimum value of $f(x)$ is bracketed by the endpoints of the new interval. For example, if $x_{\min} > x_1$ ($x_{\min}$ lies between x_1 and x_2) and $f(x_{\min}) < f(x_1)$, then we would choose the three points x_1, $x_{\min}$, and x_2. On the other hand, if $f(x_{\min}) > f(x_1)$ and $x_{\min} > x_1$, then we would choose the points x_0, x_1, and $x_{\min}$. The new set of three points provides the next parabolic approximation of the function near or at the minimum. In this manner, each newly constructed parabola is defined over an interval that always contains the minimum value of the function. The interval on which the parabolic interpolation is defined shrinks in size with each iteration.

To calculate $x_{\min}$, we must express c_1 and c_2 in Equation (8.3) in terms of the three points x_0, x_1, and x_2 and their function values. We use the second-order Lagrange interpolation formula given by Equation (6.14) (reproduced below) to construct a second-degree polynomial approximation to the function:

$$p(x) = \frac{(x - x_1)(x - x_2)}{(x_0 - x_1)(x_0 - x_2)} f(x_0) + \frac{(x - x_0)(x - x_2)}{(x_1 - x_0)(x_1 - x_2)} f(x_1) + \frac{(x - x_0)(x - x_1)}{(x_2 - x_0)(x_2 - x_1)} f(x_2).$$

Differentiating the above expression and equating it to zero, we obtain

$$\frac{(x - x_1) + (x - x_2)}{(x_0 - x_1)(x_0 - x_2)} f(x_0) + \frac{(x - x_0) + (x - x_2)}{(x_1 - x_0)(x_1 - x_2)} f(x_1) + \frac{(x - x_0) + (x - x_1)}{(x_2 - x_0)(x_2 - x_1)} f(x_2) = 0$$

Applying a common denominator to each term, the equation reduces to

$$(2x - x_1 - x_2)(x_2 - x_1)f(x_0) - (2x - x_0 - x_2)(x_2 - x_0)f(x_1) + (2x - x_0 - x_1)(x_1 - x_0)f(x_2) = 0$$

Rearranging, we get

$$x_{\min} = \frac{1}{2}\frac{(x_2^2 - x_1^2)f(x_0) - (x_2^2 - x_0^2)f(x_1) + (x_1^2 - x_0^2)f(x_2)}{(x_2 - x_1)f(x_0) - (x_2 - x_0)f(x_1) + (x_1 - x_0)f(x_2)}. \tag{8.4}$$

We now summarize the algorithm.

Algorithm for successive parabolic approximation

(1) Establish an interval within which a minimum point of the unimodal function lies.
(2) Select the two endpoints x_0 and x_2 of the interval and one other point x_1 in the interior to form a set of three points used to construct a parabolic approximation of the function.
(3) Evaluate the function at these three points.
(4) Calculate the minimum value of the parabola x_{min} using Equation (8.4).
(5) Select three out of the four points such that the new interval brackets the minimum of the unimodal function.
   (a) If $x_{min} < x_1$ then
      (i) if $f(x_{min}) < f(x_1)$ choose x_0, x_{min}, x_1;
      (ii) if $f(x_{min}) > f(x_1)$ choose x_{min}, x_1, x_2.
   (b) If $x_{min} > x_1$ then
      (i) if $f(x_{min}) < f(x_1)$ choose x_1, x_{min}, x_2;
      (ii) if $f(x_{min}) > f(x_1)$ choose x_0, x_1, x_{min}.
(6) Repeat steps (4) and (5) until the interval size is less than the tolerance specification.

Note that only function evaluations (but not derivative evaluations) are required, making the parabolic interpolation useful for functions that are difficult to differentiate. A cubic approximation of the function may or may not require evaluation of the function derivative at each iteration depending on the chosen algorithm. A discussion of the non-derivative-based cubic approximation method can be found in Rao (2002). The convergence rate of the successive parabolic interpolation method is superlinear and is intermediate to that of Newton's method discussed in the previous section and of the golden section search method to be discussed in Section 8.2.3.

Box 8.2B Optimization of a fermentation process: maximization of profit

We solve the optimization problem discussed in Box 8.2A using the method of successive parabolic interpolation. MATLAB program 8.3 is a function m-file that minimizes a unimodal single-variable function using the method of parabolic interpolation. MATLAB program 8.4 calculates the negative of the profit function (Equation (8.2)).

MATLAB program 8.3

```
function parabolicinterpolation(func, ab, tolx)
% The successive parabolic interpolation method is used to find the
% minimum of a unimodal function.

% Input variables
% func: nonlinear function to be minimized
% ab : bracketing interval [a, b]
% tolx: tolerance in estimating the minimum

% Other variables
k = 0; % counter

% Set of three points to construct the parabola
x0 = ab(1);
```

```
x2 = ab(2);
x1 = (x0 + x2)/2;

% Function values at the three points
fx0 = feval(func, x0);
fx1 = feval(func, x1);
fx2 = feval(func, x2);

% Iterative solution
while (x2 - x0) > tolx % (x2 – x0) is always positive
    % Calculate minimum point of parabola
    numerator = (x2^2 - x1^2)*fx0 - (x2^2 - x0^2)*fx1 + ...
        (x1^2 - x0^2) *fx2;
    denominator = 2*((x2 - x1)*fx0 - (x2 - x0)*fx1 + (x1 - x0)*fx2);
    xmin = numerator/denominator;

    % Function value at xmin
    fxmin = feval(func, xmin);

    % Select the next set of three points to construct new parabola
    if xmin < x1
        if fxmin < fx1
            x2 = x1;
            fx2 = fx1;
            x1 = xmin;
            fx1 = fxmin;
        else
            x0 = xmin;
            fx0 = fxmin;
        end
    else
        if fxmin < fx1
            x0 = x1;
            fx0 = fx1;
            x1 = xmin;
            fx1 = fxmin;
        else
            x2 = xmin;
            fx2 = fxmin;
        end
    end
    k = k + 1;
end
fprintf('x0 = %7.6f x2 = %7.6f f(x0) = %7.6f f(x2) = %7.6f \n', ...
    x0, x2, fx0, fx2)
fprintf('x_min = %7.6f f(x_min) = %7.6f \n',xmin, fxmin)
fprintf('number of iterations = %2d \n', k)
```

MATLAB program 8.4

```
function f = profitfunction(Q)
% Calculates the objective function to be minimized
```

```
ain = 10e-6;        % feed concentration of A (kmol/m^3)
p = 400;            % price in US dollars per kg of product
c = 15;             % operating cost per kg of product
k = 2;              % forward rate constant (1/hr)
V = 12;             % fermentor volume (m^3)
rho = 2.5;          % density (kg/m^3)
MWB = 70000;        % molecular weight (kg/kmol)

f = -((p - c)*Q*ain*MWB*(1 - 1/(1 + k*V/Q))*8000 - ...
    4000*(8000*rho*Q)^0.6);
```

We call the optimization function from the command line:

```
>> parabolicinterpolation('profitfunction', [50 70], 0.01)
x0 = 59.456686 x2 = 59.456692 f(x0) = -19195305.998460 f(x2) =
  -19195305.998460
x_min = 59.456686 f(x_min) = -19195305.998460
number of iterations = 12
```

8.2.3 Golden section search method

Bracketing methods for performing minimization are analogous to the bisection method for nonlinear root finding. Bracketing methods search for the solution by retaining only a fraction of the interval at each iteration, after ensuring that the selected subinterval contains the solution. A bracketing method that searches for an extreme point must calculate the function value at two different points in the interior of the interval, while the bisection method and regula-falsi method analyze only one interior point at each iteration. An important assumption of any bracketing method of minimization is that a unique minimum point is located within the interval, i.e. the function is unimodal. If the function has multiple minima located within the interval, this method may fail to find the minimum.

First, an interval that contains the minimum x^* of the unimodal function must be selected. Then

$$f'(x) = \begin{cases} <0 & x<x^* \\ >0 & x>x^*, \end{cases}$$

for all x that lie in the interval. If a and b are the endpoints of the interval, then the function value at any point x_1 in the interior will necessarily be smaller than at least one of the function values at the endpoints, i.e.

$$f(x_1) < \max(f(a), f(b)).$$

Within the interval $[a, b]$, two points $x_1 < x_2$ are selected according to a formula that is dependent upon the bracketing search method chosen.

The function is evaluated at the endpoints of the interval as well as at these two interior points. This is illustrated in Figure 8.5. If

$$f(x_1) > f(x_2)$$

then the minimum must lie either in the interval $[x_1, x_2]$ or in the interval $[x_2, b]$. Therefore, we select the subinterval $[x_1, b]$ for further consideration and discard the subinterval $[a, x_1]$. On the other hand if

$$f(x_1) < f(x_2)$$

Figure 8.5

Choosing a subinterval for a one-dimensional bracketing search method

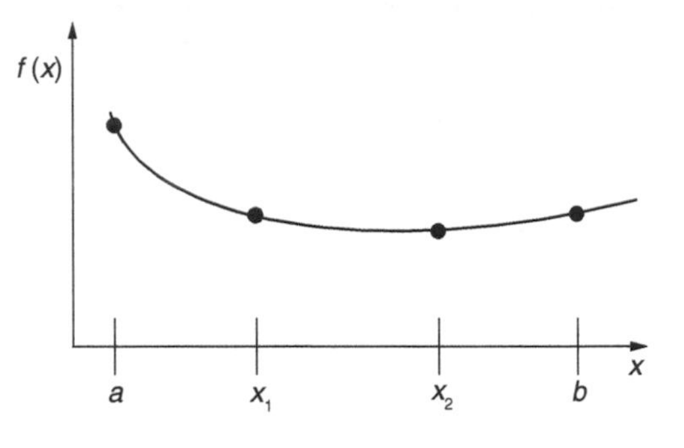

Figure 8.6

Dividing a line by the golden ratio

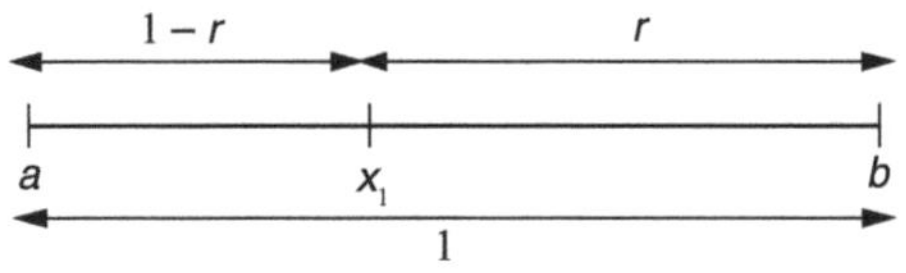

then the minimum must lie either in the interval $[a, x_1]$ or in the interval $[x_1, x_2]$. Therefore, we select the subinterval $[a, x_2]$ and discard the subinterval $[x_2, b]$.

An efficient bracketing search method that requires the calculation of only one new interior point (and therefore only one function evaluation) at each iterative step is the **golden section search method**. This method is based on dividing the interval into segments using the **golden ratio**. Any line that is divided into two parts according to the golden ratio is called a **golden section**. The golden ratio is a special number that divides a line into two segments such that the ratio of the line width to the width of the larger segment is equal to the ratio of the larger segment to the smaller segment width.

In Figure 8.6, the point x_1 divides the line into two segments. If the width of the line is equal to unity and the width of the larger segment $x_1b = r$, then the width of the smaller segment $ax_1 = 1 - r$. If the division of the line is made according to the following rule:

$$\frac{ab}{x_1b} = \frac{x_1b}{ax_1}$$

or

$$\frac{1}{r} = \frac{r}{1-r} \tag{8.5}$$

then the line ab is a golden section and the golden ratio is given by $1/r$.

Solving the quadratic equation $r^2 + r - 1 = 0$ (Equation (8.5)), we arrive at $r = (\sqrt{5} - 1)/2$ or $r = 0.618034$, and $1 - r = 0.381966$. The golden ratio is given by the inverse of r, and is equal to the irrational number $(\sqrt{5} + 1)/2 = 1.618034$. The golden ratio is a special number and was known to be investigated by ancient Greek mathematicians such as Euclid and Pythagoras. Its use has been found in architectural design and artistic works. It is also arises in natural patterns such as the spiral of a seashell.

Figure 8.7

First iteration of the golden section search method

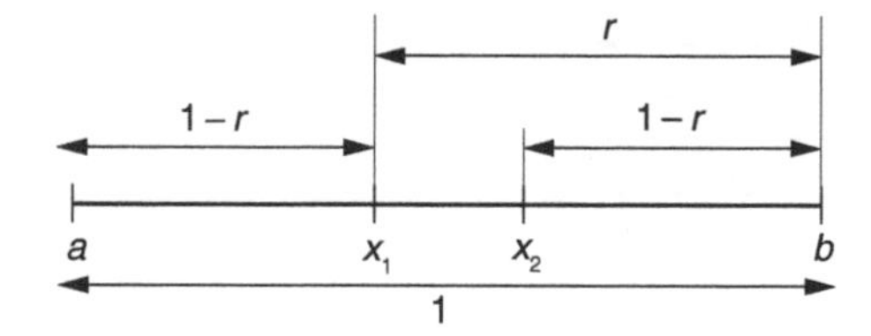

In the golden section search method, once a bracketing interval is identified (for now, assume that the initial bracketing interval has length *ab* equal to one), two points x_1 and x_2 in the interval are chosen such that they are positioned at a distance equal to $(1 - r)$ from the two ends. Thus, the two interior points are symmetrically located in the interval (see Figure 8.7).

Because this is the first iteration, the function is evaluated at both interior points. A subinterval that brackets the minimum is selected using the algorithm discussed above. If $f(x_1) > f(x_2)$, interval $[x_1, b]$ is selected. On the other hand, if $f(x_1) < f(x_2)$, interval $[a, x_1]$ is selected. Regardless of the subinterval selected, the width of the new bracket is r. Thus, the interval has been reduced by the fraction r.

The single interior point in the chosen subinterval divides the new bracket according to the golden ratio. This is illustrated next. Suppose the subinterval $[x_1, b]$ is selected. Then x_2 divides the new bracket according to the ratio

$$\frac{x_1 b}{x_2 b} = \frac{r}{1 - r},$$

which is equal to the golden ratio (see Equation (8.5)). Note that $x_2 b$ is the longer segment of the line $x_1 b$, and $x_1 b$ is a golden section. Therefore, we have

$$\frac{x_1 b}{x_2 b} = \frac{x_2 b}{x_1 x_2}.$$

Therefore, one interior point that divides the interval according to the golden ratio is already determined. For the next step, we need to find only one other point x_3 in the interval $[x_1, b]$ such that

$$\frac{x_1 b}{x_1 x_3} = \frac{r}{1 - r} = \frac{x_1 x_3}{x_3 b}.$$

This is illustrated in Figure 8.8.

The function is evaluated at x_3 and compared with $f(x_2)$. Depending on the outcome of the comparison, either segment $x_1 x_2$ or segment $x_3 b$ is discarded. The width of the new interval is $1 - r$, which is equal to r^2. Thus, the interval $[x_1, b]$ of length r has been reduced to a smaller interval by the fraction r. Thus, at every step, the interval size that brackets the minimum is reduced by the fraction $r = 0.618$. Since each iteration discards 38.2% of the current interval, 61.8% of the error in the estimation of the minimum is retained in each iteration. The ratio of the errors of two consecutive iterations is

$$\frac{|\varepsilon_{i+1}|}{|\varepsilon_i|} = 0.618.$$

Figure 8.8

Second iteration of the golden section search method

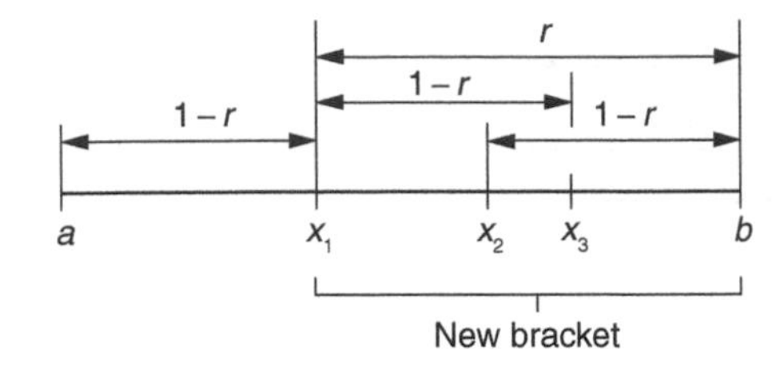

Comparing the above expression with Equation (5.10), we find $r = 1$ and $C = 0.618$. The convergence rate of this method is therefore linear. The interval reduction step is repeated until the width of the bracket falls below the tolerance limit.

Algorithm for the golden section search method

(1) Choose an initial interval $[a, b]$ that brackets the minimum point of the unimodal function and evaluate $f(a)$ and $f(b)$.
(2) Calculate two points x_1 and x_2 such that $x_1 = (1 - r)(b - a) + a$ and $x_2 = r(b - a) + a$, where $r = (\sqrt{5} - 1)/2$ and $b - a$ is the width of the interval.
(3) Evaluate $f(x_1)$ and $f(x_2)$.
(4) The function values are compared to determine which subinterval to retain.
  (a) If $f(x_1) > f(x_2)$, then the subinterval $[x_1, b]$ is chosen. Set $a = x_1$; $x_1 = x_2$; $f(a) = f(x_1)$; $f(x_1) = f(x_2)$; and calculate $x_2 = r(b - a) + a$ and $f(x_2)$.
  (b) If $f(x_1) < f(x_2)$, then the subinterval $[a, x_2]$ is chosen. Set $b = x_2$; $x_2 = x_1$; $f(b) = f(x_2)$; $f(x_2) = f(x_1)$; and find $x_1 = (1 - r)(b - a) + a$ and $f(x_1)$.
(5) Repeat step (4) until the size of the interval is less than the tolerance specification.

Another bracketing method that requires the determination of only one new interior point at each step is the Fibonacci search method. In this method, the fraction by which the interval is reduced varies at each stage. The advantage of bracketing methods is that they are much less likely to fail compared to derivative-based methods, when the derivative of the function does not exist at one or more points at or near the extreme point. Also, if there are points of inflection located within the interval, in addition to the single extremum, these will not influence the search result of a bracketing method. On the other hand, Newton's method is likely to converge to a point of inflection, if it is located near the initial guess value.

Using MATLAB

The MATLAB software contains the function `fminbnd`, which performs minimization of a nonlinear function in a single variable within a user-specified interval (also termed as **bound**). In other words, `fminbnd` carries out bounded single-variable optimization by minimizing the objective function supplied in the function call. The minimization algorithm is a combination of the golden section search method and the parabolic interpolation method. The syntax is

```
x = fminbnd(func, x1, x2)
```

or

```
[x, f] = fminbnd(func, x1, x2)
```

`func` is the handle to the function that calculates the value of the objective function at a single value of x; x_1 and x_2 specify the endpoints of the interval within which the optimum value of x is sought. The function output is the value of x that minimizes `func`, and, optionally, the value of the objective function at x. An alternate syntax is

```
x = fminbnd(func, x1, x2, options)
```

`options` is a structure that is created using the `optimset` function. You can set the tolerance for x, the maximum number of function evaluations, the maximum number of iterations, and other features of the optimization procedure using `optimset`. The default tolerance limit is 0.0001 on a change in x and $f(x)$. You can use `optimset` to choose the display mode of the result. If you want to display the optimization result of each iteration, create the `options` structure using the syntax below.

```
options = optimset('Display', 'iter')
```

Box 8.2C Optimization of a fermentation process: maximization of profit

We solve the optimization problem discussed in Box 8.2A using the golden section search method. MATLAB program 8.5 (listed below) minimizes a single-variable objective function using this method.

MATLAB program 8.5

```
function goldensectionsearch(func, ab, tolx)
% The golden section search method is used to find the minimum of a
% unimodal function.

% Input variables
% func: nonlinear function to be minimized
% ab : bracketing interval [a, b]
% tolx: tolerance for estimating the minimum

% Other variables
r = (sqrt(5) - 1) /2; % interval reduction ratio
k = 0; % counter

% Bracketing interval
a = ab(1);
b = ab(2);
fa = feval(func, a);
fb = feval(func, b);

% Interior points
x1 = (1 - r)*(b - a) + a;
```

```
x2 = r* (b - a) + a;
fx1 = feval(func, x1);
fx2 = feval(func, x2);

% Iterative solution
while (b - a) > tolx
    if fx1 > fx2
        a = x1; % shifting interval left end-point to the right
        fa = fx1;
        x1 = x2;
        fx1 = fx2;
        x2 = r* (b - a) + a;
        fx2 = feval(func, x2);
    else
        b = x2; % shifting interval right end-point to the left
        fb = fx2;
        x2 = x1;
        fx2 = fx1;
        x1 = (1 - r) * (b - a) + a;
        fx1 = feval(func, x1);
    end
    k = k + 1;
end
fprintf('a = %7.6f b = %7.6f f(a) = %7.6f f(b) = %7.6f \n', a, b, fa, fb)
fprintf('number of iterations = %2d \n', k)
```

```
>> goldensectionsearch('profitfunction', [50 70], 0.01)
a = 59.451781 b = 59.460843 f(a) = -19195305.961470 f(b) =
  -19195305.971921
number of iterations = 16
```

Finally, we demonstrate the use of `fminbnd` to determine the optimum value of the flowrate:

```
x = fminbnd('profitfunction', 50, 70,optimset('Display','iter'))
Func-count          x            f(x)            Procedure
1              57.6393       -1.91901e+007       initial
2              62.3607       -1.91828e+007       golden
3              54.7214       -1.91585e+007       golden
4              59.5251       -1.91953e+007       parabolic
5              59.4921       -1.91953e+007       parabolic
6              59.458        -1.91953e+007       parabolic
7              59.4567       -1.91953e+007       parabolic
8              59.4567       -1.91953e+007       parabolic
9              59.4567       -1.91953e+007       parabolic
Optimization terminated:
the current x satisfies the termination criteria using OPTIONS.TolX of
    1.000000e-004
x =
    59.4567
```

8.3 Unconstrained multivariable optimization

In this section, we focus on the minimization of an objective function $f(\mathbf{x})$ of n variables, where $\mathbf{x} = [x_1, x_2, \ldots, x_n]$. If $f(\mathbf{x})$ is a smooth and continuous function such that its first derivatives with respect to each of the n variables exist, then the **gradient** of the function, $\nabla f(\boldsymbol{x})$, exists and is equal to a vector of partial derivatives as follows:

$$\nabla f(\boldsymbol{x}) = \begin{bmatrix} \frac{\partial f}{\partial x_1} \\ \frac{\partial f}{\partial x_2} \\ \cdot \\ \cdot \\ \cdot \\ \frac{\partial f}{\partial x_n} \end{bmatrix}. \tag{8.6}$$

The gradient vector evaluated at $\boldsymbol{x}_0$, points in the direction of the largest rate of increase in the function at that point. The negative of the gradient vector $-\nabla f(\boldsymbol{x}_0)$ points in the direction of the greatest rate of decrease in the function at $\boldsymbol{x}_0$. The gradient vector plays a very important role in derivative-based minimization algorithms since it provides a search direction at each iteration for finding the next approximation of the minimum.

If the gradient of the function $f(\boldsymbol{x})$ at any point $\boldsymbol{x}^*$ is a zero vector (all components of the gradient vector are equal to zero), then $\boldsymbol{x}^*$ is a critical point. If $f(\boldsymbol{x}) \geq f(\boldsymbol{x}^*)$ for all $\boldsymbol{x}$ in the vicinity of $\boldsymbol{x}^*$, then $\boldsymbol{x}^*$ is a local minimum. If $f(\boldsymbol{x}) \leq f(\boldsymbol{x}^*)$, for all $\boldsymbol{x}$ in the neighborhood of $\boldsymbol{x}^*$, then $\boldsymbol{x}^*$ is a local maximum. If $\boldsymbol{x}^*$ is neither a minimum nor a maximum point, then it is called a **saddle point** (similar to the inflection point of a single-variable nonlinear function).

The nature of the critical point can be established by evaluating the matrix of second partial derivatives of the nonlinear objective function, which is called the **Hessian matrix**. If the second partial derivatives $\partial^2 f/\partial x_i \partial x_j$ of the function exist for all paired combinations i, j of the variables, where $i, j = 1, 2, \ldots, n$, then the Hessian matrix of the function is defined as

$$\boldsymbol{H}(\boldsymbol{x}) = \begin{bmatrix} \frac{\partial^2 f}{\partial x_1^2} & \frac{\partial^2 f}{\partial x_1 \partial x_2} & \cdots & \frac{\partial^2 f}{\partial x_1 \partial x_n} \\ \frac{\partial^2 f}{\partial x_2 \partial x_1} & \frac{\partial^2 f}{\partial x_2^2} & \cdots & \frac{\partial^2 f}{\partial x_2 \partial x_n} \\ & \cdot & & \\ & \cdot & & \\ & \cdot & & \\ \frac{\partial^2 f}{\partial x_n \partial x_1} & \frac{\partial^2 f}{\partial x_n \partial x_2} & \cdots & \frac{\partial^2 f}{\partial x_n^2} \end{bmatrix}. \tag{8.7}$$

The Hessian matrix of a multivariable function is analogous to the second derivative of a single-variable function. It can also be viewed as the Jacobian of the gradient vector $\nabla f(\boldsymbol{x})$. (See Section 5.7 for a definition of the Jacobian of a system of functions.) $\boldsymbol{H}(\boldsymbol{x})$ is a symmetric matrix, i.e. $a_{ij} = a_{ji}$, where a_{ij} is the element from the ith row and the jth column of the matrix.

A critical point $\boldsymbol{x}^*$ is a local minimum if $\boldsymbol{H}(\boldsymbol{x}^*)$ is positive definite. A matrix $\boldsymbol{A}$ is said to be **positive definite** if $\boldsymbol{A}$ is symmetric and $\boldsymbol{x}^{\mathrm{T}}\boldsymbol{A}\boldsymbol{x} > 0$ for all non-zero vectors $\boldsymbol{x}$.[2] Note the following properties of a positive definite matrix.

(1) The eigenvalues λ_i ($i = 1, 2, \ldots, n$) of a positive definite matrix are all positive. Remember that an eigenvalue is a scalar quantity that satisfies the matrix equation $\boldsymbol{A} - \lambda \boldsymbol{I} = 0$. A positive definite matrix of size $n \times n$ has n distinct real eigenvalues.

(2) All elements along the main diagonal of a positive definite matrix are positive.

[2] The T superscript represents a transpose operation on the vector or matrix.

(3) The determinant of each **leading principal minor** is positive. A **principal minor** of order j is a minor obtained by deleting from an $n \times n$ square matrix, $n-j$ pairs of rows and columns that intersect on the main diagonal (see the definition of a minor in Section 2.2.6). In other words, the index numbers of the rows deleted are the same as the index numbers of the columns deleted. A leading principal minor of order j is a principal minor of order j that contains the first j rows and j columns of the $n \times n$ square matrix. A 3×3 matrix has three principal minors of order 2:

$$\begin{vmatrix} a_{11} & a_{12} \\ a_{21} & a_{22} \end{vmatrix}, \begin{vmatrix} a_{11} & a_{13} \\ a_{31} & a_{33} \end{vmatrix}, \begin{vmatrix} a_{22} & a_{23} \\ a_{32} & a_{33} \end{vmatrix}.$$

Of these, the first principal minor is the leading principal minor of order 2 since it contains the first two rows and columns of the matrix. An $n \times n$ square matrix has n leading principal minors. Can you list the three leading principal minors of a 3×3 matrix?

To ascertain the positive definiteness of a matrix, one needs to show that either the first property in the list above is true or that both properties (2) and (3) hold true simultaneously.

A critical point $\boldsymbol{x}^*$ is a local maximum if $\boldsymbol{H}(\boldsymbol{x}^*)$ is negative definite. A matrix $\boldsymbol{A}$ is said to be **negative definite** if $\boldsymbol{A}$ is symmetric and $\boldsymbol{x}^{\mathrm{T}}\boldsymbol{A}\boldsymbol{x} < 0$ for all non-zero vectors $\boldsymbol{x}$. A negative definite matrix has negative eigenvalues, diagonal elements that are all less than zero, and leading principal minors that all have a negative determinant. In contrast, at a saddle point, the Hessian matrix is neither positive definite nor negative definite.

A variety of search methods for unconstrained multidimensional optimization problems have been developed. The general procedure of most multivariable optimization methods is to (1) determine a search direction and (2) find the minimum value of the function along the line that passes through the current point and runs parallel to the search direction. Derivative-based methods for multivariable minimization are called **indirect search methods** because the analytical form (or numerical form) of the partial derivatives of the function must be determined and evaluated at the initial guessed values before the numerical step can be applied. In this section we present two indirect methods of optimization: the steepest descent method (Section 8.3.1) and Newton's method for multidimensional search (Section 8.3.2). The latter algorithm requires calculation of the first and second partial derivatives of the objective function at each step, while the former requires evaluation of only the first partial derivatives of the function at the approximation $\boldsymbol{x}_i$ to provide a new search direction at each step. In Section 8.3.3, the simplex method based on the Nelder–Mead algorithm is discussed. This is a **direct search method** that evaluates only the function to proceed towards the minimum.

8.3.1 Steepest descent or gradient method

In the steepest descent method, the negative of the gradient vector evaluated at $\boldsymbol{x}$ provides the search direction at $\boldsymbol{x}$. The negative of the gradient vector points in the direction of the largest rate of decrease of $f(\boldsymbol{x})$. If the first guessed point is $\boldsymbol{x}^{(0)}$, the next point $\boldsymbol{x}^{(1)}$ lies on the line $\boldsymbol{x}^{(0)} - \alpha\nabla f(\boldsymbol{x}^{(0)})$, where α is the step size taken in the search direction from $\boldsymbol{x}^{(0)}$. A minimization search is carried out to locate a point on this line where the function has the least value. This type of one-dimensional search is called a **line search** and involves the optimization of a single variable α. The next point is given by

$$x^{(1)} = x^{(0)} - \alpha \nabla f\left(x^{(0)}\right). \tag{8.8}$$

In Equation (8.8) the second term on the right-hand side is the incremental step taken, or $\Delta x = -\alpha \nabla f\left(x^{(0)}\right)$. Once the new point $x^{(1)}$ is obtained, a new search direction is calculated by evaluating the gradient vector at $x^{(1)}$. The search progresses along this new direction, and stops when α_{opt} is found, which minimizes the function value along the second search line. The function value at the new point should be less than the function value at the previous point, or $f\left(x^{(2)}\right) < f\left(x^{(1)}\right)$ and $f\left(x^{(1)}\right) < f\left(x^{(0)}\right)$. This procedure continues iteratively until the partial derivatives of the function are all very close to zero and/or the distance between two consecutive points is less than the tolerance limit. When these criteria are fulfilled, a local minimum of the function has been reached.

Algorithm for the method of steepest descent

(1) Choose an initial point $x^{(k)}$. In the first iteration, the initial guessed point is $x^{(0)}$.
(2) Determine the analytical form (or numerical value using finite differences, a topic that is briefly discussed in Section 1.6.3) of the partial derivatives of the function.
(3) Evaluate $-\nabla f(x)$ at $x^{(k)}$. This provides the search direction $s^{(k)}$.
(4) Perform a single-variable optimization to minimize $f\left(x^{(k)} + \alpha s^{(k)}\right)$ using any of the one-dimensional minimization methods discussed in Section 8.2.
(5) Calculate the new point $x^{(k+1)} = x^{(k)} + \alpha s^{(k)}$.
(6) Repeat steps (3)–(5) until the procedure has converged upon the minimum value.

Since the method searches for a critical point, it is possible that the iterations may converge upon a saddle point. Therefore, it is important to verify the nature of the critical point by confirming whether the Hessian matrix evaluated at the final point is positive or negative definite.

An advantage of the steepest descent method is that a good initial approximation to the minimum is not required for this method to proceed towards a critical point. However, a drawback of this method is that the step size in each subsequent search direction becomes smaller as the minimum is approached. As a result, many steps are usually required to reach the minimum, and the rate of convergence slows down as the minimum point nears.

There are a number of variants to the steepest descent method that differ in the technique employed to obtain the search direction, such as the conjugate gradient method and Powell's method. These methods are discussed in Edgar and Himmelblau (1988).

Example 8.1

Using the steepest descent method, find the values of x and y that minimize the function

$$f(x, y) = x \cdot e^{x+y} + y^2. \tag{8.9}$$

To get a clearer picture of the behavior of the function, we draw a contour plot in MATLAB. This two-dimensional graph displays a number of contour levels of the function on an x–y plot. Each contour line is assigned a color according to a coloring scheme such that one end of the color spectrum (red) is associated with large values of the function and the other end of the color spectrum (blue) is associated with the smallest values of the function observed in the plot. A contour plot is a convenient tool that can be used to locate valleys (minima), peaks (maxima), and saddle points in the variable space of a two-dimensional nonlinear function.

To make a contour plot, you first need to create two matrices that contain the values of each of the two independent variables to be plotted. To construct these matrices, you use the `meshgrid` function, which creates two matrices, one for x and one for y, such that the corresponding elements of each matrix represent one point on the two-dimensional map. The MATLAB function `contour` plots the contour map of the function. Program 8.6 shows how to create a contour plot for the function given in Equation (8.9).

MATLAB program 8.6

```
% Plots the contour map of the function x*exp(x + y) + y^2
[x, y] = meshgrid(-3:0.1:0, -2:0.1:4);
z = x.*exp(x + y) + y.^2;
contour(x, y, z, 100);
set(gca, 'Line Width',2,'Font Size',16)
xlabel('x','Fontsize',16)
ylabel('y','Fontsize',16)
```

Figure 8.9 is the contour plot created by Program 8.6, with text labels added. In the variable space that we have chosen, there are two critical points, of which one is a saddle point. The other critical point is a minimum. We can label the contours using the `clabel` function. Accordingly, Program 8.6 is modified slightly, as shown below in Program 8.7.

MATLAB program 8.7

```
% Plots a contour plot of the function x*exp(x + y) + y^2 with labels.
[x, y] = meshgrid(-3:0.1:0, -1:0.1:1.5);
z = x.*exp(x + y) + y.^2;
[c, h] = contour(x, y, z, 15);
clabel(c, h)
set(gca, 'Line Width',2,'Font Size',16)
xlabel('x','Fontsize',16)
ylabel('y','Fontsize',16)
```

Figure 8.9

Contour plot of the function given in Equation (8.9)

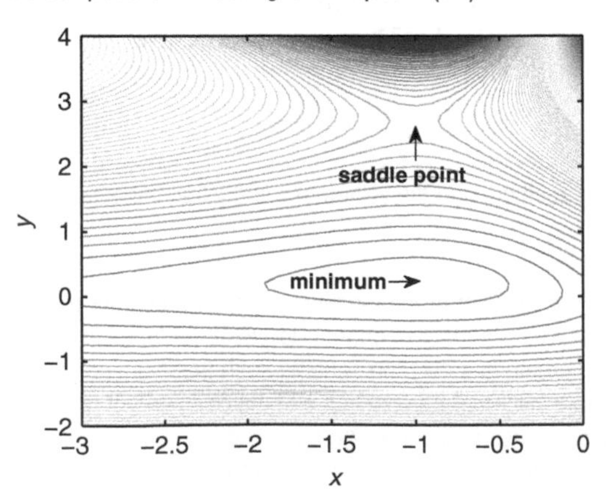

Figure 8.10

Contour plot of the function given by Equation (8.9) showing the function values at different contour levels near the minimum point

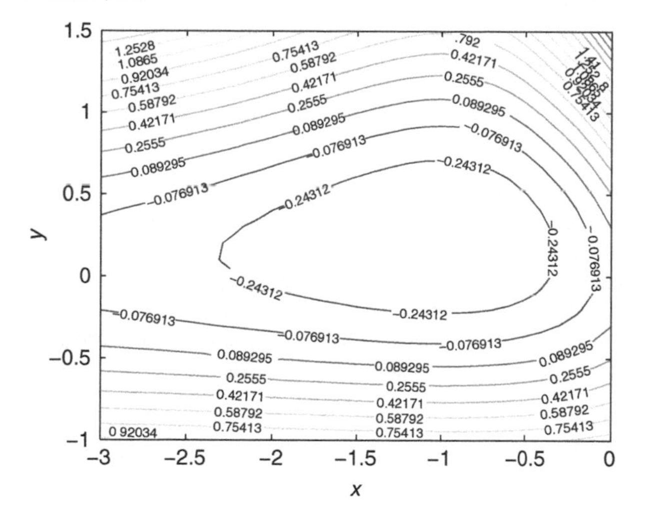

Figure 8.11

The behavior of the function at the saddle point

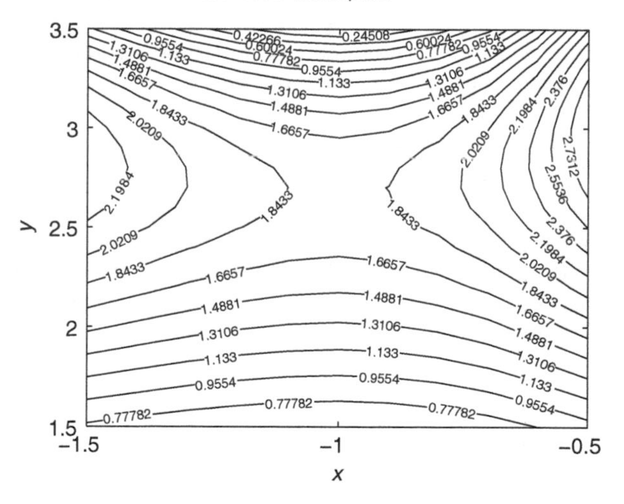

The resulting plot is shown in Figure 8.10. The second critical point is a minimum. A labeled contour plot that shows the behavior of the function at the saddle point is shown in Figure 8.11. Study the shape of the function near the saddle point. Can you explain how this type of critical point got the name "saddle"?[3]

Taking the first partial derivatives of the function with respect to x and y, we obtain the following analytical expression for the gradient:

$$\nabla f(x,y) = \begin{bmatrix} (1+x)e^{x+y} \\ xe^{x+y} + 2y \end{bmatrix}.$$

We begin with an initial guess value of $\mathbf{x}^{(0)} = (-1, 0)$:

[3] Use the command `surf(x, y, z)` in **MATLAB** to reveal the shape of this function in two variables. This might help you to answer the question.

$$\nabla f(-1,0) = \begin{bmatrix} 0 \\ -1/e \end{bmatrix}$$

Next we minimize the function:

$$f\left(\boldsymbol{x}^{(0)} - a\nabla f\left(\boldsymbol{x}^{(0)}\right)\right) = f\left(-1 - 0 \cdot a, 0 - \frac{a}{e}\right)$$
$$= -e^{-\left(1+\frac{a}{e}\right)} + \left(\frac{a}{e}\right)^2.$$

The function to be minimized is reduced to a single-variable function. Any one-dimensional minimization method, such as successive parabolic interpolation, golden section search, or Newton's method can be used to obtain the value of a that minimizes the above expression.

We use successive parabolic interpolation to perform the single-variable minimization. For this purpose, we create a MATLAB function program that calculates the value of the function $f(a)$ at any value of a. To allow the same function m-file to be used for subsequent iterations, we generalize it by passing the values of the current guess point and the gradient vector to the function.

MATLAB program 8.8

```
function f = xyfunctionalpha(alpha, x, df)
% Calculates the function f(x-alpha*df)

% Input variables
% alpha : scalar value of the step size
% x     : vector that specifies the current guess point
% df    : gradient vector of the function at x

f = (x(1)-df(1)*alpha)*exp(x(1)-df(1)*alpha + x(2)-df(2)*alpha) + ...
    (x(2) - df(2)*alpha)^2;
```

Program 8.3, which lists the function m-file that performs a one-dimensional minimization using the parabolic interpolation method, is slightly modified so that it can be used to perform our line search. In this user-defined function, the guessed point and the gradient vector are added as input parameters and the function is renamed as `steepestparabolicinterpolation` (see Program 8.10). We type the following in the Command Window:

```
>> steepestparabolicinterpolation('xyfunctionalpha', [0 10],0.01,
   [-1, 0], [0, -exp(-1)])
```

The output is

```
x0 = 0.630535 x2 = 0.630535 f(x0) = -0.410116 f(x2) = -0.410116
x_min = 0.630535 f(x_min) = -0.410116
number of iterations = 13
```

The optimum value of a is found to be 0.631, at which the function value is –0.410. The new point $\boldsymbol{x}^{(1)}$ is (–1, 0.232). The gradient at this point is calculated as

$$\nabla f(-1, 0.232) = \begin{bmatrix} 0 \\ 6 \times 10^{-5} \end{bmatrix}.$$

The gradient is very close to zero. We conclude that we have reached the minimum point in a single iteration. Note that the excellent initial guessed point and the behavior of the gradient at the initial guess contributed to the rapid convergence of the algorithm.

We can automate the steepest descent algorithm by creating an m-file that iteratively calculates the gradient of the function and then passes this value along with the guess point to `steepestparabolicinterpolation`. Program 8.9 is a function that minimizes the multivariable function using the line search method of steepest descent. It calls Program 8.10 to perform one-dimensional minimization of the function along the search direction by optimizing α.

MATLAB program 8.9

```
function steepestdescent(func, funcfalpha, x0, toldf)
% Performs unconstrained multivariable optimization using the steepest
% descent method

% Input variables
% func: calculates the gradient vector at x
% funcfalpha: calculates the value of f(x-alpha*df)
% x0: initial guess point
% toldf: tolerance for error in gradient

% Other variables
maxloops = 20;
ab = [0 10]; % smallest and largest values of alpha
tolalpha = 0.001; % tolerance for error in alpha
df = feval(func, x0);

% Miminization algorithm
for i = 1:maxloops
    alpha = steepestparabolicinterpolation(funcfalpha, ab, ...
        tolalpha, x0, df);
    x1 = x0 - alpha*df;
    x0 = x1;
    df = feval(func, x0);
    fprintf('at step %2d, norm of gradient is %5.4f \n',i, norm(df))
    if norm(df) < toldf
      break
    end
end
fprintf('number of steepest descent iterations is %2d \n',i)
fprintf('minimum point is %5.4f, %5.4f \n',x1)
```

MATLAB program 8.10

```
function alpha = steepestparabolicinterpolation(func, ab, tolx, x, df)
% Successive parabolic interpolation is used to find the minimum
% of a unimodal function.

% Input variables
% func: nonlinear function to be minimized
% ab : initial bracketing interval [a, b]
% tolx: tolerance in estimating the minimum
% x : guess point
% df : gradient vector evaluated at x
```

```
% Other variables
k = 0; % counter
maxloops = 20;

% Set of three points to construct the parabola
x0 = ab(1);
x2 = ab(2);
x1 = (x0 + x2)/2;

% Function values at the three points
fx0 = feval(func, x0, x, df);
fx1 = feval(func, x1, x, df);
fx2 = feval(func, x2, x, df);

% Iterative solution
while (x2 - x0) > tolx
    % Calculate minimum point of parabola
    numerator = (x2^2 - x1^2)*fx0 - (x2^2 - x0^2)*fx1 + (x1^2 - x0^2)*fx2;
    denominator = 2*((x2 - x1)*fx0 - (x2 - x0)*fx1 + (x1 - x0)*fx2);
    xmin = numerator/denominator;

    % Function value at xmin
    fxmin = feval(func, xmin, x, df);

    % Select set of three points to construct new parabola
    if xmin < x1
      if fxmin < fx1
          x2 = x1;
          fx2 = fx1;
          x1 = xmin;
          fx1 = fxmin;
      else
          x0 = xmin;
          fx0 = fxmin;
      end
    else
      if fxmin < fx1
          x0 = x1;
          fx0 = fx1;
          x1 = xmin;
          fx1 = fxmin;
    else
          x2 = xmin;
          fx2 = fxmin;
      end
    end
    k = k + 1;
    if k > maxloops
        break
    end
end
alpha = xmin;
fprintf('x0 = %7.6f x2 = %7.6f f(x0) = %7.6f f(x2) = %7.6f \n', x0, x2, fx0, fx2)
```

```
fprintf('x_min = %7.6f f(x_min) = %7.6f \n',xmin, fxmin)
fprintf('number of parabolic interpolation iterations = %2d \n', k)
```

Using an initial guessed point of (–1, 0) and a tolerance of 0.001, we obtain the following solution:

```
>> steepestdescent('xyfunctiongrad','xyfunctionalpha', [-1, 0],
   0.001)
x0 = 0.630535 x2 = 0.630535 f(x0) = -0.410116 f(x2) = -0.410116
x_min = 0.630535 f(x_min) = -0.410116
number of parabolic interpolation iterations = 13
at step 1, norm of gradient is 0.0000
number of steepest descent iterations is 1
minimum point is -1.0000, 0.2320
```

Instead if we start from the guess point (0, 0), we obtain the output

```
x0 = 0.000000 x2 = 1.000812 f(x0) = 0.000000 f(x2) = -0.367879
x_min = 1.000520 f(x_min) = -0.367879
number of parabolic interpolation iterations = 21
at step 1, norm of gradient is 0.3679
x0 = 0.630535 x2 = 0.630535 f(x0) = -0.410116 f(x2) = -0.410116
x_min = 0.630535 f(x_min) = -0.410116
number of parabolic interpolation iterations = 13
at step 2, norm of gradient is 0.0002
number of steepest descent iterations is 2
minimum point is -1.0004, 0.2320
```

8.3.2 Multidimensional Newton's method

Newton's method approximates the multivariate function with a second-order approximation. Expanding the function about a point $\boldsymbol{x}_0$ using a multidimensional Taylor series, we have

$$f(\boldsymbol{x}) \approx f(\boldsymbol{x}_0) + \nabla^{\mathrm{T}} f(\boldsymbol{x}_0)(\boldsymbol{x} - \boldsymbol{x}_0) + \frac{1}{2}(\boldsymbol{x} - \boldsymbol{x}_0)^{\mathrm{T}} \boldsymbol{H}(\boldsymbol{x}_0)(\boldsymbol{x} - \boldsymbol{x}_0),$$

where $\nabla^{\mathrm{T}} f(\boldsymbol{x}_0)$ is the transpose of the gradient vector. The quadratic approximation is differentiated with respect to $\boldsymbol{x}$ and equated to zero. This yields the following approximation formula for the minimum:

$$\boldsymbol{H}(\boldsymbol{x}_0)\Delta\boldsymbol{x} = -\nabla f(\boldsymbol{x}_0), \tag{8.10}$$

where $\Delta\boldsymbol{x} = \boldsymbol{x} - \boldsymbol{x}_0$. This method requires the calculation of the first and second partial derivatives of the function. Note the similarity between Equations (8.10) and (5.37). The iterative technique to find the solution $\boldsymbol{x}$ that minimizes $f(\boldsymbol{x})$ has much in common with that described in Section 5.7 for solving a system of nonlinear equations using Newton's method. Equation (8.10) reduces to a set of linear equations in $\Delta\boldsymbol{x}$. Any of the methods described in Chapter 2 to solve a system of linear equations can be used to solve Equation (8.10).

Equation (8.10) exactly calculates the minimum $\boldsymbol{x}$ of a quadratic function. An iterative process is required to minimize general nonlinear functions. If the Hessian matrix is invertible then we can rearrange the terms in Equation (8.10) to obtain the iterative formula

$$\boldsymbol{x}^{(k+1)} = \boldsymbol{x}^{(k)} - \left[\boldsymbol{H}\left(\boldsymbol{x}^{(k)}\right)\right]^{-1} \nabla f\left(\boldsymbol{x}^{(k)}\right). \tag{8.11}$$

In Equation (8.11), $-[H(x_i)]^{-1}\nabla f(x_i)$ provides the search (descent) direction. This is the Newton formula for multivariable minimization. Because the formula is inexact for a non-quadratic function, we can optimize the step length by adding an additional parameter α. Equation (8.11) becomes

$$x^{(k+1)} = x^{(k)} - \alpha_k\left[H\left(x^{(k)}\right)\right]^{-1}\nabla f\left(x^{(k)}\right). \tag{8.12}$$

In Equation (8.12), the search direction is given by $\left[\boldsymbol{H}\left(\boldsymbol{x}^{(k)}\right)\right]^{-1}\nabla f\left(\boldsymbol{x}^{(k)}\right)$ and the step length is modified by α_k.

Algorithm for Newton's multidimensional minimization ($\alpha = 1$)

(1) Provide an initial estimate $\boldsymbol{x}^{(k)}$ of the minimum of the objective function $f(\boldsymbol{x})$. At the first iteration, $k = 0$.
(2) Determine the analytical or numerical form of the first and second partial derivatives of $f(\boldsymbol{x})$.
(3) Evaluate $\nabla f(x)$ and $\boldsymbol{H}(\boldsymbol{x})$ at $\Delta\boldsymbol{x}^{(k)}$.
(4) Solve the system of linear equations $\boldsymbol{H}\left(\boldsymbol{x}^{(k)}\right)\Delta\boldsymbol{x}^{(k)} = -\nabla f\left(\boldsymbol{x}^{(k)}\right)$ for $\Delta\boldsymbol{x}^{(k)}$.
(5) Calculate the next approximation of the minimum $\boldsymbol{x}^{(k+1)} = \boldsymbol{x}^{(k)} + \Delta\boldsymbol{x}^{(k)}$.
(6) Repeat steps (3)–(5) until the method has converged upon a solution.

Algorithm for modified Newton's multidimensional minimization ($\alpha \neq 1$)

(1) Provide an initial estimate $\boldsymbol{x}^{(k)}$ of the minimum of the objective function $f(\boldsymbol{x})$.
(2) Determine the analytical or numerical form of the first and second partial derivatives of $f(\boldsymbol{x})$.
(3) Calculate the search direction $\boldsymbol{s}^{(k)} = -\left[\boldsymbol{H}\left(\boldsymbol{x}^{(k)}\right)\right]^{-1}\nabla f\left(\boldsymbol{x}^{(k)}\right)$.
(4) Perform single-parameter optimization to determine the value of α_k that minimizes $f\left(\boldsymbol{x}^{(k)} + \alpha_k\boldsymbol{s}^{(k)}\right)$.
(5) Calculate the next approximation of the minimum $\boldsymbol{x}^{(k+1)} = \boldsymbol{x}^{(k)} + \alpha_k\boldsymbol{s}^{(k)}$.
(6) Repeat steps (3)–(5) until the method has converged upon a solution.

When the method has converged upon a solution, it is necessary to check if the point is a local minimum. This can be done either by determining if $\boldsymbol{H}$ evaluated at the solution is positive definite or by perturbing the solution slightly along multiple directions and observing if iterative optimization recovers the solution each time. If the point is a saddle point, perturbation of the solution and subsequent optimization steps are likely to lead to a solution corresponding to some other critical point.

Newton's method has a quadratic convergence rate and thus minimizes a function faster than other methods, when the initial guessed point is close to the minimum. Some difficulties encountered with using Newton's method are as follows.

(1) A good initial approximation close to the minimum point is required for the method to demonstrate convergence.
(2) The first and second partial derivatives of the objective function should exist. If the expressions are too complicated for evaluation, finite difference approximations may be used in their place.
(3) This method may converge to a saddle point. To ensure that the method is not moving towards a saddle point, it should be verified that $\boldsymbol{H}\left(\boldsymbol{x}^{(k)}\right)$ is positive definite at each step. If the Hessian matrix evaluated at the current point is not positive definite, the minimum of the function will not be reached. Techniques are available

to force the Hessian matrix to remain positive definite so that the search direction will lead to improvement (reduction) in the function value at each step.

The advantages and drawbacks of the method of steepest descent and Newton's method are complementary. An efficient and popular minimization algorithm is the Levenberg–Marquart method, which combines the two techniques in such a way that minimization begins with steepest descent. When the minimum point is near (e.g. Hessian is positive definite), convergence slows down, and the iterative routine switches to Newton's method, producing a more rapid convergence.

Using MATLAB

Optimization Toolbox contains the built-in function `fminunc` that searches for the local minimum of a multivariable objective function. This function operates on two scales. It has a large-scale algorithm that is suited for an unconstrained optimization problem consisting of many variables. The medium-scale algorithm should be chosen when optimizing a few variables. The latter algorithm is a line search method that calculates the gradient to provide a search direction. It uses the quasi-Newton method to approximate and update the Hessian matrix, whose inverse is used along with the gradient to obtain the search direction. In the quasi-Newton method, the behavior of the function $f(x)$ and its gradient $\nabla f(x)$ is used to determine the form of the Hessian at $\boldsymbol{x}$. Newton's method, on the other hand, calculates $\boldsymbol{H}$ directly from the second partial derivatives. Gradient methods, and therefore the `fminunc` function, should be used when the function derivative is continuous, since they are more efficient.

The default algorithm is large-scale. The large-scale algorithm approximates the objective function with a simpler function and then establishes a trust region within which the function minimum is sought. The second partial derivatives of the objective function can be made available to `fminunc` by calculating them along with the objective function in the user-defined function file. If the `options` structure instructs `fminunc` to use the user-supplied Hessian matrix, these are used in place of the numerical approximations generated by the algorithm to approximate the Hessian. Only the large-scale algorithm can use the user-supplied Hessian matrix calculations, if provided. Note that calculating the gradient $\nabla f(\boldsymbol{x})$ along with the function $f(\boldsymbol{x})$ is optional when using the medium-scale algorithm but is required when using the large-scale algorithm.

One form of the syntax is

```
x = fminunc(func, x0, options)
```

$\boldsymbol{x}_0$ is the initial guess value and can be a scalar, vector, or matrix. Use of `optimset` to construct an `options` structure was illustrated in Section 8.2.3. `Options` can be used to set the tolerance of $\boldsymbol{x}$, to set the tolerance of the function value, to specify use of user-supplied gradient calculations, to specify use of user-supplied Hessian calculations, whether to proceed with the large-scale or medium-scale algorithm by turning "`LargeScale`" on or off, and various other specifics of the algorithm. To tell the function `fminunc` that `func` also returns the gradient and the Hessian matrix along with the scalar function value, create the following structure:

```
options = optimset('GradObj', 'on', 'Hessian', 'on')
```

to be passed to `fminunc`. When `func` supplies the first partial derivatives and the second partial derivatives at $\boldsymbol{x}$ along with the function value, it should have three output arguments in the following order: the scalar-valued function, the gradient vector, followed by the Hessian matrix (see Example 8.2). See `help fminunc` for more details.

Example 8.2

Minimize the function given in Equation (8.9) using multidimensional Newton's method.

The analytical expression for the gradient was obtained in Example 8.1 as

$$\nabla f(x,y) = \begin{bmatrix} (1+x)e^{x+y} \\ xe^{x+y} + 2y \end{bmatrix}.$$

The Hessian of the function is given by

$$H(x,y) = \begin{bmatrix} (2+x)e^{x+y} & (1+x)e^{x+y} \\ (1+x)e^{x+y} & xe^{x+y} + 2 \end{bmatrix}.$$

MATLAB Program 8.11 performs multidimensional Newton's optimization, and Program 8.12 is a function that calculates the gradient and Hessian at the guessed point.

MATLAB program 8.11

```
function newtonsmultiDoptimization(func, x0, toldf)
% Newton's method is used to minimize a multivariable objective function.

% Input variables
% func   : calculates gradient and Hessian of nonlinear function
% x0     : initial guessed value
% toldf  : tolerance for error in norm of gradient

% Other variables
maxloops = 20;
[df, H] = feval(func,x0);

% Minimization scheme
for i = 1:maxloops
    deltax = - H\df; % df must be a column vector
    x1 = x0 + deltax;
    [df, H] = feval(func,x1);
    if norm(df) < toldf
      break % Jump out of the for loop
    end
    x0 = x1;
end
fprintf('number of multi-D Newton''s iterations is %2d \n',i)
fprintf('norm of gradient at minimum point is %5.4f \n', norm(df))
fprintf('minimum point is %5.4f, %5.4f \n',x1)
```

MATLAB program 8.12

```
function [df, H] = xyfunctiongradH(x)
% Calculates the gradient and Hessian of f(x) = x*exp (x + y) + y^2

df = [(1 + x(1))*exp(x(1) + x(2)); x(1)*exp(x(1) + x(2)) + 2*x(2)];
```

```
H = [(2 + x(1))*exp(x(1) + x(2)), (1 + x(1))*exp(x(1) + x(2)); ...
(1 + x(1))*exp(x(1) + x(2)), x(1)*exp(x(1) + x(2)) + 2];
```

We choose an initial point at (0, 0), and solve the optimization problem using our optimization function developed in MATLAB:

```
>> newtonsmultiDoptimization('xyfunctiongradH', [0; 0], 0.001)
number of multi-D Newton's iterations is 4
norm of gradient at minimum point is 0.0001
minimum point is -0.9999, 0.2320
```

We can also solve this optimization problem using the `fminunc` function. We create a function that calculates the value of the function, gradient, and Hessian at any point $\mathbf{x}$.

MATLAB program 8.13

```
function [f, df, H] = xyfunctionfgradH(x)
% Calculates the valve, gradient, and Hessian of
% f(x) = x*exp(x + y) + y^2

f = x(1)*exp(x(1) + x(2)) + x(2)^2;
df = [(1 + x(1))*exp(x(1) + x(2)); x(1)*exp(x(1)+ x(2)) + 2*x(2)];
H = [(2 + x(1))*exp(x(1) + x(2)), (1 + x(1))*exp(x(1)+ x(2)); ...
    (1 + x(1))*exp(x(1) + x(2)), x(1)*exp(x(1) + x(2)) + 2];
```

Then

```
>> options = optimset('GradObj','on','Hessian','on');
>> fminunc('xyfunctionfgradH', [0; 0], options)
```

gives us the output

```
Local minimum found.
Optimization completed because the size of the gradient is less than
  the default value of the function tolerance.
ans =
    -1.0000
    0.2320
```

8.3.3 Simplex method

The simplex method is a direct method of minimization since only function evaluations are required to proceed iteratively towards the minimum point. The method creates a geometric shape called a **simplex**. The simplex has $n + 1$ vertices, where n is the number of variables to be optimized. At each iteration, one point of the simplex is discarded and a new point is added such that the simplex gradually moves in the variable space towards the minimum. The original simplex algorithm was devised by Spendley, Hext, and Himsworth, according to which the simplex was a regular geometric structure (with all sides of equal lengths). Nelder and Mead developed a more efficient simplex algorithm in which the location of the new vertex, and the location of the other vertices, may be adjusted at each iteration, permitting the simplex shape to expand or contract in the variable space. The Nelder–Mead simplex structure is not necessarily regular after each iteration. For a two-dimensional optimization problem, the simplex has the shape of a triangle. The simplex takes the shape of a tetrahedron or a pyramid for a three-parameter optimization problem.

We illustrate the simplex algorithm for the two-dimensional optimization problem. Three points in the variable space are chosen such that the lines connecting the points are equal in length, to generate an equilateral triangle. We label the three vertices according to the function value at each vertex. The vertex with the smallest function value is labeled **G** (for good); the vertex with an intermediate function value is labeled **A** (for average); and the vertex with the largest function value is labeled **B** (for bad).

The goal of the simplex method, when performing a minimization, is to move downhill, or away, from the region where the function values are large and towards the region where the function values are smaller. The function values decrease in the direction from **B** to **G** and from **B** to **A**. Therefore, it is likely that even smaller values of the function will be encountered if one proceeds in this direction. To advance in this direction, we locate the mirror image of point **B** across the line **GA**. To do this the midpoint **M** on the line **GA** is located. A line is drawn that connects **M** with **B**, which has a width w. This line is extended past **M** to a point **S** such that the width of the line **BS** is $2w$. See Figure 8.12.

The function value at **S** is computed. If $f(\mathbf{S}) < f(\mathbf{G})$, then the point **S** is a significant improvement over the point **B**. It may be possible that the minimum lies further beyond **S**. The line **BS** is extended further to **T** such that the width of **MT** is $2w$ and **BT** is $3w$ (see Figure 8.13). If $f(\mathbf{T}) < f(\mathbf{S})$, then further improvement has been made in the selection of the new vertex, and **T** is the new vertex that replaces **B**. If $f(\mathbf{T}) > f(\mathbf{S})$, then point **S** is retained and **T** is discarded.

On the other hand, if $f(\mathbf{S}) \geq f(\mathbf{B})$, then no improvement has resulted from the reflection of **B** across **M**. The point **S** is discarded and another point **C**, located at a distance of $w/2$ from **M** along the line **BS**, is selected and tested. If $f(\mathbf{C}) \geq f(\mathbf{B})$, then point **C** is discarded. Now the original simplex shape must be modified in order to make progress. The lengths of lines **GB** and **GA** contract as the points **B** and **A** are retracted towards **G**. After shrinking the triangle, the algorithm is retried to find a new vertex.

A new simplex is generated at each step such that there is an improvement (decrease) in the value of the function at one vertex point. When the minimum is approached, in order to fine-tune the estimate of its location, the size of the simplex (triangle in two-dimensional space) must be reduced.

Figure 8.12

Construction of a new triangle **GSA** (simplex). The original triangle is shaded in gray. The line **BM** is extended beyond **M** by a distance equal to w

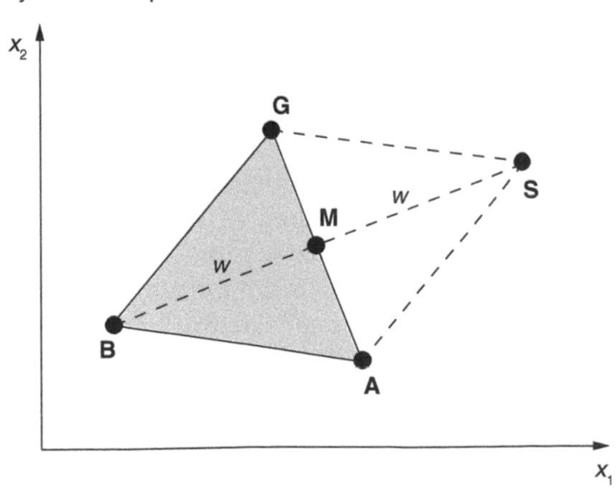

Figure 8.13

Nelder–Mead's algorithm to search for a new vertex. The point **T** may be an improvement over **S**

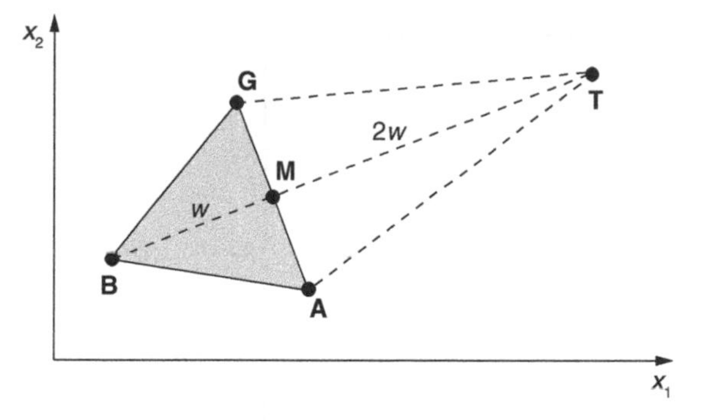

Because the simplex method only performs function evaluations, it does not make any assumptions regarding the smoothness of the function and its differentiability with respect to each variable. Therefore, this method is preferred for use with discontinuous or non-smooth functions. Moreover, since partial derivative calculations are not required at any step, direct methods are relatively easy to use.

Using MATLAB

MATLAB's built-in function `fminsearch` uses the Nelder–Mead implementation of the simplex method to search for the unconstrained minimum of a function. This function is well suited for optimizing multivariable problems whose objective function is not smooth or is discontinuous. This is because the `fminsearch` routine only evaluates the objective function and does not calculate the gradient of the function. The syntax is

```
x = fminsearch(func, x0)
```

or

```
[x, f] = fminsearch(func, x0, options)
```

The algorithm uses the initial guessed value $\mathbf{x}_0$ to construct n additional points, and 5% of each element of $\mathbf{x}_0$ is added to $\mathbf{x}_0$ to derive the $n + 1$ vertices of the simplex. Note that `func` is the handle to the function that calculates the scalar value of the objective function at $\mathbf{x}$. The function output is the value of $\mathbf{x}$ that minimizes `func`, and, optionally, the value of `func` at $\mathbf{x}$.

Box 8.3 Nonlinear regression of pixel intensity data

We wish to study the morphology of the cell synapse in a cultured aggregate of cells using immunofluorescence microscopy. Cells are fixed and labeled with a fluorescent antibody against a protein known to localize at the boundary of cells. When the pixel intensities along a line crossing the cell synapse are measured, the data in Figure 8.14 are produced.

We see a peak in the intensity near $x = 15$. The intensity attenuates to some baseline value ($I \sim 110$) in either direction. It was decided to fit these data to a Gaussian function of the following form:

$$f(x) = \frac{a_1}{\sqrt{2\pi\sigma^2}} \exp\left(-\frac{(x-\mu)^2}{2\sigma^2}\right) + a_2,$$

Figure 8.14

Pixel intensities exhibited by a cell synapse, where x specifies the pixel number along the cell synapse

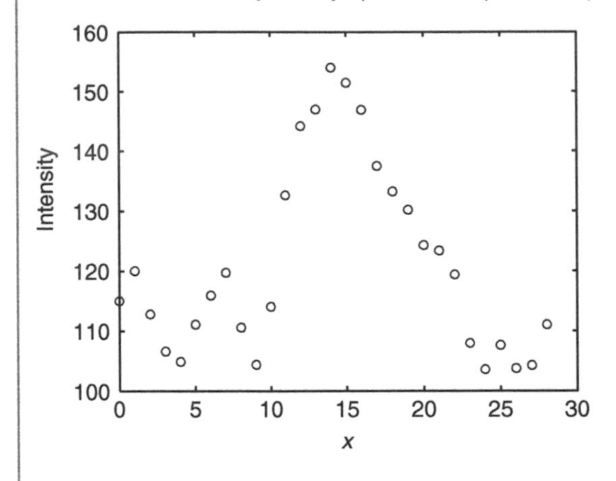

where a_1 is an arbitrary scaling factor, a_2 represents the baseline intensity, and the other two model parameters are σ and μ of the Gaussian distribution. In this analysis, the most important parameter is σ, as this is a measure of the width of the cell synapse and can potentially be used to detect the cell membranes separating and widening in response to exogenous stimulation. The following MATLAB program 8.14 and user-defined function were written to perform a nonlinear least squares regression of this model to the acquired image data using the built-in `fminsearch` function.

MATLAB program 8.14

```
% This program uses fminsearch to minimize the error of nonlinear
% least-squares regression of the unimodal Gaussian model to the pixel
% intensities data at a cell synapse
% To execute this program place the data files into the same directory in
% which this m-file is located.

% Variables
global data % matrix containing pixel intensity data
Nfiles = 4; % number of data files to be analyzed, maximum allowed is 999

for i = 1:Nfiles
    num = num2str(i);
    if length(num) == 1
        filename = ['data00',num,'.dat'];
    elseif length(num) == 2
        filename = ['data0',num,'.dat'];
    else
        filename = ['data',num,'.dat'];
    end
    data = load(filename);
    p = fminsearch('PixelSSE', [15,5,150,110]);
```

```
    % Compute model
    x = 1:.1:length(data);
    y = p(3)/sqrt(2*pi*p(2)^2)*...
        exp(-(x - p(1)).^2/(2*p(2)^2)) + p (4);

    % Compare data with the Gaussian model
    figure(i)
    plot(data(:,1),data(:,2),'ko',x,y,'k-')
    xlabel('{\itx}', 'FontSize',24)
    ylabel('Intensity', 'FontSize',24)
    set(gca, 'LineWidth',2, 'FontSize',20)
end
```

MATLAB program 8.15

```
function SSE = PixelSSE(p)
% Calculates the SSE between the observed pixel intensity and the
% Gaussian function
% f = a1/sqrt(2*pi)/sig*exp(-(x-mu)^2/(2*sig^2)) + a2

global data

% model parameters
mu = p(1);   % mean
sig = p(2);  % standard deviation
a1 = p(3);   % rescales the distribution
a2 = p(4);   % baseline

SSE = sum((data(:,2) - a1./sqrt(2*pi)./sig.*...
    exp(-(data(:,1) - mu).^2/(2*sig.^2)) - a2).^2);
```

Note the use of string variables (and string concatenation) to generate filenames; together with the `load` function this is used to automatically analyze an entire directory of up to 999 input files. Also, this program provides an example of the use of the `global` declaration to pass a data file between a function and the main program. Figure 8.15 shows the good fits that were obtained using Program 8.14 to analyze four representative data files.

This program works well, and reliably finds a best-fit Gaussian model, but what happens when the cell synapse begins to separate and show a distinct bimodal appearance such as shown in Figure 8.16?

It seems more appropriate to fit data such as this to a function comprising two Gaussian distributions, rather than a single Gaussian. The following function is defined:

$$f(x) = a_1 \cdot \left(\frac{1}{\sqrt{2\pi\sigma^2}} \exp\left(-\frac{(x-\mu_1)^2}{2\sigma^2} \right) + \frac{1}{\sqrt{2\pi\sigma^2}} \exp\left(-\frac{(x-\mu_2)^2}{2\sigma^2} \right) \right) + a_2.$$

Figure 8.15

Observed pixel intensities at four cell synapses and the fitted Gaussian models

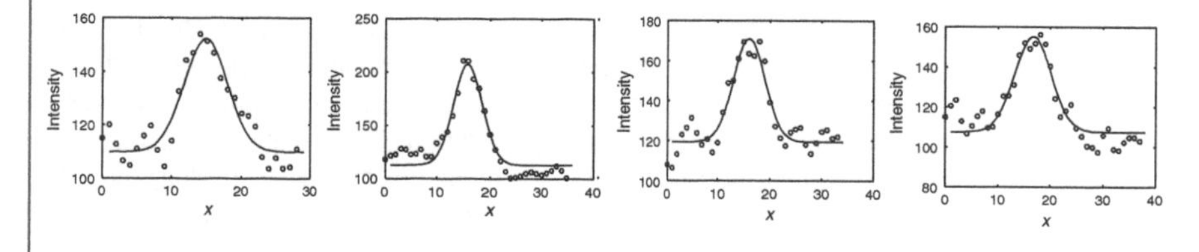

Figure 8.16

Pixel intensities exhibited by separating cell synapse

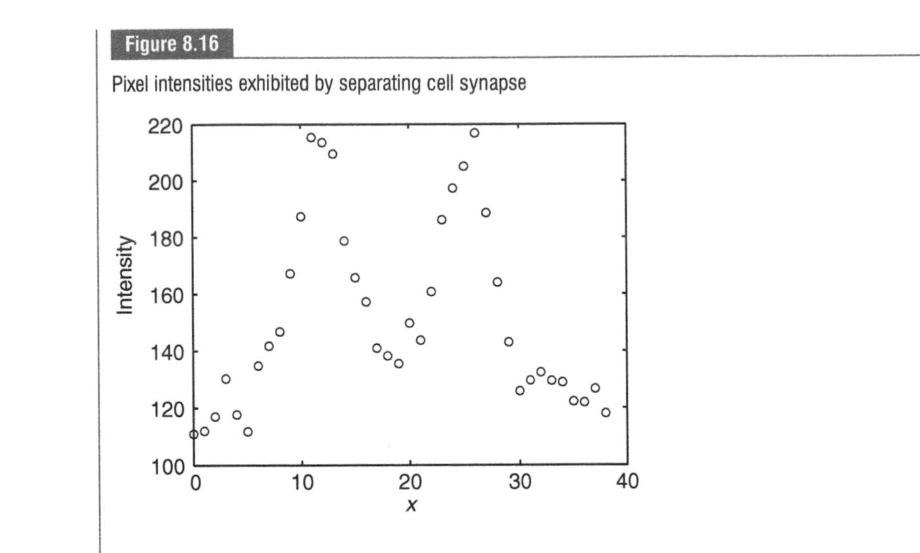

Figure 8.17

Observed pixel intensities at seven cell synapses and the fitted bimodal Gaussian models

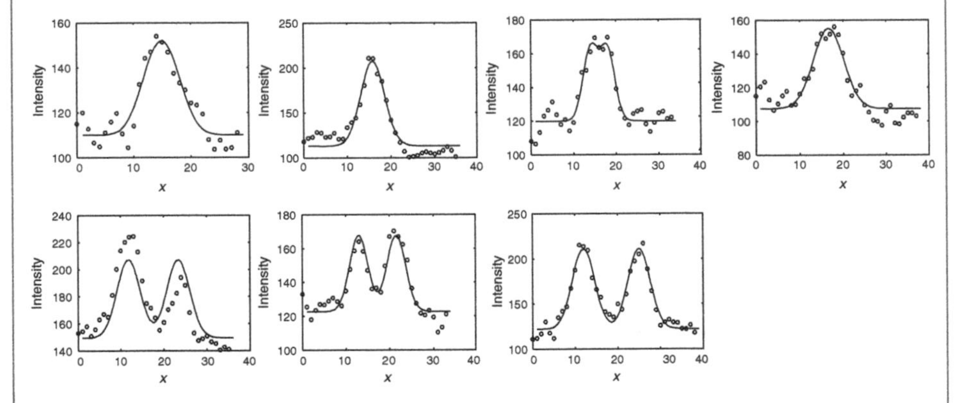

Note that we have reduced the degrees of freedom by making the two Gaussian curves share the same scaling factor a_1 and the same σ; the rationale for this restriction is discussed below. It is desirable that the new bimodal analysis program be robust enough to converge on both unimodal and bimodal data sets, to avoid the need to check each curve visually and make a (potentially subjective) human decision on each data set ... as this would defeat the purpose of our file input commands capable of analyzing up to 999 files in one keystroke! The improved, bimodal algorithm is implemented in Programs 8.16 and 8.17 given below. The relevant metric to characterize the width of the bimodal curve is now $|\mu_1 - \mu_2|$ added to some multiple of σ. As shown in Figure 8.17, Program 8.16 reliably fits both unimodal and bimodal pixel intensity curves.

The images in the first row of Figure 8.17 show the apparently unimodal images that were previously analyzed using Program 8.15, and those in the second row show obviously bimodal images; Program 8.17 can efficiently characterize both types. Interestingly, initially the third image appeared unimodal, but upon closer inspection we see a dip in the middle of the pixel range that the bimodal regression algorithm correctly detects as nearly (but not exactly) overlapping Gaussians. It was determined that if additional parameters are added to the model: (i) two, distinct values for σ, or (ii) individual scaling factors for each peak, then the least-squares regression will attempt to create an artificially low, or wide, peak, or a peak that is outside of the pixel range, in an attempt to deal with an uneven baseline at the two edges of the data range.

MATLAB program 8.16

```
% This program uses fminsearch to minimize the error of nonlinear
% least-squares regression of the bimodal Gaussian model to the pixel
% intensities data at a cell synapse
% To execute this program place the data files into the same directory in
% which this m-file is located.

% Variables
global data % matrix containing pixel intensity data
Nfiles = 7; % number of data files to be analyzed, maximum allowed is 999

for i = 1:Nfiles
    num = num2str(i);
    if length(num) == 1
        filename= ['data00',num,'.dat'];
    elseif length(num) == 2
        filename = ['data0',num,'.dat'];
    else
        filename = ['data',num,'.dat'];
    end
    data = load(filename);

    p = fminsearch('PixelSSE2', [10,25,5,220,150]);

    % Compute model
    x = 1:.1:length(data);
    y = p(4).*...
      (1/sqrt(2*pi*p(3).^2)*exp(-(x - p(1)).^2/(2*p(3).^2)) + ...
      1/sqrt(2*pi*p(3).^2)*exp(-(x - p(2)).^2/(2*p(3).^2))) + p(5);

    % Compare observations with the Gaussian model
    figure(i)
    plot(data(:,1),data(:,2),'ko',x,y,'k-')
    xlabel('[\itx]', 'FontSize',24)
    ylabel('Intensity', 'FontSize',24)
    set(gca, 'LineWidth',2, 'FontSize',20)
end
```

MATLAB program 8.17

```
function SSE = PixelSSE2(p)
% Calculates the SSE between the observed pixel intensity and the
% Gaussian function
% f = a1*(1/sqrt(2*pi*sig^2)*exp(-(x-mu1)^2/(2*sig^2))+...
% 1/sqrt(2*pi*sig^2)*exp(-(x-mu2)^2/(2*sig^2))) + a2

global data

% model parameters
mu1 = p(1); % mean of first peak
mu2 = p(2); % mean of second peak
```

```
sig = p(3);   % standard deviation
a1 = p(4);    % rescales the distribution
a2 = p(5);    % baseline

SSE = sum((data(:,2)-a1.* ...
    (1/sqrt(2*pi*sig.^2)*exp(-(data(:,1)-mu1).^2/(2*sig.^2))+...
1/sqrt(2*pi*sig.^2)*exp(-(data(:,1)-mu2).^2/(2*sig.^2)))-a2).^2);
```

Box 8.4A Kidney functioning in human leptin metabolism

The rate of leptin uptake by the kidney was modeled using Michaelis–Menten kinetics in Box 3.9A in Chapter 3:

$$R = -\frac{dS}{dt} = \frac{R_{max}S}{K_m + S}.$$

In Box 3.9A, the rate equation was linearized to obtain the Lineweaver–Burk equation. Linear least-squares regression was used to estimate the value of the two model parameters: R_{max} and K_m. Here, we will use nonlinear regression to obtain estimates of the model parameters.

The best-fit parameters obtained from the linear regression analysis are used as the initial guessed values for the nonlinear minimization problem:

$$K_m = 10.87, \qquad R_{max} = 1.732.$$

Our goal is to minimize the sum of the squared errors (residuals). Program 8.18 calculates the sum of the squared residuals as

$$SSE = \sum_{i=1}^{m} (y_i - \hat{y}_i)^2,$$

where

$$\hat{y}_i = \frac{R_{max}S}{K_m + S}.$$

We use `fminsearch` (or the simplex algorithm) to locate the optimal values of R_{max} and K_m.

```
>> p = fminsearch('SSEleptinuptake', [1.732, 10.87])
```

The `fminsearch` algorithm finds the best-fit values as $R_{max} = 4.194$ and $K_m = 26.960$. These numbers are very different from the numbers obtained using the Lineweaver–Burk analysis.

We minimize again, this time using the `fminunc` function (line search method). Since we do not provide a gradient function, the medium-scale optimization is chosen automatically by `fminunc`:

```
>> p2 = fminunc('SSEleptinuptake', [1.732, 10.87])
Warning: Gradient must be provided for trust-region method;
using line-search method instead.
Local minimum found.
Optimization completed because the size of the gradient is less than
the default value of the function tolerance.
p2 =
4.1938 26.9603
```

Both nonlinear search methods obtain the same minimum point of the SSE function. A plot of the data superimposed with the kinetic model of leptin uptake fitted by (1) the line search method and (2) the

Figure 8.18

Observed rates of renal leptin uptake and the Michaelis–Menten kinetic model with different sets of best-fit parameters

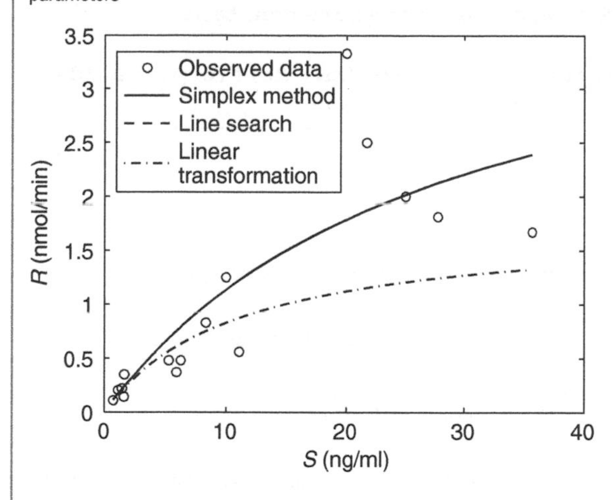

simplex method is shown in Figure 8.18. Included in this plot is the Michaelis–Menten model with best-fit parameters obtained by *linear* least-squares regression. From Figure 8.18, you can see that the fit of the model to the data when linear regression of the transformed equation is used is inferior to the model fitted using nonlinear regression. Transformation of the nonlinear equation distorts the actual scatter of the data points along the direction of the y-axis and produces a different set of best-fit model parameters. In this example, the variability in the data is large, which makes it difficult to obtain estimates of model parameters with good accuracy (small confidence intervals). For such experiments, it is necessary to obtain many data points so that the process is well characterized.

MATLAB program 8.18

```
function SSE = SSEleptinuptake(p)
% Data
% Plasma leptin concentration
S= [0.75; 1.18; 1.47; 1.61; 1.64; 5.26; 5.88; 6.25; 8.33; 10.0; 11.11; ...
   20.0; 21.74; 25.0; 27.77; 35.71];
% Renal Leptin Uptake
R = [0.11; 0.204; 0.22; 0.143; 0.35; 0.48; 0.37; 0.48; 0.83; 1.25; ...
    0.56; 3.33; 2.5; 2.0; 1.81; 1.67];

% model variables
Rmax = p(1);
Km = p(2);

SSE = sum((R - Rmax*S./(Km + S)).^2);
```

MATLAB program 8.19

```
function SSE = SSEpharmacokineticsofAZTm(p)
global data
```

```
% model variables
Im = p(1);
km = p(2); % drug absorption rate constant
ke = p(3); % drug elimination rate constant

SSE = sum((data(:,2) - Im*km/(km - ke)*(exp(-ke.*data(:,1))- ...
    exp(-km.*data(:,1)))).^2);
```

Box 8.1B Pharmacokinetic and toxicity studies of AZT

The pharmacokinetic model for the maternal compartment is given by

$$C_m(t) = I_m\left(\frac{k_m}{k_m - k_e}\right)\left(e^{-k_e t} - e^{-k_m t}\right), \tag{8.13}$$

where k_m is the absorption rate constant for the maternal compartment, and the pharmacokinetic model for the fetal compartment is given by

$$C_f(t) = I_f\left(1 - e^{-k_f t}\right), \tag{8.14}$$

where k_f is the absorption rate constant for the fetal compartment. Nonlinear regression is performed for each compartment using `fminsearch`. Program 8.19 is a function m-file that computes the SSE for the maternal compartment model. The data can be entered into another m-file that calls `fminsearch`, or can be typed into the Command Window.

```
% Data
global data
time = [0; 1; 3; 5; 10; 20; 40; 50; 60; 90; 120; 150; 180; 210; 240];
AZTconc = [0; 1.4; 4.1; 4.5; 3.5; 3.0; 2.75; 2.65; 2.4; 2.2; 2.15; 2.1;
  2.15; 1.8; 2.0];
data = [time, AZTconc];
```

Figure 8.19

Drug concentration profile in maternal compartment

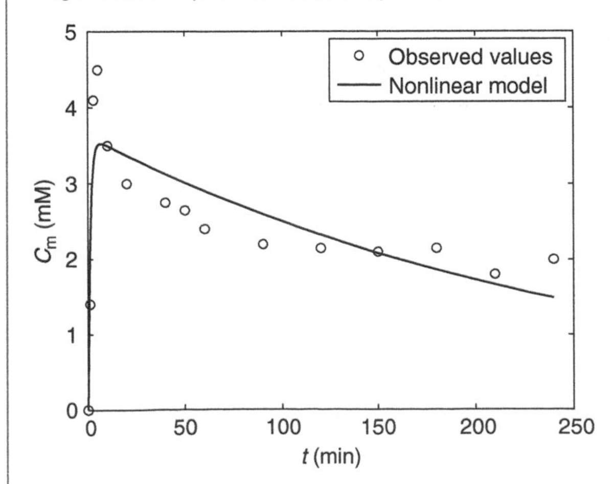

Figure 8.20

Drug concentration profile in fetal compartment

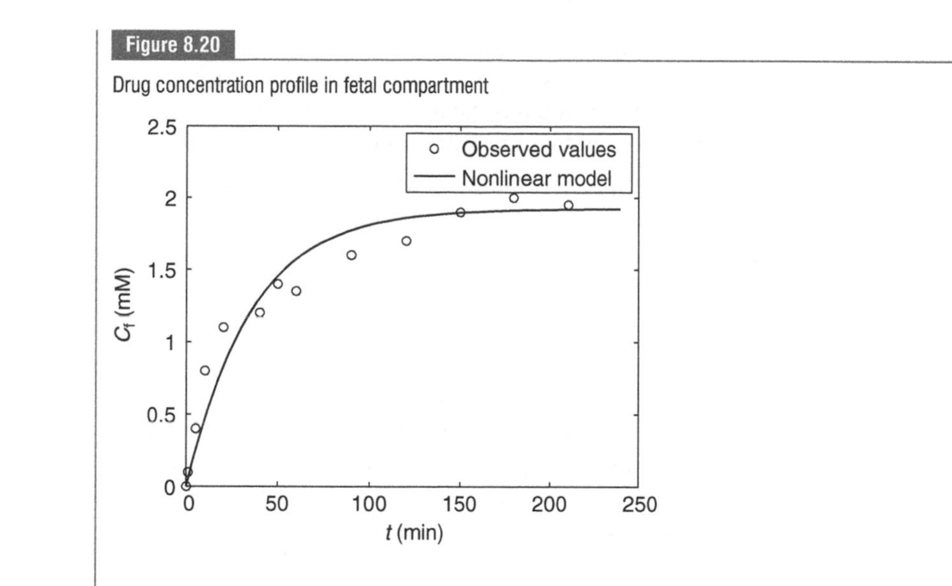

It is recommended that you create an m-file that loads the data into memory, calls the optimization function, and plots the model along with the data. This m-file (`pharmacokineticsofAZT.m`) is available on the book's website.

The values of the parameters for the maternal compartment model are determined to be

$I_m = 4.612$ mM,
$k_m = 0.833$/min,
$k_e = 0.004$/min.

The best-fit values for the fetal compartment model are

$I_f = 1.925$ mM,
$k_f = 0.028$/min.

The details of obtaining the best-fit parameters for the fetal compartment model are left as an exercise for the reader.

Figures 8.19 and 8.20 demonstrate the fit between the models (Equations (8.13) and (8.14)) and the data.

What if we had used the first-order absorption and elimination model for the fetal compartment? What values do you get for the model parameters?

Another way to analyze this problem is to couple the compartments such that

$$k_e = k_f = k_{mf},$$

where k_{mf} is the rate constant for transfer of drug from the maternal to the fetal compartment. In this case you would have two simultaneous nonlinear models to fit to the data. How would you perform the minimization of two sums of the squared errors?

8.4 Constrained nonlinear optimization

Most optimization problems encountered in practice have constraints imposed on the set of feasible solutions. As discussed in Section 8.1, a constraint placed on an objective function $f(\boldsymbol{x})$ can belong to one of two categories:

(1) equality constraints $h_j(\boldsymbol{x}) = 0,\ j = 1, 2, \ldots, m$, and
(2) inequality constraints $g_k(\boldsymbol{x}) \leq 0, k = 1, 2, \ldots, p$.

The objective function can be subjected to either one type of constraint or a mix of both types of constraints. Optimization problems can have up to m equality constraints, where $m < n$ and n is the number of variables to be optimized. An equality or inequality constraint can be further classified as follows:

(1) a linear function in two or more variables (up to n variables),
(2) a nonlinear function in two or more variables (up to n variables),
(3) an assignment function (e.g. $y = 10$) in the case of an equality constraint, or
(4) a bounding function (e.g. $y \geq 0, z \leq 2$) in the case of an inequality constraint.

In a one-dimensional optimization problem, only inequality constraints can be imposed. An equality constraint placed on the objective function would immediately solve the optimization problem. An inequality constraint(s), such as $x \geq 0$ and $x \leq c$, where c is some constant, specifies the search interval. Minimization (or maximization) of the objective function is performed to locate the optimum x^* within the interior of the search interval. The function is then evaluated at the computed optimum x^*, and compared to the value of the function at the interval endpoints. If the function has the least value at an endpoint (on the boundary of the variable space) then this point is chosen as the feasible solution; otherwise the interior optimum, x^*, is the answer.

For a multidimensional variable space, several methodologies have been devised to solve the constrained optimization problem. When only equality constraints have been imposed, and m, the number of equality constraints, is small, the best way to handle the optimization problem is to use the m constraints to reduce the number of variables in the objective function to $(n - m)$. Example 8.3 illustrates how an n-dimensional optimization problem with m equality constraints can be reduced to an unconstrained optimization problem in $(n - m)$ variables.

Example 8.3

Suppose we have a sheet of cardboard with surface area A. The entire cardboard surface is to be used to construct a box. Maximize the volume of the open box with a square base that can be made out of this cardboard.

Suppose each side of the base is of length x and the box height is h. The volume of the box is

$$V = x^2 h.$$

Our objective function to be maximized is

$$f(x, h) = x^2 h.$$

The surface area of the box is

$$A = x^2 + 4xh.$$

We have one nonlinear equality constraint,

$$h(x, h) = x^2 + 4xh - A = 0.$$

We can eliminate the variable h from the objective function by using the equality constraint to obtain an expression for h as a function of x. Substituting

$$h = \frac{A - x^2}{4x}$$

into the objective function, we obtain

$$f(x) = x^2 \frac{A - x^2}{4x}.$$

We have reduced the optimization problem in two variables subject to one equality constraint to an unconstrained optimization problem in one variable. To find the value of x that maximizes V, we take the derivative of $f(x)$ and equate this to zero. The resulting nonlinear equation is solved analytically to obtain $x = \sqrt{A/3}$, and $h = 0.5\sqrt{A/3}$. The maximum volume of the box[4] is calculated as $A^{3/2}/6\sqrt{3}$.

In general, optimization problems with $n > 2$ variables subjected to $m > 1$ equality constraints may not be reducible to an unconstrained optimization problem in $(n - m)$ variables. Unless the equality constraints are simple assignment functions or linear equations, the complexity of the nonlinear equality constraints makes it difficult to express m variables in terms of the $(n - m)$ variables.

A classical approach to solving a constrained problem is the **Lagrange multiplier method**. In a constrained problem, the variables of the objective function are not independent of each other, but are linked by the m equality constraints and some or all of the p inequality constraints. Therefore, the partial derivatives of the objective function with respect to each of the n variables cannot all be independently set to zero. Consider an objective function $f(x)$ in n variables that is subjected to m equality constraints $h_j(x)$. The following simultaneous conditions must be met at the constrained optimum point:

$$df = \sum_{i=1}^{n} \frac{\partial f}{\partial x_i} dx_i = 0$$

and

$$dh_j = \sum_{i=1}^{n} \frac{\partial h_j}{\partial x_i} dx_i = 0.$$

Let's consider the simplest case where $n = 2$ and $m = 1$. The trivial solution of the two simultaneous linear equations is $dx_1 = dx_2 = 0$, which is not meaningful. To ensure that a non-trivial solution exists, the determinant of the 2×2 coefficient matrix

$$\begin{vmatrix} \frac{\partial f}{\partial x_1} & \frac{\partial f}{\partial x_2} \\ \frac{\partial h}{\partial x_1} & \frac{\partial h}{\partial x_2} \end{vmatrix}$$

must equal zero. This is possible only if

$$\frac{\partial f}{\partial x_1} = -\lambda \frac{\partial h}{\partial x_1}$$

and

$$\frac{\partial f}{\partial x_2} = -\lambda \frac{\partial h}{\partial x_2},$$

where λ is called the **Lagrange multiplier**. To convert the constrained minimization problem into an unconstrained problem such that we can use the techniques discussed in Section 8.3 to optimize the variables, we construct an augmented objective function called the **Lagrangian function**:

$$L(x_1, x_2, \lambda) = f(x_1, x_2) + \lambda h(x_1, x_2).$$

If we take the partial derivatives of $L(x_1, x_2, \lambda)$ with respect to the two variables and the Lagrange multiplier, we obtain three simultaneous equations in three variables whose solution yields the optimum, as follows:

[4] Note that the solution of this problem is theoretical and not necessarily the most practical since a box of these dimensions may not in general be cut from a single sheet of cardboard without waste pieces.

$$\frac{\partial L}{\partial x_1} = \frac{\partial f}{\partial x_1} + \lambda \frac{\partial h}{\partial x_1} = 0,$$
$$\frac{\partial L}{\partial x_2} = \frac{\partial f}{\partial x_2} + \lambda \frac{\partial h}{\partial x_2} = 0,$$
$$\frac{\partial L}{\partial \lambda} = h(x_1, x_2) = 0.$$

For an objective function in n variables subjected to m equality constraints, the Lagrangian function is given by

$$L(\boldsymbol{x}, \lambda) = f(\boldsymbol{x}) + \sum_{j=1}^{m} \lambda_j h_j(\boldsymbol{x}), \tag{8.15}$$

where $\lambda = [\lambda_1, \lambda_2, \ldots, \lambda_m]$. The constrained optimization problem in n variables has been converted to an unconstrained problem in $n + m$ variables. For $\boldsymbol{x}^*$ to be a critical point of the constrained problem, it must satisfy the following $n + m$ conditions:

$$\frac{\partial L(\boldsymbol{x}, \lambda)}{\partial x_i} = \frac{\partial f(\boldsymbol{x}, \lambda)}{\partial x_i} + \sum_{j=1}^{m} \lambda_j \frac{\partial h_j(\boldsymbol{x})}{\partial x_i} = 0; \qquad i = 1, 2, \ldots, n,$$

and

$$\frac{\partial L(\boldsymbol{x}, \lambda)}{\partial \lambda_j} = h_j(\boldsymbol{x}) = 0; \qquad j = 1, 2, \ldots, m.$$

To find the optimum, one can solve either Equation (8.15) using unconstrained optimization techniques discussed in Section 8.3 or the set of $n + m$ equations given by the necessary conditions. In other words, while the optimized parameters of an unconstrained problem are obtained by solving the set of equations given by

$$\nabla f = 0,$$

for a constrained problem we must solve the augmented set of equations given by

$$\nabla^* L = 0,$$

where

$$\nabla^* = \begin{pmatrix} \nabla_x \\ \nabla_\lambda \end{pmatrix}.$$

How does one incorporate p inequality constraints into the optimization analysis? This is done by converting the inequality constraints into equality constraints using slack variables:

$$g_k(\boldsymbol{x}) + \sigma_k^2 = 0; \quad k = 1, 2, \ldots, p,$$

where σ_k^2 are called **slack variables**. The term is squared so that the slack variable term will always be either equal to or greater than zero. If a slack variable σ_k is zero at any point $\boldsymbol{x}$, then the kth inequality constraint is said to be **active** and $\boldsymbol{x}$ lies on the boundary specified by the constraint $g_k(\boldsymbol{x})$. If $\sigma_k \neq 0$, then the point $\boldsymbol{x}$ lies in the interior of the boundary and $g_k(\boldsymbol{x})$ is said to be an **inactive** inequality constraint. If p inequality constraints are imposed on the objective function, one must add p Lagrange multipliers to the set of variables to be optimized and p slack variables.

The Lagrangian function for a minimization problem subjected to m equality constraints and p inequality constraints is

$$L(\boldsymbol{x}, \lambda) = f(\boldsymbol{x}) + \sum_{j=1}^{m} \lambda_j h_j(\boldsymbol{x}) + \sum_{k=1}^{p} \lambda_{m+k}\left[g_k(\boldsymbol{x}) + \sigma_k^2\right]. \tag{8.16}$$

Equation (8.16) is the augmented objective function in $n + m + 2p$ variables with no constraints. At the optimum point,

$$h_j(\boldsymbol{x}^*) = 0,$$
$$g_k(\boldsymbol{x}^*) + \sigma_k^2 = 0,$$

and

$$L(\boldsymbol{x}^*, \lambda) = f(\boldsymbol{x}^*).$$

A solution is a feasible critical point if it satisfies the following conditions:

$$\frac{\partial L(\boldsymbol{x}, \boldsymbol{\lambda})}{\partial x_i} = 0; \qquad i = 1, 2, \ldots, n,$$
$$\frac{\partial L(\boldsymbol{x}, \boldsymbol{\lambda})}{\partial \lambda_j} = 0; \qquad j = 1, 2, \ldots, m + p,$$
$$\frac{\partial L(\boldsymbol{x}, \boldsymbol{\lambda})}{\partial \sigma_k} = 2\lambda_{m+k}\sigma_k = 0; \qquad k = 1, 2, \ldots, p.$$

Note that when $\sigma_k \neq 0$, the kth inequality constraint is inactive and $\lambda_{m+k} = 0$, since this inequality constraint does not affect the search for the optimum.

Example 8.4

Minimize the function

$$f(x_1, x_2) = (x_1 + 1)^2 - x_1 x_2, \tag{8.17}$$

subject to the constraint

$$g(x_1, x_2) = -x_1 - x_2 \leq 4.$$

The Lagrangian function for this constrained problem is given by

$$L\left(x_1, x_2, \lambda, \sigma^2\right) = (x_1 + 1)^2 - x_1 x_2 + \lambda\left(-x_1 - x_2 - 4 + \sigma^2\right).$$

A critical point of L satisfies the following conditions:

$$\frac{\partial L}{\partial x_1} = 2(x_1 + 1) - x_2 - \lambda = 0,$$
$$\frac{\partial L}{\partial x_2} = -x_1 - \lambda = 0,$$
$$\frac{\partial L}{\partial \lambda} = \left(-x_1 - x_2 - 4 + \sigma^2\right) = 0,$$
$$\frac{\partial L}{\partial \sigma} = 2\lambda\sigma = 0.$$

The last equation states that if the inequality constraint is inactive ($\sigma^2 > 0$) then $\lambda = 0$, and if the inequality constraint is active ($\sigma^2 = 0$) then $\lambda \neq 0$. In other words, both σ and λ cannot be non-zero simultaneously.

When $\lambda = 0$, only the first two conditions apply. In this case the critical point is given by

$$x_1 = 0; \ x_2 = 2.$$

Figure 8.21

Contour plot of the objective function with the inequality constraint shown by the dashed line. The feasible region is located above this line

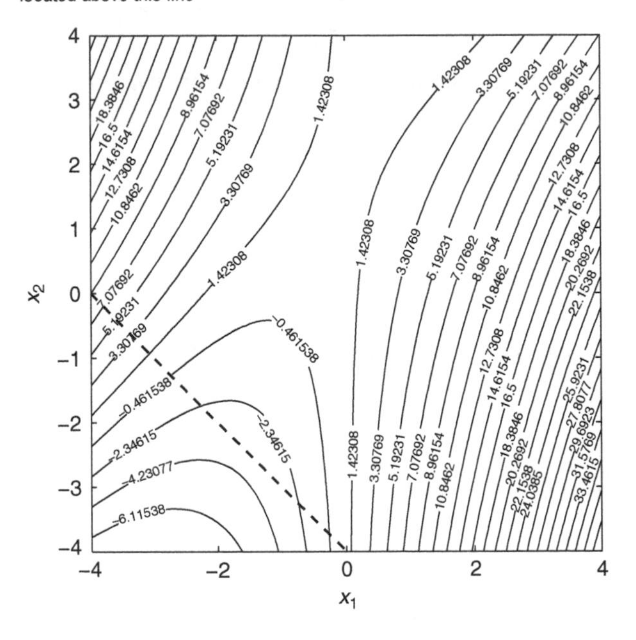

This is a saddle point, as is evident from the contour plot shown in Figure 8.21. The Hessian of the objective function at point (0, 2) is given by

$$H = \begin{bmatrix} 2 & -1 \\ -1 & 0 \end{bmatrix},$$

and $|H| = -1 < 0$. Therefore, the Hessian is not positive definite at this critical point, and (0, 2) is not a local minimum. Note that the Hessian is also not negative definite, indicating that this point is also not a local maximum.

When $\lambda \neq 0$, the first three conditions reduce to a system of three simultaneous linear equations whose solution is

$$x_1 = -\frac{3}{2};\ x_2 = -\frac{5}{2};\ \lambda = \frac{3}{2}.$$

Upon inspection of Figure 8.21, we find that this is indeed a local minimum of the constrained function.

Another technique used to solve constrained optimization problems is the **penalty function method**, in which a penalty function is constructed. The penalty function is a parameterized objective function that is not subject to any constraints. The penalty function can be minimized using any of the techniques discussed in Section 8.3. If the constrained optimization problem is to minimize $f(\mathbf{x})$, subject to m equality constraints $h_j(\mathbf{x}) = 0;\ j = 1, 2, \ldots, m$, then the penalty function is defined as[5]

$$P(\mathbf{x}) = f(\mathbf{x}) + r\sum_{j=1}^{m}\left(h_j(\mathbf{x})\right)^2, \tag{8.18}$$

[5] The penalty function can be constructed in several different ways.

where the parameter r is a positive number. The penalty function method is an iterative minimization scheme in which the value of r is gradually increased at each iteration, and Equation (8.18) is successively minimized to obtain a series of optimal points. The penalty function is first minimized using a small value of r, say 1.0. The optimum value of $\boldsymbol{x}$ obtained at this iteration is used as the starting guess value for the next iteration, in which the value of r is increased. The increase in r with each subsequent iteration shifts the optimum point of Equation (8.18) towards the optimum $\boldsymbol{x}^*$ of $f(\boldsymbol{x})$. When $r \to \infty$, $\| \boldsymbol{h}(\boldsymbol{x}) \| \to 0$, and the optimum of the penalty function approaches the optimum of the objective function. Inequality constraints can also be incorporated into a penalty function through the use of slack variables, as demonstrated earlier in the Lagrange multiplier method.

Using MATLAB

The `fmincon` function in Optimization Toolbox performs constrained minimization of a function of several variables. The constraints can be of the following types:

(1) linear inequality and equality constraints: $\boldsymbol{Ax} \leq \boldsymbol{b}$; $\boldsymbol{A}_{\text{eq}}\boldsymbol{x} = \boldsymbol{b}_{\text{eq}}$,
(2) nonlinear inequality and equality constraints: $\boldsymbol{Cx} \leq 0$; $\boldsymbol{C}_{\text{eq}}\boldsymbol{x} = 0$,
(3) bounds: $\boldsymbol{l} \leq \boldsymbol{x} \leq \boldsymbol{u}$, where $\boldsymbol{l}$ is the lower bound vector and $\boldsymbol{u}$ is the upper bound vector.

The syntax for `fmincon` is

```
x = fmincon(func, x0, A, b, Aeq, beq, lb, ub, confunc, options)
```

`func` returns the value of the objective function at any point $\boldsymbol{x}$; `x0` is the starting guess point, `A` is the linear inequality coefficient matrix, `b` is the linear inequality constants vector, `Aeq` is the linear equality coefficient matrix, and `beq` is the linear equality constants vector. If there are no linear inequality constraints, assign a null vector `[]` to `A` and `b`. Similarly, if no linear equality constraints exist, set `Aeq = []` and `beq = []`. Note that `lb` and `ub` are vectors that define the upper and lower bounds of the variables. If no bounds exist, use a null vector as a place holder for `lb` and `ub`. Also, `confunc` is a function that defines the nonlinear inequality and equality constraints. It calculates the values of $\boldsymbol{Cx}$ and $\boldsymbol{C}_{\text{eq}}\boldsymbol{x}$. The objective function `func` is minimized such that $\boldsymbol{Cx} \leq 0$ and $\boldsymbol{C}_{\text{eq}}\boldsymbol{x} = 0$. If nonlinear constraints do not exist for the problem, then the non-existent function should be represented by a null vector `[ ]`.

There are several different ways in which the function can be called. See `help fmincon` for more information on its syntax.

Example 8.5

Use `fmincon` to solve the constrained optimization problem described in Example 8.4.

We create two functions m-files, one to evaluate the objective function, and the other to evaluate the nonlinear constraint.

MATLAB program 8.20

```
function f = func(x)
f = (x(1) + 1)^2 - x(1)*x(2);
```

In MATLAB, we type the following. The output is also displayed below.

```
>> A = [-1 -1]; b = 4;
>> fmincon('func', [0,1], A, b)
```

```
Warning: Trust-region-reflective method does not currently solve this
  type
of problem, using active-set (line search) instead.
> In fmincon at 439
Local minimum found that satisfies the constraints.
Optimization completed because the objective function is non-
  decreasing in
feasible directions, to within the default value of the function
  tolerance,
and constraints were satisfied to within the default value of the
constraint tolerance.
<stopping criteria details>
Active inequalities (to within options.Tolcon = 1e-006) :
lower           upper           ineqlin         ineqnonlin
                                   1
ans =
   -1.5000 -2.5000"
```

What happens if you provide the initial guess point as (0, 2), which is the saddle point of the objective function?

8.5 Nonlinear error analysis

Nonlinear regression gives us best-fit values of the parameters of the nonlinear model. It is very important that we also assess our confidence in the computed values and future model predictions. The confidence interval for each parameter conveys how precise our estimate of the true value of the parameter is. If the confidence intervals are very wide, then the fitted model is of little use. The true parametric values could lie anywhere within large confidence intervals, giving us little confidence in the best-fit values of our model. In Chapter 3, we demonstrated how to obtain the covariance matrix Σ_x^2 of the linear model parameters $\boldsymbol{x}$:

$$\Sigma_x^2 = E\left((\boldsymbol{x} - \boldsymbol{\mu}_x)(\boldsymbol{x} - \boldsymbol{\mu}_x)^{\mathrm{T}}\right).$$

Along the diagonal of the covariance matrix are the variances associated with each best-fit parameter, which are used to estimate the confidence intervals.

Two popular techniques for estimating the error associated with the fitted nonlinear model parameters are the undersampling method and the bootstrap method. In both methods, several data sets are generated by reproducing values from the original data set. The mathematical model is then fitted to each of the data sets derived from the master data set. This produces several sets of parameters and therefore a distribution of values for each parameter. The distributions are used to calculate the variances and covariances for each pair of parameters.

(1) **The method of undersampling** This method is very useful when a large data set is available. Suppose N is the number of data points in the data set. If every mth point in the data set is picked and used to create a new, smaller data set, then this new data set will contain N/m points, and m distinct data sets can be derived this way. In this method each data point in the original data set is used only once to create the m derived sets. The model is fitted to each of the m data sets or "samples" to produce m estimates of the model parameters. The standard deviation of each parameter can be

calculated from their respective distributions. Alternatively, the covariance matrix can be estimated using the following formula:

$$\Sigma_{ij}^2 = s_{x_i x_j}^2 = \frac{1}{m-1}\left(\overline{x_i x_j} - \overline{x}_i \overline{x}_j\right). \tag{8.19}$$

Equation (8.19) is the estimated covariance between the ith and jth model parameter, where $i, j = 1, 2, \ldots, n$. Two limitations of this method are (i) N must be large and (ii) the nonlinear regression must be performed m times, which can be a time-consuming exercise.

(2) **The bootstrap method** This method is appropriate for large as well as moderately sized data sets. The N data points in the data set serve as a representative sample of the infinite number of observable data points. Another data set of size N can be created by performing random sampling with replacement (via a Monte Carlo simulation) from the N data points of the original data set. In this manner, m data sets are constructed from the original data set. Nonlinear regression is then performed m times (once for each newly constructed data set) to obtain m sets of the parameter values. The covariance between each pair of parameters is calculated using Equation (8.19). You may ask, How many N sized data sets should be produced from the original data? Note that m should be large enough to reduce sampling bias from the bootstrap procedure, which can influence the covariance estimate.

Box 8.4B Kidney functioning in human leptin metabolism

Estimate the variance in the model parameters K_m and R_{max} using the bootstrap method.

The data set has $N = 16$ points. We generate $m = 10$ data sets each with 16 data points, by replicating the data points in the original data set. Program 8.22 performs the bootstrap routine and calls `fminunc` to perform a nonlinear regression to obtain best-fit parameter values for all ten data sets. Note the use of the MATLAB function `cov(X)` in Program 8.21. This function calculates the covariance between the n variables that are represented by the n columns in the $m \times n$ matrix $\mathbf{X}$, where each row is a different observation.

A nonlinear optimization algorithm may not always converge to the correct optimal solution. Be aware that, occasionally, automated optimization procedures can produce strange results. This is why good starting guesses as well as good choices for the options available in minimizing a function are critical to the success of the technique. We present here the results of a successful run. The ten sets of best-fit values for the two parameters of the model were obtained as follows:

```
pdistribution =
4.4690      29.7062
7.1541      70.2818
4.3558      32.4130
3.2768      18.2124
4.4772      21.9166
8.1250      50.4744
3.2075      26.1971
4.1410      22.5261
7.0033      60.6923
3.5784      21.1609
```

The mean of each distribution is

$$\overline{R_{max}} = 4.979,$$

$$\overline{K}_m = 35.358.$$

The covariance matrix is computed as

$$\Sigma = \begin{bmatrix} 3.147 & 29.146 \\ 29.146 & 339.49 \end{bmatrix}.$$

From the covariance matrix we obtain the standard deviations of the distribution of each parameter as

$$s_{R_{\max}} = 1.774$$

and

$$s_{K_m} = 18.42.$$

MATLAB program 8.21

```
% Bootstrap routine is performed to estimate the covariance matrix
clear all
global S
global R

% Data
% Plasma leptin concentration
Sorig = [0.75; 1.18; 1.47; 1.61; 1.64; 5.26; 5.88; 6.25; 8.33; ...
    10.0;11.11; 20.0; 21.74; 25.0; 27.77; 35.71];
% Renal Leptin Uptake
Rorig = [0.11; 0.204; 0.22; 0.143; 0.35; 0.48; 0.37; 0.48; 0.83; ...
    1.25; 0.56; 3.33; 2.5; 2.0; 1.81; 1.67];

% Other variables
N = length(Sorig); % size of original data set
m = 10; % number of derived data sets of size N

% Bootstrap to generate m data sets
dataset(1:N,2,m) = 0;
for i = 1:m
    for j = 1:N
        randomno = floor(rand()*N + 1);
        dataset(j,:,i) = [Sorig(randomno) Rorig(randomno)];
    end
end

% Perform m nonlinear regressions
pdistribution(1:m, 1:2) = 0; % preallocating m sets of parameter
  values
for i = 1:m
S = dataset(:,1,i);
R = dataset(:,2,i);
p0 = [1.732, 10.87]; % initial guess value obtained from Box 3.9A
p1 = fminunc('SSEleptinuptake', p0);
pdistribution(i,:) = p1;
end

% Calculate the covariance matrix
sigma = cov(pdistribution);
muRmax = mean(pdistribution(:,1));
muKm = mean(pdistribution(:,2));
```

8.6 End of Chapter 8: key points to consider

(1) The optimal value of the parameters of a nonlinear model are obtained by performing **nonlinear least-squares regression** or minimization of

$$\mathrm{SSE} = \sum_{i=1}^{m} (y_i - \hat{y}_i)^2.$$

Least-squares regression provides the "best-fit" for the model parameters if the following assumptions are true:
(a) variability is dominant in the values of the dependent variable y (i.e. the independent variables are precisely known),
(b) the observations are normally distributed about the y-axis,
(c) all observations are independent of each other, and
(d) the data exhibit homoscedasticity.

(2) A point at which the gradient (or first derivative) of a function is zero is called a **critical point**. This point can either be a minimum, maximum, or saddle point of the function.

(3) Unconstrained minimization of an objective function in one variable can be carried out using Newton's method, successive parabolic interpolation, or the golden section search method.
(a) **Newton's method** has a second-order rate of convergence and is preferred when the first and second derivatives of the objective function are well-defined and easy to calculate.
(b) The **golden section search method** has a first-order rate of convergence and is preferred when the derivative of the function does not exist at one or more points near or at the minimum, or when the derivative is difficult to calculate.
(c) The **parabolic interpolation method** is a convenient method to use since it does not compute the function derivative and has superlinear convergence.

(4) For a single-variable optimization problem, the nature of the critical point can be tested using the second derivative test. For a multivariable problem, the nature of the critical point can be established by evaluating the matrix of second partial derivatives of the objective function, called the **Hessian matrix**. A critical point $\boldsymbol{x}^*$ is a local minimum if $\boldsymbol{H}(\boldsymbol{x}^*)$ is positive definite, and is a local maximum if $\boldsymbol{H}(\boldsymbol{x}^*)$ is negative definite.

(5) Unconstrained multidimensional optimization can be carried out using the method of steepest descent, Newton's multidimensional method, or the simplex method.
(a) The **steepest descent method** evaluates the gradient of the objective function to determine a search direction. An advantage of this method is that the starting guess values do not need to be close to the minimum to attain convergence.
(b) **Newton's multidimensional method** evaluates the gradient vector and the Hessian matrix of the function at each step. This method has a fast rate of convergence, but convergence is guaranteed only for good initial guess values.
(c) The **simplex method** performs function evaluations (but not derivative evaluations) at multiple points in the variable space. It draws a geometrical structure called a simplex that gradually shifts towards the minimum.

(6) The variables of an objective function $f(\boldsymbol{x})$ can have constraints placed on them. Constraints can be of two types:
(a) equality constraints $h_j(\boldsymbol{x}) = 0;\ \ j = 1, 2, \ldots, m$, or
(b) inequality constraints $g_k(\boldsymbol{x}) \leq 0;\ k = 1, 2, \ldots, p$.

(7) The **Lagrange multiplier method** performs constrained minimization by converting an optimization problem in n variables and $m + p$ constraints into an unconstrained problem in $n + m + 2p$ variables. The inequality constraints are converted to equality constraints with the use of **slack variables.**

(8) **Bootstrapping** and **undersampling** are used to estimate the covariance matrix, and thereby the confidence intervals, of model parameters whose values are obtained using nonlinear regression.

8.7 Problems

8.1. **Nonlinear regression of a pharmacokinetic model** The following equation describes the plasma drug concentration for a one-compartment pharmacokinetic model with first-order drug absorption:

$$C(t) = C_0[\exp(-k_1 t) - \exp(-k_2 t)].$$

Such a model has been successfully used to describe oral drug delivery. Given a set of experimental data (`tdat,Cdat`), write a MATLAB function that can be fed into the built-in function `fminsearch` to perform a nonlinear regression for the unknown model parameters C_0, k_1, and k_2. Try your function on the data set given in Table P8.1.

8.2. A classic test example for multidimensional minimization is the Rosenbrock banana function:

$$f(\boldsymbol{x}) = 100\left(x_2 - x_1^2\right)^2 + (1 - x_1)^2.$$

The traditional starting point is $(-1.2, 1)$. Start by defining a function m-file that takes in x_1 and x_2 as a single vector $\boldsymbol{x}$ and outputs the value of $f(x)$. Next, use the built-in MATLAB function `fminsearch` to solve for the minimum of the above equation, starting from the specified point of $(-1.2,1)$. Use the optional two-output-argument syntax of `fminsearch` so that the minimum x and the function value are reported. Use your results to answer the following "riddle": when is a *root-finding* problem the same as a *minimization* problem?

8.3. **Leukocyte migration in a capillary** Many times in (biomedical) engineering, after an external stimulus or signal is applied, there is a *time delay* where nothing happens ... and then a linear response is observed. One example might be a leukocyte (white blood cell) that is squeezing through a small capillary. A positive pressure is applied at one end of the capillary, and, following a short delay, the leukocyte begins to move slowly through the capillary at a constant velocity.

Table P8.1. *Time course of drug concentration in plasma*

Time (h)	0	1	2	3	4	5	6	7	8	9	10	11	12
Plasma conc. (mM)	0.004	0.020	0.028	0.037	0.040	0.036	0.028	0.021	0.014	0.009	0.007	0.004	0.002

Table P8.2. *Time-dependent position of a leukocyte in a capillary subjected to a pressure gradient*

Time, t(s)	Position, x(μm)
0.0	0.981
0.3	1.01
0.6	1.12
0.9	1.13
1.2	2.98
1.5	2.58
1.8	3.10
2.1	5.89

Figure P8.1

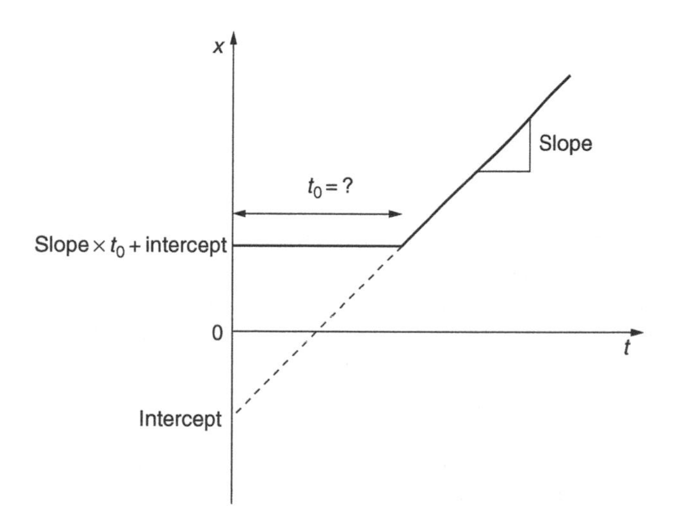

Table P8.2 provides an example of such a data set, which exhibits some random error inherent in experimental measurements.

If you plot the data in Table 8.3 you will see that the position of the cell is relatively constant around $x = 1$ μm until $t \sim 1$ s when it begins to move in a linearly time-dependent fashion. We want to model this data set as a line after the (unknown) time delay. Thus, our theoretical curve is a *piecewise* linear function (see Figure P8.1).

To regress our data correctly to such a model we must use *nonlinear* regression techniques. Here's how it's done.

(a) Create a function that takes in the three regression parameters (`slope, intercept, t0`) as a vector called `params`, and outputs the sum of the squared residuals (`SSE`) between the data and the model. The first two lines of your function should look like:

```
function SSE = piecewisefunc(params)
global t x
```

Note that we are passing the data to the function as two global variables, so the same `global` command should appear at the beginning of your main program. Next, your function must calculate the SSE (sum of the squared error). Build a `for` loop that executes once for each element of t. Inside your `for` loop, test to see whether `t(i)` is greater than `t0` (the time delay). If it is, then the SSE is calculated in the usual way as:

```
SSE = SSE + (x(i) - (slope*t(i) + intercept))^2;
```

If `t(i)` is lower than `t0`, then the SSE is defined as the squared difference between the data and the constant value of the model before `t0`. This looks like:

```
SSE = SSE + (x(i) - (slope*t0 + intercept))^2;
```

where `slope*t0 + intercept` is just the constant value of x before the linear regime of behavior. Remember to initialize SSE at zero before the `for` loop.

(b) In your main program, you must enter the data, and declare the data as global so that your function can access these values. Then, you may use the `fminsearch` function to perform the nonlinear regression and determine the optimum values of the `params` vector. This is accomplished with the statement:

```
paramsopt = fminsearch('piecewisefunc', ...
            [slopeguess,interceptguess,t0guess])
```

Note that to obtain the correct solution we must feed a good initial guess to the program. You can come up with suitable guesses by looking at your plot of the data.

(c) In your main program, output the optimum parameter values to the Command Window. Also, prepare a plot comparing the data and best-fit model.

8.4. **Optimizing drug dosage** A patient is being treated for a certain disease simultaneously with two different drugs: x units of the first drug and y units of the second drug are administered to the patient. The body's response R to the drugs is measured and is modeled as a function of x and y as

$$f(x, y) = x^3 y(15 - x - y).$$

Find the values of x and y that maximize R using the method of steepest *ascent*. Since you are *maximizing* an objective function, you will want solve the one-dimensional problem given by

$$\max_{\alpha} f(\boldsymbol{x}_0 + \alpha \nabla f(\boldsymbol{x}_0))$$

to yield the iterative algorithm

$$\boldsymbol{x}^{(k+1)} = \boldsymbol{x}^{(k)} + \alpha_{\text{opt}} \nabla f\left(\boldsymbol{x}^{(k)}\right).$$

Carry out the line search (i.e. find α_{opt}) using successive parabolic interpolation. To maximize $f(x, y)$, at each iteration you will need to minimize

$$h(\alpha) = -f\left(\boldsymbol{x}^{(k)} + \alpha \nabla f\left(\boldsymbol{x}^{(k)}\right)\right).$$

Use an initial guess of $\boldsymbol{x}_0 = (x_0, y_0) = (8, 2)$.

(a) To ensure that the minimization is successful, you will need to choose the maximum allowable value of the step size α carefully. When the magnitude of the gradient is large, even a small step size can move the solution very far from the original starting point. To find an acceptable interval for α, compute $h(\alpha)$ for

Figure P8.2

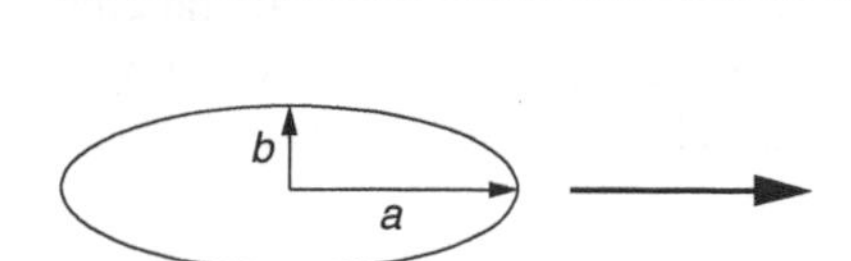

a few different values of α. Choose α_{max} such that zero and α_{max} bracket the minimum along the direction of the search. Perform this exercise using the starting point (8, 2) and use your chosen value of α_{max} for all iterations.

(b) Write a function m-file that performs the multidimensional minimization using steepest ascent. You can use the `steepestparabolicinterpolation` function listed in Program 8.10 to perform the one-dimensional minimization.

(c) Perform three iterations of steepest ascent (`maxloops = 3`). Use the following tolerances: tolerance for $\alpha = 0.0001$; tolerance for $\| \nabla f \| = 0.01$.

8.5. **Friction coefficient of an ellipsoid moving through a fluid** Many microorganisms, such as bacteria, have an elongated, ellipsoidal shape. The friction coefficient ξ relates the force on a particle and its velocity when moving through a viscous fluid – thus, it is an important parameter in centrifugation assays. A lower friction coefficient indicates that a particle moves "easier" through a fluid. The friction coefficient for an ellipsoid moving in the direction of its long axis, as shown in Figure P8.2, is given by

$$\xi = \frac{4\pi\mu a}{\ln(2a/b) - 1/2},$$

where μ is the fluid viscosity (for water, this is equal to 1 μg/(s μm)) (Saltzman, 2001). For a short axis of $b = 1$ μm, find the value of a that corresponds to a *minimum* in the friction coefficient in water. Start with an initial guess of $a = 1.5$ μm, and show the first three iterations.

(a) Use the golden section search method to search for the optimum.

(b) Check your answer by using Newton's method to find the optimum.

8.6. **Nonlinear regression of chemical reaction data** An experimenter examines the reaction kinetics of an NaOH:phenolphthalein reaction. The extent of reaction is measured using a colorimeter, where the absorbance is proportional to the concentration of the phenolphthalein. The purpose is to measure the reaction rate constant k at different temperatures and NaOH concentrations to learn about the reaction mechanism. In the experiment, the absorbances A_i are measured at times t_i. The absorbance should fit an exponential rate law:

$$A = C_1 + C_2 e^{-kt},$$

where C_1, C_2, and k are unknown constants.

(a) Write a MATLAB function that can be fed into the built-in function `fminsearch` to perform a nonlinear regression for the unknown constants.

(b) Now suppose that you obtain a solution, but you suspect that the `fminsearch` routine may be getting stuck in a local minimum that is not the most accurate solution. How would you test this?

8.7. **Nonlinear regression of microfluidic velocity data** In the entrance region of a microfluidic channel, some distance is required for the flow to adjust from upstream conditions to the fully developed flow pattern. This distance depends on the flow conditions upstream and on the Reynolds number ($\mathrm{Re} = Dv\rho/\mu$), which is a dimensionless group that characterizes the relative importance of inertial fluid forces to viscous fluid forces. For a uniform velocity profile at the channel entrance, the computed length in laminar flow (entrance length relative to hydraulic diameter = $L_{\mathrm{ent}}/D_{\mathrm{h}}$) required for the centerline velocity to reach 99% of its fully developed value is given by

$$\frac{L_{\mathrm{ent}}}{D_h} = a \cdot \exp(-b \cdot \mathrm{Re}) + c \cdot \mathrm{Re} + d.$$

Suppose that you have used micro-particle image velocimetry (μ-PIV) to measure this entrance length as a function of Reynolds number in your microfluidic system, and would like to perform a nonlinear regression on the data to determine the four model parameters a, b, c, and d. You would like to perform nonlinear regression using the *method of steepest descent*.

(a) What is the objective function to be minimized?
(b) Find the gradient vector.
(c) Describe the minimization procedure you will use to find the optimal values of a, b, c, and d.

8.8. **Using hemoglobin as a blood substitute: hemoglobin–oxygen binding** Find the best-fit values for the Hill parameters P_{50} and n in Equation 2.27.

(a) Use the `fminsearch` function to find optimal values that correspond to the least-squared error.
(b) Compare your optimal values with those obtained using linear regression of the transformed equation.
(c) Use the bootstrap method to calculate the covariance matrix of the model parameters. Create ten data sets from the original data set. Calculate the standard deviation for each parameter.

References

Boal, J. H., Plessinger, M. A., van den Reydt, C., and Miller, R. K. (1997) Pharmacokinetic and Toxicity Studies of AZT (Zidovudine) Following Perfusion of Human Term Placenta for 14 Hours. *Toxicol. Appl. Pharmacol.*, **143**, 13–21.

Edgar, T. F., and Himmelblau, D. M. (1988) *Optimization of Chemical Processes* (New York: McGraw-Hill).

Fournier, R. L. (2007) *Basic Transport Phenomena in Biomedical Engineering* (New York: Taylor & Francis).

Nauman, B. E. (2002) *Chemical Reactor Design, Optimization, and Scaleup* (New York: McGraw-Hill).

Rao, S. S. (2002) *Applied Numerical Methods for Engineers and Scientists* (Upper Saddle River, NJ: Prentice Hall).

Saltzman, W. M. (2001) *Drug Delivery: Engineering Principles for Drug Therapy* (New York: Oxford University Press).

9 Basic algorithms of bioinformatics

9.1 Introduction

The primary goal of the field of **bioinformatics** is to examine the biologically important information that is stored, used, and transferred by living things, and how this information acts to control the chemical environment within living organisms. This work has led to technological successes such as the rapid development of potent HIV-1 proteinase inhibitors, the development of new hybrid seeds or genetic variations for improved agriculture, and even to new understanding of pre-historical patterns of human migration. Before discussing useful algorithms for bioinformatic analysis, we must first cover some preliminary concepts that will allow us to speak a common language.

Deoxyribonucleic acid (DNA) is the genetic material that is passed down from parent to offspring. DNA is stored in the nuclei of cells and forms a complete set of instructions for the growth, development, and functioning of a living organism. As a macromolecule, DNA can contain a vast amount of information through a specific sequence of bases: guanine (G), adenine (A), thymine (T), and cytosine (C). Each base is attached to a phosphate group and a deoxyribose sugar to form a **nucleotide** unit. The four different nucleotides are then strung into long polynucleotide chains, which comprise genes that are thousands of bases long, and ultimately into chromosomes. In humans, the molecule of DNA in a single chromosome ranges from 50×10^6 nucleotide pairs in the smallest chromosome, up to 250×10^6 in the largest. If stretched end-to-end, these DNA chains would extend 1.7 cm to 8.5 cm, respectively! The directional orientation of the DNA sequence has great biological significance in the duplication and translation processes. DNA sequences proceed from the 5′ carbon phosphate group end to the 3′ carbon of the deoxyribose sugar end.

As presented in a landmark 1953 paper, Watson and Crick discovered, from X-ray crystallographic data combined with theoretical model building, that the four bases of DNA undergo complementary hydrogen binding with the appropriate base on a neighboring strand, which causes complementary sequences of DNA to assemble into a double helical structure. The four bases, i.e. the basic building blocks of DNA, are classified as either purines (A and G) or pyrimidines (T and C), which comprise either a double or single ring structure (respectively) of carbon, nitrogen, hydrogen, and oxygen. A purine can form hydrogen bonds only with its complementary pyrimidine base pair. Specifically, when in the proper orientation and separated by a distance of 11 Å, guanine (G) will share three hydrogen bonds with cytosine (C), whereas adenine (A) will share two hydrogen bonds with thymine (T). This phenomenon is called **complementary base pairing**. The two paired strands that form the DNA double helix run anti-parallel to each other, i.e. the 5′ end of one strand faces the 3′ end of the complementary strand. One DNA chain can serve as a template for the assembly of a second, complementary DNA strand, which is precisely how the genome is replicated prior to cell division. Complementary DNA will spontaneously pair up (or "**hybridize**") at room or body temperature, and can be caused to separate or "melt" at elevated temperature, which was later exploited in the widely used polymerase chain reaction (PCR) method of DNA amplification in the laboratory.

So how does the genome, encoded in the sequence of DNA, become the thousands of proteins within the cell that interact and participate in chemical reactions? It has been determined that each gene corresponds in general to one protein. The process by which the genes encoded in DNA are made into the myriad of proteins found in the cell is referred to as the "**central dogma**" of molecular biology. Information contained in DNA is "**transcribed**" into the single-stranded polynucleotide ribonucleic acid (RNA) by RNA polymerases. The nucleotide sequences in DNA are copied directly into RNA, except for thymine (T), which is substituted with uracil (U, a pyrimidine) in the new RNA molecule. The RNA gene sequences are then "**translated**" into the amino acid sequences of different proteins by protein/RNA complexes known as ribosomes. Groups of three nucleotides, each called a "**codon**" ($N = 4^3 = 64$ combinations), encode the 20 different amino acids that are the building blocks of proteins. Table 9.1 shows the 64 codons and the corresponding amino acid or start/stop signal.

The three stop codons, UAA, UAG, and UGA, which instruct ribosomes to terminate the translation, have been given the names "ochre," "amber," and "opal," respectively. The AUG codon, encoding for methionine, represents the translation start signal. Not all of the nucleotides in DNA encode for proteins, however. Large stretches of the genome, called **introns**, are **spliced** out of the sequence by enzyme complexes which recognize the proper splicing signals, and the remaining **exons** are joined together to form the protein-encoding portions. Major sequence repositories are curated by the National Center for Biotechnology Information (NCBI) in the United States, the European Molecular Biology Laboratory (EMBL), and the DNA Data Bank of Japan.

9.2 Sequence alignment and database searches

The simplest method for identifying regions of similarity between two sequences is to produce a graphical **dot plot**. A two-dimensional graph is generated, and a dot is placed at each position where the two compared sequences are identical. The word size can be specified, as in MATLAB program 9.1, to reduce the noisiness produced by many very short (length = 1–2) regions of similarity. Identity runs along the main diagonal, and common subsequences including possible **translocations** are seen as shorter off-diagonal lines. **Inversions**, where a region of the gene runs in the opposite direction, appear as lines perpendicular to the main diagonal, while **deletions** appear as interruptions in the lines. Figure 9.1 shows the sequence of human P-selectin, an adhesion protein important in inflammation, compared against itself. In the P-selectin secondary structure, it is known that nine consensus repeat domains[1] exist, and these can be seen as

[1] A consensus repeat domain is a sequence of amino acids that occurs with high frequency in a polypeptide.

Table 9.1. *Genetic code for translation from mRNA to amino acids*

	2nd base			
	U	C	A	G
1st base				
U	UUU: phenylalanine (Phe/F)	UCU: serine (Ser/S)	UAU: tyrosine (Tyr/Y)	UGU: cysteine (Cys/C)
	UUC: phenylalanine (Phe/F)	UCC: serine (Ser/S)	UAC: tyrosine (Tyr/Y)	UGC: cysteine (Cys/C)
	UUA: leucine (Leu/L)	UCA: serine (Ser/S)	UAA: stop (“ochre”)	UGA: stop (“opal”)
	UUG: leucine (Leu/L)	UCG: serine (Ser/S)	UAG: stop (“amber”)	UGG: tryptophan (Trp/W)
C	CUU: leucine (Leu/L)	CCU: proline (Pro/P)	CAU: histidine (His/H)	CGU: arginine (Arg/R)
	CUC: leucine (Leu/L)	CCC: proline (Pro/P)	CAC: histidine (His/H)	CGC: arginine (Arg/R)
	CUA: leucine (Leu/L)	CCA: proline (Pro/P)	CAA: glutamine (Gln/Q)	CGA: arginine (Arg/R)
	CUG: leucine (Leu/L)	CCG: proline (Pro/P)	CAG: glutamine (Gln/Q)	CGG: arginine (Arg/R)
A	AUU: isoleucine (Ile/I)	ACU: threonine (Thr/T)	AAU: asparagine (Asn/N)	AGU: serine (Ser/S)
	AUC: isoleucine (Ile/I)	ACC: threonine (Thr/T)	AAC: asparagine (Asn/N)	AGC: serine (Ser/S)
	AUA: isoleucine (Ile/I)	ACA: threonine (Thr/T)	AAA: lysine (Lys/K)	AGA: arginine (Arg/R)
	AUG: start, methionine (Met/M)	ACG: threonine (Thr/T)	AAG: lysine (Lys/K)	AGG: arginine (Arg/R)
G	GUU: valine (Val/V)	GCU: alanine (Ala/A)	GAU: aspartic acid (Asp/D)	GGU: glycine (Gly/G)
	GUC: valine (Val/V)	GCC: alanine (Ala/A)	GAC: aspartic acid (Asp/D)	GGC: glycine (Gly/G)
	GUA: valine (Val/V)	GCA: alanine (Ala/A)	GAA: glutamic acid (Glu/E)	GGA: glycine (Gly/G)
	GUG: valine (Val/V)	GCG: alanine (Ala/A)	GAG: glutamic acid (Glu/E)	GGG: glycine (Gly/G)

Box 9.1 Sequence formats

When retrieving genetic sequences from online databases for analysis using commonly available web tools, it is important to work in specific sequence formats. One of the most common genetic sequence formats is called Fasta. Let's find the amino acid sequence for ADAM metallopeptidase domain 17 (ADAM17), a transmembrane protein responsible for the rapid cleavage of L-selectin from the leukocyte surface (see Problem 1.14 for a brief introduction to selectin-mediated binding of flowing cells to the vascular endothelium), in the Fasta format. First we go to the Entrez cross-database search page at the NCBI website (www.ncbi.nlm.nih.gov/sites/gquery), a powerful starting point for accessing the wealth of genomic data available online. By clicking on the "Protein" link, we will restrict our search to the protein sequence database. Entering "human ADAM17" in the search window produces 71 hits, with the first being accession number AAI46659 for ADAM17 protein [Homo sapiens]. Clicking on the AAI46659 link takes us to a page containing information on this entry, including the journal citation in which the sequence was first published. Going to the "Display" pulldown menu near the top of the page, we select the "FASTA" option and the following text is displayed on the screen:

```
>gi|148922164|gb|AAI46659.1| ADAM17 protein [Homo sapiens]
MRQSLLFLTSVVPFVLAPRPPDDPGFGPHQRLEKLDSLLSDYDILSLSNI
QQHSVRKRDLQTSTHVETLLTFSALKRHFKLYLTSSTERFSQNF
KVVVVDGKNESEYTVKWQDFFTGHVVGEPDSRVLAHIRDDDVIIRI
NTDGAEYNIEPLWRFVNDTKDKRMLVYKSEDIKNVSRLQSPKVCGYLKVDN
EELLPKGLVDREPPEELVHRVKRRADPDPMKNTCKLLVVADHRF
YRYMGRGEESTTTNYLIHTDRAN
```

This is the amino acid sequence of human ADAM17, displayed in Fasta format. It is composed of a greater than symbol, no space, then a description line, followed by a carriage return and the sequence in single-letter code. The sequence can be cut and pasted from the browser window, or output to a text file. We may also choose to output a specified range of amino acids rather than the entire sequence. If we instead select the "GenPept" option from the Display pulldown menu (called "GenBank" format for DNA sequences) then we see the original default listing for this accession number. At the bottom of the screen, after the citation information, is the same ADAM17 sequence in the GenPept/GenBank format:

```
ORIGIN
        1 mrqsllflts vvpfvlaprp pddpgfgphq rlekldslls dydilslsni
qqhsvrkrdl
       61 qtsthvetll tfsalkrhfk lyltssterf sqnfkvvvvd gkneseytvk
wqdfftghvv
      121 gepdsrvlah irdddviiri ntdgaeynie plwrfvndtk dkrmlvykse
diknvsrlqs
      181 pkvcgylkvd neellpkglv dreppeelvh rvkrradpdp mkntckllvv
adhrfyrymg
      241 rgeestttny lihtdran
//
```

Note the formatting differences between the GenBank and Fasta sequence formats. Other sequence formats used for various applications include Raw, DDBJ, Ensembl, ASN.1, Graphics, and XML. Some of these are output options on the NCBI website, and programs exist (READSEQ; SEQIO) to convert from one sequence format to another.

off-diagonal parallel lines in Figure 9.1(b) once the spurious dots are filtered out by specifying a word size of 3. Website tools are available (European Molecular Biology Open Software Suite, EMBOSS) to make more sophisticated dot plot comparisons.

Figure 9.1

Dot plot output from MATLAB program 9.1. The amino acid sequence of human adhesion protein P-selectin was obtained from the NCBI Entrez Protein database (www.ncbi.nlm.nih.gov/sites/entrez?db=protein) and compared against itself while recording single residue matches (a) or two out of three matches (b).

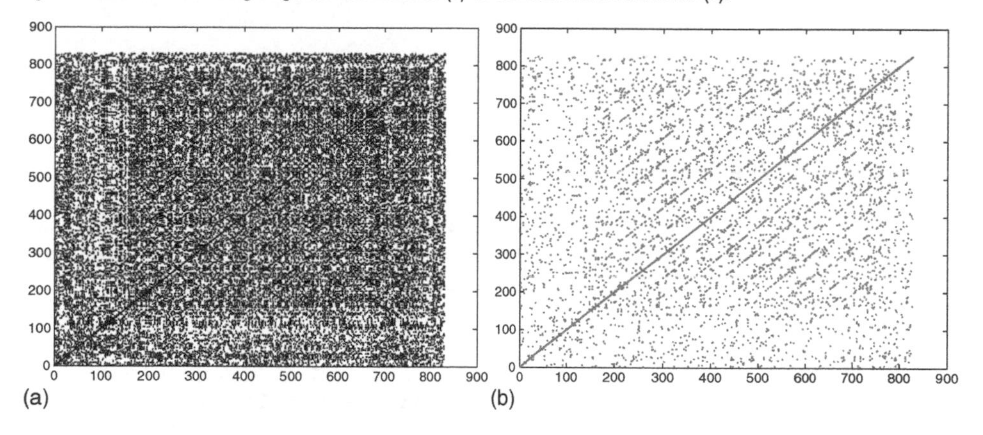

MATLAB program 9.1

```
% Pseldotplot.m

% This m-file compares a protein or DNA sequence with itself, and produces a
% dot plot to display sequence similarities.

% P-selectin protein amino acid sequence pasted directly into m-file

S='MANCQIAILYQRFQRVVFGISQLLCFSALISELTNQKEVAAWTYHYSTKAYSWNISRKYCQN
   RYTDLVAIQNKNEIDYLNKVLPYYSSYYWIGIRKNNKTWTWVGTKKALTNEAENWADNEPNNK
   RNNEDCVEIYIKSPSAPGKWNDEHCLKKKHALCYTASCQDMSCSKQGECLETIGNYTCSCYPG
   FYGPECEYVRECGELELPQHVLMNCSHPLGNFSFNSQCSFHCTDGYQVNGPSKLECLASGIWT
   NKPPQCLAAQCPPLKIPERGNMTCLHSAKAFQHQSSCSFSCEEGFALVGPEVVQCTASGVWTA
   PAPVCKAVQCQHLEAPSEGTMDCVHPLTAFAYGSSCKFECQPGYRVRGLDMLRCIDSGHWSAP
   LPTCEAISCEPLESPVHGSMDCSPSLRAFQYDTNCSFRCAEGFMLRGADIVRCDNLGQWTAPA
   PVCQALQCQDLPVPNEARVNCSHPFGAFRYQSVCSFTCNEGLLLVGASVLQCLATGNWNSVPP
   ECQAIPCTPLLSPQNGTMTCVQPLGSSSYKSTCQFICDEGYSLSGPERLDCTRSGRWTDSPPM
   CEAIKCPELFAPEQGSLDCSDTRGEFNVGSTCHFSCDNGFKLEGPNNVECTTSGRWSATPPTC
   KGIASLPTPGVQCPALTTPGQGTMYCRHHPGTFGFNTTCYFGCNAGFTLIGDSTLSCRPSGQW
   TAVTPACRAVKCSELHVNKPIAMNCSNLWGNFSYGSICSFHCLEGQLLNGSAQTACQENGHWS
   TTVPTCQAGPLTIQEALTYFGGAVASTIGLIMGGTLLALLRKRFRQKDDGKCPLNPHSHLGTY
   GVFTNAAFDPSP';

n=length(S);

figure(1)

% For first figure, individual residues in the sequence are compared
% against the same sequence. If the residue is a "match", then a dot is
% plotted into that i, j position in figure 1.
```

```
for i=1:n
    for j=1:n
        if S(i)==S(j)
            plot(i,j,'.','MarkerSize',4)
            hold on
        end
    end
end

figure(2)

% For the second figure, a more stringent matching condition is imposed. To
% produce a dot at position (i, j), an amino acid must match AND one of the
% next two residues must match as well.

for i=1:n-3
    for j=1:n-3
        if S(i)==S(j) & (S(i+1)==S(j+1) | S(i+2)==S(j+2))
            plot(i,j,'.','MarkerSize',4)
            hold on
        end
    end
end
```

The **alignment** of two or more sequences can reveal evolutionary relationships and suggest functions for undescribed genes. For sequences that share a common ancestor, there are three mechanisms that can account for a character difference at any position: (1) a **mutation** replaces one letter for another; (2) an **insertion** adds one or more letters to one of the two sequences; or (3) a **deletion** removes one or more letters from one of the sequences. In sequence alignment algorithms, insertions and deletions are dealt with by introducing a **gap** in the shorter sequence. To determine a numerically "optimal" alignment between two sequences, one must first define a match score to reward an aligned pair of identical residues, a mismatch score to penalize an aligned pair of non-identical residues, and a gap penalty to weigh the placement of gaps at potential insertion/deletion sites. Box 9.2 shows a simple example of the bookkeeping involved in scoring proposed sequence alignments. The scoring of gaps can be further refined by the introduction of **origination penalties** that penalize the initiation of a gap, and a **length penalty** which grows as the length of a gap increases. The proper weighting of these two scores is used to favor a single multi-residue insertion/deletion event over multiple single-residue gaps as more likely from an evolutionary point of view.

Alignment algorithms need not weigh each potential residue mismatch as equally likely. **Scoring matrices** can be used to incorporate the relative likelihood of different base or amino acid substitutions based on mutation data. For instance, the transition transversion matrix (Table 9.2) penalizes an A↔G or C↔T substitution less severely because **transitions** of one purine to another purine or one pyrimadine to another pyrimadine (respectively) is deemed more likely than **transversions** in which the basic chemical group is altered.

While common sequence alignment algorithms (e.g. BLAST) for nucleotides utilize simple scoring matrices such as that in Table 9.2, amino acid sequence alignments utilize more elaborate scoring matrices based on the relative frequency of substitution rates observed experimentally. For instance, in the PAM (Point/percent Accepted Mutation) matrix, one PAM unit represents an average of 1% change in all amino acid positions.

Table 9.2. *Transition transversion matrix*

	A	C	G	T
A	1	–5	–1	–5
C	–5	1	–5	–1
G	–1	–5	1	–5
T	–5	–1	–5	1

Box 9.2 Short sequence alignment

The Kozak consensus sequence is a sequence which occurs on eukaryotic mRNA, consisting of the consensus GCC, followed by a purine (adenine or guanine), three bases upstream of the start codon AUG, and then followed by another G. This sequence plays a major role in the initiation of the translation process. Let's now consider potential alignments of the Kozak-like sequences in fluit fly, CAAAATG, and the corresponding sequence in terrestrial plants, AACAATGGC.[2] Clearly, the best possible alignment with no insertion/deletion gaps is:

```
CAAAATG
AACAATGGC     (alignment 1)
```

which shows five nucleotide matches, one C–A mismatch and one A–C mismatch, for a total of two mismatches. If we place one single-nucleotide gap in the fluit fly sequence to represent a "C" insertion in the plant sequence, we get

```
CAA-AATG
 AACAATGGC     (alignment 2)
```

which shows six matches, no mismatches, and one gap of length 1. Alternatively, if we introduce a gap of length 2 in the second sequence, we obtain the following alignment:

```
  CAAAATG
AACA--ATGGC     (alignment 3)
```

which results in five matches, no mismatches, and one gap of length 2. Note that we could have placed this gap one position to the left or right and we would have obtained the same score. For a match score of +1, a mismatch score of –1, and a gap penalty of –1, then the three proposed alignments produce total scores of 3, 5, and 3, respectively. Thus, for this scoring system we conclude that alignment 2 represents the optimal alignment of these two sequences.

The PAM1 matrix was generated from 71 protein sequence groups with at least 85% identity (similarity) among the sequences. Mutation probabilities are converted to a log odds scoring matrix (logarithm of the ratio of the likelihoods), and rescaled into integer values. Similarly, the PAM250 matrix represents 250 substitutions in 100 amino acids. Another commonly used scoring matrix is BLOSUM, or "BLOCK Substitution Matrix." BLOCKS is a database of about 3000 "blocks" or short, continuous multiple alignments. The BLOSUM matrix is generated by again calculating the substitution frequency (relative frequency of a mismatch) and converting this to a log odds score. The BLOSUM62 matrix (Table 9.3) has 62% similarity, meaning that 62% of residues

[2] Note that we are considering the DNA equivalent of the mRNA sequence, thus the unmethylated version T (thymine) appears instead of U (uracil).

Table 9.3. *BLOSUM62 scoring matrix*

	C	S	T	P	A	G	N	D	E	Q	H	R	K	M	I	L	V	F	Y	W
C	**9**	–1	–1	–3	0	–3	–3	–3	–4	–3	–3	–3	–3	–1	–1	–1	–1	–2	–2	–2
S	–1	**4**	1	–1	1	0	1	0	0	0	–1	–1	0	–1	–2	–2	–2	–2	–2	–3
T	–1	1	**4**	1	–1	1	0	1	0	0	0	–1	0	–1	–2	–2	–2	–2	–2	–3
P	–3	–1	1	**7**	–1	–2	–1	–1	–1	–1	–2	–2	–1	–2	–3	–3	–2	–4	–3	–4
A	0	1	–1	–1	**4**	0	–1	–2	–1	–1	–2	–1	–1	–1	–1	–1	–2	–2	–2	–3
G	–3	0	1	–2	0	**6**	–2	–1	–2	–2	–2	–2	–2	–3	–4	–4	0	–3	–3	–2
N	–3	1	0	–2	–2	0	**6**	1	0	0	–1	0	0	–2	–3	–3	–3	–3	–2	–4
D	–3	0	1	–1	–2	–1	1	**6**	2	0	–1	–2	–1	–3	–3	–4	–3	–3	–3	–4
E	–4	0	0	–1	–1	–2	0	2	**5**	2	0	0	1	–2	–3	–3	–3	–3	–2	–3
Q	–3	0	0	–1	–1	–2	0	0	2	**5**	0	1	1	0	–3	–2	–2	–3	–1	–2
H	–3	–1	0	–2	–2	–2	1	1	0	0	**8**	0	–1	–2	–3	–3	–2	–1	2	–2
R	–3	–1	–1	–2	–1	–2	0	–2	0	1	0	**5**	2	–1	–3	–2	–3	–3	–2	–3
K	–3	0	0	–1	–1	–2	0	–1	1	1	–1	2	**5**	–1	–3	–2	–3	–3	–2	–3
M	–1	–1	–1	–2	–1	–3	–2	–3	–2	0	–2	–1	–1	**5**	1	2	–2	0	–1	–1
I	–1	–2	–2	–3	–1	–4	–3	–3	–3	–3	–3	–3	–3	1	**4**	2	1	0	–1	–3
L	–1	–2	–2	–3	–1	–4	–3	–4	–3	–2	–3	–2	–2	2	2	**4**	3	0	–1	–2
V	–1	–2	–2	–2	0	–3	–3	–3	–2	–2	–3	–3	–2	1	3	1	**4**	–1	–1	–3
F	–2	–2	–2	–4	–2	–3	–3	–3	–3	–3	–1	–3	–3	0	0	0	–1	**6**	3	1
Y	–2	–2	–2	–3	–2	–3	–2	–3	–2	–1	2	–2	–2	–1	–1	–1	–1	3	**7**	2
W	–2	–3	–3	–4	–3	–2	–4	–4	–3	–2	–2	–3	–3	–1	–3	–2	–3	1	2	**11**

match perfectly. It is similar in magnitude to PAM250, but has been determined to be more reliable for protein sequences.

When applying a scoring matrix to determine an optimal alignment between two sequences, a brute-force search of all possible alignments would in most cases be prohibitive.

For instance, suppose one has two sequences of length n. If we add some number of gaps such that (i) each of the two sequences has the same length and (ii) a gap cannot align with another gap, it can be shown that, as $n \to \infty$, the number of possible alignments approaches

$$C_n^{2n} \approx \frac{2^{2n}}{\sqrt{n\pi}}$$

which for $n = 100$ takes on the astronomical sum of $\sim 10^{59}$!

To address this, Needleman and Wunsch (1970) first applied the **dynamic programming** concept from computer science to the problem of aligning two protein sequences. They defined a maximum match as the largest number of amino acids of one protein (X) that can be matched with those of another protein (Y) while allowing for all possible deletions. The maximum match is found with a matrix in which all possible pair combinations that can be formed from the amino acid sequences are represented. The rows and columns of this matrix are associated with the ith and jth amino acids of proteins X and Y, respectively. All of the possible comparisons are then represented by pathways through the two-dimensional matrix. Every i or j can only occur once in a pathway because a particular amino acid cannot occupy multiple positions at once. A pathway can be denoted as a line connecting elements of the matrix, with complete diagonal pathways comprising no gaps.

As first proposed by Needleman and Wunsch, each viable pathway through the matrix begins at an element in the first column or row. Either i or j is increased by exactly one, while the other index is increased by one or more; in other words, only diagonal (not horizontal or vertical) moves are permitted through the matrix. In this manner, the matrix elements are connected in a pathway that leads to the final element in which either i or j (or both simultaneously) reach their final values. One can define various schemes for rewarding amino acid matches or penalizing mismatches. Perhaps the simplest method is to assign a value of 1 for each match and a value of 0 for each mismatch. More sophisticated scoring methods can include the influence of neighboring residues, and a **penalty factor** for each gap allowed. This approach to sequence alignment can be generalized to the simultaneous comparison of several proteins, by extending the two-dimensional matrix to a multidimensional array.

The maximum-match operation is depicted in Figure 9.2, where 12 amino acids within the hydrophobic domain of the protein β-synuclein (along the vertical axis, i) are compared against 13 amino acids of the α-synuclein sequence (horizontal axis, j). Note that α-synuclein (SNCA) is primarily found in neural tissue, and is a major component of pathological regions associated with Parkinson's and Alzheimer's diseases (Giasson *et al.*, 2001). An easy way to complete the depicted matrix is to start in the last row, at the last column position. The number within each element represents the maximum total score when starting the alignment with these two amino acids and proceeding in the positive i, j directions. Since, in this final position A = A, we assign a value of 1. Now, staying in the final row and moving towards the left (decreasing the j index), we see that starting the final amino acid of β-syn at any

Figure 9.2.

Comparison of two sequences using the Needleman–Wunsch method.

	P	S	E	E	G	Y	C	D	Y	E	P	E	A
P	10	9	8	7	7	6	5	5	4	3	3	1	0
Q	9	9	8	7	7	6	5	5	4	3	2	1	0
E	8	8	9	8	7	6	5	5	4	4	2	2	0
E	7	7	8	8	7	6	5	5	4	4	2	1	0
Y	7	7	7	6	6	7	6	5	5	3	2	1	0
C	6	6	6	5	5	5	6	5	4	3	2	1	0
E	5	5	6	6	5	5	5	5	4	4	2	2	0
Y	4	4	4	4	4	5	4	4	5	3	2	1	0
E	3	3	4	4	3	3	3	3	3	4	2	2	0
P	3	2	2	2	2	2	2	2	2	2	3	1	0
E	1	1	2	2	1	1	1	1	1	2	1	2	0
A	0	0	0	0	0	0	0	0	0	0	0	0	1

other position of α-syn results in no possible matches, and so the rest of this row fills out as zeros. Moving up one row to the $i = 11$ position of β-syn, and proceeding from right to left, we see that the first partial alignment that results in matches is EA aligned with EA, earning a score of 2. Moving from right to left in the decreasing j direction, we see the maximum score involving only the last two amino acids of β-syn vary between 1 and 2. Moving up the rows, we may continue completing the matrix by assigning a score in each element corresponding to the maximum score obtained when starting at that position and proceeding in the positive i and j directions while allowing for gaps. The remainder of the matrix can be calculated in a similar fashion. The overall maximum match is then initiated at the greatest score found in the first row or column, and a pathway to form this maximum match is traced out as depicted in Figure 9.2. As stated previously, no horizontal or vertical moves are permitted. Although not depicted here, it is sometimes possible to find alternate pathways with the same overall score. In this example, the maximum match alignment is determined to be as follows:

α-syn: **P S E E G Y C D Y E P E A**
β-syn: **P Q E E – Y C E Y E P E A**
***   &   ***

where the asterisks indicate mismatches, and a gap is marked with an ampersand.

A powerful and popular algorithm for sequence comparison based on dynamic programming is called the **Basic Local Alignment Search Tool**, or **BLAST**. The first step in a BLAST search is called "seeding," in which for each word of length W in the query, a list of all possible words scoring within a threshold T are generated. Next, in an "extension" step, dynamic programming is used to extend the word hits until the score drops by a value of X. Finally, an "evaluation" step is performed to weigh the significance of the extended hits, and only those above a predefined threshold are reported. A variation on the BLAST algorithm, which includes the possibility of two aligned sequences connected with a gap, is called BLAST2 or Gapped BLAST. It finds two non-overlapping hits with a score of at least T and a maximum distance d from one another. An ungapped extension is performed and if the generated highest-scoring segment pairs (HSP) have sufficiently high scores when normalized appropriately by the segment length, then a gapped extension is initiated. Results are reported for alignments showing sufficiently low ***E*-value**. The E-value is a statistic which gives a measure of whether potential matches might have arisen by chance. An E-value of 10 signifies that ten matches would be expected to occur by chance. The value 10 as a cutoff is often used as a default, and $E < 0.05$ is generally taken as being significant. Generally speaking, long regions of moderate similarity are more significant than short regions of high identity. Some of the specific BLAST programs available at the NIH National Center for Biotechnology Information (NCBI, http://blast.ncbi.nlm.nih.gov/Blast.cgi) include: **blastn**, for submitting nucleotide queries to a nucleotide database; **blastp**, for protein queries to the protein database; **blastx**, for searching protein databases using a translated nucleotide query; **tblastn**, for searching translated nucleotide databases using a protein query; and **tblastx**, for searching translated nucleotide databases using a translated nucleotide query.

Further variations on the original BLAST algorithm are available at the NCBI Blast query page, and can be selected as options. **PSI-BLAST**, or Position-Specific Iterated BLAST, allows the user to build a position-specific scoring matrix using the results of an initial (default option) blastp query. **PHI-BLAST**, or Pattern Hit Initiated BLAST, finds proteins which contain the pattern and similarity within the region of the pattern, and is integrated with PSI-BLAST (Altschul *et al.*, 1997).

There are many situations where it is desirable to perform sequence comparison between three or more proteins or nucleic acids. Such comparisons can help to identify functionally important sites, predict protein structure, or even start to reconstruct the evolutionary history of the gene. A number of algorithms for achieving **multiple sequence alignment** are available, falling into the categories of dynamic programming, **progressive alignment**, or **iterative search methods**. Dynamic programming methods proceed as sketched out above. The classic Needleman–Wunsch method is followed with the difference that a higher-dimensional array is used in place of the two-dimensional matrix. The number of comparisons increases exponentially with the number of sequences; in practice this is dealt with by constraining the number of letters or words that must be explicitly examined (Carillo and Lipman, 1988). In the progressive alignment method, we start by obtaining an optimal pairwise alignment between the two most similar sequences among the query group. New, less related sequences are then added one at a time to the first pairwise alignment. There is no guarantee following this method that the optimal alignment will be found, and the process tends to be very sensitive to the initial alignments. The **ClustalW** ("Weighted" Clustal; freely available at www.ebi.ac.uk/clustalw or align.genome.jp among other sites) algorithm is a general purpose multiple sequence alignment program for DNA or proteins. It performs pairwise alignments of all submitted sequences (maximum sequence no. = 30; maximum length = 10 000) and then produces a phylogenetic tree (see Section 9.3) for comparing evolutionary relationships (Thompson *et al.*, 1994). Sequences are submitted in FASTA format, and the newer versions of ClustalW can produce graphical output of the results. **T-Coffee** (www.ebi.ac.uk/t-coffee or www.tcoffee.org) is another multiple sequence alignment program, a progressive method which combines information from both local and global alignments (Notredame *et al.*, 2000). This helps to minimize the sensitivity to the first

Box 9.3 BLAST search of a short protein sequence

L-selectin is an adhesion receptor which is important in the trafficking of leukocytes in the bloodstream. L-selectin has a short cytoplasmic tail, which is believed to perform intracellular signaling functions and help tether the molecule to the cytoskeleton (Green *et al.*, 2004; Ivetic *et al.*, 2004). Let's perform a simple BLAST search on the 17 amino acid sequence encoding the human L-selectin cytoplasmic tail, RRLKKGKKSKRSMNDPY, to see if any unexpected connections to other similar protein sequences can be revealed. Over the past ten years, performing such searches has become increasingly user friendly and simple. From the NCBI BLAST website, we choose the "protein blast" link. Within the query box, which requests an accession number or FASTA sequence, we simply paste the 17-AA code and then click the BLAST button to submit our query. Some of the default settings which can be viewed are: blastp algorithm, word size = 3, Expect threshold = 10, BLOSUM62 matrix, and gap costs of −11 for gap creation and −1 for gap extension. Within seconds we are taken to a query results page showing 130 Blast hits along with the message that our search parameters were adjusted to search for a short input sequence. This automatic parameter adjustment can be disabled by unchecking a box on the query page. All non-redundant GenBank CDS translations + PDB + SwissProt + PIR + PRF excluding environmental samples from WGS projects have been queried.[3] Table 9.4 presents the first 38 hits, possessing an *E*-value less than 20.

The first group of letters and numbers is the Genbank ID and accession number (a unique number identified for sequence entries), and in the Blast application these are hyperlinks that take you to the database entry with sequence data, source references, and annotation into the protein's known functions. The first 26 hits are all for the L-selectin protein (LECAM-1 was an early proposed name), with the species given in square brackets. The "PREDICTED" label indicates a genomic sequence that is believed to encode a protein based on comparison with known data. Based on the low *E*-value scores of the first 20+ hits, the sequence of the L-selectin cytoplasmic tail is found to be highly conserved among species, perhaps suggesting its evolutionary importance. Later we will explore a species-to-species comparison of the full L-selectin protein. The first non-L-selectin hit is for the amphioxus dachshund protein in a small worm-like creature called the Florida lancelet. Clicking on the amphioxus dachshund score of 33.3 takes us to more specific information on this alignment:

```
>gb|AAQ11368.1| amphioxus dachshund [Branchiostoma floridae]
Length = 360
Score = 33.3 bits (71), Expect = 2.8
Identities = 10/12 (83%), Positives = 11/12 (91%), Gaps = 0/12 (0%)
Query 2      RLKKGKKSKRSM   13
             RLKKGKK+KR M
Sbjct 283    RLKKGKKAKRKM   294
```

Note that the matched sequence only involves 12 of the 17 amino acids from the L-selectin cytoplasmic tail, from positions 2 to 13. In the middle row of the alignment display, only matching letters are shown and mismatches are left blank. Conservative substitutions are shown by a + symbol. Although no gaps were inserted into this particular alignment, in general these would be indicated by a dash. The nuclear factor dachshund (dac) is a key regulator of eye and leg development in Drosophila. In man, it is a retinal determination protein. However, based on the relatively large *E*-score and short length for this and other non-L-selectin hits, in the absence of any other compelling information it would be unwise to overinterpret this particular match.

alignments. Another benefit of the T-Coffee program is that it can combine partial results obtained from several different alignment methods, such as one alignment from ClustalW, another from a program called Dialign, etc., and will combine this information to produce a new multiple sequence alignment that integrates this

[3] These terms refer to the names of individual databases containing protein sequence data.

Table 9.4. *Output data from BLAST search on L-selectin subsequence*

<table>
<tr><th></th><th>Score</th><th>E-value</th></tr>
<tr><td>dbj|BAG60862.1| unnamed protein product [Homo sapiens]</td><td>57.9</td><td>1e–07</td></tr>
<tr><td>emb|CAB55488.1| L-selectin [Homo sapiens] >emb|CAI19356.1| se ...</td><td>57.9</td><td>1e–07</td></tr>
<tr><td>gb|AAC63053.1| lymph node homing receptor precursor [Homo sap ...</td><td>57.9</td><td>1e–07</td></tr>
<tr><td>emb|CAA34203.1| pln homing receptor [Homo sapiens]</td><td>57.9</td><td>1e–07</td></tr>
<tr><td>emb|CAB43536.1| unnamed protein product [Homo sapiens] >prf|| ...</td><td>57.9</td><td>1e–07</td></tr>
<tr><td>ref|NP_000646.1| selectin L precursor [Homo sapiens] >sp|P141 ...</td><td>57.9</td><td>1e–07</td></tr>
<tr><td>emb|CAA34275.1| unnamed protein product [Homo sapiens]</td><td>57.9</td><td>1e–07</td></tr>
<tr><td>ref|NP_001009074.1| selectin L [Pan troglodytes] >sp|Q95237.1 ...</td><td>55.4</td><td>6e–07</td></tr>
<tr><td>ref|NP_001106096.1| selectin L [Papio anubis] >sp|Q28768.1|LY ...</td><td>52.0</td><td>7e–06</td></tr>
<tr><td>ref|NP_001036228.1| selectin L [Macaca mulatta] >sp|Q95198.1| ...</td><td>52.0</td><td>7e–06</td></tr>
<tr><td>sp|Q95235.1|LYAM1_PONPY RecName: Full=L-selectin; AltName: Fu ...</td><td>52.0</td><td>7e–06</td></tr>
<tr><td>ref|NP_001106148.1| selectin L [Sus scrofa] >dbj|BAF91498.1| ...</td><td>47.7</td><td>1e–04</td></tr>
<tr><td>ref|XP_537201.2| PREDICTED: similar to L-selectin precursor (...</td><td>44.8</td><td>0.001</td></tr>
<tr><td>ref|XP_862397.1| PREDICTED: similar to selectin, lymphocyte i ...</td><td>44.8</td><td>0.001</td></tr>
<tr><td>ref|XP_001514463.1| PREDICTED: similar to L-selectin [Ornitho ...</td><td>43.9</td><td>0.002</td></tr>
<tr><td>gb|EDM09364.1| selectin, lymphocyte, isoform CRA_b [Rattus no ...</td><td>40.1</td><td>0.026</td></tr>
<tr><td>ref|NP_001082779.1| selectin L [Felis catus] >dbj|BAF46391.1| ...</td><td>40.1</td><td>0.026</td></tr>
<tr><td>ref|NP_062050.3| selectin, lymphocyte [Rattus norvegicus] >gb ...</td><td>40.1</td><td>0.026</td></tr>
<tr><td>dbj|BAE42834.1| unnamed protein product [Mus musculus]</td><td>40.1</td><td>0.026</td></tr>
<tr><td>dbj|BAE42004.1| unnamed protein product [Mus musculus]</td><td>40.1</td><td>0.026</td></tr>
<tr><td>dbj|BAE37180.1| unnamed protein product [Mus musculus]</td><td>40.1</td><td>0.026</td></tr>
<tr><td>dbj|BAE36559.1| unnamed protein product [Mus musculus]</td><td>40.1</td><td>0.026</td></tr>
<tr><td>sp|P30836.1|LYAM1_RAT RecName: Full=L-selectin; AltName: Full ...</td><td>40.1</td><td>0.026</td></tr>
<tr><td>ref|NP_035476.1| selectin, lymphocyte [Mus musculus] >sp|P183 ...</td><td>40.1</td><td>0.026</td></tr>
<tr><td>gb|AAN87893.1| LECAM-1 [Sigmodon hispidus]</td><td>39.7</td><td>0.034</td></tr>
<tr><td>ref|NP_001075821.1| selectin L [Oryctolagus cuniculus] >gb|AA ...</td><td>34.6</td><td>1.2</td></tr>
<tr><td>gb|AAQ11368.1| amphioxus dachshund [Branchiostoma floridae]</td><td>33.3</td><td>2.8</td></tr>
<tr><td>ref|XP_001491605.2| PREDICTED: similar to L-selectin [Equus c ...</td><td>32.0</td><td>6.8</td></tr>
<tr><td>ref|YP_001379300.1| alpha-L-glutamate ligase [Anaeromyxobacte ...</td><td>31.2</td><td>12</td></tr>
<tr><td>ref|NP_499490.1| hypothetical protein Y66D12A.12 [Caenorhabdi ...</td><td>31.2</td><td>12</td></tr>
<tr><td>ref|XP_002132434.1| GA25185 [Drosophila pseudoobscura pseudoo ...</td><td>30.8</td><td>16</td></tr>
<tr><td>ref|YP_001088484.1| ABC transporter, ATP-binding protein [Clo ...</td><td>30.8</td><td>16</td></tr>
<tr><td>ref|ZP_01802998.1| hypothetical protein CdifQ_04002281 [Clost ...</td><td>30.8</td><td>16</td></tr>
<tr><td>gb|AAF34661.1|AF221715_1 split ends long isoform [Drosophila ...</td><td>30.8</td><td>16</td></tr>
<tr><td>gb|AAF13218.1|AF188205_1 Spen RNP motif protein long isoform ...</td><td>30.8</td><td>16</td></tr>
<tr><td>ref|NP_722616.1| split ends, isoform C [Drosophila melanogast ...</td><td>30.8</td><td>16</td></tr>
<tr><td>ref|NP_524718.2| split ends, isoform B [Drosophila melanogast ...</td><td>30.8</td><td>16</td></tr>
<tr><td>ref|NP_722615.1| split ends, isoform A [Drosophila melanogast ...</td><td>30.8</td><td>16</td></tr>
</table>

Figure 9.3.

Multiple sequence alignment of L-selectin protein sequence from eight different species, generated using the ClustalW program. Columns with consensus residues are shaded in black, with residues similar to consensus shaded in gray. The consensus sequence shows which residues are most abundant in the alignment at each position.

King and Mody Numerical and Statistical Methods for Bioengineering © 2009

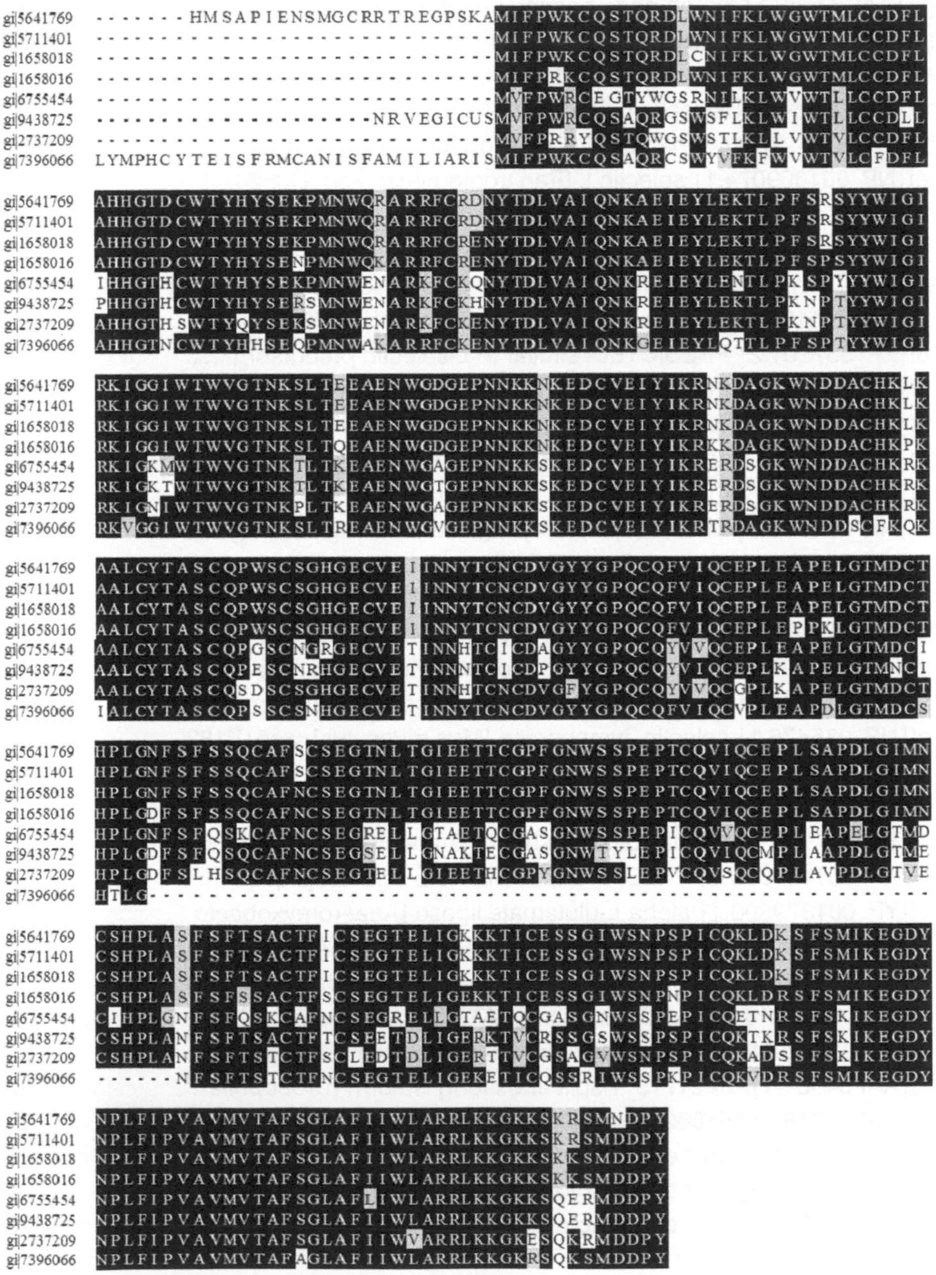

```
gi|5641769  -------HMSAPIENSMGCRRTREGPSKAMIFPWKCQSTQRDLWNIFKLWGWTMLCCDFL
gi|5711401  -----------------------------MIFPWKCQSTQRDLWNIFKLWGWTMLCCDFL
gi|1658018  -----------------------------MIFPWKCQSTQRDLCNIFKLWGWTMLCCDFL
gi|1658016  -----------------------------MIFPRKCQSTQRDLWNIFKLWGWTMLCCDFL
gi|6755454  -----------------------------MVFPWRCEGTYWGSRNILKLWVWTLLCCDFL
gi|9438725  --------------------NRVEGICUSMVFPWRCQSAQRGSWSFLKLWIWTLLCCDLL
gi|2737209  -----------------------------MVFPRRYQSTQWGSWSTLKLLVWTVLCCDFL
gi|7396066  LYMPHCYTEISFRMCANISFAMILIARISMIFPWKCQSAQRCSWYVFKFWVWTVLCFDFL

gi|5641769  AHHGTDCWTYHYSEKPMNWQRARRFCRDNYTDLVAIQNKAEIEYLEKTLPFSRSYYWIGI
gi|5711401  AHHGTDCWTYHYSEKPMNWQRARRFCRDNYTDLVAIQNKAEIEYLEKTLPFSRSYYWIGI
gi|1658018  AHHGTDCWTYHYSEKPMNWQRARRFCRENYTDLVAIQNKAEIEYLEKTLPFSRSYYWIGI
gi|1658016  AHHGTDCWTYHYSENPMNWQKARRFCRENYTDLVAIQNKAEIEYLEKTLPFSPSYYWIGI
gi|6755454  IHHGTHCWTYHYSEKPMNWENARKFCKQNYTDLVAIQNKREIEYLENTLPKSPYYYWIGI
gi|9438725  PHHGTHCWTYHYSERSMNWENARKFCKHNYTDLVAIQNKREIEYLEKTLPKNPTYYWIGI
gi|2737209  AHHGTHSWTYQYSEKSMNWENARKFCKENYTDLVAIQNKREIEYLEKTLPKNPTYYWIGI
gi|7396066  AHHGTNCWTYHHSEQPMNWAKARRFCKENYTDLVAIQNKGEIEYLQTTLPFSPTYYWIGI

gi|5641769  RKIGGIWTWVGTNKSLTEEAENWGDGEPNNKKNKEDCVEIYIKRNKDAGKWNDDACHKLK
gi|5711401  RKIGGIWTWVGTNKSLTEEAENWGDGEPNNKKNKEDCVEIYIKRNKDAGKWNDDACHKLK
gi|1658018  RKIGGIWTWVGTNKSLTEEAENWGDGEPNNKKNKEDCVEIYIKRNKDAGKWNDDACHKLK
gi|1658016  RKIGGIWTWVGTNKSLTQEAENWGDGEPNNKKNKEDCVEIYIKRKKDAGKWNDDACHKPK
gi|6755454  RKIGKMWTWVGTNKTLTKEAENWGAGEPNNKKSKEDCVEIYIKRERDSGKWNDDACHKRK
gi|9438725  RKIGKTWTWVGTNKTLTKEAENWGTGEPNNKKSKEDCVEIYIKRERDSGKWNDDACHKRK
gi|2737209  RKIGNIWTWVGTNKPLTKEAENWGAGEPNNKKSKEDCVEIYIKRERDSGKWNDDACHKRK
gi|7396066  RKVGGIWTWVGTNKSLTREAENWGVGEPNNKKSKEDCVEIYIKRTRDAGKWNDDSCFKQK

gi|5641769  AALCYTASCQPWSCSGHGECVEIINNYTCNCDVGYYGPQCQFVIQCEPLEAPELGTMDCT
gi|5711401  AALCYTASCQPWSCSGHGECVEIINNYTCNCDVGYYGPQCQFVIQCEPLEAPELGTMDCT
gi|1658018  AALCYTASCQPWSCSGHGECVEIINNYTCNCDVGYYGPQCQFVIQCEPLEAPELGTMDCT
gi|1658016  AALCYTASCQPWSCSGHGECVEIINNYTCNCDVGYYGPQCQFVIQCEPLEPPKLGTMDCT
gi|6755454  AALCYTASCQPGSCNGRGECVETINNHTCICDAGYYGPQCQYVVQCEPLEAPELGTMDCI
gi|9438725  AALCYTASCQPESCNRHGECVETINNNTCICDPGYYGPQCQYVIQCEPLKAPELGTMNCI
gi|2737209  AALCYTASCQSDSCSGHGECVETINNHTCNCDVGFYGPQCQYVVQCGPLKAPELGTMDCT
gi|7396066  IALCYTASCQPSSCSNHGECVETINNYTCNCDVGYYGPQCQFVIQCVPLEAPDLGTMDCS

gi|5641769  HPLGNFSFSSQCAFSCSEGTNLTGIEETTCGPFGNWSSPEPTCQVIQCEPLSAPDLGIMN
gi|5711401  HPLGNFSFSSQCAFSCSEGTNLTGIEETTCGPFGNWSSPEPTCQVIQCEPLSAPDLGIMN
gi|1658018  HPLGNFSFSSQCAFNCSEGTNLTGIEETTCGPFGNWSSPEPTCQVIQCEPLSAPDLGIMN
gi|1658016  HPLGDFSFSSQCAFNCSEGTNLTGIEETTCGPFGNWSSPEPTCQVIQCEPLSAPDLGIMN
gi|6755454  HPLGNFSFQSKCAFNCSEGRELLGTAETQCGASGNWSSPEPICQVVQCEPLEAPELGTMD
gi|9438725  HPLGDFSFQSQCAFNCSEGSELLGNAKTECGASGNWTYLEPICQVIQCMPLAAPDLGTME
gi|2737209  HPLGDFSLHSQCAFNCSEGTELLGIEETHCGPYGNWSSLEPVCQVSQCQPLAVPDLGTVE
gi|7396066  HTLG--------------------------------------------------------

gi|5641769  CSHPLASFSFTSACTFICSEGTELIGKKKTICESSGIWSNPSPICQKLDKSFSMIKEGDY
gi|5711401  CSHPLASFSFTSACTFICSEGTELIGKKKTICESSGIWSNPSPICQKLDKSFSMIKEGDY
gi|1658018  CSHPLASFSFTSACTFICSEGTELIGKKKTICESSGIWSNPSPICQKLDKSFSMIKEGDY
gi|1658016  CSHPLASFSFSSACTFSCSEGTELIGEKKTICESSGIWSNPNPICQKLDRSFSMIKEGDY
gi|6755454  CIHPLGNFSFQSKCAFNCSEGRELLGTAETQCGASGNWSSPEPICQETNRSFSKIKEGDY
gi|9438725  CSHPLANFSFTSACTFTCSEETDLIGERKTVCRSSGSWSSPSPICQKTKRSFSKIKEGDY
gi|2737209  CSHPLANFSFTSTCTFSCLEDTDLIGERTTVCGSAGVWSNPSPICQKADSSFSKIKEGDY
gi|7396066  ------NFSFTSTCTFNCSEGTELIGEKETICQSSRIWSSPKPICQKVDRSFSMIKEGDY

gi|5641769  NPLFIPVAVMVTAFSGLAFIIWLARRLKKGKKSKRSMNDPY
gi|5711401  NPLFIPVAVMVTAFSGLAFIIWLARRLKKGKKSKRSMDDPY
gi|1658018  NPLFIPVAVMVTAFSGLAFIIWLARRLKKGKKSKKSMDDPY
gi|1658016  NPLFIPVAVMVTAFSGLAFIIWLARRLKKGKKSKKSMDDPY
gi|6755454  NPLFIPVAVMVTAFSGLAFLIWLARRLKKGKKSQERMDDPY
gi|9438725  NPLFIPVAVMVTAFSGLAFIIWLARRLKKGKKSQERMDDPY
gi|2737209  NPLFIPVAVMVTAFSGLAFIIWVARRLKKGKESQKRMDDPY
gi|7396066  NPLFIPVAVMVTAFAGLAFIIWLARRLKKGKRSQKSMDDPY
```

information. T-Coffee tends to obtain better results than ClustalW for sequences with less than 30% identity, but is slower. Figure 9.3 shows a multiple sequence alignment using ClustalW of eight species (human, chimpanzee, orangutan, rhesus monkey, domestic dog, house mouse, Norway/brown rat, and hispid cotton rat) of

Figure 9.4.

Multiple sequence alignment of L-selectin protein sequence from seven different species, generated using the TCoffee program. Consensus regions are shaded in grey and marked with asterisks.

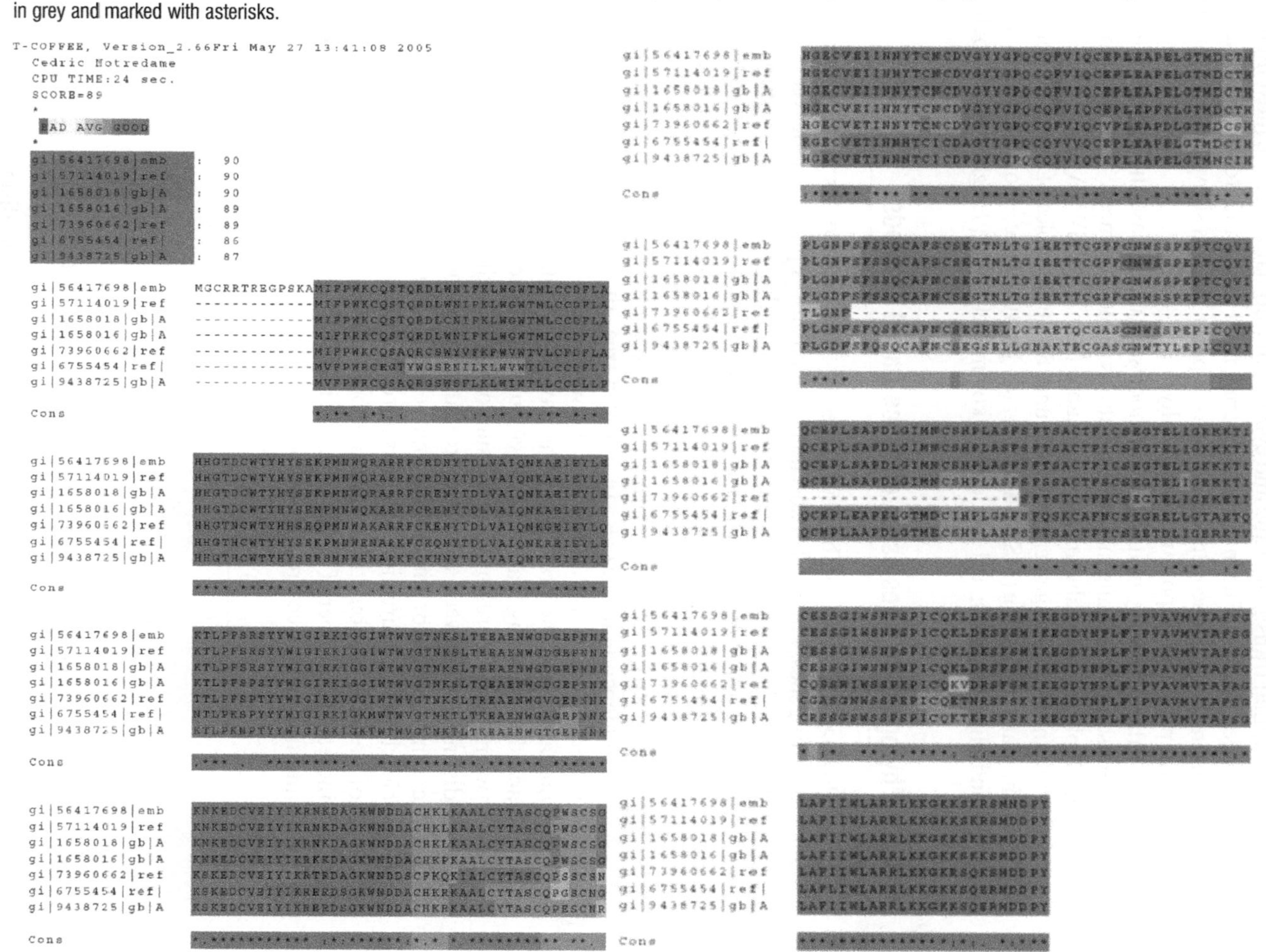

L-selectin protein, processed using BoxShade (a commonly used simple plotting program). ClustalW and BoxShade programs were accessed at the Institute Pasteur website. Figure 9.4 shows a multiple sequence alignment performed via the T-Coffee algorithm at the Swiss EMBnet website.

Figures 9.3 and 9.4 demonstrate that the N-terminal binding domain, and the C-terminal transmembrane and cytoplasmic domains, are the most highly conserved between species. The intermediate epidermal growth factor-like domain, two short consensus repeat sequences, and short spacer regions show more variability and are perhaps not as important biologically.

A third category of iterative methods seek to increase the multiple sequence alignment score by making random alterations to the alignment. The first step is to obtain a multiple sequence alignment, for instance from ClustalW. Again, there is no guarantee that an optimal alignment will be found. Some examples of iterative methods include: simulated annealing (MSASA; Kim *et al.*, 1994); genetic algorithm (SAGA; Notredame and Higgins, 1996); and hidden Markov model (SAM; Hughey and Krogh, 1996).

9.3 Phylogenetic trees using distance-based methods

Phylogenetic trees can be used to reconstruct genealogies from molecular data, not just for genes, but also for species of organisms. A phylogenetic tree, or dendrogram, is an acyclic two-dimensional graph of the evolutionary relationship among three or more genes or organisms. A phylogenetic tree is composed of **nodes** and **branches** connecting them. Terminal nodes (filled circles in Figure 9.5), located at the ends of branches, represent genes or organisms for which data are available, whereas internal nodes (open circles) represent inferred common ancestors that have given rise to two independent lineages. In a **scaled tree**, the lengths of the branches are made proportional to the differences between pairs of neighboring nodes, whereas in **unscaled trees** the branch lengths have no significance. Some scaled trees are also **additive**, meaning that the relative number of accumulated differences between any two nodes is represented by the sum of the branch lengths. Figure 9.5 also shows the distinction between a **rooted tree** and an **unrooted tree**. Rooted trees possess a common ancestor (iii) with a unique path to any "leaf", and time progresses in a clearly defined direction, whereas unrooted trees are undirected with a fewer number of possible combinations. The number of possible trees (N) increases sharply with increasing number of species (n), and is given by the following expressions for rooted and unrooted trees, respectively:

$$N_R = \frac{(2n-3)!}{2^{n-2}(n-2)!},$$

$$N_U = \frac{(2n-5)!}{2^{n-3}(n-3)!}.$$

Table 9.5 shows the number of possible phylogenetic trees arising from just 2–16 data sets.

Clustering algorithms are algorithms that use the distance (number of mutation events) to calculate phylogenetic trees, with the trees based on the relative numbers of similarities and differences between sequences. Distance matrices can be constructed by computing pairwise distances for all sequences. **Sequential clustering** is a technique in which the data sets are then linked to successively more distant taxa (groups of related

Table 9.5. *Number of possible rooted and unrooted phylogenetic trees as a function of the number of data sets*

No. of data sets, n	No. of rooted trees, N_R	No. of unrooted trees, N_U
2	1	1
4	15	3
8	135 135	10 395
16	6.19×10^{15}	2.13×10^{14}

Figure 9.5.

Examples of a rooted and unrooted phylogenetic tree. Terminal nodes are represented as filled circles and (inferred) internal nodes are represented as open circles.

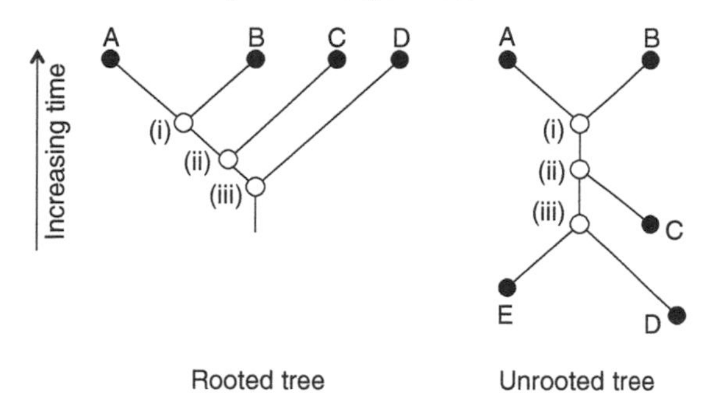

organisms). Distances between pairs of DNA sequences are relatively simple to compute, and are given by the sum of all base pair differences between the two sequences. The pairs of sequences to be compared must be similar enough to be aligned. All base changes are treated equally, and insertions/deletions are usually weighted more heavily than replacements of a single base, via a gap penalty. Distance values are often normalized as the "number of changes per 100 nucleotides." In comparison, amino acid distances are more challenging to compute. Different substitutions have varying effects on structure and function, and some amino acid substitutions of course require greater than one DNA mutation (see Table 9.1). Replacement frequencies are provided by the PAM and BLOSUM matrices, as discussed previously.

The originally proposed distance matrix method, which is simple to implement, is called the unweighted-pair-group method with arithmetic mean (UPGMA; Sneath & Sokal, 1973). UPGMA and other distance-based methods use a measure of all pairwise genetic distances between the taxa being considered. We begin by clustering the two species with the smallest distance separating them into a new composite group. For instance, if comparing species A through E, suppose that the closest pair is DE. After the first cluster is formed, then a new distance matrix is formed between the remaining species and the cluster DE. As an example, the distance between A and the cluster DE would be calculated as the arithmetic mean $d_{A(DE)} = (d_{AD} + d_{AE})/2$. The species closest together in the new distance matrix are then clustered together to form a new composite species. This process is continued until all species to be compared are grouped. Groupings are then graphically represented by the phylogenetic tree, and in some cases the branch length is used to show the relative distance between groupings. Box 9.4 gives an illustration of this sequential process.

Box 9.4 Phylogenetic tree generation using the UPGMA method

Consider the alignment of five different DNA sequences of length 40 shown in Table 9.6. We first generate a pairwise distance matrix to summarize the number of non-matching nucleotides between the five sequences:

Species	A	B	C	D
B	13	–	–	–
C	3	15	–	–
D	8	15	9	–
E	7	14	4	11

The unambiguous closest species pair is AC, with a distance of 3. Thus, our tree takes the following form:

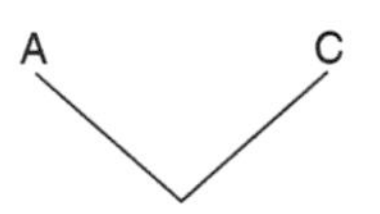

We now generate a new pairwise distance matrix, treating the AC cluster as a single entity:

Species	AC	B	D
B	14	–	–
D	8.5	15	–
E	5.5	14	11

Thus, the next closest connection is between the AC cluster and E, with a distance of 5.5. Our tree then becomes:

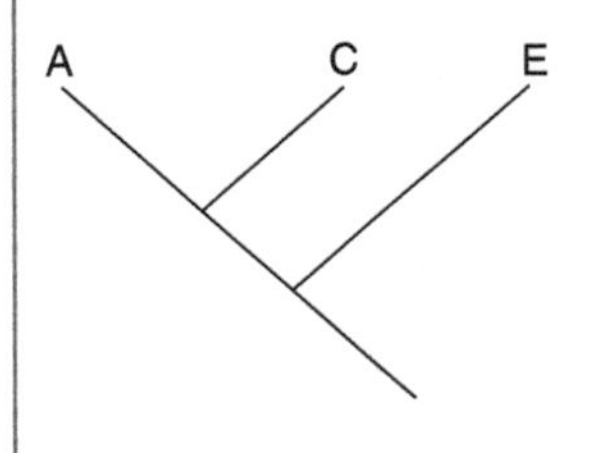

Table 9.6. *Pairwise distance matrix from alignment of five sequences*

	10	20	30	40
A:	TGCGCAAAAA	CATTGCCCCC	TCACAATAGA	AATAAAGTGC
B:	TGCGCAAACA	CATTTCCTAA	GCACGCTTGA	ACGAAGGGGC
C:	TGCGCAAAAC	CATTGCCACC	TCACAATAGA	AATAAATTGC
D:	TGCCCAAAAC	CCTTGCGCCG	TCACGATAGA	AAAAAAGGGC
E:	TGCGCAAAAC	CATTTCCACG	ACACAATCGA	AATAAATTGC

Note that this tree is unscaled. Similarly, the updated pairwise distance matrix is:

Species	(AC)E	B
B	14	–
D	9.75	15

As shown in the matrix, D shares more in common with the (AC)E cluster than with species B, and so our final, unambiguous phylogenetic tree is:

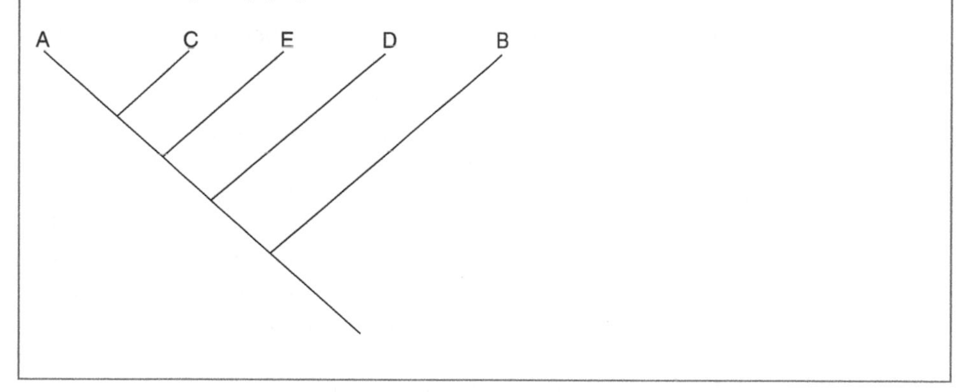

9.4 End of Chapter 9: key points to consider

(1) The process by which the genes encoded in DNA are made into the myriad of proteins found in the cell is referred to as the "**central dogma**" of molecular biology. A group of three nucleotides called a "**codon**" ($N = 4^3 = 64$ combinations) encodes the 20 different amino acids that are the building blocks of proteins.

(2) The **alignment** of two or more sequences can reveal evolutionary relationships and suggest functions for undescribed genes. For sequences that share a common ancestor, there are three mechanisms that can account for a character difference at any position: (1) a **mutation** replaces one letter for another; (2) an **insertion** adds one or more letters to one of the two sequences; (3) a **deletion** removes one or more letters from one of the sequences.

(3) When applying a scoring matrix to determine an optimal alignment between two sequences, a brute-force search of all possible alignments would in most cases be prohibitive. A powerful and popular algorithm for sequence comparison based on dynamic programming is called **Basic Local Alignment Search Tool**, or **BLAST**.

(4) Some of the specific BLAST programs available at the NIH National Center for Biotechnology Information (NCBI, http://blast.ncbi.nlm.nih.gov/Blast.cgi) include: **blastn**, for submitting nucleotide queries to a nucleotide database; **blastp**, for protein queries to the protein database; **blastx**, for searching protein databases using a translated nucleotide query; **tblastn**, for searching translated nucleotide databases using a protein query; and **tblastx**, for searching translated nucleotide databases using a translated nucleotide query.

(5) There are many situations where it is desirable to perform sequence comparison between three or more proteins or nucleic acids. Such comparisons can help to identify functionally important sites, predict protein structure, or reconstruct the evolutionary history of the gene. A number of algorithms for achieving **multiple sequence alignment** are available such as **ClustalW** and **T-Coffee**.

(6) **Phylogenetic trees** can be used to reconstruct genealogies from molecular data, not just for genes, but also for species of organisms. A phylogenetic tree, or dendrogram, is an acyclic two-dimensional graph of the evolutionary relationship among three or more genes or organisms. A phylogenetic tree is composed of **nodes** and **branches** connecting them.

9.5 Problems

9.1. **Genome access and multiple sequence alignment** Tumor necrosis factor (TNF)-related apoptosis inducing ligand (TRAIL) is a naturally occurring protein that specifically induces apoptosis in many types of cancer cells via binding to "death receptors." Obtain the amino acid sequences for the TRAIL protein for human, mouse, and three other species from the Entrez cross-database search at the National Center for Biotechnology Information (NCBI) website (www.ncbi.nlm.nih.gov/sites/gquery). Perform a multiple sequence alignment using the ClustalW and TCoffee algorithms; these can be accessed from many different online resources. What regions of the TRAIL protein are most highly conserved between different species?

9.2. **Dot plot comparison of two genomic sequences** Compare the human TRAIL protein sequence from Problem 9.1 to the sequence for human tumor necrosis factor (TNF)-α. TNF-α is an important inflammatory cytokine with cytotoxic effects at sufficient concentration. Generate a dot plot and interpret the diagram in terms of any common domains that may be revealed.

9.3. **Phylogenetics** The following six amino acid sequences are related by a common ancestor:

EYGINEVV, EVGINAER, EVGINEYR, ENGINRNR, ENLYNEYR, ENGINRYI

Construct a rooted phylogenetic tree for these related sequences, identify the sequence of the common ancestor, and calculate the distances between the lineages.

References

Altschul, S. F., Madden, T. L., Schaffer, A. A., Zhang, J., Zhang, Z., Miller, W., and Lipman, D. J. (1997) Gapped BLAST and PSI-BLAST: A New Generation of Protein Database Search Programs. *Nucleic Acids Res.*, **25**, 3389–402.

Carrillo, H. and Lipman, D. (1988) The Multiple Sequence Alignment Problem in Biology. *SIAM J. Appl. Math.*, **48**, 1073–82.

Giasson, B. I., Murray, I. V. J., Trojanowski, J. Q., and Lee, V. M.-Y. (2001) A Hydrophobic Stretch of 12 Amino Acid Residues in the Middle of a-Synuclein Is Essential for Filament Assembly. *J. Biol. Chem.*, **276**, 2380–6.

Green, C. E., Pearson, D. N., Camphausen, R. T., Staunton, D. E., and Simon, S. I. (2004) Shear-dependent Capping of L-selectin and P-selectin Glycoprotein Ligand 1 by E-selectin Signals Activation of High-avidity Beta 2-integrin on Neutrophils. *J. Immunol.*, **284**, C705–17.

Hughey, R. and Krogh, A. (1996) Hidden Markov Models for Sequence Analysis: Extension and Analysis of the Basic Method. *CABIOS*, **12**, 95–107.

Ivetic, A., Florey, O., Deka, J., Haskard, D. O., Ager, A., and Ridley, A. J. (2004) Mutagenesis of the Ezrin-Radixin-Moesin Binding Domain of L-selectin Tail Affects Shedding, Microvillar Positioning, and Leukocyte Tethering. *J. Biol. Chem.*, **279**, 33 263–72.

Kim, J., Pramanik, S., and Chung, M. J. (1994) Multiple Sequence Alignment Using Simulated Annealing. *Bioinformatics*, **10**, 419–26.

Needleman, S. B. and Wunsch, C. D. (1970) A General Method Applicable to the Search for Similarities in the Amino Acid Sequence of Two Proteins. *J. Mol. Biol.*, **48**, 443–53.

Notredame, C. and Higgins, D. (1996) SAGA: Sequence Alignment by Genetic Algorithm. *Nucleic Acids Res.*, **24**, 1515–24.

Notredame, C., Higgins, D., and Heringa, J. (2000) T-Coffee: A Novel Method for Multiple Sequence Alignments. *J. Mol. Biol.*, **302**, 205–17.

Sneath, P. H. A. and Sokal, R. R. (1973) *Numerical Taxonomy* (San Francisco, CA: W. H. Freeman and Company), pp 230–4.

Thompson, J. D., Higgins, D. G., and Gibson, T. J. (1994) CLUSTAL W: Improving the Sensibility of Progressive Multiple Sequence Alignment Through Sequence Weighting, Positions-Specific Gap Penalties and Weight Matrix Choice. *Nucleic Acids Res.*, **22**, 4673–80.

Appendix A Introduction to MATLAB

MATLAB is a technical computing software that consists of programming commands, functions, an editor/debugger, and plotting tools for performing math, modeling, and data analysis. **Toolboxes** that provide access to a specialized set of MATLAB functions are available as add-on applications to the MATLAB software. A Toolbox addresses a particular field of applied science or engineering, such as optimization, statistics, signal processing, neural networks, image processing, simulation, bioinformatics, parallel computing, partial differential equations, and several others.

The MATLAB environment includes customizable desktop tools for managing files and programming variables. You can minimize, rearrange, and resize the windows that appear on the MATLAB desktop. The **Command Window** is where you can enter commands sequentially at the MATLAB prompt "≫." After each command is entered, MATLAB processes the instructions and displays the results immediately following the executed statement. A semicolon placed at the end of the command will suppress MATLAB-generated text output (but not graphical output).

The **Command History** window is a desktop tool that lists previously run commands. If you double click on any of these statements, it will be displayed in the Command Window and executed. The **Current Directory** browser is another desktop tool that allows you to view directories, open MATLAB files, and perform basic file operations. The **Workspace** browser displays the name, data type, and size of all variables that are defined in the **base workspace**. The base workspace is a portion of the memory that contains all variables that can be viewed and manipulated within the Command Window.

In the **Editor/Debugger** window you can create, edit, debug, save, and run m-files; **m-files** are programs that contain code written in the MATLAB language. Multiple m-files can be opened and edited simultaneously within the Editor window. The document bar located at the bottom of the Editor window has as many tabs as there are open m-files (documents). Desktop tools (windows) can be docked to and undocked from the MATLAB desktop. An undocked window can be resized or placed anywhere on the screen. Undocked tools can be docked at any time by clicking the dock button or using menu commands. See MATLAB help for an illustration of how to use MATLAB desktop tools.

A1.1 Matrix operations

MATLAB stands for "MATrix LABoratory." A matrix is a rectangular (two-dimensional) array. MATLAB treats each variable as an array. A scalar value is treated as a 1×1 matrix and a vector is viewed as a row or column matrix. (See Section 2.2 for an introduction to vectors and matrices.) Dimensioning of a variable is not required in MATLAB. In programming languages, such as Fortran and C, all arrays must be explicitly dimensioned and initialized. Most of the examples and problems in this book involve manipulating and operating on matrices, i.e. two-dimensional arrays. Operations involving multidimensional (three- or higher-dimensional) data structures are slightly more complex. In this section, we confine our discussion to matrix operations in MATLAB.

The square brackets `[]` is the matrix operator used to construct a vector or matrix. Elements of the matrix are entered within square brackets. A comma or space separates one element from the next in the same row. A semicolon separates one row from the next. Note that the function performed by the semicolon operator placed inside a set of square brackets is different from that when placed at the end of a statement. We create a 4×3 matrix by entering the following statement at the prompt:

```
>> A = [1 2 3; 4 5 6; 7 8 9; 10 11 12]
A =
     1    2    3
     4    5    6
     7    8    9
    10   11   12
```

MATLAB generates the output immediately below the statement. Section 2.2.1 demonstrates with examples of how to create a matrix and how to access elements of a matrix stored in memory. MATLAB provides several functions that create special matrices. For example, the `ones` function creates a matrix of ones; the `zeros` function creates a matrix of zeros, the `eye` function creates an identity matrix, and the `diag` function creates a diagonal matrix from a vector that is passed to the function. These functions are also discussed in Section 2.2.1. Other MATLAB matrix operators that should be committed to memory are listed below.

(1) The MATLAB colon operator `(:)` represents a series of numbers between the first and last value in the expression (*`first:last`*). This operator is used to construct a vector whose elements follow an increasing or decreasing arithmetic sequence; for example,

```
>> a = 1:5;
```

creates the 1×5 vector $\boldsymbol{a} = (1, 2, 3, 4, 5)$. The default increment value is 1. Any other step value can be specified as (*`first:step:last`*), e.g.

```
>> a = 0:0.5:1
a =
     0      0.5000   1.0000
```

Note that use of the matrix constructor operator `[ ]` with the colon operator when creating a vector is unnecessary.

(2) The MATLAB transpose operator `( ' )` rearranges the elements of a matrix by reflecting the matrix elements across the diagonal. This has the effect of converting the matrix rows into corresponding columns (or matrix columns into corresponding rows), or converts a row vector to a column vector and vice versa.

Two or more matrices can be concatenated using the same operators (square brackets operator and semicolon operator) used to construct matrices. Two matrices $\boldsymbol{A}$ and $\boldsymbol{B}$ can be concatenated horizontally as follows:

```
C = [A B];
```

Because the result ***C*** must be a rectangular array, i.e. a matrix, this operation is permitted only if both matrices have the same number of rows. To concatenate two matrices vertically, the semicolon operator is used as demonstrated below:

```
C = [A; B];
```

To perform this operation, the two matrices ***A*** and ***B*** must have the same number of columns.

Section 2.2.1 shows how to access elements of a matrix. Here, we introduce another method to extract elements from a matrix. One method is called **array indexing**. In this method, a matrix (array) can be used to point to elements within another matrix (array). More simply stated, the numeric values contained in the **indexing array *B*** determine which elements are chosen from the array ***A***:

```
>> A = [1:0.5:5]
A =
    1.0000   1.5000   2.0000   2.5000   3.0000   3.5000
    4.0000   4.5000   5.0000
>> B = [2 3; 8 9];
>> C = A(B)
C =
    1.5000   2.0000
    4.5000   5.0000
```

A special type of array indexing is **logical indexing**. The indexing array ***B*** is first generated using a logical operation on ***A***; ***B*** will have the same size as ***A*** but its elements will either be ones or zeros. The positions of the ones will correspond to the positions of those elements in ***A*** that satisfy the logical operation, i.e. positions at which the operation holds true. The positions of the zeros will correspond to the positions of those elements in ***A*** that failed to satisfy the logical operation:

```
>> A = [2 -1 3; -1 -2 1];
>> B = A < 0
B =
    0   1   0
    1   1   0
```

Note that ***B*** is a logical array. The elements are not of numeric data type (e.g. `double`, which is the default numeric data type in MATLAB), but of `logical` (non-numeric) data type. The positions of the ones in ***B*** will locate the elements in ***A*** that need to be processed:

```
>> A(B) = 0
A =
    2   0   3
    0   0   1
```

An empty matrix is any matrix that has either zero rows or zero columns. The simplest empty matrix is the 0×0 matrix, which is created using a set of square brackets with no elements specified within, i.e. `A = [ ]` is an empty 0×0 matrix.

A1.2 Programming in MATLAB

An m-file contains programming code written in the MATLAB language. All m-files have the `.m` extension. There are two types of m-files: **functions** and **scripts**. A script contains a set of executable and non-executable (commented) statements. It does not accept input parameters or return output variables. It also does not have its own designated workspace (memory space). If called from the command line (or run by clicking the run button from the Editor toolbar), the script directly accesses the base workspace. It can create and modify variables in the base workspace. Variables created by a script persist in the base workspace even after the script has completed its run. However, if called from within a function, a script can access only the workspace of the calling function.

A function can accept input parameters and can return output variables to the calling program. A function begins with a function description line. This line contains the name of the function, the number of input parameters (or arguments), and the number of output arguments. One example of a function statement is

```
function f = func1(parameter1, parameter2, parameter3)
```

The first line of a function begins with the keyword[1] `function`. The name of this function is `func1`. A function name must always start with a letter and may also contain numbers and the underscore character. This function has three input parameters and returns one output variable `f`. The name of the m-file should be the same as the name of the function. Therefore this function should be saved with the name `func1.m`; `f` is assigned a value within the function and is returned by the function to the calling program. A function does not need an `end` statement as the last line of the program code unless it is nested inside another function. If the function returns more than one output value, the output variables should be enclosed within square brackets:

```
function [f1, f2, f3] = func1(parameter1, parameter2)
```

If the function does not return any output values, then the starting line of the function does not include an equals sign, as shown below:

```
function func1(parameter1, parameter2, parameter3)
```

If there are no parameters to be passed to the function, simply place a pair of empty parentheses immediately after the function name:

```
function f = func1()
```

To pass a string as an argument to a function, you must enclose the string with single quotes.

Every function is allotted its own **function workspace**. Variables created within a function are placed in this workspace, and are not shared with the base workspace or with any other function workspaces. Therefore, variables created within a function cannot be accessed from outside the function. Variables defined within a function are called **local variables**. The input and output arguments are not local variables since they can be accessed by the workspace of the calling program. The **scope** of any variable that is defined within a function is said to be *local*. When the function execution terminates, local variables are lost from the memory, i.e. the function

[1] A keyword is a reserved word in MATLAB that has a special meaning. You should not use keywords for any other purpose than what they are intended for. Examples of keywords are `if`, `elseif`, `else`, `for`, `end`, `break`, `switch`, `case`, `while`, `global`, and `otherwise`.

workspace is emptied. The values of local variables are therefore not passed to subsequent function calls.

Variables can be shared between a function and the calling program workspace by passing the variables as arguments into the function and returning them to the calling function as output arguments.[2] This is the most secure way of sharing variables. In this way, a function cannot modify any of the variables in the base workspace except those values that are returned by the function. Similarly, the function' s local variables are not influenced by the values of the variables existing in the base workspace.

Another way of extending the scope of a variable created inside a function is to declare the variable as `global`. The value of a **global variable** will persist in the function workspace and will be made available to all subsequent function calls. All other functions that declare this variable as `global` will have access to and be able to modify the variable. If the same variable is made `global` in the base workspace, then the value of this variable is available from the command line, to all scripts, and to all functions containing statements that declare the variable as `global`.

Example A1.1

Suppose a variable called `data` is defined and assigned values either in a script file or at the command line. (This example is taken from Box 8.1B.)

```
% Data
global data
time = [0; 1; 3; 5; 10; 20; 40; 50; 60; 90; 120; 150; 180; 210; 240];
AZTconc = [0; 1.4; 4.1; 4.5; 3.5; 3.0; 2.75; 2.65; 2.4; 2.2; 2.15; 2.1;
2.15; 1.8; 2.0];
data = [time, AZTconc];
```

`data` is a 15 × 2 matrix. Alternatively, the values of `data` can be stored in a file and loaded into the memory using the `load` command (see Section A1.4).

The function m-file `SSEpharmcokineticsofAZTm.m` is created to calculate the error in a proposed nonlinear model. To do this, it must access the values stored in `data`. Therefore, in this function, `data` is declared as `global`. The first two lines of the function are

```
function SSE = SSEpharmacokineticsof AZTm(p)
global data
```

By making two global declarations, one at the command line and one inside the function, the function can now access the variable `data` that is stored in the base workspace.

At various places in an m-file you will want to explain the purpose of the program and the action of the executable statements. You can create comment lines anywhere in the code by using the percent sign. All text following `%` on the same line is marked as a comment and is ignored by the compiler.

Example A1.2

The following function is written to calculate the sum of the first *n* natural numbers, where *n* is an input parameter. There is one output value, which is the sum calculated by the function. Note the use of comments.

[2] If the function modifies the input arguments in any way, the updated values will not be reflected in the calling program's workspace. Any values that are needed from the function should be returned as output arguments by the function.

```
function s = first_n_numbers(n)
% This function calculates the sum of the first n natural numbers.
% n: the last number to be added to the sequence
s = n*(n + 1)/2; % s is the desired sum.
```

In this function there are no local variables created within the function workspace. The input variable *n* and the output variable *s* are accessible from outside the function and their scope is therefore not local.

Suppose we want to know the sum of the first ten natural numbers. To call the function from the command line, we type

```
>> n_sum = first_n_numbers(10)
n_sum =
      55
```

Variables in MATLAB do not need to be declared before use. An assignment statement can be used to create a new variable, e.g. `a = 2` creates the variable *a*, in the case that it has not been previously defined. Any variable that is created at the command prompt is immediately stored in the base workspace.

If a statement in an m-file is too long to fit on one line in the viewing area of the Editor, you can continue the statement onto the next line by typing three consecutive periods on the first line and pressing Enter. This is illustrated below (this statement is in Program 7.13):

```
f = [-k*y(1)*y(2) + beta*(y(1)+y(3)) - mu*y(1); ...
    k*y(1)*y(2) - 1/gamma*y(2); ...
    1/gamma*y(2) - mu*y(3)];
```

Here, *f* is a 3×1 column vector.

A1.2.1 Operators

You are familiar with the arithmetic operators +, −, *, /, and ^ that act on scalar quantities. In MATLAB these operators can also function as a matrix operator on vectors and matrices. If $\boldsymbol{A}$ and $\boldsymbol{B}$ are two matrices, then `A*B` signifies a matrix multiplication operation; `A^2` represents the `A*A` matrix operation; and `A/B` means (right) division[3] of $\boldsymbol{A}$ by $\boldsymbol{B}$ or `A*inv(B)`. Matrix operations are performed according to the rules of linear algebra and are discussed in detail in Chapter 2.

Often, we will want to carry out element-by-element operations (array operations) as opposed to combining the matrices as per linear algebra rules. To perform array operations on matrices, we include the dot operator, which precedes the arithmetic operator. The matrices that are combined arithmetically (element-by-element) must be of the same size. Arithmetic operators for element-by-element operations are listed below:

`.*` performs element-by-element multiplication;
`./` performs element-by-element division;
`.^` performs element-by-element exponentiation.

The statement `C = A.*B` produces a matrix $\boldsymbol{C}$ such that $\boldsymbol{C}(i,j) = \boldsymbol{A}(i,j) * \boldsymbol{B}(i,j)$. The sizes of $\boldsymbol{A}$, $\boldsymbol{B}$, and $\boldsymbol{C}$ must be the same for this operation to be valid. The statement `B = A.^2` raises each element of $\boldsymbol{A}$ to the power 2.

[3] Left division is explained in Chapter 2.

Note that + and − are element-wise operators already and therefore do not need a preceding dot operator to distinguish between a matrix operation and an array operation.

For addition, subtraction, multiplication, or division of a matrix A with any scalar quantity, the arithmetic operation is identically performed on each element of the matrix A. The dot operator is not required for specifying scalar–matrix operations. This is illustrated in the following:

```
>> A = [10 20 30; 40 50 60];
>> B = A - 10
B =
    0   10   20
   30   40   50
>> C = 2*B
C =
    0   20    40
   60   80   100
```

MATLAB provides a set of mathematical functions such as `exp`, `cos`, `sin`, `log`, `log10`, `arcsin`, etc. When these functions operate on a matrix, the operation is performed element-wise.

The **relational operators** available in MATLAB are listed in Table A.1. They can operate on both scalars and arrays.

Comparison of arrays is done on an element-by-element basis. Two arrays being compared must be of the same size. If the result of the comparison is true, a "1" is generated for that element position. If the result of the element-by-element comparison is false, a "0" is generated. The output of the comparison is an array of the same size as the arrays being compared. See the example below.

```
>> A = [2.5 7 6; 4 7 2; 4 2.2 5];
>> B = [6 8 2; 4 2 6.3; 5 1 3];
>> A >= B
ans =
     0   0   1
     0   1   0
     0   1   1
```

One category of **logical operators** in MATLAB operates element-wise on logical arrays, while another category operates only on scalar logical expressions. Note that `logical` is a data type and has a value of 1, which stands for `true`, or has a value of

Table A.1. *Relational operators*

Operator	Definition
<	less than
<=	less than or equal to
>	greater than
>=	greater than or equal to
==	equal to
~=	not equal to

Table A.2. *Short-circuit logical operators*

Operator	Definition
&&	AND
||	OR

Table A.3. *Element-wise logical operators*

Operator	Definition
&	AND
|	OR
~	NOT

0, which represents `false`. The elements of a logical array are all of the `logical` data type and have values equal to either 1 or 0.

Short-circuit logical operators operate on scalar `logical` values (see Table A.2). The `&&` operator returns 1 (true) if both logical expressions evaluate as true, and returns 0 (false) if any one or both of the logical expressions evaluate as false. The `||` operator returns true if any one or both of the logical expressions are true.

These operators are called short-circuit operators because the result of the logical operation is determined by evaluating only the first logical expression. If the result of the logical operation still remains unclear, then the second logical expression is evaluated.

Element-wise logical operators compare logical arrays element-wise and return a logical array that contains the result (see Table A.3).

A1.2.2 Program control statements

Programs contain not only operators that operate on data, but also control statements. Two categories of control that we discuss in this section are conditional control and loop control.

Conditional control

`if-else-end`

The most basic form of the `if` statement is

```
if logical expression
    statement 1
    statement 2
    .
    .
    .
    statement n
end
```

The `if` statement evaluates a logical expression. If the expression is true, the set of statements contained within the `if-end` block are executed. If the logical expression is found to be false, control of the program transfers directly to the line immediately after the `end` statement. The statements contained within the `if-end` block are not executed. Good coding practice includes indenting statements located within a block as shown above. Indentation improves readability of the code.

Another syntax for the `if` control block is as follows:

```
if logical expression #1
    set of statements #1
elseif logical expression #2
    set of statements #2
else
    set of statements #3
end
```

In this control statement, if logical expression #1 turns out to be false, logical expression #2 is evaluated. If this evaluates to true, then statement set #2 is executed, otherwise statement set #3 is executed. Note the following points.

(1) The `elseif` sub-block as well as the `else` sub-block are optional.
(2) There can be more than one `elseif` statement specified within the `if-end` block.
(3) If none of the logical expressions hold true, the statements specified between the `else` line and the `end` line are executed.

`switch-case`

When the decision to execute a set of statements rests on the value of an expression, the switch-case control block should be used. The syntax is

```
switch expression
    case value_1
        set of statements #1
    case value_2
        set of statements #2
    .
    .
    .
    case value_n
        set of statements #n
    otherwise
        set of statements #(n + 1)
end
```

The value of the expression is evaluated. If its value matches the value(s) given in the first `case` statement, then statement set #1 is executed. If not, the next `case` value(s) is checked. Once a matching value is found and the corresponding statements for that `case` are executed, control of the program is immediately transferred to the line following the `end` statement, i.e. the values listed in subsequent `case` statements are not checked. Optionally, one can add an `otherwise` statement to the `switch-case` block. If none of the `case` values match the expression value, the statements in the `otherwise` sub-block are executed.

Loop control

for loop

The for loop is used to execute a set of statements a specified number of times. The syntax is

```
for index = first:step:last
     statements
end
```

If the value of *index* is inadvertently changed within the loop, this will not modify the number of loops made.

while loop

The while loop executes a series of statements over and over again as long as a specified condition holds true. Once the condition becomes false, control of the program is transferred to the line following the while-end block. The syntax is

```
while logical expression
     statements
end
```

The break statement is used to "break" out of a for-end or while-end loop. Usually, the break statement is contained within an if-end block placed within the loop. A particular condition is tested by the if-end statement. If the logical expression evaluated by the if statement is found to be true, the break command is encountered and looping terminates. Control of the program is transferred to the next line after the end statement of the loop.

Example A1.3

This example is adapted from Program 5.5. The following function m-file uses the fixed-point iterative method to solve a nonlinear equation in a single variable:

```
x1 = function fixedpointmethod(gfunc, x0, tolx)
% Fixed-Point Iteration used to solve a nonlinear equation in x

% Input variables
% gfunc : nonlinear function g(x) whose fixed-point we seek
% x0 : initial guess value
% tolx : tolerance for error in estimating root

maxloops = 50;
for i = 1:maxloops
    x1 = feval(gfunc, x0);
    if (abs(x1 - x0) <= tolx)
       break % Jump out of the for loop
    end
    x0 = x1;
end
```

To perform fixed-point iteration we must first convert the nonlinear equation $f(x) = 0$ whose root we seek into the form $x = g(x)$, where $g(x)$ is evaluated iteratively to obtain a sufficiently accurate value of x that satisfies $x = g(x)$. Once the difference in two consecutive values of x falls below a threshold called the tolerance limit, the program exits out of the loop.

MATLAB has defined certain parameters that return special values. Some important functions that are discussed in this book are:

- `ans`: Any time you evaluate an expression at the command line, but have not assigned a variable to store the value of the expression, MATLAB places the value into a function named `ans`. The most recent answer not assigned to a particular variable is stored in `ans`.
- `pi`: This function stores the value of the irrational number $\pi = 3.141592653589793$.
- `inf`: This function represents infinity. Numerical calculations that involve division by zero generate the result `inf`.
- `eps`: Gap in the floating-point number line. This function is explained in Chapter 1.
- `realmax`: The largest floating-point number that is recognized by the machine. This function is explained in Chapter 1.
- `realmin`: The smallest floating point number that is recognized by the machine. This function is explained in Chapter 1.

A1.3 Two-dimensional plotting

MATLAB provides a variety of plotting commands and tools to visualize data. The plotting function of particular interest in this book is `plot`, which draws a two-dimensional graph with linear-scale x- and y-axes. Before plotting any data, it is recommended that you first create a figure window in which the graph axes and line plot will appear.[4] This is done by typing the statement

```
figure
```

This precedes the `plot` statement. The `plot` function has many syntax forms. We start with a basic example,

```
plot(x, y)
```

where $\boldsymbol{x}$ and $\boldsymbol{y}$ are two vectors that must be of the same length. This statement plots a line using data points $(\boldsymbol{x}, \boldsymbol{y})$ such that the elements of $\boldsymbol{y}$ are paired with the corresponding elements of $\boldsymbol{x}$. You can also specify the line style and color of the plot, e.g.

```
plot(x,y,'-k')
```

plots the points as a continuous black line.

The color codes and line style codes are given in Table A.4. You can also mark the location of data points with a marker shape of any color (see Table A.5); e.g.

```
plot(x,y,'sb')
```

plots a blue square where each data point is located.

You can combine line styles and marker styles to distinguish one line from another when producing black and white plots. Suppose we plot two sets of data on the same axes. The following statement is an example of how you can differentiate the two lines simply using different markers:

```
plot(x1, y1, '-xk', x2, y2, '-ok')
```

Note that a `plot` function can accept multiple data sets.

[4] If you do not do this, you run the risk of generating a new plot in the current figure window, thereby erasing or modifying a previously created figure.

Table A.4. *Color and line style codes for MATLAB plotting function*

Letter	Color	Character	Line style
k	black	-	continuous line
b	blue	- -	dashed line
y	yellow	:	dotted line
m	magenta	-.	dash–dot line
g	green		
w	white		
r	red		
c	cyan		

Table A.5. *Marker style codes for MATLAB plotting functions*

Character	Marker
^	upward-pointing triangle
v	downward-pointing triangle
d	diamond
s	square
o	circle
*	star
x	×
+	plus sign

You can specify the width of the line by providing a value for the `LineWidth` property, equal to the number of points:

```
plot (x, y, '-k', 'LineWidth' ,2)
```

You can set a range of properties of the plot axes using the `set` function. To set the properties of an existing figure axes, you will need to obtain a handle of the axes object. The handle of the current axis is returned by the function `gca`, which stands for "get current axes."

Suppose you wish to change the width of the axes lines and the size of the tick label font:

```
set(gca, 'LineWidth' , 2, 'FontSize' , 16)
```

You can specify the axes limits using the `axis` function, whose syntax is

```
axis([xmin xmax ymin ymax])
```

If the data units must be equal in length for both the x- and y-axis, for instance to make a plot of a circle appear as a circle rather than an ellipse, the command

Figure A1.1

Plot of sin *x*.

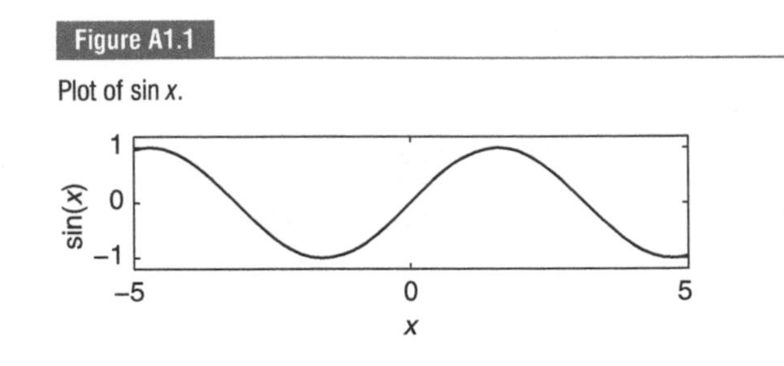


```
axis equal
```

will equalize the unit length on both axes.

You can provide labels[5] for the *x*-axis and *y*-axis using the `xlabel` and `ylabel` functions, respectively. The `title` function is used to give the figure a title.

Example A1.4

The following script plots the sin *x* function for a range of values of *x*:

```
x = -5:0.01:5;
y = sin(x);
plot (x, y, '-k','LineWidth',2)
set(gca, 'LineWidth',2, 'FontSize',16)
xlabel('x')
ylabel('sin(x)')
axis equal
axis([-5, 5, -1.2, 1.2])
```

The output of this script is shown in Figure A1.1.

If multiple plots are to be displayed within a single figure window, use the `subplot` function. The syntax is

```
subplot(no_of_rows, no_of_columns, figure_position)
```

Suppose there are six plots in a figure arranged in three rows and two columns. Each plot is positioned in the figure using the following statements:

```
figure
% first plot
subplot(3, 2, 1)
      plot(x1, y1)
% second plot
subplot(3,2,2)
      plot(x2, y2)
.
.
.
% sixth plot
subplot(3,2,6)
      plot(x6, y6)
```

[5] Formatting of text such as italics, superscripts, subscripts, etc. can be done using TeX commands. See MATLAB help for details.

A1.4 Data import and export

A1.4.1 Importing data

There are two ways to load text data (ASCII format) from a file into the workspace. One method is to use the Import Wizard that is accessible from the file menu (File ≫ Import Data). When the Import Wizard starts, it will ask you to select the file that contains the data you wish to import. Next, you will be asked to specify the delimiting character that separates adjacent data values. Often, the delimiter will be the space character, although a comma, semicolon, or tab character can also be used as a delimiter. The data extracted from the file are displayed in rows and columns in the same window for your verification. The Wizard creates a single variable named after the filename and stores all of the numerical data in that variable. It also creates another variable for alphabetical data. If row headers or column headers are present in the data, the Wizard extracts the headers and creates a separate variable for them. In the next window, you are asked to choose which variables to save. When row (or column) headers are available, MATLAB will ask if you want to save each row (or column) as a separate variable. In this case, each row variable is named after its corresponding header.

Import of numerical data from a file can be automated within an m-file using the `load` function. Consider the following statement:

```
load a.dat
```

This is the command syntax for `load`. The `load` function will treat the file `a.dat` as a text (ASCII) file and will save the numerical data contained in the file to a single variable of type double. The variable saved in the workspace is a two-dimensional array and is named `a`, after the filename. For the `load` function to execute without error, the data stored in the file must be in rectangular form, i.e. the number of elements in each row must be the same. Also, within each row the row elements should be separated by one or more spaces. The functional form of the syntax for `load` is

```
b = load(a.dat);
```

Using the functional form you can specify the name of the variable that will contain the data extracted from `a.dat`.

A1.4.2 Exporting data

Data generated within a function or script can be output to the screen or to a file using the `fprintf` function. This function outputs the data as a character string. You can specify how the numerical data should appear in the output using *conversion specifiers*. The syntax for this function is

```
fprintf(fid, format, a, b, ...)
```

where `fid` is a file identifier that specifies which file the data should be written to; `fid` is an integer and is obtained from the function `fopen`, e.g.

```
fid = fopen('myfile.txt', 'w')
```

Table A.6. *Conversion specifiers for the* `fprintf` *function*

Specifier	Purpose
%e	exponential notation
%f	fixed-point notation
%d	integer notation

To specify, output to screen fid = 1. If fid is not specified, output is sent to screen. Note that `format` is a string consisting of text and conversion specifiers. The numerical data (`a, b, ...`) to be written is provided after the `format` string in sequential order of appearance.

A conversion specifier always begins with a percent sign (%). Several conversion specifiers are presented in Table A.6. Note that %d can be used with integers and with non-integral values that have been converted to integers using the MATLAB functions fix, floor, `ceil`, or `round`. The minimum width of numeric data (number of significant digits) and the precision (number of significant digits to the right of the decimal point) can be optionally specified with the conversion specifier. See MATLAB `help` for more information on `fprintf` and `fopen`.

Example A1.5

This example illustrates the use of `fprintf`. The function `squareroot` calculates the square root of each element in vector ***x***. It then prints to the screen the formatted result:

```
function squareroot (x)
y = sqrt (x) ;
z = [x; y] ; % x and y are row vectors
fprintf( 'Square root of %6.2f is %6.2f.\n' , z)
```

Note that `'\n'` creates a new line. At the command prompt, we type the following:

```
>> x = 0:0.1:1;
>> squareroot (x)
Square root of   0.00 is   0.00.
Square root of   0.10 is   0.32.
Square root of   0.20 is   0.45.
Square root of   0.30 is   0.55.
Square root of   0.40 is   0.63.
Square root of   0.50 is   0.71.
Square root of   0.60 is   0.77.
Square root of   0.70 is   0.84.
Square root of   0.80 is   0.89.
Square root of   0.90 is   0.95.
Square root of   1.00 is   1.00.
```

To save data that does not require special formatting, use the `save` command. You can save the entire workspace or specific workspace variables to a file using `save`. By itself, the `save` command creates a binary file with a `.mat` extension called `matlab.mat` which contains all the workspace variables. You can specify the name of the file using the following syntax:

```
save filename.mat
```

The variables from this file can be loaded into the memory using the `load` command. You can choose which variables to save by specifying them in the `save` statement:

```
save filename.mat x y z
```

You can also save the file as an ASCII text file (8 digit form) using the `-ascii` option. This creates a text file that can be input into other programs such as Excel and Word. To save the numeric data in 16 digit form, include the `-double` option:

```
save filename.dat x y z -ascii - double
```

Appendix B Location of nodes for Gauss–Legendre quadrature

Here we derive the optimal location of the nodes for an n-point Gaussian quadrature. The truncation error of the Gaussian quadrature formula is the integral of Equation (6.17), i.e. the truncation error of the polynomial interpolant,

$$E(x) = \int_a^b \frac{f^{n+1}(\xi(x))}{(n+1)!}(x - x_0)(x - x_1)\cdots(x - x_n)dx.$$

The truncation error can be rewritten as a function of the integration variable z that is defined on the interval $[-1, 1]$ as follows:

$$E(z) = \left(\frac{b-a}{2}\right)^{n+2}\int_{-1}^{1}\frac{f^{n+1}\left(\xi\left(\frac{b-a}{2}z + \frac{a+b}{2}\right)\right)}{(n+1)!}(z - z_0)(z - z_1)\cdots(z - z_n)dz.$$

The interpolating polynomial function $p(x)$ is of degree n and is constructed from function values obtained at the $n + 1$ nodes, $x_0, x_1, \ldots, x_n$ whose locations we seek. We define $\psi(z) = (z - z_0)(z - z_1)\cdots(z - z_n)$, where $\psi(z)$ is a polynomial of degree $n + 1$. Suppose $f(x)$ is a polynomial of degree $\leq 2n + 1$, then its $(n + 1)$th derivative is a polynomial of degree $\leq n$. Let

$$\omega(x) = \frac{f^{n+1}(\xi(x))}{(n+1)!},$$

where $\omega(x)$ is a polynomial of degree $\leq n$. Then, $\omega(z)$ is also a polynomial of degree $\leq n$. For any polynomial $f(x)$ of degree $\leq 2n + 1$, the Gaussian quadrature formula will be exact if the truncation error, $E(z) = 0$, i.e.

$$\int_{-1}^{1}\omega(z)\psi(z)dz = 0. \qquad \text{(B.1)}$$

The above integral is the inner product of two polynomial functions.

The **inner product** of two continuous functions $f(x)$ and $g(x)$ is defined as

$$\langle f, g\rangle = \int_a^b f(x)g(x)w(x)dx, \qquad \text{(B.2)}$$

where $w(x)$ is a weighting function and $w(x) > 0$ for $a \leq x \leq b$. Just as the dot product of any non-zero vector with itself is a positive number, the inner product of any function with itself is a positive number, i.e.

$$\langle f, f\rangle = \int_a^b f(x)f(x)w(x)dx > 0. \qquad \text{(B.3)}$$

Equation (B.1) has a weighting function equal to one. If two continuous functions are orthogonal to each other, their inner product is equal to zero. (Section 2.2 discusses orthogonality (normality) of vectors in Euclidean vector space.) Let $\Phi_n(x)$ be any **orthogonal polynomial function**, where the subscript n is the degree of the polynomial. The inner product of two orthogonal polynomial functions is defined as

$$\langle \Phi_n(x), \Phi_m(x) \rangle = \int_a^b \Phi_n(x)\Phi_m(x)w(x)dx = c_{nm}\delta_{mn}, \tag{B.4}$$

where δ is the identity matrix I such that $\delta_{mn} = 1$ for $m = n$ and $\delta_{mn} = 0$ for $m \neq n$; c_{nm} is a positive constant.[1] A class of orthogonal polynomials called **Legendre polynomials** is defined for a weighting factor $w(x) = 1$. *Their properties of orthogonality exist only on the interval* $[-1, 1]$. The inner product of two Legendre polynomials is given by

$$\langle \Phi_n(z), \Phi_m(z) \rangle = \int_{-1}^{1} \Phi_n(z)\Phi_m(z)dx = c_{nm}\delta_{mn}.$$

If the roots of $\psi(z)$ are made equal to the roots z_i $(i = 0, 2, \ldots, n)$ of the $(n + 1)$th-degree Legendre polynomial $\Phi_{n+1}(z)$, then $\psi(z)$ is a Legendre polynomial of degree $n + 1$. Any polynomial of degree n can be expressed as a linear combination of Legendre polynomials of degree up to n. Therefore, the polynomial function $\omega(z)$ can be written as a linear combination of the first $n + 1$ Legendre polynomials:

$$\omega(z) = a_0\Phi_0(z) + a_1\Phi_1(z) + a_2\Phi_2(z) + \cdots + a_n\Phi_n(z).$$

The inner product of a Legendre polynomial of degree $n + 1$ with a function that is a linear combination of Legendre polynomials of up to degree n is zero. There is no term involving the inner product of two orthogonal polynomials of the same degree as long as $f(x)$ is a polynomial of degree $\leq 2n + 1$. If the degree of $f(x)$ is equal to $2n + 2$, then $\omega(x)$ will be of degree $n + 1$, the same as that of $\psi(x)$ and according to Equation (B.3), a non-zero inner product produces the non-zero truncation error.

[1] If $c_{nm} = 1$ for $n = m$ and $c_{nm} = 0$ for $n \neq m$, then the polynomials are said to be orthonormal.

Index for MATLAB commands

anova1, MATLAB function 267

backslash operator, MATLAB function 96, 118
bar, MATLAB function 157
binopdf, MATLAB function 168

chi2cdf, MATLAB function 276
chi2gof, MATLAB function 281
chi2pdf, MATLAB function 276
cond, MATLAB function 100

det, MATLAB function 74
diag, MATLAB function 55
double, MATLAB data type 16

eps, MATLAB function 13
errorbar, MATLAB function 272
eye, MATLAB function 55

fminbnd, MATLAB function 498
fmincon, MATLAB function 529
fminsearch, MATLAB function 515
fminunc, MATLAB function 511
format, MATLAB function 16, 21
fprintf, MATLAB function 572

global, MATLAB keyword 563

hist, MATLAB function 157

inv, MATLAB function 75

lu, MATLAB function 95

mean, MATLAB function 146
mldivide, MATLAB function
  see backslash operator
multcompare, MATLAB function 267

norm, MATLAB function 65
normcdf, MATLAB function 176
norminv, MATLAB function 177
normpdf, MATLAB function 169
normplot, MATLAB function 234

ODE solvers, MATLAB function 445–447
ones, MATLAB function 54

poisspdf, MATLAB function 169
polyfit, MATLAB function 124
polyval, MATLAB function 110

quad, MATLAB function 175, 387
quadl, MATLAB function 402

range, MATLAB function 147
rank, MATLAB function 72
ranksum, MATLAB function 298
realmax, MATLAB function 11
realmin, MATLAB function 13
round, MATLAB function 7

signrank, MATLAB function 295
signtest, MATLAB function 292
single, MATLAB data type 16
single, MATLAB function 17
size, MATLAB function 56
std, MATAB function 147
structure, MATLAB data type 446

tcdf, MATLAB function 182
tinv, MATLAB function 182
ttest, MATLAB function 244, 249
ttest2, MATLAB function 251

var, MATLAB function 147

zeros, MATLAB function 54
ztest, MATLAB function 236

Index

absolute error 16
adaptive quadrature 385
array indexing 561

backward substitution 76, 81, 82, 85
banded matrices 87
base-2 number system 7
basis vectors 69, 70
Bernoulli, Jacob 148, 159
Bernoulli, Jacques
  see Bernoulli, Jacob
Bernoulli, James
  see Bernoulli, Jacob
Bernoulli trial 159, 253
Bernoulli process 159
bimodal 145
binomial distribution 160
binomial variable 161, 254
Bonferroni adjustment 266
box plot 267

cardinal variable 229
case-control study
  see retrospective study
categorical variable
  see nominal variable
chemostat 415
chi-square statistic 274
coefficient of determination 115
cofactor 74
cohort study
  see prospective study
colon operator 55, 58
column space 69, 70
column vector 53
complementary event 149
completely randomized design 211
condition number 99–100
confounding variable 213
contingency table 281, 285
control group 211
convergence
  of a series, 32
  criteria, 39
covariance 187
covariance matrix
  of model parameters 198
  of *y* 199
  of model predictions 199
cubic spline 363
cumulative distribution function (cdf) 176

density curve
  see probability density curve
density function 168
descriptive statistics 142
determinant of a matrix 73
  calculation of 74,
direct methods 76
Dirichlet boundary condition 462
dot plot 540
dot product 59
  operator, 59
double precision 11
dynamic programming 547
  BLAST 548
  Needleman and Wunsch 547
eigenvalue problem 458
eigenvalues 458
eigenvectors 458
elementary row transformation 78
empirical model 480
enzymes 334
epsilon
  see machine epsilon
equivalent matrix 78
event 149
expectation
  see expected value
expected value 162, 170
exponent 6, 11
exposure variable 211

F statistic 264
false negative 156, 227
false positive 156, 226
finite difference formula 33–36
  first backward finite difference, 36
  first central finite difference, 36
  first forward finite difference, 36
Fisher, R. A. 261, 264
fixed-point 320
floating-point numbers 6, 10
  discontinuity in floating-point number line 13
flops 88
forward substitution method 88
function workspace 562

Gauss, Carl Friedrich 167
Gaussian elimination 76,
  using Thomas method 87, 133–134
  with complete pivoting 87
  with partial pivoting 82, 85,
  with scaled partial pivoting 87
  without pivoting 80, 82

global minimum 484
Gombauld, Antoine 147
Gosset, William Sealy 181

Hessian matrix 501
histogram 143, 157

ill-conditioning (instability) 26
independent events 151
inferential statistics 142, 144, 178
inflection point 487
inline function 347
inner product 575
interval estimate 178
inverse of a matrix 74

ligand 331
linear operator 65
linearly dependent 68
linearly independent 68
linear transformation 65, 66
Lineweaver–Burk analysis 194
local minimum 483, 484
logical indexing 561
lower triangular matrix 78
LU decomposition method
  see LU factorization method
LU factorization method 88
  using Doolittle method 89

machine epsilon 13
matrix constructor operator 54
matrix–matrix multiplication
    61–63
matrix–vector multiplication 61
mean 145
mean value theorem 419
mechanistic model 480
median 145–146
m-file
  function 562
  script 561–562
Michaelis–Menten kinetics
    334, 413
minor 73
  principal 502
  leading principal 502
mixed boundary condition 462
mode 146
Moivre, Abraham De 166
Monod equation 37
multimodal function 484
multiple root 313
mutually exclusive events 149

Neumann boundary condition 462
nominal variable 229
norm 64
  matrix norm, 65
  p-norm, 64
  vector norm, 64
normal equations 109,
    112, 114
normal probability plot 232

Ockham, William 101
Ockham's razor 101
ODE solvers, MATLAB functions
    445–447
order of convergence
  see rate of convergence
order of magnitude 6, 28
ordinal variables 229
overflow error 12
oxygen transport
  in the body, 3
  in skeletal tissue, 3–4

p value 220
Pascal, Blaise 148
permutation matrices 87, 93–94
pharmacodynamics 104
pharmacokinetic models 48
pharmacokinetics 104
phylogenetic tree 553
pivot column 78
pivot element 78
pivot row 78
point estimate 178
population parameter 145
power of a hypothesis test 227
  of a one-sample *z* test 236–240
  of a two sample *z* test 243
precision 5
probability density curve 167
probability density function (pdf)
  see density function
probability mass function 169
prospective study 215

quantile 232

random sample 149
random variable 157
range 146
rank
  deficiency of, 71
  of a coefficient matrix, 71
  of a matrix, 71
rate of convergence 316
receptor 331
relative error 16
representative sample 104, 149
  see also random sample
residual 98, 107
  variance 193
response variable 211, 212
retrospective study 215
row space 69, 70
row vector 53

saddle point 501
sampling 149
  without replacement 154
  error 178
sampling distribution of the
    mean 178
scoring matrices 544
  PAM matrix 544
  BLOSUM matrix 545
significance level 221
significant digits

loss of, 22
number of, 18–20
single precision 11
sparse matrices 87
SSR 109
standard deviation 147
statistic 141, 267
steady state process 51
subtractive cancellation 22

Taylor series 26–27
derivation of finite difference formulas, 33–36
tolerance
absolute, 39
relative, 39
transpose 57–58
transpose operation, 58
tridiagonal matrices 87
see also Gaussian elimination using Thomas method
unbiased estimate 171
underflow error 13
unimodal function 484
upper triangular matrix 78

Vandermonde matrix 364
variable scope
local 562
global 563
variance 146–147, 162, 170
vector space 69
basis for, 69
dimension of, 69
spanned by, 69, 70

Lightning Source UK Ltd.
Milton Keynes UK
UKHW031239220319
339642UK00015B/308/P

9 780521 871587